Flora of Florida, Volume VII

UNIVERSITY PRESS OF FLORIDA

Florida A&M University, Tallahassee
Florida Atlantic University, Boca Raton
Florida Gulf Coast University, Ft. Myers
Florida International University, Miami
Florida State University, Tallahassee
New College of Florida, Sarasota
University of Central Florida, Orlando
University of Florida, Gainesville
University of North Florida, Jacksonville
University of South Florida, Tampa
University of West Florida, Pensacola

Flora of Florida

VOLUME VII

DICOTYLEDONS, OROBANCHACEAE THROUGH ASTERACEAE

Richard P. Wunderlin, Bruce F. Hansen, and Alan R. Franck

University Press of Florida
Gainesville · Tallahassee · Tampa · Boca Raton
Pensacola · Orlando · Miami · Jacksonville · Ft. Myers · Sarasota

Published in the United States of America

25 24 23 22 21 20 6 5 4 3 2 1

Library of Congress Cataloging-in-Publication Data
Names: Wunderlin, Richard P., 1939– author. | Hansen, Bruce F., author. | Franck, Alan R., author.
Title: Flora of Florida. Vol. 7 : Dicotyledons, Orobanchaceae through Asteraceae / Richard P. Wunderlin, Bruce F. Hansen, and Alan R. Franck.
Description: Gainesville : University Press of Florida, 2020. | Includes bibliographical references and index.
Identifiers: LCCN 2017032670 | ISBN 978-0-8130-6626-4 (cloth : alk. paper)
Subjects: LCSH: Plants—Florida—Classification. | Plants—Florida—Identification.
Classification: LCC QK154 .W852 2016 | DDC 581.9759—dc23
LC record available at https://lccn.loc.gov/2017032670

The University Press of Florida is the scholarly publishing agency for the State University System of Florida, comprising Florida A&M University, Florida Atlantic University, Florida Gulf Coast University, Florida International University, Florida State University, New College of Florida, University of Central Florida, University of Florida, University of North Florida, University of South Florida, and University of West Florida.

University Press of Florida
2046 NE Waldo Road
Suite 2100
Gainesville, FL 32609
http://upress.ufl.edu

Dedicated to the many botanists whose works on the Asteraceae resulted in a much better understanding of this large and difficult family of the Florida flora.

Contents

Acknowledgments ix

Introduction xi

Orobanchaceae 1

Aquifoliaceae 26

Campanulaceae 34

Menyanthaceae 49

Goodeniaceae 53

Calyceraceae 55

Adoxaceae 56

Caprifoliaceae 61

Pittosporaceae 66

Araliaceae 67

Apiaceae 77

Asteraceae 115

Literature Cited 425

Index to Common Names 439

Index to Scientific Names 451

Acknowledgments

The facilities and collections of many herbaria were utilized in preparing this volume. The courtesies extended and the loan of specimens by the curators are gratefully appreciated. These include the Florida Museum of Natural History (FLAS), Florida State University (FSU), Harvard Herbaria (A, GH), Marie Selby Botanical Garden (SEL), New York Botanical Garden (NY), University of Central Florida (FTU), University of North Carolina, Chapel Hill (NCU), and University of West Florida (UWFP). We are especially grateful to Kent Perkins (FLAS), Loran Anderson (FSU), and Austin Mast (FSU) for their continuous support.

The *Flora of Florida* project has been strongly supported by the University of South Florida Institute for Systematic Botany.

Introduction

Volume 1 of the *Flora of Florida* provides background information on the physical setting, vegetation, history of botanical exploration, and systematic treatments of the pteridophytes and gymnosperms. Volumes 2 through 7 will contain the dicotyledons and volumes 8 through 10, the monocotyledons.

This volume contains the taxonomic treatments of 12 families.

ORGANIZATION OF THE FLORA

Taxa Included

Florida, with more than 4,400 taxa, has the third most diverse vascular plant flora of any state in the United States. The *Flora of Florida* is a treatment of all indigenous and naturalized vascular plant taxa currently known to occur in the state. Naturalized is defined as those nonindigenous taxa growing outside of cultivation and naturally reproducing. This includes plants that have escaped from cultivation as well as those that were intentionally or accidentally introduced by human activities in post-Columbian times. Taxa that have not been recently recollected and may no longer exist in the wild in Florida are formally treated both for historical completeness and on the premise that they may be rediscovered in the future.

A taxon is formally treated in this flora if (1) an herbarium specimen has been seen to document its occurrence in Florida, or (2) a specimen is cited from Florida in a monograph or revision whose treatment is considered sound.

Taxa Excluded

Literature reports of taxa attributed to Florida that are considered to be erroneous or highly questionable and therefore to be excluded from this flora are listed following the treatment for the genus, or in the case of genera not otherwise treated, at the end of the family. The reason for exclusion is given in each case. Most commonly, the taxon is excluded because it is based on a misidentified specimen(s) or lack of documentation by means of a specimen, or it is based on a misapplied name, that is, a name correctly applied to a plant not found in Florida.

Systematic Arrangement

Recent studies have demonstrated that the traditional dicotyledons are paraphyletic and that the monophyletic monocotyledons are derived from within the dicotyledons. We believe that the arrangement as proposed by the Angiosperm Phylogeny Group (Stevens, 2019) has merit and is followed in this work with slight modifications. For convenience, the genera and species within each family are arranged alphabetically.

Descriptions

Descriptions are based on Florida material and are given for each family, genus, species, and infraspecific taxon.

Common Names

Non-Latinized names given for the taxa are derived from published sources as well as from our own experience. No attempt is made to list all names that have been applied to a taxon, to standardize names with a specific source, or to supply a name for species where one is not in general usage. For plants lacking a common name, the generic name may be used as is the usual practice.

Derivation of Scientific Names

The derivation of the generic name and that of each specific and infraspecific epithet is given.

Synonymy

A literature citation is given for each species, infraspecific taxon, and synonym. Synonyms listed are only those that have been cited for Florida in manuals, monographic treatments, and technical papers. Also included is the basionym and all homotypic synonyms of a name introduced into synonymy. The homotypic synonyms are listed in chronological order in a single paragraph, and the paragraphs of synonyms are put in chronological order according to the basionym of each. If the type of a taxon is a Florida collection and is known, this information is given. We do not attempt to lectotypify the numerous Florida taxa needing lectotypification in the belief that this is best left to monographers.

For families and genera, only the author and date of publication are given. Family and generic synonyms listed are those that have been used in the major publications pertinent to the Florida flora.

Citation of periodical literature conforms to that cited in Lawrence et al. (1968) and Bridson and Smith (1991). Other literature citations conform to those cited in Stafleu and Cowan (1976 et seq.). Author abbreviations are those listed in Brummitt and Powell (1992).

Habitat

The terminology used for plant communities generally follows that of Myers and Ewel (1990) but may vary.

Distribution

The global distribution is given for each family and genus where native and naturalized. Relative abundance in Florida (ranked as common, frequent, occasional, or rare) and the distribution are given for each species and infraspecific taxon. The format for distribution of species and infraspecific taxa is: Florida; North America (continental United States, Canada, and Greenland); tropical America (West Indies, Mexico, Central America, and South America); Old World (Europe, Africa, Asia, Australia, and Pacific Islands). For taxa occurring in all of these areas, the phrase "nearly cosmopolitan" is used. For taxa of limited distribution in Florida, range statements by county are usually given. For taxa of wide distribution in Florida, the range is given in general terms: *panhandle*—from the Suwannee River west to Escambia County; *peninsula*—east of the Suwannee River and south of the Georgia line southward through the Florida Keys. Because of the vast floristic differences in peninsular Florida, this region is often further subdivided into northern, central, and southern regions and the keys. The northern region is east of the Suwannee River and south of the Georgia line southward through Gilchrist, Alachua, Putnam, and Flagler Counties. The central region extends from Levy, Marion, and Volusia Counties southward through Lee, Hendry, and Palm Beach Counties. The southern peninsula consists of the southernmost four counties (Collier, Broward, Monroe, and Miami-Dade). The Florida Keys consist of the chain of islands from Key Largo to the Marquesas Keys and the Dry Tortugas. Politically, they are part of Monroe County. The panhandle is subdivided into eastern, central, and western regions. The eastern region consists of the counties west of the Suwannee River west through Jefferson County, the central region extends from Leon and Wakulla Counties west through Holmes, Washington, and Bay Counties, and the western region consists of the westernmost four counties (Walton, Okaloosa, Santa Rosa, and Escambia).

Since the species distribution may change as new data are added, please refer to the *Atlas of Florida Plants* website (Wunderlin et al., 2019) for current information.

Endemic or Exotic Status

Endemic taxa are those whose global distribution is confined to the political boundary of Florida. If a taxon is a non-native, the region of nativity is given. Non-native taxa are those that are known to have become part of the flora following the arrival of Ponce de Leon in 1513. Admittedly, this is an arbitrary starting point as several species are believed to have been introduced by Paleo-Indians before 1513. Technically, these are considered as native. Another problem in interpretation arises when propagules arrive after 1513 by some means other than human activity (that is, hurricanes, storms, sea-drift, or animals) and the species becomes established. Again technically, these are considered as non-natives. It is sometimes difficult to determine whether a widespread species is native or exotic, and our opinion may differ from that of others.

Reproductive Season

The sexual reproductive (flowering) season for each species and infraspecific taxon is given. The reproductive seasons are broadly defined as follows: spring—March through May; summer—June through September; fall—October through November; winter—December through February. Species "flowering out of season" are sometimes encountered.

Hybrids

Named hybrids are listed along with the putative parents, nomenclature, usually with comment concerning distribution in Florida.

References

Major monographs, revisions, and other pertinent literature, other than those cited in the nomenclature, are cited where appropriate in the text and listed at the end of the volume.

TAXONOMIC CONCEPTS

Taxonomic interpretations and nomenclature are generally in accord with recent monographs or revisions for the various groups except where it is believed that recent evidence necessitates a change. Citation of a monograph or revision in the text implies consideration of the work during the preparation of the treatment, but not necessarily acceptance. Where a difference of opinion exists among published treatments or the treatment in this work deviates from that of the reference cited, a discussion of alternative opinions is often provided.

Species, subspecies, and varieties are considered as entities with a high degree of population integrity. Color forms and minor morphotypes that occur within a species and that may be formally recognized as *forma* by other authors are accorded no formal recognition in this work.

No nomenclatural innovations are intentionally published in the *Flora*.

LITERATURE CITED

Bridson, G.D.R., and E. R. Smith. 1991. Botanico-Periodicum-Huntianum/Supplementum. Pittsburgh: Hunt Botanical Library.

Brummitt, R. K., and C. E. Powell. 1992. Authors of Plant Names. Royal Botanical Gardens, Kew. Basildon: Her Majesty's Stationery Office.

Lawrence, G.H.M., A.F.G. Buchheim, G. S. Daniels, and H. Dolezal. 1968. Botanico-Periodicum-Huntianum. Pittsburgh: Hunt Botanical Library.

Myers, R. I., and J. J. Ewel, eds. 1990. Ecosystems of Florida. Gainesville: University Press of Florida.

Stafleu, F. A., and R. S. Cowan. 1976 et seq. Taxonomic Literature. Edition 2. Utrecht: Bohn, Scheltema, and Holkema.

Stevens, P. F. 2019. Angiosperm Phylogeny Website, Version 14. (http://www.mobot.org/MOBOT/research/APweb/).

Wunderlin, R. P., B. F. Hansen, A. R. Franck, and F. B. Essig. 2019. Atlas of Florida Plants. (http://florida.plantatlas.usf.edu/).

Systematic Treatments

Keys to Major Vascular Plant Groups

1. Plant reproducing by spores .... PTERIDOPHYTES (volume I)
1. Plant reproducing by seeds.
   2. Leaves with a single midvein or with simple or sometimes dichotomously branched veins, these closely parallel and lacking secondary interconnecting cross-veinlets; seeds borne on the surface of specialized bract-scale structures aggregated into woody or fleshy cones or a single seed partly or wholly surrounded by a fleshy aril and drupelike or berrylike; perianth lacking .... GYMNOSPERMS (volume I)
   2. Leaves with parallel veins with secondary interconnecting cross-veinlets or with reticulate veins; seeds borne enclosed within specialized structures (carpels); perianth usually present.
      3. Vascular bundles occurring in a ring or in concentric cylinders; cotyledons 2; flower parts usually in other than whorls of 3 or multiples thereof; leaves usually reticulate-veined.. DICOTYLEDONS (volumes II–VII)
      3. Vascular bundles scattered (or rarely single); cotyledon 1; flower parts often in whorls of 3 or multiples thereof; leaves usually parallel-veined (sometimes with midvein only) (some plants diminutive, floating aquatics, the plant body thalloid, not differentiated into stems and leaves, rootless or with 1–few unbranched roots or plants partly or wholly submersed aquatics, the leaves, flowers, and fruits often much reduced)........................... MONOCOTYLEDONS (volumes VIII–X)

Dicotyledons

OROBANCHACEAE Vent. 1799. BROOMRAPE FAMILY

Holoparasitic or hemiparasitic herbs or shrubs. Leaves alternate or opposite, simple, pinnate-veined, petiolate or epetiolate, estipulate. Flowers in terminal racemes or spikes, bisexual, bracteate, bracteolate or ebracteolate; sepals 2–5, basally connate, actinomorphic or zygomorphic; petals 5, basally connate, zygomorphic; stamens 4, of equal length or didynamous, epipetalous; staminodes 1 or absent; ovary superior, 2-carpellate, 1- or 2-loculate, the style 1. Fruit a capsule; seeds numerous.

A family of about 90 genera and about 2,060 species; nearly cosmopolitan. Many of the genera were formerly placed in the Scrophulariaceae.

Rhinanthaceae Vent., 1799.

Selected reference: Bennett and Mathews (2006).

1. Plants lacking green color, drying tan or brown.
    2. Stem much branched; flowers remote .... **Epifagus**
    2. Stem simple or few branched at the base; flowers congested.
        3. Stem thick; scalelike leaves overlapping; stamens exserted .... **Conopholis**
        3. Stem slender; scalelike leaves not overlapping; stamens included.
            4. Flowers several in spikes or spikelike racemes........................ **Orobanche**
            4. Flowers solitary on long peduncles .... **Aphyllon**
1. Plants with green color (this sometimes obscured by other pigments), drying black or sometimes green.
    5. Corolla salverform.
        6. Plant much-branched, red-brown, an evident root-parasite; corolla tube sharply curved .... **Striga**
        6. Plant few-branched, green, not an evident root-parasite; corolla tube straight or only slightly curved.. **Buchnera**
    5. Corolla other than salverform.
        7. Anther theca with rigid basal spurs (reduced and fleshy in *A. maritima*).
            8. Stamens all of equal length, exserted from the corolla tube at anthesis; corolla orange .... **Macranthera**
            8. Stamens didynamous, included within the corolla tube at anthesis; corolla pink to purple or yellow.

9. Leaves filiform or scalelike; corolla pink to purple; sepals scalelike .... **Agalinis**
9. Leaves lanceolate; corolla yellow; sepals well developed .... **Aureolaria**
7. Anther theca not spurred at the base.
10. Cauline leaves opposite .... **Seymeria**
10. Cauline leaves alternate.
11. Leaves deeply lobed or divided .... **Pedicularis**
11. Leaves entire or merely toothed.
12. Flowers subtended by bright yellow, orange, or red bracts; leaves 1-veined .... **Castilleja**
12. Flowers subtended by green bracts; leaves 3-veined .... **Schwalbea**

Agalinis Raf. 1836. FALSE FOXGLOVE

Annual or perennial hemiparasitic herbs. Leaves opposite, subopposite, or alternate, simple, pinnate-veined, epetiolate. Flowers axillary, bracteate; sepals 5, basally connate, subequal; petals 5, basally connate, zygomorphic, the upper lip 2-lobed, the lower lip 3-lobed; stamens 4, didynamous, epipetalous; staminode absent; ovary superior, 2-carpellate, 2-loculate, the style and stigma 1. Fruit a capsule; seeds numerous.

A genus of about 70 species; North America, West Indies, Mexico, Central America, and South America. [From the Greek *aga,* wonder, and *linum,* flax, in reference to its resemblance to *Linum* (Linaceae).]

Gerardia Benth. 1846.

Selected reference: Hays (2002, 2010); Pennell (1929).

1. Leaves subulate or linear-filiform, 1–20 mm long, appressed to stiffly spreading; axillary fascicles absent; inflorescence with at least some short lateral branches bearing terminal flowers.
2. Stem wiry, very flexuous, laxly branched, often supported by adjacent vegetation, terete or nearly so and only obscurely angled; leaves subulate, appressed, 1–3 mm long; corolla glabrous, the 2 upper lobes arched over the stamens; inflorescence with a single, apparently terminal flower per branch .... **A. filicaulis**
2. Stem stiffly erect and brittle, conspicuously angled; leaves subulate to linear, appressed to stiffly spreading, 1–20 mm long; corolla externally pubescent on the throat and internally across the base of the 2 upper, reflexed lobes; inflorescence appearing racemose or paniculate.
3. Stem ridged on the angles and faces; leaves subulate, 1–3 mm long, appressed, the margin glabrous or glabrate; pedicel 1–3 mm long; plant gray-green .... **A. aphylla**
3. Stem ridged on the angles only; leaves linear-filiform or linear and at least some widened distally, 5–20 mm long, ascending to spreading, the margin scabrous; pedicel 5–26 mm long; plant yellowish green.
4. Corolla 12–17 mm long, pink, with 2 yellow guidelines present; capsule globose, rounded at the summit, the surface smooth at the apex; lower branches more or less upright .... **A. obtusifolia**
4. Corolla 8–11 mm long, rose-purple, yellow guidelines absent; capsule ovoid to oblong, the apex truncate, with raised veins giving the appearance of tubercles; lower branches long decumbent .... **A. flexicaulis**

1. Leaves filiform, linear, or lance-elliptic (in *A. georgiana* widening distally), 1–5 cm long, spreading; axillary fascicles present or absent; inflorescence racemose and lacking short lateral branches, or if these present, the stem terete and the plant rhizomatous.
   5. Plant perennial, sparsely branched, rhizomatous; main stem and branches terete, glabrous; midstem leaves lanceolate-elliptic or linear; corolla throat and the 3 lower lobes externally pubescent; plant of wet to saturated soils .... **A. linifolia**
   5. Plant annual, sparsely to much branched, not rhizomatous; main stem (at least distally) and branches angular and/or ridged, glabrous to scabrous; midstem leaves linear to filiform; corolla throat and the 3 lower lobes pubescent or glabrous; plant of dry to moist saline soils.
      6. Plant sprawling, widely branched, the branches spreading or laxly ascending, the main stems 0.6–3 mm wide; leaves filiform, 0.2–1 mm wide; bracts much reduced upward; pedicel very slender, spreading, 1.5–2 cm long.
         7. Corolla pubescent at the throat and the lower 3 corolla lobes, pubescent internally across the base of the upper 2 reflexed lobes; calyx lobes erect-ascending; leaves spreading; inflorescence paniculiform, the flowers often 1 per node and/or appearing terminal on the lateral branches .... **A. laxa**
         7. Corolla glabrous externally and internally, the upper lobes flattened over the stamens; calyx lobes divergent-spreading; leaves horizontal, those of the subtending branches reflexed; inflorescence racemiform, the flowers mostly 2 per node .... **A. divaricata**
      6. Plant erect, the branches ascending to spreading or arching, the main stem 1–10 mm wide or wider; leaves filiform to linear, 0.5–6 mm wide; bracts gradually or little reduced upward; pedicel slender to stout, erect-ascending, 0.2–3 cm long.
         8. Leaves linear; corolla internally glabrous across the base and sinus of the upper lobes; upper corolla lobes arched forward; upper corolla throat wall conspicuously shorter than the lower corolla throat wall; pedicel not conspicuously longer or shorter than the bracts; stem angles narrowly winged .... **A. tenuifolia**
         8. Leaves filiform to narrowly linear; corolla internally pubescent across the base of the upper lobes or at the lobe sinus; upper corolla lobes reflexed; upper and lower corolla throat walls subequal; pedicel conspicuously longer or shorter than the bracts; stem angles not winged.
            9. Pedicel longer than the subtending bracts, 1–4 cm long in flower, elongating in fruit.
               10. Pedicel glabrous or minutely puberulent at the base; fascicled leaves absent or few .... **A. setacea**
               10. Pedicel scabrous or minutely puberulent at the base; fascicled leaves abundant.
                   11. Leaves alternate, somewhat fleshy, involute; stem internodes scabridulous; plant densely leafy in appearance; calyx and capsule not glaucous; corolla lobes pubescent on the outer surface and often on the inner surface, the margin short-ciliate .... **A. filifolia**
                   11. Leaves opposite, not fleshy, revolute; stem internodes scabrous; plant not densely leafy in appearance; calyx and capsule glaucous; corolla lobes glabrous, the margin long-ciliate .... **A. pulchella**
            9. Pedicel shorter than the subtending bracts, 2–10 mm long in flower, little elongating in fruit.
               12. Leaves succulent when fresh, wrinkled when dried, flat, glabrous; calyx lobes, obtuse to somewhat acute; corolla glabrous internally across the base of the upper lobes, often pubescent at the sinus; inflorescence of the main stem mostly extending well beyond the lateral branches; plant of saline habitats .... **A. maritima**

12. Leaves not succulent when fresh, flat to somewhat folded, scabrous or glabrous; calyx lobes subulate to triangular, acute to acuminate; corolla pubescent internally across the base of the upper lobes; inflorescence of the main stem not conspicuously extended above the lateral branches; plant of various habitats, not restricted to saline soils.
    13. Plant bushy, profusely branched; stem faces glabrous or nearly so; leaves filiform, 0.5–1 mm wide, glabrous or with minute trichomes; axillary fascicles absent; bracts glabrous or nearly so, equaling or twice as long as the pedicel; sinuses of the calyx flat and nearly straight .... **A. plukenetii**
    13. Plant widely or sparingly and strictly branched; stem faces scabrid or scabridulous, at least around the nodes; leaves linear, sometimes widening distally, 0.5–5 mm wide, scabrous; axillary fascicles abundant to few; bracts scabrous, more than twice as long as the pedicel; sinuses of the calyx U- or V-shaped.
        14. Stem internodes scabridulous or glabrous; axillary fascicles absent or few and shorter than the subtending leaves.
            15. Plant widely and often much-branched; leaves linear, spreading, not fleshy unless in saline soils, 1–5 mm wide; pedicel 2–5 mm long in flower; calyx base and pedicel continuous, not separated by a transverse line; corolla (2)2.5–3.5 cm long; seeds brown to blackish, 0.9–1.4 mm long .... **A. purpurea**
            15. Plant with relatively few erect branches; leaves narrowly linear, ascending, somewhat fleshy, 0.5–1.5 mm wide; pedicel 1–3 mm long in flower; calyx base and pedicel separated by a transverse line or crease; corolla 1.8–2.5 cm long; seeds jet-black, 0.5–0.8 mm long .... **A. harperi**
        14. Stem internodes scabrous or scabridulous; axillary fascicles abundant.
            16. Plant dark green, to 2.5 m tall; stem scabrous; leaves and axillary fascicles not widened distally; calyx lobes scabrous on the midvein; corolla deep pink with red spots and yellow lines in the throat, 1.5–3 cm long; style and stigma exserted .... **A. fasciculata**
            16. Plant light green, to 6 dm tall; stem scabridulous; leaves and axillary fascicles widened distally; calyx lobes glabrous on the midvein; corolla pale pink without markings in the throat, 1.4–1.8 cm long; style and stigma not exserted .... **A. georgiana**

Agalinis aphylla (Nutt.) Raf. [From the Greek *a,* without, and *phyllon,* leaf, in reference to the stem lacking evident leaves.] SCALELEAF FALSE FOXGLOVE.

Gerardia aphylla Nuttall, Gen. N. Amer. Pl. 2: 47. 1818. *Agalinis aphylla* (Nuttall) Rafinesque, New Fl. 2: 65. 1837 ("1836"). TYPE: FLORIDA: s.d., *Baldwin s.n.* (holotype: PH).

Agalinis microphylla Rafinesque, New Fl. 2: 65. 1837 ("1836"). TYPE: FLORIDA: Duval Co.: near Jacksonville, 1 Oct 1894, *Curtiss 5272* (neotype: FSU; isoneotypes: GA, GH, MIN, PH). Neotypified by Canne-Hilliker and Hays 2010: 678).

Erect annual herb, to 1 m; stem 4-angled, striate, hispidulous on the ridges; axillary fascicles absent. Leaves opposite, subopposite or alternate distally, appressed, the blade triangular-subulate, 1–2 mm long, 0.2–0.8 mm wide, the apex acute, the base cuneate, sessile, the margin entire, the upper surface glabrous, the lower surface sparsely pubescent, those (1–6) at the base spreading, the blade ovate, 0.5–1 cm, 2–6 mm wide, the apex acute, the base cuneate, the upper and lower surfaces pubescent to glabrate. Flowers in an elongated racemiform inflorescence, the pedicel 1–2 mm long in flower, 1.5–3 mm long in fruit, ascending, clavate; bracts subulate, minute; calyx tube hemispheric, 1–2 mm long, the lobes minute, subulate, the inner surface of the tube distally and the lobes sparsely granular-puberulent; corolla pink, with 2 yellow lines in the throat, 1.5–2 cm long, the tube 1–1.4 cm long, slightly upcurved, the lobes 5–6 mm long, the apex rounded to emarginate, spreading, the outer surface sparsely pubescent, the inner surface pubescent below the sinus and over the base of the upper lobes. Fruit depressed globose to globose, 3–4 mm long, dark brown; seeds 3- to 4-angled, ca. 1 mm long, pale to dark brown, reticulate.

Bogs and mesic to wet pine savannas and prairies. Occasional; northern peninsula, central and western panhandle. North Carolina south to Florida, west to Louisiana. Summer–fall.

Agalinis divaricata (Chapm.) Pennell [*Di,* two, and *varicare,* straddle, in reference to the stem branching system.] PINELAND FALSE FOXGLOVE.

Gerardia divaricata Chapman, Fl. South. U.S. 299. 1860. *Agalinis divaricata* (Chapman) Pennell, Bull. Torrey Bot. Club 40: 437. 1913. TYPE: FLORIDA: Franklin Co.: Apalachicola, s.d., *Chapman s.n.* (lectotype: US). Lectotypified by Pennell (1929: 239.)

Gerardia mettaueri A. W. Wood, Class-Book Bot., ed. 1861. 530. 1861. TYPE: FLORIDA: 1855, *Mettauer s.n.* (holotype: NY).

Gerardia mettaueri A. W. Wood var. *clausa* A. W. Wood, Class-Book Bot., ed. 1861. 530. 1861. TYPE: FLORIDA.

Erect annual herb, to 8 dm; stem 4-angled, striate, minutely scabrous or glabrous; axillary fascicles scarcely developed or absent, if present, then much shorter than the subtending leaves. Leaves opposite, spreading or slightly reflexed, the blade filiform, 1.5–2.5 cm long, 0.2–0.5 mm wide, the apex acute, the base cuneate, the margin entire, the upper and lower surfaces minutely scabrous, sessile. Flowers 6–12 in an elongated racemiform inflorescence, (1)2(3) flowers per node, the pedicel 1–3 cm long in flower, 2–3 mm long in fruit, slightly clavate, glabrous; bracts filiform, shorter than the pedicel; calyx tube campanulate-globose, 1.5–2 mm long, truncate, the lobes 0.1–0.3 mm long, subulate, the inner surface of the tube distally and the lobes finely granular-puberulent; corolla purple, faintly red-purple spotted in the throat of the lower lobe, 1–1.5 cm long, the tube 1–1.2 cm long, straight, the upper lobes 2–3 mm long, slightly arched, the apex rounded to slightly retuse, the lower lobes 5–6 mm long, spreading, the inner and outer surfaces glabrous, the lobe margin ciliate. Fruit globose, 2–3 mm long, dark brown.

Dry pine-oak savannas, sandhills, and mesic bog margins. Common; northern and central peninsula, central and western panhandle. Georgia, Alabama, and Florida. Summer–fall.

Agalinis fasciculata (Elliott) Raf. [In clusters, in reference to the leaves.] BEACH FALSE FOXGLOVE.

Gerardia fasciculata Elliott, Sketch Bot. S. Carolina 2: 115. 1822. *Agalinis fasciculata* (Elliott) Rafinesque, New Fl. 2: 63. 1837 ("1836"). *Gerardia purpurea* Linnaeus var. *fasciculata* (Elliott) Chapman,

Fl. South. U.S. 300. 1860. *Gerardia fasciculata* Elliott subsp. *typica* Pennell, Acad. Nat. Sci. Philadelphia Monogr. 1: 441. 1935, nom. inadmiss.

Agalinis fasciculata (Elliott) Rafinesque var. *peninsularis* Pennell, Proc. Acad. Nat. Sci. Philadelphia 81: 178. 1929. *Gerardia fasciculata* Elliott subsp. *peninsularis* (Pennell) Pennell, Acad. Nat. Sci. Philadelphia Monogr. 1: 444. 1935. TYPE: FLORIDA: Miami-Dade Co.: between Homestead and Camp Jackson, 11 Apr 1904, *Small & Wilson 1579* (holotype: NY).

Erect annual herb, to 1 m; stem 4-angled, striate, scabrous; axillary fascicles abundant, subequaling the subtending leaves. Leaves opposite or subopposite distally, spreading, linear, 1.5–4 cm long, 0.7–2.5(3) cm wide, the apex acute, the base cuneate, the margin entire, the upper and lower surfaces scabrous, sessile. Flowers 12–30 in a racemiform inflorescence, the pedicel 2–5(8) mm long, ascending, clavate, scabrellous to glabrescent; bracts longer than the pedicel. Calyx tube hemispheric, (1)3–4 mm long, in flower, 3–6 mm long in fruit, subtruncate, the lobes triangular-lanceolate to subulate, 0.5–2 mm long, throat of the tube granular, the lobes finely puberulent on the inner surface; corolla pink with 2 yellow lines and many diffused red-purple spots within the throat, 2.5–3.5 cm long, the tube 1.8–2.5 cm long, slightly upcurved, the lobes 7–10 mm long, rounded to truncate, spreading, the outer surface minutely pubescent, the inner surface slightly pubescent in the throat and over the base of the upper lobes, the lobe margins ciliate. Fruit ovoid-globose, 5–6 mm long, brown; seeds 0.5–0.8 mm long, irregularly lunate-triangular, black, reticulate.

Sandhills, coastal scrub, margin of tidal wetlands, and pinelands. Frequent; nearly throughout. New York south to Florida, west to Kansas, Oklahoma, and Texas; West Indies. Summer–fall or all year in southern peninsula.

Agalinis filicaulis (Benth.) Pennell [*Filum,* thread, and *caulis,* stem, in reference to the slender stems.] JACKSON FALSE FOXGLOVE.

Gerardia aphylla Nuttall var. *filicaulis* Bentham, in Hooker, Companion Bot. Mag. 1: 174. 1836. *Gerardia filicaulis* (Bentham) Chapman, Fl. South. U.S. 299. 1860. *Agalinis filicaulis* (Bentham) Pennell, Bull. Torrey Bot. Club 40: 438. 1913.

Gerardia mettaueri A. W. Wood var. *nuda* A. W. Wood, Class-Book Bot., ed. 1861. 530. 1861. *Gerardia nuda* (A. W. Wood) A. W. Wood, Amer. Bot. Fl. 231. 1870. TYPE: FLORIDA.

Erect annual herb, to 5 dm; stem terete, slightly striate, with laxly ascending branches, minutely scabrellous to glabrate, glaucous; axillary fascicles absent. Leaves opposite, appressed, the blade triangular-subulate, 1–2 mm long, 0.3–0.5 mm wide, the apex acuminate, the base cuneate, the margin entire, the upper surface glabrous, the lower surface minutely scabrous, sessile. Flowers few, 1(2) at a node, the pedicel 5–8 mm long in flower, 6–10 mm long in fruit, ascending, clavate, glabrous; bracts subulate, minute; calyx subglobose, the tube ca. 2 mm long, truncate, the lobes triangular-subulate, minute, the inner surface of the tube distally and the lobes minutely granular-puberulent; corolla lavender pink, 1–1.3 cm long, the tube 6–8 mm long, straight, the upper lobes 2–2.5 mm long, slightly rounded to truncate, the lower lobes 4–5 mm long, rounded to truncate, spreading, the inner and outer surfaces glabrous, lightly arched, the lobe margins ciliate. Fruit subglobose, ca. 3 mm long, dark brown; seeds irregularly lunate-triangular, ca. 0.5 mm long, brown, reticulate.

Bogs and moist to wet savannas and prairies. Occasional; northern and central peninsula, central and western panhandle. Georgia south to Florida, west to Louisiana. Fall.

Agalinis filifolia (Nutt.) Raf. [*Filum,* thread, and *folium* leaf, in reference to the very narrow leaves.] SEMINOLE FALSE FOXGLOVE.

Agalinis filifolia (Nuttall) Rafinesque, New Fl. 2: 65. 1837 ("1836"). TYPE: FLORIDA: s.d., *Baldwin s.n.* (holotype: BM?). FLORIDA: Bay Co.: C-388, 4.4 km E of the junction with FL 79, just N of West Bay, 30 Sep 1999, *Canne-Hilliker 3491* (neotype: OAC; isoneotypes: MO, NY). Neotypified by Canne-Hilliker and Hays (2010: 680).

Erect annual herb, to 8 dm; stem terete, striate-angled distally, ascending, glabrous or glabrate; axillary fascicles abundant, subequaling the subtending leaves. Leaves subopposite or alternate, narrowly linear or filiform, 1–2 cm long, 0.3–0.9 mm wide, the apex acute, the base cuneate, the margin entire, the upper surface sparingly scabrous, the lower surface glabrous, sessile. Flowers 2–12 in a racemiform inflorescence, the pedicel 1–3 cm long in flower, 2–3.5 mm long in fruit, ascending spreading; bracts filiform, to 1 cm long; calyx tube campanulate-hemispheric, ca. 4 mm long, the lobes linear-subulate, 0.5–1 mm long, the inner surface of the tube distally and the lobes sparsely granular-puberulent to glabrate; corolla rose-purple, with 2 yellow lines and diffused purple spots within the throat on the lower surface, 2.5–3.5 cm long, the tube 1.5–2 cm long, slightly upcurved, the lobes 7–9 mm long, rounded to truncate, spreading, the outer surface minutely pubescent, the inner surface pubescent below the sinus and over the base of the upper lobes, the lobe margins ciliate. Fruit ovoid-globose, 4–5 mm long, brown; seeds irregularly tetrahedral, ca. 0.5 mm long, black, reticulate.

Sandhills and coastal scrub. Common; nearly throughout. Georgia, Alabama, and Florida. Summer–fall.

Agalinis flexicaulis Hays [*Flexilis,* flexible, and *caulis,* in reference to the sprawling, decumbent branches of the secondary and tertiary branches of the larger specimens.] HAMPTON FALSE FOXGLOVE; SPRAWLING FALSE FOXGLOVE.

Agalinis flexicaulis Hays, J. Bot. Res. Inst. Texas 4: 1, f. 1. 2010. TYPE: FLORIDA: Bradford Co.: E side of FL 100, 2.7 mi. SSE of Starke, ca. 14 Oct 2009, *Hays 3452* (holotype: NY; isotypes: BRIT, MO, OAC).

Erect annual herb, to 9 dm; stem with the secondary and tertiary branches sprawling-decumbent, 4-angled distally, the angles glabrous to minutely scabridulous; axillary fascicles absent. Leaves opposite, subopposite, or alternate, the blade linear to spatulate, 7–13 mm long, 0.8–1.5 mm wide, the apex obtuse to acute, the base cuneate, the margin entire, the upper surface scabrous, the lower surface glabrous, sessile. Flowers 3–8 in a racemiform inflorescence, solitary at the nodes, the pedicel 4–12 mm long in flower, to 2 cm long in fruit, clavate, spreading; bracts linear-spatulate, 0.5–4 mm long; calyx tube campanulate, 2.2–2.7 mm long, the apex truncate, the lobes deltoid to subulate, 0.2–0.5 mm long, the inner surface of the tube distally and the lobes puberulent; corolla rose-purple, the inner surface of the lower lobe sometimes with a few rose-purple spots, 1–1.2 mm long, the tube 4–6 mm long, the outer surface and the throat pubescent, the lobes 3–4 mm long, truncate, emarginate, sometimes erose-cordate, the margin ciliate, the upper lobes reflexed, the lower lobes spreading. Fruit obovoid, 4–6 mm long, golden-brown; seeds triangular to quadrate, ca. 0.5 mm long, reticulate, golden-brown.

Pine flatwoods. Rare; Bradford County. Endemic. Fall.

Agalinis georgiana (C. L. Boynton) Pennell [Of Georgia.] BOYNTON'S FALSE FOXGLOVE.

Gerardia georgiana C. L. Boynton, Biltmore Bot. Stud. 1: 148. 1902. *Agalinis georgiana* (C. L. Boynton) Pennell, Bull. Torrey Bot. Club 40: 427. 1913.

Erect annual herb, to 8 dm; stem 4-angled, striate, sparsely scabrellous or glabrous; axillary fascicles abundant, subequaling the subtending leaves. Leaves opposite or subopposite distally, spreading, the blade linear 1–1.5 cm long, 0.5–1 mm wide, the apex acute, the base cuneate, the margin entire, the upper surface sparsely scabrous, the lower surface glabrous, sessile. Flowers 10–20 in a racemiform inflorescence, the pedicel ascending, 0.5–2 mm long in flower, 2–4 mm long in fruit, glabrous; bracts filiform, 5–10 mm long, exceeding the pedicel; calyx hemispheric, subtruncate, the tube 2–3 mm long, the lobes triangular-lanceolate, 0.5–1 mm long, the inner surface of the tube distally and the lobes sparsely granular-puberulent; corolla lavender-pink. 1.5–1.8 cm long, the tube 1–1.2 cm long, slightly upcurved, the lobes 5–6 mm long, the apex rounded to retuse, spreading, the outer surface minutely pubescent, the inner surface pubescent below the sinus and over the width of the base of the upper lobes, the lobe margins ciliate. Fruit ovoid-globose, 3.5–5 mm long, brown; seeds lunate-triangular to quadrangular, ca. 0.5 mm long, brown, reticulate.

Pine savannas, pine-oak woods, and bog margins. Occasional; central and western panhandle. Georgia, Alabama, and Florida. Summer–fall.

Agalinis georgiana is listed as endangered in Florida (Florida Administrative Code, Chapter 5B-40).

Agalinis harperi Pennell [Commemorates Roland McMillan Harper (1878–1966), staff botanist for the Geological Survey of Alabama.] HARPERS FALSE FOXGLOVE.

Agalinis harperi Pennell, in Small, Fl. Miami 167, 200. 1913. *Gerardia harperi* (Pennell) Pennell, Acad. Nat. Sci. Philadelphia Monogr. 1: 441. 1935. TYPE: FLORIDA: Wakulla Co.: St. Marks, 25 Sep 1912, *Pennell 4707* (holotype: PH).

Agalinis delicatula Pennell, Bull. Torrey Bot. Club 40: 425. 1913. *Agalinis pinetorum* Pennell var. *delicatula* (Pennell) Pennell, Proc. Acad. Nat. Sci. Philadelphia 81: 174. 1929. *Gerardia pulchella* Pennell var. *delicatula* (Pennell) Pennell, Acad. Nat. Sci. Philadelphia Monogr. 1: 441. 1935. TYPE: FLORIDA: Holmes Co.: Ponce de Leon, 17 Sep 1912, *Pennell 4661* (holotype: PH).

Agalinis pinetorum Pennell, Bull. Torrey Bot. Club 40: 424. 1913. *Gerardia pinetorum* Pennell, Acad. Nat. Sci. Philadelphia Monogr. 1: 440. 1935. TYPE: FLORIDA: Wakulla Co.: St. Marks, 26 Sep 1916, *Pennell 4708* (holotype: PH).

Erect annual herb, to 8 dm; stem 4-angled, striate, minutely scabrous or glabrous; axillary fascicles absent or present and much shorter than the subtending leaves. Leaves opposite or sometime subopposite distally, spreading, the blade linear or filiform, 1–3.5 cm long, 0.5–1 mm wide, the apex acute, the base cuneate, the margin entire, the upper surface scabrous, the lower surface glabrous, sessile. Flowers 8–20 in a racemiform inflorescence, the pedicel ascending, clavate, 0.5–2 mm long in flower, 2–3 mm long in fruit; calyx tube 1.5–3 mm long, hemispheric, the apex subtruncate, the lobes triangular-ovate to subulate, 1–2 mm long, the inner surface of the tube distally and the inner surface of the lobes sparsely granular-puberulent; corolla pink, with 2 yellow lines and sometimes with diffuse purple spots in the throat on the lower surface, 1.5–2.5 cm long, the tube 1–1.4 cm long, straight or slightly upcurved, the lobes 5–8 mm long, slightly rounded to truncate or emarginate, spreading, the outer surface pubescent, the inner surface pubescent below the sinus and over the base of the upper lobes, the lobe margins ciliate. Fruit subglobose, 4–5 mm long, dark brown; seeds irregularly lunate-triangular, ca. 1 mm long, reticulate.

Bogs, mesic to wet savannas and prairies, and interdunal swales. Occasional; nearly throughout. South Carolina south to Florida, west to Louisiana; West Indies. Summer–fall.

Agalinis laxa Pennell [Loose, in reference to the branches in a lax and open arrangement.] TWOLINE FALSE FOXGLOVE.

Agalinis laxa Pennell, Bull. Torrey Bot. Club 40: 431. 1913. *Gerardia laxa* (Pennell) Pennell, Acad. Nat. Sci. Philadelphia Monogr. 1: 449. 1935.

Erect annual herb, to 1 m; stem 4-angled and striate distally, glabrous; axillary fascicles absent or if present, then shorter than the subtending leaves. Leaves opposite or subopposite distally, spreading, the blade linear to filiform, 2–3 cm long, 0.5–0.9 mm wide, the apex acute, the base cuneate, the margin entire, the upper surface scarious or glabrous, the lower surface glabrous, sessile. Flowers 3–8 in a racemiform inflorescence, 1 or 2 per node, the pedicel spreading or slightly ascending, 1.5–3 cm long in flower, 2.5–5 cm long in fruit, glabrous; bracts filiform, to 5 mm long; calyx tube urceolate-hemispheric, 2.5–3 mm long, the apex truncate, the lobes subulate, 0.2–0.5 mm long, the inner surface of the tube distally and the lobes sparsely glandular-puberulent; corolla pink with 2 yellow lines and small red-purple spots in the throat of the lower lobe, 1.5–1.8 cm long, the tube 1.1–1.3 cm long, slightly upcurved, the lobes 4–5 mm long, rounded, spreading, the outer surface minutely pubescent, the inner surface pubescent below the sinus and over the width of the base of the upper lobes, the lobe margins ciliate. Fruit ovoid-globose, 4–5 mm long, brown; seeds irregularly narrowly triangular to quadrate-lunate, 0.3–0.5 mm long, brown to black, reticulate.

Pinelands and sandhills. Rare; Duval, Madison, and Hernando Counties. South Carolina, Georgia, and Florida. Summer–fall.

Agalinis linifolia (Nutt.) Britton [With linear leaves.] FLAXLEAF FALSE FOXGLOVE.

Gerardia linifolia Nuttall, Gen. N. Amer. Pl. 47. 1818. *Agalinis perennis* Rafinesque, New Fl. 2: 73. 1837 ("1836"), nom. illegit. *Agalinis linifolia* (Nuttall) Britton, in Britton & A. Brown, Ill. Fl. N. U.S., ed. 2. 3: 209. 1913.

Erect perennial herb, to 1.5 m; stem terete, glabrous; axillary fascicles absent. Leaves opposite, erect, the blade linear, 3–5 cm long, 1–3 mm wide, the apex acuminate, the base cuneate, the margin entire, the upper and lower surfaces glabrous, sessile. Flowers 8–20 in an elongate racemiform inflorescence, 1 or 2 flowers per node, the pedicel erect, 0.5–2 cm long in flower, 1–2.5 cm long in fruit, clavate, glabrous; bracts 3–10 mm long, glabrous; calyx tube subhemispheric, 4–4.5 mm long, the apex truncate, the lobes subulate, minute, the inner surface of the tube distally and the lobes granular-puberulent; corolla pink with diffuse red-purple spots in the throat of the lower lobe, the inner and outer surfaces glabrous. Fruit globose to globose-ovoid, 6–8 mm long, dark brown to black; seeds triangular to rectangular, 1–1.5 mm long, pale brown, reticulate.

Bogs, pond, stream, and river margins, pineland depressions, and cypress dome and strand margins. Delaware and Maryland south to Florida, west to Louisiana; West Indies. Frequent; nearly throughout. Summer–fall.

Agalinis maritima (Raf.) Raf. var. **grandiflora** (Benth.) Pennell [Growing by the sea; large-flowered.] SALTMARSH FALSE FOXGLOVE.

Gerardia maritima Rafinesque var. *grandiflora* Bentham, in Hooker, Companion Bot. Mag. 1: 208. 1836. *Gerardia spiciflora* Engelmann, Boston J. Nat. Hist. 5: 227. 1845. *Agalinis spiciflora* (Engelmann) Pennell, Proc. Acad. Nat. Sci. Philadelphia 71: 277. 1920. *Agalinis maritima* (Rafinesque) Rafinesque var. *grandiflora* (Bentham) Pennell, Proc. Acad. Nat. Sci. Philadelphia 81: 154. 1929. *Gerardia maritima* Rafinesque subsp. *grandiflora* (Bentham) Pennell, Acad. Nat. Sci. Philadelphia Monogr. 1: 432. 1935.

Gerardia maritima Rafinesque var. *major* Chapman, Fl. South. U.S. 300. 1860. TYPE: FLORIDA: Franklin Co.: Apalachicola, s.d. *Chapman s.n.* (lectotype: NY; isolectotypes: GH, US). Lectotypified by Pennell (1935: 433).

Erect annual herb, to 6 dm; stem slightly 4-angled to subterete, glabrous; axillary fascicles absent. Leaves opposite or occasionally subopposite distally, the blade linear, 2–4 cm long, 1–3 mm wide, the apex acute, the base cuneate, the margin entire, the upper surface slightly scabrous-roughened, the lower surface glabrous, sessile. Flowers 4–10 in an elongate racemiform inflorescence, the pedicel 2–10 mm long in flower, 5–12 mm long in fruit; bracts shorter than the pedicels, rarely slightly longer; calyx tube hemispheric-campanulate, 3–4 mm long, the apex subtruncate, the lobes triangular, 0.5–1.5 mm long, the apex obtuse to acute, the inner surface of the tube and the lobes slightly granular-puberulent; corolla pink with 2 yellow lines and small purple-red spots in the throat on the lower lip, 1.5–2 cm long, straight or slightly upcurved, the lobes 4–5 mm long, rounded to slightly emarginated, spreading, the outer surface pubescent distally, the inner surface pubescent below the sinus and over most of the base of the upper lobes, the upper and lower lobe margins ciliate. Fruit globose to globose ovoid, 5–6 mm long, dark brown; seeds irregularly oblong-angulate, 0.8–1 mm long, reticulate, dark brown.

Brackish to saline marshes and tidal wetlands. Frequent; nearly throughout in coastal counties. Virginia south to Florida, west to Texas; West Indies and Mexico. Summer–fall.

Agalinis obtusifolia Raf. [With the leaf apex blunt.] TENLOBE FALSE FOXGLOVE.

Agalinis obtusifolia Rafinesque, New Fl. 2: 64. 1837 ("1836"). *Gerardia obtusifolia* (Rafinesque) Pennell, Acad. Nat. Sci. Philadelphia Monogr. 1: 471. 1935. TYPE: FLORIDA: Santa Rosa Co.: along Fl 4, 0.7 km W of the Okaloosa County line, E of Munson, 25 Sep 1999, *Canne-Hilliker & Hays 3427* (neotype: OAC; isoneotypes: GH, MO, NY, PH). Neotypified by Canne-Hilliker & Hays (20: 678).
Agalinis setacea J. F. Gmelin var. *parvifolia* Bentham, in Hooker, Companion Bot. Mag. 1: 174. 1836. *Gerardia parvifolia* (Bentham) Chapman, Fl. South. U.S. 300. 1860. *Agalinis parvifolia* (Bentham) Small ex Britton, in Britton & A. Brown, Ill. Fl. N. U.S., ed. 2. 3: 212. 1913.
Agalinis tenella Pennell, Bull. Torrey Bot. Club 40: 434. 1913. *Gerardia tenella* (Pennell) Pennell, Acad. Nat. Sci. Philadelphia Monogr. 1: 471. 1935.

Erect annual herb, to 8 dm; stem 4-angled, striate, minutely scabrellous on the ridges or glabrate; axillary fascicles absent. Leaves opposite or subopposite distally, spreading to ascending, the blade linear or linear-spatulate, 1–1.5 cm long, 0.5–1.5 mm wide, the apex acute to obtuse, the base cuneate, the margin entire, the upper surface scabrous, the lower surface glabrous, sessile. Flowers 6–14 in a racemiform inflorescence, the pedicel 5–15 mm long in flower, 8–25 mm long in fruit, ascending-spreading; bracts to 5 mm long; calyx tube hemispheric, ca. 2 mm long, the apex truncate, the lobes subulate, minute, the inner surface of the tube distally and the lobes minutely granular-puberulent; corolla pink, sometimes with 2 faint yellow lines and red-purple spots in the throat on the lower lip, 1–1.5 cm long, the tube 9–12 mm long, straight, the lobes 3–4 mm long, round to retuse, spreading, the outer surface sparsely pubescent, especially distally, the inner surface pubescent below the sinus and over the base of the upper lobes, the lobe margins ciliate. Fruit globose-ovoid to -ellipsoid, 3–4 mm long, pale brown; seeds irregularly lunate-triangular to -oblong 0.7–0.9 mm long, yellow-brown, reticulate.

Pine savannas, flatwoods, and bog margins. Frequent; peninsula, central and western panhandle; Pennsylvania south to Florida, west to Kentucky, Tennessee, Mississippi, and Louisiana. Summer–fall.

Agalinis plukenetii (Elliott) Raf. [Commemorates Leonard Plukenet (1641–1706), English botanist, Royal Professor of Botany, and gardener to Queen Mary.] PLUKENET'S FALSE FOXGLOVE.

Gerardia plukenetii Elliott, Sketch Bot. S. Carolina 2: 114. 1822. *Agalinis plukenetii* (Elliott) Rafinesque, New Fl. 2: 63. 1837 ("1836").
Agalinis keyensis Pennell, Proc. Acad. Nat. Sci. Philadelphia 71: 282. 1920. *Gerardia keyensis* (Pennell) Pennell, Acad. Nat. Sci. Philadelphia Monogr. 1: 453. 1935. TYPE: FLORIDA: Monroe Co.: Big Pine Key, s.d., *Blodgett s.n.* (holotype: GH; isotype: NY).

Erect annual herb, to 8 dm; stem 4-angled, striate, glabrous or sparsely scabrellous; axillary fascicles absent or if present, then shorter than the subtending leaves. Leaves opposite or subopposite distally, filiform, 2–3.5 cm long, 0.2–0.8 cm wide, the apex acuminate, the base cuneate, the margin entire, the upper surface sparsely scabrous, the lower surface glabrous, sessile. Flowers 1–6 in a racemiform inflorescence, the pedicel 5–10 mm long in flower, 8–15 mm long in fruit, ascending, clavate; bracts subequaling or slightly longer than the pedicel; calyx tube hemispheric, 3–4 mm long, the apex truncate, the lobes triangular-subulate, 0.3–0.5 mm long, the inner surface distally and the lobes granular-puberulent; corolla rose-pink, the inner surface 2 yellow lines and diffuse red-purple spots in the throat of the lower lobe, 2–3 cm long, the upper lobes reflexed, the lower lobes spreading, the inner surface pubescent across the base of the upper lobes. Fruit globose, 4–5 mm long, brown; seeds irregularly triangular-quadrate, dark brown, reticulate, 0.6–0.8 mm long.

Savannas. Frequent; northern counties, central peninsula. Tennessee and South Carolina south to Florida, west to Louisiana. Fall.

Agalinis pulchella Pennell [Beautiful and little.] CHATTAHOOCHEE FALSE FOXGLOVE.

Agalinis pulchella Pennell, Bull. Torrey Bot. Club 40: 428. 1913. *Gerardia pulcherrima* Pennell, Acad. Nat. Sci. Philadelphia Monogr. 1: 448. 1935.

Erect annual herb, to 1 m; stem 4-angled, striate, scabrous; the axillary fascicles abundant, subequaling the subtending leaves. Leaves opposite or subopposite distally, spreading, the blade linear, 2–3 cm long, 0.4–0.9 mm wide, the apex acute, the base cuneate, the margin entire, the upper surface scabrous, the lower surface glabrous, sessile. Flowers 4–6 in a racemiform inflorescence, the pedicel ascending-spreading, 1.5–3 cm long in flower, 2.5–4 cm long in fruit; bracts 8–12 mm long, the upper surface scabrous; calyx tube hemispheric, 3–4 mm long, the apex truncate, the lobes subulate, to 0.5 mm long, the inner surface of the tube sparsely puberulent distally; corolla rose-purple with 2 yellow lines and relatively large red-purple spots in the throat of the lower lobe. Fruit subglobose, 5–6 mm long, brown; seeds irregularly broadly triangular to quadrangular, 0.6–0.7 mm long, pale to dark brown, reticulate.

Pine savannas and sandhills. Occasional; central and western panhandle. Georgia south to Florida, west to Texas. Fall.

Agalinis purpurea (L.) Pennell [Purple, in reference to the corolla color.] PURPLE FALSE FOXGLOVE.

Gerardia purpurea Linnaeus, Sp. Pl. 610. 1753. *Agalinis palustris* Rafinesque, New Fl. 2: 62. 1837 ("1836"), nom. illegit. *Agalinis purpurea* (Linnaeus) Pennell, Bull. Torrey Bot. Club: 126. 1913. *Aureolaria purpurea* (Linnaeus) Farwell, Rep. (Annual) Michigan Acad. Sci. 20: 189. 1918.

Gerardia erecta J. F. Gmelin, Syst. Nat. 2: 928. 1792. *Agalinis erecta* (J. F. Gmelin) Pennell, in Small, Fl. Florida Keys 133, 155. 1913.

Agalinis corymbosa Rafinesque, New Fl. 2: 63. 1837 ("1836"). TYPE: FLORIDA: Santa Rosa Co.: T1S, R28W, Sec 24, SE of SW, 10 Oct 1998, *Hays 2005* (neotype: NLU; isoneotype: FLAS). Neotypified by Canne-Hilliker and Hays (2010: 677).

Agalinis purpurea (Linnaeus) Pennell var. *carteri* Pennell, Proc. Acad. Nat. Sci. Philadelphia 81: 169. 1929. *Gerardia purpurea* Linnaeus var. *carteri* (Pennell) Pennell, Acad. Nat. Sci. Philadelphia Monogr. 1: 439. 1935. TYPE: FLORIDA: Miami-Dade Co.: between Long Prairie and Camp Longview, 31 Oct 1906, *Small & Carter* 2692 (holotype: NY).

Erect annual herb, to 1.2 m; stem 4-angled, striate, the angles winged, with short, broad-based trichomes on the angles or glabrous; axillary fascicles absent or weakly developed. Leaves opposite or occasionally subopposite distally, the blade linear to linear-subulate, 1.5–4 cm long, ca. 1 mm wide, the apex acute, the base cuneate, the margin entire, sometimes revolute, the upper and lower surfaces short-scabrid with broad-based trichomes, sessile. Flowers 6–14 in a racemiform inflorescence, 1 or 2 at a node, the pedicel 2–6 mm long in flower, 3–6 mm long in fruit, clavate; bracts linear, 6–8 mm long; calyx hemispheric, 3–4 mm long, the lobes ca. 1 mm long, triangular-subulate, the inner surface of the tube distally and the lobes slightly granular-puberulent; corolla pink to purple, with 2 yellow lines and purple spots in the throat of the lower lobe, 2–3.8 mm long, the tube 1.5–2 mm long, slightly upward curved, the lobes 5–9 mm long, rounded to truncate, spreading, the outer surface pubescent, the inner surface pubescent below the sinus and over the base of the lower lobes. Fruit globose or globose-ovoid, 5–7 mm long, dark brown; seeds irregularly lunate-triangular, 1–1.5 mm long, brown, reticulate.

Bogs, moist to wet savannas, and interdunal swales. Frequent; nearly throughout. Quebec south to Florida, west to Minnesota, Nebraska, Kansas, and Texas. Summer–fall.

Agalinis setacea (J. F. Gmel.) Raf. [Bristlelike, in reference to the leaves.] THREADLEAF FALSE FOXGLOVE.

Gerardia setacea J. F. Gmelin, Syst. Nat. 2: 928. 1792. *Agalinis setacea* (J. F. Gmelin) Rafinesque, New Fl. 2: 64. 1837 ("1836").

Gerardia aphylla Nuttall var. *grandiflora* Bentham, in Hooker, Companion Bot. Mag. 1: 174. 1836. *Gerardia plukenetii* Elliott var. *microphylla* A. Gray, Syn. Fl. N. Amer. 2(1): 293. 1878, nom. illegit. *Gerardia microphylla* Small, Fl. S.E. U.S. 1077, 1338. 1903. *Agalinis oligophylla* Pennell, Bull. Torrey Bot. Club 40: 432. 1913.

Gerardia holmiana Greene, Pittonia 4: 56, pl. 10. 1899. *Agalinis holmiana* (Greene) Pennell, Bull. Torrey Bot. Club 40: 429. 1913.

Agalinis stenophylla Pennell, Proc. Acad. Nat. Sci. Philadelphia 71: 281. 1920. *Gerardia stenophylla* (Pennell) Pennell, Acad. Nat. Sci. Philadelphia Monogr. 1: 452. 1935. TYPE: FLORIDA: Hillsborough Co.: Tampa, Oct 1877, *Garber 281* (holotype: PH; isotype: F).

Erect annual herb, to 8 dm; stem 4-angled, striate, glabrous or sparsely scabrellous; axillary fascicles absent or only slightly developed, shorter than the subtending leaves. Leaves opposite or subopposite distally, spreading, the blade setaceous-filiform, 2–3.5 cm long, 0.2–0.8 mm wide, the apex acuminate, the base cuneate, the margin entire, the upper surface slightly scabrous, the lower surface glabrous, sessile. Flowers 1–6 in a racemiform inflorescence, 1(2) per node, the pedicel ascending, clavate, (2)5–10 mm long in flower, 5–10(15) mm long in fruit; bracts to 1 cm long; calyx tube hemispheric, the apex truncate, the lobes triangular-subulate to subulate, 0.3–0.5 mm long, 3–4 mm long, the inner surface of the tube distally and the lobes sparsely granular-puberulent; corolla rose-pink with 2 yellow lines and numerous small diffused purple spots in the throat of the lower lobe, the outer and inner surfaces glabrous. Fruit globose, 4–5 mm long, brown; seeds irregularly triangular to quadrangular, 0.6–0.8 mm long, dark brown to black, reticulate.

Sandhills and scrub. Occasional; northern and central peninsula, central panhandle. New York south to Florida, west to Arkansas. Summer–fall.

Agalinis tenuifolia (Vahl) Raf. [*Tenuis,* slender, and *folius* leaved.] SLENDERLEAF FALSE FOXGLOVE.

Gerardia tenuifolia Vahl, Symb. Bot. 3: 79. 1804. *Agalinis tenuifolia* (Vahl) Rafinesque, New Fl. 2: 64. 1837 ("1836"). *Aureolaria tenuifolia* (Vahl) Farwell, Rep. (Annual) Michigan Acad. Sci. 20: 189. 1918. *Gerardia tenuifolia* Vahl subsp. *typica* Pennell, Acad. Nat. Sci. Philadelphia Monogr. 1: 460. 1935, nom. inadmiss.

Gerardia leucanthera Rafinesque, Fl. Ludov. 50: 1817. *Agalinis tenuifolia* (Vahl) Rafinesque var. *leucanthera* (Rafinesque) Pennell, Proc. Acad. Nat. Sci. Philadelphia 71: 286. 1920. *Gerardia tenuifolia* Vahl subsp. *leucanthera* (Rafinesque) Pennell, Acad. Nat. Sci. Philadelphia Monogr. 1: 461. 1935. *Gerardia tenuifolia* Vahl var. *leucanthera* (Rafinesque) Shinners, Field & Lab. 18: 130. 1950.

Gerardia tenuifolia Vahl var. *filiformis* Chapman, Fl. South. U.S. 300. 1860; non Bentham, 1820. TYPE: "Florida to Mississippi, and northward."

Erect annual herb, to 1 m; stem 4-angled, striate, glabrate; axillary fascicles absent or only slightly developed and shorter than the subtending leaves. Leaves opposite or subopposite distally, spreading, the blade linear, 2–5 cm long, 1–3.5 mm wide, the margin acuminate, the base cuneate, the margin entire, the upper surface slightly scabrous, the lower surface glabrous, sessile. Flowers 6–23 in a racemiform inflorescence, flowers solitary at a node, the pedicel 1–2 cm long, ascending-spreading, slightly clavate; bracts shorter than the leaves; calyx tube hemispheric, 2.5–3 mm long in flower and fruit, the apex truncate, the lobes triangular-subulate, 0.2–1 mm long, the inner surface of the tube distally and the lobes slightly granular-puberulent; corolla purple with 2 yellow lines and small diffused red-purple spots in the throat of the lower lobe, 1.5–2.5 cm long, the tube 6–10 mm long, straight, the lobes 4–5 mm long, rounded, the upper broadly arched over the stamens and style, the lower spreading, the outer surface minutely pubescent, the inner surface pubescent near the base of the filaments, the lobe margins ciliate. Fruit globose, 5–7 mm long, brown; seeds irregularly oblong-ovoid, 0.6–0.9 mm long, brown, reticulate.

Moist to dry savannas and bluffs. Occasional; northern counties, central peninsula. Quebec south to Florida, west to Manitoba, North Dakota, South Dakota, Wyoming, Colorado, and New Mexico. Summer–fall.

EXCLUDED TAXA

Agalinis maritima (Rafinesque) Rafinesque—Because infraspecific categories were not recognized, the typical variety was reported for Florida by implication by Chapman (1860, 1883, 1897, all as *Gerardia maritima*), Small (1903, as *Gerardia maritima*), Radford et al. (1964, 1968), Correll and Johnston (1970), Long and Lakela (1971), Godfrey and Wooten (1981), Correll and Correll (1982), Clewell (1985), Wunderlin (1982, 1998), and Wunderlin and Hansen (2003, 2011). All Florida plants are var. *grandiflora*.

Agalinis skinneriana (A. W. Wood) Britton—This northern species was reported for Florida by Chapman (1897, as *Gerardia skinneriana* A. W. Wood) and Small (1903, as *Gerardia skinneriana* A. W. Wood), the name misapplied to Florida material of *A. obtusifolia*.

Aphyllon Mitch. 1769. BROOMRAPE

Annual holoparasitic herbs. Leaves alternate, simple, pinnate-veined, petiolate. Flowers solitary, bracteate, ebracteolate; sepals 4 or 5, basally connate, zygomorphic; petals 5, basally connate, zygomorphic; stamens 4, didynamous, epipetalous; staminode absent; ovary superior, 2-carpellate, 1-loculate, the style 1. Fruit a capsule, dehiscence loculicidal; seeds numerous.

A genus of about 22 species; North America, Mexico, South America, Europe, Africa, Asia, Australia, and Pacific Islands.

The genus *Aphyllon*, previously included in *Orobanche* by authors, was recently resurrected by Schneider (2016).

Selected reference: Schneider (2016).

Thelesia Raf. ex Britton, 1894.

Aphyllon uniflorum (L.) Torr. & A. Gray ex A. Gray [Flowers solitary.] ONEFLOWERED BROOMRAPE.

Orobanche uniflora Linnaeus, Sp. Pl. 633. 1753. *Phelypaea biflora* Sprengel, Syst. Veg. 2: 818. 1825, nom. illegit. *Aphyllon uniflorum* (Linnaeus) Torrey & A. Gray ex A. Gray, Manual 290. 1848. *Thalesia uniflora* (Linnaeus) Britton, Mem. Torrey Bot. Club 5: 298. 1894. *Orobanche uniflora* Linnaeus var. *typica* Achey, Bull. Torrey Bot. Club 60: 443. 1933, nom. inadmiss.

Erect annual herb, to 1.5 dm; stem unbranched or few-branched basally, yellow, glabrous. Leaves with the blade obovate to ovate, 5–10 mm long, the apex acuminate, the base clasping, the margin entire, the upper and lower surfaces glabrous, sessile. Flowers solitary, the pedicel 3–10(15) cm long, glandular-pubescent; bracts like the leaves but smaller; calyx yellow, 4–12 mm long, the lobes lanceolate-ovate, subequaling the tube; corolla white to pale yellow or purplish-tinged with light purple veins, 1.5–2.5 mm long, the lobes elliptic to obovate, 3–6(9) mm long. Fruit ovoid, ca. 1 cm long; seeds minute, brown.

Mesic hammocks; parasitic on the roots of various herbaceous and woody species. Rare; northern and central peninsula, central panhandle. Nearly throughout North America; Mexico; Europe. Native to North America and Mexico. Spring.

Aureolaria Raf. 1837. YELLOW FALSE FOXGLOVE

Annual or perennial hemiparasitic herbs. Leaves opposite, simple, pinnate-veined, petiolate. Flowers in racemes, bracteate, ebracteolate; sepals 5, basally connate, actinomorphic; petals 5, basally connate, zygomorphic; stamens 4, didynamous, epipetalous; staminode absent; ovary superior, 2-carpellate and -loculate, the style 1. Fruit a capsule, dehiscence loculicidal; seeds numerous.

A genus of 8 species; North America and Mexico. [*Aureolus*, golden, in reference to corolla color.]

Selected reference: Pennell (1928).

1. Plants glabrous .... **A. flava**
1. Plants pubescent.
   2. Pubescence of single, eglandular trichomes; lower leaves with 1–2 pairs of large lobes below the middle .... **A. virginica**
   2. Pubescence, at least on the upper part of the stem, of stipitate-glandular trichomes; lower leaves pinnatifid with 5–8 pairs of pinnae .... **A. pectinata**

Aureolaria flava (L.) Farw. [*Flavus,* yellow, in reference to the flower color.] SMOOTH YELLOW FALSE FOXGLOVE.

Gerardia flava Linnaeus, Sp. Pl. 610. 1753. *Aureolaria villosa* Rafinesque, New Fl. 2: 59. 1837 ("1836"), nom. illegit. *Dasistoma flavum* (Linnaeus) A. W. Wood, Amer. Bot. Fl. 230. 1870. *Aureolaria flava* (Linnaeus) Farwell, Rep. (Annual) Michigan Acad. Sci. 20: 188. 1918. *Agalinis flava* (Linnaeus) B. Boivin, Naturaliste Canad. 94: 644. 1967.

Aureolaria reticulata Rafinesque, New Fl. 2: 59. 1837 ("1836"). *Aureolaria flava* (Linnaeus) Farwell var. *reticulata* (Rafinesque) Pennell, Proc. Acad. Nat. Sci. Philadelphia 71: 272. 1920. *Aureolaria flava* (Linnaeus) Farwell subsp. *reticulata* (Rafinesque) Pennell, Acad. Nat. Sci. Philadelphia Monogr. 1: 395. 1935. *Gerardia flava* Linnaeus var. *reticulata* (Rafinesque) Cory, Rhodora 38: 407. 1936. TYPE: "Florida and Alabama."

Dasistoma bignoniiflorum Small, Bull. New York Bot. Gard. 1: 285. 1899. TYPE: FLORIDA: Hillsborough Co.: Tampa Bay, 1834, *Burrows s.n.* (holotype: NY).

Erect perennial herb, to 2 m; stem glabrous. Leaves with the blade lanceolate, 5–15 cm long, 1.2–6 cm wide, the apex acute, the base cuneate, the margin shallowly to deeply pinnatifid, the basal sometimes 2-pinnatifid, rarely entire, the upper and lower surfaces glabrous, the petiole 0.5–2.5 cm long. Flowers in a terminal raceme, the pedicel 0.5–1 cm long; bracts linear or slightly pinnatifid, leaflike, 1–4.5 cm long, 2–5 mm wide; calyx campanulate, the tube 4–8 mm long, the lobes linear to narrowly deltate, 4–7 mm long, glabrous; corolla yellow, campanulate, the tube 2.5–3.5 cm long, glabrous, the lobes 5–9 mm long, 6–16 mm wide, the margin glabrous or ciliate. Fruit ovoid to pyriform, ca. 1 cm long, glabrous; seeds winged.

Dry hammocks. Occasional; northern and central counties. Maine south to Florida, west to Ontario, Michigan, Illinois, Missouri, Arkansas, and Texas. Summer–fall.

Aureolaria pectinata (Nutt.) Pennell [With narrow close-set divisions like a comb.] FERNLEAF YELLOW FALSE FOXGLOVE.

Gerardia pedicularia Linnaeus var. *pectinata* Nuttall, Gen. N. Amer. Pl. 2: 48. 1818. *Gerardia pectinata* (Nuttall) Torrey ex Bentham, in Hooker, Companion Bot. Mag. 1: 206. 1836. *Dasistoma pectinatum* (Nuttall) Bentham, in de Candolle, Prodr. 10: 521. 1846. *Aureolaria pectinata* (Nuttall) Pennell, Bull. Torrey Bot. Club 40: 414. 1913. *Agalinis pedicularia* (Linnaeus) S. F. Blake var. *pectinata* (Nuttall) S. F. Blake, Rhodora 20: 70. 1918. *Aureolaria pectinata* (Nuttall) Pennell subsp. *typica* Pennell, Acad. Nat. Sci. Philadelphia Monogr. 1: 403. 1935, nom. inadmiss.

Aureolaria pectinata (Nuttall) Pennell subsp. *floridana* Pennell, Bull. Torrey Bot. Club 40: 414. 1913. TYPE: FLORIDA: Franklin Co.: Fort Gadsden, 20 Sep 1912, *Pennell 4683* (holotype: PH).

Erect annual herb, to 1 m; stem villous and stipitate-glandular. Leaves with the blade lanceolate, (0.5)1–5 cm long, 0.5–2.5 cm wide, the apex acute or obtuse, the base cuneate, the margin pinnatifid with 5–8 pairs of pinnae, these usually pinnate-margined, the upper and lower surfaces villous and stipitate-glandular, the petiole 1–8 mm long. Flowers in a terminal raceme, the pedicel 0.5–2 cm long, villous and stipitate-glandular; bracts leaflike, 1–1.5 cm long, 0.5–1 cm wide; calyx campanulate, the tube 3–8 mm long, stipitate-glandular, the lobes linear, pinnatifid, 0.7–1.5 cm long; corolla campanulate, the tube 2.5–4 cm long, gibbous, the outer surface floccose, the lobes 0.7–1 cm long, 0.6–1.5 cm wide. Fruit pyriform, 1–1.5 cm long, hispid; seeds winged.

Sandhills and scrub. Occasional; nearly throughout. Virginia south to Florida, west to Missouri, Oklahoma, and Texas. Summer–fall.

Aureolaria virginica (L.) Pennell [Of Virginia.] DOWNY YELLOW FALSE FOXGLOVE.

Rhinanthus virginicus Linnaeus, Sp. Pl. 603. 1753. *Gerardia quercifolia* Pursh, Fl. Amer. Sept. 423. 1814, nom. illegit. *Dasistoma quercifolium* Bentham, in de Candolle, Prodr. 10: 520. 1846, nom. illegit. *Aureolaria virginica* (Linnaeus) Pennell, Bull. Torrey Bot. Club 40: 409. 1913. *Agalinis virginica* (Linnaeus) S. F. Blake, Rhodora 20: 71. 1918.

Aureolaria microcarpa Pennell, Proc. Acad. Nat. Sci. Philadelphia 71: 270. 1920.

Erect perennial herb, to 1 m; stem brown-pubescent. Leaves with the blade lanceolate to oblong, the apex acute, the base cuneate to subtruncate, the margin serrate to pinnatifid with 1–2 pairs of lobes below the middle, the upper and lower surfaces pubescent, the petiole 2–5 mm long. Flowers in a terminal raceme, the pedicel 3–5 mm long, pubescent; bracts leaflike, 1–3 cm long, 0.5–1.5 cm wide, the margin entire to crenate; calyx campanulate, the tube 5–6 mm long, the lobes narrowly deltate, 0.5–1 cm long, pubescent; corolla campanulate, the tube 2.5–3.5 cm long, 1–1.5 cm wide, the margin glabrous or ciliate. Fruit pyriform, 1–1.5 cm long, densely pubescent; seeds winged.

Dry hammocks. Occasional; Clay County, central and western panhandle. New Hampshire and Vermont south to Florida, west to Ontario, Michigan, Indiana, Kentucky, Tennessee, Mississippi, Louisiana, and Texas. Spring–summer.

Buchnera L. 1753. BLUEHEARTS

Perennial hemiparasitic herbs. Leaves opposite, simple, pinnate-veined, epetiolate. Flowers in terminal spikes, bracteate, bracteolate; sepals 5, basally connate, zygomorphic; petals 5, basally connate, zygomorphic; stamens 4, didynamous, epipetalous; staminode absent; ovary superior, 2-carpellate and -loculate, the style 1. Fruit a capsule, dehiscence loculicidal; seeds numerous.

A genus of about 100 species; North America, West Indies, Mexico, Central America, South America, Africa, Asia, Australia, and Pacific Islands. [Commemorates Andreas Elias von Büchner (1701–1769), German physician.]

Selected reference: Philcox (1965).

Buchnera americana L. [Of America.] AMERICAN BLUEHEARTS.

Buchnera americana Linnaeus, Sp. Pl. 630. 1753. *Erinus americanus* (Linnaeus) Miller, Gard. Dict., ed. 8. 1768.

Buchnera levicaulis Rafinesque, New Fl. 2: 39. 1837 ("1836"). TYPE: FLORIDA.
Buchnera floridana Gandoger, Bull. Soc. Bot. France 66: 217. 1919. TYPE: FLORIDA: Polk Co.: s.d., *Ohlinger 463* (holotype: ?; isotyes: BM, F, MO).
Buchnera elongata Swartz var. *obtusa* Pennell, Proc. Acad. Nat. Sci. Philadelphia 71: 288. 1920.
Buchnera brevifolia Pennell, in Small, Man. S.E. Fl. 1223, 1508. 1933. TYPE: FLORIDA: Santa Rosa Co.(?): Robinson Island, 22 May 1901, *Tracy 7605* (holotype: NY).

Erect biennial herb, stem to 9 dm, spreading hirsute proximally, appressed hirsute or glabrate distally. Leaves with the blade of the larger leaves narrowly ovate to lanceolate, 2.5–6.5 cm long, 0.5–1.5 cm wide, much reduced distally, the apex acute or obtuse, the base cuneate, the margin entire or irregularly dentate, the upper surface glabrous, the lower surface short-hispid, sessile or subsessile. Flowers in an irregular spike, the pedicel 1–2 mm long; bract 1, ovate-lanceolate, 4–6 mm long; bracteoles linear-subulate, 2–3 mm long; calyx tubular, 5–8 mm long, slightly zygomorphic, the tube obscurely 10-nerved, ascending-appressed hispid, the lobes narrowly triangular; corolla blue or violet, rarely white, salverform, 1–2 cm long, slightly bilabiate, the outer surface glabrate, the inner surface pilose; stamen filaments pilose. Fruit ovoid, 5–8 mm long, black; seeds numerous, 0.5–1 mm long.

Flatwoods and disturbed sites. Common; nearly throughout. New York and Ontario south to Florida, west to Michigan, Illinois, Kansas, Oklahoma, and Texas; West Indies, Mexico, and Central America. Spring–fall.

EXCLUDED TAXA

Buchnera longifolia Kunth—This tropical American species was reported for Florida by Liogier (1995), the name misapplied to Florida material of *B. americana*.
Buchnera palustris (Aublet) Sprengel—Reported for Florida by Chapman (1860, 1883, 1897, all as *B. elongata* Swartz), Small (1903, 1913a, 1913b, 1913c, 1933, all as B. *elongata* Swartz), and Long and Lakela (1971, as *B. elongata* Swartz), the name misapplied to Florida material of *B. americana*.

Castilleja Mutis ex L. f. 1781. INDIAN PAINTBRUSH

Annual hemiparasitic herbs. Leaves alternate, simple, pinnate-veined, epetiolate. Flowers in spikes, bracteate, ebracteolate; sepals 4, basally connate, zygomorphic; petals 5, basally connate, zygomorphic; stamens 4, didynamous, epipetalous; staminode absent; ovary superior, 2-carpellate and -loculate, the stigma 1. Fruit a capsule, dehiscence loculicidal; seeds numerous.

A genus of about 200 species; North America, Mexico, Central America, South America, and Asia. [Commemorates Domingo Castillejo (1744–1793), Spanish botanist, surgeon, and professor.]

Selected reference: Nesom and Egger (2014).

Castilleja indivisa Engelm. [Undivided, in reference to the leaves.] ENTIRELEAF INDIAN PAINTBRUSH.

Castilleja indivisa Engelmann, Boston J. Nat. Hist. 5: 255. 1845.

Erect annual herb, to 4.5 dm; stem with fine eglandular and much shorter glandular trichomes. Leaves with the blade linear-lanceolate, 2–8(9) cm long, 2–8 mm wide, the margin entire or with 1–2(3) approximately paired linear to filiform lobes 1–20 mm long, the apex acute, the base truncate, the upper and lower surfaces hirsutulous-puberulent, sessile, often subclasping. Flowers in spike 2–6(20) cm long, villous with fine eglandular trichomes; bracts lanceolate, 0.5–2.5 cm long, the apex acute, green, distally becoming shorter and obovate with an obtuse to rounded apex, with the basal ⅔ scarlet to salmon-pink, rarely white or yellowish; calyx 1.6–2.2 cm long, the primary lobes 6–9 mm long, distally red or pink, rarely white or yellowish; corolla (1.5)1.7–2.4 mm long, equaling the calyx or exserted 1–6 mm, the upper lobe 4–9 mm long, ca. ⅕ the tube length, hooded, erect and straight, enclosing the anthers, dorsally yellow to pink, glandular-puberulent, the lower lip 1–2 mm, green, white, or yellowish; stamens with the anther theca unequally attached, 1 attached at the middle and the other suspended by its apex and smaller; stigma exserted. Fruit ovate, 1–1.5 cm long.

Dry, disturbed sites. Occasional; eastern and central panhandle, northern peninsula south to Hillsborough County. Florida, Alabama, Arkansas, Louisiana, Oklahoma, and Texas; Mexico. Native to Arkansas, Louisiana, Oklahoma, Texas, and Mexico. Spring.

EXCLUDED TAXON

Castilleja coccinea (Linnaeus) Sprengel—This widespread northern species was reported for Florida by Radford et al. (1964, 1968). No Florida specimens seen.

Conopholis Wallr. 1825. SQUAWROOT

Perennial holoparasitic herbs. Leaves alternate, simple, pinnate-veined, epetiolate. Flowers in terminal spikes, bracteate, bracteolate; sepals 4–5, basally connate, zygomorphic; petals 5, basally connate, zygomorphic; stamens 4, didynamous, epipetalous; staminode absent; ovary superior, 2-carpellate, 1-loculate, the style 1. Fruit a capsule, dehiscence loculicidal; seeds numerous.

A genus of 3 species; North America, Mexico, and Central America. [From the Greek *conos,* cone, and *pholis,* scale, in reference to the cone-like inflorescence.]

Selected reference: Rodrigues et al. (2013).

Conopholis americana (L.) Wallr. [Of America.] AMERICAN SQUAWROOT; CANCERROOT.

Orobanche americana Linnaeus, Mant. Pl. 88. 1767. *Conopholis americana* (Linnaeus) Wallroth, Orobanches Gen. Diask. 78. 1825.

Erect perennial herb, to 2 dm. Stem yellow or brown, glabrous. Leaves with the blade ovate to ovate-oblong, 0.5–1.8 cm long, 0.5–1 cm wide, the apex acute, the base truncate, the margin entire, glabrous or minutely pubescent, sessile. Flowers in a terminal raceme, the pedicel to 4(6) mm long; bracts ovate to ovate-oblong, 1–2 cm long, 4–8 mm wide, the apex acuminate, the margin entire or minutely erose; bracteoles 2, minute; calyx cylindric, 3–8 mm long, the lobes deltate, the margin entire or erose, glabrous; corolla cream-colored, tubular, 8–14 mm long, 2-lipped, arching outward, glabrous. Fruit ovoid, 0.5–1.5 cm long, the style persistent, glabrous; seeds irregularly ovate, triangular, or quadrangular, ca. 1 mm long.

Mesic hammocks; parasitic of roots of various woody species. Occasional; northern counties, central peninsula. Quebec south to Florida, west to Ontario, Wisconsin, Illinois, Kentucky, Tennessee, and Alabama; Mexico. Spring.

Epifagus Nutt. 1818. BEECHDROPS

Annual holoparasitic herbs. Leaves alternate, simple, pinnate-veined, epetiolate. Flowers in terminal racemes or panicles, chasmogamous or cleistogamous, bracteate, bracteolate; sepals 5, basally connate, zygomorphic; petals 5, basally connate, zygomorphic; stamens 4, didynamous, epipetalous; staminode absent; ovary 2-carpellate, 1-loculate, the style 1. Fruit a capsule, dehiscence loculicidal; seeds numerous.

A monotypic genus; North America and Mexico. [From the Greek *epi,* upon, and *Fagus,* beech, in reference to the host plant.]

Leptamnium Raf., 1819.

Selected reference: Thieret (1969).

Epifagus virginiana (L.) W.P.C. Barton [Of Virginia.] BEECHDROPS.

Orobanche virginiana Linnaeus, Sp. Pl. 633. 1753. *Epifagus virginiana* (Linnaeus) W.P.C. Barton, Comp. Fl. Philadelph. 2: 50. 1818. *Epifagus americana* Nuttall, Gen. N. Amer. Pl. 2: 60. 1818, nom. illegit. *Mylanche virginiana* (Linnaeus) Wallroth, Orobanches Gen. Diask. 76. 1825. *Leptamnium virginianum* (Linnaeus) Rafinesque, Med. Fl. 2: 237. 1830.

Erect annual herb. Stem an ovoid underground corm with scalelike leaves and short adventitious yellow to brownish orange coralloid roots, glabrous. Leaves triangular-ovate, ca. 2 mm long, the apex acute, the base clasping, the margin entire, the upper and lower surfaces glabrous. Flowers in an erect or ascending raceme or panicle of racemes (3.5)15–50 cm long arising from the underground corm, yellow, yellow-brown, blackish purple, or purplish brown, the pedicel 1–3 mm long; bracts triangular-ovate, 2–5 mm long; bracteoles triangular-ovate, 1–2 mm long, adnate to the calyx base; chasmogamous flowers with the calyx campanulate, 2–3 mm long, oblique, the lobes 5, ca. 1 mm long; corolla tubular, 8–12 mm long, laterally compressed, the upper lobe erect, slightly incurved apically, the lower lip with the lobes erect to spreading; cleistogamous flowers with the calyx as in the chasmogamous; corolla 2–3 mm long; stamens with the anthers adnate to the stigma. Fruit obliquely ovoid, 3–4 mm long, laterally compressed; seeds ovoid to oblong, ca. 1 mm long.

Mesic hammocks; parasitic on roots of *Fagus grandifolia* (Fagaceae). Occasional; northern counties. Quebec south to Florida, west to Ontario, Wisconsin, Illinois, Missouri, Oklahoma, and Texas; Mexico. Fall.

Macranthera Nutt. ex Benth. 1835.

Biennial or perennial hemiparasitic herbs. Leaves opposite, simple, pinnate-veined, petiolate. Flowers in terminal racemes, bracteate, ebracteolate; sepals 5, basally connate, actinomorphic; petals 5, basally connate, zygomorphic; stamens 4, epipetalous; staminode absent; ovary superior, 2-carpellate and -loculate, the style 1. Fruit a capsule, dehiscence loculicidal; seeds numerous.

A monotypic genus; North America. [From the Greek *makros,* large, and *anthera,* anther, in reference to the long anthers.]

Selected reference: Alford and Anderson (2002).

Macranthera flammea (W. Bartram) Pennell [*Flammeus,* flame-colored, fiery red, scarlet, in reference to the flower color.] FLAMEFLOWER; HUMMINGBIRD-FLOWER.

Gerardia flammea W. Bartram, Travels Carolina 412. 1791. *Flamaria coccinea* Rafinesque, New Fl. 2: 71. 1837 ("1836"), nom. illegit. *Russelia flammea* (W. Bartram) Rafinesque, New Flora 2: 71. 1837 ("1838"). *Macranthera flammea* (W. Bartram) Pennell, Bull. Torrey Bot. Club 40: 124. 1913.

Macranthera lecontei Torrey, Ann. Lyceum Nat. Hist. New York 4: 80, pl. 4. 1837. *Toxopus gymnanthes* Rafinesque, New Fl. 2: 72. 1837 ("1836"), nom. illegit. *Macranthera fuschioides* (Nuttall) Leconte ex Bentham var. *lecontei* (Torrey) Chapman, Fl. South. U.S. 297. 1860. *Conradia lecontei* (Torrey) Kuntze, Revis. Gen. Pl. 2: 459. 1891.

Erect biennial or perennial herb, to 3.5 m; stem 4-angled, retrorsely puberulent. Leaves with the blade lanceolate to narrowly ovate, 8–15 cm long, 2–6 cm wide, reduced in size distally and becoming bract-like, the apex acute to acuminate, the base cuneate, decurrent on the petiole, the margin deeply pinnatifid proximally, toothed distally, the upper and lower surfaces glabrate, the petiole 0.5–1.5 cm long. Flowers terminal and on the larger branches, in a raceme 8–36 cm long, the pedicel 1–1.5 cm long, deflexed-spreading, strongly recurved distally; calyx campanulate, the tube 2–4 mm long, the lobes linear, 8–15 mm long, the outer surface retrorsely puberulent; corolla bright orange, bilabiate, 2–2.5 cm long, the lobes 3–4 mm long, subequal in length, the upper 2 lobes erect, the lower 3 lobes reflexed spreading, the outer surface densely glandular-mealy; stamens to 4.5 cm long, long-exserted, puberulent, the filaments orange; style exserted, slightly exceeding the stamens. Fruit ovoid, 1–1.3 cm long, brown, densely puberulent, the style somewhat persistent; seeds irregularly triangular, 2–3 mm long, with 2–3(5) membranous wings, brownish black.

Bogs and swamps. Occasional; central and western panhandle. Georgia south to Florida, west to Louisiana. Summer.

Macranthera flammea is listed as endangered in Florida (Florida Administrative Code, Chapter 5B-40).

Orobanche L. 1753 BROOMRAPE

Annual holoparasitic herbs. Leaves alternate, simple, pinnate-veined, epetiolate. Flowers in spikes or spikelike racemes, bracteate, ebracteolate; sepals 4 or 5, basally connate, zygomorphic; petals 5, basally connate, zygomorphic; stamens 4, didynamous, epipetalous; staminode absent; ovary 2-carpellate, 1-loculate, the style 1. Fruit a capsule, dehiscence loculicidal; seeds numerous.

A genus of about 125 species; North America, Central America, South America, Europe, Africa, Asia, Australia, and Pacific Islands. [From the Greek *orobos,* vetch, in reference to the host plant, and *ancho,* strangle, in reference to the parasitic habit.]

Orobanche minor Sm. HELLROOT.

Orobanche minor Smith, Engl. Bot. 6: t. 422. 1797.

Erect annual herb, to 7 dm; stem reddish brown, purple, or yellow, glandular-pubescent. Leaves with the blade lanceolate to oblong-ovate or triangular-ovate, 0.6–2 cm long, the apex acute to acuminate, the base clasping, the margin entire, the upper surface glabrous, the lower surface glandular-pubescent, sessile. Flowers in a spike or spikelike raceme, sessile or the pedicel to 1(3) mm long; bracts narrowly lanceolate, 6–17 mm long, the apex attenuate to subulate, slightly reflexed, the upper surface glabrous, lower surface glandular-pubescent; calyx yellow, brownish red, or brownish purple, (6)8–12 mm long, the lobes lanceolate to subulate, the outer surface glandular-villous; corolla white or yellow, sometimes purplish tinged or veined distally, 1–2 cm long, the tube slightly curved, the outer surface glandular-puberulent, the lips 3–5 mm long, spreading, the margin erose-crenulate; stamens with the filaments sparsely pubescent, the anthers tomentose or glabrous. Fruit oblong-ovoid, 5–9 mm long; seeds minute, brown.

Mesic hammocks; parasitic on the roots of various herbaceous species, primarily Asteraceae and Fabaceae. Rare; northern and central peninsula. Vermont and New York south to Florida; Washington and Oregon; Central America and South America; Europe, Africa, Asia, Australia, and Pacific Islands. Native to Europe, Africa, and Asia. Spring.

Pedicularis L. 1753. LOUSEWORT

Perennial hemiparasitic herbs. Leaves alternate, simple, pinnate-veined, petiolate. Flowers in racemes, bracteate, ebracteolate; sepals 4, basally connate, zygomorphic; petals 5, basally connate, zygomorphic; stamens 4, didynamous, epipetalous; ovary superior, 2-carpellate and -loculate, the style 1. Fruit a capsule, dehiscence loculicidal; seeds numerous.

A genus of about 600; North America, Mexico, Central America, South America, Europe, Africa, and Asia. [*Pediculus,* a louse, in reference to the early European belief that cattle feeding where *P. palustris* abounded became covered with lice.]

Selected references: Robart et al. (2015).

Pedicularis canadensis L. [Of Canada.] CANADIAN LOUSEWORT.

Pedicularis canadensis Linnaeus, Mant. Pl. 86. 1767. *Pedicularis canadensis* Linnaeus forma *typica* Farwell, Amer. Midl. Naturalist 11: 68. 1928, nom. inadmiss. *Pedicularis canadensis* Linnaeus var. *dobbsii* Fernald, Rhodora 48: 59, t. 1009(2–3), 1010. 1946.

Erect perennial herb, to 3 dm; stem rhizomatous, villous. Leaves with the blade oblanceolate to narrowly oblong, 4–15 cm long, 0.8–2.5 m wide, the apex acute to obtuse, the base cuneate, decurrent on the petiole, the margin pinnatifid or bipinnatifid, the ultimate lobes irregularly crenate or entire, the upper and lower surfaces sparsely hirsute, the petiole 1–10 cm long, the upper leaves sessile. Flowers in a short, dense, spiciform raceme; bracts leaflike, oblanceolate, the margin crenate to toothed, 8–12 mm long, the upper and lower surfaces sparsely hirsute; calyx 7–9 mm long, the tube oblique; corolla yellow, sometimes the upper lip purplish or purplish throughout, the upper lip 11–15 cm long, arched and decurved, narrowed to a truncate apex, with 2 slender teeth just below the apex, the lower lip 7–10 mm long, with 3 widely spreading lobes. Fruit narrowly oblong, 1.0–1.5 mm long; seeds 1–2 mm long, striate.

Bluffs and calcareous hammocks. Rare; Clay County, central and western panhandle. Quebec south to Florida, west to Manitoba, North Dakota, South Dakota, Nebraska, Colorado, and New Mexico. Spring.

Schwalbea L. 1753.

Perennial hemiparasitic herbs. Leaves alternate, simple, pinnate-veined, epetiolate. Flowers in racemes, ebracteate, bracteolate; sepals 5, basally connate, zygomorphic; petals 5, basally connate, zygomorphic; stamens 4, epipetalous; staminode absent; ovary superior, 2-carpellate and -loculate, the style 1. Fruit a capsule, dehiscence septicidal and loculicidal; seeds numerous.

A monotypic genus; North America. [Commemorates Christian Georg Schwalbe, early eighteenth-century medical botanist.]

Schwalbea americana L. [Of America.] CHAFFSEED.

Schwalbea americana Linnaeus, Sp. Pl. 606. 1953.
Schwalbea australis Pennell, Proc. Acad. Nat. Sci. Philadelphia 71: 289. 1920. *Schwalbea americana* Linnaeus var. *australis* (Pennell) Reveal & C. R. Broome, Castanea 46: 75. 1981.

Erect perennial herb, to 8 dm; stem densely pubescent. Leaves with the blade lanceolate to narrowly elliptic, 2–5 cm long, 6–10 mm wide, gradually reduced distally, the apex acute to obtuse, the base cuneate, the margin entire, the upper and lower surfaces densely pubescent. Flowers in a terminal raceme, the pedicel 1–5 mm long; bracts absent; bracteoles 2 just below the calyx, linear, 6–12 mm long; calyx tubular, 1.5–2.2 cm long, bilabiate, 10- to 12-ribbed, the upper lobes short-triangular, the lower lobes larger, with 2 lateral lobes and a larger central lobe 3-notched at the apex, the outer surface glandular-pubescent; petals pale yellow suffused with purple in the distal ½ and on the veins, 2.2–3.5 cm long, the upper limb slightly galeate, the outer surface glandular-pubescent; stamen filaments pubescent. Fruit ellipsoid to ovoid, 10–12 mm long, glabrous; seeds linear, flattened, 2–3 mm long, winged, yellow-brown.

Open hammocks and flatwoods. Occasional; northern and central peninsula, central and western panhandle. New York and Massachusetts south to Florida, west to Kentucky, Tennessee, Mississippi, and Texas. Spring–summer.

Schwalbea americana is listed as endangered in Florida (Florida Administrative Code, Chapter 5B-40) and in the United States (U.S. Fish and Wildlife Service, 50 CFR 23).

Seymeria Pursh, nom. cons. 1814. BLACKSENNA

Annual hemiparasitic herbs. Leaves opposite, simple, pinnate-veined, epetiolate. Flowers in racemes, bracteate, ebracteolate; sepals 5, basally connate, actinomorphic; petals 5, zygomorphic; stamens 4, epipetalous; ovary superior, 2-carpellate and -loculate, the stigma 1. Fruit a capsule, dehiscence locicidal; seeds numerous.

A genus of about 27 species; North America, West Indies, and Mexico. [Commemorates Henry Seymer (1745–1800), English naturalist.]

Afzelia J. F. Gmel., nom. rej. 1791.

Selected reference: Pennell (1925).

1. Leaf segments filiform; calyx glabrous or nearly so; capsule glabrous .... **S. cassioides**
1. Leaf segments lanceolate; calyx glandular-pubescent; capsule brown-tomentose .... **S. pectinata**

Seymeria cassioides (J. F. Gmel.) S. F. Blake [Resemble *Cassia* (Fabaceae).] YOUPON BLACKSENNA.

Afzelia cassioides J. F. Gmelin, Syst. Nat. 2: 927. 1792. *Gerardia afzelia* Michaux, Fl. Bor.-Amer. 2: 20. 1803, nom. illegit. *Gerardia cassioides* (J. F. Gmelin) Persoon, Syn. Pl. 2: 154. 1806. *Seymeria tenuifolia* Pursh, Pl. Amer. Sept. 737. 1814, nom. illegit. *Seymeria cassioides* (J. F. Gmelin) S. F. Blake, Rhodora 17: 134. 1915.

Erect annual herb, to 1 m; stem glandular hirsute to sparsely pubescent with ascending incurved trichomes. Leaves with the blade ovate, the larger ones 1–1.5 cm long, 0.8–1.5 cm wide, the apex acute, the base broad, the margin deeply bipinnatifid, the segments and pinnules filiform, the upper surface glandular hirsute to sparingly hirtellous, channeled, the lower surface glabrous, the petiole 1–3 mm long. Flowers in a raceme, the pedicel 3–6 mm long; calyx 3–4 mm long, the lobes linear, 2–3 mm long, the outer surface glabrous or sparsely glandular-hirtellous; corolla pale yellow, often with purple in the throat and at the base of the lobes, ca. 9 mm long, the tube ca. 3 mm long, the lobes ca. 6 mm long, ovate-lanceolate, widely spreading, the 2 upper ones united nearly ½ their length, the margin sparsely ciliolate, the outer surface of the tube and lobes glabrous, the inner surface of the tube with a ring of short trichomes at the point of attachment of the stamens; stamens glabrous. Fruit ovate, 4–5 mm long, laterally compressed, glabrous, black; seeds oblong-lanceolate, ca. 1 mm long, the surface furrowed, yellowish brown.

Sandhills and flatwoods. Occasional; northern counties, central peninsula. Virginia south to Florida, west to Texas; West Indies. Fall.

Seymeria pectinata Pursh [With narrow close divisions like a comb, in reference to the leaves.] PIEDMONT BLACKSENNA.

Seymeria pectinata Pursh, Fl. Amer. Sept. 737. 1814. *Afzelia pectinata* (Pursh) Kunth, Revis. Gen. Pl. 2: 457. 1891. *Seymeria pectinata* Pursh subsp. *typica* Pennell, Acad. Nat. Sci. Philadelphia Monogr. 1: 409. 1935.

Afzelia pectinata (Pursh) Kuntze var. *peninsularis* Pennell, Proc. Acad. Nat. Sci. Philadelphia 71: 265. 1920. *Seymeria pectinata* Pursh subsp. *peninsularis* (Pennell) Pennell, Acad. Nat. Sci. Philadelphia Monogr. 1: 411. 1935. TYPE: FLORIDA: Collier Co.: Marco, Jul–Aug. 1900, *Hitchcock 254* (holotype: US).

Erect annual herb, to 6 dm; stem pubescent with ascending incurved glandular and eglandular trichomes and sessile glands. Leaves with the blade triangular-ovate, the larger ones 1.5–3 cm long, 1–2.5 cm wide, the apex acute, the base broad, the margin pinnatifid or sometimes the lower ones bipinnatifid, the segments lanceolate, the upper and lower surfaces glandular-pubescent and with sessile glands, the petiole 4–7 mm long, sometimes slightly winged. Flowers in a raceme, the pedicel 3–5 mm long; calyx 5–6 mm long, the lobes lanceolate, 3–4 mm long, the outer surface glandular-pubescent and with sessile glands; corolla golden yellow, somewhat purple within at the throat and at the base of the lobes, ca. 10 mm long, the tube ca. 4 mm long, the lobes ca. 6 mm long, ovate, widely spreading, the 2 upper ones united nearly ½ their length, the margin long-ciliolate, the outer surface of the tube and lobes pubescent, the inner surface of the tube with a ring of short trichomes at the point of attachment of the stamens; stamens pubescent. Fruit ovate, 5–7 mm long, laterally compressed, brown-tomentose; seeds angular ovate, ca. 1 mm long, with 3- to 4-winged, the surface alveolate, pale brown.

Scrub and sandhills. Frequent; nearly throughout. North Carolina south to Florida, west to Louisiana. Fall.

Striga Lour. 1790. WITCHWEED

Annual hemiparasitic herbs. Leaves opposite or subopposite, simple, pinnate-veined, epetiolate. Flowers in spikes, bracteate, bracteolate; sepals 5, basally connate, zygomorphic; petals 5, basally connate, zygomorphic; stamens 4, didynamous, epipetalous; staminode absent; ovary superior, 2-carpellate and -loculate, the style 1. Fruit a capsule, dehiscence loculicidal; seeds numerous.

A genus of about 40 species; North America, Africa, Asia, and Australia. [Evil or a witch, in reference to rendering its victims prematurely aged and weak.]

Striga gesnerioides (Willd.) Vatke ex Engl. COWPEA WITCHWEED.

Buchnera gesnerioides Willdenow, Sp. Pl. 3: 338. 1899. *Striga gesnerioides* (Willdenow) Vatke ex Engler, Oesterr. Bot. Z. 25: 11. 1879.

Erect annual herb, to 3 dm; stem puberulent or glabrous. Leaves with the blade lanceolate, scale-like, appressed, 3–7 mm long, ca. 2 mm wide, the apex acute to acuminate, the base broadly cuneate, the margin entire, ciliate, the upper and lower surfaces puberulent, sessile. Flowers in a spike, opposite or alternate; bracts lanceolate, 5–7 mm long, 1–2 mm wide, glabrous or puberulent, the margin ciliate; bracteoles subulate, 2–3 mm long, the margin ciliate; calyx 5–7 mm long, ribbed, pilose or with short trichomes, scarious between the ribs, the margin ciliate; corolla brownish red or purple, rarely white, 8–12 mm long, the tube curved, the lobes obovate, the outer surface sparsely pubescent or glabrous. Fruit oblong to ovoid, 4–5 mm long, glabrous; seeds ovoid, minute, brown or black,

Disturbed sites; parasitic on *Indigofera* and *Alysicarpus ovalifolius* (Fabaceae). Occasional; central Florida. Florida; Africa and Asia. Native to Africa and Asia. Fall.

In Africa, *S. gesnerioides* is parasitic on several agricultural crops, particularly on cowpea (*Vigna unguiculata*) (Botanga and Timko, 2005).

AQUIFOLIACEAE Bercht. & J. Presl, 1825. HOLLY FAMILY

Shrubs or trees. Leaves alternate, simple, pinnate-veined, petiolate, stipulate. Flowers in axillary cymes or solitary, bisexual or unisexual (plants dioecious or polygamodioecious), bracteate; sepals 4–9, basally connate, actinomorphic; petals 4–9, basally connate, actinomorphic; stamens 4–9, epipetalous; ovary superior, 4- to 9-carpellate and -loculate, the style 1. Fruit a drupe.

A monogeneric family of about 400 species; nearly cosmopolitan.

Selected reference: Wunderlin and Poppleton (1977).

Ilex L. 1753. HOLLY

Shrubs or trees. Leaves alternate, simple, pinnate-veined, petiolate, stipulate. Flowers in axillary cymes or solitary, bisexual or unisexual (plants dioecious or polygamodioecious), bracteate; sepals 4–9, basally connate, actinomorphic; petals 4–9, basally connate, actinomorphic; stamens 4–9, epipetalous; ovary superior, 4- to 9-carpellate and -loculate, the style 1. Fruit a berrylike drupe; seeds (pyrenes) 1 per locule.

A genus of about 400 species; nearly cosmopolitan. [The ancient Latin vernacular name for the holly-oak or holm-oak (*Quercus ilex*), holly, in reference to the similar appearance of the leaves to that species.]

1. Leaves chartaceous to subcoriaceous, deciduous.
    2. Leaves with a crenate margin, the blade tapering to a narrow cuneate base .... **I. decidua**
    2. Leaves with a serrate margin, the blade tapering to a broad cuneate base.
        3. Fruit pyrene with a soft, smooth endocarp .... **I. verticillata**
        3. Fruit pyrene with a hard, ribbed endocarp.
            4. Leaves with the lower surface conspicuously reticulate-veined .... **I. amelanchier**
            4. Leaves with the lower surface only slightly or obscurely reticulate-veined.
                5. Fruiting pedicels 3–8 mm long .... **I. ambigua**
                5. Fruiting pedicels 10–23 mm long .... **I. longipes**
1. Leaves coriaceous, evergreen.
    6. Leaves with a crenate margin throughout their length .... **I. vomitoria**

6. Leaves with a dentate, serrate, or entire margin, if crenate, then only toward the apex.
 7. Leaf margin remotely crenate toward the apex .... **I. glabra**
 7. Leaf margin serrate, dentate, or entire.
  8. Leaves with the margin spinulose-dentate or entire, the teeth, or at least the leaf apex, armed with a rigid spine 1 mm long or longer .... **I. opaca**
  8. Leaves with the margin spinulose-serrate or entire, the teeth (if present) and the leaf apex unarmed or with a spine less than 1 mm long.
   9. Leaves with minute black punctuations on the lower surface .... **I. coriacea**
   9. Leaves without minute black punctuations on the lower surface.
    10. Inflorescence pedunculate; fruit red or yellow .... **I. cassine**
    10. Inflorescence not pedunculate; fruit black or purple .... **I. krugiana**

Ilex ambigua (Michx.) Torr. [Doubtful, in reference to its identity.] CAROLINA HOLLY; SAND HOLLY.

Cassine caroliniana Walter, Fl. Carol. 242. 1788; non Lamarck, 1785. *Synstima caroliniana* Rafinesque, Sylva Tellur. 49. 1838. *Ilex caroliniana* (Rafinesque) Trelease, Trans. St. Louis Acad. Sci. 5:347. 1889; non Miller, 1768.

Prinos ambiguus Michaux, Fl. Bor.-Amer. 2: 236. 1803. *Ilex ambigua* (Michaux) Torrey, Fl. New York 2: 2. 1843. *Nemopanthus ambiguus* (Michaux) A. W. Wood, Class-Book Bot., ed. 1861. 497. 1861.

Ilex ambigua (Michaux) Torrey var. *coriacea* Trelease, Trans. St. Louis Acad. Sci. 5:347. 1889. TYPE: FLORIDA: Hillsborough Co.: Tampa, Oct 1877, *Garber s.n.* (holotype: MO).

Ilex dubia (G. Don) Britton et al. forma *pseudoambigua* Loesener, Nova Acta Acad. Caes. Leop.-Carol. Germ. Nat. Cur. 78: 486. 1901. SYNTYPE: FLORIDA.

Ilex buswellii Small, Bull. Torrey Bot. Club 51: 382. 1924. TYPE: FLORIDA: Lee Co.: near the Caloosahatchee above Fort Myers, 2 Oct 1923, *Buswell s.n.* (holotype: NY).

Ilex caroliniana (Rafinesque) Trelease var. *jejuna* McFarlin, Rhodora 34: 236, pl. 231. 1932. TYPE: FLORIDA: Polk Co.: Lake Marion, s.d. *McFarlin 5550* (holotype: MICH).

Tree or shrub, to 6 m; branchlets green or brownish, glabrous or short-pubescent when young, the older ones gray or brownish. Leaves deciduous, the blade elliptic, ovate, or obovate, 2–10 cm long, 1–4(6) cm wide, the apex acute to obtuse, the base cuneate to rounded, the margin entire or appressed crenate-serrate from near the middle upward, the upper and lower surfaces glabrous or short-pubescent, the petiole 0.3–1 cm long, short-pubescent. Plants dioecious, the staminate flowers few to many in an axillary cluster, the carpellate flowers solitary or few in an axillary cluster, the pedicel 1–4 mm long; calyx 4- to 5-lobed, the lobes short-triangular, the margin ciliate; corolla white, 4- to 5-lobed, the lobes oblong to obovate. Fruit subglobose to elliptic, 5–7(10) mm long; pyrenes 4–5, ca. 5 mm long, furrowed on the rounded side.

Hammocks, scrub, sandhills, and dunes. Common; northern counties, central peninsula. Virginia south to Florida, west to Oklahoma and Texas. Winter–spring.

Ilex amelanchier M. A. Curtis ex Chapm. [Named for the genus *Amelanchier,* which it resembles.] SARVIS HOLLY; SERVICEBERRY HOLLY.

Ilex amelanchier M. A. Curtis ex Chapman, Fl. South. U.S. 270. 1860.

Ilex dubia (G. Don) Britton et al. forma *chapmaniana* Loesener, Nova Acta Acad. Caes. Leop.-Carol. Germ. Nat. Cur. 78: 487. 1901. TYPE: FLORIDA: Liberty Co.: near Bristol, s.d., *Chapman s.n.* (Holotype: ?).

Shrub or tree, to 5 m; branchlets grayish or brown when young, the older ones powdery pubescent or sparsely short-pubescent, branches smooth, grayish brown, the lenticels elliptic, pale buff-colored. Leaves deciduous, the blades oblong, oblong-obovate, or elliptic, the larger ones 5–9 cm long, 3–4 cm wide, the apex round or obtuse, rarely abruptly short-acuminate, the base rounded or short-cuneate, the margin entire or with a few small teeth, the upper surface glabrous, the lower surface shaggy pubescent, the petiole 3–15 mm long, shaggy pubescent. Plants dioecious, the staminate flowers in axillary fascicles, the carpellate flowers solitary or few in the leaf axils or on leafless portions of the branchlets of the previous season; calyx 4-lobed, the lobes narrowly triangular; corolla white or yellowish, 4-lobed, the lobes oblong to oblong-elliptic. Fruit subglobose to oblate, 5–10 mm long, red; pyrenes 4, the rounded side with 2 broad longitudinal grooves, sometimes with a rib connecting 1 or both grooves.

Swamps. Rare; central and western panhandle. North Carolina south to Florida, west to Louisiana. Spring.

Ilex amelanchier is listed as threatened in Florida (Florida Administrative Code, Chapter 5B-40).

Ilex cassine L. [Derived from the Timucua name for *I. vomitoria.*] DAHOON.

Shrub or tree to 10 m; branchlets usually short-pilose, minutely puberulent, or occasionally glabrous when young, the older ones glabrous, the bark gray, thin, and smooth. Leaves persistent, the blade oblanceolate, subspatulate, oblong, ovate, or subovate, 2–8(14) cm long, 0.8–4.5 cm wide, the apex acute, obtuse, or sometimes rounded, usually tipped with a mucro, the base broadly to narrowly cuneate, the margin entire or with a few short, sharp teeth, the upper surface sparsely pubescent along the midrib, otherwise glabrous, the lower surface pubescent along the midrib and with straight trichomes elsewhere or wholly glabrous, the petiole 0.5–1.5 cm long, usually pubescent. Plants dioecious, the staminate flowers in an axillary paniculiform cyme ca. 5 cm long, the carpellate flowers solitary or 2–4 in an axillary cyme or sometimes in a paniculiform cyme like the staminate, the peduncle ca. 5 mm long, pubescent, the pedicel 2–3 mm long, pubescent; calyx 4-lobed, the lobes triangular, the margin minutely toothed or erose; the corolla white, 4-lobed, the lobes oblong, the apex broad, slightly cupped. Fruit glabrous, red, orangish red, yellowish, or yellow, 5–8 mm long; pyrenes 4, ca. 4 mm long, irregularly ribbed on the rounded side.

1. Leaves more than 1 cm wide; branchlets short-pilose or occasionally glabrous; secondary branches at an angle of less than 45° from the main branch .... var. **cassine**
1. Leaves less than 1 cm wide; branchlets minutely strigose-puberulent; secondary branches at an angle greater than 45° from the main branch .... var. **myrtifolia**

Ilex cassine var. **cassine** DAHOON.

Ilex cassine Linnaeus, Sp. Pl. 125. 1753. *Ilex caroliniana* Miller, Gard. Dict., ed. 8. 1768. *Ilex cassine* Linnaeus var. *latifolia* Aiton, Hort. Kew. 1: 170. 1789, nom. inadmiss. *Ilex angustifolia* Salisbury, Prodr. Stirp. Chap. Allerton 70: 1796, nom. illegit.

Ilex dahoon Walter, Fl. Carol. 241. 1788. *Ageria palustris* Rafinesque, Sylva Tellur. 47. 1838, nom. illegit. *Ilex dahoon* Walter var. *grandifolia* K. Koch, Dendrologie 2(1): 224. 1869, nom. inadmiss.

Ilex cassine Linnaeus var. *angustifolia* Aiton, Hort. Kew. 1: 170. 1789. *Ilex angustifolia* (Aiton) Willdenow, Enum. Pl. 172. 1809; non Salisbury, 1796. *Ilex dahoon* Walter var. *angustifolia* (Aiton) Torrey & A. Gray ex S. Watson, Bibl. Index N. Amer. Bot. 158. 1878. TYPE: "Carolina & Florida."
Ilex floridana Lamarck, Tabl. Encycl. 1: 356. 1792. TYPE: FLORIDA.
Ilex ramulosa Rafinesque, Fl. Ludov. 110. 1817. "on the seashore near Pensacola, and at Opelousas."
Ilex laurifolia Nuttall, Amer. J. Sci. Arts 5: 289. 1822. *Ilex dahoon* Walter var. *laurifolia* (Nuttall) de Candolle, Prodr. 2: 14. 1825. TYPE: FLORIDA: east Florida, s.d., *Ware s.n.* (holotype: PH).
Ageria heterophylla Rafinesque, Sylva Tellur. 48. 1838. TYPE: "Florida and Alabama."
Ageria obovata Rafinesque, Sylva Tellur. 48. 1838. TYPE: FLORIDA.

Flatwoods, wet hammocks, swamps, and pond margins. North Carolina south to Florida, west to Texas; West Indies and Mexico. Common; nearly throughout. All year.

Ilex cassine var. *cassine* hybridizes in Florida with *Ilex opaca* var. *opaca* (*I.* ×*attenuata*).

Ilex cassine var. **myrtifolia** (Walter) Sarg. [With leaves like *Myrtus* (Myrtaceae).] MYRTLE DAHOON.

Ilex myrtifolia Walter, Fl. Carol. 241. 1788. *Ilex dahoon* Walter var. *myrtifolia* (Walter) Chapman, Fl. South. U.S. 269. 1860. *Ilex dahoon* Walter var. *parvifolia* K. Koch, Dendrologie 2(1): 225. 1869. *Ilex cassine* Linnaeus var. *myrtifolia* (Walter) Sargent, Gard. & Forest 2: 616. 1889. *Ilex cassine* Linnaeus var. *parvifolia* C. K. Schneider, Handb. Laubholzk. 2: 160. 1907, nom. illegit. *Ilex cassine* Linnaeus subsp. *myrtifolia* (Walter) E. Murray, Kalmia 13: 8. 1983.

Acid swamps and coastal swales. Frequent; northern counties, Orange County. North Carolina south to Florida, west to Texas. Spring.

Often treated as a species by authors, hybridization frequently occurs with var. *cassine* where the two overlap and it seems that varietal status is warranted.

Ilex coriacea (Pursh) Chapm. [Leathery, in reference to the leaves.] LARGE GALLBERRY; SWEET GALLBERRY.

Prinos coriaceus Pursh, Fl. Amer. Sept. 221. 1814. *Ennepta coriacea* (Pursh) Rafinesque, Sylva Tellur. 52. 1838. *Ilex coriacea* (Pursh) Chapman, Fl. South. U.S. 270. 1860.

Shrub or tree, to 5 m; branchlets of the season dark brown, finely short-pubescent, the older branches grayish to tan, the lenticels circular with a longitudinal slit. Leaves persistent, the blade elliptic, ovate, lanceolate, or subobovate, 3.5–9 cm long, 1.5–4 mm wide, the apex acute or short-acuminate, sometimes rounded or notched, the base cuneate, the margin entire or with a few nearly bristlelike, usually spreading teeth, the upper surface sublustrous, short-pubescent along the midrib, the lower surface paler, glabrous or pubescent, with scattered punctate glands, the petiole 5–10 mm long, short-pubescent. Plants dioecious, the staminate flowers in axillary clusters or solitary on year-old branchlets or solitary at the base of a shoot of the season and upward axillary to the developing leaves, the carpellate flowers solitary or in clusters axillary to the leaves of the season; calyx 5- to 9-lobed, the outer surface somewhat tuberculate, the lobes triangular, the margin entire, shallowly erose, or fine-toothed; corolla white, 5- to 9-lobed, the lobes oblong. Fruit globose to oblate, 6–8 mm long, black, lustrous; pyrenes 5–9, smooth on the rounded side.

Flatwoods, swamps, and bogs. Occasional; northern counties, central peninsula. Virginia south to Florida, west to Texas. Spring.

Ilex decidua Walter [Deciduous, in reference to the leaves.] POSSUMHAW.

Ilex decidua Walter, Fl. Carol. 241. 1788. *Prinos deciduus* (Walter) de Candolle, Prodr. 2: 16. 1825.
Ilex decidua Walter var. *curtissii* Fernald, Bot. Gaz. 33: 155. 1902. *Ilex curtissii* (Fernald) Small, Man. S.E. Fl. 815. 1933. TYPE: FLORIDA: Suwannee Co.: along the Suwannee River near Branford, 24 Oct 1900, *Curtiss 6736* (holotype: GH; isotype: NY).
Ilex cuthbertii Small, Man. S.E. Fl. 815. 1933. TYPE: "FL., and GA."

Shrub or tree, to 10 m; branchlets of the season greenish brown, glabrous, the older ones gray with nearly circular lenticels, the bark thin, gray, smooth or slightly roughened. Leaves deciduous, the blade oblanceolate, spatulate, obovate or elliptic, 1–6(8) cm long, 0.8–3(4.5) cm wide, the apex rounded, obtuse, or short-acuminate, the base cuneate, the margin appressed crenate, the teeth gland-lipped, the upper surface sometimes sparsely pubescent near or on the midrib when young, the lower surface shaggy-pubescent on the midrib, usually on the main lateral veins, and sometimes between the veins, the petiole 0.2–1.5 cm long, glabrous to sparsely short-pubescent, rarely densely so. Plants dioecious, the staminate flowers in fascicles at the junction of the spur shoots of the previous season, on the branchlets of the current season, and solitary in the axil of developing leaves, the pedicel 0.5–1(2) cm long, the carpellate flowers solitary or 2–3 in the leaf axils or the nodes on branchlets of the previous season, the pedicel 3–5(8) mm long; calyx 4- to 5-lobed, the lobes deltoid; corolla white or yellowish green, 4- to 5-lobed, the lobes oblong to oblanceolate. Fruit globose or subglobose, 6–8(10) mm long, red, orangish red, or yellow; pyrenes 5, 4–5 mm long, irregularly ridged on the rounded side.

Floodplain forests. Occasional; northern counties, central peninsula. Maryland south to Florida, west to Kansas, Oklahoma, and Texas. Spring.

Ilex glabra (L.) A. Gray [Smooth, in reference the leaf surfaces.] INKBERRY; GALLBERRY.

Prinos glaber Linnaeus, Sp. Pl. 330. 1753. *Winterlia trifolia* Moench, Methodus 74. 1794, nom. illegit. *Ennepta myricoides* Rafinesque, Sylva Tellur. 52. 1838, nom. illegit. *Ilex glabra* (Linnaeus) A. Gray, Manual, ed. 2. 264. 1856.
Ilex glabra (Linnaeus) A. Gray forma *leucocarpa* F. W. Woods, Rhodora 58: 25, t. 1222. 1956. TYPE: FLORIDA: Jackson Co.: ca. 4 mi. S of Mariana, 20 Jan 1955, *Woods C.E.F. 2* (holotype: US; isotypes: F, FLAS, FSU, GA, GH, MO, NY, TENN, UNA).

Shrub, to 3 m; branchlets of the season green, finely powdery pubescent, the older ones grayish brown, glabrous, the lenticels with vertical slits. Leaves persistent, the blade elliptic to oblanceolate or oblanceolate, 2–5 cm long, 1.5–2.5 cm wide, the upper surface glabrous, sublustrous, the lower surface pale, dull green, glabrous, with scattered punctate glands, the apex obtuse, with a small mucro, the base cuneate, the margin entire or with 1–2(3) small, appressed teeth on a side distally, the petiole 3–8 mm long. Plants dioecious or polygamodioecious, the staminate flowers in an axillary pedunculate cyme, the peduncle 6–10 mm long, the pedicel 6–10 mm long, the carpellate flowers solitary or 2–3 in the leaf axil, the peduncle 2–5 mm long, the pedicel 2–5 mm long; calyx 5- to 8-lobed, the lobes short, triangular, the outer surface glabrous; corolla white, 5- to 8-lobed, the lobes oblong. Fruit globose, 5–7 mm long, black or rarely white, dull or sublustrous; pyrenes 5–8, 3–4 mm long, smooth on the rounded side.

Flatwoods, bogs, and coastal swales. Common; nearly throughout. Nova Scotia and Maine south to Florida, west to Texas. Winter–spring.

Ilex krugiana Loes. [Commemorates Karl Wilhelm Leopold Krug (1833–1898), German businessman, naturalist, and supporter of scientific collections.] TAWNYBERRY HOLLY; KRUG'S HOLLY.

Ilex krugiana Loesener, in Urban, Bot. Jahrb. Syst. 15: 317. 1892.

Shrub or tree, to 15 m; branchlets gray, glabrous. Leaves persistent, the blade elliptic to ovate, 5–7(11) cm long, (2)2.5–4(5) cm wide, the apex acuminate to acute, the base rounded to broadly cuneate, the margin entire, the upper and lower surfaces glabrous, the upper one lustrous, the petiole (0.5)1–1.5(2) cm long. Plants dioecious, the staminate flowers solitary or in a 2- to 3-flowered cyme in fascicles in the axil of the leaves, the carpellate flowers solitary in fascicles in the axil of the leaves, the peduncle ca. 1 mm long, the pedicel 5–8 mm long; calyx 4-lobed, the lobes triangular; corolla white, 4-lobed, the lobes ovate. Fruit subglobose, 3–4 mm long, dark purple or black; pyrenes 4, ca. 3 mm long, the rounded side smooth or furrowed.

Rockland hammocks and pine rocklands. Florida; West Indies. Rare; Miami-Dade County. Spring.

Ilex krugiana is listed as threatened in Florida (Florida Administrative Code, Chapter 5B-40).

Ilex longipes Chapm. ex Trel. [Long-footed, in reference to the long carpellate flower or fruit pedicel.] GEORGIA HOLLY.

Ilex longipes Chapman ex Trelease, Trans. St. Louis Acad. Sci. 5: 346. 1889. *Ilex decidua* Walter var. *longipes* (Chapman ex Trelease) Ahles, J. Elisha Mitchell Sci. Soc. 80: 173. 1964. *Ilex decidua* walter subsp. *longipes* (Chapman ex Trelease) E. Murray, Kalmia 13: 8: 1983.

Shrub or tree, to 7 m; branchlets of the season greenish brown, glabrous, the older ones gray, with nearly circular lenticels, the bark thin, grayish, smooth or slightly roughened. Leaves deciduous, the blade elliptic to obovate, 4–6 cm long, 2–3 cm wide, the apex short-acuminate to subobtuse, the base cuneate, the margin crenate-serrate, the teeth tipped with a spinelike bristle, the upper surface glabrous, the lower surface glabrous or sparsely hirtellous, mainly along the midrib, the petiole 3–10 mm long. Plants dioecious, the staminate flowers in fascicles at the junction of the spur shoot of the previous season, on the branches of current season, and solitary in the axil of the developing leaves, the pedicel 0.5–1(2) cm long, the carpellate flowers solitary or 2–3 in the leaf axil or the node on the branches of the previous season, the pedicel 1–2.3 cm long; calyx 4- to 5-lobed, the lobes deltoid, the corolla white, 4- to 5-lobed, the lobes oblong to lanceolate. Fruit subglobose, 7–10 mm long, red, orange-red, or yellow; pyrenes 5, 4–5 mm long, irregularly ridged on the rounded side.

Mesic hammocks. Rare; central and western panhandle. Virginia south to Florida, west to Texas. Spring.

Ilex opaca Aiton [Opaque, in reference to the thick leaves.] AMERICAN HOLLY.

Tree, to 15 m; branchlets of the season sparsely short-pubescent, becoming glabrous, the older branchlets brown, roughened, with circular raised lenticels, the bark light gray, slightly roughened with wart-like processes. Leaves persistent, the blade elliptic, ovate, or obovate, 3–10(12) cm long, 1–5.5 cm wide, stiff-leathery, the apex acute, spine-tipped, the base narrowly to broadly cuneate, the margin entire or with spine-tipped teeth, these mostly distal, the upper surface of the young leaves short-pubescent along the midrib, becoming glabrate or glabrous, the lower surface glabrous or with a few short trichomes along the midrib, the petiole 5–12(18) cm long, short-pubescent. Plants dioecious, the staminate flowers in cymes on the leafless part of the branchlet, in the axil of a developing leaf, or in the axil of a leaf of the previous season, the peduncle 2–5 mm long, the pedicel 2–5 mm long; carpellate flowers solitary and disposed as the staminate, the peduncle ca. 1 mm long, the pedicel 3–4 mm long; calyx 4-lobed, the lobes triangular, the margin somewhat erose; corolla white or cream-colored, 4-lobed, the lobes elliptic. Fruit globose or ovate, red, orangish red, or yellow, 7–19(12) mm long; pyrenes 4, 6–8 mm long, irregularly ribbed on the rounded side.

1. Leaves, 3.5–5.5 cm wide, the upper surface dark green, the margin flat or only slightly revolute .... var. **opaca**
1. Leaves 1–2.5 cm wide, the upper surface light green or yellowish, the margin distinctly revolute .... var. **arenicola**

Ilex opaca var. **opaca** AMERICAN HOLLY.

Ilex opaca Aiton, Hort. Kew. 1: 169. 1789. *Ageria opaca* (Aiton) Rafinesque, Sylva Tellur. 47. 1838.
Ilex opaca Aiton forma *subintegra* Weatherby, Rhodora 23: 119. 1921. *Ilex opaca* Aiton var. *subintegra* (Weatherby) Rehder, Man. Cult. Trees 543. 1927.
Ilex opaca Aiton var. *xanthocarpa* Rehder, Mitt. Deutsch. Dendrol. Ges. 1907: 73. 1807.

Leaf blade 3.5–5.5 cm wide, the upper surface dark green, the margin flat or only slightly revolute; fruit red or yellow.

Mesic hammocks. Occasional; northern counties, central peninsula. New York and Massachusetts south to Florida, west to Michigan, Illinois, Missouri, Oklahoma, and Texas. Spring.

Ilex opaca var. **arenicola** (Ashe) Ashe [*Arena,* sandy place, and *icola,* to dwell, in reference to its habitat.] SCRUB HOLLY.

Ilex arenicola Ashe, J. Elisha Mitchell Sci. Soc. 40: 44. 1924. *Ilex opaca* Aiton var. *arenicola* (Ashe) Ashe, Quart. Charleston Mus. 1(2): 31. 1925. *Ilex opaca* Aiton subsp. *arenicola* (Ashe) E. Murray, Kalmia 12: 21. 1982. TYPE: FLORIDA: Lake Co.: near Astor Park, Apr 1923, *Ashe s.n.* (holotype: NCU?; isotype: NY).
Ilex cumulicola Small, Bull. Torrey Bot. Club 51: 382. 1924. TYPE: FLORIDA: Highlands Co.: Lake Nancesowe (Lake Jackson), 25 Dec 1922, *Small et al. 10752* (syntype: NY).

Ilex pygmaea McFarlin, Rhodora 34: 17. 1932. TYPE: FLORIDA: Polk Co.: Lake Marion, s.d., *McFarlin 4508* (holotype: MICH).
Ilex arenicola Ashe forma *sebringensis* McFarlin, Rhodora 34: 18. 1932. TYPE: FLORIDA: Highlands Co.: Sebring, 9 Jun 1931, *McFarlin 5714* (Holotype: MICH).
Ilex arenicola Ashe forma *oblanceolata* McFarlin, Rhodora 34: 234. 1932. TYPE: FLORIDA: Polk Co.: Lake Marion, s.d., *Poole 2* (holotype: MICH).
Ilex arenicola Ashe var. *obovata* McFarlin, Rhodora 34: 234, Pl. 226. 1932. TYPE: FLORIDA: Polk Co.: Lake Marion, s.d. *McFarlin 4507* (holotype: MICH).
Ilex arenicola Ashe var. *paucidens* McFarlin, Rhodora 34: 235, pl. 227. 1932. TYPE: FLORIDA: Lake Marion, s.d. *McFarlin 4501* (holotype: MICH).
Ilex arenicola Ashe var. *transiens* McFarlin, Rhodora 34: 235, pl. 228. 1932. TYPE: FLORIDA: Polk Co.: Deer Lake, Winter Haven, s.d., *McFarlin 5174* (holotype: MICH).
Ilex pygmaea McFarlin var. *subedentata* McFarlin, Rhodora 34: 235, pl. 230. 1932. TYPE: FLORIDA: Polk Co.: Lake Marion, s.d., *Poole 5* (holotype: MICH).

Leaf blade 1–2.5 cm wide, the upper surface green or yellow, the margin distinctly revolute; fruit orangish red.

Scrub. Occasional; northern and central peninsula. Endemic. Spring.

Ilex verticillata (L.) A. Gray [Verticillate, in a whorled manner, in reference to the flowers and the fruits.] COMMON WINTERBERRY.

Prinos verticillata Linnaeus, Sp. Pl. 330. 1753. *Prinos confertus* Moench, Methodus 481. 1794, nom. illegit. *Prinos gronovii* Michaux, Fl. Bor.-Amer. 2: 236. 1803, nom. illegit. *Ilex verticillata* (Linnaeus) A. Gray, Manual, ed. 2. 264. 1856.

Shrub or small tree, to 8 m; branchlets sparsely pubescent, becoming glabrous, greenish, the older branches grayish brown, with circular tan lenticels, the bark smooth, thin, brown to dark gray. Leaves deciduous, with the blade elliptic, ovate, or obovate, 2–10 cm long, 1–5.5 cm wide, the apex acuminate, the base cuneate, the margin appressed-serrate, the upper surface sparsely short pubescent, mainly along the midrib, becoming glabrous, the lower surface pubescent, sometimes only on the veins. Plants dioecious, the staminate flowers usually in short verticels or rarely sessile in the leaf axils, the peduncle 2–6 mm long, the pedicel 2–5 mm long, both slightly pubescent, the pistillate flowers solitary or 2–4 in short-pedunculate or sessile verticels in the leaf axils; calyx 5- to 7(8)-lobed, the outer surface pubescent, the lobes triangular, the apex obtuse, the margin ciliate; corolla white, 5- to 7(8)-lobed, the lobes oblong. Fruit globose, 5–7 mm long, red or sometimes yellow; pyrenes 5 or 10, 3–4 mm long, smooth on the rounded side.

Swamps. Occasional; central and western panhandle. Quebec and Newfoundland south to Florida, west to Ontario, Minnesota, Iowa, Missouri, Arkansas, and Texas. Spring.

Ilex vomitoria Aiton [To cause emesis, in reference to the effect when drunk.] YAUPON.

Ilex vomitoria Aiton, Hort. Kew. 1: 170. 1798. TYPE: FLORIDA: without data (lectotype: Catesby, Nat. Hist. Carolina 2: t. 57. 1736). Lectotypified by González-Gutiérrez and Sierra-Calzado, in Greuter and Rankin Rodríguez (2004: 12).

Ilex cassena Michaux, Fl. Bor.-Amer. 2: 229. 1803. *Ageria cassena* (Michaux) Rafinesque, Sylva Tellur. 47. 1838. TYPE: "a Carolina ad Floridam."

Shrub or tree, to 8 m; branchlets sparsely to densely short-pubescent when young, the older ones becoming glabrous with a thin waxy gray coating that breaks into an interlacing pattern, the bark thin, gray, smooth. Leaves persistent, the blade coriaceous, ovate, elliptic, oblong-elliptic, or lanceolate, 0.5–3 cm long, 0.5–2.5 cm wide, the apex obtuse to rounded, the base rounded, the margin appressed-crenate. With small glandular mucros in the sinuses and in a notch at the apex, the upper surface lustrous, short pubescent only along the midrib when young, becoming glabrous, the lower surface paler and dull green, glabrous, the petiole 2–3 mm long, short-pubescent. Flowers in short-pedunculate or sessile fascicles in the axil of the leaves at leafless nodes, plants dioecious or sometimes with a few bisexual flowers with the staminate, the pedicel 2–3 mm long, short-pubescent; calyx 4-lobed, the outer surface short-pubescent, the lobes short-triangular, the apex obtuse to rounded; corolla white or yellowish, 4-lobed, the lobes oblong, usually with the wide tip curled inward. Fruit globose, 4–8 mm long, red or sometimes yellow; pyrenes 4, 3–4 mm long, shallowly ribbed on the rounded side.

Hammocks, swamps, flatwoods, and dunes. Frequent; northern counties, central peninsula. West Virginia south to Florida, west to Oklahoma and Texas; West Indies and Mexico. Native to North America and Mexico. Spring.

HYBRID TAXON

Ilex ×attenuata Ashe [*I. cassine* × *I. opaca*] [In reference to the leaf base.] EAST PALATKA HOLLY; TOPAL HOLLY.

Ilex ×attenuata Ashe, J. Elisha Mitchell Sci. Soc. 40: 44. 1924, pro sp. TYPE: FLORIDA: Walton Co.: near Mossyhead P.O., Apr & Oct 1923, *Ashe s.n.* (holotype: NCUP?; isotype: NY). Spring.

Hammocks and disturbed sites. Rare; Walton County. North Carolina, South Carolina, Florida, and California (escaped from cultivation). Spring.

EXCLUDED TAXA

Ilex lucida (Aiton) Torrey & A. Gray ex S. Watson—Reported by Chapman (1897) and Small (1903), who misapplied the name to Florida material of *I. coriacea.*

Ilex montana Torrey & A. Gray—This mainly Appalachian species was reported for Florida by Godfrey (1988), Wunderlin (1998), and Wunderlin and Hansen (2003, 2011), all as *I. ambigua* (Michaux) Torrey var. *monticola* (A. W. Wood) Wunderlin & Poppleton, the name misapplied to *I. ambigua.*

CAMPANULACEAE Juss. 1789. BELLFLOWER FAMILY

Herbs. Leaves alternate, simple, pinnate-veined, petiolate or epetiolate, estipulate. Flowers in spikes, racemes, or solitary in the leaf axils, bisexual, bracteolate or ebracteolate; sepals 5, basally connate, actinomorphic or zygomorphic; petals 5, basally connate, actinomorphic or zygomorphic; stamens 5, free or epipetalous; ovary superior or inferior, 2-carpellate, 2- to 5-loculate, the style 1. Fruit a capsule; seeds numerous.

A family of about 84 genera and about 2,380 species; nearly cosmopolitan.

Circumscription of the family has long been controversial with some authors recognizing Campanulaceae s.l. and others the segregate Lobeliaceae.

Lobeliaceae Juss. ex Bonpl. 1813.

1. Corolla zygomorphic, longitudinally split on the upper side .... **Lobelia**
1. Corolla actinomorphic, not split on the upper side.
   2. Corolla tube 8–16 cm long; anthers and filaments connate into a tube around the style .... **Hippobroma**
   2. Corolla tube to 1 cm long; anthers and filaments free.
      3. Flowers sessile .... **Triodanis**
      3. Flowers pedicellate.
         4. Corolla funnelform to campanulate; capsule opening by 2–3 apical pores; stem leaves reduced upward .... **Wahlenbergia**
         4. Corolla rotate; capsule opening by 3–5 lateral pores; stem leaves not much reduced upward .... **Campanula**

Campanula L. 1753. BELLFLOWER

Annual or perennial herbs. Leaves alternate, simple, pinnate-veined, petiolate or epetiolate. Flowers in spikes, racemes, or solitary or in a cyme in the leaf axils, bisexual, bracteolate; sepals 5, basally connate, actinomorphic; petals 5, basally connate, actinomorphic; stamens 5, free; ovary superior, 2-carpellate, 3- to 5-loculate, the style 1. Fruit a capsule, dehiscence by lateral pores; seeds numerous.

A genus of about 240 species; North America, Europe, Africa, and Asia. [Diminutive of *campana,* bell, in reference to the corolla shape.]

The circumscription of *Campanula* is controversial and requires further study (Roquet et al., 2008).

Campanulastrum Small, 1903; *Rotantha* Small, 1933.

1. Plant erect; leaf margin conspicuously serrate-crenate .... **C. americana**
1. Plant sprawling; leaf margin entire or obscurely toothed.
   2. Corolla 14–18 mm wide; sepals 6–9 mm long .... **C. floridana**
   2. Corolla 7–8 mm wide; sepals 1–3 mm long .... **C. robinsiae**

Campanula americana L. [Of America.] AMERICAN BELLFLOWER.

Campanula americana Linnaeus, Sp. Pl. 154. 1753. *Campanulastrum americanum* (Linnaeus) Small, Fl. S.E. U.S. 1141, 1338. 1903.

Erect annual herb, to 2 m; stem glabrous. Leaves with the blade elliptic, ovate, or ovate-oblong, 5–15 cm long, 2–6 cm wide, the apex acuminate, the base broadly cuneate, decurrent on the petiole, the margin serrate, the upper surface glabrous, the lower surface sparsely scaly-pubescent on the principal veins, the petiole to 2.5 cm long. Flowers in a terminal raceme, sometimes terminating short branchlets, solitary or 2–4 in an axillary cyme, the proximal flowers on a pedicel to 1 cm long, the upper sessile or subsessile; bracteole subulate, minute; hypanthium hemispheric, 3–5 mm long, calyx lobe linear-subulate, 5–12 mm long; corolla light blue, rotate, the lobes elliptic-oblong, 6–14 mm long. Fruit obconic, 7–12 mm long, dehiscing by 3–5 pores near the apex.

Floodplain and bluff forests. Rare; Gadsden, Liberty, and Jackson Counties. Ontario south to Florida, west to Minnesota, South Dakota, Nebraska, Kansas, Oklahoma, and Louisiana. Summer–fall.

Campanula floridana S. Watson ex A. Gray [Of Florida.] FLORIDA BELLFLOWER.

Campanula floridana S. Watson ex A. Gray, Fl. N. Amer. 2(2): 13. 1878. *Rotantha floridana* (S. Watson ex A. Gray) Small, Man. S.E. Fl. 1289, 1508. 1933. TYPE: FLORIDA.

Weakly ascending to reclining perennial herb, to 4 dm; stem rhizomatous, glabrous or short-pubescent. Leaves with the blade lanceolate, elliptic-lanceolate, linear-lanceolate, or linear, 1–4 cm long, 2–5 mm wide, the apex acute, the base cuneate, usually decurrent on the petiole, the margin finely serrate, the upper surface glabrous, the lower surface glabrous or sparsely short-pubescent on the veins, upper sessile, the lower short-petiolate. Flowers mostly solitary terminating a short branchlet, the pedicel to 2 mm long; bracteoles minute; hypanthium hemispheric, ca. 3 mm long; calyx lobes 4–6 mm long, subulate, equaling or a little longer than the corolla; corolla violet, rotate, the lobes oblong-elliptic, 4–9 mm long. Fruit narrowly obconic, 3–4 mm long, dehiscing by pores near the middle.

Cypress ponds, marshes, and pond margins. Common; peninsula west to central panhandle. Alabama and Florida. Collections made in 2013 from a cemetery in Mobile County, Alabama, are considered to be introduced there. Spring.

Campanula robinsiae Small [Commemorates Margaret Dreier Robins (1868–1945), wife of Raymond Robins (1873–1954), of Chinsegut Hill.] ROBINS' BELLFLOWER; CHINSEGUT BELLFLOWER.

Campanula robinsiae Small, Torreya 26: 35. 1926. *Rotantha robinsiae* (Small) Small, Man. S.E. Fl. 1289, 1508. 1933. TYPE: FLORIDA: Hernando Co.: Chinsegut Hill near Brooksville, *Small & Robins s.n.* (holotype: NY).

Weakly ascending to reclining annual herb, to 15 cm; stem glabrous. Leaves with the blade ovate, elliptic-ovate, elliptic, lanceolate, or linear-lanceolate, 6–12 mm long, 2–6 mm wide, the apex acute to obtuse, the base narrowly to broadly cuneate, decurrent on the petiole, the margin entire or serrate, the upper and lower surfaces glabrous, sessile or the petiole to 4 mm long. Flowers solitary in the axil of the leaves, the pedicel 2–6 mm long; hypanthium hemispheric, ca. 2 mm long; sepal lobes lanceolate or subulate-lanceolate, 1–2 mm long; corolla pale blue, rotate, the lobes elliptic-ovate to elliptic-lanceolate, longer than the tube. Fruit subglobose, ca. 2 mm long, dehiscent by 3 basal pores.

Seepage areas on slopes and pond margins. Rare; Hernando and Hillsborough Counties. Endemic. Spring.

Campanula robinsiae was known only from Hernando County until discovered in Hillsborough County by Carmel vanHoek in 2006.

Campanula robinsiae is listed as endangered in Florida (Florida Administrative Code, Chapter 5B-40) and in the United States (U.S. Fish and Wildlife Service, 50 CFR 23).

Hippobroma G. Don 1834.

Perennial herbs. Leaves alternate, simple, pinnate-veined, petiolate or epetiolate. Flowers solitary in the leaf axils, bracteolate; sepals 5, basally connate, actinomorphic; petals 5, basally connate, actinomorphic; stamens 5, epipetalous, the filaments and anthers connate into a tube; ovary inferior, 2-carpellate and -loculate, the stigma 1. Fruit a capsule, dehiscence loculicidal; seeds numerous.

A monotypic genus; North America, West Indies, Mexico, Central America, South America, Africa, Asia, and Pacific Islands. [From the Greek *hippos,* horse, and *bromos,* poison, the species considered to be a strong poison fatal to horses.]

Hippobroma is sometimes placed in Lobeliaceae by authors. Also, depending on the circumscription of *Lobelia,* our species may be placed in that genus (Antonelli, 2008; Chen et al., 2016; Lagomarsino et al., 2014).

Hippobroma longiflora (L.) G. Don [With long flowers, in reference to floral tube.] MADAMFATE.

Lobelia longiflora Linnaeus, Sp. Pl. 930. 1753. *Rapuntium longiflorum* (Linnaeus) Miller, Gard. Dict., ed. 8. 1768. *Hippobroma longiflora* (Linnaeus) G. Don, Gen. Hist. 3: 717. 1834. *Isotoma longiflora* (Linnaeus) C. Presl, Prodr. Monogr. Lobel. 42. 1836. *Laurentia longiflora* (Linnaeus) Petermann, Pflanzenreich 444. 1845.

Erect perennial herb, to 3.5 dm; stem villous, sometimes glabrous or glabrate in age proximally. Leaves with the blade oblanceolate or elliptic, 7–15 cm long, 1–3.5 cm wide, the apex acute to acuminate, the base cuneate, the upper and lower surfaces densely to sparsely villous, sometimes sparsely so or glabrous, sessile or short-petiolate. Flowers solitary, axillary, the pedicel 3–10 mm long; bracteoles 2, filiform; hypanthium campanulate, obconic or ellipsoid, 6–8 mm long, villous; calyx lobes linear, 8–18 mm long, villous, the margin denticulate; corolla white, the tube 6.5–10 cm long, the lobes elliptic, 1.8–2.5 cm long, the outer surface villous; stamens with the filament tube adnate to the corolla tube, the anther tube ca. 7 mm long, exserted, the anthers with apical tufts of stiff trichomes. Fruit obconic, broadly ellipsoid, or obovoid, 1–1.5 cm long, densely villous, dehiscence by 2 apical valves.

Disturbed sites. Rare; Miami-Dade County. Florida; West Indies, Mexico, Central America, and South America; Africa, Asia, and Pacific Islands. Native to West Indies. All year.

Lobelia L. 1753.

Annual or perennial herbs. Leaves alternate, simple, pinnate-veined, petiolate or epetiolate. Flowers in spikes or racemes, bracteolate or ebracteolate; sepals 5, basally connate, zygomorphic;

petals 5, basally connate, zygomorphic; stamens 5, the filaments free at the base, connate into a tube distally, the anthers connate into a tube; ovary inferior, 2-carpellate and -loculate, the stigma 1. Fruit a capsule, dehiscence apical and loculicidal; seeds numerous.

A genus of about 415 species; nearly cosmopolitan. [Commemorates Mathias de L'Obel (1538–1616), Flemish physician and botanist.]

Lobelia is sometimes placed in the Lobeliaceae by authors.

Selected references: McVaugh (1943); Spaulding and Barger (2016).

1. Corolla red, rarely white, 3–4.5 cm long .... **L. cardinalis**
1. Corolla, blue, lavender, or white, to 2 cm long.
    2. Plant subscapose (upper leaves much reduced); basal leaves elongate and long-tapering to the margined petiole, these present at flowering time.
        3. Calyx lobe sinuses without auricles; corolla tube fenestrate (with lateral slits near the base between the calyx lobes); stamen filament tube 3–5 mm long .... **L. paludosa**
        3. Calyx lobe sinuses with small auricles; corolla not fenestrate; stamen filament tube 6–11 mm long .... **L. floridana**
    2. Plant not subscapose (upper leaves not much reduced); basal leaves not long-tapering to the margined petiole, these usually absent at flowering time.
        4. Stem leaves subulate, less than 1 mm wide .... **L. boykinii**
        4. Stem leaves 1 mm wide or wider.
            5. Stamen filament tube ca. 3 mm long.
                6. Leaves with the margin irregularly crenate or incised-dentate; capsule 5–7 mm long .... **L. homophylla**
                6. Leaves with the margin entire or shallowly crenate; capsule 2–4 mm long.
                    7. Plant with the lower stem usually trailing, rooting at the nodes, leafy, mat-forming .... **L. feayana**
                    7. Plant with the stem erect, not rooting at the nodes, with few leaves, not mat-forming .... **L. nuttallii**
            5. Stamen filament tube 5–16 mm long.
                8. Larger leaves to 3 cm long; calyx lobes with prominent basal auricles extending to the base of the floral tube .... **L. brevifolia**
                8. Larger leaves usually longer than 3 cm; calyx lobes lacking or with small basal auricles (these sometimes extending halfway to the base of the calyx tube in *L. rogersii*).
                    9. Principal stem leaves usually linear or lanceolate, less than 1 cm wide.
                        10. Larger leaves linear-lanceolate, usually less than 5 mm wide .... **L. glandulosa**
                        10. Larger leaves lanceolate to oblong-lanceolate, usually more than 5 mm wide .... **L. flaccidifolia**
                    9. Principal stem leaves elliptic, ovate, or oblanceolate, more than 1 cm wide.
                        11. Calyx sinuses with prominent drooping auricles often covering ½ of the calyx tube .... **L. rogersii**
                        11. Calyx sinuses lacking auricles or these very small and inconspicuous.
                            12. Stem densely pubescent throughout; corolla tube densely pubescent on the outer surface .... **L. puberula**
                            12. Stem sparsely pubescent or glabrous (more pubescent at the base); corolla tube glabrous or sparsely pubescent in lines on the outer surface.

13. Lower lip of the corolla villous on the upper surface near the tube throat ... . **L. apalachicolensis**
13. Lower lip of the corolla glabrous on the upper surface near the tube throat.
  14. Calyx lobe margins irregularly dentate with callose glandular teeth, sometimes only remotely toothed near the base .... **L. georgiana**
  14. Calyx lobe margins entire.. **L. amoena**

Lobelia amoena McVaugh [*Amoenus*, beautiful.] SOUTHERN LOBELIA.

Lobelia amoena Michaux, Fl. Bor.-Amer. 2: 153. 1803. *Rapuntium amoenum* (Michaux) C. Presl, Prodr. Monogr. Lobel. 23: 1836. *Dortmanna amoena* (Michaux) Kuntze, Revis. Pl. 972. 1891.

Erect perennial herb, to 1 m; stem sparsely pubescent or glabrous. Leaves with the blade lanceolate, elliptic, or ovate, 2–10 cm long, 1–4 cm wide, the apex acute, the base rounded to narrowly cuneate, the margin irregularly and finely dentate, undulate, or entire, the upper and lower surfaces glabrous, the petiole of the lower leaves to 2 cm long, the upper ones sessile or subsessile. Flowers in a loose raceme, the pedicel 3–4 m long; bracteoles linear-subulate, minute; hypanthium campanulate-hemispheric, ca. 2 mm long, glabrous; calyx lobes 6–8 mm long, ciliate basally, subulate distally, the margin entire, the basal auricles absent; corolla blue, 1.5–2 cm long, the lower lip glabrous on the inner surface, with 2 basal tubercles; stamen filament tube 6–8 mm long, the anther tube ca. 3 mm long. Fruit hemispheric, ca. 5 mm long.

Floodplain forests. Rare; Jackson County. Virginia south to Florida, west to Louisiana. Summer–fall.

Lobelia apalachicolensis D. D. Spaulding et al. [Of Apalachicola). APALACHICOLA LOBELIA.

Lobelia apalachicolensis D. D. Spaulding et al., Phytoneuron 2016-63: 1. 2016. TYPE: FLORIDA: Liberty Co.: FL 20 E of Hosford, between FL 65 and FL 267, 18 Oct 2014, *Spaulding & Taylor 14223* (holotype: AMAL; isotypes: ALNHS, AUA, FSU, JSU, UNA, NCU, USF, UWAL).

Erect, perennial herb, to 1.9 m; stem sparsely hirtellous or glabrate, usually more pubescent proximally. Leaves with the blade elliptic to ovate, 2.5–10 cm long, 1–2.5 cm wide, the apex acute to obtuse, the base cuneate, the margin with callose-glandular teeth, the upper surface sparsely hirtellous or glabrate, the lower surface moderately to densely hirtellous, sessile. Flowers in a raceme; bracteoles 1–2 mm long; hypanthium campanulate-hemispheric, densely or sparsely chaffy-hirsute or glabrous, the surface smooth or warty; calyx lobes linear-subulate, 6–8 mm long, the margin with callose-glandular teeth, the basal auricles small and inconspicuous or absent; corolla purplish to blue, ca. 2 cm long, the outer surface glabrous or sparsely pubescent in lines, the lower lip villous at the base on the inner surface, fenestrate; stamen filament tube 6–8 mm long, the anther tube ca. 3 mm long. Fruit hemispheric, ca. 6 mm long.

Wet pine flatwoods, bogs, and roadside depressions. Rare; central panhandle. Endemic. Summer–fall.

Lobelia boykinii Torr. & A. Gray ex A. DC. [Commemorates Samuel Boykin (1786–1848), physician and botanist in Georgia and Alabama.] BOYKIN'S LOBELIA.

Lobelia boykinii Torrey & A. Gray ex A. de Candolle, in de Candolle, Prodr. 7: 374. 1839. *Dortmanna boykinii* (Torrey & A. Gray ex A. de Candolle) Kuntze, Revis. Gen. Pl. 2: 972. 1891.

Erect perennial herb, to 8.5 dm; stem glabrous. Leaves with the blade linear or subulate, 0.5–1.5 cm long, 0.5 mm or less wide, the apex acuminate, the base cuneate, the margin entire or with a few minute callose-glandular teeth, sessile. Flowers in a loose raceme, the pedicel ca. 1 cm long; bracteoles absent; hypanthium campanulate, ca. 2 mm long, glabrous; calyx lobes linear-subulate, entire, without basal auricles; corolla blue with a white eye in the throat, 10–12 mm long, the lower lip usually pubescent on the inner surface, with 2 basal tubercles; stamen filament tube 3–5 mm long, the anther tube ca. 2 mm long. Fruit campanulate-hemispheric, ca. 3 mm long.

Cypress pond margins, bogs, and wet flatwoods. Rare; Gadsden, Jackson, Okaloosa, and Santa Rosa Counties. Delaware and Maryland, North Carolina south to Florida, west to Mississippi. Summer.

Lobelia boykinii is listed as endangered in Florida (Florida Administrative Code, Chapter 5B-40).

Lobelia brevifolia Nutt. ex A. DC. [With short leaves.] SHORTLEAF LOBELIA.

Lobelia brevifolia Nuttall ex A. de Candolle, in de Candolle, Prodr. 7: 377. 1839. *Dortmanna brevifolia* (Nuttall ex A. de Candolle) Kuntze, Revis. Gen. Pl. 2: 972. 1891.

Erect perennial herb, to 9 dm; stem glabrous or sparsely short-pubescent. Leaves with the blade spatulate-oblanceolate, oblong, or linear-oblong, 0.5–1.5(3) cm long, 2–5 mm wide, the apex obtuse to rounded, the base narrowly cuneate, the margin serrate, with callus-glandular teeth, the upper surface glabrous, the lower surface glabrous or sparsely short-pubescent, sessile or subpetiolate. Flowers in a terminal raceme, the pedicel 5–10 mm long, with retrorse spicule-like trichomes; bracteoles minute; hypanthium campanulate, 4–6 mm long, the outer surface with retrorse spicule-like trichomes; calyx segments triangular, with callus-glandular teeth, the basal auricles subequaling the hypanthium; corolla pale blue, grayish blue, azure blue, or purple, 1.5–2 cm long, the outer surface pubescent, usually fenestrate, the lower lip glabrous or pubescent on the inner surface near the base; stamen filament tube 6–11 mm long, the anther tube 3–4 mm long. Fruit campanulate-hemispheric, 5–6 mm long.

Wet flatwoods, bogs, and cypress pond margins. Frequent; central and western panhandle. Alabama, Florida, Mississippi, and Louisiana. Summer–fall.

Lobelia cardinalis L. [Red, like the robe worn by a Catholic cardinal, in reference to the flowers.] CARDINALFLOWER.

Lobelia cardinalis Linnaeus, Sp. Pl. 930. 1753. *Rapuntium cardinale* (Linnaeus) Miller, Gard. Dict., ed. 8. 1768. *Rapuntium coccineum* Moench, Suppl. Meth. 277. 1802, nom. illegit. *Lobelia coccinea*

Stokes, Bot. Mat. Med. 1: 343. 1812, nom. illegit. *Dortmanna cardinalis* (Linnaeus) Kuntze, Revis. Gen. Pl. 2: 380. 1891.
Lobelia cardinalis Linnaeus var. *meridionalis* Bowden, Canad. J. Genet. Cytol. 2: 241. 1960.

Erect perennial herb, to 2 m; stem glabrous or sparsely short-pubescent. Leaves with the blade lanceolate or elliptic, to 18 cm long, to 5 cm wide, the apex acute to acuminate, the base cuneate, the margin with small irregularly spaced teeth or shallowly dentate-serrate, these callose-tipped, the upper and lower surfaces glabrous or short-pubescent mainly on or near the veins, the lower leaves with petiole margined, to 3 cm long, becoming sessile upward. Flowers in a terminal raceme, the pedicel short-pubescent; bracteoles linear-subulate or narrow-lanceolate, longer than the pedicel, the margin callus-toothed; hypanthium glabrous or sometimes bristly- or scaly-pubescent; calyx lobes linear-subulate, flaring at the base, much shorter than the corolla, glabrous to sparsely pubescent, the basal auricles usually absent; corolla red or rarely white, 3–4.5 cm long, the tube fenestrate below; stamen filament tube 1.8–3.5 cm long, long-exserted from the corolla and about the same color, the anther tube 4–6 mm long, bluish gray. Fruit broadly campanulate to oblate, 8–10 mm long.

Floodplain forests and spring runs. Occasional; northern counties, central peninsula. Quebec south to Florida, west to Minnesota, Nebraska, and California; Mexico, Central America, and South America. Summer–fall.

Lobelia cardinalis is listed as threatened in Florida (Florida Administrative Code, Chapter 5B-40).

Lobelia feayana A. Gray [Commemorates William T. Féay (1803–1876), physician and botanist from Savannah, Georgia.] BAY LOBELIA.

Lobelia feayana A. Gray, Proc. Amer. Arts 12: 60. 1877. *Dortmanna feayana* (A. Gray) Kuntze, Revis. Gen. Pl. 2: 972. 1891. TYPE: FLORIDA: s.d. *Feay s.n.* (lectotype: GH). Lectotypified by Bowden, 1959: 56).

Erect or sprawling perennial herb, to 4 dm; stem decumbent at the base, often rooting at the nodes and mat forming, 1–3 cm long, leafy, the flowering portion weakly erect, 5–30 cm long, with 2 to several widely spaced leaves below the raceme, glabrous. Lower leaves with the blade orbicular to ovate, 1–1.5 cm long and wide, the apex rounded, the base subcordate to truncate, the margin crenate with minute reddish callosities in shallow indentations, the petiole 0.5–2 cm long, the upper leaves with the blade 0.5–1.5 cm long, 4–8 mm wide, lanceolate, ovate or oblanceolate, the apex acute, the base cuneate, the margin dentate, with callose-tipped teeth, sessile or short-petiolate. Flowers in a lax terminal raceme, the pedicel 2–5 mm long; bracteoles linear-subulate; hypanthium glabrous; calyx lobes triangular-subulate, the margin entire or with a few minute callose teeth distally, the basal auricles absent; corolla blue or purple with a white eye, 5–8 mm long, with 2 greenish tubercles at the base of the lower lip on the inner surface; stamen filament tube ca. 3 mm long, the anther tube 1–2 mm long. Fruit campanulate to oblate, 3–4 mm long.

Open, moist, grassy areas. Common; peninsula west to eastern panhandle. Florida and Georgia. Spring.

Lobelia flaccidifolia Small [With flaccid leaves, not able to hold up its own weight.] FOLDEAR LOBELIA.

Lobelia flaccidifolia Small, Bull. Torrey Bot. Club 24: 338. 1897.

Erect annual herb, to 7 dm; stem glabrous. Leaves with the blade lanceolate to oblong-lanceolate, 5–8 cm long, 5–8 mm wide, the upper ones reduced, the apex acute or obtuse, the base cuneate, the margin undulate, with minute sunken callose-glandular teeth, the upper and lower surfaces glabrous, the lower ones short petiolate, the upper sessile. Flowers in a loose raceme, the pedicel 4–11 mm long; bracteoles linear; hypanthium rough-pubescent; calyx lobes triangular-subulate, the outer surface pubescent, the basal auricles sagittate; corolla blue, lavender, or nearly white with a white eye, 1.5–2 cm long, the outer surface glabrous or rough pubescent, sometimes fenestrate, the lower lip somewhat pubescent near the base on the inner surface, with 2 basal tubercles; stamen filament tube 5–8 mm long, the anther tube 2–3 mm long. Fruit hemispheric, 4–6 mm long.

Floodplain forests. Rare; panhandle. Georgia and Florida. Georgia, Alabama, and Florida, also Louisiana and Texas. Summer–fall.

Lobelia floridana Chapm. [Of Florida.] FLORIDA LOBELIA.

Lobelia floridana Chapman, Bot. Gaz. 3: 9. 1878. *Lobelia paludosa* Nuttall var. *floridana* (Chapman) A. Gray, Syn. Fl. N. Amer. 2(1): 393. 1878. TYPE: FLORIDA.

Erect perennial herb, to 1 m; stem glabrous. Basal leaves with the blade narrowly oblanceolate to linear, 1–3 dm long, 0.5–1.5 cm wide, the apex acute to obtuse, the base cuneate, tapering to a winged petiole, the margin entire or crenate, with callose teeth, the upper and lower surfaces glabrous, the upper ones similar, remotely spaced, and reduced, 1–5 cm long, 3–8 mm wide. Flowers in a raceme, the pedicel 3–8 mm long; bracteoles subulate, slightly exceeding the pedicel, the margin with callose teeth; hypanthium campanulate-hemispheric, ca. 3 mm long; calyx lobes triangular-subulate, 2–6 mm long, the margin with fine callose teeth, the basal auricles small; corolla pale blue to nearly white, sometimes with a pinkish tinge, 1.3–2 cm long, the outer surface pubescent, the lower lip pubescent at the base on the lower surface, the tube not fenestrate; stamen filament tube 6–11 mm long, the anther tube ca. 3 mm long. Fruit campanulate-hemispheric, 4–5 mm long.

Wet flatwoods, cypress pond margins, and bogs. Frequent; central and western panhandle. North Carolina and Georgia south to Florida, west to Texas. Summer–fall.

Lobelia georgiana McVaugh [Of Georgia.] SOUTHERN LOBELIA.

Lobelia amoena Michaux var. *glandulifera* A. Gray, Syn. Fl. N. Amer. 2(1): 4: 1878. *Lobelia glandulifera* (A. Gray) Small, Fl. S.E. U.S. 1144, 1338. 1903; non (de Candolle) Kuntze, 1891. *Lobelia georgiana* McVaugh, Bull. Torrey Bot. Club 67: 144. 1940. TYPE: FLORIDA.

Lobelia amoena Michaux var. *obtusata* A. Gray, Syn. Fl. N. Amer. 2(1): 4. 1878. TYPE: FLORIDA: s.d., *Chapman s.n.* (holotype: GH).

Erect to sprawling perennial herb, to 1 m; stem sparsely pubescent to glabrous. Leaves with the blade lanceolate, elliptic, or ovate, 2–10 cm long, 1–4 cm wide, the apex acute, the base rounded to narrowly cuneate, the margin irregularly and finely dentate, undulate, or entire, the upper and lower surfaces glabrous, the petiole of the lower leaves to 2 cm long, the upper ones sessile or subsessile. Flowers in a loose raceme, the pedicel 3–4 cm long; bracteoles linear-subulate, minute; hypanthium campanulate-hemispheric, ca. 2 mm long, glabrous; calyx lobes 6–8 mm long, ciliate basally, subulate distally, the margin irregularly dentate with callose-glandular teeth, the basal auricles absent; corolla blue, 1.5–2 cm long, the lower lip glabrous on the inner surface, with 2 basal tubercles; stamen filament tube 6–8 mm long, the anther tube ca. 3 mm long. Fruit hemispheric, ca. 5 mm long.

Floodplain forests. Occasional; northern and central peninsula, central and western panhandle. Virginia south to Florida, west to Alabama. Summer–fall.

Lobelia glandulosa Walter [Gland-bearing, in reference to the glands on the leaves and calyx lobe margins.] GLADE LOBELIA.

Lobelia glandulosa Walter, Fl. Carol. 218. 1788. *Rapuntium glandulosum* (Walter) C. Presl, Prodr. Monogr. Lobel. 21. 1836. *Dortmanna glandulosa* (Walter) Kuntze, Revis. Gen. Pl. 2: 972. 1891.

Lobelia crassiuscula Michaux, Fl. Bor.-Amer. 2: 152. 1803. TYPE: "Carolinae maritimae usque ad Floridam."

Lobelia glandulosa Walter var. *laevicalyx* Fernald, Rhodora 49: 186. 1947.

Erect perennial herb, to 1 m; stem glabrous. Leaves with the blade linear to narrowly oblanceolate, 8–15 cm long, 3–7 mm wide, the apex acute, the base cuneate, the margin callose-glandular or entire and undulate, often with short appressed trichomes on or near the edge, sessile or subsessile. Flowers in a raceme, the pedicel 5–12 mm long, glabrous or with short, knobby, whitish trichomes; bracteoles minute; hypanthium campanulate, 3–4 mm long, glabrous or with short, knobby, whitish trichomes; calyx lobes linear-subulate, subequaling the pedicel, the margin with callose-glandular teeth, the basal auricles or minute; corolla blue, 2–3 cm long, the lower lip pubescent at the base on the inner surface, fenestrate; stamen filament tube 7–8 mm long, the anther tube 3–4 mm long. Fruit hemispheric, 6–8 mm long.

Wet flatwoods, swamps, and bogs. Common; nearly throughout. Delaware and Maryland south to Florida, west to Alabama. Summer–fall.

Lobelia homophylla E. Wimm. [With leaves all the same.] PINELAND LOBELIA.

Lobelia homophylla E. Wimmer, Repert. Spec. Nov. Regni Veg. 22: 194, pl. 32(12). 1926. TYPE: Florida: St. Johns Co.: St. Augustine, s.d., *Curtiss 6363* (holotype: ?; isotype: US).

Erect annual herb, to 9 dm; stem glabrous. Leaves with the blade ovate to suborbicular, 1.5–3 cm long, 1.5–2.5 cm wide, the apex obtuse to rounded, the base truncate to subcordate, the margin irregularly crenate, dentate, or dentate-serrate, the upper and lower surfaces glabrous, the petiole 0.5–2 cm long, the uppermost leaves sessile or subsessile. Flowers in a loose raceme, the pedicel ca. 1 cm long, filiform; ebracteolate; hypanthium campanulate, 1–2 cm long; calyx lobes subulate, 2–3 mm long, glabrous or ciliate with few short, stiff trichomes, the basal auricles absent; corolla blue, 6–8 mm long, not fenestrate, the lower lip pubescent at the base on the lower surface. Fruit ovate, 5–7 mm long.

Open wet hammocks. Frequent; peninsula. Endemic. Spring–fall.

Lobelia nuttallii Schult. [Commemorates Thomas Nuttall (1786–1859), English botanist and zoologist who worked in America from 1808 to 1844).] NUTTALL'S LOBELIA.

Lobelia nuttallii Schultes, in Roemer & Schultes, Syst. Veg. 5: 39. 1819. *Rapuntium nuttallianum* C. Presl, Prodr. Monogr. Lobel. 23. 1836, nom. illegit. *Dortmanna nuttallii* (Schultes) Kuntz, Revis. Gen. Pl. 2: 973. 1891.

Erect perennial herb, to 7 dm; stem glabrous, sometimes slightly pubescent proximally. Basal leaves (usually absent at anthesis) with the blade ovate to broadly elliptic or oblanceolate, 1.5–2 cm long, 1–1.5 cm wide, the apex obtuse to rounded, the base decurrent on the petiole, the margin entire or with a few obscure callose teeth, the upper and lower surfaces glabrous or nearly so, the petiole to 1 cm long, the upper leaves with the blade lanceolate to linear, 1–2.5 cm long, 2–4(5) mm wide, the apex acute to obtuse, the base cuneate, the margin with a few callose teeth, the upper and lower surfaces glabrous, sessile. Flowers in a raceme, the pedicel ca. 5 mm long, filiform; bracteoles minute; hypanthium hemispheric, ca. 2 mm long; calyx lobes linear-subulate, ca. 2 mm long; corolla blue with a white eye at the throat, 5–7 mm long, the lower lip glabrous at the base on the inner surface, with 2 tubercles; stamen filament tube ca. 3 mm long, the anther tube 1–2 mm long. Fruit hemispheric, 2–3 mm long.

Wet flatwoods, bogs, and seepage slopes. Occasional; central and western panhandle. New York south to Florida, west to Kentucky, Tennessee, and Alabama. Spring–fall.

Lobelia paludosa Nutt. [*Paludosa*, swampy, in reference to the habitat.] WHITE LOBELIA.

Lobelia paludosa Nuttall, Gen. N. Amer. Pl. 2: 75. 1818. *Dortmanna paludosa* (Nuttall) G. Don, Gen. Hist. 3: 715. 1834. *Rapuntium paludosum* (Nuttall) C. Presl, Prod. Monogr. Lobel. 18. 1836.
Lobelia nudicaulis Rafinesque, Atl. J. 147. 1832. TYPE: FLORIDA.

Erect perennial herb, to 6 m; stem glabrous. Basal leaves with the blade narrowly oblanceolate to linear, 5–12 cm long, 0.5–1 cm wide, the apex acute to obtuse, the base cuneate, tapering to a winged petiole, the margin entire or crenate, with callose teeth, the upper and lower surfaces glabrous, the upper ones similar, remotely spaced, and reduced, 1–5 cm long, 3–8 mm wide. Flowers in a raceme, the pedicel 3–8 mm long; bracteoles minute or absent; hypanthium campanulate-hemispheric, ca. 3 mm long; calyx lobes triangular-subulate, 2–3 mm long, the margin with fine callose teeth, the basal auricles absent; corolla pale blue to nearly white, ca. 1.5 cm long, the outer surface pubescent, the lower lip pubescent at the base on the lower surface, the tube fenestrate; stamen filament tube 3–5 mm long, the anther tube 2–3 mm long. Fruit campanulate-hemispheric, 4–5 mm long.

Wet flatwoods. Common; nearly throughout. Georgia and Florida. Spring–fall.

Lobelia puberula Michx. [With short, soft trichomes.] DOWNY LOBELIA.

Lobelia puberula Michaux, Fl. Bor.-Amer. 2: 152. 1803. *Rapuntium puberulum* (Michaux) C. Presl, Prodr. Monogr. Lobel. 23. 1836. *Dortmanna puberula* (Michaux) Kuntze, Revis. Gen. Pl. 2: 973. 1891.

Lobelia puberula Michaux var. *glabella* Hooker, Bot. Mag. 61: pl. 3292. 1834; non Elliott, 1821. *Rapuntium puberulum* (Michaux) C. Presl var. *glabellum* C. Presl, Prodr. Monogr. Lobel. 23: 1836. *Lobelia puberula* Michaux var. *laeviuscula* C. Mohr, Contr. U.S. Natl. Herb. 6: 750. 1901.

Lobelia puberula Michaux var. *simulans* Fernald, Rhodora 49: 184. 1947. *Lobelia puberula* Michaux forma *simulans* (Fernald) Bowden, Bull. Torrey Bot. Club 86: 98. 1956.

Erect perennial herb to 1.5 m; stem densely puberulent to short-hirsute, sometimes becoming glabrate in age. Leaves with the blade oblanceolate, obovate, elliptic, or lanceolate, the lower 3–12 cm long, 1–4 cm wide, the lower ones the largest and becoming gradually reduced upward, the apex acute, the base cuneate, the margin coarsely to finely serrate or dentate, with callose teeth, sometimes with the margin entire and with the callosities along the edge, the upper and lower surfaces puberulent, rarely becoming glabrate, sessile. Flowers in a terminal raceme, the pedicel 3–5 mm long; bracteoles small, basal; hypanthium hemispheric, 2–3 mm long, the outer surface puberulent; calyx lobes subulate, 5–7 mm long, the margin with callose teeth, the outer surface puberulent, the base with small auricles; corolla blue or purple with a white eye in the throat, 1.5–2 cm long, the outer surface puberulent, the lower lip glabrous at the base on the inner surface; stamen filament tube 6–15 mm long, the anther tube 3–6 mm long. Fruit hemispheric, 5–7 mm long.

Open wet hammocks and swamps. Occasional; northern counties, central peninsula. Pennsylvania and New Jersey south to Florida, west to Illinois, Missouri, Oklahoma, and Texas. Summer–fall.

Lobelia rogersii Bowden [Commemorates Rogers McVaugh (1909–2009), botanist at the University of Michigan and authority on Asteraceae, Myrtaceae, Campanulaceae, Rosaceae, and the flora of Mexico.] MCVAUGH'S LOBELIA.

Lobelia rogersii Bowden, Canad. J. Bot. 39: 1688. 1961, pro. hybrid.

Erect perennial herb, to 1 m; stem densely short-hirsute, sometimes sparsely so or glabrate. Leaves with the blade lanceolate, elliptic, or oblong, 1–9 cm long, 0.5–2 cm wide, the apex acute to obtuse, the base broadly to narrowly cuneate, the margin dentate-serrate, with callose teeth, the upper and lower surfaces short-hirsute, the upper becoming glabrate in age, sessile or subpetiolate. Flowers in a terminal raceme, the pedicel 2–3 mm long, short-hirsute; bracteoles minute; hypanthium campanulate-hemispheric, 3–5 mm long, the surface with retrorse spicule-like hirsute trichomes; calyx lobes triangular-lanceolate, with callose teeth (rarely entire), the basal auricles large, drooping, subequaling the hypanthium; corolla blue with a white eye, 1.6–2 cm long, the outer surface pubescent, fenestrate, the lower lip pubescent on the inner surface near the base; stamen filament tube 6–11 mm long, the anther tube 3–4 mm long. Fruit campanulate-hemispheric, 5–6 mm long.

Wet flatwoods, bogs, and seeps. Occasional; central and western panhandle. Georgia south to Florida, west to Mississippi. Summer–fall.

Lobelia rogersii, originally described as a hybrid of *L. brevifolia* and L. *puberula*, was treated as such by Wunderlin and Hansen (2003, 2011) but is now considered to be a species of hybrid origin as proposed by Spaulding and Barger (2016) as followed here. It occurs where the ranges of *L. brevifolia* and *L. puberula* overlap in the south and also in the north where only *L. puberula* occurs.

EXCLUDED TAXA

Lobelia cliffortiana Linnaeus—Reported for Florida by Small (1903, 1913a, 1933), based on a misapplication of the name to our material of *L. homophylla.*

Lobelia elongata Small—This Atlantic coast species was reported for Florida by Small (1903, 1913a, 1933), the name misapplied to our material of *L. puberula.*

Lobelia spicata Lamarck—Reported for Florida by Small (1903). This species is known only from the north of Florida.

Lobelia xalapensis Kunth—Reported for Florida by Chapman (1878) and Small (1903), the name misapplied to Florida material of *L. homophylla.*

Triodanis Raf. ex Greene 1838. VENUS' LOOKING-GLASS

Annual herbs. Leaves alternate, simple, pinnate-veined, petiolate or epetiolate. Flowers axillary in a spiciform inflorescence, dimorphic, bisexual, bracteate; sepals 5, basally connate, actinomorphic; petals 5, basally connate, actinomorphic; stamens 5, free; ovary inferior, 2-carpellate and 2(3)-loculate, the stigma 1. Fruit a capsule, dehiscence by pores; seeds numerous.

A genus of 7 species; North America, Mexico, Central America, South America, Asia, and Australia. [From the Greek *tri,* three, and *odontos,* toothed, in reference to the 3 toothlike lobes in the cleistogamous flowers of some species.]

Selected reference: McVaugh (1945).

1. Pores of the capsule near the top; leaves longer than wide; chasmogamous (purple) flowers 1 or rarely 2 at the stem apex .... **T. biflora**
1. Pores of the capsule near or below the middle; leaves wider than long; chasmogamous flowers (purple) several in the upper leaf axils .... **T. perfoliata**

Triodanis biflora (Ruiz & Pav.) Greene [Two-flowers, in reference to the chasmogamous and cleistogamous flowers.] SMALL VENUS' LOOKING-GLASS.

Campanula biflora Ruiz López & Pavón, Fl. Peruv. 2: 55, pl. 200(b). 1799. *Specularia biflora* (Ruiz López & Pavón) C. A. Meyer, Index Sem. Hort. Petrop. 1: 17. 1836. *Pentagonia biflora* (Ruiz López & Pavón) Kuntze, Revis. Gen. Pl. 2: 381. 1891. *Legouzia biflora* (Ruiz López & Pavón) Britton, Mem. Torrey Bot. Club 5: 309. 1894. *Triodanis biflora* (Ruiz López & Pavón) Greene, Man. Bot. San Francisco 230. 1894. *Triodanis perfoliata* (Linnaeus) Nieuwland var. *biflora* (Ruiz López & Pavón) T. R. Bradley, Brittonia 27: 114. 1975. *Triodanis perfoliata* (Linnaeus) Nieuwland subsp. *biflora* (Ruiz López & Pavón) Lammers, Novon 16: 72. 2006.

Erect or reclining annual herb, to 4.5 dm; stem usually 5-angled, hirsute proximally, retrorsely hispid to scabrous distally. Leaves with the blade elliptic or ovate, 1–3 cm long, 6–10 mm wide, those subtending the flowers much reduced and bract-like, the apex of the lower ones obtuse, that of the upper acute, the base broadly cuneate, the margin crenate, sometimes with a few basal callose teeth, the upper and lower surfaces sparsely hispid to glabrous, the lower ones short petiolate, the upper sessile. Flowers 1(2–3) in the axil of the upper leaves and forming a spiciform inflorescence, dimorphic, the inflorescence with 1, rarely 2, terminal chasmogamous, the others cleistogamous; chasmogamous flowers with the hypanthium 3–6 mm long, glabrous to sparsely hispid; calyx lobes triangular to lanceolate, 1–2 mm long, erect to spreading, glabrous to sparsely hispid; corolla purple, sometimes with white lines, the tube 1–2 mm long, the lobes 5–10 mm long, the outer surface ciliate near the apex; cleistogamous flowers with the hypanthium and calyx lobes as in the chasmogamous. Fruit clavate to ovoid, 5–8 mm long, glabrous to sparsely hispid, the pores near the apex, ovate to round.

Disturbed sites. Occasional; northern counties, central peninsula. New York south to Florida, west to California; Mexico and South America; Asia. Native to North America. Spring–fall.

Triodanis perfoliata (L.) Nieul. [With the leaves meeting around the stem and the stem appearing to be growing through the leaves.] CLASPING VENUS' LOOKING-GLASS.

Campanula perfoliata Linnaeus, Sp. Pl. 169. 1753. *Campanula amplexicaulis* Michaux, Fl. Bor.-Amer. 1: 108. 1803, nom. illegit. *Prismatocarpus perfoliatus* (Linnaeus) Sweet, Hort. Brit. 251. 1826. *Specularia perfoliata* (Linnaeus) A. de Candolle, Monogr. Campan. 351. 1830. *Dysmicodon perfoliatum* (Linnaeus) Nuttall, Trans. Amer. Philos. Soc., ser. 2. 8: 256. 1843. *Pentagonia perfoliata* (Linnaeus) Kuntze, Revis. Gen. Pl. 2: 381. 1891. *Legouzia perfoliata* (Linnaeus) Britton, Mem. Torrey Bot. Club 5: 309. 1894. *Triodanis perfoliata* (Linnaeus) Nieuwland, Amer. Midl. Naturalist 3: 192. 1914.

Erect or decumbent annual herb; to 5(10) dm; stem usually 5-angled, hispid. Leaves with the blade cordate, 0.5–2(2.5) cm long and wide, slightly reduced upward, the apex acute, obtuse, or rounded, the base cordate, the margin crenate to serrate, the upper surface glabrous, the lower surface hispid or scabrous, sessile. Flowers 1–3 in the axil of most of the leaves and forming a spiciform inflorescence, dimorphic, primarily cleistogamous, but usually a few of the upper nodes chasmogamous; chasmogamous flowers with the hypanthium 3–5 mm long, slightly inflated, glabrous; calyx lobes triangular to lanceolate, 4–9 mm long, erect to spreading, glabrous or ciliate; corolla purple, sometimes with white lines or entirely white, the tube 1–2 mm long, the lobes triangular to elliptic, 5–7 mm long, the outer surface ciliate near the apex; cleistogamous flowers with the calyx lobes narrowly triangular, 2–3 mm long, otherwise as in the chasmogamous. Fruit ovoid, 4–7 mm long, glabrous to sparsely hispid, the pores near or below the middle, elliptic.

Disturbed sites. Frequent; northern counties, central peninsula. Nearly throughout North America. North America; West Indies, Mexico, Central America, and South America; Asia and Australia. Native to North America and Mexico. Spring–fall.

Wahlenbergia Schrad. & Roth, nom. cons. 1821.

Perennial herbs. Leaves alternate, simple, epetiolate. Flowers terminal or solitary in the leaf axils, ebracteate, ebracteolate; sepals 5, basally connate, actinomorphic; petals 5, basally connate, actinomorphic; stamens 5, free; ovary inferior, 3-carpellate and -loculate, the style 1. Fruit a capsule, dehiscence by pores.

A genus of about 250 species; nearly cosmopolitan. [Commemorates Georg (Göran) Wahlenberg (1780–1851), Swedish physician and naturalist.]

1. Hypanthium narrowly obconic to subcylindric, 3.5–4.5 mm long at anthesis, to 12 mm long in fruit; fruit opening by 2 terminal pores; plant perennial .... **W. linarioides**
1. Hypanthium ellipsoid or ovoid, 1.5–3 mm long at anthesis, 7.5 mm long in fruit; fruit opening by 3 terminal pores; plant annual .... **W. marginata**

Wahlenbergia linarioides (Lam.) A. DC. [Resembling *Linaria* (Plantaginaceae).] TUFFYBELLS.

Campanula linarioides Lamarck, Encycl. 1: 580. 1785. *Wahlenbergia linarioides* (Lamarck) A. de Candolle, Monogr. Campan. 158. 1830. *Campanopsis linarioides* (Lamarck) Kuntze, Revis. Gen. Pl. 2: 379. 1891.

Erect perennial herb, to 5 dm; stem glabrous. Leaves linear, (0.5)1–2.5 cm long, 3–4 mm wide, the apex acuminate to obtuse, the base cuneate, the margin entire or irregularly denticulate, the upper and lower surfaces glabrous, sessile. Flowers in a paniculiform inflorescence; pedicel 7–10 mm long; ebracteolate; hypanthium obconic, 4–6 mm long, the outer surface glabrous; calyx lobes triangular-lanceolate, 3–5 mm long; corolla blue-white, campanulate, 5–6 mm long, the lobes ovate-lanceolate. Fruit obconic, 10-veined, 8–10 mm long.

Disturbed sites. Occasional; Clay and Pasco Counties, central and western panhandle. Alabama and Florida; Mexico and South America. Native to South America. Spring.

Wahlenbergia marginata (Thunb.) A. DC. [In reference to the slightly recurved leaf margin.] SOUTHERN ROCKBELL.

Campanula marginata Thunberg, Fl. Jap. 89. 1784. *Campanopsis marginata* (Thunberg) Kuntze, Revis. Gen. Pl. 2: 378. 1891. *Wahlenbergia marginata* (Thunberg) A. de Candolle, Monogr. Campan. 143. 1830.

Erect perennial herb, to 6 dm; stem glabrous or puberulent. Leaves mostly basal, the blade oblanceolate, elliptic, or linear, 1–4 cm long, 2–3 mm wide, the apex acute, the base narrowly cuneate, the margin entire, sinuate, or sparsely serrulate, the upper and lower surfaces sparsely hirsute or glabrous, sessile. Flowers terminal or solitary in the leaf axil, the pedicel to 1–5 cm long; hypanthium obovoid or conic, glabrous; calyx lobes triangular or subulate, 2–3 mm long; corolla blue to violet, 5–8 mm long, the lobes shorter than the tube; the lobes obovate or oblong. Fruit turbinate to hemispheric, 3–6 mm long, 3-locular; seeds oblong, slightly compressed, ca. 0.5 mm long, smooth, brown.

Disturbed sites. Frequent; northern counties, central peninsula. North Carolina south to

Florida, west to Texas; Mexico; Asia, Australia, and Pacific Islands. Native to Asia, Australia, and Pacific Islands. Spring–fall.

MENYANTHACEAE Dumort. 1829. BOGBEAN FAMILY

Herbs. Leaves alternate, simple, palmate-veined, petiolate, estipulate. Flowers in axillary sub-umbellate clusters, ebracteate, ebracteolate, bisexual; sepals 5, basally connate, actinomorphic; petals 5, basally connate, actinomorphic; stamens 5, epipetalous; ovary superior, 2-carpellate, 1-loculate, the style 1. Fruit a capsule, indehiscent.

A family of 5 genera; nearly cosmopolitan.

Nymphoides Ség. 1754. FLOATINGHHEART

Perennial herbs. Leaves alternate, simple, palmate-veined, petiolate. Flowers in axillary sub-umbellate clusters, ebracteate, ebracteolate, bisexual; sepals 5, basally connate, actinomorphic; petals 5, basally connate, actinomorphic; stamens 5, epipetalous; ovary superior, 2-carpellate, 1-loculate, the stigma 1. Fruit a capsule, indehiscent.

A genus of about 50 species; nearly cosmopolitan. [To resemble *Nymphaea* (Nymphaeaceae).] *Nymphoides* is sometimes placed in the Gentianaceae by authors.

Selected references: Middleton et al. (2018); Tippery and Les (2011); Tippery et al. (2015).

1. Corolla conspicuously fimbriate on the inner surface.
   2. Corolla all yellow .... **N. grayana**
   2. Corolla white with a yellow center.
      3. Seed surface with tubercles .... **N. indica**
      3. Seed surface without tubercles.......... **N. humboldtiana**
1. Corolla glabrous on the inner surface.
   4. Corolla yellow.......... **N. peltata**
   4. Corolla white
      5. Corolla lobes with a median longitudinal crest on the inner surface.......... **N. cristata**
      5. Corolla lobes lacking a median longitudinal crest on the inner surface.
         6. Floating leaves ovate to reniform, the lower surface conspicuously roughened; fruit much exceeding the persistent calyx.......... **N. aquatica**
         6. Floating leaves cordate-ovate, the lower surface smooth or only slightly roughened; fruit only slightly exceeding the persistent calyx .... **N. cordata**

Nymphoides aquatica (J. F. Gmel.) Kuntze [*Aqua,* water, in reference to its aquatic habitat.] BIG FLOATINGHEART.

Villarsia aquatica J. F. Gmelin, Syst. Nat. 2: 447. 1791. *Menyanthes trachysperma* Michaux, Fl. Bor.-Amer. 1: 126. 1803, nom. illegit. *Villarsia lacunosa* Ventenat, Choix. Pl. t. 9. 1803, nom. illegit. *Trachysperma natans* Rafinesque, Med. Repos., ser. 2. 5: 352. 1808, nom. illegit. *Villarsia trachysperma* Elliott, Sketch Bot. S. Carolina 1: 230. 1817, nom. illegit. *Limnanthemum lacunosum* Grisebach, Gen. Sp. Gent. 347. 1838, nom. illegit. *Limnanthemum trachyspermum* A. Gray, Manual, ed. 5. 390. 1867, nom. illegit. *Limnanthemum aquaticum* (J. F. Gmelin) Britton, Trans. New York Acad. Sci.

9: 12. 1889. *Nymphoides aquatica* (J. F. Gmelin) Kuntze, Revis. Gen. Pl. 2: 429. 1891. *Nymphoides lacunosa* Kuntze, Revis. Gen. Pl. 2: 429. 1891, nom. illegit. *Trachysperma aquaticum* (J. F. Gmelin) House, New York State Mus. Bull. 233–34: 65. 1922. *Trachysperma lacunosum* House, New York State Mus. Bull. 233–34: 66. 1922.

Perennial aquatic herb. Stem rhizomatous in the submerged substrate with several slender petiole-like, slightly roughened, purple-punctate ones bearing 1–several leaves. Floating leaves with the blade ovate to reniform, 5–15 cm long and wide, the apex rounded or emarginate, the base deeply cordate, the margin entire, undulate, or irregularly toothed, the upper surface green or sometimes with purple markings, especially around the edge, the lower surface purple, glandular-punctate and roughened, the petiole 3–5(8) cm long. Flowers 1–several in a subumbelliform cluster, sometimes admixed or subtended with a cluster of fleshy, blunt-tipped, tuber-like roots, the pedicel to 8 cm long, purple glandular-punctate; calyx 4–5 mm long, the lobes elliptic-oblong, purple glandular-punctate, the apex rounded to obtuse; corolla white, subrotate, 1–2 cm wide, lobed nearly to the base, the lobes obovate, the margin membranous, irregularly laciniate-undulate. Fruit ovate, 10–14 mm long, much exceeding the persistent calyx; seeds globose, papillate.

Ponds and swamps. Frequent; nearly throughout. Delaware and Maryland south to Florida, west to Texas. Spring–summer.

Nymphoides cordata (Elliott) Fernald [Heart-shaped, in reference to the floating leaves.] LITTLE FLOATING HEART.

Villarsia cordata Elliott, Sketch Bot. S. Carolina 1: 230. 1817. *Nymphoides cordata* (Elliott) Fernald, Rhodora 40: 338. 1938.

Perennial aquatic herb. Stem rhizomatous in the submerged substrate with several slender petiole-like green or purple-punctate ones bearing 1(2) leaves. Floating leaves with the blade ovate to cordate, 3–7 cm long and wide, the apex rounded, the base deeply cordate, the margin entire or undulate, the upper surface green or variegated with purple, the lower surface purple, glandular-punctate, the petiole to 2–3(5) cm long. Flowers 1–several in a subumbelliform cluster, sometimes admixed or subtended with a cluster of fleshy, blunt-tipped tuber-like roots, the pedicel to 3 cm long, green or rarely purple-punctate; calyx 2–4 mm long, green or rarely purple-punctate, the lobes elliptic-oblong, the apex acute; corolla white or cream-colored, subrotate, 0.5–1 cm wide, lobed nearly to the base, the lobes obovate, the margin membranous, irregularly laciniate-undulate. Fruit ovate to subglobose, ca. 4 mm long, only slightly exceeding the persistent calyx; seeds subglobose, smooth or sparsely papillate.

Ponds and lakes. Occasional; central and western panhandle. Newfoundland, Quebec, and Ontario south to Florida, west to Louisiana. Summer–fall.

Nymphoides cristata (Roxb.) Kuntze [*Crista*, crest, in reference to the median longitudinal crest on the inner surface of the corolla lobes.] CRESTED FLOATINGHEART.

Menyanthes cristata Roxburgh, Pl. Coromandel 2: 3, pl. 105. 1799. *Villarsia cristata* (Roxburgh) Sprengel, Syst. Veg. 1: 582. 1824. *Limnanthemum cristatum* (Roxburgh) Grisebach, Gen. Sp. Gent. 342. 1838. *Nymphoides cristata* (Roxburgh) Kuntze, Revis. Gen. Pl. 2: 429. 1891.

Perennial aquatic herb. Stem a rhizome in the submerged substrate with several slender petiole-like purple-punctate ones bearing 1–several leaves. Floating leaves with the blade ovate-orbicular, 3–10 cm long and wide, the apex rounded to obtuse, the base deeply cordate, the margin entire or slightly undulate, the upper surface green or sometimes with irregular purple markings, especially around the margin, the lower surface purple, glandular-punctate, the petiole 1–2 cm long. Flowers 1–several in a subumbellatiform cluster, sometimes admixed with a cluster of fleshy, blunt-tipped tuber-like roots, the pedicel 3–4.5 cm long, green; calyx 3–6 mm long, lobed nearly to the base, the lobes ovate-elliptic or ovate, the apex obtuse; corolla white with a yellow center, subrotate, 1–2 cm wide, lobed nearly to the base, the lobes ovate, the inner surface with a median longitudinal crest, the margin undulate. Fruit subglose, 3–5 mm long, exceeding the persistent calyx; seeds globose, smooth or scabrous.

Ponds and canals. Occasional; peninsula. Escaped from cultivation. South Carolina and Florida; Asia. Native to Asia. Summer–fall.

Nymphoides cristata is listed as a Category I invasive species by the Florida Exotic Pest Plant Council (FLEPPC, 2017).

Nymphoides grayana (Griseb.) Kuntze [Commemorates Asa Gray (1810–1888), American botanist at Harvard University.] GRAY'S FLOATINGHEART.

Limnanthemum grayanum Grisebach, Cat. Pl. Cub. 181. 1866. *Nymphoides grayana* (Grisebach) Kuntze, Revis. Gen. Pl. 2: 429. 1891. *Trachysperma grayanum* (Grisebach) House, New York State Mus. Bull. 233–34: 66. 1921.

Perennial aquatic herb. Stem rhizomatous in the submerged substrate with several slender petiole-like, slightly roughened purple-punctate ones bearing 1–several leaves. Floating leaves with the blade subrotund, 5–12 cm long and wide, the apex rounded, the base deeply cordate, the margin entire or slightly undulate, the upper surface green, the lower surface purple, glandular-punctate and roughened, the petiole ca. 3 mm long. Flowers several in a subumbelliform cluster, sometimes admixed with a cluster of fleshy, blunt-tipped, tuber-like roots, the pedicel 4–8 cm long; calyx 6–8 mm long, lobed nearly to the base, the lobes elliptic-oblong, the apex acute; corolla yellow, subrotate, 1.5–2 cm wide, lobed nearly to the base, the lobes obovate, the inner surface fimbriate, the margin undulate. Fruit not seen.

Ponds and shallow lakes. Rare; central peninsula. Florida; West Indies. Summer–fall.

Nymphoides grayana was first reported for Florida by Tippery et al. (2015) as a natural range extension possibly by migratory waterfowl because of its proximity to native populations in the West Indies and the absence of the species in the water plant trade.

Nymphoides humboldtiana (Kunth) Kuntze [Commemorates Friedrich Heinrich Alexander von Humboldt (1769–1859), German polymath, geographer, naturalist, and explorer.] FLOATINGHEART.

Villarsia humboldtiana Kunth, in Humboldt et al., Nov. Gen. Sp. 3: 187. 1819. ("1818"). *Limnanthemum humboldtianum* (Kunth) Grisebach, Gen. Sp. Gent. 347. 1838 ("1839"). *Nymphoides humboldtiana* (Kunth) Kuntze, Revis. Gen. Pl. 2: 429. 1891. *Trachysperma humboldtianum* (Kunth) House, New York State Mus. Bull. 233–34: 66: 1921 ("1920").

Perennial aquatic herb. Stem rhizomatous in the submerged substrate with several slender petiole-like purple-punctate ones bearing 1–several leaves. Floating leaves with the blade ovate to subcordate, 3–18 cm long and wide, the apex rounded, the base deeply cordate, the margin entire or slightly undulate, the upper surface green, the lower surface green or purplish, glandular-punctate, the petiole 3–6 cm long. Flowers 1–several in a subumbelliform cluster, sometimes admixed with a cluster of fleshy, blunt-tipped tuber-like roots, the pedicel 3–5 cm long; calyx 3–6 mm long, lobed nearly to the base, the lobes lanceolate to elliptic, the apex obtuse; corolla white with a yellow center, subrotate, 7–12 mm long, lobed nearly to the base, the lobes ovate-elliptic, the inner surface densely fimbriate, the margin undulate. Fruit elliptic, 3–5 mm long, subequaling the persistent calyx; seeds globose, smooth.

Rivers, lakes, and ponds. Rare; central peninsula. Florida and Texas; Mexico, Central America, and South America. Summer–fall.

Nymphoides humboldtiana has long been misidentified as the non-native *N. indica,* which it closely resembles and is frequently cultivated in Florida. It can be distinguished from *N. indica* by its seeds, which lack tubercles while *N. indica* has seeds with tubercles (Middleton et al., 2018).

Nymphoides indica (L.) Kuntze [Of India.] WATER SNOWFLAKE.

Menyanthes indica Linnaeus, Sp. Pl. 145. 1753. *Villarsia indica* (Linnaeus) Ventenat, Choix Pl. t. 9. 1803. *Limnanthemum indicum* (Linnaeus) Grisebach, Gen. Sp. Gent. 343. 1838. *Nymphoides indica* (Linnaeus) Kuntze, Revis. Gen. Pl. 2: 429. 1891.

Perennial aquatic herb. Stem rhizomatous in the submerged substrate with several slender petiole-like purple-punctate ones bearing 1–several leaves. Floating leaves with the blade ovate to subcordate, 3–18 cm long and wide, the apex rounded, the base deeply cordate, the margin entire or slightly undulate, the upper surface green, the lower surface green or purplish, glandular-punctate, the petiole 3–6 cm long. Flowers 1–several in a subumbelliform cluster, sometimes admixed with a cluster of fleshy, blunt-tipped tuber-like roots, the pedicel 3–5 cm long; calyx 3–6 mm long, lobed nearly to the base, the lobes lanceolate to elliptic, the apex obtuse; corolla white with a yellow center, subrotate, 7–12 mm long, lobed nearly to the base, the lobes ovate-elliptic, the inner surface densely fimbriate, the margin undulate. Fruit elliptic, 3–5 mm long, subequaling the persistent calyx; seeds globose, smooth.

Rivers, lakes, and ponds. Rare; central peninsula. Escaped from cultivation. Florida; Africa, Asia, and Australia. Native to Africa, Asia, and Australia. Summer–fall.

Nymphoides peltata (S. G. Gmel.) Kuntze [Shield-shaped, in reference to the floating leaves.] YELLOW FLOATINGHEART.

Limnanthemum peltatum S. G. Gmelin, Novi Comment. Acad. Sci. Imp. Petrop. 14(1): 527, t. 17(2). 1770 ("1769"). *Nymphoides peltata* (S. G. Gmelin) Kuntze, Revis. Gen. Pl. 2: 429. 1891.

Perennial aquatic herb. Stem rhizomatous in the submerged substrate, with several slender petiole-like purple-punctate ones bearing 1–several leaves. Floating leaves with the blade ovate to suborbicular, 1.5–8 cm long and wide, the apex obtuse to rounded, the base deeply cordate, the margin entire or slightly undulate, the upper surface glabrous, the lower surface purple, densely glandular-punctate, the petiole 5–10 cm long, widening at the base into an amplexicaul sheath. Flowers several in a subumbelliform cluster, sometimes admixed or subtended with a cluster of fleshy blunt-tipped roots, the pedicel 3–7 cm long; calyx 7–9 mm long, lobed to near the base, the lobes elliptic-lanceolate, the apex obtuse; corolla yellow, rotate, 2.5–3 cm wide, lobed nearly to the base, the lobes obovate, the margin membranous, irregularly laciniate-undulate, the apex rounded to emarginate. Fruit elliptic, 1.7–2.5 cm long, greatly exceed the persistent calyx; seeds elliptic, compressed, with hooked cilia on the margin.

Ponds. Rare; Orange County. Escaped from cultivation. Scattered areas nearly throughout North America; Europe and Asia. Native to Europe and Asia. Summer–fall.

GOODENIACEAE R. Br., nom. cons. 1810. GOODENIA FAMILY

Shrubs or trees. Leaves alternate, simple, petiolate or epetiolate, estipulate. Flowers in axillary cymes, bracteate, bracteolate, bisexual; sepals 5, basally connate, actinomorphic; petals 5, basally connate, zygomorphic; stamens 5, free; ovary inferior, 2-carpellate and -loculate, the style 1. Fruit a drupe.

A family of 12 genera and about 400 species; North America, West Indies, Mexico, Central America, South America, Africa, Asia, Australia, and Pacific Islands.

Scaevola L. 1771. NAUPAKA

Shrubs or trees. Leaves alternate, simple, petiolate or epetiolate. Flowers in axillary cymes, bracteate, bracteolate, bisexual; sepals 5, basally connate, actinomorphic; petals 5, basally connate, zygomorphic, the posterior side divided to the base; stamens 5, free; ovary inferior, 2-carpellate and -loculate, the style 1. Fruit a drupe.

A genus of about 130 species; North America, West Indies, Mexico, Central America, South America, Africa, Asia, Australia, and Pacific Islands. [Latin for "left-handed," in reference to Gaius Mucius Scaevola, fifth-century BC Roman legendary assassin, who is said to have burned his right hand away as a show of bravery during the early years of the Roman Republic, this in reference to the zygomorphic flower appearing "one-sided."]

Selected reference: Howarth et al. (2003).

1. Cymes with the central flower sessile, the lateral flowers pedicellate; calyx lobes to 1 mm long; fruit black .... **S. plumieri**
1. Cymes with all flowers pedicellate; calyx lobes 3.5–5 mm long; fruit white or yellowish white .... **S. taccada**

Scaevola plumieri (L.) Vahl [Commemorates Charles Plumier (1646–1704), French botanist who collected in the West Indies.] BEACHBERRY; INKBERRY; GULLFEED.

Lobelia plumieri Linnaeus, Sp. Pl. 929. 1753. *Lobelia frutescens* Miller, Gard. Dict., ed. 9. 1768, nom. illegit. *Scaevola lobelia* Murray, Syst. Veg., ed. 13. 178. 1774, nom. illegit. *Scaevola plumieri* (Linnaeus) Vahl, Symb. Bot. 2: 36. 1791.

Shrub or small tree, to 3 m; branchlets glabrous, with a tuft of white trichomes in the leaf axils. Leaves often spirally arranged at the branch apex, the blade obovate, 8–150 cm long, 4–8 cm wide, slightly succulent, the apex rounded, the base cuneate, the margin entire or slightly undulate, the upper and lower surfaces glabrous, sessile or short-petiolate. Flowers in an axillary cyme, the peduncle 1.5–2 cm long, the pedicel 2–4 mm long, bracts and bracteoles lanceolate, 2–3 mm long, glabrous; calyx 4–5 mm long, the outer surface glabrous, the hypanthium ovoid, ca. 2 mm long, the calyx lobes ovate to rounded, 1–2 mm long; corolla white or pinkish, ca. 2 cm long, the outer surface glabrous, the inner surface of the tube densely white-villous, the tube 1–1.5 cm long, the lower side divided to the base, the lobes linear-lanceolate, the margin membranous, undulate. Fruit ovoid, 1–1.5 cm long, 8-ribbed, black; pyrenes 2.

Coastal strands. Frequent; central and southern peninsula. Florida, Louisiana, and Texas; West Indies, Mexico, Central America, and South America; Africa and Asia. All year.

Scaevola plumieri is listed as threatened in Florida (Florida Administrative Code, Chapter 5B-40).

Scaevola taccada (Gaertn.) Roxb. [Latinized vernacular name "takkada" used in Ceylon.] BEACH NAUPAKA.

Lobelia taccada Gaertner, Fruct. Sem. Pl. 1: 119, t. 25(5). 1788. *Scaevola taccada* (Gaertner) Roxburgh, Hort. Beng. 15: 1814. *Romeria lobelia* Dennstedt, Schlüssel Hortus Malab. 10, 24, 30. 1818, nom. illegit. *Scaevola bela-modagam* Schultes, in Roemer & Schultes, Syst. Veg. 5: 163. 1819, nom. illegit. *Scaevola sericea* Vahl var. *taccada* (Gaertner) Makino, Bot. Mag. (Tokyo) 18(208): 68. 1904.

Scaevola sericea Vahl, Symb. Bot. 2: 37. 1791. *Lobelia sericea* (Vahl) Kuntze, Revis. Gen. Pl. 2: 377. 1891. *Lobelia sericea* (Vahl) Kuntze var. *typica* Kuntze, Revis. Gen. Pl. 2: 377. 1891, nom. inadmiss. *Scaevola frutescens* K. Krause, in Engler, Pflanzenr. 4(Heft 54): 125. 1912, nom. illegit. *Scaevola frutescens* K. Krause var. *sericea* (Vahl) Merrill, Philipp. J. Sci. 7: 353. 1912, nom. inadmiss. *Scaevola taccada* (Gaertner) Roxburgh, var. *sericea* (Vahl) H. St. John, Taxon 9: 205. 1960.

Shrub or small tree, to 3 m; branchlets pubescent or glabrate, with a dense tuft of white trichomes in the leaf axils. Leaves often spirally arranged at the branch apex, the blade spatulate or obovate, 10–20 cm long, 4–8 cm wide, slightly succulent, the apex rounded, truncate, or emarginated, the base cuneate, the margin entire or undulate, the upper and lower surfaces glabrous or sericeous, sessile or short-petiolate. Flowers in an axillary cyme, the peduncle ca. 1 cm long, the pedicel 4–5 mm long; bracts and bracteoles lanceolate, 2–3 mm long, with a tuft of white trichomes in the axils; calyx 6–8 mm long, the outer surface glabrous, the hypanthium obovoid, 3–4 m long, the lobes lanceolate, 3.5–5 mm long; corolla white, pale yellow, or pinkish, ca. 2 cm long, the outer surface glabrous, the inner surface of the tube densely white-villous, the tube 1–1.5 cm long, lower side divided to the base, the lobes lanceolate, 5–8 mm long, the margin membranous, undulate. Fruit ovoid to subglobose, 8-ribbed, white or yellowish white, 7–10 mm long; pyrenes 2.

Coastal strands. Occasional; central and southern peninsula. Escaped from cultivation. Florida; West Indies; Africa, Asia, Australia, and Pacific Islands. Native to Africa, Asia, Australia, and Pacific Islands. All year.

Scaevola taccada is listed as a Category I invasive species by the Florida Exotic Pest Plant Council (FLEPPC, 2017).

CALYCERACEAE CALYCERA FAMILY

Herbs. Leaves alternate, simple, pinnate-veined, epetiolate, estipulate. Flowers in terminal or axillary capitula surrounded by a bracteate involucre, bisexual or unisexual; sepals 5, basally connate, actinomorphic; petals 5, basally connate, actinomorphic; stamens 5, epipetalous; ovary superior, 2-carpellate, 1-loculate, style 1. Fruit an achene.

A family of 4 genera and about 60 species; North America, South America, and Pacific Islands.

Selected reference: DeVore (1994).

Acicarpha Juss. 1803.

Annual herbs. Leaves alternate, simple, pinnate-veined, epetiolate. Flowers in terminal or axillary capitula surrounded by a bracteate involucre, bisexual or unisexual; sepals 5, basally connate, actinomorphic; petals 5, basally connate, actinomorphic; stamens 5, epipetalous; ovary superior, 2-carpellate, 1-loculate, style 1. Fruit an achene.

A genus of 5 species; North America and South America. [From the Greek *acis,* pointed, and *carphos,* small dry body, in reference to the fruit with spines.]

Selected reference: DeVore (1991).

Acicarpha tribuloides Juss. [To resemble *Tribulus* (Zygophyllaceae), in reference to the fruit, which resembles a Roman tribulus, a device once used to impede cavalry.] MADAM GORGON.

Acicarpha tribuloides Jussieu, Ann. Mus. Natl. Hist. Nat. 2: 348, t. 58. 1803. *Acicarpha tribuloides* Jussieu var. *dentata* Kuntze, Revis. Gen. Pl. 3(2): 126. 1898, nom. illegit.

Erect annual herb, to 5 dm; stem glabrous. Leaves 5–16 cm long, 1.5–6 cm wide, the apex acute, the base amplexicaul, the margin irregularly serrate-lobed, the upper and lower surfaces glabrous, sessile. Flowers in a terminal or axillary capitulum 3–8 mm wide, often surrounded by several reduced involucrate leaves, the pedicel 2–3 mm long; calyx ca. 1 mm long, 5-lobed, adnate to the fruit, becoming hard and spinescent; corolla white, 2–3 mm long, the tube 1–2 mm long, the lobes ca. 1 mm long. Fruit enclosed by the adherent, hard, and spinescent calyx, this often remaining attached to the receptacle, the achenes and receptacle fused together and forming an ovate, rigid, spiny disseminule that detaches from the spindly peduncle.

Dry disturbed sites. Rare; Walton and Escambia Counties. Not collected since 1885 (see DeVore, 1991). New Jersey, Pennsylvania, North Carolina, Florida, and Alabama; South America; Pacific Islands. Native to South America. Spring–fall.

ADOXACEAE E. Mey. 1839. MOSHATEL FAMILY

Shrubs or trees. Leaves opposite, simple or odd-pinnately or bipinnately compound, the simple ones pinnate- or palmate-veined, petiolate or epetiolate, stipulate or estipulate. Flowers in terminal compound cymes, ebracteate, ebracteolate, bisexual; sepals 5, basally connate, actinomorphic; petals 5, basally connate, actinomorphic; stamens 5, epipetalous; ovary inferior, 3- to 5-carpellate and -loculate, the style 1 or the stigma sessile. Fruit a drupe.

A family of 5 genera and about 200 species; nearly cosmopolitan.

1. Leaves irregularly pinnately and bipinnately compound .... **Sambucus**
1. Leaves simple .... **Viburnum**

Sambucus L. 1753. ELDER

Shrubs. Leaves opposite, odd-pinnately or bipinnately compound, the leaflets pinnate-veined, petiolate, estipulate. Flowers in terminal compound cymes, ebracteate, ebracteolate, bisexual; sepals 5, basally connate, actinomorphic; petals 5, basally connate, actinomorphic; stamens 5, epipetalous; ovary inferior, 3- to 5-carpellate and -loculate, the stigma sessile. Fruit a drupe.

A genus of about 30 species; nearly cosmopolitan. [Sambuca, an ancient Greek string instrument and a wind instrument made with tubes of the wood.]

Sambucus is sometimes placed in the Caprifoliaceae by authors.

Selected references: Applequist (2015); Bolli (1994); Eriksson and Donoghue (1997).

Sambucus nigra L. subsp. **canadensis** (L.) R. Bolli [*Niger*, black, in reference to the fruit; of Canada.] AMERICAN ELDER; ELDERBERRY.

Sambucus canadensis Linnaeus, Sp. Pl. 269. 1753. *Sambucus bipinnata* Moench, Methodus 506. 1794, nom. illegit. *Sambucus canadensis* Linnaeus forma *typica* Schwerin, Mit. Deutsch Dendrol. Ges. 29: 215. 1920, nom. inadmiss. *Sambucus canadensis* Linnaeus var. *glabra* Schwerin, Mit. Deutsch Dendrol. Ges. 29: 215. 1920, nom. inadmiss. *Sambucus nigra* Linnaeus subsp. *canadensis* (Linnaeus) Bolli, Diss. Bot. 223: 168: 1994. *Sambucus nigra* Linnaeus var. *canadensis* (Linnaeus) B. L. Turner, in B. L. Turner et al. Sida, Bot. Misc. 24: 6. 2003.

Sambucus intermedia Carrière, Rev. Hort. 1876: 400. 1876.

Sambucus canadensis Linnaeus var. *laciniata* A. Gray, Syn. Fl. N. Amer. 1(2): 9. 1884. *Sambucus canadensis* Linnaeus subsp. *laciniata* (A. Gray) Murray, Kalmia 13: 31. 1983. TYPE: FLORIDA: Brevard Co.: Indian River, s.d., *Palmer s.n.* (holotype: GH).

Sambucus simpsonii Rehder, in Sargent, Trees & Shrubs 2: 187, t. 175. 1911. TYPE: FLORIDA: Manatee Co.: Bradenton, Jun 1910, *Simpson s.n.* (holotype: A).

Shrub, to 4 m; branchlets glabrous with prominent lenticels, the bark grayish brown. Leaves with the blade irregularly odd-pinnately and partially bipinnately compound, the leaflets usually 5–11, lanceolate, ovate, or elliptic, 5–15(18) cm long, 2–6(8) cm wide, the apex acuminate, the base rounded to cuneate, the upper surface short pubescent on the midrib and main veins, sometimes sparsely so between the veins, the lower surface pubescent or sometimes so only on the midrib and main veins, sessile or the petiolule to ca. 1 cm long, the leaflets sometimes subtended by a subulate stipel or stipitate-gland, the petiole 3–10 cm long, pubescent, sometimes with a line of short bristlelike trichomes or a gland-like protuberance at the junction with the leaf axil between the petioles. Flowers in a flat-topped to broadly rounded compound terminal cyme to 4 dm wide, the branches mealy and puberulent; hypanthium campanulate, ca. 1 mm long; calyx lobes short-triangular; corolla white, lobed nearly to the base, the lobes ovate, ca. 2 mm long, spreading. Fruit subglobose, 4–6 mm long, purplish black; pyrenes 3–5.

Open hammocks, floodplain forests, swamps, and disturbed sites. Common; nearly throughout. Nearly throughout North America; West Indies, Mexico, and Central America. Native to North America, Mexico, and Central America. Spring–fall.

EXCLUDED TAXON

Sambucus canadensis L. var. *canadensis*—The typical variety was reported for Florida by Clewell (1985), Godfrey (1988), Small (1933), and Wunderlin (1982, 1998), the name misapplied to Florida material of *S. nigra* subsp. *canadensis*.

Viburnum L. 1753.

Shrubs or trees. Leaves opposite, simple, pinnate- or palmate-veined, petiolate or epetiolate, stipulate or estipulate. Flowers in terminal compound umbelliform cymes or pyramidal paniculiform arrays, bracteate, ebractiolate, bisexual; sepals 5, basally connate, actinomorphic; petals 5, basally connate, actinomorphic; stamens 5, epipetalous; ovary 3-carpellate and -loculate, the style 1. Fruit a drupe.

A genus of about 175 species; nearly cosmopolitan. [Etruscan vernacular name for the wayfaring tree (*Viburnum lantana*).]

Viburnum is sometimes placed in the Caprifoliaceae by authors.

Selected reference: McAtee (1956).

1. Flowers in a pyramidal paniculiform array .... **V. odoratissimum**
1. Flowers in a terminal compound umbelliform array.
    2. Leaves with coarsely dentate margins, the base rounded or cordate.
        3. Leaf blade shallowly 3-lobed .... **V. acerifolium**
        3. Leaf blade unlobed .... **V. dentatum**
    2. Leaves with the margin finely crenate, undulate, or entire, the base cuneate.
        4. Leaves spatulate, short-petiolate (less than 5 mm long) or sessile .... **V. obovatum**
        4. Leaves elliptic to obovate, distinctly petiolate.
            5. Leaves with the margin finely serrate; peduncle less than 0.5 cm long .... **V. rufidulum**
            5. Leaves with the margin crenate, undulate, or entire; peduncle 1–4 cm long .... **V. nudum**

Viburnum acerifolium L. [With leaves like *Acer* (Aceraceae).] MAPLELEAF VIBURNUM.

Viburnum acerifolium Linnaeus, Sp. Pl. 268. 1753.
Viburnum involucratum Chapman, Bot. Gaz. 3: 5. 1878; non Wallich ex de Candolle, 1830. *Viburnum densiflorum* Chapman, Fl. South. U.S., ed. 2. 1883. *Viburnum acerifolium* Linnaeus var. *densiflorum* (Chapman) McAtee, Rev. Nearctic Viburnum 79. 1956. TYPE: FLORIDA: s.d., *Chapman s.n.* (holotype: ?; isotypes: GH, NY, US).

Shrub, to 2 m; branchlets pubescent with long simple and short stellate trichomes, glabrous in age, the older stems brown, smooth. Leaves with the blade ovate, usually palmately 3-veined and -lobed, sometimes some or rarely all pinnate-veined, 4–9 cm long and wide, the apex and lobes obtuse to acuminate, the base subcordate or truncate, the margin irregularly blunt-toothed or undulate, the upper surface with a few straight trichomes, the lower surface pubescent with long simple and short stellate trichomes, the petiole 0.8–1.5 cm long; stipules acicular, 2–3 mm long, inserted at and adnate to the petiole base. Flowers in a terminal compound cyme 2–6 cm wide, the peduncle 2–4 cm long, pubescent, with 5–7 primary branches, the pedicel ca. 5 mm long; bracts minute; hypanthium campanulate, ca. 1 m long, glabrate; calyx lobes rounded, minute; corolla white or whitish, lobed about ½ way to the base, the lobes subrotund to reniform, ca. 1 mm long. Fruit ellipsoid, 8–10 mm long, black or purple-black.

Bluff forests. Occasional; central and western panhandle. Quebec south to Florida, west to Ontario, Wisconsin, Illinois, Arkansas, and Texas. Spring.

Viburnum dentatum L. [Toothed, in reference to the leaf margin.] SOUTHERN ARROWWOOD.

Viburnum dentatum Linnaeus, Sp. Pl. 268. 1753.
Viburnum dentatum Linnaeus var. *semitomentosum* Michaux, Fl. Bor.-Amer. 1: 179. 1803. *Viburnum tomentosum* Rafinesque, Med. Repos., ser. 2. 5: 354. 1808; non Lamarck, 1779; nec. Thunberg, 1784. *Viburnum semitomentosum* (Michaux) Rehder, Rhodora 6: 59. 1904.
Viburnum dentatum Linnaeus var. *scabrellum* Torrey & A. Gray, Fl. N. Amer. 2: 16. 1841. *Viburnum scabrellum* (Torrey & A. Gray) Torrey & A. Gray ex Chapman, Fl. South. U.S. 172. 1860. *Viburnum molle* Michaux var. *tomentosum* Chapman, Fl. South. U.S., ed. 3. 190. 1897, nom. illegit. SYNTYPE: FLORIDA.

Shrub or small tree, to 3 m; branchlets stellate pubescent, glabrescent in age, with wart-like lenticels. Leaves with the blade ovate, ovate-elliptic, lanceolate, or obovate, 3–12 cm long, 2.5–8 cm wide, the apex acute to obtuse, the base rounded, truncate, or subcordate, the margin dentate-serrate, the leaf veins ending in a tooth, the upper and lower surfaces sparsely stellate-pubescent, the petiole 1–3 cm long, stellate pubescent. Flowers in a terminal compound cyme 4–10 cm wide, the peduncle 1.5–2 cm long, stellate-pubescent, with 6–8 primary branches, the pedicel ca. 1 mm long; bracts lanceolate, ca. 2 mm long; hypanthium campanulate, ca. 1 mm long, sparsely stellate pubescent; calyx lobes lanceolate, minute; corolla white or whitish lobed ca. ½ way to the base, the lobes broadly ovate, ca. 1 mm long. Fruit ellipsoid or globose, 5–8 mm long, blue-black.

Moist hammocks and swamps. Frequent; northern counties, central peninsula. Maine, New York, and Massachusetts south to Florida, west to Iowa, Missouri, Arkansas, and Texas. Spring.

Viburnum nudum L. [Naked, in reference to the evident, long-pedunculate inflorescence.] POSSUMHAW.

Viburnum nudum Linnaeus, Sp. Pl. 268. 1753. *Viburnum laurifolium* Rafinesque, Alsogr. Amer. 52. 1838, nom. illegit. *Viburnum nudum* Linnaeus var. *claytonii* Torrey & A. Gray, Fl. N. Amer. 2: 14. 1841, nom. inadmiss.

Viburnum nitidum Aiton, Hort. Kew. 1: 371. 1789. *Viburnum nudum* Linnaeus var. *nitidum* (Aiton) Zabel, Gartenflora 38: 463. 1889. *Viburnum cassinoides* Linnaeus var. *nitidum* (Aiton) McAtee, Rev. Nearctic Viburnum 34. 1956.

Viburnum nudum Linnaeus var. *angustifolium* Torrey & A. Gray, Fl. N. Amer. 2: 14. 1841.

Viburnum nudum Linnaeus var. *serotinum* Ravenel ex Chapman, Fl. South. U.S., ed. 2. 624. 1883.

Viburnum nudum Linnaeus var. *grandiflorum* A. Gray, Syn. Fl. N. Amer. 1(2): 12. 1884. TYPE: FLORIDA: s.d., *Treat s.n.* (holotype: GH).

Shrub or small tree, to 5 m; branchlets with rusty-scurfy scales, glabrescent in age, with warty lenticels. Leaves with the blade elliptic, oblong, lanceolate, ovate, or obovate, the apex acute, obtuse, or rounded, the base cuneate to rounded, the margin entire, undulate, or crenate-serrate, slightly revolute, the upper surface sparsely and finely glandular-dotted, the lower surface with rusty-scurfy scales, mainly along the midrib, the petiole 0.5–2 cm long, with rusty-scurfy scales. Flowers in a terminal compound cyme, the peduncle 1–4 cm long, with rusty-scurfy scales, with 4–5 primary branches, subsessile or the pedicel ca. 1 mm long; bracts ovate-lanceolate, minute; hypanthium elliptic, 1–2 mm long, with rusty-scurfy scales; calyx lobes lanceolate, minute; corolla white or whitish, lobed nearly to the base, the lobes suborbicular, ca. 1 mm long. Fruit ellipsoid-ovoid, 6–10 mm long, blue, glaucous.

Wet flatwoods, bogs, and swamps. Frequent; northern counties, central peninsula. Newfoundland and Quebec south to Florida, west to Ontario, Wisconsin, Illinois, Arkansas, and Texas. Spring.

Viburnum obovatum Walter [Obovate, in reference to the leaf shape.] WALTER'S VIBURNUM; SMALL-LEAF VIBURNUM.

Viburnum obovatum Walter, Fl. Carol. 116. 1788.

Viburnum capitatum Rafinesque, Alsogr. Amer. 56. 1838. TYPE: "Florida or Georgia."

Viburnum nashii Small, Fl. S.E. U.S. 1123, 1338. 1903. TYPE: FLORIDA: Gadsden Co.: River Junction, 5 Sep 1895, *Nash 2593* (holotype: NY).

Shrub or small tree, to 5 m; branchlets with rusty-scurfy scales, usually narrowly 2-winged, becoming gray in age. Leaves with the blade oblanceolate or spatulate, 2–5 cm long, 1–3 cm wide, the apex obtuse or rounded, the base narrowly cuneate, the margin entire or finely and irregularly serrate or crenate-dentate from about the middle, revolute, the upper surface glabrous, the lower surface glandular-dotted, sessile or with the petiole, narrowly winged and to 6 mm long. Flowers in a terminal compound cyme 4–6 cm wide, the peduncle 3–6 mm long, with small rusty-scurfy scales, with 2–5 primary branches, the pedicel 3–4 mm long; bracts lanceolate, minute; hypanthium elliptic-campanulate, ca. 2 mm long, with small rusty-scurfy scales; calyx lobes lanceolate, minute; corolla white or pinkish, lobed nearly to the base, the lobes broadly elliptic, 1–2 mm long. Fruit ellipsoid, ovate, or subglobose, 6–10 mm long, black.

Floodplain forests. Common; nearly throughout. South Carolina, Georgia, Alabama, and Florida. Spring.

Viburnum odoratissimum Ker Gawl. [Very sweet-smelling.] SWEET VIBURNUM.

Viburnum odoratissimum Ker Gawler, Bot. Reg. 6: pl. 456. 1820. *Thyrsosma chinensis* Rafinesque, Sylva Tellur. 130. 1838, nom. illegit. *Microtinus odoratissimus* (Ker Gawler) Oersted, Vidensk. Meddel. Dansk Naturhist Foren. Kjoebenhaven 12: 294. 1860.

Shrub or small tree, to 10(15) m; branchlets glabrous, those of the previous year with small raised lenticels. Leaves with the blade elliptic, oblong, oblong-obovate, or obovate, 7–20 cm long, (3)4–9 cm wide, leathery, the apex acute, mucronate, sometimes obtuse or rounded, the base broadly cuneate or rounded, the margin undulate or entire, rarely irregularly serrate apically, the upper and lower surfaces glabrous, sometimes sparsely stellate-pubescent on the veins, the lower surface sparsely minutely gland-dotted, the petiole 1–2(3) cm long, glabrous or sparsely stellate-pubescent. Flowers in a terminal, pyramidal, paniculiform array, the peduncle 4–10 mm long, glabrous or sparsely stellate-pubescent, the pedicel 1–3 mm long; bracts ovate or lanceolate, ca. 1 cm long, foliose; bracteoles similar and reduced; hypanthium elliptic, ca. 2 mm long, glabrous or sparsely stellate-pubescent; calyx lobes triangular, ca. 1 mm long, glabrous; corolla white or whitish, lobed about ½ way to the base, the lobes suborbicular, 2–3 mm long. Fruit ovoid-elliptic, ca. 8 mm long, blue-black, glabrous.

Disturbed woods. Rare; Leon County. Escaped from cultivation. Florida; Asia. Native to Asia. Spring.

Viburnum rufidulum Raf. [Rusty-colored, in reference to the reddish trichomes on the young branchlets, leaves, and inflorescences.] RUSTY BLACKHAW.

Viburnum rufidulum Rafinesque, Alsogr. Amer. 56. 1838.

Viburnum prunifolium Linnaeus var. *ferrugineum* Torrey & A. Gray, Fl. N. Amer. 2: 15. 1841. *Viburnum ferrugineum* (Torrey & A. Gray) Small, in Small & Vail, Mem. Torrey Bot. Club 4: 123. 1893; non Rafinesque, 1838; nec (Oersted) Donnell Smith, 1889. *Viburnum rufotomentosum* Small, Bull. Torrey Bot. Club 23: 410. 1896.

Viburnum rufidulum Rafinesque var. *floridanum* Ashe, J. Elisha Mitchell Sci. Soc. 40: 47. 1924. TYPE: FLORIDA: Walton Co.

Shrub or small tree, to 5 m; branchlets rusty-pubescent with stellate and simple trichomes, the older ones grayish with a few scattered lenticels, the bark with a block-like pattern. Leaves with the blades ovate, elliptic, obovate, or suborbicular, the apex obtuse or rounded, sometimes emarginate, the base obtuse or rounded, the margin finely serrate, the upper surface glabrous, the lower surface patchy rusty-pubescent, especially along the veins, the petiole 0.8–1.5 cm long, narrowly winged, the margin finely serrate, the upper surface purplish and glabrous, the lower surface densely rusty-pubescent. Flowers in a terminal compound cyme 5–10 cm wide, sessile, with 2–5 primary branches, these densely rusty-pubescent, the pedicel 2–3 mm long;

bracts lanceolate, minute; hypanthium elliptic-campanulate, ca. 2 mm long, rusty pubescent; calyx lobes lanceolate, minute; corolla cream-colored, lobed nearly to the base, the lobes suborbicular, 1–2 mm long. Fruit ellipsoid, 1–1.5 cm long, blue, glaucous.

Moist calcareous hammocks. Occasional; northern counties, central peninsula. Virginia and Ohio south to Florida, west to Illinois, Missouri, Kansas, Oklahoma, and Texas. Spring.

EXCLUDED TAXA

Viburnum cassinoides Linnaeus—Reported for Florida by Kurz and Godfrey (1962) and Small (1903, 1913a, 1913d, 1933), the name misapplied to Florida material of *V. nudum*.

Viburnum corymbosum (Miller) Rehder—Reported for Florida by West and Arnold (1946), misapplied to Florida material of *V. obovatum*. This name is untypified, but probably is a Rhamnaceae or Aquifoliaceae (fide McAtee (1956).

Viburnum molle Michaux—Reported for Florida by Small (1903). No Florida specimens seen.

Viburnum prunifolium Linnaeus—Reported for Florida by Chapman (1897) and by Correll and Johnston (1970). No Florida specimens seen.

CAPRIFOLIACEAE Juss. 1789. HONEYSUCKLE FAMILY

Herbs, vines, shrubs, or trees. Leaves opposite, simple, pinnate-veined, petiolate or epetiolate, estipulate. Flowers in axillary or terminal spikes or cymes, bracteate, bisexual or unisexual (plants gynodioecious); sepals 5, basally connate, actinomorphic or zygomorphic; petals 5, basally connate, zygomorphic; stamens 5, epipetalous; ovary inferior, 2- to 3-carpellate and -loculate. Fruit a berry, drupe, or achene.

A family of 42 genera and about 890 species; nearly cosmopolitan.

Valerianaceae Batsch 1802.

Selected reference: Ferguson (1965).

1. Herbs .... **Valerianella**
1. Vines, shrubs, or trees.
    2. Herbaceous vines.
        3. Corolla to 2 mm long; fruit an achene with a persistent plumose-like calyx .... **Valeriana**
        3. Corolla 3–5 cm long; fruit a berry .... **Lonicera**
    2. Shrubs or trees.
        4. Flowers solitary or few-flowered in axillary or terminal cymes; sepals much enlarged .... **Abelia**
        4. Flowers in congested axillary spikes; sepal lobes not much enlarged .... **Symphoricarpos**

Abelia R. Br. 1818.

Shrubs. Leaves opposite, simple, pinnate-veined, petiolate. Flowers axillary, bracteate, bisexual; sepals 5, basally connate, actinomorphic; petals 5, basally connate, zygomorphic; stamens 5, epipetalous; ovary inferior, 3-carpellate and -loculate, 1 fertile, 2 sterile, the style 1. Fruit an achene.

A genus of about 30 species; North America, Mexico, and Asia. [Commemorates Clarke Abel (1780–1826), British physician and naturalist.]

Abelia ×grandiflora (Rovelli ex André) Rehder (*A. chinensis* R. Brown × *A. uniflora* R. Brown ex Wallich) [Large-flowered.] LARGEFLOWER ABELIA; GLOSSY ABELIA.

Abelia rupestris Lindley var. *grandiflora* Rovelli ex André, Rév. Hort. 1886: 488. 1886. *Abelia ×grandiflora* (Rovelli ex André) Rehder, in L. H. Bailey, Cycl. Amer. Hort. 1: 1900.

Shrub, to 2(3) m; branchlets red, pubescent when young, glabrous in age, the bark light brown, shredding. Leaves ovate, 2–6 cm long, 0.5–2.5 cm wide, the apex acute, mucronulate, the base cuneate, the margin irregularly crenate-serrate, the upper surface glabrous, the lower surface glabrous or sparsely pubescent on the veins or in the vein axils, the petiole 1–3 mm long, slightly winged. Flowers solitary or few in an axillary or terminal cyme, the pedicel 2–4 mm long; bracts lanceolate, ca. 1 mm long; hypanthium elliptic, 3–4 mm long, the outer surface glabrous or sparsely pubescent; calyx lobes elliptic-oblanceolate, 6–8 mm long, reddish; corolla white tinged with pink, campanulate, ca. 2 cm long, gibbous at the base, the lower lip hirsute on the inner surface. Fruit oblong, 8–10 mm long, sparsely pilose or glabrous, with the persistent sepals at the apex.

Disturbed sites. Rare; Franklin and Escambia Counties. Escaped from cultivation. North Carolina, Alabama, and Florida; Europe, Africa, Asia, and Australia. Native to Asia. Summer–fall.

Lonicera L. 1753. HONEYSUCKLE

Woody vines. Leaves opposite, simple, pinnate-veined, petiolate or epetiolate, estipulate. Flowers in axillary or terminal spikes, bracteate, bisexual; sepals 5, basally connate, actinomorphic; petals 5, basally connate, zygomorphic; stamens 5, epipetalous; ovary inferior, 2- to 3-carpellate and -loculate, the style 1. Fruit a berry.

A genus of about 180 species; nearly cosmopolitan. [Commemorates Adam Lonicer (1528–1586), German physician and botanist.]

Nintooa Sweet, nom. nud., 1830; *Phenianthus* Raf., 1820.

1. Corolla white, yellow, or pinkish, the lobes very unequal, nearly as long as the tube; leaves below the inflorescence petiolate .... **L. japonica**
1. Corolla red, the lobes subequal, much shorter than the tube; leaves below the inflorescence connate-perfoliate .... **L. sempervirens**

Lonicera japonica Thunb. [Of Japan.] JAPANESE HONEYSUCKLE.

Lonicera japonica Thunberg, in Murray, Syst. Veg., ed. 14. 216. 1784. *Caprifolium japonicum* (Thunberg) Dumont d'Urville, Bot. Cult., ed. 2. 7: 209. 1814. *Nintooa japonica* (Thunberg) Sweet, Hort. Brit., ed. 2. 258. 1839. *Lonicera acuminata* Wallich var. *japonica* (Thunberg) Miquel, Ann. Mus. Bot. Lugduno-Batavum 2: 270. 1866. *Lonicera japonica* Thunberg forma *typica* Zabel, in Beissner et al., Handb. Laubholzben. 451. 1903, nom. inadmiss.

Climbing or trailing woody vine; stem pubescent when young, glabrous in age, the bark reddish brown, sometimes shredding. Leaves with the blade ovate, elliptic, or oblong, 3–8 cm long, 1.5–4 cm wide, the apex acute to obtuse or rounded and apiculate, the base broadly cuneate, rounded, truncate, or subcordate, often slightly unequal, the margin entire or undulate, usually ciliate, the young shoots sometimes pinnately lobed, the upper and lower surfaces sparsely pubescent, at least on the midrib, the petiole 2–8(11) mm long, pubescent. Flowers solitary or in pairs in an axillary or terminal cluster, subtended by a pair of small bract-like leaves; bracts 2, lanceolate, minute, pubescent; hypanthium elliptic, ca. 1 mm long, glandular-pubescent; calyx lobes narrowly triangular, ca. 1 mm long, pubescent; corolla white or cream-colored, occasionally pinkish or purplish tinged, yellowish in age, 3–5 cm long, the tube narrow funnelform, abruptly expanded and bilabiate, the upper lip 4-lobed, the lower 1-lobed, the outer surface sparsely glandular-pubescent, the inner surface sparsely pubescent; stamens and style exserted from the throat. Fruit subglobose, 5–8 mm long, black; seeds 2–3 mm long, lengthwise ridged, finely reticulate.

Disturbed sites. Frequent; nearly throughout. Escaped from cultivation. Nearly throughout North America; South America; Europe, Africa, Asia, Australia, and Pacific Islands. Native to Asia. Spring–summer.

Lonicera japonica is listed as a Category I invasive species by the Florida Exotic Pest Plant Council (FLEPPC, 2017).

Lonicera sempervirens L. [Evergreen.] CORAL HONEYSUCKLE; TRUMPET HONEYSUCKLE.

Lonicera sempervirens Linnaeus, Sp. Pl. 173. 1753. *Periclymenum sempervirens* (Linnaeus) Miller, Gard. Dict., ed. 8. 1768. *Lonicera sempervirens* Linnaeus var. *major* Aiton, Hort. Kew. 1: 230. 1789, nom. inadmiss. *Lonicera flammea* Salisbury, Prodr. Stirp. Chap. Allerton 138. 1796, nom. illegit. *Lonicera sempervirens* Linnaeus var. *ovata* Veillard, in Duhamel du Monceau, Traite Arbr. Arbust. 1: 48. 1801, nom. inadmiss. *Caprifolium sempervirens* (Linnaeus) Moench, Suppl. Meth. 194. 1802. *Periclymenum sempervirens* (Linnaeus) Miller var. *latifolium* spach, Hist. Nat. Vég. 8: 345. 1839, nom. inadmiss. *Phenianthus sempervirens* (Linnaeus) Rafinesque ex Small, Man. S.E. Fl. 1274. 1933.

Lonicera angustifolia Rafinesque, New Fl. 3: 19. 1838 ("1836"); non Wallich ex de Candolle, 1830; nec Wenderoth, 1831. *Kantemon angustifolium* Rafinesque, New. Fl. 3: 19. 1838 ("1836"). TYPE: FLORIDA.

Climbing or trailing woody vine, to 5 m; stem glabrous, the bark reddish gray, shredding. Leaves opposite, the uppermost 1–2 pairs often connate-perfoliate, the blade oblong, elliptic, obovate, or suborbicular, 3–7 mm long, 1–4 cm wide, the apex obtuse or rounded, sometimes mucronate, the base narrowly cuneate or rounded, the margin entire or undulate, slightly revolute, the upper surface glabrous, the lower surface glabrous, glaucous, sessile or the petiole to 4 mm long, slightly winged. Flowers 2–4 at a node in a terminal interrupted spike, each cluster subtended by a pair of greatly reduced bract-like leaves; bracts 2, minute; hypanthium 2–3 mm long, elliptic, glabrous, glaucous; calyx lobes subrotund, minute; corolla red on the outer surface, yellow on the inner, 4–4.5 cm long, narrowly funnelform, abruptly expanded with the lobes ovate, subequal, slightly bilabiate, the outer surface glabrous, the inner surface of the tube pubescent; stamens and style only slightly exserted from the throat. Fruit subglobose 9–12 mm long, orange-red; seeds ca. 5 mm long, brown, alveolate-reticulate.

Floodplain forests and hammocks. Common; northern counties; central peninsula. Quebec south to Florida, west to Michigan, Iowa, Missouri, Kansas, Oklahoma, and Texas. Spring–summer.

Symphoricarpos Duham. 1755. SNOWBERRY

Woody vines. Leaves opposite, simple, pinnate-veined, petiolate or epetiolate. Flowers in axillary spikes, bracteate, bisexual; sepals 5, basally connate, actinomorphic; petals 5, basally connate, zygomorphic; stamens 5, epipetalous; ovary inferior, 2-carpellate and 4-loculate, the style 1. Fruit a drupe.

A genus of about 15 species; North America, Mexico, Central America, and Asia. [From the Greek *symphorein,* to bear together, and *karpos,* fruit, in reference to closely packed fruits.]

Selected reference: Jones (1940).

Symphoricarpos orbiculatus Moench [Rounded, in reference to the sometimes suborbicular leaves.] CORALBERRY; INDIAN-CURRANT.

Lonicera symphoricarpos Linnaeus, Sp. Pl. 178. 1753. *Symphoricarpos orbiculatus* Moench, Methodus 503. 1794. *Symphoricarpos parviflorus* Dumont de Courset, Bot. Cult. 2: 573. 1802, nom. illegit. *Symphoria glomerata* Pursh, Fl. Amer. Sept. 162. 1814, nom. illegit. *Symphoricarpos symphoricarpos* (Linnaeus) MacMillan, Bull. Torrey Bot. Club 19: 15. 1892, nom. inadmiss.
Symphoricarpos vulgaris Michaux, Fl. Bor.-Amer. 1: 105. 1803.

Shrub, to 1(2) m; branchlets brown to purplish, gray-pubescent, the older ones light brown, the bark shredding. Leaves with the blade ovate, elliptic, or suborbicular, 1.5–5 cm long, 1–3 mm wide, the apex rounded, obtuse, or sometimes acute, the base rounded to broadly cuneate, the margin entire or undulate, occasional with a few blunt, irregular teeth, the upper surface glabrous, the lower surface shaggy short-pubescent, glaucous, the petiole 1–3 mm long, pubescent. Flowers in a short, compact, axillary spike; bracts short, ovate; hypanthium urceolate, 1–2 mm long, pubescent; calyx lobes deltoid, minute; corolla pinkish, campanulate, 2–3 mm long, the lobes ovate-oblong, erect to slightly spreading, the tube villous on the inner surface. Fruit ellipsoid or subglobose, often oblique, 4–5 mm long, red or pink, sometimes tinged with purple, the apex with the persistent calyx; pyrenes 2, ovoid or ellipsoid, flattened on 1 side, ca. 3 mm long, smooth.

Calcareous hammocks. Rare; Levy and Jackson Counties. Ontario south to Florida, west to Minnesota, South Dakota, Nebraska, Kansas, Oklahoma, and Texas, also Utah; Mexico. Fall.

Symphoricarpos orbiculatus is listed as endangered in Florida (Florida Administrative Code, Chapter 5B-40).

Valeriana L. 1753. VALERIAN

Perennial herbs. Leaves opposite, simple, pinnate-veined, petiolate. Flowers in axillary and terminal compound cymes, bracteate, gynodioecious; sepals 5, basally connate, 11- to 15-fid, actinomorphic; petals 5, basally connate, zygomorphic; stamens 5, epipetalous; ovary inferior, 3-carpellate and -loculate, 1 fertile, 2 sterile, the style 1. Fruit an achene. [Commemorates Publius Licinius Valerianus (253–260 AD), Roman emperor.]

A genus of about 270 species; North America, West Indies, Mexico, Central America, South America, Europe, Africa, and Asia.

Valeriana is sometimes placed in the Valerianaceae by authors.

Selected reference: Meyer (1951).

Valeriana scandens L. [Climbing.] FLORIDA VALERIAN.

Valeriana scandens Linnaeus, Sp. Pl., ed. 2. 1762. *Valeriana scandens* Linnaeus var. *genuina* K.A.E. Müller, in Martius, Fl. Bras. 6(4): 344. 1885, nom. inadmiss.

Perennial herbaceous vine. Stem glabrous or sparsely pilose. Leaves with the blade ovate-cordate, undivided or 3-parted, the undivided ones 3–11 cm long, 2–8 cm wide, the terminal lobe of the divided ones 2–11 cm long, 1–6 cm wide, ovate-cordate, the lateral lobes smaller, often somewhat oblique, the apex acute to acuminate, the base cuneate, rounded, truncate, or subcordate, the margin serrate, crenate, dentate, or merely undulate to nearly entire, the upper surface glabrous or sparsely pilose, the lower surface glabrous, the petiole 1–9 cm long. Flowers in a terminal or axillary compound cyme, the inflorescence to 4 dm long, glabrous or pilosulous; bracts lanceolate, 1–2 mm long, glabrous or pilosulous; hypanthium ca. 2 mm long, elliptic-ovate, the outer glabrous; calyx lobes minute; corolla white, campanulate, 1–2 mm long on the perfect flowers, that of the carpellate to 1 mm long, the lobes less than ½ the length of the gibbous tube, the outer surface glabrous, the inner surface glabrous or sparsely pilosulous at the tube throat. Fruit oblong to ovate, 1–3 mm long, tawny, glabrous, with the persistent and enlarged calyx lobes ca. 5 mm long, 11- to 15-fid, setose.

Floodplain forests. Occasional; northern and central peninsula. Florida; West Indies, Mexico, Central America, and South America. Spring–summer.

Valerianella Mill. 1754. CORNSALAD

Herbs. Leaves opposite, simple, pinnate-veined, epetiolate. Flowers in terminal spikes, bracteate, bisexual; sepals 5, basally connate, actinomorphic; petals 5, basally connate, zygomorphic; stamens 5, epipetalous; ovary inferior, 3-carpellate and -loculate, 1 fertile, 2 sterile, the style 1. Fruit an achene.

A genus of about 65 species; North America, Mexico, Europe, Africa, and Asia. [*Valeriana* and *ellus*, diminutive, resembling the genus *Valeriana*, but smaller in size.]

Valerianella is sometimes placed in the Valerianaceae by authors.

Selected reference: Eggers (1969).

Valerianella radiata (L.) Dufr. [Spreading outward, in reference to the calyx lobes.] BEAKED CORNSALAD.

Valeriana locusta Linnaeus var. *radiata* Linnaeus, Sp. Pl. 34. 1753. *Valeriana radiata* (Linnaeus) Willdenow, Sp. Pl. 1: 184. 1797. *Fredia radiata* (Linnaeus) Michaux, Fl. Bor.-Amer. 1: 18. 1803. *Valerianella radiata* (Linnaeus) Dufresne, Hist. Nat. Valer. 57. 1811.

Annual or short-lived perennial herb, to 6 dm; stem glabrous or pubescent along the angles. Leaves with the blade of the lower ones oblong-spatulate, the base amplexicaule and joined to that of the opposite one, the margin entire, sparsely ciliate, the upper surface glabrous, the lower sparsely pubescent along the midrib, the blade of the upper ones oblong ovate, the apex obtuse to rounded, the base not connate, the margin coarsely toothed or lobed at the base. Flowers in a terminal cyme; bracts lanceolate or oblanceolate, the margin slightly ciliate or glabrous; hypanthium elliptic, ca. 1 mm long, sparsely pubescent; calyx lobes ca. 1 mm long; corolla white, funnelform, ca. 2 mm long, the tube shorter than the limb, with a saccate gibbosity at the base of the throat on the lower side, the lobes spreading, subequal. Fruit ovoid, ca. 2 mm long, 2 locules sterile, the remaining 1-seeded, grooved between the locules, yellowish, glabrous or sparsely pubescent.

Wet disturbed sites. Occasional; northern counties, Levy County. Connecticut south to Florida, west to Kansas, Oklahoma, and Texas. Spring.

PITTOSPORACEAE R. Br., nom. cons. 1814. PITTOSPORUM FAMILY

Shrubs or trees. Leaves alternate, simple, petiolate, estipulate. Flowers in terminal or axillary subterminal cymes; bracteate, bracteolate, bisexual; sepals 5, free, actinomorphic; petals 5, free, actinomorphic; stamens 5, free; ovary superior, 3-carpellate and -loculate, the style 1. Fruit a capsule, dehiscence septicidal.

A family of about 7 genera and about 200 species; North America, Africa, Asia, Australia, and Pacific Islands.

Pittosporum Banks ex Gaertn., nom. cons. 1788. CHEESEWOOD

Shrubs or trees. Leaves alternate, simple, pinnate-veined, petiolate. Flowers in terminal or axillary subterminal cymes; bracteate, bracteolate, bisexual; sepals 5, free, actinomorphic; petals 5, free, actinomorphic; stamens 5, free; ovary superior, 3-carpellate and -loculate, the style 1. Fruit a capsule, dehiscence septicidal.

A genus of about 140 species; North America, Africa, Asia, Australia, and Pacific Islands. [From the Greek, *pitta,* pitch, and *sporos,* seed, in reference to resinous pulp around the seeds.]

1. Leaves with the blade elliptic to oblong, rarely obovate, the margin flat; fruit dehiscent by 2 valves .... **P. pentandrum**
1. Leaves with the blade obovate, the margin revolute; fruit dehiscent by 3 valves .... **P. tobira**

Pittosporum pentandrum (Blanco) Merr. [From the Greek *pentos,* five, and *andros,* male, in reference to the five stamens.] TAIWANESE CHEESEWOOD.

Aquilaria pentandra Blanco, Fl. Filip. 373. 1837. *Pittosporum pentandrum* (Blanco) Merrill, Publ. Bur. Gov. Lab. Philipp. 27: 19. 1905.

Shrub or tree, to 12 m; branchlets glabrous, lenticellate. Leaves with the blade elliptic or oblong, rarely obovate, 4–10 cm long, 2–5 cm wide, the apex acute or obtuse, sometimes rounded, the base cuneate, slightly decurrent on the petiole, the margin entire or crenate-undulate, the upper and lower surfaces glabrous, the petiole 0.5–1.5 cm long. Flowers in a terminal paniculiform inflorescence of cymes, to 10 cm long, the pedicel 2–5 mm long; bracts lanceolate, ca. 2 mm long; bracteoles ovate-lanceolate, ca. 2 mm long; sepals ovate-lanceolate, ca. 2 mm long; petals white, elliptic, 5–7 mm long. Fruit subglobose, 6–8 mm long, slightly compressed, glabrous, orange, dehiscent by 2 valves; seeds angular, ca. 3 mm long, red.

Disturbed sites. Rare; Miami-Dade County. Escaped from cultivation. Florida; Asia. Native to Asia. Summer.

Pittosporum pentandrum is listed as a Category II invasive species by the Florida Exotic Pest Plant Council (FLEPPC, 2017).

Pittosporum tobira (Thunb.) Aiton [Japanese vernacular name for the plant.] JAPANESE CHEESEWOOD.

Euonymus tobira Thunberg, Fl. Jap. 99. 1784. *Pittosporum tobira* (Thunberg) Aiton, Hort. Kew., ed. 2. 2: 27. 1811.

Shrubs or small trees, to 6 m; branchlets puberulent, lenticellate, gray and glabrous in age. Leaves with the blade obovate, green or sometimes irregularly white-variegated along the margin, the apex rounded or obtuse, usually emarginate, the base cuneate, the margin entire, revolute, the upper surface glabrous, the lower surface puberulent, the petiole 5–10 mm long, puberulent. Flowers in a terminal or subterminal axillary cyme, the pedicel 0.5–1 cm long, puberulent; bracts lanceolate, 4–5 mm long, puberulent; bracteoles 2–3 mm long, puberulent; sepals lanceolate, 3–4 mm long; petals white, becoming yellow, oblanceolate, 6–8 mm long. Fruit subglobose to subelliptic, ca. 1.2 cm long, slightly angular, puberulent, yellow-brown, dehiscent by 3 valves; seeds angular, ca. 4 mm long, red.

Disturbed sites. Rare; Gadsden County. Escaped from cultivation. North Carolina, Florida, Alabama, and California. Native to Asia. Summer.

ARALIACEAE Juss., nom. cons. 1789. GINGSENG FAMILY

Herbs, vines, shrubs, or trees. Leaves alternate, simple, palmately compound, or pinnately compound, petiolate, stipulate or estipulate. Flowers in simple or compound umbels, bisexual or unisexual (plants polygamomonoecious); sepals 4–5, basally connate or free, actinomorphic; petals 4–12, basally connate or free, actinomorphic; stamens 4–12, free; ovary inferior, 2- to 5- to 12-carpellate and -loculate, styles 1–5 or absent. Fruit a drupe or schizocarp.

A family of about 43 genera and about 1,450 species; nearly cosmopolitan.

Hederaceae Giseke, 1792.

Selected references: Graham (1966); Lowry et al. (2004); Mathias and Constance (1944b, 1945).

1. Leaves simple.
  2. Creeping herb with an elongate, rhizomatous, horizontal stem; inflorescence arising along the stem at the nodes .... **Hydrocotyle**
  2. Erect herb, subshrub, or vine; inflorescence arising terminally or subterminally.
    3. Plant an erect herb or subshrub; leaves peltate.......... **Tetrapanax**
    3. Plant a vine; leaves with the base cordate or truncate .... **Hedera**
1. Leaves palmately or pinnately compound.
  4. Leaves palmately compound.
    5. Leaves 3-foliolate; branches and petioles armed with recurved prickles.......... **Eleutherococcus**
    5. Leaves 5- to 7-foliolate; branches and petioles without prickles.......... **Schefflera**
  4. Leaves pinnately compound.
    6. Stem with coarse prickles.......... **Aralia**
    6. Stem without prickles .... **Polyscias**

Aralia L. 1753. SPIKENARD

Shrub or tree. Leaves alternate, 2- or 3-pinnately compound, petiolate, stipulate. Flowers in terminal compound umbels, bracteate, bracteolate, bisexual or unisexual (plants polygamo-monoecious); sepals 5, basally connate, actinomorphic; petals 5, free, actinomorphic; stamens 5; ovary inferior, 2- to 5-carpellate and -loculate, the styles 2–5, basally connate. Fruit a drupe.

A genus of about 68 species; North America, West Indies, Mexico, Central America, South America, and Asia. [From the French-Canadian vernacular name "aralie," perhaps derived from a Native American name, the original specimens sent with that name to Tournefort by the Quebec physician, Dr. Sarrazin (1730–1740).]

Aralia spinosa L. [With spines.] DEVIL'S WALKINGSTICK.

Aralia spinosa Linnaeus, Sp. Pl. 273. 1753. *Angelica spinosa* (Linnaeus) Shecut, Fl. Carol. 167. 1806.

Shrub or small tree, to 8 m; stem armed with stiff, sharp, straight or curved prickles, sometimes sloughing off on older stems, the leafless nodes with a raised narrow leaf scar nearly obliquely encircling the stem. Leaves with the blade 2- or 3-pinnately compound, 3–12 dm long and wide, the leaflet blade ovate, 3–10 cm long, 1.5–8 cm wide, pinnate-veined, the apex acute or acuminate, the base subcordate, truncate, rounded, or cuneate, the margin serrate, the upper and lower surfaces glabrous or sometimes with a few stiff trichomes on 1 or both surfaces, sometimes the midrib on the lower surface with a few hooked prickles, the petiole to 1.5–3 dm long, with prickles or unarmed, the base obliquely clasping the stem; stipules reniform, partly adnate to the petiole. Flowers in a terminal compound umbel to 12 dm long and sometimes as wide, pubescent with short, stiff trichomes; bracts and bracteoles 1–8 mm long; hypanthium cupuliform, ca. 1 mm long, glabrous; calyx lobes minute; corolla white, the lobes elliptic-lanceolate, ca. 2 mm long. Fruit subglobose, 5–8 mm long, purplish black, the axes reddish purple, with apex with the persistent styles; pyrenes 5.

Hammocks. Occasional; northern counties, central peninsula. Maine south to Florida, west to Illinois, Missouri, Okalahoma, and Texas. Summer.

Eleutherococcus Maxim. 1859.

Scandent shrubs or climbers. Leaves alternate, trifoliolate, petiolate, estipulate. Flowers in terminal racemes of umbels or compound umbels, bisexual, bracteate, bracteolate; sepals 5, basally connate, actinomorphic; petals 5, free, actinomorphic; stamens 5, free; ovary inferior, 2-carpellate and -loculate, the styles 2, united to about the middle. Fruit a drupe.

A genus of about 40 species; North America and Asia. [From the Greek *eleuthero,* free, and *koccos,* seed, in reference to the separated pyrenes.]

Eleutherococcus trifoliatus (L.) S. Y. Hu [With three leaves (in this case, having three leaflets).] CLIMBING GINSENG.

Zanthoxylum trifoliatum Linnaeus, Sp. Pl. 270. 1753. *Panax aculeatus* Aiton, Hort. Kew. 4: 448. 1789, nom. illegit. *Aralia trifoliata* (Linnaeus) Meyen, Observ. Bot. 2: 332. 1835. *Acanthopanax aculeatus* Witte, Ann. Hort. Bot. 4: 89. 1861, nom. Illegit. *Acanthopanax trifoliatus* (Linnaeus) Voss, Vilm. Blumengärtn., ed. 3. 1: 406. 1894. *Eleutherococcus trifoliatus* (Linnaeus) S. Y. Hu, J. Arnold Arbor. 61: 110. 1980.

Scandent shrubs, to 7 m; branchlets glabrous, with scattered recurved prickles. Leaves 3-foliolate, the leaflets elliptic-ovate, 4–10 cm long, 2–4.5 cm wide, the apex acute to acuminate, the base cuneate, usually oblique, the margin irregularly serrate, the upper surface glabrous, the lower surface glabrous or sparsely setose on midvein and principal veins, the petiolule 2–8 mm long, the petiole 2–6 cm long, glabrous, with scattered prickles. Flowers in a terminal raceme of 3–10 umbels or a compound umbel, the peduncle 2–7 cm long, the pedicel 1–2 cm long; bracts and bracteoles ovate-lanceolate, to 3 mm long; hypanthium conical, ca. 2 mm long; calyx lobes minute; corolla white, the lobes ovate-lanceolate, ca. 2 mm long, reflexed. Fruit subglobose, laterally compressed, 3–4 mm long, the apex with the persistent styles, glabrous.

Disturbed sites. Rare; Alachua County. Escaped from cultivation. Florida; Asia. Native to Asia. Summer–fall.

Hedera L. 1753. IVY

Woody vine. Leaves alternate, simple, palmate-veined, petiolate, estipulate. Flowers in terminal simple or compound racemes of umbels, bracteate, bracteolate, bisexual; sepals 5, basally connate, actinomorphic; petals 5, free, actinomorphic; stamens 5, free; ovary inferior, 2- to 5-carpellate and -loculate, the style 1. Fruit a drupe.

A genus of about 15 species; North America, Europe, Africa, Asia, Australia, and Pacific Islands. [*Prehendere,* to seize, in reference to its climbing by prehensile aerial roots.]

Selected reference: Ackerfield and Wen (2002).

1. Young branchlets and leaves with reddish brown scalelike trichomes less than 0.5 mm long .... **H. algeriensis**
1. Young branchlets and leaves with white stellate trichomes 0.5–1.0 mm long .... **H. helix**

Hedera algeriensis Hibberd [Of Algeria.] ALGERIAN IVY.

Hedera algeriensis Hibberd, Fl. World 7: 57. 1864.

Woody vines, to 20 m or more; branchlets attaching to the substrate with adventitious aerial roots, with reddish brown scalelike trichomes less than 0.5 mm long, usually with 10–15 rays. Leaves with the blade cordate or suborbicular, 5–20 cm long, 5–12 cm wide, the apex acute or rounded, the base rounded or cordate, the margin entire, undulate, or palmately 3- to 5-lobed, the upper and lower surface with scalelike stellate pubescence, the trichomes less than 0.5 mm long, with 10–15 lobes, glabrous or glabrate in age, the petiole 2–15 cm long. Flowers in a terminal simple or compound raceme of umbels, the peduncle 2–8 cm long, stellate-pubescent, the pedicel ca. 1 cm long; hypanthium conic-cupulate, 1–2 mm long, stellate pubescent; sepal lobes minute; petals greenish yellow, elliptic-ovate, 2–4 mm long, spreading-reflexed; fruit subglobose to obovoid, 7–10 mm long, black; pyrenes 2–5.

Disturbed sites. Rare; Jefferson County, central panhandle. Escaped from cultivation. Florida; Europe and Africa. Native to Africa. Summer.

Hedera helix L. [From the Greek, coiled, spirally twisted, in reference to the stems.] ENGLISH IVY.

Hedera helix Linnaeus, Sp. Pl. 202. 1753. *Hedera poetica* Salisbury, Prodr. Stirp. Chap. Allerton 143. 1796, nom. illegit. *Hedera diversifolia* Stokes, Bot. Med. 1: 456. 1812, nom. illegit. *Hedera communis* Gray, Nat. Arr. Brit. Pl. 2: 491. 1821, nom. illegit. *Hedera poetarum* Bertoloni, Prael. Rei Herb. 78. 1827, nom. illegit. *Hedera helix* Linnaeus var. *vulgaris* de Candolle, Prodr. 4: 261. 1830, nom. inadmiss. *Hedera arborea* Carrière, Rev. Hort. 1890; non. Walter, 1788; nec. Swartz, 1797; nec Röhling, 1812. *Hedera helix* Linnaeus var. *europaea* Voss, Vilm. Blumengärtn., ed. 3. 1: 407. 1894, nom. inadmiss. *Hedera helix* Linnaeus var. *typica* Schelle, in Beissner et al., Handb. Laubholzben. 363. 1903, nom. inadmiss. *Hedera helix* Linnaeus subsp. *euhelix* Hayek, Repert. Spec. Nov. Regni Veg. Beih 30(1): 955. 1926, nom. inadmiss.

Woody vines, to 20 m or more; branchlets attaching to the substrate with adventitious aerial roots, with white stellate trichomes 0.5–1 mm long, with less than 10 rays. Leaves with the blade ovate, cordate, or suborbicular, 2–12 cm long, and wide, the apex acute or rounded, the base rounded or cordate, the margin entire, undulate, or palmately 3- to 5-lobed, the upper and lower surfaces stellate pubescent, the trichomes 0.5–1 mm long, with 10 or fewer lobes, glabrous or glabrate in age, the petiole 1.5–8 cm long. Flowers in a terminal simple or compound raceme of umbels, the peduncle 2–8 cm long, stellate-pubescent, the pedicel ca. 1 cm long; hypanthium conic-cupulate, 1–2 mm long, stellate pubescent; sepal lobes minute; petals greenish yellow, elliptic-ovate, 2–4 mm long, spreading-reflexed; fruit subglobose to obovoid, 7–10 mm long, black; pyrenes 2–5.

Disturbed sites. Occasional; northern counties. Escaped from cultivation. Nova Scotia, Massachusetts, and New York south to Florida, west to Wisconsin, Illinois, Nebraska, Missouri,

Arkansas, and Arizona, also British Columbia south to California; Europe, Africa, Asia, Australia, and Pacific Islands. Native to Europe, Africa, and Asia. Summer.

Hydrocotyle L. 1753. MARSHPENNYWORT

Perennial herbs. Leaves alternate, simple, palmate-veined, petiolate, estipulate. Flowers in axillary simple or compound umbels; sepals 5, free, actinomorphic; petals 5, free, actinomorphic; ovary 2-carpellate and -loculate, styles 2. Fruit a schizocarp.

A genus of about 130 species; nearly cosmopolitan. [From the Greek *hydro,* water, and *cotyle,* cup, in reference to the leaves of some species being somewhat cup-shaped.]

Hydrocotyle has been previously placed in the Apiaceae by various workers.

Mathias and Constance (1944a).

1. Leaves peltate.
    2. Inflorescence a whorled spike or raceme .... **H. verticillata**
    2. Inflorescence an umbel.
        3. Floral umbels simple (rarely weakly proliferating) .... **H. umbellata**
        3. Floral umbels compound and proliferous .... **H. bonariensis**
1. Leaves not peltate, sometimes with a deep basal sinus, but this not extending completely to the petiole.
    4. Stems and leaves hirsute; fruit hispidulous .... **H. bowlesioides**
    4. Stems and leaves glabrous; fruit glabrous.
        5. Plant robust; petiole 2–3 mm thick; fruit distinctly pedicellate .... **H. ranunculoides**
        5. Plant delicate; petiole less than 1 mm thick; fruit sessile or subsessile .... **H. sibthorpioides**

Hydrocotyle bonariensis Comm. ex Lam. [Of Buenos Aires, Argentina.] LARGELEAF MARSHPENNYWORT.

Hydrocotyle bonariensis Commerson ex Lamarck, Encycl. 3: 153, 1789. *Hydrocotyle umbellata* Linnaeus var. *bonariensis* (Commerson ex Lamarck) Spegazzini, Anales Soc. Ci. Argent. 48: 54. 1899.

Perennial herb; stem stoloniferous, creeping, rooting at the nodes, glabrous. Leaves with the blade peltate, broader than long or orbicular, 3–5(8) cm long, 4–8(12) cm wide, the margin crenate or shallowly and irregularly crenate-lobed, often more deeply so at the apex and especially at the base, the upper and lower surfaces glabrous, the petiole to 3 dm long. Flowers at first in simple globose umbel and soon in a proliferous, obpyramid, compound umbel, the peduncle to 3 dm long, longer than the leaves, the branches 7–10(20) cm long; involucrate bracts lanceolate, inconspicuous; calyx lobes minute or obsolete; petals white or yellow, ovate, ca. 1 mm long. Fruit broadly ellipsoid to reniform, laterally compressed, 1–2 mm long, 2–4 mm wide, the ribs evident, glabrous.

Moist depressions of beaches and dunes, ponds, and canals. Frequent; nearly throughout. Virginia south to Florida, west to Texas; West Indies, Mexico, Central America, and South America; Africa and Australia. Native to North America, tropical America, and Africa. Spring–summer.

Hydrocotyle bowlesioides Mathias & Constance [Resembling *Bowlesia* (Apiaceae), in reference to the leaves.] LARGELEAF MARSHPENNYWORT.

Hydrocotyle bowlesioides Mathias & Constance, Bull. Torrey Bot. Club 69: 151. 1942.

Perennial herb; stem stoloniferous, creeping, rooting at the nodes, hirsute. Leaves with the blade suborbicular or somewhat reniform, 1.5–3 cm long and wide, the margin shallowly 5-lobed, the lobes triangular, the lobe apex obtuse or rounded, the base with a cordate or subtruncate sinus extending to the petiole, the margin crenate, the upper and lower surfaces hirsute, the petiole 1–12 cm long, reflexed hirsute. Flowers in a simple umbel, the peduncle 2–12 mm long, much shorter than the leaves, sessile; involucral bracts lanceolate, inconspicuous; calyx lobes minute or obsolete; petals white or yellow, ovate, ca. 1 mm long. Fruit ellipsoid, laterally compressed, ca. 1 mm long, 1–2 mm wide, the ribs evident, hispidulous.

Wet disturbed sites. Rare; Leon and Orange Counties. North Carolina, South Carolina, Georgia, Alabama, Florida, and Louisiana; West Indies, Central America, and South America; Pacific Islands. Native to Central America. Spring–summer.

Hydrocotyle ranunculoides L. f. [Resembling *Ranunculus* (Ranunculaceae), in reference to the leaves.] FLOATING MARSHPENNYWORT.

Hydrocotyle ranunculoides Linnaeus f., suppl. Pl. 177. 1782 ("1781"). *Hydrocotyle ranunculoides* Linnaeus f. var. *lobata* Urban, in Martius, Fl. Bras. 11(1): 283. 1879, nom. inadmiss. *Hydrocotyle ranunculoides* Linnaeus f. forma *genuina* Urban, in Martius, Fl. Bras. 11(1): 284. 1879, nom. inadmiss.

Perennial herb; stem stoloniferous, floating or creeping, rooting at the nodes, glabrous. Leaves with the blade suborbicular or reniform, 0.5–8 cm long and wide, with a deep basal sinus extending nearly to the petiole, the margin deeply crenate-lobed, 5–6-lobed to about the middle, the upper and lower surfaces glabrous, the petiole 1–30 cm long, often thickened to 2–3 mm. Flowers in a simple umbel, the peduncle 0.5–6 cm long, shorter than the leaves, the pedicel 1–3 mm long. Fruit suborbicular, 1–3 mm long, 2–3 mm wide, the ribs obsolete, glabrous.

Pond and lake margins and riverbanks. Frequent; nearly throughout. New York south to Florida, west to Kansas, Oklahoma, and Texas, also British Columbia south to California and Arizona; West Indies, Mexico, Central America, and South America; Africa, Asia, Australia, and Pacific Islands. Native to North America, tropical America, Africa, and Asia. Spring–summer.

Hydrocotyle sibthorpioides Lam. [Resembling *Sibthorpia* (Plantaginaceae), in reference to the leaves.] LAWN MARSHPENNYWORT.

Hydrocotyle sibthorpioides Lamarck, Encycl. 3: 153. 1789. *Chondrocarpus sibthorpioides* (Lamarck) Sweet, Hort. Brit. 185. 1826.

Perennial herb; stem creeping, rooting at the nodes, glabrous. Leaves with the blade suborbicular, 0.5–1.5 cm long and wide, the base with a deep sinus extending to the petiole, the margin crenate-lobed, the upper and lower surfaces glabrous, the petiole 0.5–2 cm long. Flowers in a simple umbel, sessile or subsessile, the peduncle 0.5–2 cm long, subequaling the leaves. Fruit suborbicular, 1–2 mm long and wide, laterally compressed, the ribs evident, glabrous.

Pond margins. Rare; Baker, Leon, and Jackson Counties. Pennsylvania and New Jersey south to Florida, west to Arkansas and Louisiana, also California; Africa, Asia, and Australia. Native to Africa and Asia. Spring–summer.

Hydrocotyle umbellata L. [In an umbel, in reference to the flowers.] MANYFLOWER MARSHPENNYWORT.

Hydrocotyle umbellata Linnaeus, Sp. Pl. 234. 1753. *Hydrocotyle umbellulata* Michaux, Fl. Bor.-Amer. 1: 161. 1803, nom. illegit. *Hydrocotyle umbellata* Linnaeus var. *umbellulata* de Candolle, Prodr. 4: 60. 1830, nom. inadmiss.

Perennial herb; stem stoloniferous, floating or creeping, rooting at the nodes, glabrous. Leaves with the blade peltate, orbicular, 5–7.5 mm long and wide, the margin crenate-lobed, the upper and lower surfaces glabrous, the petiole 0.5–40 cm long. Flowers in a simple umbel, rarely weakly proliferating and compound, the pedicel 2–25 mm long, the peduncle 0.5–40 cm long, subequaling the leaves; involucrate bracts lanceolate, inconspicuous; calyx lobes minute or obsolete; petals white or yellow, ovate, ca. 1 mm long. Fruit suborbicular to elliptical, laterally compressed, 1–2 mm long, 2–3 mm wide, the ribs evident, glabrous.

Marshes and pond margins. Frequent; nearly throughout. Nova Scotia and New York south to Florida, west to Minnesota, Oklahoma, and Texas, also Oregon and California; West Indies, Mexico, Central America, and South America; Africa. Spring–summer.

Hydrocotyle verticillata Thunb. [Whorled, in reference to the flowers.] WHORLED MARSHPENNYWORT.

Perennial herb; stem stoloniferous, creeping, rooting at the nodes, glabrous. Leaves with the blade peltate, orbicular, 0.5–6 cm long and wide, the margin crenate-lobed, the upper and lower surfaces glabrous, the petiole 0.5–26 cm long. Flowers in simple, interrupted few-flowered verticels 1.5–17 cm long, this once or twice 2-furcate, rarely 3- or 4-furcate, the flowers with a pedicel to 10 mm long or sessile; involucrate bracts lanceolate, inconspicuous; calyx lobes minute or obsolete; petals white or yellow, ovate, ca. 1 mm long. Fruit ellipsoid, laterally compressed, 1–3 mm long, 2–4 mm wide, the ribs evident, glabrous.

1. Flowers or fruits sessile or subsessile .... var. **verticillata**
1. Flowers or fruits with an evident pedicel to 10 mm long .... var. **triradiata**

Hydrocotyle verticillata var. **verticillata**

Hydrocotyle verticillata Thunberg, Hydrocotyle 2, 5, t. s.n.[2]. 1798. *Hydrocotyle vulgaris* Linnaeus var. *verticillata* (Thunberg) Persoon, syn. Pl. 1: 301. 1805.
Hydrocotyle interrupta Muhlenberg ex Elliott, Sketch Bot. S. Carolina 1: 345. 1817.

Flowers or fruits sessile or subsessile.

Floodplain forests, pond margins, and marshes. Frequent; nearly throughout. New York and Massachusetts south to Florida, west to Texas; West Indies and Mexico. Spring–summer.

Hydrocotyle verticillata var. **triradiata** (A. Rich.) Fernald

Hydrocotyle tribotrys Ruiz López & Pavón, Fl. Peruv. 3: 24, pl. 246(b). 1802. *Hydrocotyle bonariensis* Commerson ex Lamarck var. *tribotrys* (Ruiz López & Pavón), G. Don, Gen. Hist. 3: 249. 1834, nom. illegit.
Hydrocotyle polystachya A. Richard var. *triradiata* A. Richard, Ann. Gen. Sci. Phys. 4: 171. 1820. *Hydrocotyle verticillata* Thunberg var. *triradiata* (A. Richard) Fernald, Rhodora 41: 437. 1939.
Hydrocotyle umbellata Linnaeus var. *ambigua* A. Gray, Manual, ed. 5. 190. 1867. *Hydrocotyle canbyi* J. M. Coulter & Rose, Bot. Gaz. 12: 103. 1887. *Hydrocotyle ambigua* (A. Gray) Britton et al., Prelim. Cat. 21. 1888, nom. illegit; non Pursh, 1814.
Hydrocotyle australis J. M. Coulter & Rose, Contr. U.S. Natl. Herb. 7: 28. 1900. TYPE: FLORIDA: Palm Beach Co.: Lake Worth, 8 May 1895, *Curtiss 5376* (holotype: US).

Flowers or fruits with an evident pedicel to 10 mm long.

Floodplain forests, pond margins, and marshes. Occasional; nearly throughout. Massachusetts south to Florida, west to California; West Indies, Mexico, Central America, and South America. Spring–summer.

Polyscias J. R. Forst. & G. Forst. 1775. ARALIA

Shrubs or trees. Leaves alternate, odd-pinnately compound, petiolate, stipulate. Flowers in compound umbels in a paniculiform inflorescence, bisexual or unisexual (plants andromonoecious), bracteate; sepals 5, basally connate, actinomorphic; petals 5, free, actinomorphic; stamens 5, free; ovary inferior, 3–4(5)-carpellate and -loculate, the styles 3–4(5). Fruit a drupe.

A genus of about 160 species; North America, Central America, Africa, Asia, Australia, and Pacific Islands. [From the Greek *poly,* many, and *skia,* shadow, in reference to the variable foliage.]

Polyscias guilfoylei (W. Bull) L. H. Bailey [Commemorates William Robert Guilfoyle (1840–1912), English-born Australian botanist.] FROSTED ARALIA.

Aralia guilfoylei W. Bull, Cat. 83: 4. 1873. *Northopanax guilfoylei* (W. Bull), Merrill, Philipp. J. Sci. 7: 242. 1912. *Polyscias guilfoylei* (W. Bull) L. H. Bailey, Rhodora 18: 153. 1916.

Shrub or tree, to 5 m; branchlets glabrous, the bark roughened, gray. Leaves with the blade odd-pinnately compound, (2.5)3.5–5.5 dm long, the leaflets (5)7–9, elliptic, ovate, or obovate, 5–15(20) cm long, 2.5–12 cm wide, the apex obtuse, acute, or acuminate, the base cuneate, often somewhat oblique, the margin irregularly shallowly, spinulose-serrate, variegated pale yellow or whitish, especially along the margin, the petiolules 1–3.5 cm long, the petiole 7–18 cm long, the base clasping the stem, slightly membranous winged for 1.5–3 cm; stipules adnate to the petiole. Flowers in a terminal panicle of umbels, bisexual or unisexual (plants andromonoecious); with the terminal umbel bisexual and often 1 or 2 lateral umbels staminate), the pedicel 4–10 mm long; bracts minute; hypanthium turbinate, 1–2 mm long; calyx cupuliform, minute, the rim undulate and inconspicuously 5-toothed, hyaline; petals greenish white, 5, deltoid-oblong, ca. 2 mm long; styles free nearly to the base. Fruit subglobose, 4–5 mm long, ribbed; pyrenes 3–4(5).

Disturbed sites. Rare; Monroe County keys. Escaped from cultivation. Florida; West Indies, Central America; Africa, Asia, Australia, and Pacific Islands. Commonly cultivated in tropical America and possible also naturalized there. Native to Asia. Summer–fall.

Schefflera J. R. Forst. & G. Forst. 1775.

Shrubs or trees. Leaves alternate, palmately compound, petiolate, stipulate. Flowers in a compound panicle of umbels, bisexual or unisexual (plants andromonoecius), bracteate; sepals 5, basally connate, actinomorphic; petals 5–12, free; stamens 5–12, free; ovary inferior or half-inferior, 5- to 12-carpellate and -loculate, the styles absent. Fruit a drupe.

A genus of about 900 species; North America, West Indies, Central America, South America, Africa, Asia, Australia, and Pacific Islands. [Commemorates Johann Peter Ernst von Scheffler (1739–1809), Polish physician and botanist.]

Several recent phylogenetic studies have shown that *Schefflera* is polyphyletic (Plunkett et al., 2005). The genus will ultimately have to be restricted to a small group of species from the Pacific Islands. The other species currently placed in it will be transferred to other genera, including the two taxa naturalized in Florida.

1. Leaflets (including the petiolule) ca. 30 cm long, ca. 10 cm wide; inflorescence a compound panicle, the flowers in short stalked heads, pink to red; ovary 10- to 12-carpellate, red ripening black................ **S. actinophylla**
1. Leaflets (including the petiolule) ca. 12 cm long, ca. 4 cm wide; inflorescence a compound panicle of short-stalked umbels, greenish white; ovary 5- to 7-carpellate, orange ripening black................ **S. arboricola**

Schefflera actinophylla (Endl.) Harms [Starlike, radiating from the center, in reference to the leaves.] AUSTRALIAN UMBRELLA TREE; OCTOPUS TREE.

Brassaia actinophylla Endlicher, in Endlicher & Fenzl, Nov. Stirp. Dec. 89. 1839. *Schefflera actinophylla* (Endlicher) Harms, in Engler & Prantl, Nat. Pflanzenfam. 3(8): 36. 1894.

Shrub or tree, to 12 m; branchlets glabrous, the bark gray, fissured. Leaves with (5)7–16 leaflets, the blade oblong, 8–20 cm long, 5–10 cm wide, the apex acuminate to obtuse, the base cuneate to rounded, the margin entire or few toothed at the apex, the upper and lower surfaces glabrous, the petiolule 5–10 cm long, the petiole 20–40 cm long. Flowers in a terminal panicle of 6–13 primary rays to 8 dm long, each raceme-like with stalked head-like umbels, these pink to red; calyx pink or red, obscure, 5-lobed; corolla pink or red, the petals (7)12(18), 3–5 mm long; stamens as many as the petals; stigmas sessile. Fruit subglobose, 6–12 mm long, red ripening dark purple; pyrenes 10–12.

Disturbed sites; often epiphytic when young. Occasional; central and southern peninsula. Escaped from cultivation. Florida; West Indies, Mexico, Central America, and South America; Africa, Asia, Australia, and Pacific Islands. Native to Asia and Australia. Summer–fall.

Schefflera arboricola (Hayata) Merr. [*Arbor,* tree, and *-cola,* dweller, in reference to its sometimes epiphytic habit.] DWARF SCHEFFLERA.

Heptapleurum arboricola Hayata, Icon. Pl. Formosan. 6: 23, t.4. 1916. *Schefflera arboricola* (Hayata) Merrill, Lingnan Sci. J. 5: 139. 1929.

Shrubs, to 4 m; branchlets glabrous. Leaves with the blade palmately compound, the leaflets (5)7–9(10), obovate, oblong, or elliptic, 6–12 cm long, 1.5–3.5(4.5) cm wide, the apex obtuse or acute, the base cuneate, the margin entire, the upper and lower surfaces glabrous, the petiolule (0.5)1–3 cm long, the petiole (6)10–12 cm long; stipules adnate to the petiole. Flowers in a compound panicle of short-stalked umbels, the primary axis 3–8 cm long, the secondary axis to 10 cm long, bisexual or unisexual (plants andromonoecious), the pedicel less than 3 mm long; calyx 5-lobed, subentire, corolla greenish white, the petals 5; ovary 5- to 6-carpellate, the stigmas sessile. Fruit subglobose, ca. 5 mm long, 5- to 6-ribbed, orange ripening purple-black; pyrenes 5–6.

Disturbed sites; often epiphytic when young. Rare; central and southern counties. Escaped from cultivation. Florida; Africa, Asia, Australia, and Pacific Islands. Native to Asia. Summer–fall.

Tetrapanax (K. Koch) K. Koch 1859.

Shrubs or trees. Leaves alternate, simple, palmate-veined, petiolate, stipulate. Flowers in a terminal compound panicle of umbels, bisexual, bracteate; sepals 4(5), basally connate, actinomorphic; petals 4(5), free, actinomorphic; stamens 4(4); ovary inferior, 2-carpellate and -loculate, the styles 2. Fruit a drupe.

A monotypic genus; North America and Asia. [*Tetra,* four, and the genus *Panax,* in reference to the *Panax* species with a 4-part floral structure.]

Tetrapanax papyrifer (Hook.) K. Koch [Paper-bearing, in reference to the pith of the stem that was used to make paper.] RICEPAPER PLANT.

Aralia papyrifer Hooker, Hooker's J. Bot. Kew Gard. Misc. 4: 53, pl. 1–2. 1852. *Didymopanax papyrifer* (Hooker) K. Koch, Wochenschr. Gaertneri Pflanzenk. 2: 69. 1859. *Tetrapanax papyrifer* (Hooker) K. Koch, Wochenschr. Gaertneri Pflanzenk. 2: 371. 1859. *Fatsia papyrifer* (Hooker) Miquel ex Witte, Fl. Jardins Pays-Bas. 4: 87. 1861. *Panax papyrifer* (Hooker) F. Mueller, Frag. 4: 122. 1864. *Echinopanax papyrifer* (Hooker) Kuntze, Revis. Gen. Pl. 1: 271. 1891.

Shrub or small tree, to 3.5 m; stem densely ferruginous or pale brown stellate-tomentose. Leaves with the blade ovate-oblong or suborbicular, 0.5–5(7) dm long and wide, 7- to 12-lobed, the base cordate, the margin coarsely and irregularly serrate, the apex of the lobes acute or acuminate, the upper surface glabrous, the lower surface ferruginous or pale brown stellate tomentose, the petiole to 2–4 dm long, glabrous; stipules awl-shaped, 7–8 mm long. Flowers in a terminal compound panicle of subglobose many-flowered umbels 1–2 cm long, ferruginous or pale brown stellate-tomentose, the pedicel 4–6 mm long, stellate-tomentose; bracts linear-lanceolate, 1–3 mm long, the outer surface stellate-tomentose; hypanthium ca. 1 mm long, stellate-tomentose; calyx minute, 4(5)-lobed, stellate-tomentose; corolla yellowish white, the petals 4(5), oblong, ca. 2 mm long; stamens 4(5). Fruit subglobose, slightly laterally compressed, dark purple, the apex with the 2 persistent recurved styles, glabrous; pyrenes 2.

Disturbed hammocks. Occasional northern and central peninsula, central and western panhandle. Escaped from cultivation. Alabama and Florida; Asia, Australia, and Pacific Islands. Native to Asia. Summer.

EXCLUDED TAXON

Panax quinquefolius Linnaeus—Reported for Florida by Small (1903, 1913a, 1933). No Florida specimens seen.

APIACEAE Lindl., nom. cons. 1826. CARROT FAMILY

Herbs. Leaves alternate or opposite, simple or compound, pinnate-, palmate-, or parallel-veined, petiolate or epetiolate, estipulate. Flowers in terminal and axillary simple or compound umbels, bracteate or ebracteate, bracteolate or ebracteolate, bisexual or unisexual; sepals 5, actinomorphic or zygomorphic, or absent; petals 5, free, actinomorphic; stamens 5, free; ovary inferior, 2-carpellate and -loculate. Fruit a schizocarp, mericarps 2.

A family of about 434 genera and about 3,780 species; nearly cosmopolitan.

Ammiaceae Bercht. & C. Presl, 1820; *Umbelliferae* Juss., nom alt., 1789.

Many Apiaceae contain aromatic oils and emit a characteristic odor when crushed. Some species are used as food or flavoring such as *Anethum graveolens* (dill), *Apium graveolens* (celery), *Coriandrum sativum* (coriander), *Daucus carota* (carrot), *Foeniculum vulgarum* (fennel), and *Petroselinum crispum* (parsley). However, others contain polyacetylene compounds that are extremely poisonous when ingested (*Cicuta maculata*) or furanocoumarins, which cause photodermatitis (*Ammi* spp., *Angelica* spp., and *Daucus* spp.).

Selected references: Coulter and Rose (1900); Mathias and Constance (1944b, 1945).

1. Flowers in dense involucrate heads .... **Eryngium**
1. Flowers in umbels.
    2. Leaves all simple.
        3. Leaves linear or terete, septate.
            4. Leaves flat; stem stoloniferous .... **Lilaeopsis**
            4. Leaves terete; stem erect .... **Tiedemannia**
        3. Leaves cordate or suborbicular, not septate.
            5. Leaves sessile, the upper ones perfoliate .... **Bupleurum**
            5. Leaves petiolate.
                6. Leaf blades suborbicular to reniform, usually deeply lobed; trichomes stellate; stem decumbent but not stoloniferous .... **Bowlesia**
                6. Leaf blades ovate to oblong, not deeply lobed; trichomes simple; stem stoloniferous .... **Centella**
    2. Leaves (at least some) compound or dissected nearly to the base and appearing compound.

7. Fruit echinate, tuberculate, or hispid.
8. Leaves palmately dissected into 3–5 lobes .... **Sanicula**
8. Leaves pinnately compound or dissected.
9. Ovary and/or fruit 5–10 times as long as wide .... **Scandix**
9. Ovary and/or fruit less than 2 times as long as wide.
10. Ovary and/or fruit tuberculate .... **Spermolepis**
10. Ovary and/or fruit hispid or bristly.
11. Fruit winged, hispid .... **Angelica**
11. Fruit not winged, bristly.
12. Involucre of several pinnate bracts .... **Daucus**
12. Involucre absent or of 1(2) linear bract(s) .... **Torilis**
7. Fruit ribbed, winged, or smooth.
13. Fruit more than 2 times as long as wide.
14. Leaves trifoliolate .... **Cryptotaenia**
14. Leaves 2- to 3-pinnately compound.
15. Sepals minute but evident at 10x magnification .... **Trepocarpus**
15. Sepals absent .... **Chaerophyllum**
13. Fruit 2 times as long as wide or less.
16. Involucral bracts present, conspicuous on the primary umbel (1–3 small linear ones on *Oxypolis*).
17. Fruit with prominently winged lateral ribs, the dorsal and lateral ones low and blunt .... **Oxypolis**
17. Fruit with subequal ribs.
18. Fruit with winged ribs .... **Sium**
18. Fruit without winged ribs.
19. Fruit with lateral ribs conspicuously wider than the dorsal ones .... **Ptilimnium**
19. Fruit with the dorsal ribs and lateral ribs subequal .... **Ammi**
16. Involucral bracts absent or greatly reduced and inconspicuous on the primary umbel (sometimes the primary umbels sessile or subsessile, appearing as 1–3 simple umbels).
20. Margin of the leaf segments entire.
21. Corolla white or greenish white; compound umbels sessile or subsessile, appearing as 1–3 simple umbels .... **Cyclospermum**
21. Corolla yellow; compound umbels distinctly pedunculate.
22. Fruit subterete, the lateral ribs not winged .... **Foeniculum**
22. Fruit dorsally flattened, the lateral ribs winged .... **Anethum**
20. Margin of the leaf segments serrate-dentate.
23. Compound umbels sessile or subsessile, appearing as 1–3 simple umbels .... **Apium**
23. Compound umbels distinctly pedunculate.
24. Corolla yellow, greenish yellow, or purple.
25. Central flower or fruit sessile or subsessile .... **Zizia**
25. Central flower or fruit pedicellate.
26. Sepals evident at 10x magnification .... **Thaspium**
26. Sepals minute, not evident at 10x magnification .... **Petroselinum**
24. Corolla white or greenish white.

27. Fruit subglobose .... **Coriandrum**
27. Fruit laterally flattened.
    28. Fruit with the ribs winged.. **Ligusticum**
    28. Fruit with the ribs rounded.
        29. Plant delicate, to 1 m tall.. **Angelica**
        29. Plant robust, over 1 m tall.. **Cicuta**

Ammi L. 1753.

Herbs. Leaves alternate, pinnately compound, petiolate. Flowers in terminal and axillary compound umbels, bracteate, bracteolate, bisexual; sepals 5, basally connate, actinomorphic; petals 5, free; stamens 5, free, actinomorphic; ovary inferior, 2-carpellate and -loculate; fruit a schizocarp, mericarps 2.

A genus of 6 species; North America, Europe, Africa, Asia, Australia, and Pacific Islands. [From the Greek *amos,* sand, in reference to the habitat.]

1. Leaf segments linear or filiform, the margins entire; rays of the umbel with the bases forming a discoid structure (plexus) .... **A. visnaga**
1. Leaf segments elliptic, the margins finely serrate; rays of the umbel with the bases not forming a discoid structure (plexus) .... **A. majus**

Ammi majus L. [Large, in reference to the size.] LARGE BULLWORT.

Ammi majus Linnaeus, Sp. Pl. 243. 1753. *Apium ammi* Crantz, Stirp. Austriac. Fasc. 1(3): 109. 1767, nom. illegit. *Apium ammi-majus* Crantz, Cl. Umbell. Emend. 103. 1767, nom. illegit. *Ammi elatum* Salisbury, Prod. Stirp. Chap. Allerton 162. 1796, nom. illegit. *Ammi diversifolium* Noulet, Fl. Bras. Sous-Pyren. 279. 1837, nom. illegit. *Ammi diversifolium* Noulet var. *latifolium* Noulet, Fl. Bras. Sous-Pyren. 280. 1837, nom. inadmiss. *Sison majus* (Linnaeus) Eaton & Wright, Man. Bot., ed. 8. 429. 1840. *Visnaga vulgaris* Bubani, Fl. Pyren. 2: 350. 1900, nom. illegit. *Selinum ammoides* E.H.L. Krause, in Sturm, Deutschl. Fl., ed 2. 12: 43. 1904, nom. illegit. *Carum majus* (Linnaeus) Koso-Poljansky, Bull. Soc. Imp. Naturalistes Moscou, ser. 2. 29: 198. 1916.

Erect or ascending annual herb, to 8 dm; stem striate, glabrous. Leaves with the blade oblong to broadly triangular-ovate, ternately or pinnately compound, 2–20 cm long and wide, the leaflets elliptic-lanceolate, 2.5–4 cm long, 1–2.5 cm wide, the apex acute, the base cuneate, the margin finely serrate, the upper and lower surfaces glabrous, the petiole ca. 1 cm long, the base sheathing. Flowers in a terminal or axillary compound umbel, the peduncle 8–12 cm long, the rays numerous, 2–7 cm long, the base not forming a discoid plexus, the flowers numerous per umbellet, the pedicel 3–12 mm long; bracts numerous, 1- or 2-pinnately dissected, subequaling or longer than the rays; bracteoles numerous, entire, subequaling or slightly shorter than the pedicel; calyx lobes triangular, minute; corolla white, the petals ovate to obovate, ca. 1 mm long, the apex rounded or emarginate. Fruit oblong-elliptic, laterally flattened, 1.5–2 mm long, each mericarp with 5 shallow ribs, dark brown.

Disturbed sites. Rare; Leon, Alachua, and Palm Beach Counties. New Jersey, South Carolina south to Florida, west to Texas, also Oregon, California, and Arizona; Europe, Africa, and Asia.

Frequently cultivated in other areas where possibly naturalized. Native to Europe, Africa, and Asia. Spring.

Ammi visnaga (L.) Lam. [European vernacular name for the plant.] TOOTHPICKWEED.

Daucus visnaga Linnaeus, Sp. Pl. 242. 1753. *Apium visnaga* (Linnaeus) Crantz, Cl. Umbell. Emend. 104. 1767. *Ammi visnaga* (Linnaeus) Lamarck, Fl. Franc. 3: 426. 1779 ("1778"). *Visnaga daucoides* Gaertner, Fruct. Sem. Pl. 1: 92. 1788. *Daucus laevis* Salisbury, Prodr. Stirp. Chap. Allerton 162. 1796, nom. illegit. *Sium visnaga* (Linnaeus) Stokes, Bot. Mat. Med. 2: 106. 1812. *Selinum visnaga* (Linnaeus) E.H.L. Krause, in Sturm, Deutschl. Fl., ed. 2. 12: 44. 1904.

Erect or ascending annual herb, to 1 m; stem striate, glabrous. Leaves with the blade oblong to broadly triangular or ovate, ternately to pinnately compound, 2–20 cm long and wide, the ultimate segments linear or filiform, 1–2 cm long, 1–2 mm wide, the apex acute, the margin entire, the upper and lower surfaces glabrous, the petiole ca. 1 cm long, the base sheathing. Flowers in a terminal or axillary compound umbel, the peduncle 6–8 cm long, the rays numerous, 2–4 cm long, the bases forming a discoid plexus, the flowers numerous per umbellet, the pedicel 3–5 mm long, unequal; bracts numerous, 1- or 2-pinnately dissected, subequaling or longer than rays; bracteoles, lanceolate, entire, subequaling the pedicel; calyx lobes triangular, minute; corolla white, ca. 1 mm long; the apex rounded or emarginate. Fruit ellipsoid, ca. 2 mm long, laterally flattened, each mericarp with 5 shallow ribs, brown.

Disturbed sites. Rare; Escambia County. Not recently collected. Pennsylvania, North Carolina, Florida, Alabama, Texas, and California; Europe, Africa, and Asia. Frequently cultivated in other areas where possibly naturalized. Native to Europe, Africa, and Asia. Spring.

Anethum L. 1753. DILL

Herbs. Leaves alternate, pinnately compound, petiolate, estipulate. Flowers in terminal and axillary compound umbels, ebracteate, ebracteolate, bisexual; sepals absent; petals 5, free, actinomorphic; stamens 5, free; ovary inferior, 2-carpellate and -loculate; fruit a schizocarp, mericarps 2.

A monotypic genus; nearly cosmopolitan. [From the Greek *aitein,* blaze, in reference to the pungent seeds.]

Anethum graveolens L. [*Gravis* heavy, strong, and *olens,* smelling.] DILL.

Anethum graveolens Linnaeus, Sp. Pl. 263. 1753. *Selinum anethum* Crantz, Cl. Umbell. Emend. 63. 1767, nom. illegit. *Anethum arvense* Salisbury, Prodr. Stirp. Chap. Allerton 168. 1796, nom. illegit. *Pastinaca graveolens* (Linnaeus) Bernhardi, Syst. Verz. 171. 1800. *Selinum graveolens* (Linnaeus) Vest, Man. Bot. 501. 1805. *Ferula graveolens* (Linnaeus) Sprengel, Neue Schriften Naturf. Ges. Halle 2: 14. 1813. *Pastinaca anethum* Sprengel, in Roemer & Schultes, Syst. Veg. 6: 587. 1820, nom. illegit. *Peucedanum graveolens* (Linnaeus) C. B. Clark, in Hooker f., Fl. Brit. India 2: 709. 1879. *Peucedanum anethum* Baillon, Traité Bot. Méd. Phan. 1045. 1883, nom. illegit.

Erect or ascending annual herb, to 1.5 m; stem striate, glabrous. Leaves with the blade ovate, 3- or more pinnately dissected, 4–35 cm long and wide, the ultimate segments narrowly linear to threadlike, 0.5–2 cm long, the apex acuminate, the upper and lower surfaces glabrous, the petiole 1–3 cm long, the base sheathing. Flowers in a terminal or axillary compound umbel, the peduncle 6–12 cm long, the rays 10–many, 3–10 cm long, the flowers many per umbellet, the pedicel 3–10 mm long; calyx absent; corolla yellow, the petals ovate, ca. 1 mm long, the apex rounded. Fruit ovate-elliptic, laterally flattened, each mericarp with 5 ribs, the lateral and sometimes the dorsal and intermediate ones winged.

Disturbed sites. Rare; Miami-Dade County. Escaped from cultivation. Nearly throughout North America; nearly cosmopolitan. Native to Asia. Summer.

Angelica L., nom. cons. 1753.

Herbs. Leaves alternate, pinnately compound, petiolate. Flowers in terminal and axillary compound umbels, ebracteate, bracteolate, bisexual; sepals 5, basally connate, actinomorphic, or absent; petals 5, free, actinomorphic; stamens 5, free; ovary inferior, 2-carpellate and -loculate; fruit a schizocarp, the mericarps 2.

A genus of about 60 species; [From the Greek *angelikos*, messenger, in reference to the medical properties.]

1. Leaf segments coarsely and irregularly toothed; inflorescence glabrous; fruit glabrous........ **A. dentata**
1. Leaf segments finely and evenly toothed; inflorescence pubescent; fruit hispid.................. **A. venenosa**

Angelica dentata (Chapm. ex Torr. & A. Gray) M. Coult. & Rose [With sharp outward pointing teeth, in reference to the leaf margin.] COASTALPLAIN ANGELICA.

Archangelica dentata Chapman ex Torrey & A. Gray, Fl. N. Amer. 1: 622. 1840. *Angelica dentata* (Chapman ex Torrey & A. Gray) J. M. Coulter & Rose, Bot. Gaz. 12: 61. 1887. TYPE: FLORIDA: Gadsden Co.: s.d., *Chapman s.n.* (holotype: NY; isotype: NY).

Erect or ascending perennial herb, to 1 m; stem striate, glabrous. Leaves with the blade triangular-ovate, pinnately or 2- to 3-ternately compound, (5)10–25 cm long, 5–20 cm wide, the ultimate segments lanceolate, 1–3 cm long, 3–8 mm wide, occasionally with 1 or 2 basal lobes, the apex acute, the base cuneate, the margin coarsely and irregularly toothed, the upper and lower surfaces glabrous, the petiole 3–10 cm long, the base sheathing, the uppermost reduced to bladeless sheaths. Flowers in a terminal or axillary compound umbel, the inflorescence glabrous, the peduncle 3–5 cm long, the rays 5–12, 1–4 cm long, the flowers 5–many per umbellet, the pedicel 2–5 mm long; bracteoles minute; calyx lobes triangular, minute, or absent; corolla white, the petals obovate, ca. 1 mm long, the apex rounded or emarginate. Fruit oblong-ovate, 5–6 mm long, laterally flattened, glabrous, brown, each mericarp with 5 ribs, the dorsal and intermediate ribs sometimes narrowly winged, the lateral ones with broad papery wings.

Sandhills and flatwoods. Occasional; central panhandle. Georgia and Florida. Summer–fall.

Angelica venenosa (Greenway) Fernald [Very poisonous.] HAIRY ANGELICA.

Ferula villosa Walter, Fl. Carol. 115. 1788. *Angelica villosa* (Walter) Britton et al., Prelim. Cat. 22. 1888; non Lagasca, 1816. *Archangelica villosa* (Walter) Kuntze, Revis. Gen. Pl. 1: 265. 1891.
Cicuta venenosa Greenway, Trans. Amer. Philos. Soc. 3: 235. 1793. *Angelica venenosa* (Greenway) Fernald, Rhodora 45: 301. 1943.

Erect or ascending perennial herb, to 1.5 m; stem striate, glabrous. Leaves with the blade triangular-ovate, pinnately or 2- to 3-ternately compound, (4)10–25 cm long and wide, the ultimate segments ovate to elliptic, 1.5–5 cm long, 1 cm or more wide, occasionally with 1 or 2 basal lobes, the apex acute or obtuse, the base cuneate, the margin finely serrate, the upper and lower surfaces puberulent or glabrate, the petiole 4–20 cm long, the base sheathing, the uppermost reduced to bladeless sheaths. Flowers in a terminal or axillary compound umbel, the inflorescence pubescent, the peduncle 2–10 cm long, the rays numerous, 1–4 cm long, the flowers 8–many per umbellet, the pedicel 2–10 mm long; bracteoles minute; calyx lobes triangular, minute, or absent; corolla white, the petals obovate, ca. 1 mm long, the apex rounded or emarginate. Fruit oblong-elliptic, laterally flattened, 4–6 mm long, sparsely pubescent, dark brown, each mericarp with 5 ribs, the dorsal and intermediate ribs sometimes narrowly winged, the lateral ones with broad papery wings.

Hammocks. Rare; Leon, Jackson, Santa Rosa, and Escambia Counties. New York and Ontario, south to Florida, west to Illinois, Missouri, Oklahoma, and Louisiana. Summer.

EXCLUDED TAXON

Angelica triquinata Michaux—This northern species was reported for Florida by Chapman (1860, as *Angelica hirsuta* Muhlenberg; 1897, as *Archangelica hirsuta* Torrey & A. Gray, a misapplication of the name to Florida material of *A. venenosa*.)

Apium L. 1753. CELERY

Herbs. Leaves alternate, pinnately compound, petiolate. Flowers in terminal and axillary compound umbels, ebracteate, ebracteolate, bisexual; sepals 5, basally connate, actinomorphic; petals 5, free, actinomorphic; stamens 5, free; ovary inferior, 2-carpellate and -loculate. Fruit a schizocarp, mericarps 2.

A genus of about 20 species; nearly cosmopolitan. [Ancient Greek vernacular name for the plant.]

Celeri Adans., 1763.

Apium graveolens L. [*Gravis*, heavy, strong, and *olens*, smelling.] WILD CELERY.

Apium graveolens Linnaeus, Sp. Pl. 264. 1753. *Seseli graveolens* (Linnaeus) Scopoli, Fl. Carniol., ed. 2. 1: 215. 1772. *Apium celleri* Gaertner, Fruct. Sem. Pl. 1: 99. 1788, nom. illegit. *Sium apium* Roth, Tent. Fl. Germ. 1: 128. 1788, nom. illegit. *Apium maritimum* Salisbury, Prodr. Stirp. Chap. Allerton 169. 1796, nom. illegit. *Sium graveolens* (Linnaeus) Vest, Man. Bot. 517. 1805. *Apium vulgare* Bubani, Fl. Pyren. 2: 344. 1899, nom. illegit. *Selinum graveolens* (Linnaeus) E.H.L. Krause, in sturm, Deutschl.

Fl., ed. 2. 12: 38. 1904; non (Linnaeus) Vest, 1805. *Celeri graveolens* (Linnaeus) Britton, in Britton & A. Brown, Ill. Fl. N. U.S., ed. 2. 2: 660. 1913. *Carum graveolens* (Linnaeus) Koso-Poljansky, Bull. Soc. Imp. Naturalistes Moscou, ser. 2. 29: 199. 1916.

Erect or ascending perennial herb, to 1 m; stem striate, glabrous. Leaves with the blades oblong to obovate, pinnately compound, the leaflets 3–9, ovate, 0.8–5 cm long and wide, the apex rounded, the base cuneate, the margin 1- or 2-ternately lobed or toothed, the upper and lower surfaces glabrous, the petiole to 3 dm long, the base sheathing, the uppermost leaves simple and appearing as an involucre below the sessile inflorescence. Flowers in a terminal or axillary compound umbel, this sessile or subsessile, the rays 7–15, 0.7–2.5 cm long, flowers 7–17 per umbellet, the pedicel 1–3 mm long; sepals triangular, minute; corolla white, the petals broadly ovate, ca. 1 mm long, the apex rounded. Fruit oblong-elliptic to suborbicular, laterally flattened, 1–2 mm long, each mericarp with 5 narrow, rounded ribs, brown.

Disturbed sites. Occasional; central and southern peninsula, central panhandle. Escaped from cultivation. Nearly cosmopolitan. Native to Europe, Africa, and Asia. Spring–summer.

Bowlesia Ruiz & Pav. 1794.

Herbs. Leaves opposite, simple, palmate-veined, petiolate. Flowers in simple axillary umbels, bracteate, ebracteolate, bisexual; sepals 5, basally connate, actinomorphic; petals 5, free, actinomorphic; stamens 5, free; ovary inferior, 2-carpellate and -loculate. Fruit a schizocarp, mericarps 2.

A genus of about 15 species; North America, West Indies, Mexico, Central America, South America, Europe, and Africa. [Commemorates William Bowles (1705–1780), Irish naturalist.]

Selected reference: Mathias and Constance (1965).

Bowlesia incana Ruiz & Pav. [Hoary, white, in reference to its general appearance.] HOARY BOWLESIA.

Bowlesia incana Ruiz López & Pavón, Fl. Peruv. 3: 28, t. 268(a). 1802. *Bowlesia incana* Ruiz López & Pavón forma *crassifolia* Urban, in Martius, Fl. Bras. 11(1): 292. 1879, nom. inadmiss.

Decumbent annual herb; stem to 6 dm long, striate, stellate-pubescent. Leaves with the blade suborbicular or reniform, 0.5–4.5 cm long and wide, palmate-veined, the apex rounded, the base truncate or cordate, the margin irregularly palmately lobed, the lobes obtuse, the upper and lower surfaces stellate-pubescent, the petiole 0.8–3 cm long. Flowers in a simple, axillary, 2- to 5-flowered umbel, this sessile or subsessile, the pedicel 1–3 m long; bracts few, lanceolate, scarious, white; calyx lobes triangular, to 1 mm long, hyaline; corolla greenish white to purplish, the petals ovate to suborbicular, 1–2 mm long, the apex rounded. Fruit ellipsoid to globose, ca. 2 mm long, stellate-pubescent, each mericarp with 5 inconspicuous ribs.

Open wet hammocks. Occasional; northern peninsula, Hillsborough County, central and western panhandle. Florida west to California; West Indies, Mexico, Central America, and South America. Native to South America. Spring.

Bupleurum L. 1753.

Herbs. Leaves alternate, simple, pinnate-veined, epetiolate. Flowers in terminal and axillary compound umbels, ebracteate, bracteolate, bisexual; sepals absent; petals 5, free, actinomorphic; stamens 5, free; ovary inferior, 2-carpellate and -loculate. Fruit a schizocarp, mericarps 2.

A genus of about 150 species; North America, Europe, Africa, and Asia. [From the Greek *Bous,* oxen, and *pleuron,* side, in reference to the supposed swelling in oxen when they feed on the plant.]

Selected reference: Neves and Watson (2004).

Bupleurum lancifolium Hornem. [With lance-shaped leaves.] HARE'S-EAR.

Bupleurum lancifolium Hornemann, Hort. Bot. Hafn. 1: 267. 1813.

Erect or ascending annual herb, to 5 dm; stem striate, glabrous, glaucous. Leaves with the blade ovate to elliptic-ovate, 1.5–8 cm long, 1.5–2.5 cm wide, pinnate-veined, the apex rounded to obtuse, the basal and lower ones clasping the stem with nearly rounded auricles, the median and upper ones perfoliolate, the margin entire. Flowers in a terminal or axillary compound umbel, the peduncle 1.5–2 cm long, the rays 4–10, 0.5–1.5 cm long, flowers 10–12 per umbellet, the pedicel 1–3 mm long, giving the umbellets a dense head-like appearance; bracteoles 5, 0.5–1.5 cm long, fused at the base, ovate, the apex acute; sepals absent; corolla yellow or greenish, the petals broadly oblong to suborbicular, ca. 1 mm long, the apex rounded. Fruit oblong, 2–3 mm long, slightly laterally flattened, glabrous, purplish brown to black, each mericarp with 5 inconspicuous ribs.

Disturbed sites. Rare; Hillsborough and Pinellas Counties. Massachusetts, Connecticut, Pennsylvania, West Virginia, Ohio, and Florida, also Texas, Arizona, and California; Europe, Africa and Asia. Native to Europe, Africa, and Asia. Spring.

EXCLUDED TAXON

Bupleurum rotundifolium Linnaeus—Reported for Florida by Wunderlin (1982, 1998) and Wunderlin and Hansen (2003), the name misapplied to Florida material of *B. lancifolium*).

Centella L. 1763.

Herbs. Leaves alternate, simple, palmate-veined, petiolate. Flowers in simple axillary umbels, bracteate, ebracteolate, bisexual; sepals absent; petals 5, free, actinomorphic; stamens 5, free; ovary inferior, 2-carpellate and -loculate. Fruit a schizocarp, mericarps 2.

A genus of about 45 species; nearly cosmopolitan. [From the Greek *centron,* a sharp point, prickle, and the Latin diminutive *ella,* in reference to the prickles on the type species.]

Centella is sometimes placed in the Araliaceae by various authors.

Selected reference: Schubert (2014).

Centella asiatica (L.) Urb. [Of Asia.] SPADELEAF.

Hydrocotyle asiatica Linnaeus, Sp. Pl. 234. 1753. *Centella asiatica* (Linnaeus) Urban, in Martius, Fl. Bras. 11(1): 287. 1879. *Hydrocotyle sarmentosa* Salisbury, Prodr. Stirp. Chap. Allerton 159. 1796, nom. illegit.

Hydrocotyle erecta Linnaeus f., Suppl. Pl. 177. 1782 ("1781"). *Centella erecta* (Linnaeus f.) Fernald, Rhodora 42: 295. 1940.

Hydrocotyle asiatica Linnaeus var. *floridana* J. M. Coulter & Rose, Rev. N. Amer. Umbell. 136. 1888. *Centella asiatica* (Linnaeus) Urban var. *floridana* (J. M. Coulter & Rose) J. M. Coulter & Rose, Contr. U.S. Natl. Herb. 7: 30. 1900. *Centella repanda* (Persoon) Small forma *floridana* (J. M. Coulter & Rose) Small, Fl. S.E. U.S. 859, 1336. 1903. *Centella floridana* (J. M. Coulter & Rose) Nannfeldt, Svensk. Bot. Tidskr. 18: 411. 1924. TYPE: FLORIDA: Duval Co.: near Jacksonville, s.d. *Curtiss 988* (lectotype: US). Lectotypified by Coulter and Rose (1900: 30).

Hydrocotyle ficaroides Michaux, Fl. Bor.-Amer. 1: 161. 1803; non Lamarck, 1789. *Hydrocotyle repanda* Persoon, Syn. Pl. 1: 302. 1805. *Glyceria repanda* (Persoon) Nuttall, Gen. N. Amer. Pl. 1: 177. 1818. *Centella repanda* (Persoon) Small, Fl. S.E. U.S. 859, 1336. 1903. *Centella asiatica* (Linnaeus) Urban var. *repanda* (Persoon) Domin, Bot. Jahrb. Syst. 41: 159. 1908.

Prostrate perennial herb; stem horizontal, rooting at the nodes, striate, glabrous or tomentose. Leaves arising at the nodes of the horizontal stem, the blade ovate to orbicular-ovate, 1.5–5 cm long and wide, palmate-veined, the apex rounded, the base deeply cordate to truncate, the margin denticulate, dentate, or slightly sinuate, the upper and lower surfaces glabrous or tomentose, the petiole 1–30 cm long, glabrous or tomentose, the base sheathing. Flowers in a simple, axillary, loose or subcapitate umbel, 1–5 per node, the peduncle shorter than the leaf petiole from the same node, the flowers 2–4 per umbel, subsessile or the pedicel to 4 mm long; bracts 2, oblanceolate; sepals absent; corolla white or reddish tinged, the petals oblanceolate, ca. 1 mm long, glabrous or pubescent on the lower surface. Fruit suborbicular, 3–4 mm long, laterally flattened, each mericarp with 5 ribs, prominently nerved with slightly raised reticulate venation between the ribs, slightly pubescent.

Marshes, wet hammocks, floodplain forests, and wet disturbed sites. Common; nearly throughout. New Jersey south to Florida, west to Texas, also Ohio and Oregon; nearly cosmopolitan. Summer.

Chaerophyllum L. 1753. CHERVIL

Herbs. Leaves alternate, pinnately compound, petiolate. Flowers in terminal and axillary compound umbels, ebracteate, bracteolate, bisexual; sepals absent; petals 5, free, actinomorphic; stamens 5, free; ovary inferior, 2-carpellate and -loculate. Fruit a schizocarp, mericarps 2.

A genus of about 35 species; North America, Europe, Africa, and Asia. [From the Greek, *chairo,* to rejoice, to please, and *phyllon,* leaf, in reference to the fragrant foliage.]

1. Fruit elliptic-fusiform, widest near the middle, the ribs narrower than the intervals .... **C. procumbens**
1. Fruit lanceolate, widest at the base, the ribs wider than the intervals .... **C. tainturieri**

Chaerophyllum procumbens (L.) Crantz [Prostrate.] SPREADING CHERVIL.

Scandix procumbens Linnaeus, Sp. Pl. 257. 1753. *Chaerophyllum procumbens* (Linnaeus) Crantz, Cl. Umbell. Emend. 77. 1767. *Myrrhis procumbens* (Linnaeus) Sprengel, Pl. Min. Cogn. Pug. 2: 56. 1815.

Erect or ascending annual herb, to 6 dm; stem striate, glabrous or sparsely pubescent toward the base and sometimes at the nodes. Leaves with the blade ovate to oblong-ovate, pinnately or ternately and then pinnately 3 times compound, 1–12 cm long and wide, the ultimate segments entire, few-toothed, or pinnately lobed or dissected, 1–8 mm long, 1–4 mm wide, the apex obtuse or rounded, the upper surface glabrous, the lower surface glabrous or sparsely pubescent along the veins, the petiole 2–4 cm, reduced upward, the uppermost ones subsessile, the base sheathing. Flowers in a terminal or axillary compound umbel, the axillary umbel sessile, the terminal with a peduncle to 3 cm long, the rays 1–5, 0.5–1.5 cm long at flowering, elongating to 5.5 cm at fruiting, the flowers 2–15 per umbellet, sessile or the pedicel to 2 mm long at flowering, unequal, elongating to 1 cm in fruit; bracteoles 4–6, elliptic ovate to oblong-ovate, ca. 2 mm long, usually fused at the base, usually pubescent along the margin; sepals absent; corolla white, the petals obovate, ca. 1 mm long, the apex rounded or shallowly notched. Fruit elliptic-fusiform, widest near the middle, 5–10 mm long, laterally flattened, dark brown, each mericarp with 5 low, blunt ribs, these narrower than the intervals.

Wet hammocks. Rare; Alachua County, central panhandle. New York south to Florida, west to Ontario, Wisconsin, Iowa, Nebraska, Kansas, Oklahoma, and Alabama. Spring.

Chaerophyllum tainturieri Hook. [Commemorates Louis François Tainturier (fl. 1825–1840), a market-gardener in Louisiana who collected plants and sent them to Hooker.] HAIRYFRUIT CHERVIL.

Chaerophyllum tainturieri Hooker, Companion Bot. Mag. 1: 47. 1835. *Chaerophyllum procumbens* (Linnaeus) Crantz var. *tainturieri* (Hooker) J. M. Coulter & Rose, Bot. Gaz. 12: 160. 1887.
Chaerophyllum tainturieri Hooker var. *floridanum* J. M. Coulter & Rose, Contr. U.S. Natl. Herb. 7: 60. 1900. *Chaerophyllum floridanum* (J. M. Coulter & Rose) Bush, Trans. Acad. Sci. St. Louis 12: 62. 1902. TYPE: FLORIDA: Duval Co.: Sister Islands, St. Johns River, Mar 1880, *Curtis 1040* (holotype: US).

Erect or ascending annual herb, to 7 dm; stem striate, sparsely pubescent. Leaves with the blade ovate to oblong-ovate, pinnately or ternately and then pinnately 3 times compound, 1–12 cm long and wide, the ultimate segments entire, few toothed, or pinnately lobed or dissected, 1–6 mm long, 1–3 mm wide, the apex obtuse or rounded, the upper surface glabrous, the lower surface glabrous or sparsely pubescent to moderately pubescent, the petiole 2–4 cm, reduced upward, the uppermost ones subsessile, the base sheathing. Flowers in a terminal or axillary compound umbel, the axillary umbel sessile, the terminal with a peduncle to 4 cm long, the rays 1–5, 0.5–1.5 cm long at flowering, elongating to 7 cm at fruiting, the flowers 2–15 per umbellet, sessile or the pedicel to 2 mm long at flowering, unequal, elongating to 1 cm in fruit; bracteoles 4–6, elliptic ovate to oblong-ovate, ca. 2 mm long, usually fused at the base, usually pubescent along the margin; sepals absent; corolla white, the petals obovate, ca. 1 mm long, the apex rounded or shallowly notched. Fruit lanceolate, widest at the base, 4–8 mm long, laterally flattened, dark brown, each mericarp with 5 low, blunt ribs, these wider than the intervals.

Wet hammocks. Occasional; northern counties, central peninsula. Delaware and Maryland south to Florida, west to Nebraska, Kansas, Oklahoma, and Arizona. Spring.

Cicuta L. 1753. WATER HEMLOCK

Herbs. Leaves alternate, pinnately compound, petiolate. Flowers in terminal and axillary compound umbels, bracteate or ebracteate, bracteolate or ebracteolate, bisexual; sepals 5, basally connate, actinomorphic; petals 5, free, actinomorphic; stamens 5, free; ovary inferior, 2-carpellate and -loculate. Fruit a schizocarp, mericarps 2.

Selected reference: Mulligan (1980).

A genus of 4 species; North America and Europe. [Ancient Latin vernacular name for the Old World poisonous hemlock.]

Cicuta maculata L. [Spotted, in reference to the stem, which is sometimes purple mottled.] SPOTTED WATER HEMLOCK.

Cicuta maculata Linnaeus, Sp. Pl. 256. 1753. *Cicutaria maculata* (Linnaeus) Lamarck, Encycl. 2: 2. 1786. *Cicuta virosa* Linnaeus var. *maculata* (Linnaeus) J. M. Coulter & Rose, Rev. N. Amer. Umbell. 130. 1888. *Coriandrum maculatum* (Linnaeus) Roth, Tent. Fl. Germ. 1: 130. 1788.

Cicuta curtissii J. M. Coulter & J. M. Rose, Contr. U.S. Natl. Herb. 7: 97. 1900. *Cicuta maculata* Linnaeus var. *curtissii* (J. M. Coulter & Rose) Fernald, Rhodora 41: 439. 1939. TYPE: FLORIDA: s.d., Jun 1880, *Curtiss 1030* (holotype: US; isotypes: NY, US).

Cicuta mexicana J. M. Coulter & Rose, Proc. Wash. Acad. Sci. 1: 145. 1900.

Perennial herb, to 2 m; stem striate and sometimes slightly angled, glabrous, glaucous, sometimes purple-spotted or -mottled toward the base. Leaves with the blade ovate to triangular-ovate, 2–40 cm long, the basal and lower ones 2- to 3-pinnately compound, the median and upper progressively reduced and 1- to 2-pinnately compound, the uppermost ones sometimes simple, the apex acute, the base unequally rounded or broadly cuneate, the margin finely to coarsely serrate, sometimes unevenly lobed, the upper and lower surfaces glabrous, the petiole 1–3 dm long, the base sheathing. Flowers in a terminal or axillary compound umbel, the peduncle 1–2 cm long, the rays numerous, 1.5–6.5 cm long, often unequal in length; flowers usually numerous in an umbellet, the pedicel 2–10 mm long; bracts 1–4, linear, shorter than the rays, or absent, the bracteoles 3–7 or absent, mostly shorter than the pedicel, linear to lanceolate, with a thin, white, papery margin, the apex acute or acuminate; sepals triangular, minute; corolla white, the petals obovate, 1–2 mm long, the apex narrowed to a slender tip. Fruit oblong-elliptic, 2–4 mm long, laterally flattened, dark brown to reddish brown, each mericarp with 5 pale, blunt and somewhat corky ribs.

Swamps, marshes, and riverbanks. Frequent; nearly throughout. Nearly throughout North America; Mexico. Summer-fall.

EXCLUDED TAXA

Cicuta bulbifera Linnaeus—Reported for Florida by Mulligan (1980). No Florida specimens seen.
Cicuta virosa Linnaeus—Reported for Florida by Hiroe (1979), the name misapplied to Florida material of *C. maculata.*

Coriandrum L. 1753. CORIANDER

Herbs. Leaves alternate, pinnately compound, petiolate or epetiolate. Flowers in terminal and axillary compound umbels, bracteate, bracteolate, bisexual; sepals 5, basally connate, actinomorphic; petals 5, free, actinomorphic; stamens 5, free; ovary inferior, 2-carpellate and -loculate. Fruit a schizocarp, mericarps 2.

A genus of 2 species; nearly cosmopolitan. [From the Greek *koriannon,* this partly derived from *koris,* "bedbug," in reference to the fetid bedbug-like smell of the unripe fruit.]

Coriandrum sativum L. [Planted, cultivated.] CORIANDER.

Coriandrum sativum Linnaeus, Sp. Pl. 265. 1753. *Coriandrum majus* Gouan, Hortus Monsp. 145. 1762, nom. illegit. *Coriandrum globosum* Salisbury, Prodr. Stirp. Chap. Allerton 166. 1796, nom. illegit. *Coriandrum sativum* Linnaeus var. *vulgare* Alefeld, Landw. Fl. 165. 1866, nom. inadmiss. *Selinum coriandrum* E.H.L. Krause, in Sturm, Deutschl. Fl., ed. 2. 12: 163. 1904.

Erect or ascending annual herb, to 7 dm; stem striate, glabrous. Leaves with the blade ovate, 1- to 2-pinnately compound, 3–15 cm long and wide, upward progressively more divided and 2- to 3-pinnately dissected, the leaflets broadly oblanceolate, 1–2 cm long, the ultimate segments linear, entire, or with a few teeth or lobes, the apex acute or blunt, the base cuneate, the upper and lower surfaces glabrous. Flowers in a terminal or axillary compound umbel, the peduncle 4–6 cm long, the rays 2–8, 1–2.5 cm long, unequal, the flowers numerous per umbellet, the pedicel 2–5 mm long, unequal; bract 1, linear, shorter than the rays; bracteoles 3–5, shorter than or pedicel; sepals triangular, minute, those of the outermost flowers ovate, to 1 mm long; corolla white to pale pink, the petals obovate, the apex rounded or shallowly notched, some or all of the outermost ones narrowly obovate, to 4 mm long, the apex rounded. Fruit oblong-elliptic, to subglobose, 2–3 mm long, not laterally flattened, brown, each mericarp with 5 pale, low, blunt ribs, the mericarps not separating or tardily so.

Disturbed sites. Rare; Volusia and Highlands Counties. Escaped from cultivation. Nearly through North America; nearly cosmopolitan. Native to Europe and Asia. Summer.

Cryptotaenia DC., nom. cons. 1829. HONEWORT

Herbs. Leaves alternate, pinnately compound, petiolate. Flowers in terminal and axillary compound umbels, bracteate, bracteolate or ebracteolate, bisexual; calyx 5, basally connate or absent, actinomorphic; petals 5, free, actinomorphic; stamens 5, free; ovary inferior, 2-carpellate and -loculate. Fruit a schizocarp, mericarps 2.

A genus of 7 species; North America, Europe, Africa, Asia, and Pacific Islands. [From the Greek *cryptos,* hidden, and *tainia,* a ribbon, in reference to the concealed oil tubes.]

Cryptotaenia canadensis (L.) DC. [Of Canada.] CANADIAN HONEWORT.

Sison canadensis Linnaeus, Sp. Pl. 252. 1753. *Chaerophyllum canadense* (Linnaeus) Crantz, Cl. Umbell. Emend. 79. 1767. *Sium canadense* (Linnaeus) Lamarck, Encycl. 1: 407. 1785. *Myrrhis canadensis* (Linnaeus) Gaertner, Fruct. Sem. Pl. 1: 109. 1788. *Scandix ternata* Moench, Methodus 101. 1794, nom. illegit. *Conopodium canadense* (Linnaeus) W.D.J. Koch, Nova Acta Phys.-Med. Acad. Caes. Leop.-Carol. Nat. Cur. 12: 119. 1825. *Cryptotaenia canadensis* (Linnaeus) de Candolle, Prodr. 4: 119. 1830. *Deringa canadensis* (Linnaeus) Kuntze, Revis. Gen. Pl. 1: 266. 1891.

Erect or ascending perennial herb, to 1 m; stem striate, glabrous. Leaves pinnately compound, broadly ovate, 3–13 cm long and wide, the leaflets 3, oblong-lanceolate to elliptic or obovate, 3–15 cm long, 2–5 cm wide, the central leaflet sometimes with a pair of deep basal lobes, 1 or both of the lateral ones sometimes with a single basal lobe, the apex acuminate, the base cuneate, long-tapered, subsessile, the margin irregularly coarsely to finely serrate, the petiole to 10 cm long, reduced upward and becoming subsessile, the base sheathing. Flowers in a terminal or axillary compound umbel, these often becoming a paniculiform inflorescence, the peduncle 1–2 cm long, the rays 2–7, 0.5–5 cm long, unequal, the flowers 2–10 per umbellet, the pedicel 2–30 mm long, unequal; bract 1, linear, shorter than the rays, or absent; bracteoles 1 or 2, linear, shorter than the pedicel; sepals minute or absent; corolla white, the petals obovate, ca. 1 mm long, the apex rounded or with an abrupt tip. Fruit narrowly oblong-elliptic, 4–7(8) mm long, laterally flattened, arched or curved, each mericarp with 5 narrow, blunt ribs, dark brown, the ribs lighter greenish yellow, glabrous.

Floodplain forests. Rare; Gadsden, Liberty, and Jackson Counties. Quebec south to Florida, west to Manitoba, North Dakota, South Dakota, Nebraska, Kansas, Oklahoma, and Texas. Spring.

Cryptotaenia canadensis is listed as endangered in Florida (Florida Administrative Code, Chapter 5B-40).

Cyclospermum Lag. 1821. MARSH PARSLEY

Herbs. Leaves alternate, pinnately compound, petiolate. Flowers in terminal and axillary simple or compound umbels, ebracteate, ebracteolate, bisexual; sepals absent; petals 5, free, actinomorphic; stamens 5, free; ovary inferior, 2-carpellate and -loculate. Fruit a schizocarp, mericarps 2.

A genus of 3 species; nearly cosmopolitan. [From the Greek *kyclos,* circle, and *sperma,* seed, in reference to the shape of the fruit and seed.]

Cyclospermum leptophyllum (Pers.) Sprague ex Britton & P. Wilson [With narrow leaves, in reference to the narrow leaf segments.] MARSH PARSLEY.

Pimpinella leptophylla Persoon, Syn. Pl. 1: 324. 1805. *Aethusa leptophylla* (Persoon) Sprengel, Pl. Umbell. Prodr. 22. 1813. *Helosciadium leptophyllum* (Persoon) de Candolle, Mém. Soc. Phys. Genève

4: 493. 1829. *Apium leptophyllum* (Persoon) F. Mueller ex Bentham, Fl. Austral. 3: 372. 1867. *Apium ammi* (Savi) Urban var. *leptophyllum* (Persoon) Kuntze, Revis. Gen. Pl. 3(2): 111. 1898. *Selinum leptophyllum* (Persoon) E.H.L. Krause, in Sturm, Deutschl. Fl., ed. 2. 12: 28. 1904. *Cyclospermum leptophyllum* (Persoon) Sprague ex Britton & P. Wilson, Bot. Porto Rico 6: 52. 1925.

Sesili ammi Savi, Due Cent. Piante 71. 1804. *Apium ammi* Urban, in Martius, Fl. Bras. 11(1): 341. 1879; non Crantz, 1767. *Apium ammi* (Savi) Urban var. *genuinum* H. Wolff, in Engler, Pflanzenr. 4(Heft 90): 54. 1927, nom. inadmiss.

Erect annual herb, to 6 dm; stem striate, glabrous. Leaves with the blade oblong-ovate, 3- to 4-pinnately compound, 3.5–10 cm long, 3.5–8 cm wide, the ultimate divisions filiform, 1.5–7 mm long, the lower ones sometimes with lanceolate or linear segments, the apex acute, margin entire, the petiole 2.5–11 cm long, the base sheathing. Flowers in a terminal or axillary compound umbel, sessile or subsessile, these sometimes with a single ray and appearing as a simple, pedunculate umbel, the rays 1–5, 5–8 mm long, the flowers 8–15 per umbellet, the pedicel 2–4 mm long; calyx absent; corolla white, the petals ovate, ca. 1 mm long. Fruit ovoid, 2–3 mm long, laterally compressed, each mericarp with 5 rounded ribs, glabrous.

Disturbed sites. Frequent; nearly throughout. New York south to Florida, west through the lower southern states to California and north to Oregon; nearly cosmopolitan. Native to tropical America. Spring–summer.

Daucus L. 1753. WILD CARROT

Herbs. Leaves alternate, pinnately compound, petiolate. Flowers in terminal and axillary compound umbels, bracteate, bracteolate, bisexual; sepals 5, basal connate, actinomorphic, or absent; petals 5, free, actinomorphic; stamens 5, free; ovary inferior, 2-carpellate and -loculate. Fruit a schizocarp, mericarps 2.

A genus of about 25 species; nearly cosmopolitan. [From the Greek *daukos*, ancient vernacular name for *D. carota*].

1. Involucral bracts scarious-margined, spreading or reflexed in fruit; spines on the fruit not barbed .... **D. carota**
1. Involucral bracts not scarious-margined, appressed to the umbel in fruit; spines on the fruit apically barbed .... **D. pusillus**

Daucus carota L. [European vernacular name for the plant.] QUEEN ANNE'S LACE.

Daucus carota Linnaeus, Sp. Pl. 242. 1753. *Caucalis carota* (Linnaeus) Crantz, Cl. Umbell. Emend. 113. 1767. *Caucalis daucus* Crantz, Stirp. Austriac. Fasc. 3: 125. 1767, nom. illegit. *Daucus vulgaris* Necker, Delic. Gall-Belg. 139. 1768, nom. illegit. *Daucus carota* Linnaeus var. *sylvestris* Hoffmann, Deutschl. Fl. 1: 91. 1791, nom. inadmiss. *Daucus esculentus* Salisbury, Prodr. Stirp. Chap. Allerton 162. 1796, nom. illegit. *Carota sylvestris* Ruprecht, Fl. Ingr. 468. 1860. *Daucus carota* Linnaeus var. *typicus* Pospichal, Fl. Oesterr. Kuestenl. 2: 200. 1898, nom. inadmiss. *Daucus communis* Rouy & Camus, Fl. France 7: 231. 1901, nom. illegit. *Daucus carota* Linnaeus var. *genuinus* Burnat, in Burnat et al., Fl. Alpes Marit. 4: 245. 1906, nom. inadmiss. *Daucus carota* Linnaeus forma *normalis* Thellung, in Hegi, Ill. Fl. Mitt.-Eur. 5: 1514. 1926. *Daucus carota* Linnaeus subvar. *typicus* Thellung, in Hegi, Ill. Fl. Mitt.-Eur. 5: 1514. 1926, nom. inadmiss. *Daucus carota* Linnaeus subsp. *mediterraneus* Zagorodskich, Dokl. Akad. Nauk. SSSR 25: 522–55. 1932, nom. inadmiss.

Annual or biennial herb, to 1.5 m; stem striate, sparsely pubescent with spreading or recurved broad-based trichomes. Leaves with the blade oblong to triangular-ovate, 2- to 4-pinnately compound or dissected, 5–20 cm long, the ultimate segments linear or lanceolate, 2–12 mm long, 1–2 mm wide, the apex acuminate, the margin entire or few-toothed or -lobed, sparsely to moderately pubescent, especially along the margins and veins, the petiole 0.5–2 cm long, the base sheathing. Flowers in a terminal or axillary compound umbel, the peduncle to 15 cm long, the rays numerous, 1.5–3 cm long, unequal, spreading and loosely ascending at flowering, curving upward and inward at fruiting, the flowers numerous per umbellet, the central flower sessile, the pedicel of the others 1–8 mm long, unequal; bracts 4–15, like the leaves, 0.5–4 cm long, pinnately 1- to 2-dissected, scarious-margined, spreading to loosely ascending at flowering, spreading to loosely reflexed in fruit; bracteoles 5–13, shorter than to slightly longer than the pedicel, the margin entire or sometimes pinnately few-lobed distally; calyx triangular, minute, or absent; corolla white, the sessile central one of each umbellet usually purple or pink, the petal obovate, ca. 1–2 mm long, those of some of the outer ones somewhat larger than the inner ones, the apex notched into 2 unequal lobes. Fruit oblong-elliptic, 3–4 mm long, laterally compressed, each mericarp with 5 inconspicuous primary ribs and 4 secondary ribs between the primary ones, these slender, winged, the margin with a row of flattened bristles pointed at the tip.

Disturbed sites. Occasional; northern counties, central peninsula. Nearly throughout North America; nearly cosmopolitan. Native to Europe, Africa, and Asia. Spring.

Daucus pusillus Michx. [Very small, in reference to the plant size.] AMERICAN WILD CARROT.

Daucus pusillus Michaux, Fl Bor.-Amer. 1: 164. 1803.
Daucus scadiophylus Rafinesque, New Fl. 4: 24. 1838 ("1836"). SYNTYPE: FLORIDA.

Annual or biennial herb, to 6(8) dm, stem striate, sparsely pubescent with spreading or recurved broad-based trichomes. Leaves with the blade oblong to triangular-ovate, 2- to 4-pinnately compound or dissected, 3–14 cm long, the ultimate segments linear or lanceolate, 1–6 mm long, to 1 mm wide, the apex acuminate, the margin entire or few-toothed or -lobed, sparsely to moderately pubescent, especially along the margins and veins, the petiole 0.5–2 cm long, the base sheathing. Flowers in a terminal or axillary compound umbel, the peduncle to 15 cm long, the rays numerous, 0.5–4 cm long, unequal, spreading and loosely ascending at flowering, curving upward and inward in fruit, the flowers numerous per umbellet, the central flower sessile, the pedicel of the others 1–9 mm long, unequal; bracts 4–15, like the leaves, 0.5–4 cm long, pinnately 1- to 2-dissected, not scarious-margined, spreading to loosely ascending at flowering, appressed to the umbel in fruit; bracteoles 5–13, shorter than to slightly longer than the pedicel, the margin entire or sometimes pinnately few-lobed distally; calyx triangular, minute, or absent; corolla white, the sessile central one of each umbellet usually purple or pink, the petals obovate, ca. 1–2 mm long, those of some of the outer ones somewhat larger than the inner ones, the apex notched into 2 unequal lobes. Fruit oblong-elliptic, 3–6 mm long, laterally compressed, each mericarp with 5 inconspicuous primary ribs and 4 secondary ribs between the primary ones, these slender, winged, the margin with a row of flattened bristles minutely barbed at the tip.

Dry disturbed sites. Occasional; northern counties, central peninsula. Virginia south to Florida, west to California, north to British Columbia; Mexico and South America. Native to North America and Mexico. Spring.

Eryngium L. 1753. ERYNGO

Herbs. Leaves alternate or opposite, simple, pinnate-, pinnipalmate-, or parallel-veined, petiolate or epetiolate, estipulate. Flowers in terminal and axillary, cymose, head-like umbels, bracteate, bracteolate, bisexual; sepal lobes 5, basally connate, actinomorphic; petals 5, free, actinomorphic; stamens 5, free; ovary inferior, 2-carpellate and -loculate. Fruit a schizocarp, mericarps 2.

A genus of about 250 species; nearly cosmopolitan. [From the Greek *eryngion,* eructate, belching, in reference to the use of the plant to reduce flatulence, also from the Latin vernacular name *eryngion,* this from a diminutive of *erungos,* the vernacular name for "sea holly."]

Selected reference: Mathias and Constance (1941).

1. Leaf blades parallel-veined and grasslike .... **E. yuccifolium**
1. Leaf blades reticulate-veined, not parallel-veined and grasslike.
    2. Leaves callous-margined.
        3. Plants diffuse, the branches decumbent at the base; leaves densely and uniformly disposed on the stem .... **E. aromaticum**
        3. Plant erect; leaves reduced in number upward on the stem .... **E. cuneifolium**
    2. Leaves not callous-margined.
        4. Stem prostrate or weakly ascending; heads solitary in the leaf axils.
            5. Leaves pinnatifid or pinnatisect; corolla greenish white .... **E. divaricatum**
            5. Leaves simple to ternate- or palmate-lobed or -dissected; corolla blue.
                6. Flowering heads cylindrical; flowers subequaling their subtending bracteoles; involucral bracts spreading well beyond the basal flowers to ca. ⅔ the length of the head .... **E. prostratum**
                6. Flowering heads subglobose; flowers conspicuously exceeded by their subtending bracteoles; involucral bracts not or barely extending beyond the basal flowers .... **E. baldwinii**
        4. Stems erect; heads in cymes.
            7. Leaves divided or lobed, the segments spine-tipped .... **E. foetidum**
            7. Leaves entire or merely toothed on the margin, lacking spines.
                8. Larger leaves oblong-lanceolate to ovate, 2–10 cm long, the margin crenate .... **E. integrifolium**
                8. Larger leaves linear or narrowly lanceolate, to 5 dm long, the margin entire or serrate-dentate .... **E. aquaticum**

Eryngium aquaticum L. [Growing in water.] RATTLESNAKEMASTER.

Eryngium aquaticum Linnaeus, Sp. Pl. 232. 1753. *Eryngium aquaticum* Linnaeus var. *normale* H. Wolff, in Engler, Pflanzenr. 4(Heft 61): 240. 1913, nom. inadmiss.
Eryngium virginianum Lamarck, Encycl. 4: 759. 1797.
Eryngium ravenelii A. Gray, Boston J. Nat. Hist. 6: 209. 1850. *Eryngium aquaticum* Linnaeus var. *ravenelii* (A. Gray) Mathias & Constance, Amer. Midl. Naturalist 25: 382. 1941.
Eryngium mettaueri A. W. Wood, Class-Book Bot., ed. 1861. 379. 1861. TYPE: FLORIDA: Wakulla Co.: near Newport, s.d., *Mettauer s.n.* (holotype: NY?).
Eryngium floridamum J. M. Coulter & Rose, Bot. Gaz. 13: 142. 1888. *Eryngium aquaticum* Linnaeus var. *floridanum* (J. M. Coulter & Rose) Mathias and Constance, Amer. Midl. Naturalist 25: 382. 1941. TYPE: FLORIDA: Duval Co.: near Jacksonville, s.d., *Curtiss s.n.* (lectotype: US). Lectotypified by Mathias and Constance (1941: 382).

Erect perennial herb, to 1.8 m; stem, striate, glabrous. Leaves alternate or opposite, the lower leaves with a long broadly winged, septate, entire or serrate-dentate petiole sheathing at the base, the blade linear, linear-lanceolate, or -elliptic, the leaves (including the petiole) to 5 dm long, the blade to 15 cm long, to 5 cm wide, the apex acute to rounded, the base narrowly cuneate to subcordate, the margin serrate or dentate, the upper and lower surfaces glabrous, the leaves reduced upward and mostly without blades. Flowers in a terminal or axillary, cymose, ovoid head, 1.5–3 cm long; bracts 6–10, linear-attenuate, reflexed, commonly blue, the margin entire or with 2–several teeth or lobes; bracteoles oblong, sometimes with 3 terminal lobes, the central one longer than the lateral, extending beyond the flowers and fruits, the heads thus bristly; calyx lobes lanceolate, 2–3 mm long, the apex acuminate; corolla blue, ovate, ca. 2 mm long, the apex acuminate. Fruit oblong, 2–3 mm long, the angles with flat scales.

Marshes, cypress ponds, and creek swamps. Occasional; northern counties, central peninsula. New York south to Florida, West to Mississippi, also Ontario and British Columbia. Summer.

Eryngium aromaticum Baldwin [With an odor.] FRAGRANT ERYNGO.

Eryngium aromaticum Baldwin, in Elliott, Sketch Bot. S. Carolina 1: 344. 1817. TYPE: FLORIDA.

Erect perennial herb, to 6 dm; stem basally decumbent, striate, glabrous. Leaves alternate or rarely opposite, densely and uniformly disposed on the stem, the blade elliptic-spatulate, 3–4 cm long, 1–1.5 cm wide, the apex acute, the base cuneate, the margin pinnate-lobed, the upper lobes lanceolate to linear-lanceolate, the lower linear to filiform, the margin callous and pale, the apex and lobes acuminate and bristle-tipped, the broader upper part of the leaf often deciduous in age and the lower bristlelike lobes persistent, the upper and lower surfaces glabrous, sessile or subsessile. Flowers in an open, cymose, subglobose head, 8–10 cm long; bracts numerous, 4–6 mm long, linear-lanceolate, reflexed, the margin entire; bracteoles oblong; calyx lobes lanceolate, the apex acuminate, extending beyond the flowers and fruits, the heads thus bristly; calyx-lobes lanceolate, 2–3 mm long, the apex acuminate; corolla blue, ovate, ca. 2 mm long. Fruit oblong, ca. 2 mm long, the angles papillate.

Sandhills, flatwoods, and scrub. Common; peninsula west to central panhandle. Georgia, Alabama, and Florida. Fall.

Eryngium baldwinii Spreng. [Commemorates William Baldwin (1779–1819), American physician and botanist.] BALDWIN'S ERYNGO.

Eryngium gracile Baldwin, in Elliott, Sketch Bot. S. Carolina 1: 345. 1817; non Delaroche f., 1808. *Eryngium baldwinii* Sprengel, Syst. Veg. 1: 870. 1825. *Streblanthus gracilis* Rafinesque, New Fl. 4: 36. 1838 ("1836"), nom. illegit.

Eryngium gracile Nuttall, Gen. N. Amer. Pl. 1: 175. 1818; non Delaroche f., 1808; nec Baldwin, 1817. *Atirsita pumila* Rafinesque, New Fl. 37. 1838 ("1836"). TYPE: FLORIDA: "West Florida," s.d., *Baldwin s.n.* (holotype: PH?).

Streblanthus humilis Rafinesque, New Fl. 4: 35. 1838 ("1836"). TYPE: FLORIDA: Hillsborough Co.: Tampa Bay.

Streblanthus tenuifolius Rafinesque, New Fl. 4: 36. 1838 ("1836").

Weakly ascending or prostrate perennial herb; stem to 5 dm long, rooting at the nodes, striate, glabrous. Leaves alternate or opposite, the blade ovate to elliptic, (0.5)1–7 cm long, (0.3)0.5–2.5 cm wide, the basal ones the largest, reduced upward, simple or ternate- or palmate-lobed or dissected, the apex acute to obtuse, the base rounded to cuneate, the margin entire, dentate, or pinnatifid, the upper and lower surfaces glabrous, the petiole of the basal ones 1–3.5 cm long, reduced upward and becoming sessile. Flowers in a solitary, axillary, subglobose head 6–8 mm long, the peduncle 1–2(3.5) cm long; bracts 5–10, linear-subulate, shorter than or barely extending beyond the basal flowers, reflexed, the margin entire, the apex acuminate; bracteoles linear-lanceolate, conspicuously exceeding the flowers; calyx lobes ovate, ca. 1 mm long, scarcely extending outward from between the flowers; corolla blue, the petals obovate, ca. 1 mm long, the apex rounded or shallowly notched. Fruit subglobose to obconic, ca. 2 mm long, the angles with elongated, clavate papillae or tubercles.

Wet flatwoods, floodplain forests, and wet disturbed sites. Common; peninsula west to central panhandle. Georgia and Florida. Spring–fall.

Eryngium cuneifolium Small [With wedge-shaped leaves.] WEDGELEAF ERYNGO; SCRUB ERYNGO.

Eryngium cuneifolium Small, Man. S.E. Fl. 964, 1506. 1933. TYPE: FLORIDA: Highlands Co.: about Lake Nancesowee [Lake Jackson], Dec 1927, *Small & Mosier s.n.* (holotype: NY).

Erect perennial herb, to 5 dm; stem striate, glabrous. Leaves alternate or opposite, mostly basal, 3–4 cm long, the basal ones with the blade spatulate, pinnatifid or pinnately lobed, the apical lobe with 3–5 teeth and cuneate at the base, the lower pinna 1 or 2 pairs, linear, the lobes and pinnae spine-tipped, the margin entire, callous and pale, the petiole 2–10 mm long, winged, the base sheathing, the upper leaves similar in general appearance, but smaller, 0.5–1.5 cm long, to 1 cm wide, the apical lobes narrower, sessile or subsessile. Flowers in a terminal or axillary subglobose head 5–8 mm long, the peduncle 0.5–4 cm long; bracts linear-subulate, spine-tipped, slightly extending outward from the head base; bracteoles similar to the bracts, slightly longer than the flowers; calyx lobes lanceolate, 1–2 mm long, the apex acuminate; corolla white or blue, the petals lanceolate, 1–2 mm long. Fruit 1–2 mm long, the angles papillate.

Scrub. Rare; Highlands County. Endemic. Summer–fall.

Eryngium cuneifolium is listed as endangered in Florida (Florida Administrative Code, Chapter 5B-40) and in the United States (U.S. Fish and Wildlife Service, 50 CFR 23).

Eryngium divaricatum Hook. & Arn. [Spreading at an angle, in reference to the inflorescence branches.] BALLAST ERYNGO.

Eryngium divaricatum Hooker & Arnott, Bot. Misc. 3: 350. 1833.

Procumbent perennial herb, to 15 cm; stem to 5 dm long, striate, glabrous. Leaves alternate or opposite, the basal ones oblong, 4–20 cm long, 1–4 cm wide, pinnatifid or pinnatisect, the lobes narrowly lanceolate, the apex acute, the margin entire or irregularly serrate, the petiole short, winged, the base sheathing, the leaves greatly reduced upward on the stem. Flowers in a solitary, terminal or axillary, cylindric to ovoid head 6–10 mm long, the peduncle ca. 1 cm long; bracts 6–7, linear-subulate, ca. 10 mm long, much longer than flowers, the margin entire, extending outward from the head; bracteoles linear-subulate, slightly longer than the flowers; calyx lobes ovate, ca. 0.5 mm long, the apex obtuse; corolla greenish-white, the petals ovate, ca. 1 mm long, the apex inflexed. Fruit subglobose, 1–2 mm long, the angles with obpyriform scales.

Disturbed sites. Rare; Escambia County. Not recently collected. North Carolina and Florida; Mexico; South America. Native to South America. Spring–fall.

Eryngium foetidum L. [Bad-smelling.] SPIRITWEED.

Eryngium foetidum Linnaeus, Sp. Pl. 232. 1753.

Erect perennial herb, to 6 dm; stem striate, glabrous. Leaves alternate or opposite, the basal ones with the blade lanceolate to oblanceolate, (3)10–16(30) cm long, (1)2.5–4(5) cm wide, the apex rounded to obtuse, the base cuneate, the margin spinulose-serrate, the upper and lower surfaces glabrous, the petiole to 3 cm long, the upper leaves usually opposite, the blade oblanceolate to obovate, 2–4 cm long, 0.5–1.5 cm wide, the apex rounded to acute, the base cuneate, the margin spinulose-serrate or lobed, the lobes spine-tipped, the upper and lower surfaces glabrous, sessile. Flowers in a terminal or axillary, cymose, cylindric head 0.7–1.5 cm long; bracts 5–6, foliaceous, lanceolate, 1–4 cm long, 2–10 mm wide, much exceeding the head, unequal, the margin entire or few-toothed or -lobed; bracteoles lanceolate, subequaling the flower, those at the head apex usually larger and forming an apical tuft; calyx lobes triangular-ovate, to 1 mm long, the apex mucronate; corolla white or yellow green, the petals obovate, ca. 1 mm long. Fruit subglobose, ca. 2 mm long, the angles conspicuously papillate.

Disturbed sites. Rare; Miami-Dade County. Georgia and Florida; nearly cosmopolitan. Native to tropical America. Summer.

Eryngium integrifolium Walter [With entire-margined leaves.] BLUEFLOWER ERYNGO.

Eryngium integrifolium Walter, Fl. Carol. 112. 1788. *Eryngium integrifolium* Walter var. *typicum* H. Wolff, in Engler, Pflanzenr. 4(Heft 61) 160. 1913, nom. inadmiss.

Erect perennial herb, to 8 dm; stem striate, glabrous. Leaves alternate, the blade of the lower ones (often not present at flowering) elliptic, ovate, or oblong, 4–8 cm long, 3–4 cm wide, the apex obtuse, the base obliquely cordate, the margin entire or crenate, the upper and lower surfaces glabrous, the petiole ca. 1 cm long, the stem leaves 2–4 cm long, 1–2 cm wide, reduced upward, the blade ovate to linear-lanceolate, the apex acute to obtuse, the base obliquely rounded to cuneate, the margin crenate to spinulose-toothed, rarely laciniate, the upper and lower surfaces glabrous, the petiole to 1 cm long, slightly winged, the base sheathing. Flowers in a cymose hemispheric to subglobose head, to ca. 1.5 cm long, the peduncle ca. 2 cm long; bracts 6–10, linear, 1–2 cm long, the margin entire or with 3–5 spinose teeth, spreading out from the head; bracteoles linear-oblong, ca. 5 mm long, with 3 spinose cusps apically, longer than the flowers or fruits; sepal lobes lance-ovate, ca. 2 mm long, the apex acuminate; corolla blue, the petals lanceolate-ovate, ca. 1 mm long. Fruit obpyramidal, ca. 2 mm long, the angles with white papillae, these becoming flat and scalelike at maturity.

Bogs and wet flatwoods. Occasional; northern counties. Virginia south to Florida, west to Oklahoma and Texas. Summer–fall.

Eryngium prostratum Nutt. ex DC. [Lying on the ground, in reference to the growth habit.] CREEPING ERYNGO.

Eryngium prostratum Nuttall ex de Candolle, Prodr. 4: 92. 1830. *Streblanthus heterophylus* Rafinesque, New Fl. 4: 35. 1838 ("1836"), nom. illegit.

Weakly ascending or prostrate perennial herb; stem to 5 dm long, rooting at the nodes, striate, glabrous. Leaves alternate or opposite, the blade ovate to elliptic, simple to ternate or palmate-lobed or dissected, (0.5)1–7 cm long, (0.3)0.5–2.5 cm wide, the basal ones the largest, reduced upward, the apex acute to obtuse, the base rounded to cuneate, the margin entire, dentate, or pinnatifid, the upper and lower surfaces glabrous, the petiole of the basal ones 1–3.5 cm long, reduced upward and becoming sessile. Flowers in a solitary, axillary, cylindric head 6–8 mm long, the peduncle 1–2(3.5) cm long; bracts 5–10, linear to narrowly oblanceolate, considerably exceeding the radius of the head to ca. ⅔ the length of the head, reflexed, the margin entire, the apex acuminate; bracteoles linear-lanceolate, subequaling the flowers; calyx lobes ovate, scarcely extending outward from between the flowers; corolla blue, the petals obovate, ca. 1 mm long, the apex rounded or shallowly notched. Fruit subglobose to obconic, ca. 2 mm long, the angles with minute tubercles.

Floodplain forests, bogs, and pond and lake margins. Frequent; northern counties, central peninsula. Delaware south to Florida, west to Illinois, Missouri, Kansas, Oklahoma, and Texas. Spring–fall.

Eryngium yuccifolium Michx. [With leaves like *Yucca* (Agavaceae).] BUTTON RATTLESNAKEMASTER; BUTTON ERYNGO.

Eryngium yuccifolium Michaux, Fl. Bor.-Amer. 1: 164. 1803.

Eryngium yuccifolium Michaux var. *synchaetum* A. Gray ex J. M. Coulter & Rose, Rev. N. Amer. Umbell. 94. 1888. *Eryngium synchaetum* (A. Gray ex J. M. Coulter & Rose) J. M. Coulter & Rose, Contr.

U.S. Natl. Herb. 7: 44. 1900. *Eryngium aquaticum* Linnaeus var. *synchaetum* (A. Gray ex J. M. Coulter & Rose) H. Wolff in Engler, Pflanzenr. 4(Heft 61): 242. 1913. TYPE: FLORIDA: Duval Co.: near Jacksonville, s.d., *Curtiss 996* (lectotype: GH). Lectotypified by J. M. Coulter and Rose (1900: 44).

Erect or ascending perennial herb, to 1.5 m; stem striate, glabrous, glaucous. Leaves alternate, the blade linear, 0.4–10 dm long, 1–4 cm wide, parallel-veined and grasslike in appearance, the apex attenuate, the base sheathing, the margin with antrorse spines or bristles, sometimes entire, the upper and lower surfaces glabrous. Flowers in a terminal or axillary, cymose, subglobose or ovoid head, 1–2.5 cm long, the peduncle ca. 2 cm long; bracts 4–10, ovate-lanceolate, 4–16 mm long, the apex subulate, the margin entire or finely spinulose-toothed, spreading or reflexed; bracteoles 6–10 mm long, similar to the bracts, extending outward from between the flowers or fruits; sepal lobes ovate, 2–3 mm long, the apex obtuse and mucronate; corolla white, the petals oblong, 2–3 mm long, the apex rounded. Fruit obpyramidal, ca. 3 mm long, the angles with ascending lanceolate, papery scales.

Bogs, wet flatwoods, and floodplain forests. Frequent; nearly throughout. Connecticut south to Florida, west to Minnesota, Nebraska, Kansas, Oklahoma, and Texas. Summer.

EXCLUDED TAXA

Eryngium cervatesii Delaroche—This Mexican species was reported for Florida by Chapman (1860, 1883), who misapplied the name to Florida material of *E. baldwinii.*

Eryngium longifolium Cavanilles—This Mexican species was reported for Florida by Chapman (1897), who misapplied the name to Florida material of *E. yuccifolium.*

Foeniculum Mill. 1754. FENNEL

Perennial herb. Leaves alternate, pinnately dissected, petiolate. Flowers in terminal and axillary compound umbels, ebracteate, ebracteolate, bisexual; sepals absent; petals 5, free, actinomorphic; stamens 5, free; ovary inferior, 2-carpellate and -loculate. Fruit a schizocarp, mericarps 2.

A genus of 3 species; nearly cosmopolitan. [*Foenum,* hay, in reference to the thread- or straw-like segments of the leaves.]

Foeniculum vulgare Mill. [Common.] SWEET FENNEL.

Anethum foeniculum Linnaeus, Sp. Pl. 263. 1753. *Ligusticum foeniculum* (Linnaeus) Crantz, Cl. Umbell. Emend. 82. 1767. *Foeniculum vulgare* Miller, Gard. Dict., ed. 8. 1768. *Foeniculum officinale* Allioni, Fl. Pedem. 2: 25. 1785, nom. illegit. *Anethum foeniculum* Linnaeus var. *vulgare* (Miller) Schkuhr, Bot. Handb. 1: 223. 1791, nom. inadmiss. *Anethum rupestre* Salisbury, Prodr. Stirp. Chap. Allerton 168. 1796, nom. illegit. *Meum foeniculum* (Linnaeus) Sprengel, in Roemer & Schultes, Syst. Veg. 6: 433. 1820. *Foeniculum foeniculum* (Linnaeus) H. Karsten, Deut. Fl. 827. 1882, nom inadmiss. *Foeniculum commune* Bubani, Fl. Pyren. 2: 372. 1900, nom. illegit. *Foeniculum vulgare* Miller forma *officinale* Fiori & Beguinot, Fl. Ital. 2: 173. 1900, nom. inadmiss. *Selinum foeniculum* (Linnaeus) E.H.L. Krause, in Sturm, Deutschl. Fl., ed. 2. 12: 115. 1904. *Seseli foeniculum* (Linnaeus) Koso-Poljansky, Bull. Soc. Imp. Naturalistes Moscou, ser. 2. 29: 183. 1916.

Erect or ascending perennial herb, to 2 m; stem striate, glabrous, glaucous. Leaves with the blade ovate to broadly triangular-ovate, 3–30 cm long and wide, the basal and median stem leaves 3- to 5-pinnately dissected, the ultimate segments linear, 0.4–4 cm, the apex short-tapered to an abrupt, sharp tip, the upper and lower surfaces glabrous, the petiole 0.5–10 cm long, winged, the base sheathing, the leaves gradually reduced, 1- to 2-pinnate or sometimes bladeless. The flowers in a terminal or axillary compound umbel, the peduncle 5–15 cm long, the rays numerous, 1–6.5 cm long, unequal, the flowers 12–many per umbellet, the pedicel 2–10 mm long, unequal; calyx absent; corolla, yellow, the petals obovate, ca. 1 mm long, the apex rounded or obtuse. Fruit oblong, slightly laterally flattened, each mericarp dark brown with 5 light yellowish ribs.

Open disturbed sites. Rare; Volusia County. Nearly throughout North America; nearly cosmopolitan. Native to Europe, Africa, and Asia. Summer.

Ligusticum L. 1753. LICORICE-ROOT

Herb. Leaves alternate, pinnately compound, petiolate. Flowers in terminal and axillary compound umbels, ebracteate, bracteolate, bisexual; sepals 5, basally connate, actinomorphic; petals 5, free, actinomorphic; stamens 5, free; ovary inferior, 2-carpellate and -loculate. Fruit a schizocarp, mericarps 2.

A genus of about 60 species; North America, Europe, and Asia. [From Ancient Greek *ligustikon,* the vernacular name for one of the species, this derived from the Italian region of Liguria where the plant occurs.]

Ligusticum canadense (L.) Vail [Of Canada.] CANADIAN LICORICE-ROOT.

Ferula canadensis Linnaeus, Sp. Pl. 247. 1753. *Ligustrum canadense* (Linnaeus) Vail, Mem. Torrey Bot. Club 4: 121. 1894.

Erect or ascending perennial herb, to 1.5 m; stem striate, glabrous. Leaves with the blade ovate to triangular-ovate, (4)10–24 cm long and wide, 3- to 4-pinnately compound, the upper leaves sometimes simple, the leaflets lanceolate to ovate or oblong-ovate, 2.5–12 cm long, 1–6 cm wide, the apex acute or acuminate, the base cuneate, the margin irregularly serrate, sometimes with 1 or 2 basal lobes, the upper and lower surfaces glabrous, sessile or subsessile, the petiole to 8 cm long, the base sheathing. The flowers in terminal and axillary compound umbels, the peduncle 8–12 cm long, the rays 6–14, 2–5 cm long, unequal, the flowers 6–14 per umbellet, the pedicel 2–4 m long; bracteoles 2–5, lanceolate, the margin entire, shorter than the pedicel; sepal lobes triangular, minute; corolla white, the petals obovate, ca. 1 mm long, the apex rounded or obtuse. Fruit ovate-elliptic, 4–7 mm long, slightly laterally flattened, each mericarp dark brown with 5 light brown narrowly winged ribs.

Mixed hardwoods. Rare; Okaloosa County. Pennsylvania and New Jersey south to Florida, west to Missouri, Arkansas, and Mississippi. Summer.

Lilaeopsis Greene 1891. GRASSWORT

Herbs. Stem rhizomatous. Leaves alternate, simple, epetiolate. Flowers in a simple axillary umbel, bracteate, ebracteolate, bisexual; sepals 5, basally connate, actinomorphic, or absent; petals 5, free, actinomorphic; stamens 5, free; ovary 2-carpellate and -loculate. Fruit a schizocarp, mericarps 2.

A genus of 14 species; North America, West Indies, Mexico, South America, Europe, Africa, Australia, and Pacific Islands. [*Lilaea* (Juncaginaceae), and the Greek *opsis,* to resemble.]

Selected reference: Affolter (1985).

1. Leaves 10–30(50) cm long; peduncles much shorter than the leaves .... **L. carolinensis**
1. Leaves 1–8(10) cm long; peduncles as long as the leaves or longer .... **L. chinensis**

Lilaeopsis carolinensis J. M. Coult. & Rose [Of Carolina.] CAROLINA GRASSWORT.

Lilaeopsis carolinensis J. M. Coulter & Rose, Bot. Gaz. 24: 48. 1897. *Crantziola carolinensis* (J. M. Coulter & Rose) Koso-Poljansky, Bull. Soc. Imp. Naturalistes Moscou, ser. 2. 29: 125. 1916. *Crantzia carolinensis* (J. M. Coulter & Rose) Chodat, Bull. Soc. Bot. Genève 12: 31. 1920.

Stem rhizomatous, rooting at the nodes, striate, glabrous. Leaves with the blade narrowly spatulate, or rarely linear, 10–30(50) cm long, 2–15 mm wide, 7- to 10(20)-septate, the apex rounded or acute, the base expanded into a scarious sheath 2–20 mm long, the margin entire, the upper and lower surfaces glabrous. Flowers (3)5–10(14) in a simple axillary umbel, these 1–2 at a node, the peduncle 1–6(9) cm long, much shorter than the leaves, the pedicel 1–10(30) mm long; bracts triangular, minute; calyx lobes minute or absent; corolla dull white or maroon-tinted, the petals lanceolate, ca. 1 mm long. Fruit ovoid or obovoid, laterally flattened, ca. 2 mm long, each mericarp with 5 winged ribs.

Freshwater marshes. Occasional; northern and central peninsula, central and western panhandle. Virginia south to Florida, west to Arkansas and Louisiana; Europe; South America. Native to South America. Spring.

Lilaeopsis chinensis (L.) Kuntze [Of China.] EASTERN GRASSWORT.

Hydrocotyle chinensis Linnaeus, Sp. Pl. 234. 1753. *Hydrocotyle sinensis* J. F. Gmelin, Syst. Nat. 2: 468. 1791, nom. illegit. *Lilaeopsis chinensis* (Linnaeus) Kuntze, Revis. Gen. Pl. 3(2): 114. 1898. *Crantzia chinensis* (Linnaeus) Druce, Bot. Exch. Soc. Club Brit. Isles 3: 416. 1914.

Hydrocotyle lineata Michaux, Fl. Bor.-Amer. 1: 162. 1803. *Crantzia lineata* (Michaux) Nuttall, Gen. N. Amer. Pl. 1: 178. 1818. *Hallomuellera lineata* (Michaux) Kuntze, Revis. Gen. Pl. 1: 267. 1891. *Lileopsis lineata* (Michaux) Greene, Pittonia 2: 192. 1891. *Crantziola lineata* (Michaux) Koso-Poljansky, Bull. Soc. Imp. Naturalistes Moscou, ser. 2. 29: 125. 1916.

Stem rhizomatous, rooting at the nodes, striate, glabrous. Leaves with the blade linear or narrowly spatulate, 1–8(10) cm long, (1)2–5 mm wide, 4- to 8(10)-septate, the apex rounded, the base expanded into a scarious sheath 1–4(7) mm long, the margin entire, the upper and lower surfaces glabrous. Flowers (3)4–8(13) in a simple axillary umbel, these 1–2 at a node, the peduncle (1)2.5–6.5(8) cm long, subequaling or longer than the leaves, the pedicel 2–9 mm long; bracts triangular, minute; calyx lobes minute or absent; corolla white or pinkish, the petals elliptic, 1–2 mm long. Fruit ellipsoid to ovoid, laterally flattened, ca. 2 mm long, the mericarp with 5 winged ribs.

Brackish marshes. Occasional; northern counties, central peninsula. Nova Scotia and Maine, south to Florida, west to Texas. Spring.

EXCLUDED TAXON

Lilaeopsis attenuata (Hooker & Arnott) Fernald—Reported for Florida by Wilhelm (1984), the name misapplied to Florida material of *L. carolinensis.*

Oxypolis Raf. 1825. COWBANE

Herbs. Leaves alternate, pinnately compound, petiolate. Flowers in terminal and axillary compound umbels, bracteate, bracteolate, bisexual; sepals 5, basally connate, actinomorphic; petals 5, free, actinomorphic; stamens 5, free; ovary inferior, 2-carpellate and -loculate. Fruit a schizocarp, mericarps 2.

A genus of 5 species; North America. [From the Greek *oxys,* sharp, and *polis,* white, in reference to subulate involucres and white petals.]

Selected reference: Feist et al. (2012).

1. Leaves pinnate-lobed, the lobes oblong-lanceolate, the veins reticulate .... **O. rigidior**
1. Leaves ternate-lobed, the lobes linear, the veins parallel .... **O. ternata**

Oxypolis rigidior (L.) Raf. [Stiff, in reference to the habit.] STIFF COWBANE.

Sium rigidius Linnaeus, Sp. Pl. 251. 1753. *Oenanthe rigidius* (Linnaeus) Crantz, Cl. Umbell. Emend. 85. 1767. *Pastinaca rigidior* (Linnaeus) Sprengel, in Roemer & Schultes, Syst. Veg. 6: 586. 1820. *Archemora rigidior* (Linnaeus) de Candolle, Prodr. 4: 188. 1830. *Oxypolis rigidior* (Linnaeus) Rafinesque, Bull. Bot. (Geneva) 1: 218. 1830. *Peucedanum rigidium* (Linnaeus) Baillon, Hisp. Pl. 7: 187. 1879; non A. W. Wood, 1870. *Tiedemannia rigidior* (Linnaeus) J. M. Coulter & Rose, Bot. Gaz. 12: 74. 1887.

Sium denticulatum Baldwin, in Elliott, Sketch Bot. S. Carolina 1: 354. 1817. *Archemora denticulata* (Baldwin) de Candolle, Prodr. 4: 188. 1830. *Oxypolis denticulata* (Baldwin) Rafinesque, Bull. Bot. (Geneva) 1: 218. 1830. *Pastinaca denticulata* (Baldwin) D. Dietrich, Syn. Pl. 2: 971. 1840.

Erect or ascending perennial herb, to 1.5 m; stem striate, glabrous, sometimes slightly glaucous. Leaves with the blade ovate to triangular, pinnately compound with (3)5–9 leaflets, 4–30 cm long, the leaflets linear-lanceolate, oblong, or narrowly obovate, 3.5–15 cm long, 0.5–4.5 cm wide, the veins reticulate, the apex acute to acuminate, with a sharp tip, the base cuneate, the margin entire or with a few teeth mostly above the middle, these narrowed to a sharp tip, the upper surface glabrous, the lower surface glabrous and glaucous, the petiole 3–8 cm long, the base sheathing, the upper leaves sessile or subsessile, sometimes reduced to a nearly bladeless sheath. Flowers in a terminal or axillary compound umbel, the peduncle 8–15 cm long, the rays 10–many, 1–8(12) cm long, unequal, the flowers 10–many per umbellet, the pedicel 4–15(20) mm long; bracts 1–3, linear, 2–20 mm long, entire; bracteoles 2–9, linear, entire, mostly shorter than the pedicel; calyx lobes triangular, minute; corolla white, the petals ovate, ca. 1 mm long, the apex narrowed to a short, slender tip. Fruit oblong-elliptic, 4–7 mm long, laterally flattened, tan, each mericarp with a dorsal and 2 intermediate ribs, the lateral ribs with broad, thin wings.

Titi and bay swamps, bogs, and marshes. Occasional; central and western panhandle. Ontario and New York south to Florida, west to Minnesota, Iowa, Missouri, Oklahoma, and Texas. Fall.

Oxypolis ternata (Nutt.) A. Heller [In threes, in reference to the leaf segments.] PIEDMONT COWBANE.

Peucedanum ternatum Nuttall, Gen. N. Amer. Pl. 1: 182. 1818. *Sataria linearis* Rafinesque, New Fl. 4: 21. 1838 ("1836"). *Tiedemannia ternata* (Nuttall) J. M. Coulter & Rose, Bot. Gaz. 12: 74. 1887. *Archemora ternata* (Nuttall) Nuttall, in Torrey & A. Gray, Fl. N. Amer. 1: 631. 1840. *Oxypolis ternata* (Nuttall) A. Heller, Cat. N. Amer. Pl. 5. 1898.

Neurophyllum longifolium Torrey & A. Gray, Fl. N. Amer. 1: 613. 1840. TYPE: FLORIDA: "Middle Florida," Sep, *Croom s.n.* (lectotype: NY). Lectotypified by Feist (2009: 662).

Erect or ascending perennial herb, to 1 m; stem striate, glabrous. Leaves with the blade of the proximal ones with (1–2)3 leaflets, the leaflets linear, 8–20(35) cm long, 2–3(6) cm wide, the veins parallel, the apex acuminate, the base narrowly cuneate, the margin entire, the petiole to 20 cm long, the base sheathing, the uppermost leaflets few, simple, reduced, sessile or subsessile. Flowers in a terminal or axillary compound umbel, the peduncle 4–15 cm long, the rays 5–11, 2–8 cm long, unequal, the flowers 8–12 per umbellet, the pedicel 2–15 mm long; bracts 2–5, linear-subulate, 4–10 mm long; bracteoles several, linear-filiform, 2–3 m long; calyx lobes triangular, minute; corolla white, the petals ovate to obovate, ca. 1 mm long, the apex short-acuminate, inflexed. Fruit elliptic to obovate, 3–5 mm long, laterally flattened, tan, each mericarp with a dorsal and 2 intermediate ribs, the lateral ribs with broad, thin wings.

Occasional; wet pine flatwoods, bogs, and seepage slopes. Eastern and central panhandle. Virginia south to Florida, also Texas. Fall.

Petroselinum Hill 1756. PARSLEY

Herbs. Leaves alternate, pinnately compound, petiolate. Flowers in terminal and axillary compound umbels, bracteate, bracteolate, bisexual; sepals 5, basally connate, actinomorphic; petals 5, free, actinomorphic; stamens 5, free; ovary inferior, 2-carpellate and -loculate. Fruit a schizocarp, mericarps 2.

A genus of 2 species; North America, Europe, and Africa. [From the Greek *petros,* rock, and *selinum,* celery, "rock-celery" the vernacular name.]

Petroselinum crispum (Mill.) Fuss [Irregularly waved and twisted, in reference to the leaves.] PARSLEY.

Apium crispum Miller, Gard. Dict., ed. 8. 1768. *Apium petroselinum* Linnaeus var. *crispum* (Miller) J. Willmott, Enum. Pl. Hort. Kew. 16. 1798. *Apium petroselinum* Linnaeus var. *crispifolium* Hayne, Getreue Darstell. Gew. 7: t. 23. 1821, nom. illegit. *Petroselinum sativum* Hoffmann ex Gaudin var. *crispum* (Miller) de Candolle, Prodr. 4: 102. 1830. *Petroselinum sativum* Hoffmann ex Gaudin subvar. *crispum* (Miller) Cosson & Germain de Saint-Pierre, Fl. Descr. Anal. Paris 206. 1945. *Petroselinum crispum* (Miller) Fuss, Fl. Transsilv. 254. 1866. *Carum petroselinum* (Linnaeus) Bentham var. *crispum* (Miller) Beck, Fl. Nied.-Oesterr. 1: 621. 1892. *Petroselinum petroselinum* (Linnaeus) H. Karsten var. *crispum* (Miller) Ascherson & Graebner, Fl. Nordostdeut. Flachl. 519. 1898. *Petroselinum hortense* Hoffmann forma *crispum* (Miller) Fiori & Beguinot, Fl. Ital. 2: 160. 1900. *Petroselinum hortense* Hoffmann var. *crispum* (Miller) Gaudin ex Schinz & Thellung, in Schinz & R. Keller, Schweiz Fl., ed. 3. 2: 262. 1914.

Erect annual or biennial herb, to 1.5 m; stem striate, glabrous or sparsely hirtellous proximally, the root fusiform. Proximal leaves with the blade ovate, 2- to 3-pinnatisect, the pinnae and pinnules petiolulate, the ultimate divisions ovate-lanceolate to broadly linear, trifid or lobulate, the margin tooth or lobed, the upper and lower surfaces glabrous or hirtellous, the petiole to ca. 10 cm long, the base sheathing, the upper leaves reduced, the ultimate divisions becoming fewer and more linear, entire or nearly so. Flowers in a terminal or lateral compound umbel, the peduncle 3–10 cm long, the rays 10–20, 0.5–1.5 cm long, unequal, the flowers 10–15 per umbellet, the pedicel 2–4 mm long; bracts 5–6, linear, shorter than the rays; bracteoles 5–6, shorter than the pedicel; calyx lobes minute; corolla yellow, the petals orbicular, ca. 1 mm long, the apex emarginate, with narrow inflexed tips. Fruit ovoid to oblong-ovoid, 2–3 mm long, laterally flattened, each mericarp brown, with 5 blunt tan ribs.

Disturbed sites. Rare; Leon County. Escaped from cultivation. Newfoundland, Ontario, and sporadic nearly throughout the United States; Europe, Africa, Asia, Australia, and Pacific Islands. Native to Europe and Africa. Summer.

Ptilimnium Raf. 1825. MOCK BISHOPWEED

Herbs. Leaves alternate, pinnately compound, petiolate. Flowers in terminal and axillary compound umbels, bracteate, bracteolate, bisexual; sepals 5, basally connate, actinomorphic; petals 5, free, actinomorphic; stamens 5, free; ovary inferior, 2-carpellate and -loculate. Fruit a schizocarp, mericarps 2.

A genus of 5 species, North America. [From the Greek *Ptilon,* feather or down, and *limne,* mud, in reference to the finely divided leaves and the habitat.]

Selected reference: Easterly (1957); Feist et al. (2012).

Ptilimnium capillaceum (Michx.) Raf. [Hairlike, in reference to the leaf segments.] MOCK BISHOPWEED; HERBWILLIAM.

Ammi capillaceum Michaux, Fl. Bor.-Amer. 1: 164. 1803. *Sison capillaceum* (Michaux) Sprengel, Syst. Veg. 1: 897. 1825. *Discopleura capillacea* (Michaux) de Candolle, Coll. Mém. 5: 38. 1829. *Ptilimnium capillaceum* (Michaux) Rafinesque, Bull. Bot. (Geneva) 1: 217. 1830.

Erect annual herb, to 8 dm; stem striate, glabrous. Leaves with the blade pinnately decompound, usually with 3 divisions per rachis node, the ultimate segments linear, 3–25 mm long, the margin entire, the upper and lower surfaces glabrous, the petiole 1–1.5 cm long, the base sheathing, the leaves reduced upward on the stem. Flowers in a terminal or axillary compound umbel, the peduncle 3–10 cm long, the rays 5–25, 1–3.5 cm long, unequal, the flowers 5–20 per umbellet, the pedicel 3–6(12) mm long; bracts 2–9, 4–12 mm long, similar to the leaves but smaller; bracteoles linear, shorter than to subequaling the pedicels; calyx lobes triangular, minute; corolla white, rarely pink, the petals obovate, ca. 1 mm long, the apex abruptly acuminate. Fruit ovate, 2–3 mm long, laterally flattened, light yellowish brown, each mericarp with 5 blunt ribs.

Swamps, ponds, pond margins, and wet ditches. Common, nearly throughout. Massachusetts and New York south to Florida, west to Missouri, Kansas, Oklahoma, and Texas, also South Dakota. All year.

EXCLUDED TAXON

Ptilimnium nuttallii (de Candolle) Britton—Reported for Florida by Chapman (1860, 1883, 1897, all as *Discopleura nuttallii* de Candolle), the name misapplied to Florida material of *P. capillaceum*.

Sanicula L. 1753. BLACKSNAKEROOT

Herbs. Leaves alternate, simple, palmate-lobed and/or -compound, petiolate. Flowers in terminal and axillary compound umbels, bracteate, bracteolate, bisexual or unisexual (plant polygamomonoecious); sepals 5, basally connate, actinomorphic; petals 5, free, actinomorphic; stamens 5, free; ovary inferior, 2-carpellate and -loculate. Fruit a schizocarp, mericarps 2.

A genus of about 40 species; nearly cosmopolitan. [*Sanare*, to heal, in reference to its medicinal use.]

Selected references: Pryer and Phillippe (1989); Shan and Constance (1951).

1. Styles longer than the bristles on the ovary or fruit; staminate flowers 10–25.
   2. Fruit evidently pedicellate; ovary or fruit bristles gradually tapering to the base, arranged in rows; petals yellowish green; calyx lobes ovate, connate in the basal ⅓ .... **S. odorata**
   2. Fruit sessile or subsessile; ovary or fruit bristles bulbous at the base, not arranged in rows; petals greenish white; calyx lobes lanceolate, nearly free .... **S. marilandica**
1. Styles equaling or shorter than the bristles on the ovary or fruit; staminate flowers 3–8.
   3. Fruit pedicellate; roots fibrous .... **S. canadensis**
   3. Fruit sessile or subsessile; roots thickened and cordlike .... **S. smallii**

Sanicula canadensis L. [Of Canada.] CANADIAN BLACKSNAKEROOT.

Sanicula canadensis Linnaeus, Sp. Pl. 235. 1753. *Caucalis canadensis* (Linnaeus) Crantz, Cl. Umbell. Emend. 110. 1767. *Sanicula marilandica* Linnaeus var. *canadensis* (Linnaeus) Torrey, Fl. N. Middle United States 302. 1824. *Sanicula canadensis* Linnaeus var. *typica* H. Wolff, in Engler, Pflanzenr. 4(Heft 61): 67. 1913, nom. inadmiss. *Sanicula canadensis* Linnaeus var. *genuina* Fernald, Rhodora 42: 467. 1940, nom. inadmiss.

Sanicula floridana E. P. Bicknell, Bull. Torrey Bot. Club 24: 581. 1897. *Sanicula canadensis* Linnaeus var. *floridana* (E. P. Bicknell) H. Wolff, in Engler, Pflanzenr. 4(Heft 61): 67. 1913. TYPE: FLORIDA: May, *Curtiss 944* (holotype: NY; isotypes: GH, MIN, MISSA, NY, P, US, VT, YU).

Erect, biennial herb, to 1 m; stem striate, glabrous, the roots fibrous. Leaves with the blade ovate to suborbicular, 2–10(14) cm long and wide, deeply palmately 3- to 5(7)-lobed and/or compound, the apex acute to obtuse, the base cuneate, the margin irregularly serrate, the upper and lower surfaces glabrous, the petiole 0.5–8 cm long. Flowers in a terminal or axillary compound umbel, the peduncle 3–8 cm long, the rays 2–3, 0.2–3 cm long, unequal, the flowers 4–6 per umbellet, the umbellet with staminate and bisexual flowers (staminate 1–7 per umbellet), the pedicel ca. 1 mm long; bracts leaflike, lanceolate, 0.8–3.5 cm long; bracteoles elliptic-ovate, minute; calyx lobes lanceolate, 0.5–1 mm long, the apex tapered to a sharp point; corolla greenish white, the petals shorter than the sepals; stamens with white anthers; styles shorter than the fruit bristles, shorter than to subequaling the sepals. Fruit oblong ovate, slightly laterally flattened, 2–5 mm long, with hooked bristles in longitudinal rows, the pedicel 1–1.5 mm long.

Moist to dry hammocks and rocky floodplain forests. Frequent; northern counties, central peninsula. Quebec south to Florida, west to Ontario, Minnesota, South Dakota, Wyoming, Nebraska, Kansas, Oklahoma, and Texas. Spring-summer.

Sanicula marilandica L. [Of Maryland.] MARYLAND BLACKSNAKEROOT.

Sanicula marilandica Linnaeus, Sp. Pl. 235. 1753. *Caucalis marilandica* (Linnaeus) Crantz, Cl. Umbell. Emend. 110. 1767. *Sanicula canadensis* Linnaeus var. *marilandica* (Linnaeus) Hitchcock, Trans. Acad. Sci. St. Louis 5: 497. 1889.

Erect perennial herb, to 1 m; stem striate, glabrous, the roots fibrous. Leaves with the blade ovate to suborbicular or reniform, 6–10 cm long, 2–15 cm wide, deeply palmately 5-lobed and/or compound, the apex acute to obtuse, the base cuneate, the margin irregularly serrate, the upper and lower surfaces glabrous, the petiole 6–10 cm long, the uppermost subsessile. Flowers in a terminal or axillary compound umbel, the peduncle 4–8 cm long, the rays 1–3, 3–8 cm long, the flowers numerous per umbellet, the umbellet with staminate and bisexual flowers (staminate 12–25 per umbellet), sessile or subsessile; bracts leaflike, lanceolate, 0.8–4 cm long; bracteoles elliptic-ovate, minute; calyx lobes lanceolate, 0.5–1 mm long, free nearly to the base, the apex tapered to a blunt point; corolla greenish white, the petals longer than the sepals; stamens with yellow anthers; styles longer than the fruit bristles. Fruit oblong ovate, slightly laterally flattened, 4–6 mm long, with hooked bristles in rows, these bulbous at the base, sessile or subsessile.

Hammocks and bluff forests. Occasional; northern counties. Nearly throughout North America except for some southwestern U.S. states. Spring.

Sanicula odorata (Raf.) Pryer & Phillippe [With an odor.] CLUSTERED BLACKSNAKEROOT.

Triclinium odoratum Rafinesque, Fl. Ludov. 80. 1817. *Sanicula triclinium* de Candolle, Prodr. 4: 85. 1830, nom. illegit. *Sanicula triclinaris* Saint-Lager, Ann. Soc. Bot. Lyon 7: 134. 1880, nom. illegit. *Sanicula odorata* (Rafinesque) Pryer & Phillippe, Canad. J. Bot. 67: 703. 1989.
Sanicula gregaria E. P. Bicknell, Bull. Torrey Bot. Club 22: 354, pl. 242. 1895.

Erect perennial herb, to 8 dm; stem striate, glabrous, the roots fibrous. Leaves with the blade ovate to suborbicular, 2–12 cm long and wide, deeply palmately 3- to 5(7)-lobed and/or compound, the apex acute to obtuse, the base cuneate, the margin irregularly serrate, the upper and lower surfaces glabrous, the petiole 0.5–5 cm long, the uppermost subsessile. Flowers in a terminal or axillary compound umbel, the peduncle 3–5 cm long, the rays 1–3, 1–6 cm long, the flowers numerous per umbellet, the umbellet with staminate and bisexual flowers (staminate 12–15 per umbellet), the pedicel to 1 mm long; bracts leaflike, lanceolate, 0.8–4 cm long; bracteoles elliptic-ovate, minute; calyx lobes lanceolate, 0.5–1 mm long, the apex tapered to a blunt point; corolla yellowish green, the petals longer than the sepals; stamens with yellow anthers; styles longer than the fruit bristles, up to 2 times as long as the sepals. Fruit oblong ovate, slightly laterally flattened, 3–5 mm long, with hooked bristles in rows, the pedicel 0.5–1 mm long.

Rocky floodplain forests and bluff forests. Occasional; central panhandle. Quebec south to Florida, west to Ontario, North Dakota, South Dakota, Nebraska, Kansas, Oklahoma, and Texas. Spring.

Sanicula smallii E. P. Bicknell [Commemorates John Kunkel Small (1869–1938), American botanist associated with the New York Botanical Garden and specializing in the southeastern U.S. flora.] SMALL'S BLACKSNAKEROOT.

Sanicula smallii E. P. Bicknell, Bull. Torrey Bot. Club 24: 578. 1897.

Erect perennial herb, to 6 dm; stem striate, glabrous, the roots thickened and cord-like. Leaves with the blade ovate to suborbicular, 1.5–12 cm long and wide, deeply palmately 3- or 5-lobed and/or compound, the apex acute to obtuse, the base cuneate, the margin irregularly serrate, the upper and lower surfaces glabrous, the petiole 5–15 cm long. Flowers in a terminal or axillary compound umbel, the peduncle 3–6 cm long, the rays 1–5, 0.2–6 cm long, the flowers 4–8 per umbellet, the umbellet with staminate and bisexual flowers (staminate 1–7 per umbellet), sessile; bracts leaflike, lanceolate, 0.6–4 cm long; bracteoles elliptic-ovate, minute; calyx lobes lanceolate, 1–1.5 mm long, the apex tapered to a sharp point; corolla greenish white, subequaling or shorter than the sepals; stamens with white anthers; styles shorter than the fruit bristles, slightly longer than the sepals. Fruit oblong ovate, slightly laterally flattened, 4–6 mm long, with hooked bristles in rows, sessile.

Calcareous hammocks and bluff forests. Rare; central and western panhandle. Virginia south to Florida, west to Illinois, Missouri, Arkansas, and Texas. Spring.

Scandix L. 1753.

Herbs. Leaves alternate, pinnately compound, petiolate. Flowers in terminal and axillary compound umbels, bracteate or ebracteate, bracteolate, bisexual; sepals 5, basally connate, actinomorphic; petals 5, free, actinomorphic or zygomorphic; stamens 5, free; ovary inferior, 2-carpellate and -loculate. Fruit a schizocarp, mericarps 2.

A genus of about 20 species; North America, South America, Europe, Africa, Asia, and Australia. [Latin vernacular name of the plant used by Pliny the Elder.]

Scandix pecten-veneris L. [*Pecten,* comb, and *veneris,* Venus.] SHEPARDSNEEDLE; VENUSCOMB.

Scandix pecten-veneris Linnaeus, Sp. Pl. 256. 1753. *Chaerophyllum pecten-veneris* (Linnaeus) Crantz, Stirp. Austr. Fasc. 2: 66. 1763. *Scandix pecten* Scopoli, Fl. Carniol., ed. 2. 1: 211. 1771, nom. illegit. *Pecten veneris* Lamarck, Fl. France 3: 437. 1779 ("1778"). *Myrrhis pecten-veneris* (Linnaeus) Allioni, Fl. Pedem. 2: 29. 1785. *Scandix rostrata* Salisbury, Prodr. Stirp. Chap. Allerton 166. 1796, nom. illegit. *Pectinaria vulgaris* Bernhardi, Syst. Verz. 168. 1800, nom. illegit. *Scandix pectenifera* Stokes, Bot. Mat. Med. 2: 122. 1812, nom. illegit. *Scandix vulgaris* Gray, Nat. Arr. Brit. Pl. 2: 503. 1821, nom. illegit. *Scandix pectiniformis* Saint-Lager, Ann. Soc. Bot. Lyon 7: 70. 1880, nom. illegit. *Scandix pecten-veneris* Linnaeus var. *genuina* Aznavour, Bull. Soc. Bot. France 46: 143. 1899, nom. inadmiss. *Wylia pecten-veneris* (Linnaeus) Bubani, Fl. Pyren. 2: 407. 1900. *Selinum pecten* E.H.L. Krause, in Sturm, Fl. Deutschl., ed. 2. 12: 75. 1904, nom. illegit. *Scandix pecten-veneris* Linnaeus subsp. *eupecten-veneris* Thellung, in Hegi, Ill. Fl. Mitt.-Eur. 5: 1039. 1926, nom. inadmiss.

Erect or ascending annual herb, to 5 dm; stem striate, hispid. Leaves (1)2- to 3-pinnate, the blade lanceolate, 2–10 cm long, 1–4 cm wide, the ultimate segments linear-lanceolate, 1–4 mm long, to 1 mm wide, the apex mucronulate, the margin hispid, the petiole 10–15 cm long, the margin ciliate or entire, the base sheathing. Flowers in a terminal or axillary compound umbel, the peduncle 1–4 cm long, the rays (1)2–3, (1)2–3(4) cm long; flowers bisexual or sometime sterile; bracts 1, similar to the leaves but reduced in size, or absent; bracteoles several, oblong, 4–7 mm long, the apex laciniate, the margin serrate or ciliate; calyx lobes minute; corolla white, the petals oblong, sometimes the outer flowers zygomorphic, the apex acuminate, inflexed. Fruit linear, 4–5 cm long, subcylindric, the margin bristly, the beak (nonseed-bearing part) 5–10 times as long as the seed-bearing part, each mericarp with 5 slender ribs.

Disturbed sites. Rare; Alachua and Jackson Counties. Nearly throughout North America; Europe, Africa, Asia, and Australia. Native to Africa, Europe, and Asia. Spring.

Sium L. 1753. WATERPARSNIP

Herbs. Leaves alternate, pinnately compound, petiolate. Flowers in terminal and axillary compound umbels, bracteate, bracteolate, bisexual; sepals 5, basally connate, actinomorphic or absent; petals 5, free, actinomorphic; stamens 5, free; ovary inferior, 2-carpellate and -loculate, mericarps 2.

A genus of about 12 species; North America, Europe, Africa, and Asia. [From the Greek *Sion,* vernacular name for a marsh plant.]

Sium suave Walter [*Suavis,* sweet, apparently in reference to its taste.] HEMLOCK WATERPARSNIP.

Sium suave Walter, Fl. Carol. 115. 1788.

Sium cicutifolium Schrank, Baier. Fl. 1: 558. 1789. *Critamus dauricus* Hoffmann, Gen. Umbell. Pl., ed. 2. 184. 1816, nom. illegit. *Falcaria dahurica* de Candolle, Prodr. 4: 110. 1830, nom. illegit. *Apium cicutifolium* (Schrank) Bentham & Hooker f. ex Forbes & Hemsley, J. Linn. Soc., Bot. 23: 328. 1887.

Sium lineare Michaux, Fl. Bor.-Amer. 1: 167. 1803. *Drepanophyllum lineare* (Michaux) Koso-Poljansky, Bull. Soc. Imp. Naturalistes Moscou, ser. 2. 28: 181. 1915. *Sium cicutifolium* Schrank var. *lineare* (Michaux) H. Wolff, in Engler, Pflanzenr. 4(Heft 90): 57. 1927.

Sium lineare Michaux var. *intermedium* Torrey & A. Gray, Fl. N. Amer. 1: 611. 1840. TYPE: FLORIDA: s.d., *Chapman s.n.* (Holotype: NY?).

Sium floridanum Small, Man. S.E. Fl. 976, 1506. 1933. *Sium suave* Walter var. *floridanum* (Small) C. F. Reed, Phytologia 63: 411. 1887. TYPE: FLORIDA: along the Chipola River, s.d., *Chapman s.n.* (holotype: NY).

Erect or ascending perennial herb, to 2 m; stem striate, glabrous. Leaves with the blade narrowly oblong to broadly ovate, 2–30 cm long, with (3)7–17 leaflets, the leaflets pinnately or dichotomously dissected in submerged leaves, the uppermost ones sometimes simple, the leaflets 1–10 cm long, those of the emergent leaves linear to lanceolate or narrowly ovate, the apex acuminate, the base rounded or cuneate, the margin finely serrate, those of the submerged leaves usually deeply and finely dissected into linear segments, the petiole of the lower ones to 15 cm long, reduced upward to 1 cm long, the base sheathing; the flowers in a terminal or axillary compound umbel, the peduncle 3–8 cm long, the rays 10–20, 1–4.5 cm long, unequal, the flowers numerous per umbellet, the pedicel 3–5 mm long, unequal; bracts 6–10, linear or lanceolate, 3–15 mm long, the wider ones with the margin irregular, white, and papery; bracteoles 4–8, linear to narrowly lanceolate, shorter or sometimes subequaling the pedicel, sometimes with a white margin; sepals triangular, minute, or absent; corolla white, the petals elliptic-obovate, ca. 1 mm long, the apex shallowly notched or tapered to a short, slender tip. Fruit broadly elliptic, 2–3 mm long, slightly laterally flattened, tan or yellowish, each mericarp with 5 prominent short, corky, winged ribs.

Swamps and spring runs. Occasional; northern counties, central peninsula. Nearly throughout North America; Asia. Spring–summer.

Spermolepis Raf. 1825. SCALESEED

Herbs. Leaves alternate, pinnately compound, petiolate. Flowers in terminal and axillary compound umbels, bracteolate, bisexual; sepals 5, basally connate, actinomorphic, or absent; petals 5, free, actinomorphic; stamens 5, free; ovary inferior, 2-carpellate and -loculate. Fruit a schizocarp, mericarps 2.

A genus of 11 species; North America, Mexico, South America, and Pacific Islands. [From the Greek *sperma,* seed, and *lepis,* scale, in reference to the scurfy or bristly fruit.]

Selected reference: Nesom (2012).

1. Fruit tuberculate .... **S. divaricata**
1. Fruit echinate with hooked bristles .... **S. echinata**

Spermolepis divaricata (Walter) Raf. [Spreading at an angle, in reference to the inflorescence branches.] ROUGHFRUIT SCALESEED.

Daucus divaricatus Walter, Fl. Carol. 114. 1788. *Ammi divaricatum* (Walter) Persoon, Syn. Pl. 1: 308. 1805. *Aethusa divaricata* (Walter) Sprengel, Pl. Umbell. Prodr. 22. 1813. *Sison divaricatum* (Walter) Sprengel, Sp. Umbel. 113. 1818. *Spermolepis divaricata* (Walter) Rafinesque, Bull. Bot. (Geneva) 1: 217. 1830. *Babiron divaricatum* (Walter) Rafinesque, New Fl. 4: 24. 1838 ("1836"). *Leptocaulis divaricata* (Walter) de Candolle, Coll. Mém. 5: 39. 1829. *Apium divaricatum* (Walter) A. W. Wood, Amer. Bot. Fl. 140. 1870.
Babiron dichotomum Rafinesque, New Fl. 4: 24. 1838 ("1836"). TYPE: FLORIDA.

Erect annual herb, to 4 dm; stem glabrous. Leaves with the blade oblong to oblong-ovate, 0.5–5 cm long, pinnately 2–4 times dissected, the ultimate segments narrowly linear or threadlike. Flowers in a terminal or axillary compound umbel, the peduncle 1–5 cm long, the rays 3–7, 5–35 mm long, equal or subequal, spreading to loosely ascending, straight or slightly curved, the flowers 1–6 per umbellet, the pedicel 2–15 mm long, the central flower pedicellate or sessile; bracteoles 1–3, linear, shorter than the pedicel; calyx lobes minute or absent; corolla white, the petals oblong to ovate, ca. 1 mm long. Fruit elliptic ovate, 1–2 mm long, laterally flattened, each mericarp with 5 inconspicuous, narrow, rounded ribs, with minute tubercles.

Disturbed sites. Frequent; nearly throughout. New Jersey south to Florida, west to Texas. Spring.

Spermolepis echinata (Nutt. ex DC.) A. Heller [With prickles, in reference to the fruit.] BRISTLY SCALESEED.

Leptocaulis echinata Nuttall ex de Candolle, Prodr. 4: 107. 1830. *Apium echinatum* (Nuttall ex de Candolle) Bentham & Hooker f. ex S. Watson, Bibl. Index N. Amer. Bot. 412. 1878. *Spermolepis echinata* (Nuttall ex de Candolle) A. Heller, Contr. Herb. Franklin Marshall Coll. 1: 73. 1895.

Ascending to erect annual herb, to 4 dm; stem glabrous. Leaves with the blade ovate to oblong-ovate, 0.7–2.5 cm long, pinnately 2–4 times dissected, the ultimate segments narrowly linear or threadlike. Flowers in a terminal or axillary compound umbel, the peduncle 0.8–6 cm long, the rays 5–15, 1–12 mm long, unequal, strongly ascending, the flowers 1–6 flowers per umbellet, the pedicel 1–7 mm long, the central umbellet of most umbels sessile or subsessile, with 1–2 flowers, these pedicellate or sessile; bracts absent; bracteoles 1–3, linear, shorter than the pedicel; corolla white, the petals oblong to ovate, ca. 1 mm long. Fruit ovate, 1–2 mm long, laterally flattened, each mericarp with 5 inconspicuous, narrow, rounded ribs, with short hooked bristles.

Disturbed sites. Occasional; northern counties, central peninsula. New York, Virginia south to Florida, west to Iowa, Kansas, Oklahoma, and Texas; Mexico. Spring.

EXCLUDED TAXON

Spermolepis inermis (Nuttall ex de Candolle) Mathias & Constance—Reported for Florida by Bell and Constance (1957), based on a misidentification of Florida material of *S. divaricata.*

Thaspium Nutt. 1818. MEADOWPARSNIP

Herbs. Leaves alternate, simple or pinnately compound, petiolate. Flowers in terminal and axillary compound umbels, bracteate, bracteolate, bisexual; sepals 5, basally connate, actinomorphic; petals 5, free, actinomorphic; stamens 5, free; ovary inferior, 2-carpellate and -loculate. Fruit a schizocarp, mericarps 2.

A genus of 5 species; North America. [Modification of *Thapsia,* a related genus known from the Thapsus peninsula, now Magnisi in modern-day Sicily.]

1. Cauline leaves 1- to 2-ternately compound, the margins coarsely toothed .... T. **barbinode**
1. Cauline leaves 3-foliolate, the margins evenly and finely toothed .... T. **trifoliatum**

Thaspium barbinode (Michx.) Nutt. [With bearded nodes.] HAIRYJOINT MEADOWPARSNIP.

Ligusticum barbinode Michaux, Fl. Bor.-Amer. 1: 167. 1803. *Thaspium barbinode* (Michaux) Nuttall, Gen. N. Amer. Pl. 1: 196. 1818.

Thaspium barbinode (Michaux) Nuttall var. *pinnatifidum* J. M. Coulter & Rose, Rev. N. Amer. Umbell. 84. 1888. *Thaspium barbinode* (Michaux) Nuttall var. *chapmanii* J. M. Coulter & Rose, Contr. U.S. Natl. Herb. 7: 148. 1900, nom. illegit. *Thaspium chapmanii* Small, Man. S.E. Fl. 980. 1933. TYPE: FLORIDA: Jackson Co.: s.d., *Curtis 1023* (lectotype: US). Lectotypified by Coulter and Rose (1900: 148).

Erect or ascending perennial herb, to 1 m; stem striate, usually pubescent with a band of short, white trichomes at the base of at least the uppermost leaf sheath, sometimes also sparsely and minutely pubescent toward the apex. Basal leaves with the blade 6–25 cm long and wide, ternately or ternately and then pinnately 2 or 3 times compound, the ultimate segments lanceolate to ovate, 2–6 cm long, 1–3.5 cm wide, the apex acute, the base cuneate to rounded, often unevenly so, the margin coarsely serrate, often with 1 or 2 lobes near the base, hispidulous, the upper and lower surfaces glabrous or sparsely white pubescent, the petiole 5–10 cm long, the base sheathing, the stem leaves gradually reduced upward in size and the amount of dissection. Flowers in a terminal or axillary compound umbel, the peduncle 3–8 cm long, the rays 5–10(15), 1–3 cm long, subequal, the flowers numerous per umbellet, the pedicel 2–5 mm long; bracts linear, 1–4 mm long; bracteoles linear, 1–4 mm long; calyx lobes minute; corolla yellow, the petals obovate, ca. 1 mm long. Fruit ovate or oblong, 3–6 mm long, each mericarp with 5 ribs, several of these winged.

Bluff forests. Rare; Jackson County. Nova Scotia, Ontario, New York south to Florida, west to Minnesota, Iowa, Kansas, Oklahoma, and Texas. Spring.

Thaspium trifoliatum (L.) A. Gray [Leaves with three leaflets.] PURPLE MEADOWPARSNIP.

Thaspia trifoliata Linnaeus, Sp. Pl. 262. 1753. *Smyrnium cordatum* Walter, Fl. Carol. 114. 1788, nom. illegit. *Zizia cordata* W.D.J. Koch ex de Candolle, Prodr. 4: 100. 1830, nom. illegit. *Thaspium cordatum* Torrey & A. Gray, Fl. N. Amer. 1: 615. 1840, nom. illegit. *Thaspium cordatum* Torrey & A. Gray var. *atropurpureum* A. W. Wood, Class-Book Bot., ed. 2. 289. 1847, nom. inadmiss. *Thaspium trifoliatum* (Linnaeus) A. Gray, Manual, ed. 2. 156. 1856. *Thaspium aureum* (Linnaeus) Nuttall var. *cordatum* Britton et al., Prelim. Cat. 22. 1888, nom. illegit. *Thaspium aureum* (Linnaeus) Nuttall var. *trifoliatum* (Linnaeus) Britton et al., Prelim. Cat. 22. 1888. *Thaspium ziziopsis* Daniels, Univ. Missouri Stud., Sci. Ser. 1: 329. 1907, nom. illegit.

Erect or ascending perennial herb, to 8 dm; stem striate, usually pubescent with a band of short white trichomes at the base of the uppermost leaf sheath, otherwise glabrous. Basal leaves with the blade 3–9 cm long, simple or ternately lobed or compound, the leaflets ovate to oblong-ovate or lanceolate, 1–7 cm long, 1–3.5 cm wide, the base broadly cuneate to cordate, often unevenly so, the margin finely crenate-dentate, sometimes with 1 or 2 lobes near the base, with a narrow, white border, the upper and lower surfaces glabrous or slightly scabrous along the main veins, the petiole 6–10 cm long, the upper leaves similar and gradually reduced upward, mostly ternately lobed or compound, less commonly simple or ternately 2 times lobed or compound, the margin finely serrate, white and glabrous. Flowers in a terminal or axillary compound umbel, the peduncle 3–10 cm, the rays 6–10(18), 0.5–3 cm long, unequal, flowers numerous per umbellet, the pedicel 1–4 mm long; bracts 4–8, linear to ovate-lanceolate, 2–4 mm long; bracteoles ovate-triangular, minute; calyx lobes minute; corolla yellow or purplish, the petals obovate, ca. 1 mm long. Fruit ovate, 3–5 mm long, each mericarp with 5 winged ribs.

Bluff forests and creek banks. Rare; central panhandle. Ontario and New York south to Florida, west to Minnesota, Illinois, Missouri, Kansas, Oklahoma, and Texas. Spring.

EXCLUDED TAXON

Thaspium trifoliatum (Linnaeus) A. Gray var. *flavum* S. F. Blake—Reported for Florida by Radford et al (1964, 1968). No Florida material of this yellow-flowered plant seen.

Tiedemannia DC. 1829. WATER COWBANE

Herbs. Leaves alternate, simple, petiolate. Flowers in terminal and axillary compound umbels, bracteate, bracteolate, bisexual; sepals 5, basally connate, actinomorphic; petals 5, free, actinomorphic; stamens 5, free; ovary inferior, 2-carpellate and -loculate. Fruit a schizocarp, mericarps 2.

A genus of 2 species; North America. [Commemorates Friedrich Tiedemann (1781–1861), German zoologist].

Selected reference: Feist et al. (2012).

Tiedemannia filiformis (Walter) Feist & S. R. Downie [Threadlike, in reference to the phyllodes.] WATER COWBANE.

Perennial herb, to 2 m; stem striate, glabrous. Leaves phyllodal, the proximal elongate tapering, to 6 cm long, septate, the base sheathing, glabrous, much reduced upward. Flowers in a terminal or axillary compound umbel, the peduncle 3–12 cm long, the rays 6–12, 2–4 cm long, subequal, the flowers 10–25 per umbellet, the pedicel 4–8 mm long, to 1 cm long in fruit; bracts 4–10, linear-lanceolate, 4–6 mm long, the apex attenuate, the margin scarious; bracteoles 6–10, linear, 1–3 mm long, the apex attenuate; calyx lobes deltoid, minute; corolla white or garnet-maroon, the petals ovate, 1–2 mm long, the apex inflexed. Fruit ovate, elliptic, or obovate, 5–8 mm long, slightly laterally flattened, each mericarp with 5 ribs, the lateral ones winged.

1. Petals garnet-maroon; phyllodes conspicuously septate, the joints brittle and easily broken .... subsp. **greenmanii**
1. Petals white; phyllodes inconspicuously septate, the joints not brittle .... subsp. **filiformis**

Tiedemannia filiformis subsp. **filiformis** WATER COWBANE.

Oenanthe filiformis Walter, Fl. Carol. 113. 1788. *Oenanthe carolinensis* Persoon, Syn. Pl. 1: 318. 1805, nom. illegit. *Oenanthe teretifolia* Muhlenberg, Cat. Pl. Amer. Sept. 32. 1813, nom. illegit. *Sium teretifolium* Elliott, Sketch Bot. S. Carolina 1: 354. 1817, nom. illegit. *Tiedemannia teretifolia* de Candolle, Coll. Mém. 5: 51, 81. 1829, nom. illegit. *Oxypolis caroliniana* Rafinesque, Bull. Bot. (Geneva) 1: 218. 1830, nom. illegit. *Peucedanum teretifolium* A. W. Wood, Amer. Bot. Fl. 136. 1870, nom. illegit. *Oxypolis filiformis* (Walter) Britton, Mem. Torrey Bot. Club 5: 239. 1894. *Tiedemannia filiformis* (Walter) Feist & S. R. Downie, in Feist et al., Taxon 61: 413. 2012.

Phyllodes green, inconspicuously septate, not readily breaking at the joints. Petals white.

Swamps and wet flatwoods. Common; nearly throughout. North Carolina south to Florida, west to Texas. Summer–fall.

Tiedemannia filiformis (Walter) Feist & S. R. Downie subsp. **greenmanii** (Mathias & Constance) Feist & S. R. Downie [Commemorates Jesse More Greenman (1867–1951), American botanist, specialist on the floras of Mexico and Central America.] GIANT WATER COWBANE.

Oxypolis greenmanii Mathias & Constance, Bull. Torrey Bot. Club 69: 152. 1942. *Oxypolis filiformis* (Walter) Britton subsp. *greenmanii* (Mathias & Constance) Judd, Rhodora 84: 277. 1982. *Tiedemannia filiformis* (Walter) Feist & S. R. Downie subsp. *greenmanii* (Mathias & Constance) Feist & S. R. Downie, Taxon 61: 414. 2012. TYPE: FLORIDA: Gulf Co.: Wewahitchka, Aug 1896, *Chapman* s.n. (holotype: MO).

Phyllodes suffused with purple, conspicuously septate, the joints brittle and easily broken. Petals garnet-maroon.

Marshes, cypress ponds, and wet flatwoods. Rare; Calhoun, Gulf, and Bay Counties. Endemic. Summer–fall.

Tiedemannia filiformis subsp. *greenmanii* (as *Oxypolis greenmanii*) is listed as endangered in Florida (Florida Administrative Code, Chapter 5B-40).

Torilis Adans. 1763. HEDGEPARSLEY

Herbs. Leaves alternate, pinnately compound, petiolate. Flowers in terminal and axillary compound cymes, bracteate, bracteolate, bisexual; sepals 5, basally connate, actinomorphic; petals 5, free, actinomorphic; stamens 5, free; ovary inferior, 2-carpellate and -loculate. Fruit a schizocarp, mericarps 2.

A genus of about 8 species; North America, Europe, Africa, Asia, and Australia. [*Toro,* to carve or emboss, in reference to the fruit covered with tubercles.]

Torilis arvensis (Huds.) Link [Pertaining to fields or cultivated land.] SPREADING HEDGEPARSLEY.

Caucalis arvensis Hudson, Fl. Angl. 98. 1762. *Torilis divaricata* Moench, Suppl. Meth. 34. 1802, nom. illegit. *Caucalis divaricata* Heller, Fl. Wirceb. 1: 168. 1810, nom. illegit. *Torilis arvensis* (Hudson) Link, Enum. Hort. Berol. Alt. 1: 265. 1821. *Anthriscus arvensis* (Hudson) Koso-Poljansky, Bull. Soc. Imp. Naturalistes Moscou, ser. 2. 29: 151. 1916 ("1915").

Erect or ascending annual herb, to 1 m; stem striate, appressed hispid. Leaves 1- to 3-pinnately compound, the blade ovate-triangular, 3–15 cm long, 2–8 cm wide, the basal ones the largest and reduced upward on the stem, the apex acuminate, the base rounded to truncate, the leaflets ovate to linear-lanceolate, 0.5–6 cm long, 2–18 mm wide, deeply pinnately dissected, the lobes with the margins, mostly sharply few-toothed, the upper and lower surfaces hirsute, the petiole 2–8 mm long, the base sheathing. Flowers in a terminal or axillary compound umbel, the peduncle 3–10 cm long, the rays 3–10, 0.5–2.5 cm long, unequal, the flowers 5–12 per umbellet, the pedicel 1–4 cm long, unequal; bracts 1–2, narrowly triangular or linear, 1–12 mm long; bracteoles 3–8, longer than the pedicel; calyx lobes triangular, minute; corolla white, the petals obovate, ca. 1 m long, the apex rounded or shallowly notched. Fruit oblong-elliptic, 3–5 mm long, slightly laterally flattened, each mericarp with 5 ribs, these obscured by a dense covering of straight or slightly arched bristles ca. 1 mm long.

Disturbed sites. Rare; Escambia County. Nearly throughout North America; Europe, Africa, Asia, and Australia. Native to Europe, Africa, and Asia. Spring.

EXCLUDED TAXA

Torilis japonica (Houttuyn) de Candolle—Reported for Florida by Fernald (1950). No Florida specimens seen.

Torilis nodosa (Linnaeus) Gaertner—Reported for Florida by Small (1903, 1913a, 1933) and Mathias and Constance (1944b) as "adventive throughout the southern United States," which would presumably include Florida. No Florida specimens seen.

Trepocarpus Nutt. ex DC. 1829.

Herbs. Leaves alternate, pinnately compound, petiolate. Flowers in terminal and axillary compound umbels, bracteate, bracteolate, bisexual; sepals 5, basally connate, actinomorphic; petals 5, free, actinomorphic; stamens 5, free; ovary inferior; 2-carpellate and -loculate. Fruit a schizocarp, mericarps 2.

A monotypic genus; North America. [From the Greek *trepo,* to turn, and *carpos,* fruit, in reference to the fruits turning outward at maturity.]

Trepocarpus aethusae Nutt. ex DC. [Resembling the genus *Aethusa* (Apiaceae).] WHITENYMPH.

Trepocarpus aethusae Nuttall ex de Candolle, Coll. Mém. 5: 56. 1829.

Erect or ascending annual herb, to 7 dm; stem striate, glabrous. Leaves 3- to 4-pinnately compound or dissected, ovate to broadly triangular, 3–12 cm long, 1–10 cm wide, the apex acute, the base rounded to subtruncate, the ultimate segments linear, 2–12 mm long, the apex short tapered to sharp point, the upper and lower surfaces glabrous, the petiole 0.5–2 cm long, the base sheathing. Flower in a terminal or axillary compound umbel, the peduncle 2–4 cm long, the rays 2–4, 0.5–1.5 cm long, the flowers 2–8 per umbellet, the pedicel absent or to 3 mm long, the umbellet sometimes appearing head-like; bracts 2–4, linear, 2–12 mm long, these sometimes with a few linear lobes; bracteoles 2–6, linear, subequaling or longer than the pedicel; calyx lobes triangular, minute, unequal; corolla white, the petals obovate, ca. 1 mm long, the apex shallowly notched or tapering to an abrupt tip. Fruit narrowly oblong, 8–10 mm long, slightly laterally flattened, light brown, the mericarps with 5 inconspicuous ribs, and 4 conspicuous, rounded, corky, secondary ribs lighter in color.

Floodplain forests. Occasional; Levy County, eastern and central panhandle. South Carolina south to Florida, west to Missouri, Oklahoma, and Texas. Spring.

Zizia W.D.J. Koch 1824. ALEXANDERS

Perennial herbs. Leaves alternate, simple or ternately compound, petiolate. Flowers in terminal or terminal and axillary compound umbels, bracteate, bracteolate, bisexual; sepals 5, basally connate; petals 5, free; stamens 5, free; ovary inferior, 2-carpellate and -loculate. Fruit a schizocarp, the mericarps 2.

A genus of 5 species; North America. [Commemorates Johann Baptist Ziz (1779–1829), German botanist.]

1. Basal leaves simple .... **Z. aptera**
1. Basal leaves 1- to 2-ternately compound.
    2. Leaf margins with the teeth averaging 5–10 per cm .... **Z. aurea**
    2. Leaf margins with the teeth averaging 2–3 per cm .... **Z. trifoliata**

Zizia aptera (A. Gray) Fernald [Without wings, in reference the fruits, which lack wings.] MEADOW ALEXANDERS.

Thaspium trifoliatum (Linnaeus) A. Gray var. *apterum* A. Gray, Manual, ed. 2. 156. 1856. *Zizia aptera* (A. Gray) Fernald, Rhodora 41: 441. 1939.

Erect perennial herb, to 7 dm; stem glabrous. Leaves with the blade cordate, the basal ones simple, the upper ones 1-ternately divided, 4–7 cm long, 3–5 cm wide, the apex acute to obtuse, the base cordate, the lobes of the upper ones lanceolate, the base rounded to cuneate, the margin coarsely serrate, sometimes lobed, the upper and lower surfaces glabrous, the petiole 5–10 cm long, much reduced upward on the stem and the leaves becoming sessile or subsessile. Flowers in a terminal or axillary compound umbel, 11–19 per umbel, the peduncle 6–10 cm long, the rays 12–18, 1–3 cm long, the central flower in the umbel sessile or subsessile, the others with the pedicel 1–4 mm long; bracts 3–9, linear, 1–2 mm long; bracteoles minute; calyx lobes triangular, minute; petals yellow, obovate, ca. 1 mm long. Fruit ovate to oblong-ovate, 2–4 cm long, laterally compressed, the mericarps reddish brown, with 5 yellowish brown ribs.

Bay swamp margins. Rare; Calhoun and Holmes Counties. Nearly throughout North America. Spring.

Zizia aurea (L.) W.D.J. Koch [*Aurum*, gold, in reference to the flower color.] GOLDEN ALEXANDERS.

Smyrnium aurea Linnaeus, Sp. Pl. 262. 1753. *Thaspium aureum* (Linnaeus) Nuttall, Gen. N. Amer. Pl. 1: 196. 1818. *Sison aureum* (Linnaeus) Sprengel, in Roemer & Schultes, Syst. Veg. 6: 410. 1820. *Zizia aurea* (Linnaeus) W.D.J. Koch, Nova Acta Phys.-Med. Acad. Caes. Leop.-Carol. Nat. Cur. 12: 129. 1825.

Erect perennial herb, to 11(15) dm; stem glabrous. Leaves with the blade cordate, 1- to 2-ternately compound, the basal ones 4–13(20) cm long, the leaflets 1–12 cm long, broadly oblong-ovate to lanceolate, the apex acute, the base rounded to cordate, often unequally so, sometimes with 1–2 lobes toward the base, the margin serrate, the teeth averaging 5–10 per cm, the leaves gradually reduced upward on the stem, becoming more finely divided with the leaflets and/or segments narrower, the uppermost ones only ternately 1- to 3-lobed, the upper and lower surfaces glabrous, the petiole 10(15) cm long, much reduced upward on the stem. Flowers in a terminal or axillary compound umbel, 11–19 per umbel, the peduncle 5–15 cm long, the rays 10–21, 0.5–5 cm long, the central flower in the umbel sessile or subsessile, the others with the pedicel 1–4 mm long; bracts 3–9, linear, 1–3 mm long; bracteoles minute, calyx lobes triangular, minute; petals yellow, obovate, ca. 1 mm long. Fruit ovate to oblong-ovate, 2–4 cm long, laterally compressed, the mericarps reddish brown with 5 yellowish brown ribs.

Hammocks and bluff forests. Rare; northern peninsula, Levy County, central and western panhandle. Quebec south to Florida, west to Manitoba, Montana, South Dakota, Nebraska, Oklahoma, and Texas. Spring.

Zizia trifoliata (Michx.) Fernald [Leaves with three leaflets.] MEADOW ALEXANDERS.

Sison trifoliatum Michaux, Fl. Bor.-Amer. 1: 168. 1803. *Sium trifoliatum* (Michaux) Poiret, in Lamarck, Encycl., Suppl. 1: 622. 1811. *Zizia trifoliata* (Michaux) Fernald, Rhodora 42: 298. 1940.
Zyzia arenicola Rose, Proc. U.S. Natl. Mus. 29: 442. 1905.
Zizia latifolia Small, Man. S.E. Fl. 982, 1506. 1933. TYPE: FLORIDA: Liberty Co.: near Bristol, 28 Aug, Curtiss s.*n*. (holotype: NY).

Erect perennial herb, to 10(15) dm; stem glabrous. Leaves with the blade cordate, 1- to 2-ternately compound, the basal ones 4–13(20) cm long, the leaflets often 3, 1–12 cm long, broadly oblong-ovate or lanceolate, the apex acute, the base rounded or cordate, often unequally so, sometimes with 1–2 lobes toward the base, the margin coarsely serrate, the teeth averaging 2–3 per cm, the leaves gradually reduced upward on the stem, becoming more finely divided with the leaflets and/or segments narrower, the uppermost ones only ternately 1- to 3-lobed, the upper and lower surfaces glabrous, the petiole 10(15) cm long, much reduced upward on the stem. Flowers in a terminal or axillary compound umbel, 4–10 per umbel, the peduncle 5–15 cm long, the rays 10–20, 1–5 cm long, the central flower in the umbel sessile or subsessile, the others with the pedicel 1–4 mm long; bracts 3–9, linear, 1–3 mm long; bracteoles minute; calyx lobes triangular, minute; petals yellow, obovate, ca. 1 mm long. Fruit ovate to suborbicular, 2–4 cm long, laterally compressed, the mericarps reddish brown with 5 yellowish brown ribs.

Floodplain forests and mesic hammocks. Rare; Levy, Leon, Gadsden, and Liberty Counties. Maryland south to Florida, west to Arkansas. Spring–summer.

EXCLUDED GENERA

Berula erecta (Hudson) Coville—Reported for Florida by Mathias and Constance (1944b) as "throughout the U.S.," which would presumably include Florida and by Correll and Johnston (1970). No Florida specimens seen.

Conium maculatum Linnaeus—Reported for Florida by Radford et al. (1964, 1968) and by Wilhelm (1984). No Florida specimens seen.

Pastinaca sativa Linnaeus—Reported by Small (1903) as "throughout the United States," which would presumably include Florida. No Florida specimens seen.

Trachyspermum ammi (Linnaeus) Sprague—Reported for Florida by Small (1903, as *Cyclospermum ammi* (Linnaeus) Lagasca y Sagura). Small was attempting to use the epithet *ammi* in the sense of Savi and Urban (*Apium ammi* (Savi) Urban, in Martius, Fl. Bras. 11(1): 341. 1879; non Crantz, 1767), to apply the name to material of *Cyclospermum leptophyllum*. However, Small's citation caused a misapplication since that name is based on *Seseli ammi* Savi, which applies to a Eurasian species, not *Sison ammi* Linnaeus, which applies to *Cyclospermum ammi*.

ASTERACEAE Bercht. & J. Pesl 1820. ASTER FAMILY

Herbs, shrubs, or vines. Leaves alternate, opposite, or whorled, simple or compound, pinnate- or palmate-veined, petiolate or epetiolate, estipulate. Flowers in terminal and axillary capitula surrounded by phyllaries, sometimes also with calycauli, paleate or epaleate, bisexual, unisexual, or sterile; petals (0)5, basally connate, actinomorphic or zygomorphic; stamens (4)5, epipetalous, the filaments connate or free, the anthers connate or free; ovary inferior, 2-carpellate, 1-loculate. Fruit a cypsela, an apical pappus (modified sepals) present or absent.

A family of about 1,500 genera and about 23,000 species. Nearly cosmopolitan.

The Asteraceae share a number of unusual morphological features to which distinct terms have been applied.

The flowers occur in a head-like structure known as a *capitulum* (pl. *capitula*), which is solitary or arranged in a variety of arrays. Each capitulum is surrounded by involucral bracts called *phyllaries*. Sometimes there may be smaller bractlets just below these called a *calyculus* (pl. *calyculi*). The flowers in the capitulum are on a *receptacle*, which may be smooth, alveolate (honeycombed), or foveolate (pitted) or have trichomes or subulate enations. The flowers on the receptacle may have a bract at their base called a *palea* (pl. *paleae*).

The corolla may be tubular and actinomorphic with 5 equal lobes (*disk flower*), zygomorphic and bilabiate, zygomorphic with a flat *ligule* with 5 teeth (*ligulate flower*), or zygomorphic with a flat *lamina* terminating in (0)2–3(4) teeth (*ray flower*).

The capitulum may be *ligulate* (all ligulate flowers), *discoid* (all disk flowers), *radiate* (outer ray flowers and inner disk flowers), *radiant* (peripherial disk flowers zygomorphic and raylike), or *trimorphic* (1–several outer rows with ray flowers, the next row with bilabiate flowers, and the inner ones with disk flowers.

The fruit is technically a *cypsela* (pl. *cypselae*), although some incorrectly refer to it as an *achene*. The cypsela is a dry, bicarpellate, uniloculate, 1-seeded, inferior ovary while an achene is a dry, 1-carpellate, 1-loculate, superior ovary.

The fruit sometimes has a *pappus* (pl. *pappi*), which is interpreted as the highly modified remains of a calyx.

The Asteraceae have long been recognized as a natural group, although some authors have divided it into three or more families. A tribal classification was established in the nineteenth century by H. Cassini and has since been much modified by others. Recent molecular phylogenetic studies have futher modified it. A recent proposal by Panero and Funk (2002) recognizes 10 subfamilies and 35 tribes.

In Florida, the 139 genera recognized are placed in 12 tribes following the system used in the *Flora of North America* (Flora of North America Editorial Committee, 2006). A key to the tribes and genera in North America is provided by Barkley et al. (2006). In our work, the genera are arranged alphabetically and not by tribes. The generic placement in tribes is listed here for reference.

Anthemideae: *Achillea, Anthemis, Artemisia, Cladanthus, Glebonis, Leucanthemum, Soliva.*

Arctotideae: *Haplocarpha.*

Astereae: *Aphanostephanus, Baccharis, Bigelowia, Boltonia, Bradburia, Brintonia, Chrysoma, Chrysopsis, Conyza, Croptilon, Doellingeria, Erigeron, Eurybia, Euthamia, Gerbera, Heterotheca, Ionactis, Oclemena, Pityopsis, Rayjacksonia, Sericocarpus, Solidago, Symphyotrichum.*

Cichorieae: *Cichorium, Crepis, Hieracium, Hypochaeris, Krigia, Lactuca, Launaea, Lygodesmia, Nabalus, Pilosella, Pyrropappus, Sonchus, Taraxacum, Youngia.*

Cynareae: *Centaurea, Cirsium, Onopordum.*

Eupatorieae: *Ageratina, Ageratum, Brickellia, Carphephorus, Chromolaena, Conoclinium, Eupatorium, Eutrochium, Fleischmannia, Garberia, Hartwrightia, Koanophyllon, Liatris, Mikania, Praxelis, Sclerolepis.*

Gnaphalieae: *Antennaria, Facelis, Filago, Gamochaeta, Pseudognaphalium.*

Heliantheae: *Acanthospermum, Acmella, Ambrosia, Arnica, Balduina, Berlandiera, Bidens, Borrichia, Calyptocarpus, Chrysogonum, Coreopsis, Cosmos, Dracopsis, Echinacea, Eclipta, Enydra, Flaveria, Gaillardia, Galinsoga, Helenium, Helianthus, Heliopsis, Hymenopappus, Iva, Lagascea, Marshallia, Melampodium, Melanthera, Palafoxia, Parthenium, Pascalia, Pectis, Phoebanthus, Polymnia, Ratibida, Rudbeckia, Silphium, Smallanthus, Sphagneticola, Synedrella, Tagetes, Tetragonotheca, Thymophylla, Tithonia, Tridax, Verbesina, Xanthium, Zinnia.*

Inuleae: *Dittrichia, Pulicaria.*

Mutisieae: *Chaptalia.*

Plucheeae: *Pluchea, Pterocaulon, Sachsia.*

Senecioneae: *Arnoglossum, Crassocephalum, Emilia, Erechtites, Euryops, Gynura, Hasteola, Packera, Pseudogynoxys, Senecio.*

Vernonieae: *Centratherum, Cyanthillium, Elephantopus, Pseudelephantopus, Stokesia, Vernonia.*

Ambrosiaceae Bercht. & J. Presl, 1820; *Carduaceae* Bercht. & J. Presl, 1820; *Cichoriaceae* Juss., 1789; *Compositae* Giseke, nom. alt., 1792.

1. Capitula ligulate and perfect; sap usually milky .... KEY 1
1. Capitula radiate, discoid, or trimorphic, these variously staminate, carpellate, or perfect; sap clear.
    2. Capitula radiate (ligule of the ray flowers sometimes small and inconspicuous) or trimorphic.
        3. Ray flowers yellow or orange (sometimes marked with purple or brown at the base) .... KEY 2
        3. Ray flowers white, pink, purple, red, or blue .... KEY 3
    2. Capitula discoid.
        4. Pappus of capillary or plumose bristles .... KEY 4
        4. Pappus of scales, a few awns, very short chaffy bristles, a mere crown, or absent .... KEY 5

KEY 1

1. Pappus consisting of scales, scales and bristles, or absent.
    2. Flowers blue (rarely white or pink); pappus of minute truncate scales .... **Cichorium**
    2. Flowers yellow or orange; pappus of scales and bristles or absent .... **Krigia**
1. Pappus consisting of capillary or plumose bristles.
    3. Pappus of plumose bristles .... **Hypochaeris**
    3. Pappus of capillary bristles.
        4. Fruit strongly laterally flattened.
            5. Flowers blue or purple, if yellow, then the fruits distinctly filiform-beaked or narrowed toward the apex .... **Lactuca**
            5. Flowers yellow; fruits not filiform-beaked or narrowed toward the apex .... **Sonchus**
        4. Fruit cylindrical, columnar, or fusiform (sometimes slightly laterally flattened).
            6. Flowers pink or lavender.
                7. Leaves greatly reduced (to 3 mm wide), mostly scalelike; capitula few, terminating the branches .... **Lygodesmia**
                7. Leaves broad (more than 3 mm wide); capitula many, in thyrsiform, paniculate, or subracemiform arrays .... **Nabalus**
            6. Flowers yellow.
                8. Fruit (at least those in the center of the disk) filiform-beaked.
                    9. Fruit beak with a ring of soft, white, reflexed trichomes at the apex .... **Pyrrhopappus**
                    9. Fruit beak lacking a ring of trichomes at the apex .... **Taraxacum**
                8. Fruit sometimes tapering above, but not filiform-beaked.
                    10. Leaves finely prickly; capitula slender pedunculate, scattered along the elongate main branches of the inflorescence; pappus bristles of 2 distinct types, a few of them coarser than the others .... **Launaea**
                    10. Leaves not finely prickly; capitula open corymbose or paniculate, terminating the ultimate branches; pappus bristles sometimes of uneven thickness, but not of 2 distinct types.
                        11. Phyllaries imbricate; plants perennial.
                            12. Cypsela 4–5 mm long; pappus bristles in 2 series .... **Hieracium**
                            12. Cypsela less than 3 mm long; pappus bristles in 1 series .... **Pilosella**

11. Phyllaries consisting of 2 series, the inner ones long and subequal, the outer ones short and unequal; plants annual.
    13. Phyllaries 8–12 mm high; stem with glandular trichomes below, glabrous ones above .... **Crepis**
    13. Phyllaries 3.5–6 mm high; stem puberulent below, glabrous above ..... **Youngia**

KEY 2

1. Receptacle epaleate.
   2. Phyllaries and leaves with evident large sessile glands.
      3. Leaves entire; phyllaries free .... **Pectis**
      3. Leaves pinnatifid; principal phyllaries connate most of their length.
         4. Pappus of 10–12 similar scales, each palmately divided into 3–5 slender bristles, the central one the longest .... **Thymophylla**
         4. Pappus of several very unequal scales not palmately divided into bristles .... **Tagetes**
   2. Phyllaries and leaves lacking evident large sessile glands (leaves sometimes finely glandular-dotted).
      5. Pappus of scales or absent.
         6. Pappus absent.
            7. Leaves opposite, the margin entire or merely toothed .... **Flaveria**
            7. Leaves alternate, bipinnatifid or nearly so .... **Glebionis**
         6. Pappus of scales.
            8. Leaves white-tomentose on the lower surface .... **Haplocarpha**
            8. Leaves glabrous or pubescent, but not white-tomentose on the lower surface.
               9. Fruit partly or wholly covered with a conspicuous basal tuft of trichomes . **Gaillardia**
               9. Fruit sometimes pubescent on the angles or ribs, but lacking a conspicuous basal tuft of trichomes .... **Helenium**
      5. Pappus (at least of the disk flowers) of capillary bristles.
         10. Lower leaves opposite, the upper ones sometimes alternate .... **Arnica**
         10. All leaves alternate.
             11. Shrubs, subshrubs, or vines.
                 12. Vines; flowers orange to orange-red; leaves toothed .... **Pseudogynoxys**
                 12. Shrubs or subshrubs; flowers yellow; leaves entire or deeply pinnatifid.
                     13. Leaves entire .... **Chrysoma**
                     13. Leaves deeply pinnatifid .... **Euryops**
             11. Herbs.
                 14. Inner phyllaries equal, the outer ones much reduced.
                     15. Cauline leaves gradually reduced upward, the segments linear-lanceolate, white-tomentose on the lower surface .... **Senecio**
                     15. Cauline leaves much reduced, most leaves basal, if gradually reduced upward, then the segments oblanceolate, glabrous on the lower surface .... **Packera**
                 14. Phyllaries imbricate.
                     16. Pappus double, the outer of short scales or bristles, the inner of long bristles.
                         17. Ray flowers epappose .... **Heterotheca**
                         17. Ray flowers with an evident pappus like that of the disk flowers.
                             18. Lower cauline leaves linear, with parallel veins and often grasslike; fruit of the disk flowers fusiform .... **Pityopsis**

18. Lower cauline leaves not as above; fruit of the disk flowers compressed-obconic.
    19. Capitula in a corymbiform, subumbelliform, or paniculiform array; pappus bristles relatively fine, uniformly whitish to stramineous, the outer series linear to narrowly triangular .... **Chrysopsis**
    19. Capitula born singly or in a loose paniculiform array; pappus bristles relatively thick and rigid, the inner series yellow to rust-colored proximally, whitish to tan distally, the outer series light tan, scalelike .... **Bradburia**

16. Pappus simple.
    20. Pappus tan or brown.
        21. Leaves with spinulose-tipped teeth; phyllaries 10–12 mm long .... **Rayjacksonia**
        21. Leaves entire, or if toothed, then the teeth not spinulose-tipped; phyllaries 5–8 mm long .... **Croptilon**
    20. Pappus white.
        22. Acaulescent herbs with a solitary capitulum on the elongate scapes............ .... **Gerbera**
        22. Caulescent herbs with numerous capitula in various arrays.
            23. Receptacle alveolate.
                24. Pappus bristles united at the base into a cup .... **Dittrichia**
                24. Pappus bristles not united at the base into a cup .... **Pulicaria**
            23. Receptacle not alveolate.
                25. Capitula in corymbose arrays .... **Euthamia**
                25. Inflorescence in paniculiform, racemose, or thyrsoid arrays .... **Solidago**

1. Receptacle paleate, or if epaleate, and then deeply alveolate with a honeycomblike surface (*Balduina*) or with hard setifom enations (*Gaillardia*).
    26. Disk flowers sterile.
        27. Inner phyllaries investing the fruit and enlarging to form a perigynium and enclosing it, conspicuously prickly .... **Acanthospermum**
        27. Inner phyllaries not as above.
            28. Ray flower fruits in 2–3 series .... **Silphium**
            28. Ray flower fruits in 1 series.
                29. Leaves alternate.. **Berlandiera**
                29. Leaves opposite.
                    30. Leaves lobed.
                        31. Larger leaves pinnately lobed; fruit prominently 4- to 6-ribbed or -angled .... **Polymnia**
                        31. Larger leaves palmately 3- to 5-lobed at the base, the lobes pinnatifid; fruit many-striate .... **Smallanthus**
                    30. Leaves entire to coarsely dentate-crenate.
                        32. Receptacle flat; ray flowers ca. 5.. **Chrysogonum**
                        32. Receptacle convex; ray flowers 8–13.. **Melampodium**
    26. Disk flowers fertile.
        33. Leaves all alternate.

34. Receptacle epaleate, with hard setiform enations .... **Gaillardia**
34. Receptacle with chaff-like paleae, lacking setiform enations.
35. Receptacle strongly conic or columnar.
36. Ray flowers subtended by paleae or by the inner series of phyllaries that resemble palea.
37. Leaves simple, basally cordate-clasping .... **Dracopis**
37. Leaves pinnatifid or bipinnatifid, not basally cordate-clasping .... **Ratibida**
36. Ray flowers not subtended by paleae.
38. Paleae subtending the disk flowers spinescent, surpassing the disk flowers .... **Echinacea**
38. Paleae not spinescent, equaling or shorter than the disk flowers (short awn-pointed and a little longer than the disk flowers in *R. triloba*) .... **Rudbeckia**
35. Receptacle merely convex or low conic.
39. Receptacle epaleate, deeply alveolate with a honeycomblike surface .... **Balduina**
39. Palea not as above.
40. Fruit flattened parallel to the phyllaries, at a right angle to the radius of the capitula; phyllaries evidently in 2 series and dimorphic .... **Coreopsis**
40. Fruit either not flattened or flattened at a right angle to the phyllaries and in alignment with the radius of the capitula; phyllaries imbricate or subequal.
41. Leaves seldom more than 5 mm wide .... **Phoebanthus**
41. Leaves more than 5 mm wide.
42. Fruit strongly flattened .... **Verbesina**
42. Fruit not strongly flattened .... **Tithonia**
33. Leaves opposite or whorled, at least the lower ones.
43. Ray flower corolla persistent on the achene and becoming papery.
44. Fruits quadrangular, little if any compressed; plants perennial .... **Heliopsis**
44. Fruits radially compressed; plants annual .... **Zinnia**
43. Ray flower corolla deciduous from the achene.
45. Pappus of 20 long, plumose bristles .... **Tridax**
45. Pappus various but not as above or absent.
46. Receptacle with bristlelike paleae .... **Flaveria**
46. Receptacles with chaff-like paleae.
47. Paleae flat or nearly so, not clasping the fruit; fruit usually flattened parallel to the phyllaries at a right angle to the radius of the capitula.
48. Phyllaries imbricate.
49. Fruit distinctly dimorphic, those of the ray flowers with a conspicuous lacerate-winged margin, those of the disk flowers wingless .... **Synedrella**
49. Fruit all alike, wingless or somewhat thick-winged distally, but not lacerate .... **Calyptocarpus**
48. Phyllaries in 2 series and dimorphic.
50. Fruit beaked, not much flattened .... **Cosmos**
50. Fruit not beaked, usually strongly flattened.
51. Fruit mostly wing-margined; pappus awns (if present) smooth or antrorsely barbed .... **Coreopsis**

51. Fruit wingless or scarcely winged; pappus awns (if present) retrorsely barbed, if antrorsely barbed, then the ray flowers much reduced............ ....**Bidens**

47. Paleae folded and clasping the fruit; fruit not flattened, or if so, then radially flattened.

52. Phyllaries 4(5)... **Tetragonotheca**

52. Phyllaries more than 5.

53. Fruit (at least of the disk flowers) strongly flattened, also often with the margins winged.

54. Receptacle flat to short-conic; style branches with an acute appendage.. .... **Verbesina**

54. Receptacle elongate, conic or subcyclindic; style branches truncate, lacking an appendage.. **Acmella**

53. Fruit thick or quadrangular to subterete, not strongly flattened or with winged margins.

55. Ray flowers sterile.

56. Leaves all 0.5–7 mm wide; pappus of 1–2 persistent scales and several bristlelike shorter scales .... **Phoebanthus**

56. Leaves always some on the plant more than 7 mm wide; pappus of 2 caducous awn scales, rarely also with a few shorter scales ....**Helianthus**

55. Ray flowers fertile.

57. Shrubs; pappus a crown around the margin of the broad-topped fruit ....**Borrichia**

57. Herbs; pappus a crown at the strongly constricted center of the top of the fruit.

58. Erect herbs; leaves entire or few-toothed; peduncle less than 3 cm long... **Pascalia**

58. Creeping herbs; leaves serrate and usually 3-lobed; peduncle 3–10 cm long.. **Sphagneticola**

KEY 3

1. Capitula with trimorphic flowers (outer 1–several rows ligulate, the next row tubular, the central ones bilabiate) .... **Chaptalia**

1. Capitula radiate.

2. Receptacle lacking paleae.

3. Pappus of scales, a few short awns, a short crown, or absent.

4. Receptacle hemispheric or evidently conic .... **Boltonia**

4. Receptacle flat to merely convex or low conic.

5. Pappus of 6–10 well-developed aristate scales .... **Gaillardia**

5. Pappus various, but not as above.

6. Leaves, stems, and phyllaries densely and softly pubescent..................**Aphanostephus**

6. Leaves, stems, and phyllaries glabrous or glabrate .... **Leucanthemum**

3. Pappus of capillary bristles, sometimes also with some short outer bristles or small scales

7. Plants annual; ligule of the ray flowers to 2 mm long.

8. Fruit 2-nerved .... **Conyza**
8. Fruit 3- or more-nerved .... **Symphyotrichum**
7. Plants perennials or with the ligule of the ray flowers more than 2 mm long (sometimes both).
9. Acaulescent herbs, the scape with a solitary capitulum .... **Gerbera**
9. Caulescent herbs, the scape with 2–many capitula.
10. Phyllaries subequal to subimbricate, often green in part, but not leafy or with a chartaceous base and an herbaceous green tip .... **Erigeron**
10. Phyllaries either subequal and the outer leafy, or evidently imbricate, with a chartaceous base and an evident green tip.
11. Pappus of 2 types, the inner ones longer than the outer.
12. Leaves to 0.5 cm wide, 1-nerved; pappus with both the inner and outer ones capillary .... **Ionactis**
12. Leaves 1.5 cm wide or wider, reticulate-nerved; pappus with the inner ones clavate-tipped (sometimes only slightly so in *Oclemena*), the outer ones capillary.
13. Fruit fusiform, with nonresinous nerves, the surface covered with small sessile glands; outer pappus bristles much shorter than the inner ones .... **Oclemena**
13. Fruit compressd-obconic, with orange-resinous nerves, the outer surface glabrous or strigose, but eglandular; outer pappus bristles subequaling the inner ones .... **Doellingeria**
11. Pappus one type.
14. Leaves basally disposed, linear and grasslike; pappus bristles coarse, the larger ones somewhat flattened and apically clavate .... **Eurybia**
14. Basal leaves absent at flowering time, or if present, then not linear and grasslike, or if so, then the pappus bristles soft and fine, terete and not apically clavate.
15. Ray and disk flowers white; fruit sericeous; capitula cylindric .... **Sericocarpus**
15. Ray flowers white or purple, the disk flowers yellow or red-purple; fruit glabrous or pubescent, but not sericeous; capitula campanulate .... **Symphyotrichum**
2. Receptacle with paleae.
16. Leaves opposite (at least the basal ones).
17. Pappus of several well-developed scales or long-plumose bristles.
18. Pappus plumose; phyllaries 7–10 mm long .... **Tridax**
18. Pappus scales often somewhat fimbriate, but not plumose; phyllaries to 4 mm long .... **Galinsoga**
17. Pappus of a few awns or teeth, a short crown, or absent.
19. Disk flowers sterile .... **Polymnia**
19. Disk flowers fertile.
20. Corolla persistent on the fruit and becoming papery .... **Zinnia**
20. Corolla deciduous.
21. Ray flowers ferile, the ligule ca. 1 mm long; disk flowers white .... **Eclipta**
21. Ray flowers sterile, the ligule much more than 1 mm long; disk flowers yellow or red.

22. Fruits distinctly beaked .... **Cosmos**
22. Fruits sometimes narrowed, but not beaked .... **Bidens**
16. Leaves all alternate.
23. Ray flowers 1–5, the ligule less than 1 cm long.
24. Disk flowers sterile .... **Parthenium**
24. Disk flowers fertile.
25. Leaves pinnately dissected .... **Achillea**
25. Leaves merely toothed .... **Verbesina**
23. Ray flowers either more than 5 or the ligule more than 1 cm long, or both.
26. Pappus of 6- to 10-awned scales; receptacle epaleate, with setiform enations .... **Gaillardia**
26. Pappus not as above; receptacle with chaff-like paleae.
27. Leaves pinnately or bipinnately dissected.
28. Ray flowers brown-purple .... **Ratibida**
28. Ray flowers white or white and yellow.
29. Ray flowers white with a yellow base; leaves pinnatisect; corolla of the disk flowers saccate or spurred at the base .... **Cladanthus**
29. Ray flowers entirely white; leaves 2- to 3-pinnatisect; corolla of the disk flowers not saccate or spurred at the base .... **Anthemis**
27. Leaves entire or merely toothed.
30. Fruits strongly flattened parallel to the phyllaries .... **Coreopsis**
30. Fruits not flattened.
31. Ligule of the ray flowers 2–8 cm long .... **Echinacea**
31. Ligule of the ray flowers 1–1.5 cm long .... **Rudbeckia**

KEY 4

1. Capitula falsely radiate, the enlarged corollas of the marginal flowers irregularly 5-lobed above the tubular base.
2. Phyllaries with pectinate or lacerate tips .... **Centaurea**
2. Phyllaries without pectinate or lacerate tips .... **Stokesia**
1. Capitula not falsely radiate.
3. Leaves spine-margined.
4. Pappus of plumose bristles .... **Cirsium**
4. Pappus of unequal, capillary or somewhat chaffy-flattened, merely barbellate bristles .... **Onopordum**
3. Leaves not spine-margined.
5. Receptacle with at least some paleae .... **Carphephorus**
5. Receptacle lacking paleae (paleae present but deciduous in *Filago*).
6. Outer flowers or all flowers carpellate.
7. Pappus plumose .... **Facelis**
7. Pappus not plumose.
8. Leaves mainly or wholly basal.
9. Capitula with marginal carpellate flowers and central staminate flowers .... **Sachsia**
9. Capitula unisexual (plants dioecious) .... **Antennaria**

8. Leaves mainly cauline.
 10. Phyllaries 1–4 or absent; paleae present, but deciduous............................ **Filago**
 10. Phyllaries more than 4; paleae absent.
  11. Stem conspicuously winged with the decurrent leaf bases.
   12. Capitula in dense spikes..**Pterocaulon**
   12. Capitula in broad corymbiform cymes...**Pluchea**
  11. Stem not conspicuously winged with the decurrent leaf bases.
   13. Phyllaries scarious or hyaline.
    14. Capitula array a cyme or paniculate cluster; pappus bristles distinct, individually deciduous ....**Pseudognaphalium**
    14. Capitula array a narrow spikelike panicle; pappus bristles united in a ring and deciduous as a whole.. **Gamochaeta**
   13. Phyllaries herbaceous.
    15. Phyllaries in 1 series; calyculi often present..........................**Erechtites**
    15. Phyllaries imbricate; calyculi never present.
     16. Capitula unisexual (plants dioecious)............................. **Baccharis**
     16. Capitula with the marginal flowers carpellate, the central ones bisexual..**Pluchea**
6. Flowers all bisexual and fertile.
 17. Phyllaries in 1 series; calyculi sometimes present.
  18. Plants with conspicuous purple trichomes...**Gynura**
  18. Plants glabrous, or if with trichomes, these never purple.
   19. Flowers bright yellow...**Senecio**
   19. Flowers various-colored, but not bright yellow.
    20. Plants annual; fruits with short trichomes on the angles.
     21. Involucre with several small bracts at the base; fruits 8- to 10-ribbed...... ....**Crassocephalum**
     21. Involucre lacking small bracts at the base; fruits 5-ribbed............ **Emilia**
    20. Plants perennial; fruits glabrous.
     22. Phyllaries 10–15; flowers 20–40 .... **Hasteola**
     22. Phyllaries 5; flowers 5 .... **Arnoglossum**
 17. Phyllaries imbricate; calyculi absent.
  23. Flowers bright yellow.
   24. Capitula array corymbose...**Bigelowia**
   24. Capitula array paniculate..**Solidago**
  23. Flowers cream, white, purple, or lavender.
   25. Pappus distinctly double, the inner of long bristles, the outer of very short bristles.
    26. Phyllaries linear-subulate, less than 5 mm long; fruits terete, lacking ribs or faintly ribbed; plants annual...**Cyanthillium**
    26. Phyllaries ovate-elliptic to lanceolate, more than 5 mm long; fruits ribbed or furrowed; plants perennial .... **Vernonia**
   25. Pappus of similar bristles.
    27. Vines..**Mikania**
    27. Herbs or shrubs.
     28. Phyllaries arranged in 5 vertical ranks of 3.................................**Garberia**

28. Phyllaries not arranged in ranks.
    29. Phyllaries (at least the outer ones) evidently narrowly few-striate longitudinally .... **Brickellia**
    29. Phyllaries not striate longitudinally.
        30. Leaves all alternate.
            31. Capitula array racemose or spicate, if corymbose, then the phyllaries 1.7–2.3 cm high .... **Liatris**
            31. Capitula array corymbose or thyrsoid-paniculate.
                32. Leaves with the margin entire .... **Carphephorus**
                32. Leaves with the margin toothed (at the least the lower ones) .... **Brintonia**
        30. Leaves opposite or whorled, sometimes the distal alternate.
            33. Receptacle conspicuously conic.
                34. Phyllaries ca. 1 mm wide or wider, with conspicuous dark striations; petiole less than ¼ the blade length; capitula cylindric-campanulate, 6–8 mm long, 3–4 mm wide .... **Praxelis**
                34. Phyllaries less than 1 mm wide, lacking striations; petiole ¼–½ the blade length; capitula subglobose, 3–4 mm long, 4–5 mm wide .... **Conoclinium**
            33. Receptacle flat or merely convex.
                35. Phyllaries deciduous .... **Chromolaena**
                35. Phyllaries persistent.
                    36. Shrubs .... **Koanophyllon**
                    36. Herbs (sometimes suffrutescent).
                        37. Leaves whorled .... **Eutrochium**
                        37. Leaves opposite or alternate.
                            38. Flowers pink, purplish, or whitish with pink or lilac lobes; stems decumbent, rooting at the nodes. .... **Fleischmannia**
                            38. Flowers white; stems erect.
                                39. Phyllaries subequal .... **Ageratina**
                                39. Phyllaries unequal .... **Eupatorium**

KEY 5

1. Capitula unisexual, the carpellate ones burr-like.
    2. Phyllaries of the carpellate capitula with hooked spines .... **Xanthium**
    2. Phyllaries of the carpellate capitula tuberculate or with straight spines .... **Ambrosia**
1. Capitula with bisexual flowers or with both carpellate and staminate flowers in the same capitulum, the capitula not burr-like.
    3. Receptacle with paleae (sometimes with only a few near the margin in *Hartwrightia*).
        4. Pappus of 2 series, the outer of 10 awns, these alternating with much shorter, minutely pubescent, sparsely pectinate inner ones .... **Centaurea**
        4. Pappus various, but not as above.
            5. Phyllaries 4; paleae tightly enclosing the fruit .... **Enydra**

5. Phyllaries more than 4; paleae not enclosing the fruit.
6. Phyllaries in 2 series and evidently dimorphic.. **Bidens**
6. Phyllaries in 1–several series, but not evidently dimorphic.
7. Capitula inconspicuously radiate.. **Eclipta**
7. Capitula discoid.
8. Capitula with the central flowers staminate, the marginal ones carpellate.......... **Iva**
8. Capitula with the flowers all bisexual.
9. Flowers 7–10 in each capitulum .... **Hartwrightia**
9. Flowers more than 20 in each capitulum.
10. Pappus of 5 short scales.. **Marshallia**
10. Pappus of 1-several awns, awn-scales, bristles, or absent.
11. Fruits strongly flattened, often wing-margined .... **Verbesina**
11. Fruits not strongly flattened or wing-margined.
12. Plants scapose, the basal leaves well developed; corolla deep purple.. .... **Helianthus**
12. Plants with well-developed cauline leaves, the basal leaves few or absent at flowering; corolla white or yellow to orange-yellow.
13. Receptacle convex; corolla white .... **Melanthera**
13. Receptacle conic to subcylindric; corolla yellow to orange............. .... **Acmella**
3. Receptacle lacking paleae.
14. Leaves, or most of them, opposite or whorled.
15. Leaves whorled.. **Sclerolepis**
15. Leaves opposite.
16. Capitula 1-flowered, aggregated into axillary or terminal glomerules................. **Lagascea**
16. Heads many-flowered, in a corymbose array... **Ageratum**
14. Leaves, or most of them, alternate or all basal.
17. Leaves entire or merely toothed or lobed.
18. Capitula 2- to 5-flowered, aggregated into dense clusters subtended by several foliaceous bracts.
19. Capitula array corymbose-paniculate; pappus bristles all uniform, straight .... **Elephantopus**
19. Capitula array spicate or racemose-spicate; pappus bristles of 2 types, the 2 lateral ones longer and with an S-shaped bend at the end, the other ones straight .... **Pseudelephantopus**
18. Capitula many-flowered, or if few-flowered, then not aggregated into dense clusters subtended by several foliaceous bracts.
20. Capitula in a corymbiform or paniculifom array; corolla white or pinkish to pale lavender.. **Palafoxia**
20. Capitula solitary or occasionally of 2–3 in a cluster; corolla reddish purple to blue... .... **Centratherum**
17. Leaves, or most of them, pinnatifid or pinnately dissected.
21. Pappus of evident scales.
22. Phyllaries erect-appressed, the inner ones wide, obtuse, scarious, whitish or yellowish petaloid distally.. **Hymenopappus**
22. Phyllaries spreading, not petaloid.. **Gaillardia**

21. Pappus obsolete.
   23. Capitula solitary on a long peduncle or racemose-paniculate; fruits not spine-tipped with the persistent style .... **Artemesia**
   23. Capitula sessile in the forks of dichotomous branches; fruits spine-tipped with the persistent style .... **Soliva**

Acanthospermum Schrank, nom. cons. 1820. STARBURR

Herbs. Leaves opposite, simple, pinnate-veined, petiolate or epetiolate. Flowers in terminal capitula (appearing axillary by sympodial growth), radiate; receptacle paleate; ray flowers carpellate; petals 5, basally connate, zygomorphic; ovary inferior, 2-carpellate, 1-loculate; disk flowers staminate; petals 5, basally connate, actinomorphic; stamens 5, epipetalous, the filaments free, the anthers connate. Fruit a cypsela; pappus absent.

A genus of 6 species; nearly cosmopolitan. [From the Greek *acantha,* spine, and *sperma,* seed, in reference to the spiny fruit.]

Selected references: Blake (1921); Strother (2006a).

1. Fruits with hooked prickles only .... **A. australe**
1. Fruits with hooked prickles below and 2 large spines at the apex.
   2. Leaf blade abruptly contracted to a definite, usually winged petiole .... **A. humile**
   2. Leaf blade gradually narrowed to a sessile or short petiolate base .... **A. hispidum**

Acanthospermum australe (Loefl. ex L.) Kuntze [Southern]. PARAGUAY STARBURR.

Melampodium australe Loefling ex Linnaeus, Sp. Pl., ed. 2. 1303. 1763. *Centrospermum xanthioides* Kunth, in Humboldt et al., Nov. Gen. Sp. 4: 270. 1820, nom. illegit. *Acanthospermum xanthioides* de Candolle, Prodr. 5: 521. 1836, nom. illegit. *Acanthospermum xanthioides* de Candolle var. *obtusifolium* de Candolle, Prodr. 5: 522. 1836, nom. inadmiss. *Acanthospermum australe* (Loefling ex Linnaeus) Kuntze, Revis. Gen. Pl. 1: 303. 1891.

Procumbent annual herb; stem to 6(15) dm long, striate, appressed pubescent. Leaves with the blade rhombic-ovate, 1.5–3.5 cm long, 0.8–3 cm wide, the apex acute or rounded, the base cuneate, the margin irregularly dentate-serrate, the upper and lower surfaces sparsely appressed pilosulous, gland-dotted, the petiole 3–15 mm long. Flowers in a solitary terminal capitulum (appearing axillary by sympodial growth), the peduncle to 1.5 cm long; involucre hemispheric; phyllaries in 2 series, the inner investing the fruit, enlarging, and enclosing it, the outer ovate-elliptic, 3–4 mm long, the apex obtuse, sparsely pilose; receptacle convex; paleae 2–3 mm long, scarious, glandular and ciliate, fimbriate-ciliate at the deeply emarginate apex; ray flowers 5–8; corolla yellow, the lamina ca. 1 mm long, stipitate-glandular; disk flowers ca. 12; corolla yellowish, 2–3 mm long, pilose proximally, glandular distally. Cypsela obliquely ellipsoid-fusiform, 7–9 mm long, slightly compressed, 5- to 7-ribbed, densely glandular, lacking terminal spines, the ribs with 1–2 rows of uncinate prickles to 2 mm long; pappus absent.

Disturbed sites. Frequent; northern counties, central peninsula. Massachusetts and New York south to Florida, west to Texas, also Oregon; nearly cosmopolitan. Native to South America. Spring–fall.

Acanthospermum hispidum DC. [Hispid.] HISPID STARBURR.

Acanthospermum hispidum de Candolle, Prodr. 5: 522. 1836. *Acanthospermum humile* (Swartz) de Candolle var. *hispidum* (de Candolle) Kuntze, Revis. Gen. Pl. 1: 303. 1891.

Erect annual herb, to 5 dm; stem striate, hispid-pilose. Leaves with the blade elliptic- or deltate-ovate, 2–12 cm long, 0.8–8 cm wide, the apex acute or obtuse, mucronulate, the base cuneate, the margin serrulate or subentire, the upper and lower surfaces hispid-pilose, the lower surface gland-dotted, short-petiolate or sessile. Flowers in a solitary terminal capitulum (appearing axillary by sympodial growth), the peduncle 3–15 mm long; involucre hemispheric; phyllaries in 2 series, the inner investing the fruit, enlarging, and enclosing it, the outer ovate or elliptic-ovate, 3–4 mm long, the apex acute, hispid-pilose mainly on the margin; receptacle convex; paleae 2 mm long, stipitate-glandular, lacerate-ciliate at the subtruncate apex; ray flowers 5–8; corolla pale yellow, the lamina 1–2 mm long, sparsely hispid pilose; disk flowers ca. 7; corolla pale yellow, ca. 2 mm long, sparsely hispid-pilose. Cypsela obliquely cuneate, strongly compressed, 4–5 mm long, 5-ribbed, gland-dotted, with small uncinate-hispid prickles over the body, with 2 terminal straight or slightly curved, strongly divergent, terminal spines 3–4 mm long; pappus absent.

Disturbed sites. Occasional; northern counties, central peninsula. New York, Ontario, New Jersey, Virginia, South Carolina south to Florida, west to Alabama, also Oregon; nearly cosmopolitan. Native to South America. Spring–fall.

Acanthospermum humile (Sw.) DC. [Low, in reference to the habit.] LOW STARBURR.

Melampodium humile Swartz, Prodr. 114. 1788. *Acanthospermum humile* (Swartz) de Candolle, Prodr. 5: 522. 1836. *Acanthospermum humile* (Swartz) de Candolle var. *normale* Kuntze, Revis. Gen. Pl. 1: 303. 1891, nom. inadmiss.

Erect or decumbent annual herb, to 3 dm; stem striate, puberulent and hispid-pilose. Leaves with the blade ovate or deltoid-ovate, 1–2.8 cm long, 1–3.3 cm wide, the apex acute to obtuse, the base abruptly narrowed, the margin crenate-dentate or repand-serrate, the upper and lower surfaces hispid-pilose, particularly along the veins, gland-dotted, the petiole 0.4–1.8 cm long, winged, the wing irregularly serrate, lobate, or entire. Flowers in a solitary terminal capitulum (appearing axillary by sympodial growth), sessile or the peduncle to 3 mm long; involucre hemispheric; phyllaries in 2 series, the inner investing the fruit, enlarging, and enclosing it, the outer ovate, the apex acute, hispid-pilose mainly on the margin; the receptacle convex; paleae 1–2 mm long, lacerate or emarginate; ray flowers 5–7; corolla pale yellow, the lamina ca. 1 mm long, hispid-pilose; disk flowers 5; corolla pale yellow, 1–2 mm long, stipitate-glandular and sparsely hispid-pilose. Cypsela obliquely trigonous, ca. 3 mm long, compressed, gland-dotted and pilosulous, uncinate-prickly on the ribs and apical margin, the sides with a few sparse prickles, with 2 large terminal spines (1 uncinate and 1 straight) 2–3 mm long; pappus absent.

Disturbed sites. Rare; Escambia County. Not recently collected. New York, South Carolina, Alabama, and Florida; West Indies, Central America, and South America. Native to West Indies. All year.

Achillea L. 1753. YARROW

Herbs. Leaves alternate, simple, pinnate-veined, petiolate or eptiolate. Flowers in capitula in terminal simple or compound corymbiform arrays, radiate; receptacle paleate; ray flowers carpellate; petals 5, basally connate, zygomorphic; disk flowers bisexual; petals 5, basally connate, actinomorphic; stamens 5, epipetalous, the filaments free, the anthers connate; ovary inferior, 2-carpellate, 1-loculate. Fruit a cypsela; pappus absent.

A genus of about 115 species; nearly cosmopolitan. [Named for the Greek god Achilles, who is said to have used the plants to treat his wounds.]

Selected reference: Trock (2006a).

Achillea millefolium L. [*Mille,* thousand, thousand-leaved.] COMMON YARROW.

Achillea millefolium Linnaeus, Sp. Pl. 899. 1753. *Alitubus millefolium* (Linnaeus) Dulac, Fl. Hautes-Pyrénées 500. 1867. *Chamaemelum millefolium* (Linnaeus) E.H.L. Krause, in Sturm, Deutschl. Fl., ed. 2. 13: 216. 1905.

Achillea millefolium Linnaeus var. *occidentalis* de Candolle, Prodr. 6: 24. 1838. *Achillea occidentalis* (de Candolle) Rafinesque ex Rydberg, Bull. Torrey Bot. Club 37: 456. 1910. *Achillea millefolium* Linnaeus subsp. *occidentalis* (de Candolle) Hylander, Uppsala Univ. Arsskr. 7: 314. 1945.

Erect perennial herb, to 7.5 dm; stem striate, densely lanate to glabrate. Leaves with the blade oblong to lanceolate, 3.5–35 cm long, 0.5–3.5 cm wide, 1- to 2-pinnately dissected, the ultimate lobes lanceolate, the upper and lower surfaces lanate, tomentose, or glabrate, the lower ones short-petiolate, the middle and upper ones sessile. Flowers in 10–many capitula in a terminal or axillary simple or compound corymbiform array, the peduncle 1–4 mm long; involucre campanulate; phyllaries 20–30, in 3 series, ovate to lanceolate, green to yellowish, the outer surface tomentose, gland-dotted; receptacle convex; paleae lanceolate, 2–4 mm long; ray flowers (3)5–8; corolla white or light pink to purple, the lamina broadly elliptic, 2–3 mm long and wide; disk flowers 10–20; corolla white or grayish white, 2–5 mm long, gland-dotted. Cypsela elliptical, 1–2 mm long, laterally flattened, the lateral margins winged; pappus absent.

Disturbed sites. Rare; northern counties, Volusia County. Throughout North America; Mexico, Europe, Asia, Australia, and Pacific Islands. Native to North America, Mexico, Europe, and Asia. Spring.

Acmella Rich. ex pers. 1807. SPOTFLOWER

Herbs. Leaves opposite, simple, pinnate-veined, petiolate. Flowers in terminal capitula, radiate or discoid; receptacle paleate; ray flowers carpellate; petals 5, basally connate, zygomorphic;

disk flowers bisexual; petals 5, basally connate, actinomorphic; stamens 5, epipetalous, the filaments free, the anthers connate; ovary inferior, 2-carpellate, 1-loculate. Fruit a cypsela; pappus absent.

A genus of about 30 species; nearly cosmopolitan. [Sinhalese (northern India) vernacular name for *Blainvillea acmella.*]

Selected references: Jansen (1985); Strother (2006b).

1. Leaves truncate or cordate at the base, the petiole not winged .... **A. pilosa**
1. Leaves attenuate at the base, the petiole winged.
   2. Leaves lance-ovate .... **A. oppositifolia**
   2. Leaves linear or narrowly lanceolate .... **A. pusilla**

Acmella oppositifolia (Lam.) R. K. Jansen var. **repens** (Walter) R. K. Jansen [With opposite leaves; creeping or prostrate.] OPPOSITELEAF SPOTFLOWER.

Anthemis repens Walter, Fl. Carol. 211. 1788. *Spilanthes repens* (Walter) Michaux, Fl. Bor.-Amer. 2: 131. 1803. *Acmella repens* (Walter) Persoon, Syn. Pl. 2: 473. 1807. *Ceratocephalus repens* (Walter) Kuntze, Revis. Gen. Pl. 326. 1891. *Spilanthes americana* Hieronymus var. *repens* (Walter) A. H. Moore, Proc. Amer. Acad. Arts 42: 547. 1907. *Acmella oppositifolia* (Lamarck) R. K. Jansen var. *repens* (Walter) R. K. Jansen, Syst. Bot. Monogr. 8: 34. 1985.

Acmella occidentalis Nuttall, Gen. N. Amer. Pl. 2: 171. 1818; non (Willdenow) Richard, 1807. *Acmella nuttalliana* Rafinesque, New Fl. 1: 52. 1836. *Spilanthes nuttallii* Torrey & A. Gray, Fl. N. Amer. 2: 356. 1842, nom. illegit.

Decumbent or repent perennial herb; stem 2(3) dm long, rooting at the nodes, striate, pilose or glabrate. Leaves with the blade lance-ovate, 0.8–9.5 cm long, 0.5–3 cm wide, the apex acuminate or rounded, the base cuneate, the margin denticulate or coarsely dentate, sometimes sparsely ciliate, the upper and lower surfaces pilose, strigose, or glabrous, the petiole 0.3–4 cm long, narrowly winged. Flowers in a solitary terminal capitulum, the peduncle 5–10 cm long; involucre hemispheric; phyllaries 10–16 in 2 series, lanceolate, the apex acuminate, the margin entire or sinuate, sometimes irregularly dentate, ciliate, 3–7 mm long, the outer series spreading and slightly longer than the inner; receptacle conic; paleae lanceolate, 3–5 mm long, stramineous, loosely enclosing the fruit; ray flowers 5–19; corolla orange-yellow, the lamina elliptic, 3–7 mm long, sparsely pilose; disk flowers numerous; corolla orange-yellow, 2–3 mm long, glabrous. Cypsela elliptic, compressed, 1–2 mm long, ciliate with straight-tipped trichomes; pappus absent.

Floodplain forests. Common; nearly throughout. North Carolina south to Florida, west to Oklahoma and Texas; Mexico. All year.

Acmella pilosa (Hook. & Arn.) R. K. Jansen [With pilose trichomes.] HAIRY SPOTFLOWER.

Acmella pilosa R. K. Jansen, Syst. Bot. Monogr. 8: 27. 1985.

Decumbent perennial herb; stem 2–3 dm long, rooting at the nodes, striate, pilose. Leaves with the blade ovate or deltate, the apex acute, the base truncate or cordate, the margin crenate or coarsely dentate, the upper and lower surfaces pilose, the petiole 0.5–1.5 cm long. Flowers in a solitary terminal capitulum, the peduncle 6–18 cm long; involucre hemispheric; phyllaries 15–20 in 3 series, the outer series 5, ovate to elliptic, 4–5 mm long, the apex rounded or acuminate, the margin entire or sinuate, ciliate, the middle series 5, ovate or elliptic, 4–6 mm long, the apex acute or acuminate, the margin entire or sinuate, ciliate, the inner series 5–11, lanceolate or narrowly ovate, 3–6 mm long, the apex acute or acuminate, the margin entire or sinuate, ciliate; receptacle conic; palea lanceolate, 3–4 mm long, stramineous, loosely enclosing the fruit; ray flowers 9–20; corolla yellow, the lamina 3–6 mm long, moderate to densely pilose; disk flowers numerous; corolla yellow, ca. 2 mm long, glabrous. Cypsela elliptic, compressed, 1–2 mm long, densely ciliate with straight-tipped trichomes; pappus absent.

Wet disturbed sites. Rare; Miami-Dade County. Florida; Mexico and Central America. Native to Mexico and Central America. All year.

Acmella pusilla (Hook. & Arn.) R. K. Jansen [Small, in reference to the size of the plant.] DWARF SPOTFLOWER.

Spilanthes pusilla Hooker & Arnott, J. Bot. (Hooker) 3: 317. 1841. *Spilanthes stolonifera* de Candolle var. *pusilla* (Hooker & Arnott) Baker, in Martius, Fl. Bras. 6: 235. 1884. *Acmella pusilla* (Hooker & Arnott) R. K. Jansen, Syst. Bot. Monogr. 8: 46. 1985.

Spilanthes americana Hieronymus forma *longiinternodiata* A. H. Moore, Proc. Amer. Acad. Arts 42: 549. 1907. TYPE: FLORIDA: Carrabelle, 9 Jun 1897, *Curtiss 5882* (holotype: GH; isotypes: E, F, G, HBG, MO, NY, UC).

Repent perennial herb; stem 2(3) dm long, rooting at the nodes, striate, sparsely to moderately pilose. Leaves with the blade linear or narrowly lanceolate, 1.2–4 cm long, 3–10 mm wide, the apex acute, the base cuneate, the margin sinuate-dentate or rarely entire, sparsely ciliate, the upper and lower surfaces glabrous, the petiole 2–4 mm long, narrowly winged. Flowers in a solitary terminal capitulum, the peduncle 3–10 mm long; involucre hemispheric; phyllaries 9–12 in 2 series, the outer series 5–6, narrowly or broadly ovate, 3–4 mm long, the apex acute, the margin sinuate, ciliate, the inner series 4–6, narrowly ovate, the margin sinuate or irregularly dentate, 3–4 mm long, sparsely to moderately ciliate; receptacle conic; paleae lanceolate, 3–5 mm long, stramineous; Ray flowers 6–13; corolla yellow to yellow-orange, the lamina elliptic, 3–6 mm long, pilose; disk flowers numerous; corolla yellow to yellow-orange, 2–3 mm long, glabrous. Cypsela elliptic, compressed, 1–2 mm long, moderately ciliate with recurve-tipped trichomes; pappus absent.

Disturbed sites. Rare; Santa Rosa, Franklin, and Escambia Counties. Escaped from cultivation. North Carolina south to Florida; West Indies and South America. Native to South America. Spring.

EXCLUDED TAXA

Acmella oppositifolia (Lamarck) R. K. Jansen—The typical variety was reported for Florida by Small (1933), Long and Lakela (1971), Cronquist (1980), Godfrey and Wooten (1981), and Wunderlin (1982), all as *Spilanthes americana* Linnaeus f. All Florida material is var. *repens*.

Ageratina Spach 1841. SNAKEROOT

Herbs. Leaves opposite, simple, pinnate-veined, petiolate. Flowers in capitula in terminal and axillary corymbiform arrays, discoid, bisexual; receptacle epaleate; petals 5, basally connate, actinomorphic; stamens 5, epipetalous, the filaments free, the anthers connate; ovary inferior, 2-carpellate, 1-loculate. Fruit a cypsela; pappus present.

A genus of about 250 species; North America, Mexico, Central America, and South America. [From the genus *Ageratum,* and *ina,* diminutive.]

Selected references: Clewell and Wooten (1971); Nesom (2006a).

1. Leaf blades membranaceous, the margin serrate or dentate .... **A. altissima**
1. Leaf blades chartaceous or subcoriaceous, the margin crenate or crenate-serrate.
    2. Larger leaf blades at least 5 times longer than the petiole; corolla lobes densely long-pubescent; fruit glabrous .... **A. aromatica**
    2. Larger leaf blades 1–4 times longer than the petiole; corolla lobes sparsely short-pubescent or glabrous; fruit usually short pubescent, at least at the apex .... **A. jucunda**

Ageratina altissima (L.) R. M. King & H. Rob. [Tallest.] WHITE SNAKEROOT.

Ageratum altissimum Linnaeus, Sp. Pl. 839. 1753. *Eupatorium altissimum* (Linnaeus) Linnaeus, Syst. Veg., ed. 13. 614. 1774; non Linnaeus, 1753. *Eupatorium rugosum* Houttuyn, Nat. Hist. 2(10): 558. 1779. *Eupatorium urticifolium* Reichard, Syst. Pl. 3: 719. 1780, nom. illegit. *Eupatorium ageratoides* Linnaeus f., Suppl. Pl. 355. 1782 ("1781"), nom. illegit. *Batschia nivea* Moench, Methodus 567. 1794, nom. illegit. *Kyrstenia altissima* (Linnaeus) Greene, Leafl. Bot. Observ. Crit. 1: 8. 1903. *Ageratina altissima* (L.) R. M. King & H. Robinson, Phytologia 19: 212. 1970.

Erect or ascending perennial herb, to 8(12) dm; stem sometimes basally rhizomatous, striate, puberulent. Leaves with the blade deltoid ovate, ovate, or broadly lanceolate, 4–10(13) cm long, 2.5–8(9) cm wide, membranaceous, the apex acute or acuminate, the base rounded, truncate, or cordate, the margin coarsely, often doubly serrate, or dentate, the upper and lower surfaces glabrous or puberulent, the petiole (0.5)1–3(5) mm long. Flowers in capitula in a corymbiform array, discoid, bisexual, the peduncle 1–5 mm long; involucre hemispheric; phyllaries 8–15 in 3 series, the outer series absent or of 1 short phyllary, the 2 inner series of 8–14 phyllaries, 3–7 mm long, subequal, oblong, spatulate, or linear, the apex acute or obtuse, ciliate or fimbriate, the margin often scarious, the outer surface glabrous or pubescent; receptacle convex; flowers 9–34; corolla white, 3–4 mm long, pubescent. Cypsela prismatic, 2–3 mm long, 5-ribbed, glabrous; pappus of numerous barbellate bristles in 1 series.

Open, calcareous hammocks and floodplain forests. Rare; Jefferson County, central panhandle. Quebec south to Florida, west to North Dakota, South Dakota, Nebraska, Kansas, Oklahoma, and Texas, also Saskatchewan and Northwest Territories. Summer–fall.

Ageratina altissima possibly hybridizes with *A. aromatica* in Florida where the ranges of the two species overlap (Clewell and Wooten, 1971).

Ageratina aromatica (L.) Spach [Fragrant.] LESSER SNAKEROOT.

Eupatorium aromaticum Linnaeus, Sp. Pl. 839. 1753. *Ageratina aromatica* (Linnaeus) Spach, Hist. Nat. Vég. 10: 286. 1841. *Kyrstenia aromatica* (Linnaeus) Greene, Leafl. Bot. Observ. Crit. 1: 8. 1903. *Eupatorium tracyi* Greene, Pittonia 4: 278. 1901. *Kyrstenia tracyi* (Greene) Greene, Leafl. Bot. Observ. Crit. 1: 8. 1903. TYPE: "Florida to Louisiana."

Erect perennial herb, to 8(10) dm; stem striate, villous-puberulent. Leaves with the blade deltoid, ovate, or lanceolate, 2–7(9) cm long, 1.5–4 cm wide, chartaceous or subcoriaceous, the apex acute or obtuse, the base rounded, truncate, or subcordate, the margin crenate-serrate, the upper and lower surfaces minutely pilose, the petiole 1–8(12) mm long. Flowers in terminal or axillary capitula in a corymbiform array, discoid, bisexual, the peduncle 2–9 mm long; involucre hemispheric; phyllaries 8–14 in 3 series, the outer series of 1–2 short phyllaries or absent, the 2 inner series of 8–12 phyllaries, 3–5 mm long, subequal, oblong or spatulate, the apex acute or obtuse, sometimes fimbriate, the margin often scarious, sometimes ciliate, the outer surface pubescent; receptacle convex; flowers 10–19; corolla white, 3–4 mm long, densely long-pubescent. Cypsela prismatic, 2–3 mm long, 5-ribbed, glabrous or sparsely pubescent; pappus of numerous barbellate bristles in 1 series.

Sandhills, flatwoods, and hammocks. Occasional; northern counties. New York and Massachusetts, south to Florida, west to Ohio, Kentucky, Tennessee, Mississippi, and Louisiana. Fall.

Ageratina aromatica possibly hybridizes with *A. altissima* or with *A. jucunda* in Florida where the ranges of the species overlap (Clewell and Wooten, 1971).

Ageratina jucunda (Greene) Clewell & Wooten [Pleasing.] HAMMOCK SNAKEROOT.

Eupatorium suaveolens Chapman, Bot. Gaz. 3: 5. 1878; non Kunth, 1818. *Eupatorium aromaticum* Linnaeus var. *incisum* A. Gray, Syn. Fl. N. Amer. 1(2): 101. 1884. *Eupatorium incisum* (A. Gray) Chapman, Fl. South. U.S., ed. 3. 216. 1897; non Richard, 1792; nec Grisebach, 1866. *Eupatorium jucundum* Greene, Pittonia 3: 180. 1897. *Kyrstenia jucunda* (Greene) Greene, Leafl. Bot. Observ. Crit. 1: 8. 1903. *Ageratina jucunda* (Greene) Clewell & Wooten, Brittonia 23: 142. 1971. *Ageratina aromatica* (Linnaeus) Spach var. *incisa* (A. Gray) C. F. Reed, Phytologia 63: 412. 1987. TYPE: FLORIDA: Manatee/Pinellas Co.

Erect perennial herb, to 8(10) dm; stem striate, minutely pilose. Leaves with the blade deltate or rhombic, 2–6(7) cm long, 1.5–4 cm wide, chartaceous or subcoriaceous, the apex acute or acuminate, the base cuneate, truncate, or subcordate, the margin crenate-serrate, the upper surface glabrous, the lower surface glabrous or sparsely pubescent on the veins, the petiole 7–15(22) mm long. Flowers in terminal or axillary capitula in a corymbiform array, discoid, bisexual, the peduncle 2–10 mm long; involucre hemispheric; phyllaries 8–12(14) in 3 series, the outer series absent or of 1 short phyllary, the 2 inner series of 8–11(13) phyllaries, 2–4 mm long, subequal, oblong, the apex acute to obtuse, sometimes fimbriate, the margin often scarious, sometimes fimbriate, the outer surface minutely pubescent; receptacle convex; flowers 7–18; corolla white, 3–4 mm long, sparsely pubescent or glabrous. Cypsela prismatic, 2–3 mm long, 5-ribbed, glabrous or sparsely hirtellous on the distal ⅓; pappus of numerous barbellate bristles in 1 series.

Sandhills and hammocks. Frequent; peninsula west to central panhandle. Georgia and Florida. Summer–fall.

Ageratina jucunda possibly hybridizes with *A. aromatica* in Florida where the ranges of the two species overlap (Clewell and Wooten, 1971).

Ageratum L. 1753. WHITEWEED

Herbs. Leaves opposite, simple, pinnate-veined, petiolate. Flowers in capitula in terminal and axillary cymiform or corymbiform arrays, discoid, bisexual; receptacle epaleate; petals 5, basally connate, actinomorphic; stamens 5, epipetalous, the filaments free, the anthers connate; ovary inferior, 2-carpellate, 1-loculate. Fruit a cypsela; pappus present.

Selected references: Johnson (1971); Nesom (2006b).

1. Leaves and stems glabrous (stems puberulous-pilose at the nodes) .... **A. maritimum**
1. Leaves and stems viscid pilose or stipitate-glandular.
    2. Phyllaries densely pilose, stipitate-glandular .... **A. houstonianum**
    2. Phyllaries sparsely pilose or glabrate, eglandular .... **A. conyzoides**

Ageratum conyzoides L. [Resembling the genus *Conyza.*] TROPICAL WHITEWEED.

Ageratum conyzoides Linnaeus, Sp. Pl. 839. 1753, nom. cons. *Ageratum hirtum* Lamarck, Encycl. 1: 54. 1783, nom. illegit. *Ageratum humile* Salisbury, Prodr. Stirp. Chap. Allerton 188. 1796, nom. illegit. *Ageratum odoratum* Vilmorin, Fl. Pleine Terre, ed. 2. 42. 1866, nom. illegit. *Carelia conyzoides* (Linnaeus) Kuntze, Revis. Gen. Pl. 345. 1891. *Eupatorium conyzoides* (Linnaeus) E.H.L. Krause, in Sturm, Deutschl. Fl., ed. 2. 13: 32. 1905; non Miller, 1768; nec Vahl, 1794.

Erect annual or perennial herb or subshrub, to 1.5 m; stem striate, sparsely to densely villous. Leaves with the blade ovate or elliptic-oblong, 2–8 cm long, 1–5 cm wide, the apex acute or rounded, the base broadly cuneate, the margin crenate, the upper surface sparsely pilose, especially on the veins, or glabrate, the lower surface sparsely pilose and gland-dotted, the petiole (0.5)1–3.5(7) cm long, white pilose. Flowers in capitula in a terminal or axillary cymiform or corymbiform array, the peduncle 0.5–1.5 cm long, pilose, eglandular; involucre campanulate; phyllaries 30–40 in 2–3 series, oblong-lanceolate 3–4 mm long, the apex abruptly tapering to a subulate tip to 1 mm long, the outer surface glabrous or sparsely pilose, eglandular, the margin often ciliate; receptacle conic; disc flowers numerous; corolla blue to lavender, sometimes white, 2–3 mm long, glabrous or sparsely pubescent distally. Cypsela prismatic, 1–2 mm long, 5-ribbed, black, sparsely strigose-hispidulous along the ribs; pappus of 5–6 aristate, basally free, apically setiferous scales 0.5–2(3) mm long.

Disturbed sites. Rare; St. Lucie, Broward, and Collier Counties. Escaped from cultivation. Connecticut, Maryland, Kentucky, North Carolina, Missouri, Georgia south to Florida, west to Louisiana, also California; West Indies, Mexico, Central America, and South America; Africa, Asia, Australia, and Pacific Islands. Native to South America. All year.

Ageratum houstonianum Mill. [Commemorates William Houston (1695–1733), Scottish surgeon and botanist who collected in the West Indies, Mexico, and South America.] BLUEMINK.

Ageratum houstonianum Miller, Gard. Dict., ed. 8. 1768. *Carelia houstoniana* (Miller) Kuntze, Revis. Gen. Pl. 1: 325. 1891. *Ageratum mexicanum* Sims forma *houstonianum* (Miller) Voss, Vilm. Blumengärtn., ed. 3. 1: 445. 1894. *Ageratum houstonianum* Miller forma *normale* B. L. Robinson, Contr. Gray Herb. 68: 5. 1923, nom. inadmiss. *Ageratum houstonianum* Miller var. *typicum* B. L. Robinson, Contr. Gray Herb. 68: 5: 1923, nom. inadmiss.

Erect to decumbent annual herb, to 8 dm; stem striate, sparsely to densely pilose. Leaves with the blade deltate or ovate, 3–8 cm long, 2.5–4 cm wide, the apex acute, obtuse, or rounded, the base cordate or truncate, the margin crenate, the upper and lower surfaces viscid-pilose to stipitate-glandular, the petiole 0.6–3.5 cm long, densely white-pilose. Flowers in capitula in a terminal or axillary cymiform or corymbiform array, the peduncle 0.5–1 cm long, viscid-puberulent, pilose, and stipitate-glandular, rarely glabrous; involucre campanulate; phyllaries 30–40 in 2–3 series, narrowly lanceolate, 1–2 mm long, the apex gradually tapering, indurate-subulate, the outer surface stipitate-glandular, sparsely to densely pilose, rarely glabrous, ciliate or eciliate; receptacle conic; disk flowers numerous; corolla lavender, rarely white, 2–3 mm long, the outer surface puberulent. Cypsela prismatic, 1–2 mm long, 5-ribbed, black, sparsely strigose-hispidulous along the ribs; pappus of 5 oblong, basally free, apically setiferous scales, the setae scabrous.

Wet, disturbed sites. Occasional; central and southern peninsula. Escaped from cultivation. Massachusetts and Connecticut, North Carolina south to Florida, west to Alabama, also Texas; Mexico and Central America; Pacific Islands. Native to Mexico and Central America. All year.

Ageratum maritimum Kunth [Growing by the sea.] CAPE SABLE WHITEWEED.

Ageratum maritimum Kunth, in Humboldt et al., Nov. Gen. Sp. 4: 150. 1820. *Carelia maritima* (Kunth) Kuntze, Revis. Gen. Pl. 1: 325. 1891.

Coelestina maritima Torrey & A. Gray, Fl. N. Amer. 2: 63. 1841. *Ageratum littorale* A. Gray, Proc. Amer. Acad. Arts 16: 78. 1881. *Carelia littorale* (A. Gray) Kuntze, Revis. Gen. Pl. 1: 325. 1891. TYPE: FLORIDA: Monroe Co.: Key West, s.d., *Bennett s.n.* (lectotype: GH). Lectotypified by Johnson (1971: 70).

Ageratum littorale A. Gray forma *setigerum* B. L. Robinson, Proc. Amer. Acad. Arts. 49: 468. 1913.

Ageratum littorale A. Gray forma *album* Moldenke, Amer. Midl. Naturalist 32: 562. 1944. TYPE: FLORIDA: Big Pine Key, 20 Mar 1930, *Moldenke 817a* (holotype: NY).

Decumbent or rarely erect annual or perennial herb, to 5 dm; stem rhizomatous, rooting at the nodes, striate, glabrous, puberulous-pilose at the nodes. Leaves with the blade deltate-ovate or oblong, 0.8–4 cm long, 0.5–3 cm wide, fleshy, the apex acute, the base broadly cuneate or truncate, the margin crenate, the upper and lower surfaces glabrous; the petiole 0.3–1.5(2.1) cm long, with long white trichomes along the margin. Flowers in capitula in a terminal or axillary cymbiform or corymbiform array, the peduncle 0.5–4 cm long, glabrous or glabrate; involucre campanulate; phyllaries 30–40 in 2–3 series, elliptic-lanceolate, the apex abruptly tapered or subobtuse, the outer surface glabrous or glabrate; receptacle conic; disk flowers numerous; corolla lavender, blue, or white, 2–3 mm long, glabrous or puberulent. Cypsela prismatic, 1–2 mm long, 5-ribbed, dark brown, glabrous; pappus coroniform, ca. 0.5 mm long, undulate, laciniate, or pectinate, of 1–2 setae, rarely of separate scales.

Hammocks and dunes. Rare; Monroe County keys. Florida; West Indies; Mexico, and Central America. All year.

Ageratum maritimum (as *A. littorale*) is listed as endangered in Florida (Florida Administative Code, Chapter 5B-40).

Ambrosia L. 1753. RAGWEED

Herbs or shrubs. Leaves opposite or alternate, simple, pinnate- or palmate-veined, petiolate or epetiolate. Flowers in capitula in racemiform or spiciform arrays, discoid, unisexual; receptacle epaleate; petals absent; stamens 5, the filaments connate, the anthers free; ovary inferior, 2-carpellate, 1-loculate. Fruit a cypsela; pappus absent.

A genus of about 40 species; nearly cosmopolitan. [From the Greek myths, the food or drink of the Greek gods, depicted as conferring longevity or immortality on whoever consumed it. Its application to certain plants is apparently related to the healing quality of the species.]

Selected reference: Strother (2006c).

1. Plant creeping, densely hispid; lower leaves 3-pinnatifid .... **A. hispida**
1. Plant erect, short scabrous or canescent; lower leaves palmately 3- to 5-lobed, pinnatifid, or bipinnatifid.
    2. Leaves palmately 3- to 5-lobed .... **A. trifida**
    2. Leaves pinnatifid or bipinnatifid.
        3. Leaves distinctly petiolate, the upper primarily alternate, the lower opposite, the lower ones bipinnatifid; fruiting involucres with 5–7 spines .... **A. artemisiifolia**
        3. Leaves sessile or subsessile, all opposite, the lower and upper ones pinnatifid; fruiting involucres unarmed or with blunt tubercules .... **A. psilostachya**

Ambrosia artemisiifolia L. [With leaves like *Artemisia*.] COMMON RAGWEED.

Ambrosia artemisiifolia Linnaeus, Sp. Pl. 988. 1753. *Ambrosia elatior* Linnaeus var. *artemisiifolia* (Linnaeus) Farwell, Rep. (Annual) Michigan Acad. Sci. 15: 190. 1913.

Ambrosia elatior Linnaeus, Sp. Pl. 987. 1753. *Ambrosia elata* Salisbury, Prodr. Stirp. Chap. Allerton 175. 1796, nom. illegit. *Ambrosia artemisiifolia* Linnaeus var. *elatior* (Linnaeus) Descourtilz, Fl. Med. Antilles 1: 329. 1821.

Iva monophylla Walter, Fl. Carol. 232. 1788. *Ambrosia paniculata* Michaux, Fl. Bor.-Amer. 2: 183. 1803, nom. illegit. *Ambrosia artemisiifolia* Linnaeus var. *paniculata* Blankinship, Rep. (Annual) Missouri Bot. Gard. 18: 173. 1907. *Ambrosia monophylla* (Walter) Rydberg, in Britton, N. Amer. Fl. 33: 17. 1922.

Ambrosia glandulosa Scheele, Linnaea 22: 157. 1849.

Erect annual herb, to 1(1.5) m; stem coarsely pubescent to hispidulous. Leaves opposite proximally, alternate distally, the blade ovate, 4–10 cm long, 2–7 cm wide, the upper pinnatifid, the lower bipinnatifid, the apex acute or obtuse, the base cuneate, the ultimate lobes with the margin entire, blunt-toothed, or dentate, the upper and lower surfaces pilose or strigillose, gland-dotted, the petiole 2–4 cm long, narrowly winged, the upper leaves reduced and becoming subsessile. Flowers in a terminal array, the staminate distal, the carpellate proximal; staminate capitula numerous in a racemiform array, the peduncle to 2 mm long, the flowers 12–20+; phyllaries 5–8 in 1 series, basally connate, the involucre obliquely cupulate, 2–3 mm wide, the margin crenate, the outer surface hispid to pilosulous or glabrate; carpellate capitula in axillary clusters below the staminate, the flower 1; phyllaries 8 or more in several series, obovoid, fused to the ovary, the outer surface pubescent or glabrate. Cypsela subglobose to pyriform, 2–3 mm long, the fused phyllaries with 3–5(7) straight spines or tubercles to 0.5 mm long near the middle or distal, slightly pilosulous.

Disturbed sites. Common; nearly throughout. Nearly throughout North America; nearly cosmopolitan. Summer–fall.

Ambrosia hispida Pursh [With bristly trichomes.] COASTAL RAGWEED.

Ambrosia hispida Pursh, Fl. Amer. Sept. 743. 1814.
Ambrosia crithmifolia de Candolle, Prodr. 5: 525. 1836.

Prostrate or decumbent perennial herb; stem to 5(8) dm long, hispid. Leaves opposite, the blade deltate, ovate, or elliptic, 2–3.5 cm long, 1–3 cm wide, 2- to 3-pinnately lobed, the apex acute, the base cuneate, the ultimate lobes with the margin entire or toothed, the upper and lower surfaces hispid or strigillose, the petiole 0.5–2.5 cm long. Flowers in a terminal array, the staminate distal and the carpellate proximal; staminate capitula numerous in a racemiform array, the peduncle 1–2 mm long, the flowers 5–20+; phyllaries 5–8 in 1 series, basally connate, the involucres obliquely cupulate, 2–3 mm wide, the margin crenate or undulate, the outer surface strigillose; carpellate capitula in axillary clusters below the staminate, the flower 1; phyllaries 8 or more in several series, obovoid, fused to the ovary, the outer surface strigillose. Cypsela pyriform, 1–2 mm long, the fused phyllaries smooth or with 1–5+ straight spines or tubercles to 0.5 mm long near the middle or distal, strigillose.

Beaches and dunes. Occasional; Brevard and Lee Counties, southern peninsula. Florida; West Indies, Central America, and South America. All year.

Ambrosia psilostachya DC. [From the Greek *psil,* smooth, and *stachys,* spike, in reference to inflorescence.] WESTERN RAGWEED.

Ambrosia psilostachya de Candolle, Prodr. 5: 526. 1836.
Ambrosia rugelii Rydberg, in Britton, N. Amer. Fl. 33: 19. 1922. TYPE: FLORIDA: s.d., *Rugel 508* (holotype: NY).

Erect perennial herb, to 0.6(1) m; stem hispidulous to strigose. Leaves opposite proximally, alternate distally, the blade deltate to lanceolate, 2–8(14) cm long, 0.5–3.5(5) cm wide, pinnately toothed or lobed, the apex acute, the base cuneate or truncate, the ultimate lobes linear-lanceolate, with the margin subentire or toothed, the upper and lower surfaces hirsutulous or strigose, gland-dotted, the petiole to 2.5 cm long, often somewhat winged, or sessile. Flowers in a terminal racemiform or spiciform array, the staminate distal, the carpellate proximal; staminate capitula numerous in a racemiform array, the peduncle to 2 mm long, the flowers 5–15(30+); phyllaries 5–8 in 1 series, basally connate, the involucres obliquely cupulate, 2–4(5) mm wide, the margin crenate or undulate, the outer surface hirsutulous; carpellate capitula in axillary clusters below the staminate, the flower 1; phyllaries 8 or more in several series, obovoid, fused to the ovary, the outer surface hirsutulous. Cypsela obpyramidal to subglobose, 2–3 mm long, the fused phyllaries smooth or with 1–6 straight spines or tubercles to 1 mm long, these mostly distal, hirsutulous.

Disturbed sites. Occasional; northern counties, Citrus County. Nearly throughout North America; nearly cosmopolitan. Native to North America and Mexico. Summer–fall.

Ambrosia trifida L. [Three-lobed, in reference to the frequently 3-lobed leaves.] GIANT RAGWEED.

Ambrosia trifida Linnaeus, Sp. Pl. 988. 1753. *Ambrosia trifida* Linnaeus var. *normalis* Kuntze, Revis. Gen. Pl. 1: 305. 1891, nom. inadmiss.

Erect annual herb, to 1.5(4) m; stem scabrellous. Leaves opposite, the blade rounded-deltate to ovate or elliptic, 4–15(25) cm long, 3–7(20) cm wide, palmately 3- to 5-lobed, the apex acute, the base truncate to cuneate, sometimes decurrent on the petiole, the lobe margin toothed or entire, the upper and lower surfaces scabrellous, gland-dotted, the petiole 1–3(7) cm long. Flowers in a terminal array, the staminate distal, the carpellate proximal; staminate capitula numerous in a racemiform array, the peduncle 1–3 mm long, the flowers 3–25+; phyllaries 5–8 in 1 series, basally connate, the involucre saucer-shaped, 2–4 mm wide, the margin crenate or undulate, the outer surface scabrellous, often with 1–3 black nerves, scabrellous; carpellate capitula in axillary clusters below the staminate, the flower 1; phyllaries 8 or more in several series, obovoid, fused to the ovary, the outer surface scabrellous. Cypsela pyramidal, 3–5(7) mm long, the fused phyllaries with 4–8 straight spines to 1 mm long, these distal, the outer ones with a rib or sometimes slightly winged, glabrous or glabrate.

Disturbed sites. Occasional; northern counties, Marion and Broward Counties. Nearly throughout North America; nearly cosmopolitan. Native to North America and Mexico. Summer–fall.

EXCLUDED TAXON

Ambrosia peruviana Willdenow—Reported for Florida by Wunderlin (1982, as *Ambrosia cumanensis* Kunth) and Liogier and Martorelli (1982), who misapplied the name to Florida material of *Ambrosia psilostachya*).

Antennaria Gaertn. 1791. PUSSYTOES

Perennial herbs. Leaves simple, alternate, pinnate-veined, petiolate or epetiolate. Flowers in capitula in corymbiform arrays, discoid, unisexual (plants dioecious); receptacle epaleate; petals 5; basally connate, actinomorphic; stamens 5; epipetalous, the filaments free, the anthers connate; ovary inferior, 2-carpellate, 1-loculate. Fruit a cypsela; pappus present.

A genus of about 45 species; North America, Mexico, South America, Europe, and Asia. [*Antenna*, antennae, and *aris*, belonging to, in reference to the clavate pappus bristles in the staminate flowers that resemble the antennae of some insects.]

Selected references: Bayer (2006).

Antennaria plantaginifolia (L.) Hook. [With leaves like some species of *Plantago* (Plantaginaceae).] PUSSYTOES; WOMAN'S TOBACCO.

Gnaphalium plantaginifolium Linnaeus, Sp. Pl. 850. 1753. *Gnaphalium dioicum* Linnaeus var. *plantaginifolium* (Linnaeus) Michaux, Fl. Bor.-Amer. 2: 127. 1803. *Antennaria plantaginea* Sweet, Hort. Brit. 221. 1826, nom. illegit.; non R. Brown, 1817. *Antennaria plantaginifolia* (Linnaeus) Hooker, Fl. Bor.-Amer. 1: 330. 1834.
Antennaria plantaginea R. Brown, Trans. Linn. Soc. London 12: 123. 1817.
Antennaria plantaginea R. Brown, var. *petiolata* Fernald, Proc. Boston Soc. Nat. Hist. 28: 242. 1898.

Ascending perennial herb, to 2(2.5) dm; stem gray-pubescent. Basal leaves with the blade obovate to suborbicular, 3.5–7.5 cm long, 1.5–3.5 cm wide, the apex minutely mucronate, the base cuneate, the margin entire, the upper surface green-glabrescent to gray-pubescent, the lower surface gray-tomentose, the petiole 1–2 cm long, winged, the cauline linear, 0.7–3.5 cm long, 0.5–1 cm wide, the apex acuminate, sessile. Flowers in 4–7(30) capitula in a tight corymbiform array; phyllaries in 3–6 series, those of the staminate capitula 5–7(8) mm long, those of the carpellate capitula 5–7 mm long, narrow, unequal, the proximal papery or membranous, the distal somewhat scarious, the apex acute or obtuse; receptacle convex, foveolate, epaleate; disk flowers numerous; corolla white, that of the staminate flowers 2–4 mm long, that of the carpellate flowers 3–4 mm long. Cypsela ellipsoid or ovate, 0.5–1.5 mm long, slightly papillate, glabrous; pappus of 10–20 capillary barbellate bristles, basally connate and shed together, those of the staminate flowers 3–4 mm long, clavate, those of the carpellate flowers 4–6 mm long.

Hammocks. Rare; Leon County. Nova Scotia, New Brunswick, and Quebec, south to Florida, west to Manitoba, Minnesota, Iowa, Missouri, Arkansas, and Texas. Spring.

EXCLUDED TAXA

Antennaria parlinii Fernald subsp. *fallax* (Greene) Bayer & Stebbins—Reported by Correll and Johnston (1970, as *A. fallax* Greene) as occurring in "most of e. U.S.," which presumably would include Florida. Excluded from Florida by Bayer and Stebbins (1993).
Antennaria solitaria Rydberg—Reported for Florida by Arriagada (1998). No Florida specimens seen.

Anthemis L. 1753. CHAMOMILE

Herbs. Leaves simple, alternate, pinnate-veined, petiolate or epetiolate. Flowers in solitary capitula at the branch tips, radiate; receptacle paleate; ray flowers carpellate or sterile; petals 5, basally connate, zygomorphic; disk flowers bisexual; petals 5, basally connate, actinomorphic; stamens 5, epipetalous, the filaments free, the anthers connate; ovary inferior, 2-carpellate, 1-loculate. Fruit a cypsela; pappus present or absent.

A genus of about 175 species; North America, South America, Europe, Africa, Asia, Australia, and Pacific islands. [From the Greek *anthemon,* flower.]

Selected reference: Watson (2006a).

1. Receptacle with paleae throughout; ray flowers carpellate .... **A. arvensis**
1. Receptacle with paleae only toward the middle; ray flowers sterile .... **A. cotula**

Anthemis arvensis L. [Persisting in fields and cultivated land.] CORN CHAMOMILE.

Anthemis arvensis Linnaeus, Sp. Pl. 894. 1753. *Chamaemelum arvense* (Linnaeus) Schreber, Spic. Fl. Lips. 18. 1771.

Ascending or erect annual herb, to 3(8) dm; stem sometimes decumbent and rooting at the nodes, strigose-sericeous, villous, or glabrescent. Leaves with the blades oblanceolate to oblong-elliptic or ovate, 1–4 cm long, 0.8–1.6 cm wide, deeply 1- to 2-pinnately lobed, the basal lobes sometimes appearing fascicled, the ultimate lobes triangular, narrowly elliptic, linear, 0.5–4 mm long, the apex acute, the base cuneate, the ultimate margin dentate or lobed, the upper and lower surfaces strigose-sericeous, villous, or glabrescent. Flowers in a solitary capitulum at the branch apex, the peduncle 3–15 cm long; involucre hemispheric; phyllaries imbricate, slightly pubescent; receptacle conic; paleae throughout the receptacle, lanceolate to oblanceolate, 3–4 mm long, slightly naviculate with a spinose apex; ray flowers 5–20, carpellate; corolla white, rarely tinged with pink, the lamina 5–15 mm long, sometimes inconspicuously glandular-dotted; disk flowers numerous, bisexual; corolla white, sometimes tinged with purple, 2–3(4) mm long, the lobes minutely gland-dotted. Cypsela oblong-obovoid, ca. 2 mm long, bluntly 4-angled, the apex obliquely truncate, the base truncate, the ribs (8)10, smooth or weakly tuberculate, glabrous; pappus absent or minutely coroniform.

Disturbed sites. Rare; Jackson County. Nearly throughout North America; South America; Europe, Africa, Asia, Australia, and Pacific Islands. Native to Europe, Africa, and Asia. Spring–summer.

Anthemis cotula L. [Pelinnaean vernacular name for species.] STINKING CHAMOMILE; MAYWEED.

Anthemis cotula Linnaeus, Sp. Pl. 894. 1753. *Anthemis cotula-foetida* Crantz, Inst. Rei Herb. 1: 300. 1766, nom. illegit. *Anthemis foetida* Lamarck, Fl. Franc. 2: 164. 1779 ("1778"), nom. illegit. *Chamaemelum cotula* (Linnaeus) Allioni, Fl. Pedem. 1: 186. 1785. *Chamaemelum foetidum* Baumgartner, Enum. Stirp. Trannsilv. 3: 144. 1817, nom. illegit. *Maruta foetida* Gray, Nat. Arr. Brit. 2: 456. 1821, nom. illegit. *Maruta cotula* (Linnaeus) de Candolle, Prodr. 6: 13. 1838. *Maruta vulgaris* Bluff & Fingerhut, in Bluff et al., Comp. Fl. German., ed. Alt. 2: 392. 1838, nom. illegit.

Ascending or erect annual herb, to 5(8) dm; stem puberulent, sparsely strigillose, glabrate, or glabrous, gland-dotted. Leaves with the blade oblanceolate, oblong-elliptic, or ovate, 1–6 cm long, 1.5–3 mm wide, deeply 1- to 2-pinnately lobed, the basal lobes sometimes appearing fascicled, the ultimate lobes triangular to narrowly elliptic to linear, 0.5–4 mm long, the apex acute, the base cuneate, the ultimate margin dentate or lobed, the upper and lower surfaces strigose-sericeous, villous, or glabrescent. Flowers in a solitary capitulum at the branch apex, the peduncle 3–8 cm long; involucre hemispheric; phyllaries imbricate, villosulous to arachnose; receptacle conic; paleae mostly distal on the receptacle, subulate to needlelike, 2–3 mm long, often gland-dotted; ray flowers 10–15, sterile; corolla white, the lamina 5–9 mm long, sometimes inconspicuously gland-dotted; disk flowers numerous, bisexual; corolla white, 2–3 mm long, the lobes sparsely gland-dotted. Cypsela oblong-obovoid, ca. 2 mm long, bluntly

4-angled, the apex obliquely truncate, the base truncate, the ribs 8(10), smooth or weakly tuberculate, often gland-dotted in the furrows; pappus absent.

Disturbed sites. Occasional; Alachua and Brevard Counties; central panhandle. Nearly throughout North America; South America; Europe, Africa, Asia, Australia, and Pacific Islands. Native to Europe, Africa, and Asia. Summer–fall.

Aphanostephus DC. 1836. DOZEDAISY

Herbs. Leaves alternate, simple, pinnate-veined, petiolate or epetiolate. Flowers in solitary terminal capitula, radiate; receptacle epaleate; ray flowers carpellate; petals 5, basally connate, zygomorphic; disk flowers bisexual; petals 5, basally connate, actinomorphic; stamens 5, epipetalous, the filaments free, the anthers connate; ovary inferior, 2-carpellate, 1-loculate. Fruit a cypsela; pappus present.

A genus of 4 species; North America and Mexico. [From the Greek *aphanes,* obscure, and *stephanos,* crown, in reference to the low coroniform pappus.]

Selected references: Nesom (2006c); Shinners (1946).

Aphanostephus skirrhobasis (DC.) Trel. ex Branner & Coville var. **thalassius** Shinners [From the Greek *Skirrhous,* cord-like, and *basis,* base; in reference to the many cord-like stems arising at the base; *thalassa,* sea, in reference to its coastal habitat.] LAZY DAISY; ARKANSAS DOZEDAISY.

Aphanostephus skirrhobasis (de Candolle) Trelease ex Branner & Corville var. *thalassus* Shinners, Wrightia 1: 106. 1946.

Erect, ascending, or decumbent annual herb, to 4 dm; stem puberulent-hispidulous, the trichomes spreading or deflexed. Leaves with the blade broadly oblanceolate or linear-lanceolate, 1–4(6) cm long, 0.5–1(2) cm wide, slightly thickened, the apex obtuse, the base cuneate, the margin entire or dentate or rarely shallowly pinnatifid, the upper and lower surfaces densely gray-pubescent, sessile or with a short winged petiole. Flowers in a solitary terminal capitulum, the peduncle 1–4 cm long; involucre depressed hemispheric; phyllaries numerous in 3–4 series, ovate-lanceolate, 4–5 mm long, unequal, the apex acute or acuminate, the margin scarious, ciliate, the outer surface hirsute or hispid; receptacle conic, pitted; ray flowers (10)22–44, carpellate; corolla white, the lamina 0.8–1.5 cm long, sometimes reddish, at least in streaks, on the outer surface; disk flowers numerous, bisexual; corolla yellow, 2–3 mm long. Cypsela columnar, ca. 2 mm long, usually 4-angled, with 4–12 ribs, the surface sparsely strigose; pappus coroniform, unequal, sometimes with awn-tipped or ciliate scales.

Coastal beach dunes or open disturbed sites. Rare; St. Johns, Bay, Santa Rosa, and Escambia Counties. Florida, Louisiana, and Texas; Mexico. Fall.

EXCLUDED TAXON

Aphanostephus skirrhobasis (de Candolle) Trelease ex Branner & Coville—Because infraspecific categories were not recognized, the typical variety was reported from Florida by Small (1903, 1913a, 1933). Our plants are all var. *thallasius.*

Arnica L. 1753. LEOPARDBANE

Herbs. Leaves opposite or alternate, simple, pinnate-veined, epetiolate. Flower in capitula in corymbiform arrays, radiate; receptacle epaleate; ray flowers carpellate; petals 5, basally connate, zygomorphic; disk flowers bisexual; petals 5, basally connate, actinomorphic; stamens 5, epipetalous, the filaments free, the anthers connate; ovary inferior, 2-carpellate, 1-loculate. Fruit a cypsela; pappus present.

A genus of about 29 species; North America, Mexico, Europe, and Asia. [Ancient Latin or Greek vernacular name for the plant.]

Selected references: Downie and Denford (1988); Wolf (2006).

Arnica acaulis (Walter) Britton et al. [Stemless.] COMMON LEOPARDBANE.

Doronicum acaule Walter, Fl. Carol. 205. 1788. *Arnica claytonii* Pursh, Fl. Amer. Sept. 527. 1814, nom. illegit. *Arnica acaulis* (Walter) Britton et al., Prelim. Cat. 30. 1888.

Doronicum nudicaule Michaux, Fl. Bor.-Amer. 2: 121. 1803. *Arnica nudicaulis* (Michaux) Nuttall, Gen. N. Amer. Pl. 2: 164. 1818. TYPE: "Virginia ad Floridam."

Erect herb, to 8(10) dm; stem hirsute and stipitate-glandular. Leaves with the blade of the basal ones broadly elliptic, ovate, or rhombic, 4–15 cm long, 1.5–8 cm wide, the apex obtuse, the base cuneate, the margin entire or coarsely serrate, the upper and lower surfaces hirsute and stipitate-glandular, the cauline leaves reduced, the proximal in 1 or 2 pairs, the distal alternate, more reduced, linear-lanceolate. Flowers in 3–20 capitula in a corymbiform array, the peduncle 2–10 cm long; involucre campanulate; phyllaries 12–18, narrowly elliptic or elliptic-lanceolate, to ca. 1 cm long, the outer surface hirsute and stipitate-glandular; receptacle convex; ray flowers 10–16, carpellate; corolla yellow, the lamina 1.5–2(2.5) cm long; disk flowers numerous, bisexual; corolla yellow, 6–8 mm long. Cypsela fusiform, 5–7 mm long, with 5–10 ribs, dark gray to dark brown, glabrous or sparsely stipitate-glandular; pappus with white barbellate bristles 4–6 mm long.

Flatwoods. Rare; Liberty and Jackson Counties. Pennsylvania and New Jersey south to Florida, west to Alabama. Spring.

Arnica acaulis is listed as endangered in Florida (Florida Administrative Code, Chapter 5B-40).

Arnoglossum Raf. 1817. INDIAN PLANTAIN

Herbs. Leaves alternate, simple, palmate-veined, petiolate or epetiolate. Flowers in capitula in corymbiform arrays, discoid, bisexual; receptacle epaleate; petals 5, basally connate, actinomorphic; stamens 5, epipetalous, the filaments free, the anthers connate; ovary inferior, 2-carpellate, 1-loculate. Fruit a cypsela; pappus present.

A genus of 8 species; North America. [From the Greek *Arnos,* lamb, and *glossa,* tongue, in reference to the leaves of some species resembling a lamb's tongue.]

Mesadenia Raf., 1832.

Selected references: Anderson (2006a).

1. Phyllaries lacking a median wing; stem terete or merely striate.
   2. Leaves with the margin entire or sinuate, rarely serrulate-denticulate .... **A. ovatum**
   2. Leaves with the margin shallowly lobed or dentate .... **A. atriplicifolium**
1. Phyllaries with a median wing; stem sharply angled or conspicuously grooved.
   3. Lower cauline leaves with the blade deltoid-hastate .... **A. diversifolium**
   3. Lower cauline leaves with the blade ovate.
      4. Phyllaries with the wings widest at the base; basal leaves with the lateral veins appressed to midrib for 2–4 cm, then spreading well above the base of the blade .... **A. album**
      4. Phyllaries evenly winged throughout or widest at the apex; basal leaves with the lateral veins diverging at the base of the blade.
         5. Lower cauline leaves with the margin crenate, 7- to 9-nerved .... **A. floridanum**
         5. Lower cauline leaves with the margin sinuate-dentate, 3- to 5-nerved .... **A. sulcatum**

Arnoglossum album L. C. Anderson [White, in reference to the flowers.] WHITE INDIAN PLANTAIN.

Arnoglossum album L. C. Anderson, Sida 18: 378. 1998. TYPE: FLORIDA: Bay Co.: just E of FL 77 on N side of Southport, T14S, R14W, Sec 21, NE ¼ of SE ¼, 6 Jun 1995, *Anderson 15555* (holotype: NY; isotypes: BRIT, FSU, MO, US).

Erect perennial herb, to 1 m; stem strongly ridged, glabrous. Basal leaves with the blade broadly ovate to narrowly oblong-lanceolate, 15–37 cm long, 5–10(16) cm wide, the lateral veins appressed to the midvein for 2–4 cm proximally and then spreading, the apex acute, the base cuneate, the margin usually entire, rarely serrulate-denticulate, the upper and lower surfaces glabrous, the petiole 12–18 cm long, the cauline leaves reduced upward, the proximal ones with the blade ovate, the apex acuminate, the margin coarsely serrate, petiolate, the distal ones smaller, sessile. Flowers in capitula in a corymbiform array; involucre cylindric or turbinate; phyllaries 5, ovate, 10–13 mm long, chalky white, the midvein prominently winged, widest at the base; receptacle flat or slightly convex, with a short central cusp, foveolate; corolla white, rarely pink tinged, 9–11 mm long. Cypsela cylindric or narrowly clavate, 4–5 mm long; pappus of numerous bristles 6–7 mm long.

Wet savannas. Rare; Gulf and Bay Counties. Endemic. Summer.

Arnoglossum album is listed as endangered in Florida (Florida Administrative Code, Chapter 5B-40).

Arnoglossum atriplicifolium (L.) H. Rob. [With leaves like *Atriplex* (Amaranthaceae).] PALE INDIAN PLANTAIN.

Cacalia atriplicifolia Linnaeus, Sp. Pl. 835. 1753. *Senecio atriplicifolius* (Linnaeus) Hooker, Fl. Bor.-Amer. 1: 332. 1834. *Mesadenia atriplicifolia* (Linnaeus) Rafinesque, New Fl. 4: 79. 1838 ("1836"). *Adenimesa atriplicifolia* (Linnaeus) Nieuwland, Amer. Midl. Naturalist 3: 193. 1914. *Conophora atriplicifolia* (Linnaeus) Nieuwland, Amer. Midl. Naturalist 3: 193. 1914. *Arnoglossum atriplicifolium* (Linnaeus) H. Robinson, Phytologia 28: 294. 1974.

Erect perennial herb, to 3 m; stem terete to weakly striate, glabrous, glaucous. Basal leaves with the blade ovate or ovate-cordate, to 21 cm long and wide, the lateral veins proximally spreading from the midvein, the apex round to obtuse, the base truncate or subcordate, the margin shallowly lobed or dentate, the upper and lower surfaces glabrous, the lower surface glaucous, the petiole to 10 cm long, the cauline leaves reduced upward, the proximal ones subdeltoid, petiolate, the margins coarsely dentate, the distal ones smaller, petiolate or sessile. Flowers in capitula in a corymbiform array; involucre cylindric or turbinate; phyllaries 5, oblong, 7–8(10) mm long, pale green, the midvein not winged; receptacle flat or slightly convex, with a short central cusp, foveolate; corolla white, sometimes green- or purple-tinged, 8–10(13) mm long. Cypsela fusiform or ellipsoid, 4–5 mm long, 8- to 10-ribbed, brown or purplish; pappus of numerous bristles 5–6(7) mm long, early deciduous.

Floodplain forests. Rare; Liberty, Jackson, and Washington Counties. New York and Massachusetts, south to Florida, west to Minnesota, Iowa, Nebraska, Kansas, and Oklahoma. Summer.

Arnoglossum diversifolium (Torr. & A. Gray) H. Rob. [With variously shaped leaves.] VARIABLELEAF INDIAN PLANTAIN.

Cacalia diversifolia Torrey & A. Gray, Fl. N. Amer. 2: 435. 1842. *Mesadenia diversifolia* (Torrey & A. Gray) Greene, Pittonia 3: 182. 1897. *Conophora diversifolia* (Torrey & A. Gray) Nieuwland, Amer. Midl. Naturalist 3: 193. 1914. *Arnoglossum diversifolium* (Torrey & A. Gray) H. Robinson, Phytologia 28: 294. 1974. TYPE: FLORIDA: s.d., *Chapman s.n.* (holotype: ?).
Mesadenia difformis Small, S.E. U.S. 1301, 1341. 1903. TYPE: FLORIDA: Walton Co.: 1885, *Curtiss s.n.* (holotype: NY).

Erect perennial herb, to 2(3) m; stem proximally ribbed and distally angular, glabrous. Basal leaves with the blade ovate or ovate-cordate, to 10 cm long and wide, the lateral veins proximally spreading from the midrib, the apex obtuse, the base truncate, the margin coarsely and irregularly denticulate or shallowly lobed, the upper and lower surfaces glabrous, the petiole to 20 cm long, the cauline leaves reduced upward, the proximal ones deltate-hastate, the margin dentate, petiolate, the distal ones smaller, short-petiolate or sessile, the upper and lower surfaces finely puberulous. Flowers in capitula in a corymbiform array; involucre cylindric or turbinate; phyllaries 5, lanceolate-ovate, (8)9–11 m long, the midvein winged, uniform or highest at the apex; receptacle flat or slightly convex, with a short central cusp, foveolate; corolla white or purplish, 8–10 mm long. Cypsela fusiform, ca. 3 mm long; pappus of numerous bristles (6)7–8 mm long.

Floodplain forests. Rare; Putnam and Levy Counties, central panhandle, Walton County. Georgia, Florida, and Alabama. Spring–summer.

Arnoglossum diversifolium is listed as threatened in Florida (Florida Administrative Code, Chapter 5B-40).

Arnoglossum floridanum (A. Gray) H. Rob. [Of Florida.] FLORIDA INDIAN PLANTAIN.

Cacalia floridana A. Gray, Proc. Amer. Acad. Arts 19: 52. 1883. *Mesadenia floridana* (A. Gray) Greene, Pittonia 3: 183. 1897. *Conophora floridana* (A. Gray) Nieuwland, Amer. Midl. Naturalist 3: 193. 1914. *Arnoglossum floridanum* (A. Gray) H. Robinson, Phytologia 28: 294. 1974. TYPE: FLORIDA.

Erect perennial herb, to 1 m; stem strongly ridged, glabrous. Basal leaves with the blade ovate to elliptic, to 8 cm long, the lateral veins proximally spreading from the midvein, the apex rounded, the base broadly cuneate, the margin entire, the upper and lower surfaces glabrous, the petiole to 5 cm long, the cauline leaves reduced upward, the proximal ones with the blade ovate, the margin crenulate, petiolate, the distal ones smaller, petiolate or sessile. Flowers in capitula in a corymbiform array; involucre cylindric or turbinate; phyllaries 5, ovate, light green, the midvein winged, uniform or highest at the apex; receptacle flat or slightly convex, with a short central cusp, foveolate; corolla white or light green, (9)11–12 mm long. Cypsela fusiform, ca. 5 mm long, green; pappus of numerous bristles 7–9 mm long.

Sandhills. Frequent; northern and central peninsula, eastern panhandle. Endemic. Summer.

Arnoglossum ovatum (Walter) H. Rob. [Egg-shaped, in reference to the leaves.] OVATELEAF INDIAN PLANTAIN.

Cacalia ovata Walter, Fl. Carol. 196. 1788. *Mesadenia ovata* (Walter) Rafinesque, New Fl. 4: 79. 1838 ("1836"). *Senecio ovatus* (Walter) MacMillan, Metasp. Minnesota Valley 555. 1892; non Willdenow, 1803. *Conophora ovata* (Walter) Nieuwland, Amer. Midl. Naturalist 3: 193. 1914. *Arnoglossum ovatum* (Walter) H. Robinson, Phytologia 29: 294. 1974.

Cacalia lanceolata Nuttall, Gen. N. Amer. Pl. 2: 138. 1818. *Mesadenia lanceolata* (Nuttall) Rafinesque, New Fl. 4: 79. 1838 ("1836"). *Arnoglossum ovatum* (Walter) H. Robinson var. *lanceolatum* (Nuttall) D. B. Ward, Novon 14: 367. 2004. TYPE: FLORIDA: s.d., *Baldwin s.n.* (lectotype: PH). Lectotypified by Ward (2004: 367).

Mesadenia elliottii R. M. Harper, Torreya 5: 184. 1905. *Cacalia elliottii* (R. M. Harper) Shinners, Field & Lab. 18: 80. 1950. *Cacalia lanceolata* Nuttall var. *elliottii* (R. M. Harper) Kral & R. K. Godfrey, Quart. J. Florida Acad. Sci. 21: 205. 1958. SYNTYPE: FLORIDA.

Mesadenia angustifolia Rydberg, Bull. Torrey Bot. Club 51: 378. 1924. TYPE: FLORIDA: Miami-Dade Co.: Everglades, Fort Lauderdale to Miami, Feb 1911, *Small et al. 3318* (holotype: NY).

Erect perennial herb, to 3 m; stem terete, glabrous. Basal leaves with the blade ovate, narrowly lanceolate, or linear-lanceolate, to 30 cm long, to 4 cm wide, the lateral veins proximally spreading from the midvein, the apex acute or obtuse, the base cuneate, the margin entire, the upper and lower surfaces glabrous, the petiole to 15 cm long, the cauline leaves reduced upward, the proximal ones petiolate, the distal ones smaller, sessile. Flowers in capitula in a corymbiform array; involucre cylindric or turbinate; phyllaries 5, lanceolate-ovate, the midvein not winged; receptacle flat or slightly convex, with a short central cusp, foveolate; corolla white or greenish, rarely purplish, 8–9(10) mm long. Cypsela fusiform or clavate, 4–5 mm long, 7- to 8-ribbed, dark brown; pappus of numerous bristles (5)6–8 mm long.

Wet flatwoods, marshes, and seepage slopes. Frequent; nearly throughout. North Carolina south to Florida, west to Texas. Georgia, Florida, and Alabama. Summer–fall.

Arnoglossum sulcatum (Fernald) H. Rob. [Furrowed, in reference to the lower stem.] GEORGIA INDIAN PLANTAIN.

Cacalia sulcata Fernald, Bot. Gaz. 33: 157. 1902. *Mesadenia sulcata* (Fernald) Small, Fl. S.E. U.S. 1301, 1341. 1903. *Arnoglossum sulcatum* (Fernald) H. Robinson, Phytologia 28: 295. 1974.

Erect perennial herb, to 1.4 m; stem strongly ridged, glabrous or sparsely pubescent. Basal leaves with the blade broadly ovate, or elliptic, 12–15(20) cm long, 4–10 cm wide, with 2 lateral veins proximally spreading from the midvein, the apex acute or obtuse, the margin entire, sinuate, crenulate, or denticulate, the upper and lower surfaces glabrous, the petiole 10–45 cm long, the cauline leaves reduced upward, the proximal petiolate, the distal smaller, petiolate or sessile. Flowers in capitula in a corymbiform array; involucre cylindric or turbinate; phyllaries 5, ovate, (8)10–12 mm long, pale green, the midvein winged, uniform or highest at the apex; receptacle flat or convex, with a short central cusp, foveolate; corolla white or greenish, sometimes purple-tinged, 7–8(10) mm long. Cypsela fusiform or clavate, ca. 4 mm long, 6- to 8-ribbed, dark brown; pappus of numerous bristles 5–7 mm long.

Flatwoods, swamps, and floodplain forests. Occasional; central and western panhandle. Fall.

EXCLUDED TAXON

Arnoglossum plantagineum Rafinesque—Reported by Correll and Johnston (1970, as *Cacalia plantaginea* (Rafinesque) Shinners) as occurring "widespread in e. U.S.," which would presumably include Florida. Excluded from Florida by Kral and Godfrey (1958). No Florida specimens seen.

Artemisia L. 1819. SAGEBRUSH

Herbs. Leaves alternate, simple, pinnate-veined, petiolate or epetiolate. Flowers in capitula in paniculiform, racemiform, or spiciform arrays, discoid, bisexual or unisexual; receptacle epaleate; petals 5, basally connate, actinomorphic; stamens 5, epipetalous, the filaments free, the anthers connate; ovary inferior, 2-carpellate, 1-loculate. Fruit a cypsela; pappus absent.

A genus of about 500 species; nearly cosmopolitan. [Derived from Artemis, ancient Greek goddess of the hunt and protector of young girls, aiding in childbirth, and relieving disease of women.]

Selected reference: Shultz (2006).

1. Principal leaves 2- to 3-pinnatifid or dissected, the lower surface glabrous or sparsely pubescent .... **A. campestris**
1. Principal leaves entire or pinnatifid, the segments relatively broad, the lower surface densely white tomentose.
    2. Leaves glabrous or glabrate on the upper surface .... **A. vulgaris**
    2. Leaves white tomentose or floccose on the upper surface.
        3. Leaves deeply lobed to 1/2 or more .... **A. stelleriana**
        3. Leaves entire, toothed, or lobed to ½ or less .... **A. ludoviciana**

Artemisia campestris L. subsp. **caudata** (Michx.) H. M. Hall & Clements [Pertaining to the plains; with a taillike appendage, in reference to the slender inflorescence.] FIELD WORMWOOD.

Artemisia caudata Michaux, Fl. Bor.-Amer. 2: 129. 1803. *Artemisia campestris* Linnaeus subsp. *caudata* (Michaux) H. M. Hall & Clements, Publ. Carnegie Inst. Wash. 326: 122. 1923. *Artemisia campestris* Linnaeus var. *caudata* (Michaux) Steyermark, Ann. Missouri Bot. Gard. 22: 676. 1935. *Artemisia caudata* Michaux forma *typica* J. Rousseau, Naturaliste Canad. 71: 198. 1944, nom. inadmiss. *Oligosporus campestris* (Linnaeus) Cassini subsp. *caudatus* (Michaux) W. A. Weber, Phytologia 55: 9. 1984.

Erect biennial herb, to 0.8(1.5) m; stem striate, sparsely pubescent or glabrous. Basal leaves with the blade ovate or elliptic, 4–12 cm long, gradually reduced upward, the cauline ones 2–4 cm long, 0.5–1.5 cm wide, 2- to 3-pinnately lobed, the ultimate lobes linear or narrowly oblong, the apex acute, the upper surface glabrous, the lower surface sparsely pubescent or glabrous, short-petiolate. Flowers in capitula in a paniculiform array, 8–10(12) cm long, 1–2(3) cm wide, short-pedunculate, the outer flowers of the capitulum carpellate, the inner ones staminate; involucre turbinate; phyllaries broadly elliptic, 2–3 mm long, the outer surface sparsely pubescent or glabrous; receptacle convex, glabrous; corolla pale yellow, 1–2 mm long, sparsely pubescent or glabrous. Cypsela oblong-lanceolate, ca. 1 mm long, slightly compressed, faintly nerved, glabrous; pappus absent.

Flatwoods. Rare; Lake, Orange, Polk, and Osceola Counties. Maine and Ontario south to Florida, west to Saskatchewan, Montana, Colorado, and New Mexico. Summer–fall.

Artemisia ludoviciana Nutt. [Of St. Louis, Missouri.] WHITE SAGEBRUSH.

Artemisia ludoviciana Nuttall, Gen. N. Amer. Pl. 2: 143. 1818. *Artemisia vulgaris* Linnaeus var. *ludoviciana* (Nuttall) Kuntze, Revis. Gen. Pl. 1: 309. 1891. *Artemisia vulgaris* Linnaeus subsp. *ludoviciana* (Nuttall) H. M. Hall & Clements, Publ. Carnegie Inst. Wash. 326: 76. 1923. *Artemisia ludoviciana* Nuttall var. *typica* Keck, Publ. Carnegie Inst. Wash. 520: 330. 1940, nom. inadmiss.

Erect perennial herb, to 8 dm; stem gray- to white-tomentose. Leaves with the blade linear or narrowly elliptic, the basal 3–11 cm long, gradually reduced upward, the basal ones entire or apically lobed to ⅓ the blade length, the cauline 1.5–11 cm long, 1–1.5 cm wide, the margin entire or lobed to pinnatifid, the upper and lower surfaces gray- or white-tomentose, the upper surface early deciduous, short-petiolate or sessile. Flowers in capitula in a paniculiform or racemiform array 5–30 cm long, 1–4 cm wide, sessile or the peduncle 2–5 mm long, the outer flowers carpellate, the inner bisexual; involucre turbinate; phyllaries broadly elliptic, 3–4 mm long, the outer surface gray or white tomentose; receptacle convex, glabrous; corolla yellow, 2–3 mm long, glabrous. Cypsela ellipsoid, less than 1 mm long, slightly compressed, obscurely nerved, glabrous; pappus absent.

Disturbed sites. Rare; Alachua, Manatee, and Escambia Counties. Nearly throughout North America. Escaped from cultivation. Native to the north and west of our area. Summer–fall.

Artemisia stellariana Besser [Commemorates Georg Wilhelm Steller (1709–1746), German-born Russian physician, natural historian, and explorer.] DUSTY MILLER.

Artemisia stellariana Besser, Nouv. Mém. Soc. Imp. Naturalistes Moscou 3: 79. 1884.

Erect or ascending perennial herb, to 6(7) dm; stem densely grayish white-tomentose or floccose. Basal and cauline leaves with the blade oblanceolate, the proximalmost pinnatifid, 3–10 cm long, 1–5 cm wide, the lobes broad, the lobe apex rounded, the upper and lower surfaces grayish white-tomentose, short-petiolate, the distal ones smaller. Flowers in capitula in a paniculiform, racemiform, or spiciform array 8–20 cm long, 2–4 cm wide, sessile or the peduncle to 3 cm long, the outer flowers carpellate, the inner ones bisexual; involucre campanulate; phyllaries broadly lanceolate, 5–8 mm long, the outer surface grayish white-tomentose; receptacle convex, glabrous; corolla yellow or tan, ca. 2 mm long, gland-dotted. Cypsela linear-oblong, ca. 2 mm long, slightly compressed, dark brown, glabrous; pappus absent.

Disturbed sites. Rare; Miami-Dade County. Escaped from cultivation. Newfoundland and Labrador south to North Carolina, west to Ontario and Minnesota, also Florida, Louisiana, Washington, and Alaska. Native to Europe, Asia, and Alaska. Summer.

Artemisia vulgaris L. [Common.] COMMON MUGWORT.

Artemisia vulgaris Linnaeus, Sp. Pl. 848. 1753. *Artemisia officinalis* Gaterau, Descr. Pl. Montauban 144. 1789, nom. illegit. *Artemisia ruderalis* Salisbury, Prodr. Stirp. Chap. Allerton 191. 1796, nom. illegit. *Artemisia vulgaris* Linnaeus var. *normalis* Kuntze, Revis. Gen. Pl. 1: 309. 1891, nom. inadmiss. *Artemisia vulgaris* Linnaeus subsp. *typica* H. M. Hall & Clements, Publ. Carnegie Inst. Wash. 326: 73. 1923, nom. inadmiss. *Artemisia vulgaris* Linnaeus var. *typica* H. St. John, Res. Stud. State Coll. Wash. 1: 106. 1929, nom. inadmiss.

Erect perennial herb, to 1.9 m; stem striate, sparsely pubescent or glabrous. Basal leaves with the blade ovate, lanceolate, or linear, (2)3–10(12) cm long, 1.8–8 cm wide, gradually reduced upward, pinnately dissected, the principal lobes to 2 cm wide, the apex obtuse, the margin irregularly lobed or serrate, the upper surface glabrous, the lower surface pubescent, short-petiolate. Flowers in capitula in a paniculiform or racemiform array (10)20–30(40) cm long, (5)7–15(20) cm wide, short-pedunculate or sessile, the outer flowers carpellate, the inner bisexual; involucre ovoid or campanulate; phyllaries lanceolate, 2–3(4) mm long, the outer surface sparsely pubescent or glabrescent; receptacle convex, glabrous; corolla yellowish or reddish brown, 2–3 mm long, glabrous. Cypsela ellipsoid, ca. 1 mm long, slightly compressed, obscurely nerved, glabrous, sometimes shiny; pappus absent.

Disturbed sites. Rare; Baker, Alachua, and Escambia Counties. Escaped from cultivation. Greenland south to Florida, west to British Columbia, Washington, and Ontario, also Alaska; Europe and Asia. Native to Europe and Asia. Summer–fall.

Baccharis L. 1753.

Shrubs or trees. Leaves alternate, simple, pinnate-veined, petiolate or epetiolate. Flowers in capitula in terminal corymbiform or paniculiform arrays or axillary glomerules, discoid, unisexual (plants dioecious); receptacle epaleate; petals 5, basally connate, actinomorphic; stamens 5, epipetalous, the filaments free, the stamens connate; ovary inferior, 2-carpellate, 1-loculate. Fruit a cypsela; pappus present.

A genus of about 400 speces; North America, West Indies, Mexico, Central America, South America, Europe, Asia, and Australia. [Derived from Bacchus, Roman god of agriculture, wine, and fertility.]

Selected reference: Sundberg and Bogler (2006).

1. Leaves linear, less than 5 mm wide .... **B. angustifolia**
1. Leaves ovate, spatulate, or elliptic, more than 6 mm wide.
    2. Leaves entire, the lateral veins obscure; pappus 3–5 mm long at maturity .... **B. diocia**
    2. Leaves with a few teeth, the lateral veins evident; pappus 8–12 mm long at maturity.
        3. Capitula sessile or subsessile, often in glomerules of 3 .... **B. glomeruliflora**
        3. Capitula pedunculate, in a paniculiform array .... **B. halimifolia**

Baccharis angustifolia Michx. [*Anguste,* narrow, and *folios,* leaves, in reference to the narrow leaves.] SALTWATER FALSEWILLOW.

Baccharis angustifolia Michaux, Fl. Bor.-Amer. 2: 125. 1803. TYPE: "Carolina ad Floridam."

Erect to ascending shrub, to 2(4) m; stem striate, glabrous, resinous. Leaves with the blade linear, 2–6 cm long, 1–2 mm wide, slightly succulent, 1-veined, the apex acute, the base cuneate, the margin entire, the upper and lower surfaces glabrous, sessile. Flowers 15–20 in capitula in a paniculiform array, pedunculate, discoid, unisexual (plants dioecious); involucre campanulate; phyllaries ovate to lanceolate, 2–4 mm long, the apex acute, the margin scarious, the median green or purplish; staminate corolla pale yellow, 2–3 mm long; carpellate corolla whitish, ca. 3 mm long. Cypsela obovoid to cylindric, 1–2 mm long, 10-ribbed, glabrous; pappus of minutely barbellate, apically attenuate bristles 6–8 mm long.

Brackish marshes, beaches, and coastal swales. Frequent; nearly throughout. North Carolina south to Florida, west to Louisiana. Summer–fall.

Baccharis angustifolia is reported to hybridize with *B. halimifolia* in Florida where the ranges of the two overlap (Sundberg and Bogler, 2006).

Baccharis dioica Vahl [Dioecious.] BROOMBUSH FALSEWILLOW.

Baccharis dioica Vahl, Symb. Bot. 3: 98, t. 74. 1794. *Baccharis rohriana* Sprengel, in Ersch & J. G. Gruber, Allg. Encycl. 7: 27. 1821, nom. illegit. *Baccharis vahlii* de Candolle, Prodr. 5: 411. 1836, nom. illegit.

Erect shrub, to 3 m; stem striate-angled, glabrous or slightly scurfy. Leaves with the blade obovate to stipulate, 1–3 cm long, 1–2 cm wide, slightly succulent, the lateral veins obscure, the apex obtuse, mucronulate, or retuse, the base cuneate and tapering onto the petiole, the margin entire or few-toothed, the upper and lower surfaces glabrous, gland-dotted, short-petiolate. Flowers 20–30 in capitula in a corymbiform or paniculiform array, pedunculate, discoid, unisexual (plants dioecious); involucre obconic; phyllaries ovate to lanceolate, 1–4 mm long, the apex obtuse or acuminate, the margin scarious-erose, the median green; staminate corolla pale yellow, 3–4 mm long; carpellate corolla whitish, 4–5 mm long. Cypsela obovoid to cylindric, 1–2 mm long, 8- to 10-ribbed, glabrous; pappus of minutely barbellate, apically attenuate bristles 3–5 mm long.

Coastal hammocks. Rare; Miami-Dade County. Not recently collected. Florida, West Indies, and Mexico. Fall.

Baccharis dioica is listed as endangered in Florida (Florida Administrative Code, Chapter 5B-40.)

Baccharis glomeruliflora Pers. [With flowers close together, in reference to the capitula in clusters.] SILVERLING.

Baccharis sessiliflora Michaux, Fl. Bor.-Amer. 2: 125. 1803; non Vahl, 1794. *Baccharis glomeruliflora* Persoon, Syn. Pl. 2: 423. 1807.

Erect to ascending shrub, to 3 m; stem striate-angled, glabrous or slightly scurfy. Leaves with the blade obovate, elliptic, or rhombic, 2–6 cm long, 1–4 cm wide, the apex acute, the base cuneate, the margin irregularly dentate distally, the upper and lower surfaces glabrous, the lower surface black gland-dotted, the petiole to 7 mm long, the blade of the distal leaves reduced and entire. Flowers 15–35 in capitula, the capitula solitary or (2)3–4 in an axillary glomerule, sessile or subsessile, discoid, unisexual (plants dioecious); involucre campanulate or obconic; phyllaries ovate to lanceolate, 1–4 mm long, the apex rounded or obtuse, the margin scarious, the median green, the apex sometimes purplish; staminate corolla pale yellow, 4–5 mm long; carpellate corolla whitish, 3–4 mm long. Cypsela obovoid or cylindric, ca. 2 mm long, 8- to 10-ribbed, glabrous; pappus of minutely barbellate, apically attenuate bristles 8–9 mm long.

Floodplain forests. Common; nearly throughout. North Carolina south to Florida, west to Mississippi. Fall.

Baccharis halimifolia L. [From the Greek *halos,* salt, and *folios,* leaves, in reference to the leaves similar to those of other maritime plants.] GROUNDSEL TREE; SEA MYRTLE.

Baccharis halimifolia Linnaeus, Sp. Pl. 860. 1753. *Baccharis cuneifolia* Moench, Methodus 574. 1794, nom. illegit. *Conyza halimifolia* (Linnaeus) Desfontaines, Tabl. Ecole Bot., ed. 2. 144. 1815.
Baccharis halimifolia Linnaeus var. *angustior* de Candolle, Prodr. 5: 412. 1836. *Baccharis angustior* (de Candolle) Britton ex Malagarriga, Contr. Inst. Geobiol. La Salle Canoas 2: 29. 1952.

Erect or ascending shrub or tree, to 3(6) m; stem striate-angled, glabrous or slightly scurfy, sometimes resinous. Leaves with the blade elliptic, obovate, or rhombic, 3–5(8) cm long, 1–4(6) cm wide, the apex acute or obtuse, the base cuneate, the margin irregularly serrate distally or entire, the upper and lower surfaces glabrous, gland-dotted, short-petiolate or sessile, the distal leaves reduced, entire. Flowers 20–30 in capitula in a broad paniculiform array, pedunculate, discoid, unisexual (plants dioecious); involucre campanulate; phyllaries ovate to lanceolate, 1–4 mm long, the apex obtuse or acute, the margin scarious, the median green; staminate corolla pale yellow, 3–4 mm long; carpellate corolla whitish, 3–4 mm long. Cypsela obovoid or cylindric, 1–2 mm long, 8- to 10-ribbed, glabrous; pappus of minutely barbellate, apically attenuate bristles 8–12 mm long.

Marshes and disturbed sites. Common; nearly throughout. New York and Massachusetts south to Florida, west to Oklahoma and Texas, also Nova Scotia; West Indies and Mexico; Europe, Asia, and Australia. Native to North America and West Indies. Fall.

Baccharis halimifolia is reported to hybridize with *B. angustifolia* in Florida where the ranges of the two overlap (Sundberg and Bogler, 2006).

Balduina Nutt. 1818. HONEYCOMBHEAD

Herbs. Leaves alternate, simple, pinnate-veined, epetiolate. Flowers in capitula, solitary or in corymbiform or paniculiform arrays, radiate; receptacle epaleate; ray flowers sterile; petals 5, basally connate, zygomorphic; disk flowers bisexual; petals 5, basally connate, actinomorphic; stamens 5, epipetalous, the filaments free, the anthers connate; ovary inferior, 2-carpellate, 1-loculate. Fruit a cypsela; pappus present.

A genus of 3 species; North America. [Commemorates William Bartram (1739–1823), American physician and botanist.]

Actinospermum Ell., 1823; *Endorima* Raf., 1819.

Selected reference: Keener (2006); Parker and Jones (1975).

1. Stem usually much branched; pappus less than 1 mm long; plant annual ....**B. angustifolia**
1. Stem solitary or rarely few-branched; pappus 1–2 mm long; plant perennial.
   2. Disk flowers yellow to orange-red .... **B. uniflora**
   2. Disk flowers dark purple...**B. atropurpurea**

Balduina angustifolia (Pursh) B. L. Rob. [*Anguste,* narrow, and *folius,* leaved, in reference to the narrow leaves.] COASTALPLAIN HONEYCOMBHEAD.

Buphthalmum angustifolium Pursh, Fl. Amer. Sept. 564. 1814. *Actinospermum angustifolium* (Pursh) Torrey & A. Gray, Fl. N. Amer. 2: 389. 1842. *Balduina angustifolia* (Pursh) B. L. Robinson, Proc. Amer. Acad. Arts 47: 215. 1911. TYPE: "Georgia and Florida," s.d., *Bartram s.n.* (holotype: BM).
Balduina multiflora Nuttall, Gen. N. Amer. Pl. 2: 175. 1818.

Erect annual herb, to 7 dm; stem angled, sparsely pubescent or glabrous. Basal leaves (absent at flowering) with the blade linear-spatulate, 1.5–6 cm long, 1–3 mm wide, the apex acute or obtuse, the base narrowly cuneate, the margin entire, the upper surface glabrous or sparsely pubescent, the lower surface glabrous, sessile, the proximal and mid cauline with the blade linear, 1.5–4.5 cm long, to 1 mm wide, the upper cauline similar, smaller, and reduced upward. Flowers in capitula in a corymbiform to paniculiform array, the peduncle 1–11.5 cm long, sparsely pubescent; involucre hemispheric or campanulate, 0.5–1.5 cm wide; phyllaries numerous in 3–5 series. lanceolate, 3–7 mm long, the apex acuminate-aristate, the proximal yellow, the distal green, the outer surface glabrous; receptacle slightly convex, pitted, the pit borders spinulose-cuspidate at the angles; ray flowers 5–13; corolla yellow, the lamina 0.8–1.8 cm long, the upper surface glabrous, the lower surface sparsely pubescent; disk flowers numerous; corolla yellow, 3–5 mm long, the outer surface sparsely pubescent. Cypsela obconic, 1–2 mm long, villous; pappus of obovate to suborbicular scales ca. 0.5 mm long.

Sandhills, scrub, and dunes. Common; nearly throughout. Georgia, Florida, Alabama, and Mississippi. Spring–summer.

Balduina atropurpurea R. M. Harper [Dark purple, in reference to the disk flowers.] PURPLEDISK HONEYCOMBHEAD; PURPLE BALDUINA.

Balduina atropurpurea R. M. Harper, Bull. Torrey Bot. Club 28: 483. 1901. *Endorima atropurpurea* (R. M. Harper) Small, Fl. S.E. U.S. 1283, 1340. 1903.

Erect perennial herb, to 9 dm, angled, glabrous or sparsely pubescent. Basal leaves (present or absent at flowering) with the blade linear-spatulate, 7–32 cm long, 5–10 mm wide, the apex acute to obtuse, the base narrowly cuneate, the margin entire, the upper surface glabrous, the lower surface glabrous or sparsely pubescent, sessile, the proximal and mid cauline 4–6 cm long, 2–6 mm wide, the distal cauline similar, smaller, and reduced upward. Flowers in capitula in a corymbiform to paniculiform array, the peduncle 5–20 cm long, sparsely pubescent; involucre hemispheric or campanulate, 1.5–2.5 cm wide; phyllaries numerous in 3–5 series, ovate to lanceolate, 3–8 mm long, the apex acute, the outer green to purple, the inner purple, the outer surface pubescent; receptacle slightly convex, pitted, the pit borders toothed at the angles; ray flowers 8–22; corolla yellow, the lamina 1.6–2.7 cm long, the upper surface glabrous, the lower surface pubescent; disk flowers numerous; corolla dark purple or yellow with a dark purple apex, 4–7 mm long, the outer surface sparsely villous. Cypsela obconic, ca. 2 mm long, villous; pappus of lanceolate, entire or apically laciniate scales ca. 2 mm long.

Bogs, wet flatwoods, and savannas. Rare; Nassau, Duval, Clay, and Putnam Counties. North Carolina south to Florida. Fall.

Balduina atropurpurea is listed as endangered in Florida (Florida Administrative Code, Chapeter 5B-40).

Balduina uniflora Nutt. [One-flowered, in reference to the solitary terminal flower.] ONEFLOWER HONEYCOMBHEAD.

Balduina uniflora Nuttall, Gen. N. Amer. Pl. 2: 175. 1818. *Balduina lutea* Rafinesque, New Fl. 4: 73. 1838 ("1836"), nom. illegit. *Actinospermum uniflorum* (Nuttall) Barnhart, Bull. Torrey Bot. Club 24: 411. 1897. *Endorima uniflora* (Nuttall) Barnhart ex Small, Fl. S.E. U.S. 1283, 1340. 1903.

Balduina bicolor Rafinesque, New Fl. 4: 73. 1838 ("1836"). TYPE: FLORIDA: s.d., *Leconte s.n.* (holotype: ?).

Erect perennial herb, to 1 m; stem sparsely pubescent to glabrous. Basal leaves (present or absent at flowering) with the blade spatulate, 3–10 cm long, 5–13 mm wide, the apex obtuse, the base narrowly cuneate, the margin entire, the upper surface glabrous, the lower surface sparsely pubescent or glabrous, sessile, the proximal cauline linear-spatulate, 1.5–5.5 cm long, 2–7 mm wide, the mid cauline and distal similar, smaller and reduced upward. Flowers in capitula, solitary or sometimes 2–4 per stem, the peduncle 13–24 cm long, pubescent; involucre hemispheric or campanulate, 1.5–2.5 cm wide; phyllaries numerous in 3–5 series, ovate to lanceolate, 4–11 mm long, the apex acute, the outer light green proximally, dark green distally, the outer surface pubescent; receptacle slightly convex, pitted, the pit borders toothed to subentire at the angles; ray flowers 8–22; corolla yellow, the lamina 1.2–3 cm long, the upper surface glabrous, the lower surface pubescent; disk flowers numerous; corolla yellow to orange-red, 6–8 mm long, the outer surface pubescent. Cypsela obconic, 1–2 mm long, villous; pappus of lanceolate, entire or apically laciniate scales 1–2 mm long.

Flatwoods, savannas, and bogs. Frequent; northern counties. North Carolina south to Florida, west to Mississippi. Summer–fall.

Berlandiera DC. 1836. GREENEYES

Herbs. Leaves alternate, simple, petiolate or epetiolate. Flowers in capitula, solitary or in paniculiform or corymbiform arrays, radiate; receptacle paleate; ray flowers carpellate; petals 5, basally connate, zygomorphic; disk flowers staminate; petals 5, basally connate, actinomorphic; stamens 5, epipetalous, the filaments free, the anthers connate; ovary inferior, 2-carpellate, 1-loculate. Fruit a cypsela; pappus absent.

A genus of 8 species; North America and Mexico. [Commemorates Jean-Louis Berlandier (1805–1851), French-born naturalist, physician, and anthropologist who lived and worked in Mexico and Texas.]

Selected reference: Pinkava (1967, 2006).

1. Leaves evenly distributed along the stem, the margin crenate, the upper and lower surfaces tomentose; disk flowers red to maroon .... **B. pumila**
1. Leaves mostly basal, the margin sinuate-pinnatifid, the upper and lower surfaces scabrous; disk flowers yellow .... **B. subacaulis**

Berlandiera pumila (Michx.) Nutt. [*Pumilio,* dwarf, in reference to its size.] SOFT GREENEYES.

Silphium pumulum Michaux, Fl. Bor.-Amer. 2: 146. 1803. *Berlandiera pumila* (Michaux) Nuttall, Trans. Amer. Philos. Soc., ser. 2. 7: 342. 1841. TYPE: FLORIDA: s.d., *Michaux 10* (holotype: P).
Silphium reticulatum Pursh, Fl. Amer. Sept. 579. 1814; non Moench, 1794. TYPE: FLORIDA.
Silphium tomentosum Pursh, Fl. Amer. Sept. 579. 1814. *Silphium pumilum* Michaux var. *tomentosum* (Pursh) Nuttall, Gen. N. Amer. Pl. 2: 183. 1818. *Berlandiera tomentosa* (Pursh) Nuttall, Trans. Amer. Philos. Soc., ser. 2. 7: 343. 1841.

Erect herb, to 1 m; stem striate, floccose. Leaves evenly distributed along the stem, the blade elliptic, ovate, or lanceolate, 6–13 cm long, 2–7 cm wide, the apex obtuse, the base truncate or cordate, the margin crenate, the upper and lower surfaces tomentose, the petiole of the proximal leaves ca. 2 cm long, gradually reduced upward, the distal ones becoming sessile or subsessile. Flowers in capitula in a paniculiform or corymbiform array, the peduncle 2–10 cm long, floccose; involucre campanulate or hemispheric 1.5–2 cm wide; phyllaries numerous in several series, obovate to suborbicular or elliptic, 1–1.5 cm long, the apex rounded, the surface sericeous; receptacle subtruncate; paleae linear-conduplicate, each wrapped around a disk flower, distally dilated, hirtellous; ray flowers usually 8; corolla deep yellow to orange-yellow with green veins on the lower surface of the lamina, the lamina 1.2–2 cm long; disk flowers numerous; corolla red to maroon, ca. 3 mm long. Cypsela obovate, 5–6 mm long, slightly laterally flattened, the edge thin, but not winged, black, pubescent, each adhering to and shed with 2 adjacent paleae, 2 disk flowers, and a subtending phyllary.

Sandhills. Frequent; northern counties, Marion and Volusia Counties. South Carolina to Florida, west to Texas. Spring–summer.

Berlandiera pumila hybridizes with *B. subacaulis* in Florida (*B.* ×*humilis*).

Berlandiera subacaulis (Nutt.) Nutt. [Somewhat stemless.] FLORIDA GREENEYES.

Silphium subacaule Nuttall, Amer. J. Sci. Arts 5: 301. 1821. *Silphium nuttallianum* Torrey, Ann. Lyceum Nat. Hist. New York 2: 216. 1828, nom. illegit. *Berlandiera subacaulis* (Nuttall) Nuttall, Trans. Amer. Philos. Soc., ser. 2. 7: 343. 1841. TYPE: FLORIDA: s.d., *Ware s.n.* (holotype: PH).

Erect herb, to 5 dm; stem often branched at the base, striate, scabrous. Leaves mostly basal, the blade oblanceolate to spatulate, 6–13 cm long, 2–7 cm wide, the apex obtuse, the base cuneate, the margin sinuate-pinnatifid, the margin of the ultimate lobes serrate-crenate, the upper and lower surfaces scabrous, the petiole, short-petiolate or sessile. Flowers in capitula, usually borne singly, the peduncle 3–20 cm long, scabrous; involucre campanulate or hemispheric, 1.5–2 mm wide; phyllaries numerous in several series, obovate to suborbicular or elliptic, 1–1.5 cm long, the apex rounded, the surface sericeous; receptacle subtruncate; paleae linear-conduplicate, each wrapped around a disk flower, distally dilated, hirtellous; ray flowers usually 8; corolla yellow with green veins on the lower surface of the lamina, the lamina 10–12 mm long; disk flowers numerous; corolla red to maroon, ca. 3 mm long. Cypsela obovate, 5–7 mm long, slightly laterally flattened, the edge thin, but not winged, black, pubescent, each adhering to and shed with 2 adjacent paleae, 2 disk flowers, and a subtending phyllary.

Sandhills. Common; peninsula west to central panhandle. Endemic. Spring–summer.

Berlandiera subacaulis hybridizes with *B. pumila* in Florida (*B. ×humilis*).

HYBRID

Berlandiera ×humilis Small [Low-growing.] (*B. pumila* × *B. subacaulis*).

Berlandiera humilis Small, Fl. S.E. U.S. 1246, 1340. 1903., pro. sp. TYPE: FLORIDA: St. Johns Co.: St. Augustine, Jun–Dec 1875, *Reynolds s.n.* (holotype: NY; isotype: F).

Sandhills. Rare; St. Johns County. Endemic. Spring–summer.

Bidens L. 1753. BEGGARTICKS

Herbs. Leaves opposite or whorled, simple or pinnately compound, pinnate- or palmate-veined, petiolate or epetiolate. Flowers in capitula, solitary or in corymbiform arrays, radiate, discoid, or disciform; calyculi present; receptacle paleate; ray flowers bisexual or sterile, sometimes absent; petals 5, basally connate, zygomorphic; disk flowers bisexual; petals 5, basally connate, actinomorphic; stamens 5, epipetalous, the filaments free, the anthers connate; ovary inferior, 2-carpellate, 1-loculate. Fruit a cypsela; pappus present or absent.

A genus of about 200 species; nearly cosmopolitan. [*Bis,* two, and *dens,* tooth, in reference to the 2-awned pappus of the original species.

1. Leaves simple, sessile .... **B. laevis**
1. Leaves lobed or divided (rarely simple) petiolate.
    2. Cypsela laterally obcompressed-quadrangulate.
        3. Leaves mostly 2- to 3-pinnately divided, the segments entire or few-toothed .... **B. bipinnata**
        3. Leaves pinnately divided or undivided, the segments finely serrate.

4. Heads radiate, the lamina of the ray flowers 5–16 mm long .... **B. alba**
4. Heads discoid or pseudoradiate (ray flowers tubular), 2–3 mm long.......... **B. pilosa**

2. Cypsela laterally flattened.
   5. Lamina of the ray flowers 1–3 cm long; outer phyllaries subequaling or shorter than the inner phyllaries.
      6. Cypsela narrowly cuneate (2.5–4 times as long as wide, the margin antrorsely barbed or ciliate.......... **B. trichosperma**
      6. Cypsela broadly cuneate (1.5–2.5 times as long as wide), the margin not barbed or ciliate .... **B. mitis**
   5. Lamina of the ray flowers less than 0.5 cm long or the ray flowers absent; outer phyllaries much exceeding the inner phyllaries.
      7. Outer phyllaries with a smooth margin.......... **B. discoidea**
      7. Outer phyllaries with a ciliate margin.......... **B. frondosa**

Bidens alba (L.) DC. [White, in reference to the ray flower laminas.] BEGGARTICKS; ROMERILLO.

Coreopsis alba Linnaeus, Sp. Pl. 908. 1753. *Bidens alba* (Linnaeus) de Candolle, Prodr. 5: 605. 1836. *Bidens pilosa* Linnaeus var. *alba* (Linnaeus) O. E. Schulz, in Urban, Symb. Antill. 7: 136. 1911.

Coreopsis leucanthema Linnaeus, Cent. Pl. 1: 29. 1755. *Coreopsis leucantha* Linnaeus, Sp. Pl., ed. 2. 1281. 1763, nom. illegit. *Bidens leucantha* Willdenow, Sp. Pl. 3: 1719. 1803, nom. illegit. *Kerneria leucantha* Cassini, in Curvier, Dict. Sci. Nat. 24: 398. 1822, nom. illegit. *Bidens pilosa* Linnaeus forma *radiata* Schultz Bipontinus, in Webb & Berthelot, Hist. Nat. Iles Canaries 3(2(2)): 242. 1844. *Bidens pilosa* Linnaeus var. *radiata* (Schultz Bipontinus) J. A. Schmidt, Beitr. Fl. Cap. Verd. Isl. 197. 1852. *Bidens pilosa* Linnaeus var. *leucantha* Harvey, in Harvey & Sonder, Fl. Cap. 3: 133. 1864, nom. illegit. *Kerneria pilosa* (Linnaeus) Lowe var. *radiata* (Schultz Bipontinus) Lowe, Man. Fl. Madeira 1: 474. 1868. *Bidens pilosa* Linnaeus forma *leucantha* Kuntze, Revis. Gen. Pl. 1: 322. 1891, nom. illegit. *Bidens leucanthema* (Linnaeus) E.H.L. Krause, in Sturm, Deutschl. Fl., ed. 2. 13: 159. 1905. *Bidens pilosa* Linnaeus subvar. *radiata* (Schultz Bipontinus) Pitard, in Pitard & Proust, Iles Canaries 226. 1909. *Bidens alba* (Linnaeus) de Candolle var. *radiata* (Schultz Bipontinus) R. E. Ballard ex Melchert, Phytologia 32: 295. 1975.

Erect or ascending annual herb, to 6(18) dm; stem striate-angled, sparsely pilose or glabrate. Leaves opposite, simple, the blade ovate to lanceolate, 3–7(12) cm long, 1–2(4.5) cm wide, pinnately lobed, the primary lobes 3–7, ovate or lanceolate, (1)2.5–8 cm long, (0.5)1–4 cm wide, the apex acute, the base truncate or cuneate, the ultimate margin serrate-crenate, sometimes ciliate, the upper and lower surfaces pilosulous, sparsely hirtellous, or glabrate, the petiole 1–3(7) cm long. Flowers in capitula, solitary or in an open corymbiform array, the peduncle 1–2(9) cm long; calyculus of (6)7–9(13) spatulate or linear bracts (3)4–5 mm long, subappressed, the margin ciliate, the outer surface hispidulous or puberulent; involucre turbinate or campanulate, 5–6 mm long, 6–8 mm wide; phyllaries (8)12(16), lanceolate or oblanceolate, 4–6 mm long; receptacle flat or slightly convex; paleae oblong or linear, to 5 mm long, flat or slightly navicular; ray flowers 5–8, sterile; corolla white, the lamina 5–16 mm long; disk flowers 12–many; corolla yellow, (2)3–5 mm long. Cypsela fusiform to narrowly oblanceolate, the apex truncate or slightly acute, laterally flattened, the margin antrorsely hispidulus, the face obscurely 2-grooved, tuberculate-hispidulous; pappus absent or 1–2 erect to divergent, retrorsely barbed awns 1–3 mm long.

Disturbed sites. Common; throughout. Massachusetts and Connecticut south to Florida, west to Alabama, also Missouri, Louisiana, and New Mexico. Nearly cosmopolitan. Native to North America, tropical America, and Africa. All year.

Some recent treatments subsume *B. alba* into *B. pilosa* s.l. (León de la Luz and Medel Narváez, 2013; Melchert, 2010; Strother and Weedon, 2006), while others continue to recognize *B. alba* and *B. pilosa* as distinct as we do here. Reportedly *B. alba* and *B. pilosa* are interfertile (Norton, 1991).

Bidens bipinnata L. [Twice-pinnate, in reference to the leaves.] SPANISH NEEDLES.

Bidens bipinnata Linnaeus, Sp. Pl. 832. 1753. *Bidens pilosa* Linnaeus var. *bipinnata* (Linnaeus) Hooker f., Fl. Brit. India 3: 309. 1881.

Erect annual herb, to 1(1.5) m; stem striate-angled, glabrous or sparsely pubescent. Leaves opposite, simple, the blade deltate, ovate, or lanceolate, (2)3–7 cm long, (2)3–6 cm wide, (1)2(3)-pinnisect, the ultimate lobes obovate or lanceolate, 1.5–4.5 cm long, 1–2.5 cm wide, the apex rounded, acute, or attenuate, the base truncate or cuneate, the margin irregularly and coarsely serrate or entire, sometimes ciliolate, the upper and lower surfaces glabrous or sparsely hirtellous, the petiole 2–5 cm long. Flowers in capitula, solitary or in a corymbiform array, the peduncle (1)2–5(10) cm long; calyculus of (7)8(10) linear bracts 3–5 mm long, subappressed, the margin ciliate, the outer surface glabrous or sparsely hirtellous; involucre campanulate; phyllaries 8–12, lanceolate to linear, 4–6 mm long; receptacle flat or slightly convex; paleae oblong or linear, to 5 mm long, flat to slightly navicular; ray flowers 3–5, sterile, or absent; corolla yellow or whitish, the lamina 1–2(3) mm long; disk flowers 10–many; corolla yellow or whitish, 2–3 mm long. Cypsela black, red-brown, or stramineous, the outer oblanceolate to obovate, 7–15 mm long, weakly laterally flattened, the inner linear to fusiform, 12–18 mm long, 4-angled, the faces 2-grooved, usually tuberculate-hispidulous; pappus of (2)3–4 erect to divergent barbed awns 2–4 mm long.

Bluff forests, floodplain forests, marshes, and disturbed sites. Frequent; peninsula west to central panhandle. New Brunswick south to Florida, west to Ontario, Nebraska, Kansas, Oklahoma, New Mexico, and Arizona; Mexico and South America; Africa, Europe, Asia, Australia, and Pacific Islands. Native to Asia. Summer–fall.

Bidens discoidea (Torr. & A. Gray) Britton [Disklike, in reference to the capitulum composed of only disk flowers.] SMALL BEGGARTICKS.

Coreopsis discoidea Torrey & A. Gray, Fl. N. Amer. 2: 339. 1842. *Bidens discoidea* (Torrey & A. Gray) Britton, Bull. Torrey Bot. Club 20: 281. 1893.

Erect annual herb, to 6(18) dm; stem striate-angled, glabrous. Leaves opposite, simple or pinnately compound (usually 3-foliolate), the blade deltate, lanceolate, or ovate, 3–8(10) cm long, 1–3(8) cm wide, the leaflets lanceolate or ovate, (1)2–5(10) cm long, 0.5–2(4) cm wide, the apex acuminate or attenuate, the base cuneate, the margin irregularly serrate, sometimes ciliate, the upper and lower surfaces glabrous, the petiolule to 1 cm long, the petiole 1–4(6) cm long. Flowers in capitula, solitary or 2–3 in a corymbiform array, the peduncle 1–2(5) cm long; calyculus of (3)4(5) spatulate or linear bracts (3)12–25 mm long, subappressed, sometimes foliaceous, the outer surface glabrous; involucre hemispheric; phyllaries 5–7, oblong to lanceolate, 4–6 mm long; receptacle flat or slightly convex; paleae oblong or linear, to 5 mm long, flat to slightly navicular; ray flowers absent; disk flowers (10)15–20; corolla yellow-orange, ca. 2 mm long. Cypsela linear or lanceolate, slightly laterally flattened, the apex truncate, red-brown, the outer 3–5 mm long, the inner 4–6 mm long, the surface tuberculate, antrorsely strigillose; pappus of 2 erect, antrorsely barbed or smooth awns to 3 mm long.

Marshes and floodplain forests. Rare; Nassau and Alachua Counties, central panhandle, Escambia County. New Brunswick south to Florida, west to Ontario, Minnesota, Iowa, Missouri, Oklahoma, and Texas. Summer–fall.

Bidens frondosa L. [Leafy.] DEVIL'S BEGGARTICKS.

Bidens frondosa Linnaeus, Sp. Pl. 832. 1753.
Bidens melanocarpa Wiegand, Bull. Torrey Bot. Club 26: 405. 1899. SYNTYPE: FLORIDA.

Erect annual herb, to 6(18) dm; stem striate-angled, glabrous or sparsely pubescent. Leaves opposite, pinnately compound, 3(5)-foliolate, the blade deltate, lanceolate, or ovate, 3–8(15) cm long, 2–6(10) cm wide, the leaflets lanceolate or ovate, (1.5)3.5–6(12) cm long, (0.5)1–2.5(3) cm wide, the apex acuminate, the base cuneate, the margin dentate or serrate, sometimes ciliate, the upper and lower surfaces glabrous or hirtellous, sessile or short-petiolulate, the petiole 1–4(6) cm long. Flowers in capitula, solitary or 2–3 in a corymbiform array, the peduncle 1–4(8) cm long; calyculus of (5)8(10) spatulate, oblanceolate, or linear bracts 0.5–2(6) cm long, ascending or spreading, sometimes foliaceous, the outer surface glabrous or hirtellous, the margin ciliate; involucre campanulate to hemispheric; phyllaries 6–12, oblong, ovate, or lanceolate, 5–9 mm long, the outer with a ciliate margin; receptacle flat or slightly convex; paleae oblong or linear, to 5 mm long, flat to slightly navicular; ray flowers 1–3 or absent, sterile; corolla yellow, the lamina 2–4 mm long; disk flowers numerous; corolla yellow-orange, ca. 3 mm long. Cypsela obovate the apex truncate, slightly laterally flattened, blackish-brown or stramineous, the outer 5–7 mm long, the inner 7–10 mm long, the faces 1-nerved, sometimes tuberculate, the margin antrorsely or retrorsely barbed; pappus of 2 erect or spreading, antrorsely or retrorsely barbed awns 2–5 mm long.

Marshes. Occasional; Alachua County, panhandle. Nearly throughout North America; Mexico; Europe. Native to North America and Mexico. Fall.

Bidens laevis (L.) Britton et al. [Smooth, in reference to the leaves.] BURRMARIGOLD; SMOOTH BEGGARTICKS.

Helianthus laevis Linnaeus, Sp. Pl. 906. 1753. *Heliopsis laevis* (Linnaeus) Persoon, Syn. Pl. 2: 473. 1807. *Bidens laevis* (Linnaeus) Britton et al., Prelim. Cat. 29. 1888.

Coreopsis perfoliata Walter, Fl. Carol. 214. 1788. *Bidens chrysanthemoides* Michaux, Fl. Bor.-Amer. 2: 136. 1803, nom. illegit.
Bidens nashii Small, Bull. Torrey Bot. Club 25: 481. 1898. *Bidens chrysanthemoides* Michaux var. *nashii* (Small) Jepson, Fl. W. Calif. 544. 1901. TYPE: FLORIDA/LOUISIANA.

Erect annual herb, to 6(1.2) dm; stem striate-angled, glabrous. Leaves opposite or sometimes 3–4 in a whorl, simple, the blade elliptic, lanceolate, or linear, (2)5–10(15) cm long, 1–3(4) cm wide, the apex acute or acuminate, the base cuneate or rounded, the margin coarsely dentate or serrate, sometimes ciliate, the upper and lower surfaces glabrous, sessile. Flowers in capitula, solitary or in a corymbiform array, the peduncle (1)2–6 cm long; calyculus of 5–7(10) oblanceolate, lanceolate, or linear bracts (6)10–12(20) mm long, spreading, foliaceous, the margin ciliate, the outer surface glabrous or hispidulous at the base; involucre turbinate to hemispheric; phyllaries 8–12, ovate, obovate, or oblong-lanceolate, (4)6–8(10) mm long, the apex sometimes orange or purplish; paleae flat to slightly navicular, narrowly oblong to linear, to 5 mm long, the apex sometimes orange or purplish; ray flowers 7–8, sterile, or absent; corolla orange-yellow, the lamina (1)2–3 mm long; disk flowers numerous; corolla yellow or orange-yellow, 3–6 mm long. Cypsela obovate, the apex truncate, laterally flattened or sometimes 3- to 4-angled, the outer 6–8 mm long, the inner 8–10 mm long, the margin retrorsely ciliate or barbed, the faces usually 1-nerved or striate, glabrous; pappus of 2–4 erect or slightly spreading, retrorsely barbed awns 3–5 mm long.

Marshes. Frequent; nearly throughout. Maine south to Florida, west to California; Mexico, Central America, and South America; Pacific Islands. Native to North America, Mexico, Central America, and South America. Fall.

Bidens mitis (Michx.) Sherff [Mild, innoxious, in reference to the original specimens in comparison to other related species.] SMALLFRUIT BEGGARTICKS.

Coreopsis mitis Michaux, Fl. Bor.-Amer. 2: 140. 1803. *Diodonta mitis* (Michaux) Nuttall, Trans. Amer. Philos. Soc., ser. 2. 7: 360. 1841. *Coreopsis aurea* Aiton var. *incisa* Torrey & A. Gray, Fl. N. Amer. 2: 340. 1842. *Bidens mitis* (Michaux) Sherff, Bot. Gaz. 81: 43. 1926.
Diodonta leptophylla Nuttall, Trans. Amer. Philos. Soc., ser. 2. 7: 360. 1841. *Coreopsis aurea* Aiton var. *leptophylla* (Nuttall) Torrey & A. Gray, Fl. N. Amer. 2: 339. 1842. *Bidens coronata* (Linnaeus) Fischer ex Britton var. *leptophylla* (Nuttall) C. Mohr, Contr. U.S. Nat. Herb. 6: 808. 1901. *Bidens mitis* (Michaux) Sherff var. *leptophylla* (Nuttall) Small, Man. S.E. Fl. 1454. 1933.

Erect annual herb, to 1 m; stem striate-angled, glabrous or sparsely hirtellous. Leaves opposite, the blade deltate, ovate, or lanceolate, 3–10 cm long, 1–5 cm wide, the apex acute or attenuate, the base cuneate, the margin pinnatisect, the ultimate lobes 3(7), ovate, lanceolate or linear, 1–6 cm long, 5–10 mm wide, the ultimate margin irregularly serrate, undulate, or entire, sometimes ciliolate, the upper and lower surfaces glabrous or hirtellous, the petiole 0.5–2(3) cm long. Flowers in capitula in an open corymbiform array, the peduncle 2–12 cm long; calyculus of 7–10 spatulate or linear bracts 5–10 mm long, sometimes foliaceous, the margin ciliate, the outer surface glabrous or hirtellous; involucre hemispheric; phyllaries 6–8, lanceolate, 4–5 mm long; receptacle flat or slightly convex; paleae oblong or linear, to 5 mm long, flat or slightly navicular; ray flowers 8–13, sterile; corolla yellow, the lamina 1–2.5 cm long; disk flowers 25–50; corolla yellow, 2–3 mm long. Cypsela obovate, the apex subtruncate, laterally flattened or unequally 3(4)-angled, reddish brown, the outer 3–4 mm long, the inner 4–5 mm long, the margin smooth, the faces sometimes 1-nerved, sometimes tuberculate, glabrous, or hirtellous; pappus absent or of 2 erect, antrorsely barbed awns or deltate scales to 1 mm long.

Marshes, cypress domes, and swamps. Common, nearly throughout. Virginia south to Florida, west to Missouri, Arkansas, and Texas. All year.

Bidens pilosa L. [With pilose trichomes, in reference to the leaf surface.] SPANISH NEEDLES.

Bidens pilosa Linnaeus, Sp. Pl. 832, 1753. *Kerneria tetragona* Moench, Methodus 505. 1794, nom. illegit. *Bidens pilosa* Linnaeus forma *discoidea* Schultz Bipontinus, in Webb & Berthelot, Hist. Iles Canaries 3(2(2)): 242. 1844. *Bidens leucantha* Willdenow var. *pilosa* (Linnaeus) Grisebach, Cat. Pl. Cub. 155. 1866. *Kerneria pilosa* (Linnaeus) Lowe, Man. Pl. Madeira 1: 474. 1868. *Kerneria pilosa* (Linnaeus) Lowe var. *discoidea* (Schultz Bipontinus) Lowe, Man. Fl. Madeira 1: 474. 1868, nom. inadmiss. *Bidens pilosa* Linnaeus subvar. *discoidea* (Schultz Bipontinus) Pitard & Proust, Iles Canaries 226. 1909, nom. inadmiss.

Erect or ascending annual herb, to 6(18) dm; stem striate-angled, sparsely pilose or glabrate. Leaves opposite, simple, the blade ovate to lanceolate, 3–7(12) cm long, 1–2(4.5) cm wide, pinnately lobed, the primary lobes 3–7, ovate or lanceolate, (1)2.5–8 cm long, (0.5)1–4 cm wide, the apex acute, the base truncate or cuneate, the ultimate margin serrate-crenate, sometimes ciliate, the upper and lower surfaces pilosulous, sparsely hirtellous, or glabrate, the petiole 1–3(7) cm long. Flowers in capitula, solitary or in an open corymbiform array, the peduncle 1–2(9) cm long; calyculus of (6)7–9(13) spatulate or linear bracts (3)4–5 mm long, subappressed, the margin ciliate, the outer surface hispidulous or puberulent; involucre turbinate or campanulate; phyllaries 7–10, lanceolate or oblanceolate, 4–6 mm long; receptacle flat or slightly convex; paleae oblong or linear, to 5 mm long, flat or slightly navicular; ray flowers (3)5–8, bisexual or absent; corolla white, the lamina, 7–15 mm long; disk flowers numerous; corolla yellow, (2)3–5 mm long. Cypsela fusiform to narrowly oblanceolate, the apex truncate or slightly acute, laterally flattened, the margin antrorsely hispidulous, the face obscurely 2-grooved, tuberculate-hispidulous; pappus of 3–5 erect to slightly divergent, retrorsely barbed awns 1–3 mm long.

Disturbed sites. Rare; Sarasota, Leon, and Gulf Counties. Quebec and Ontario south to Florida, west throughout the southern states to California and north to Oregon. Nearly cosmopolitan. Native to North America and tropical America. Summer–fall.

Some recent treatments subsume *B. alba* into *B. pilosa* s.l. (Strother and Weedon, 2006; Melchert, 2010; León de la Luz and Medel Narváez, 2013), while others continue to recognize *B. alba* and *B. pilosa* as distinct as we do here. Reportedly *B. alba* and *B. pilosa* are interfertile (Norton, 1991).

Bidens trichosperma (Michx.) Britton [From the Greek *trichos,* hair, and *sperma,* seed, in reference to pappus awns.] CROWNED BEGGARTICKS.

Coreopsis coronata Linnaeus, Sp. Pl., ed. 2. 1281. 1763. *Diodonta coronata* (Linnaeus) Nuttall, Trans. Amer. Philos. Soc., ser. 2. 7: 360. 1841. *Bidens coronata* (Linnaeus) Fischer ex Britton, Bull. Torrey

Bot. Club 20: 281. 1893; non Fischer ex Colla, 1834. *Bidens coronata* (Linnaeus) Fischer ex Britton var. *typica* Fernald, Rhodora 40: 349. 1838, nom. inadmiss.

Coreopsis trichosperma Michaux, Fl. Bor.-Amer. 2: 1803. *Bidens trichosperma* (Michaux) Britton, Bull. Torrey Bot. Club 20: 281. 1893. *Bidens coronata* (Linnaeus) Fischer ex Britton var. *trichosperma* (Michaux) Fernald, Rhodora 40: 350. 1938.

Erect annual herb, to 5(15) dm; stem striate-angled, glabrous or sparsely pubescent. Leaves opposite, simple, the blade deltate or ovate, 4–8(15) cm long, 1–5(13) cm wide, laciniately 1- to 2-pinnatisect, the ultimate lobes (3)5–7, oblanceolate or linear, 1–3(8) cm long, 4–8(12) mm wide, the apex acute, the base cuneate, the ultimate margin incised, dentate, serrate, or entire, the upper and lower surfaces hirtellous or strigillose, the petiole 0.5–2 cm long. Flowers in capitula in an open corymbiform array; calyculus of (6)8(10) spatulate or linear bracts 3–10(15) mm long, ascending or spreading, sometimes foliaceous, the margin sometimes ciliate, the outer surface glabrous; involucre hemispheric; phyllaries 6–8(10), oblong, 3–8 mm long; receptacle flat or slightly convex; paleae oblong or linear, to 5 mm long, flat to slightly navicular; ray flowers 7–9, sterile; corolla yellow, the lamina 1–3 cm long; disk flowers numerous, bisexual; corolla yellow, 3–5 mm long. Cypsela narrowly obovate, slightly laterally compressed, blackish brown, the apex truncate, the outer 3–6 mm long, the inner 5–9 mm long, the margin antrorsely barbed or ciliate, the faces obscurely 1-nerved, sometimes tuberculate, glabrous or sparsely hispidulous; pappus of 2 erect or slightly spreading barbed awns or scales 1–3 mm long.

Fresh and brackish marshes. Rare; Duval County. Quebec south to Florida, west to Ontario, Minnesota, South Dakota, Nebraska, Missouri, Arkansas, and Alabama. Fall.

EXCLUDED TAXA

Bidens aristosa (Michaux) Britton—Reported by Correll and Johnston (1970) as occurring in "most of the e. U.S.," which would presumably include Florida. Excluded from Florida by Cronquist (1980). No Florida specimens seen.

Bidens aurea (Aiton) Sherff—This European species was reported for Florida by Chapman (1890, 1897, both as *Coreopsis aurea* Aiton). The name misapplied to Florida material of *B. mitis*.

Bigelowia DC., nom. cons. 1836. RAYLESS GOLDENROD

Herbs. Leaves alternate, simple, pinnate-veined, epetiolate. Flowers in capitula in corymbiform arrays, discoid, bisexual; receptacle epaleate; petals 5, basally connate, actinomorphic; stamens 5, epipetalous, the filaments free, the anthers connate; ovary inferior, 2-carpellate, 1-loculate. Fruit a cypsela; pappus present.

A genus of 2 species; North America. [Commemorates Jacob Bigelow (1787–1879), American physician and botanist.]

Chondrophora Raf. ex Porter & Britton, 1894.

Selected reference: Anderson (1970).

1. Basal leaves linear-oblanceolate or narrowly spatulate, 2–14 mm wide, the base not persistent .... **B. nudata**
1. Basal leaves linear-filiform, 1–2 mm wide, the base persistent .... **B. nuttallii**

Bigelowia nudata (Michx.) DC [Naked, in reference to the sparse cauline leaves.] PINELAND RAYLESS GOLDENROD.

Erect perennial herb, to 5(7) dm long; stem striate-angled, glabrous, basally rhizomatous. Basal leaves with the blade linear-oblanceolate or narrowly spatulate, (2)3–16(22) mm long, 2–14 mm wide, the apex acute or obtuse, the base cuneate, sheathing, the margin entire, the upper and lower surfaces glabrous, sessile, the base not persistent, the cauline leaves reduced in number and size upward. Flowers in capitula in a flat-topped corymbiform array; involucre cylindric or cylindric-turbinate; phyllaries 10–20 in 3–4 series, narrowly lanceolate, 5–8 mm long, 1-nerved, the margin narrowly scarious, glabrous; receptacle flat, pitted; flowers 2–6; corolla yellow, 3–5 mm long. Cypsela turbinate or cylindric, ca. 3 mm long, slightly laterally flattened or nearly quadrate, 4- to 5-nerved, sparsely strigose; pappus of numerous tan, unequal, barbellate, apically attenuate bristles.

1. Basal leaves (2)3–8(10) cm long, (4)5–14 mm wide .... subsp. **nudata**
1. Basal leaves (5)8–16(22) cm long, 2–5 mm wide..subsp. **australis**

Bigelowia nudata subsp. **nudata**

Chrysocoma nudata Michaux, Fl. Bor.-Amer. 2: 101. 1803. *Bigelowia nudata* (Michaux) de Candolle, Prodr. 5: 329. 1836. *Bigelowia nudata* (Michaux) de Candolle var. *spathulifolia* Torrey & A. Gray, Fl. N. Amer. 2: 232. 1842, nom. inadmiss. *Bigelowia nudata* (Michaux) de Candolle forma *spathulifolia* A. Gray, Proc. Amer. Acad. Arts 8: 646. 1873, nom. inadmiss. *Aster nudatus* (Michaux) Kuntze, Revis. Gen. Pl. 1: 319. 1891. *Chondrophora nudata* (Michaux) Britton, in Britton & Porter, Mem. Torrey Bot. Club 5: 317. 1894.

Chrysocoma virgata Nuttall, Gen. N. Amer. Pl. 2: 137. 1818. *Bigelowia virgata* (Nuttall) de Candolle, Prodr. 5: 329. 1836. *Bigelowia nudata* (Michaux) de Candolle var. *virgata* (Nuttall) Torrey & A. Gray, Fl. N. Amer. 2: 232. 1942. *Bigelowia nudata* (Michaux) de Candolle forma *virgata* (Nuttall) A. Gray, Proc. Amer. Acad. Arts 8: 646. 1873. *Chondrophora nudata* (Michaux) Britton var. *virgata* (Nuttall) Britton, in Britton & Porter, Mem. Torrey Bot. Club 5: 317. 1894. *Chondrophora virgata* (Nuttall) Greene, Erythea 3: 91: 1895.

Basal leaves (2)3–8(10) cm long, (4)5–14 mm wide. Phyllaries 6–8 mm long; corolla 3–4 mm long.

Wet flatwoods and swamp margins. Frequent; northern counties and central peninsula. North Carolina south to Florida, west to Louisiana. All year.

Bigelowia nudata subsp. **australis** L. C. Anderson [Southern.]

Bigelowia nudata (Michaux) de Candolle subsp. *australis* L. C. Anderson, Sida 3: 463. 1970. *Bigelowia nudata* (Michaux) de Candolle var. *australis* (L. C. Anderson) Shinners, Sida 4: 274. 1971. TYPE: FLORIDA: Lee Co.: vicinity of Fort Myers, 20 Mar 1916, *Standley 28* (holotype: NY; isotypes, GH, MO, US).

Basal leaves (5)8–16(22) cm long, 2–5 mm wide. Phyllaries 6–8 mm long; corolla 4–5 mm long.
Moist flatwoods. Common; peninsula. Endemic. All year.

Bigelowia nuttallii L. C. Anderson [Commemorates Thomas Nuttall (1786–1859), English botanist who worked in America (1808–1841).] NUTTALL'S RAYLESS GOLDENROD.

Bigelowia nuttallii L. C. Anderson, Sida 3: 460. 1970.

Erect perennial herb, to 5(6) dm; stem striate-angled, glabrous, basally rhizomatous. Basal leaves numerous, the blade linear-filiform, (6)8–12(14) cm long, 1–2 mm wide, the apex acute, the base cuneate, sheathing, the margin entire, the upper and lower surfaces glabrous, sessile, the base persistent, the cauline leaves reduced in number and size upward. Flowers in capitula in a flat-topped corymbiform array; involucres cylindric or cylindric-turbinate; phyllaries 10–20 in 3–4 series, narrowly lanceolate, 5–8 mm long, 1-nerved, the margin narrowly scarious, glabrous; receptacle flat, pitted; flowers 2–6; corolla yellow, 4–5 mm long, glabrous. Cypsela turbinate to cylindric, ca. 3 mm long, slightly laterally flattened or nearly quadrate, 4- to 5-nerved, sparsely strigose; pappus of numerous tan, unequal, barbellate, apically attenuate bristles.

Sandstone outcrops in mesic longleaf pinelands and scrub. Rare; Washington and Pinellas Counties. Georgia, Florida, Alabama, Louisiana, and Texas. Fall–winter.

Bigelowia nuttallii is listed as endangered in Florida (Florida Administrative Code, Chapter 5B-40).

Boltonia L'Hér. 1789. DOLL'S DAISY

Herbs. Leaves alternate, simple, pinnate-veined, epetiolate. Flowers in capitula in corymbiform or paniculiform arrays, radiate; phyllaries present; receptacle epaleate; ray flowers bisexual; petals 5, basally connate, zygomorphic; disk flowers bisexual; petals 5, basally connate, actinomorphic; stamens 5, epipetalous, the filaments free, the anthers connate; ovary inferior, 2-carpellate, 1-loculate. Fruit a cypsela; pappus present.

A genus of 5 species; North America. [Commemorates James Bolton (1735–1799), English botanist and artist.]

Selected reference: Karaman-Castro and Urbatsch (2006).

1. Inflorescence narrow and few-headed .... **B. asteroides**
1. Inflorescence diffusely branched.
    2. Inflorescence usually leafy bracteate, at least some of the bracts more than 1 cm long .... **B. apalachicolensis**
    2. Inflorescence merely subulate-bracteate, the bracts usually less than 1 cm long .... **B. diffusa**

Boltonia apalachicolensis L. C. Anderson [Of Apalachicola, Florida.] APALACHICOLA DOLL'S DAISY.

Boltonia apalachicolensis L. C. Anderson, Syst. Bot. 12: 133. 1987. TYPE: FLORIDA: Franklin Co.: Forbes Island at milepost 16.8 on Apalachicola River, ca. 12.7 air mi. NNW of Apalachicola, 18 Oct 1985, *Anderson 8921* (holotype: NY; isotypes: FSUM MO).

Erect perennial herb, to 1.8 m; stem striate-angled, glabrous. Leaves with the blade linear-oblanceolate, obovate, or obliquely elliptic, 4.5–12 cm long, 8–12 mm wide, reduced upward, 1-nerved, the apex acute to acuminate, the base cuneate, the margin entire, the upper and lower surfaces glabrous, sessile. Flowers in capitula in a corymbiform array, the peduncle 0.5–3.5 cm long, with 1–3 linear-oblanceolate bracteate leaves 1–3 mm long; involucre hemispheric; phyllaries linear to subulate in 2–4 series, the outer 1–2 mm long, the inner 2–3 mm long; receptacle conic to hemispheric, obscurely pitted; ray flowers 20–35; corolla white or lilac, the lamina 5–9 mm long; disk flowers numerous; corolla white or lilac, 2–3 mm long; cypsela obovoid, 1–2 mm long, the lateral margins winged; pappus of 2–3 awns to 1 mm long.

Floodplain forests. Rare; Liberty, Franklin, Gulf, and Washington Counties. Endemic. Summer–fall.

Boltonia asteroides (L.) L'Hér. [Resembling the genus *Aster*.] WHITE DOLL'S DAISY.

Matricaria asteroides Linnaeus, Mant. Pl. 116. 1767. *Boltonia asteroides* (Linnaeus) L'Héritier de Brutelle, Sert. Angl. 27. 1789 ("1788").

Matricaria glastifolia Hill, Hort. Kew. 19, t. 3. 1769. *Boltonia glastifolia* (Hill) L'Héritier de Brutelle, Sert. Angl. 27. 1789 ("1788"). *Boltonia asteroides* (Linnaeus) L'Héritier de Brutelle var. *glastifolia* (Hill) Fernald, Rhodora 42: 486. 1940.

Erect perennial herb, to 2 m; stem striate-angled, glabrous, stoloniferous. Leaves with the blade linear or lanceolate, 2–20 cm long, 2–30 mm wide, reduced upward, 1-nerved, the apex acute or obtuse, the base cuneate, the margin entire, the upper and lower surfaces glabrous, sessile. Flowers in capitula in a corymbiform or paniculiform array, the peduncle 0.5–20 mm long, with 1–15 linear or oblanceolate bracteate leaves (0.5)1.5–9 cm long or absent; involucre hemispheric; phyllaries linear or subulate in 3–5 series, 2–4 mm long, the outer slightly smaller than the inner; receptacle conic to hemispheric, obscurely pitted; ray flowers 2–many; corolla white or lilac, the lamina 5–13 mm long; disk flowers numerous; corolla white or lilac, 2–3 mm long. Cypsela obovoid, 1–3 mm long, the lateral margins winged or wingless; pappus of 2–3 awns 1–2 mm long.

Wet hammocks. Rare; Nassau, Franklin, Gulf, and Santa Rosa Counties. New York south to Florida, west to Arkansas, Louisiana, and Texas. Summer–fall.

Boltonia diffusa Elliott [Widely spreading, in reference to the inflorescence branching.] SMALLHEAD DOLL'S DAISY.

Boltonia diffusa Elliott, Sketch Bot. S. Carolina 2: 400. 1823.

Erect perennial herb, to 2 m; stem striate-angled, glabrous, stoloniferous. Leaves with the blade linear or narrowly lanceolate, 3–11 cm long, 3–17 mm wide, reduced upward, 1-nerved, the apex acute or obtuse, the base cuneate, the margin entire, the upper and lower surfaces glabrous, sessile. Flowers in capitula in a paniculiform array, the peduncle 0.5–18 cm long, with 3–10 subulate bracteate leaves 0.5–2.5 cm long; involucre hemispheric; phyllaries subulate, in 4–6 series, 1–2 mm long, the outer slightly smaller; receptacle conic to hemispheric, obscurely pitted; ray flowers numerous; corolla white or lilac, the lamina 3–6 mm long; disk flowers numerous; corolla white or lilac, 1–2 mm long. Cypsela obovoid, 2–3 mm long, the lateral margins winged; pappus of 2–3 awns, to 1 mm long.

Marshes and wet disturbed sites. Frequent; nearly throughout. Summer–fall.

Borrichia Adans. 1763. SEASIDE OXEYE

Herbs or shrubs. Leaves opposite, simple, pinnate-veined, petiolate or epetiolate. Flowers in capitula, solitary or in cymbiform arrays, radiate; receptacle paleate; ray flowers carpellate; petals 5, basally connate, zygomorphic; disk flowers bisexual; petals 5, basally connate, actinomorphic; stamens 5, epipetalous, the filaments free, the anthers connate; ovary inferior, 2-carpellate, 1-loculate. Fruit a cypsela; pappus present.

A genus of 2 species; North America, West Indies, Mexico, and Central America. [Commemorates Ole Borch (1626–1690), Danish botanist.]

Selected references: Semple (1978, 2006a); Semple and Semple (1977).

1. Phyllaries with a prominent spreading or erect apical spine .... **B. frutescens**
1. Phyllaries without an apical spine .... **B. arborescens**

Borrichia arborescens (L.) DC. [Becoming treelike.] TREE SEASIDE OXEYE.

Buphthalmum arborescens Linnaeus, Syst. Nat., ed. 10. 1227. 1759. *Diomedea indentata* Cassini, in Cuvier, Dict. Sci. Nat. 13: 284. 1819, nom. illegit. *Diomedea glabrata* Kunth, in Humboldt et al., Nov. Gen. Sp. 4: 213. 1829, nom. illegit. *Borrichia arborescens* (Linnaeus) de Candolle, Prodr. 5: 489. 1836. *Verbesina arborescens* (Linnaeus) M. Gómez de la Maza y Jiménez, Dicc. Bot. 104. 1889.

Borrichia glabrata Small, Fl. S.E. U.S. 1263, 1340. 1903. *Borrichia arborescens* (Linnaeus) de Candolle var. *glabrata* (Small) D. B. Ward, Phytologia 94: 466. 2012. TYPE: FLORIDA: Monroe Co.: Bahia Honda Key, May *Curtiss 1412* (holotype: NY; isotypes: F, GH, MO, NY).

Decumbent to ascending shrub, to 1.5 m; stem stoloniferous, striate-angled, glabrous or sericeous. Leaves with the blade oblanceolate or linear, 5–10(15) cm long, 0.4–2 cm wide, succulent, the apex acute or obtuse, the base cuneate, the margin dentate or serrate with the teeth mostly proximal or entire, the upper and lower surfaces glabrous or sericeous, short-petiolate or sessile. Flowers in a capitulum, solitary or 3–6 in a cymbiform array, the peduncle 0.5–5 cm long; involucre ovoid or hemispheric; phyllaries 10–16 in 3–4 series, the outer ovate-elliptic, 6–9 mm long, the apex obtuse or acute, the outer surface glabrate or sericeous, the inner 6–9 mm long, the apex rounded, obtuse, or acute, the outer surface glabrous; receptacle convex; paleae lanceolate to ovate, 6–9 mm long, conduplicate and partially enclosing the cypsela, the apex acute or obtuse; ray flowers 7–15, carpellate; corolla yellow, the lamina 5–8 mm long; disk flowers 20–50, bisexual; corolla yellow, 5–6 mm long. Cypsela obpyramidal, 4–5 mm long, 4-angled, slightly laterally flattened, dark gray or black; pappus coroniform or cupular, 4-angled.

Beaches and salt marshes. Rare; Miami-Dade and Monroe Counties. Florida; West Indies and Central America. Spring–summer.

Borrichia arborescens hybridizes with *B. frutescens* (*B.* ×*cubana*) where the two are sympatric in Florida.

Borrichia frutescens (L.) DC. [Becoming shrubby.] BUSHY SEASIDE OXEYE.

Buphthalmum frutescens Linnaeus, Sp. Pl. 903. 1753. *Diomedea bidentata* Cassini, in Cuvier, Dict. Sci. Nat. 13: 283. 1819, nom. illegit. *Borrichia frutescens* (Linnaeus) de Candolle, Prodr. 5: 488. 1836.

Erect, sometimes decumbent or ascending, perennial herb or a shrub in subtropical regions, to 1(1.5) m; stem striate-angled, sericeous to puberulent or glabrate, stoloniferous. Leaves with the blade obovate, elliptic, or oblanceolate, 4.5–8(11) cm long, 1–3 cm wide, coriaceous, the apex acute or mucronate, the base cuneate, the margin dentate or serrate, usually only so proximally, often with 1–2 or more spine-tipped teeth near the base, the upper and lower surfaces villous or sericeous, rarely glabrate, short-petiolate or sessile. Flowers in capitula, terminal and solitary on the branches, the peduncle 2–6 cm long; involucre subglobose, ovoid, or hemispheric; phyllaries 20–40 in 3–4 series, the outer 2–4 mm long, with a spreading or erect apical spine, the inner 3–6 mm long, similar, the outer surface sericeous; receptacle convex; paleae oblanceolate, 4–6 mm long, with an apical spine to 3 mm long; ray flowers 15–30, carpellate; corolla yellow, the lamina 6–8 mm long; disk flowers 20–75, bisexual; corolla yellow, 5–7 mm long. Cypsela obpyramidal, 3–4(5) mm long, 3- to 4-angled, slightly flattened, dark gray or black; pappus coroniform or cupular, 3- to 4-angled.

Beaches and salt marshes. Common; nearly throughout. Virginia south to Florida, west to Texas; West Indies and Mexico. Native to North America and Mexico. Spring–summer.

Borrichia frutescens hybridizes with *B. arborescens* (*B.* ×*cubana*) where the two are sympatric in Florida.

HYBRID

Borrichia ×*cubana* Britton & S. F. Blake (*B. arborescens* × *B. frutescens*).
Borrichia cubana Britton & S. F. Blake, Mem. Torrey Bot. Club 16: 116. 1920, pro. sp.

Disturbed sites. Rare; Miami-Dade County and Monroe County keys. Florida; West Indies. All year.

Bradburia Torr. & A. Gray, nom. cons. 1842. GOLDENASTER

Herbs. Leaves alternate, simple, pinnate-veined, petiolate or epetiolate. Flowers in capitula in paniculiform arrays, radiate; receptacle epaleate; ray flowers carpellate; petals 5, basally connate, zygomorphic; disk flowers bisexual; petals 5, basally connate, actinomorphic; stamens 5, epipetalous, the filaments free, the anthers connate; ovary inferior, 2-carpellate, 1-loculate. Fruit a cypsela; pappus present.

A genus of 2 species; North America. [Commemorates John Bradbury (1768–1823), English naturalist and collector in the Missouri Territory.]

Selected reference: Semple (2006b).

Bradburia pilosa (Nutt.) Semple [With long trichomes.] SOFT GOLDENASTER.

Chrysopsis pilosa Nuttall, J. Acad. Nat. Sci. Philadelphia 7: 66. 1834. *Dipligon nuttallianum* Kuntze, Revis. Gen. Pl. 1: 334. 1891. *Chrysopsis nuttallii* Britton et al. Mem. Torrey Bot. Club 5: 316. 1894, nom. illegit. *Heterotheca pilosa* (Nuttall) Shinners, Field & Lab. 19: 68. 1951. *Bradburia pilosa* (Nuttall) Semple, Univ. Waterloo Biol. Ser. 37: 7. 1996.

Erect annual herb, to 8 dm; stem striate-angled, pilose. Basal leaves oblanceolate, 5–10 cm long, 1–2.5 cm wide, the apex acute, the base cuneate, the margin apically dentate or entire, the upper and lower surfaces pilose, the cauline leaves linear-elliptic, reduced upward to 1 cm long, the margin entire or obscurely dentate. Flowers in capitula, terminal and solitary on the branches, the peduncle 1.5–7 cm long, short-pilose, sometimes stipitate-glandular distally, with reduced linear leaves grading to scales; involucre campanulate; phyllaries 25–60 in 3–4 series, linear, 6–8 mm long, the margin scarious, the outer surface pilose, sparsely glandular; receptacle slightly convex; ray flowers (11)16(24), carpellate; corolla yellow, the lamina 7–12 mm long; disk flowers bisexual; corolla yellow, 5–6 mm long. Cypsela obconic, 2–3 mm long, laterally flattened, sometimes slightly ribbed, short-strigose, brown; pappus of numerous yellowish to rusty brown, apically attenuate bristles 5–6 mm long surrounded by an outer series of light tan flat scales to 1 mm long.

Disturbed sites. Rare; Leon County. Virginia and North Carolina, west to Kansas, Oklahoma, and Texas, also Florida. Native to the midwestern United States, from Tennessee and Mississippi westward. Spring–summer.

Brickellia Elliott, nom. cons. 1823. BRICKELLBUSH

Herbs. Leaves opposite, simple, palmate-veined, petiolate or epetiolate. Flowers in capitula in corymbiform or paniculiform arrays, discoid, bisexual; receptacle epaleate; petals 5, basally connate, actinomorphic; stamens 5, epipetalous, the filaments free, the anthers connate; ovary inferior, 2-carpellate, 1-loculate. Fruit a cypsela; pappus present.

A genus of about 100 species; North America, Mexico, and Central America.

Kuhnia L., nom. rej., 1763.

Selected reference: Scott (2006).

1. Leaves deltate-ovate or lanceolate-ovate, 2–10 cm wide; pappus bristles capillary .... **B. cordifolia**
1. Leaves rhombic-lanceolate, lanceolate, linear-lanceolate, or linear, 0.1–2 cm wide; pappus bristles plumose.

2. Leaves rhombic-lanceolate or narrowly lanceolate, 0.5–4 cm wide, the margin irregularly dentate or entire, the petiole 1–10 mm long .... **B. eupatorioides**
2. Leaves linear or narrowly oblong, 1–4 mm wide, the margin entire, sessile or the petiole to 2 mm long .... **B. mosieri**

Brickellia cordifolia Elliott [With heart-shaped leaves.] FLYR'S NEMESIS; HEART-LEAF BRICKELLBUSH.

Brickellia cordifolia Elliott, Sketch Bot. S. Carolina 2: 290. 1823. *Coleosanthes cordifolius* (Elliott) Kuntze, Revis. Gen. Pl. 1: 328. 1891.

Erect perennial herb, to 1.5 m; stem striate-angled, pubescent. Leaves with the blade deltate-ovate or lanceolate-ovate, 3–15 cm long, 2–10 cm wide, 3-nerved from the base, the apex acuminate, the base truncate or cordate, the margin coarsely to finely serrate, the upper surface sparsely pubescent, the lower surface minutely pubescent and gland-dotted, the petiole 1–4 cm long. Flowers in capitula in a corymbiform or paniculate array, the peduncle 1–2.5 cm long, glandular-pubescent; involucre broadly campanulate; phyllaries 25–35 in 4–5 series, 8–12 mm long, often purple-tinged, 4- to 7-striate, the margin scarious, ciliate, the outer lanceolate to lanceolate-ovate, the apex long-acuminate, the outer surface glandular-pubescent, the inner lanceolate, the apex acute or obtuse, recurved, the outer surface pubescent, often gland-dotted; receptacle slightly convex; flowers 35–45; corolla yellow-green, often purple-tinged, 6–7 mm long. Cypsela narrowly prismatic, 4–6 mm long, slightly laterally flattened, 10-ribbed, hispidulous or glabrate, often gland-dotted; pappus of numerous purple-tinged, barbellate bristles 6–8 mm long.

Hammocks. Rare; Alachua and Jefferson Counties, central panhandle. Georgia, Alabama, and Florida. Fall.

Brickellia cordifolia is listed as endangered in Florida (Florida Administrative Code, Chapter 5B-40).

Brickellia eupatorioides (L.) Shinners [Resembling the genus *Eupatorium*.] FALSE BONESET.

Kuhnia eupatorioides Linnaeus, Sp. Pl., ed. 2. 1662. 1763. *Critonia kuhnia* Gaertner, Fruct. Sem. Pl. 2: 411. 1791, nom. illegit. *Kuhnia glabra* Rafinesque, Herb. Raf. 29. 1833, nom. illegit. *Kuhnia kuhnia* C. Mohr, Bull. Torrey Bot. Club 24: 28. 1897, nom. illegit. et nom. inadmiss. *Brickellia eupatorioides* (Linnaeus) Shinners, Sida 4: 274. 1971.
Kuhnia paniculata Cassini, in Cuvier, Dict. Sci. Nat. 24: 516. 1822.
Kuhnia latifolia Rafinesque, New Fl. 4: 105. 1838 ("1836"). TYPE: FLORIDA.
Kuhnia virgata Rafinesque, New Fl. 4: 105. 1838 ("1836"). *Kuhnia albicaulis* Rafinesque, New Fl. 4: 105. 1838 ("1836"), nom. illegit. TYPE: "in Carolina and Florida."
Kuhnia eupatorioides Linnaeus var. *gracilis* Torrey & A. Gray, Fl. N. Amer. 2: 78. 1841.

Erect perennial herb, to 1.5 m; stem striate-angled, pubescent. Leaves with the blade rhombic-lanceolate or narrowly lanceolate, 0.5–10 cm long, 0.1–2 cm wide, 1- or 3-nerved from the base, the apex acuminate to obtuse, the base cuneate, the margin irregularly dentate or entire, the upper and lower surfaces glandular-pubescent, the petiole 1–10 mm long. Flowers in capitula in a corymbiform array, the peduncle 0.5–2 cm long, glandular-pubescent; involucre narrowly campanulate to cylindric; phyllaries 20–25 in 4–6 series, 7–11 mm long, green or yellowish, sometimes purple-tinged, 3- to 7-striate, the outer ovate or ovate-lanceolate, the apex acute or acuminate, the outer surface puberulent, slightly gland-dotted, the inner phyllaries lanceolate, the apex obtuse or acuminate, the outer surface puberulent, gland-dotted; receptacle slightly convex; flowers 6–15; corolla pale yellow or maroon, 5–7 mm long. Cypsela narrowly prismatic, 4–6 mm long, slightly laterally flattened, 10-ribbed, hispidulous, often gland-dotted; pappus of numerous white or tawny, plumose bristles 6–8 mm long.

Sandhills and hammocks. Rare; northern counties, central peninsula. Pennsylvania and New Jersey, south to Florida, west to Indiana, Kentucky, Tennessee, Arkansas, and Texas. Summer–fall.

Brickellia mosieri (Small) Shinners [Commemorates Charles A. Mosier (1871–1936), the first superintendent of Royal Palm State Park, who often collaborated with J. K. Small in his explorations of Florida.] MOSIER'S FALSE BONESET; BRICKELLBUSH.

Kuhnia mosieri Small, Man. S.E. Fl. 1229, 1508. 1933. *Brickellia mosieri* (Small) Shinners, Sida 4: 274. 1971. TYPE: FLORIDA: Miami-Dade Co.: Ross-Costello Hammock, 24 Jun 1915, *Small et al. 6544* (holotype: NY).

Kuhnia eupatorioides Linnaeus var. *floridana* R. W. Long, Rhodora 72: 39. 1970. *Brickellia eupatorioides* (Linnaeus) Shinners var. *floridana* (R. W. Long) B. L. Turner, Phytologia 67: 130. 1989. TYPE: FLORIDA: Miami-Dade Co.: Miami, 13 May 1904, *Tracy 9046* (holotype: GH).

Erect perennial herb, to 7 dm; stem striate-angled, pubescent. Leaves with the blade linear or narrowly oblong, 0.5–4 cm long, 1–4 mm wide, 1- or obscurely 3-nerved from the base, the apex acuminate or obtuse, the base cuneate, the margin entire, the upper and lower surfaces glandular-pubescent, sessile or the petiole to 2 mm long. Flowers in capitula in a corymbiform or paniculiform array, the peduncle 0.5–6 cm long, glandular-pubescent; involucre narrowly campanulate or cylindric; phyllaries 20–25 in 4–6 series, 7–11 mm long, green or yellowish, sometimes purple-tinged, 3- to 7-striate, the outer ovate or ovate-lanceolate, the apex obtuse or acuminate, the outer surface puberulent, gland-dotted; receptacle slightly convex; flowers 9–13; corolla pale yellow or maroon, 6–7 mm long. Cypsela narrowly prismatic, 4–5 mm long, slightly laterally flattened, 10-ribbed, hispidulous, often gland-dotted; pappus of numerous white or tawny, plumose bristles 6–8 mm long.

Pinelands. Rare; Miami-Dade County. Endemic. Summer–fall.

Brickellia mosieri is listed as endangered in Florida (Florida Administrative Code, Chapter 5B-40) and in the United States (U.S. Fish and Wildlife Service, 50 CFR 23).

Brintonia Greene 1895. MOCK GOLDENROD

Herbs. Leaves alternate, simple, pinnate-veined, petiolate. Flowers in capitula in paniculiform arrays, discoid, bisexual; receptacle epaleate; petals 5, basally connate, actinomorphic; stamens

5, epipetalous, the filaments free, the anthers connate; ovary inferior, 2-carpellate, 1-loculate. Fruit a cypsela; pappus present.

A monotypic genus; North America. [Commemorates Jeremiah Bernard Brinton (1835–1894), physician and botanist of Philadelphia.]

Selected reference: Semple (2006c).

Brintonia discoidea (Elliott) Greene [In reference to the capitula consisting only of disk flowers.] RAYLESS MOCK GOLDENROD.

Aster discoideus Elliott, Sketch Bot. S. Carolina 2: 358. 1822. *Solidago discoidea* (Elliott) Torrey & A. Gray, Fl. N. Amer. 2: 195. 1842. *Brintonia discoidea* (Elliott) Greene, Erythea 3: 89. 1895.

Erect perennial herb, to 1.5 m; stem striate-angled, villose, rhizomatous. Basal and proximal cauline leaves with the blade ovate, 4–10 cm long, 3–8 cm wide, the apex acute, the base cordate or cuneate, the margin serrate, the upper and lower surfaces strigose, the petiole 2–8 cm long, distally winged, villous, the mid and distal cauline reduced upward to 1 cm long, to 5 mm wide. Flowers in capitula in a paniculiform array, 1–4(10) per lateral branch, the peduncle shorter than the internode, usually subtended by a small linear, foliose, bract-like leaf; involucre narrowly campanulate; phyllaries 14–22 in 2–3 series, lanceolate, 4–6 mm long, 1-nerved, the apex acuminate, the margin scarious, the outer surface villose-strigose; receptacle flat, slightly pitted, the pits surrounded with irregular ridges; flowers 8–20; corolla white or purple-tinged, 4–5 mm long. Cypsela obconic, 3–4 mm long, 5- to 10-ribbed, strigose or glabrate, brown; pappus of numerous white or purplish, barbellate bristles in 2 series, the outer slightly shorter than the inner clavate ones.

Hammocks. Rare; Leon, Gadsden, Liberty, Santa Rosa, and Escambia Counties. Georgia south to Florida, west to Louisiana. Fall.

Calyptocarpus Less. 1832.

Herbs. Leaves opposite, simple, pinnapalmate-veined, petiolate. Flowers in capitula, solitary, terminal or axillary, radiate; receptacle paleate; ray flowers carpellate; petals 5, basally connate, zygomorphic; disk flowers bisexual; petals 5, basally connate, actinomorphic; stamens 5, epipetalous, the filaments free, the anthers connate; ovary inferior, 2-carpellate, 1-loculate. Fruit a cypsela; pappus present.

A genus of 2 species; North America, Mexico, Central America, South America, Asia, Australia, and Pacific Islands. [From the Greek *kalypto,* covered or hidden, and *karpos,* fruit.]

Selected reference: Strother (2006d).

Calyptocarpus vialis Less. [*Via,* growing along roads or paths.] STRANGLER DAISY.

Calyptocarpus vialis Lessing, Syn. Gen. Compos. 221. 1832. *Synedrella vialis* (Lessing) A. Gray, Proc. Amer. Acad. Arts 17: 217. 1882.

Prostrate or decumbent perennial herb, to 3 dm; stem, striate-angled, scabrellous. Leaves with the blade deltate, ovate, or lanceolate, 1–3(6) cm long, 0.5–1.5(3) cm wide, the apex acute or obtuse, the base truncate or cuneate, the margin serrate-crenate, the upper and lower surfaces scabrellous, the petiole 1–8 mm long. Flowers in capitula, solitary, terminal or axillary, radiate; involucre obconic; phyllaries 5 in 1(2) series, linear or lanceolate, 4–5 mm long; receptacle convex; palea scarious; ray flowers 3–8; corolla pale yellow, the lamina 2–4 mm long; disk flowers 10–20; corolla yellow, 3–4 mm long. Cypsela prismatic, 2–4 mm long, laterally flattened, sometimes tuberculate; pappus of 2(5) erect or somewhat spreading awns 1–2(3) mm long.

Disturbed sites. Occasional; nearly throughout. South Carolina west to Arizona; Mexico, Central America, and South America; Asia, Australia, and Pacific Islands. Native to Texas and tropical America. All year.

Carphephorus Cass. 1816. CHAFFLEAD

Herbs. Leaves alternate, simple, pinnate-veined, petiolate or epetiolate. Flowers in capitula in corymbiform or paniculiform arrays, discoid, bisexual; receptacle paleate or epaleate; petals 5, basally connate, actinomorphic; stamens 5, epipetalous, the filaments free, the anthers connate; ovary inferior, 2-carpellate, 1-loculate. Fruit a cypsela; pappus present.

A genus of 7 species; North America. [From the Greek *karphos,* chaff, and *phoros,* bearing, in reference to the receptacular paleae.]

Litrisa Small, 1924; *Trilisa* (Cass.) Cass., 1820.

Selected references: Correa and Wilbur (1969); Nesom (2006d).

1. Capitula in a thyrsiform array .... **C. paniculatus**
1. Capitula in a corymbiform or paniculiform array.
    2. Basal leaves acicular, 1–2(3) mm wide, involute, pilose .... **C. pseudoliatris**
    2. Basal leaves linear, elliptic, or oblanceolate, 5–25 mm wide, not involute or pilose.
        3. Heads more than 1 cm long; phyllary tips rounded, the margin erose .... **C. corymbosus**
        3. Heads less than 1 cm long; phyllary tips obtuse or apiculate, the margin slightly erose to entire.
            4. Stem pubescent; basal leaves in a depressed rosette, the blade linear or linear-lanceolate .... **C. carnosus**
            4. Stem glabrous; basal leaves not in a depressed rosette, the blade oblanceolate .... **C. odoratissimus**

Carphephorus carnosus (Small) C. W. James [*Carneus,* flesh-colored, in reference to the flowers.] PINELAND CHAFFHEAD.

Litrisa carnosa Small, Bull. Torrey Bot. Club 51: 392. 1924. *Trilisa carnosa* (Small) B. L. Robinson, Contr. Gray Herb. 104: 49. 1934. *Carphephorus carnosus* (Small) C. W. James, Rhodora 60: 120. 1958. TYPE: FLORIDA: Highlands Co.: Istokpoga, 31 Aug 1922, *Small & DeWinkeler 10658* (holotype: NY).

Erect perennial herb, to 5(9) dm; stem striate-angled, puberulent or villous-hirsute, gland-dotted. Basal rosette leaves with the blade linear-lanceolate or oblanceolate, 5–7(9) cm long, (3)4–9(11) mm wide, the apex acute, the base cuneate, the margin entire, the upper and lower surfaces puberulent or glabrate, gland-dotted, sessile, clasping, the cauline abruptly reduced upward, 5–11(20) mm long, 1–3 mm wide, the upper and lower surfaces short-pubescent, gland-dotted. Flowers in capitula in a corymbiform array, the peduncle to 5 mm long; involucre campanulate; phyllaries 5–12 in 1–2(3) series, oblong, oblong-lanceolate, or elliptic, 2–5(8) mm long, the apex acute or obtuse and mucronate, the margin entire on the outer, that of the inner pectinate-ciliate, the outer surface villous-hirsute, gland-dotted; receptacle convex; paleae 1–2 linear, 3–6 mm long, or absent; flowers numerous; corolla purple-lavender or rarely white, 3–5 mm long, the outer surface gland-dotted. Cypsela prismatic, 1–2(3) mm long, 10-ribbed, scabrellous, eglandular; pappus of numerous barbellate bristles in 1–2 series, (3)4–6(7) mm long.

Wet flatwoods. Occasional; central peninsula. Endemic. Summer–fall.

Carphephorus corymbosus (Nutt.) Torr. & A. Gray [Capitula in a corymbiform array.] COASTALPLAIN CHAFFHEAD; FLORIDA PAINTBRUSH.

Liatris corymbosa Nuttall, Gen. N. Amer. Pl. 2: 132. 1819. *Carphephorus corymbosus* (Nuttall) Torrey & A. Gray, Fl. N. Amer. 2: 67. 1841. TYPE: "From North Carolina to Florida."

Erect perennial herb, to 1(1.2) m; stem striate-angled, villous-hirsute, eglandular. Basal rosette and proximal cauline leaves with the blade oblanceolate, 6–12(20) cm long, (7)9–20(25) mm wide, the apex obtuse or rounded, the base cuneate, the margin entire, the upper and lower surfaces glabrous, gland-dotted, sessile, clasping, the cauline gradually reduced upward, the upper and lower surfaces glabrous, gland-dotted, sessile, weakly clasping. Flowers in capitula in a corymbiform array, the peduncle to 1 cm long; involucre hemispheric; phyllaries 9–20 in 3–6 series, ovate to oblong, 2–10 mm long, the apex rounded to obtuse, the margin scarious, finely erose, the outer surface glabrous, eglandular; receptacle convex; paleae linear, 6–9(11) mm long; flowers numerous; corolla purple-lavender or pinkish, 7–9 mm long, the outer surface eglandular. Cypsela prismatic, 2–4(5) mm long, 10-ribbed, scabrellous, eglandular; pappus of numerous barbellate bristles in 1–2 series, 5–10 mm long.

Sandhills. Common; peninsula west to central panhandle. South Carolina, Georgia, and Florida. Summer–fall.

Carphephorus odoratissimus (J. F. Gmel.) H. J.-C. Hebert [In reference to having a strong odor.] VANILLALEAF.

Erect perennial herb, to 1.5 m; stem striate-angled, glabrous, eglandular. Basal leaves with the blade oblanceolate, (5)9–30(40) cm long, 1–6(11) cm wide, the apex rounded to obtuse, the base cuneate, the margin entire, undulate, or remotely serrate-dentate, the upper and lower surfaces glabrous, gland-dotted, sessile, clasping, the cauline similar, reduced upward. Flowers in capitula in a corymbiform or paniculiform array, the peduncle to 1 cm long; involucre campanulate; phyllaries 5–12 in 1–3 series, oblanceolate, (1)2–5(6) mm long, the apex obtuse, the margin slightly erose or entire, the inner ones sometimes scarious, the outer surface glabrous, gland-dotted; receptacle convex; paleae 1–2, linear or linear-spatulate, 3–5(8) mm long or absent; flowers numerous; corolla purple-lavender, 3–5 mm long, the outer surface gland-dotted.

Cypsela prismatic, 2–3 mm long, 10-ribbed, pubescent near the base and along the ribs, gland-dotted; pappus of numerous barbellate bristles in 1–2 series, 2–5 mm long.

1. Inflorescence a rather narrow, congested, corymbiform array, usually longer than wide; leaves with a coumarin (vanilla-like) odor when dry, the upper cauline serrate-dentate .... var. **odoratissimus**
1. Inflorescence a diffuse, paniculiform array, usually as wide as or wider than long; leaves lacking a coumarin (vanilla-like) odor when dry, the upper cauline entire............................. var. **subtropicanus**

Carphephorus odoratissimus var. **odoratissimus** VANILLALEAF.

Chrysocoma odoratissima J. F. Gmelin, Syst. Nat. 2: 1204. 1792. *Liatris odoratissima* (J. F. Gmelin) Michaux, Fl. Bor.-Amer. 2: 93. 1803. *Trilisa odoratissima* (J. F. Gmelin) Cassini, in Cuvier, Dict. Sci. Nat. 55: 310. 1828. *Carphephorus odoratissimus* (J. F. Gmelin) H. J.-C. Hebert, Rhodora 70: 483. 1968.

Liatris amplexicaulis Rafinesque, New Fl. 4: 78. 1838 ("1836"). TYPE: "Florida & Georgia."

Leaves with a coumarin (vanilla-like) odor when dry, the basal usually more than 15 cm long and more than 5 cm wide, the upper cauline with the margin often serrate-dentate, the apex spreading away from the stem. Flowers usually 7–10 per capitulum in a narrow, congested, corymbiform array.

Wet pine savannas, wet flatwoods, seepage areas, and bogs. Frequent; northern counties, central peninsula. North Carolina south to Florida, west to Louisiana. Summer.

Carphephorus odoratissimus var. *odoratissimus* sometimes intergrades with var. *subtropicanus* where their ranges overlap in central and southern Florida (Orzell and Bridges, 2002).

Carphephorus odoratissimus var. **subtropicanus** (DeLaney et al.) Wunderlin & B. F. Hansen [In reference to its distribution.] PINELAND PURPLE; FALSE VANILLALEAF.

Carphephorus subtropicanus DeLaney et al., Bot. Explorer (Florida) 1: 2. 1999. *Carphephorus odoratissimus* (J. F. Gmelin) H. J.-C. Hebert var. *subtropicanus* (DeLaney et al.) Wunderlin & B. F. Hansen, Novon 11: 366. 2001. *Trilisa subtropicana* (DeLaney et al.) Bridges & Orzell, J. Bot. Res. Inst. Texas 11: 300. 2017. TYPE: FLORIDA: Polk Co.: 4.36 mi. E of US 27 on Avon Park Cut-Off Road, T32S, R27E, Sec. 21, SE1/4, 27°37'05"W, 9 Oct 1999, *Bissett 1001* (holotype: USF; isotypes: FLAS, FSU, FTG, GA, NY, USF).

Leaves lacking a coumarin (vanilla-like) odor when dry, the basal usually less than 15 cm long and 4 cm wide, the upper cauline with the margin entire, the apex appressed to the stem. Flowers usually 10–14 per capitulum in a diffuse paniculiform array usually wider than long.

Mesic flatwoods and dry prairies. Frequent; central and southern peninsula. Endemic. Fall.

Carphephorus odoratissimus var. *subtropicanus* sometimes intergrades with var. *odoratissimus* where their ranges overlap in central and southern Florida (Orzell and Bridges, 2002).

Carphephorus paniculatus (J. F. Gmel.) H. J.-C. Hebert [Capitula in a paniculiform array.] HAIRY CHAFFHEAD.

Chrysocoma paniculata J. F. Gmelin, Syst. Nat. 2: 1204. 1792. *Liatris paniculata* (J. F. Gmelin) Michaux, Fl. Bor.-Amer. 2: 93. 1803. *Trilisa paniculata* (J. F. Gmelin) Cassini, in Cuvier, Dict. Sci. Nat. 55: 310. 1828. *Carphephorus paniculatus* (J. F. Gmelin) H. J.-C. Hebert, Rhodora 70: 483. 1968.

Erect perennial herb, to 1(1.8) m; stem striate-angled, villous-hirsute, eglandular. Basal and proximal cauline leaves with the blade oblanceolate or narrowly elliptic, 5–35 cm long, (0.5)1–4 cm wide, the apex acute, the base cuneate, the margin entire or undulate, the upper and lower surfaces glabrous, eglandular, sessile, clasping, the cauline abruptly reduced upward, lanceolate, 1–3.5(5) cm long, (2)4–9(15) mm wide, the upper and lower surfaces glabrous or hirsute (especially at the base, margin, and midrib). Flowers in capitula in a thyrsiform array, the peduncle to 5 mm long; involucre campanulate or hemispheric; phyllaries 5–12 in 1–2(3) series, oblanceolate, 3–8 mm long, the apex acute or obtuse, the margin entire, ciliate, the outer surface short-pubescent, gland-dotted, sometimes stipitate-glandular; receptacle convex; paleae 1–2, linear, 3–6 mm long or absent; flowers numerous; corolla purple-lavender, 3–5 mm long. Cypsela prismatic, 2–3 mm long, 10-ribbed, scabrellous, eglandular; pappus of numerous barbellate bristles in 1–2 series, (2)3–4(5) mm long.

Wet flatwoods and bogs. Frequent; nearly throughout. North Carolina south to Florida, west to Alabama. Fall.

Carphephorus pseudoliatris Cass. [*Pseudo*, false, and the genus *Liatris*, in reference to its resemblance to *Liatris*.] BRISTLELEAF CHAFFHEAD.

Carphephorus pseudoliatris Cassini, Bull. Sci. Soc. Philom. Paris 1816: 198. 1816.

Erect perennial herb, to 1 m; stem striate-angled, villous-hirsute, sparsely and inconspicuously gland-dotted or eglandular. Basal and proximal cauline leaves with the blade narrowly linear, 5–35 cm long, 1–2(3) mm wide, acicular, the apex acute, the base cuneate, the margin entire, involute, the upper and lower surfaces pilose near the margin, eglandular, sessile, clasping, the mid and upper cauline much reduced upward, linear, 1.5–2.5(4) cm long, ca. 1 mm wide, appressed to the stem. Flowers in capitula in a corymbiform array, the peduncle 4–6 mm long, finely villous or strigose; involucre campanulate; phyllaries numerous in 3–5 series, triangular-lanceolate, the apex acute, the margin entire, the outer surface villous, eglandular; receptacle convex; paleae nearly throughout, linear, 5–9 mm long; flowers numerous; corolla reddish purple, 4–6 mm long, the outer surface eglandular. Cypsela prismatic, 2–3 mm long, 10-ribbed, strigose, gland-dotted or eglandular; pappus of numerous barbellate bristles in 1–2 series, 4–6 mm long.

Flatwoods and bogs. Occasional; central and western panhandle. Georgia south to Florida, west to Louisiana. Fall.

EXCLUDED TAXA

Carphephorus bellidifolius (Michaux) Torrey & A. Gray—Reported for Florida by Small (1903, 1913a, 1933). Excluded from Florida by Correa and Wilbur (1969). No Florida specimens seen.
Carphephorus tomentosus (Michaux) Torrey & A. Gray—Reported for Florida by Small (1903, 1913a, 1933). Excluded from Florida by Correa and Wilbur (1969). No Florida specimens seen.

Centaurea L., nom. cons. 1753. KNAPWEED

Herbs. Leaves alternate, simple, pinnate-veined, petiolate or epetiolate. Flowers in capitula, solitary or in corymbiform arrays, radiant (peripheral flowers zygomorphic and ray-like, the inner actinomorphic and discoid), the outer sterile, the inner bisexual; receptacle epaleate; stamens 5, epipetalous, the filaments free, the anthers connate; ovary inferior, 2-carpellate, 1-loculate. Fruit a cypsela; pappus present.

A genus of about 500 species; nearly cosmopolitan. [From the Greek *kentaurieanm,* ancient plant name associated with Chiron, a centaur famous for its knowledge of medicinal plants.]

The genus *Centaurea* has great diversity and the taxonomy has been controversial. There have been many misapplications and differing taxonomic interpretations. See Keil and Ochsmann (2006) for some details regarding our taxa. This work must be considered tentative pending further studies.

Cnicus L., 1753.

Selected reference: Keil and Ochsmann (2006).

1. Corolla yellow; phyllaries spine-tipped .... **C. benedicta**
1. Corolla blue, pink, or purple; phyllaries not spine-tipped.
   2. Principal cauline leaves entire .... **C. cyanus**
   2. Principal cauline leaves pinnatifid .... **C. stoebe**

Centaurea benedicta (L.) L. [*Benedicto,* blessing, an expression of good wishes, the plant once believed to prevent evil and was used medicinally.] BLESSED THISTLE.

Cnicus benedictus Linnaeus, Sp. Pl. 826. 1753. *Centaurea benedicta* (Linnaeus) Linnaeus, Sp. Pl., ed. 2. 1296. 1763. *Calcitrapa lanuginosa* Lamarck, Fl. Franc. 2: 35. 1779 ("1778"), nom. illegit. *Benedicta officinalis* Bernhardi, Syst. Verz. 159. 1800. *Calcitrapa benedicta* (Linnaeus) Sweet, Hort. Brit. 218. 1826.

Ascending annual herb, to 6 dm; stem striate-angled, tomentose. Leaves with the blade lanceolate or oblanceolate, 6–25 cm long, 2–4 cm wide, the apex acute or obtuse or rounded, the base cuneate, the margin coarsely dentate or pinnately lobed, the lobes and teeth with short, weak spines, the upper and lower surfaces with slender cobweb-like trichomes, gland-dotted, sessile, sometimes decurrent on the stem. Flowers in capitula, terminal and solitary, subtended by an involucre-like cluster of bracteate leaves, sessile; involucre subspheric; phyllaries numerous in several series, 2–4 mm long, the outer tightly appressed at the base with spreading spine tips, the inner lanceolate, with pinnate, spreading spine tips to 2.5 cm long; receptacle flat, bristly; flowers numerous; corolla yellow, the outer ray-like, 3-lobed, shorter than the inner, sterile, the inner discoid, 2–2.5 cm long. Cypsela cylindric, 8–11 mm long, slightly curved, 20-ribbed, the apex with a 10-dentate rim, glabrous; pappus of numerous awns in 2 series, the outer 9–10 mm long, smooth or slightly roughened, the inner 2–5 mm long, minutely barbed.

Disturbed sites. Rare; northern counties. Nearly throughout North America. North America and South America; Europe, Africa, Asia, and Australia. Native to Europe, Africa, and Asia. Spring.

Centaurea cyanus L. [From the Greek *cyanos,* dark blue, in reference to the flower color.] GARDEN CORNFLOWER.

Centaurea cyanus Linnaeus, Sp. Pl. 911. 1753. *Cyanus segetus* Hill, Veg. Syst. 4: 29. 1762. *Cyanus cyanus* (Linnaeus) Hill, Hort. Kew. 64. 1768, nom. inadmiss. *Cyanus arvensis* Moench, Methodus 561. 1794, nom. illegit. *Cyanus vulgaris* Delarbre, Fl. Aubergne, ed. 2. 203. 1800, nom. illegit. *Setachna cyanus* (Linnaeus) Dulac, Fl. Hautes-Pyrénées 520. 1867. *Leucacantha cyanus* (Linnaeus) Nieuwland & Lunell, Amer. Midl. Naturalist 5: 71. 1917.

Erect herb, to 1 m; stem striate-angled, tomentose. Leaves with the blade linear-lanceolate, the basal rarely with a few lobes, 2–10 cm long, 2–5(9) mm wide, the apex acute, the base cuneate, the margin entire, the upper and lower surfaces tomentose, sessile. Flowers in capitula in a corymbiform array, the peduncle 2–6 cm long; involucre campanulate; phyllaries numerous in several series, 2–12 mm long, the outer ovate, the inner oblong, the margin white, brown, or black, scarious, fringed with short, slender teeth, the outer surface tomentose or glabrate; receptacle flat, bristly; flowers numerous; corolla blue, purple, or white, the outer ray-like, 2–2.5 cm long, the inner discoid, 1–1.5 cm long. Cypsela cylindric, somewhat laterally flattened, 4–5 mm long, finely pubescent; pappus of numerous unequal bristles 2–4 mm long.

Disturbed sites. Rare; Leon, Escambia, Hillsborough, and Polk Counties. Escaped from cultivation. Nearly throughout North America. North America and Europe. Native to Europe. Spring–fall.

Centaurea stoebe L. subsp. **micranthos** (Gugler) Hayek. [From the Greek *stoibe,* stuffing or padding, in reference to its use for packing wine jars and making bedding.] SPOTTED KNAPWEED.

Centaurea micranthos S. G. Gmelin ex Hayek, Centaurea Österr.-Ung. 92. 1901; non *C. micrantha* Hoffmannsegg & Link, 1820. *Centaurea maculosa* Lamarck subsp. *micranthos* Gugler, Centaurea Ung. Natl.-Mus. 167. 1907. *Centaurea stoebe* Linnaeus subsp. *micranthos* (Gugler) Hayek, Repert. Spec. Nov. Regni Veg. Beih. 30(2): 766. 1931. *Acosta micranthos* (Gugler) Soják, Cas. Nár. Mus., Odd. Prír. 140: 134. 1972.

Ascending to erect perennial herb, to 1.5 m; stem striate-angled, tomentose. Leaves with the blade oblanceolate or elliptic, the basal and proximal cauline 10–15 cm long, 4–7 cm wide, 1–2 times pinnatifid, the ultimate segments linear or oblong, the apex acute or obtuse with a minute tip, the upper and lower surfaces tomentose or glabrate, gland dotted, short-petiolate, the mid and upper similar but smaller and less divided. Flowers in capitula in an open corymbiform array, short pedunculate; involucre ovoid; phyllaries numerous in several series, 10–13 mm long, the outer ovate, the inner oblong, each with several prominent parallel veins, the apex acuminate, dark brown or black, scarious, fringed with slender teeth, the outer surface finely tomentose or glabrous; receptacle flat, bristly; flowers numerous; corolla pink or purple, rarely white, the outer ray-like, 1.5–2.5 cm long, the inner discoid, 1.2–1.5 cm long. Cypsela prismatic, 3–4 mm long, somewhat laterally flattened, finely pubescent; pappus of numerous, unequal, bristles to 5 mm long.

Disturbed sites. Rare; Escambia County. Nearly throughout North America. North America; Europe, Asia, Australia, and Pacific Islands. Native to Europe. Summer–fall.

EXCLUDED TAXA

Centaurea calcitrapa Linnaeus—Reported for Florida by Keil and Ochsmann (2006), basis unknown. No other reports. No Florida specimens seen.

Centaurea maculosa Lamark—Reported for Florida by Wunderlin (1998), the name misapplied to Florida material of *C. stoebe* subsp. *micranthos.*

Centaurea nigrescens Willdenow—Reported for Florida by Keil and Ochsmann (2006), basis unknown. No other reports. No Florida specimens seen.

Centaurea phrygia Linnaeus—Reported for Florida by Keil and Ochsmann (2006), basis unknown. No other reports. No Florida specimens seen.

Centaurea solstitialis Linnaeus—Reported for Florida by Small (1933) and Keil and Ochsmann (2006). No Florida specimens seen.

Centratherum Cass. 1817.

Herbs. Leaves alternate, simple, pinnate-veined, petiolate or epetiolate. Flowers in capitula, terminal and solitary, discoid, bisexual; receptacle epaleate; petals 5, basally connate, actinomorphic; stamens 5, epipetalous, the filaments free, the anthers connate; ovary inferior, 2-carpellate, 1-loculate. Fruit a cypsela; pappus present.

A genus of 5 species; North America, West Indies, Mexico, Central America, South America, Australia, and Pacific Islands. [*Centrum,* center, and *atherum,* prickle or awn, in reference to the spine-tipped middle phyllaries.]

Selected references: Kirkman (1981); Strother (2006e).

Centratherum punctatum Cass. [Marked with dots, in reference to minute punctate glands on the leaves.] LARKDAISY.

Centratherum punctatum Cassini, in Cuvier, Dict. Sci. Nat. 7: 384. 1817. *Centratherum longispinum* Cassini, Dict. Sci. Nat. 57: 346. 1828, nom. illegit. *Baccharoides punctata* (Cassini) Kurtze, Revis. Gen. Pl. 1: 320. 1891.

Erect or ascending perennial herb, to 5 dm; stem striate-angled, sometimes rooting at the proximal nodes, strigose. Leaves with the blade obovate or lanceolate, 2–6(8) cm long, 0.8–2.5(3) cm wide, the apex acute or obtuse, the base cuneate, the margin irregularly serrate-dentate, often ciliate, the upper and lower surfaces sparsely hirtellous, strigillose, or glabrate, glandular-punctate, the petiole to 1 cm long, sometimes sessile. Flowers in capitula, terminal, solitary or sometimes 2–3 in a cluster, subtended by 3–8 leaflike bracts, the peduncle 2–7 mm long;

involucre hemispheric or cylindric-campanulate; phyllaries numerous in 4–8 series, 6–12(18) mm long, membranaceous, the outer ovate, deltate, or lanceolate, greenish, the inner oblong or lanceolate, purplish, the apex rounded or acute, apiculate or spine-tipped, the margin entire, the outer surface strigillose or glabrate, sparsely glandular-punctate distally; receptacle slightly convex; flowers numerous; corolla lavender or purplish, rarely white, 5–8(10) mm long. Cypsela columnar to clavate, 1–3 mm long, 8- to 10-ribbed, 1–3 mm long, glabrous or sometimes minutely pubescent on the ribs, gland-dotted; pappus with numerous, stramineous, linear-lanceolate bristles or subulate scales 1–3 mm long.

Disturbed sites. Occasional; central and southern peninsula. Escaped from cultivation. Florida; West Indies, Mexico, Central America, and South America. Native to West Indies, Central America, and South America. All year.

Chaptalia Vent., nom. cons. 1802. SUNBONNETS

Herbs. Leaves alternate, simple, pinnate-veined, petiolate or epetiolate. Flowers in capitula, solitary, radiate; receptacle epaleate; ray flowers carpellate; petals 5, basally connate, zygomorphic; disk flowers bisexual or staminate; petals 5, basally connate, actinomorphic; stamens 5, epipetalous, the filaments free, the anthers connate; ovary inferior, 2-carpellate, 1-loculate. Fruit a cypsela; pappus present.

A genus of about 60 species; North America, West Indies, Mexico, Central America, and South America. [Commemorates J.-A. Chaptal (1756–1832), French chemist, physician, agronomist, industrialist, statesman, educator, and philanthropist.]

Selected reference: Nesom (2006e).

1. Cypsela with a slender beak ½ to ⅔ as long as the body; disk flowers bisexual .... **C. albicans**
1. Cypsela with a short stout beak less than ½ as long as the body; disk flowers staminate .... **C. tomentosa**

Chaptalia albicans (Sw.) Vent ex B. D. Jacks. [Whitish, in reference to the flowers and the lower surface of the leaves.] WHITE SUNBONNETS.

Leontodon tomentosus Linnaeus f., Suppl. Pl. 347. 1782 ("1781"). *Tussilago albicans* Swartz, Prodr. 113. 1788. *Leria albicans* (Swartz) de Candolle, Ann. Mus. Natl. Hist. Nat. 19: 68. 1812., nom. illegit. *Gerbera albicans* (Swartz) Schultz Bipontinus, in Seemann, Bot. Voy. Herald 313. 1856. *Thyrsanthema tomentosum* (Linnaeus f.) Kuntze, Revis. Gen. Pl. 1: 369. 1891. *Chaptalia albicans* (Swartz) Ventenat ex B. D. Jackson, Index Kew. 1: 506. 1893.

Scapose perennial herb, to 3.5 dm. Leaves obovate or elliptic, 2–14 cm long, 1–1.5 cm wide, the apex acute or obtuse, the base cuneate, the margin retrorsely serrulate, denticulate, or lobed, the upper surface glabrous or glabrate, the lower surface white-tomentose, sessile or subsessile. Flowers in a solitary capitulum, erect, the peduncle to 1.5 dm long in anthesis, to 3.5 dm long in fruit, dilated distally, sparsely to densely white-tomentose; involucre campanulate; phyllaries numerous in 2–5 series, linear, 7–20 mm long, the apex acute, the margin white; receptacle slightly convex; ray flowers 10–15, carpellate; corolla creamy white or purple-tinged, the lamina

4–6 mm long; disk flowers numerous, bisexual, 7–8 mm long. Cypsela fusiform, 8–11 mm long, the beak filiform, ½ to ⅔ times as long as the body, the body slightly laterally flattened, 4- to 12-ribbed, glabrous or sparsely glandular, usually so only along the ribs; pappus of numerous barbellate bristles 5–9 mm long.

Pinelands. Rare; Miami-Dade County. Florida; West Indies, Mexico, and Central America. All year.

Chaptalia albicans is listed as threatened in Florida (Florida Administrative Code, Chapter 5B-40).

Chaptalia tomentosa Vent. [Woolly pubescent, in reference to the lower surface of the leaves.] WOOLLY SUNBONNETS; PINELAND DAISY.

Tussilago integrifolia Michaux, Fl. Bor.-Amer. 2: 121. 1801. *Chaptalia integrifolia* (Michaux) Nuttall, Gen. N. Amer. Pl. 2: 182. 1818. TYPE: "Carolina ad Floridam."

Chaptalia tomentosa Ventenat, Descr. Pl. Nouv. pl. 61. 1802. *Gerbera walteri* Schultz Bipontinus, in Seemann, Bot. Voy. Herald 313. 1856. *Thyrsanthema semiflosculare* Kuntz, Revis. Gen. Pl. 1: 369. 1891.

Scapose perennial herb, to 4 dm. Leaves elliptic or elliptic-obovate, 5–18(24) cm long, 1–3 cm wide, the apex obtuse, the base cuneate, the margin denticulate, the upper surface glabrous or glabrate, the lower surface densely white-tomentose, sessile. Flowers in a solitary capitulum, erect at anthesis, nodding in bud and fruit, the pedicel to 1 dm long at anthesis, to 4 dm in fruit, not dilated distally, densely white-tomentose; involucre campanulate; phyllaries numerous in 2–5 series, linear-lanceolate, 4–10 mm long, the apex acute or obtuse, the margin reddish; receptacle slightly convex; ray flowers 10–15, carpellate; corolla creamy white, the lamina with a purple outer surface, the lamina 4–6 mm long; disk flowers numerous, staminate, 7–8 mm long. Cypsela fusiform, 4–5 mm long, the beak stout, less than ½ as long as the body, the body slightly laterally flattened, 4-to 12-ribbed, the body glabrous, the beak with ascending, spreading, swollen-apiculate trichomes; pappus of numerous barbellate bristles 5–9 mm long.

Wet flatwoods and bogs. Frequent; nearly throughout. North Carolina south to Florida, west to Texas. Spring.

EXCLUDED TAXON

Chaptalia dentata (Linnaeus) Cassini—Reported for Florida by Small (1903, 1913a, 1913b, 1933), Long and Lakela (1971), Cronquist (1980), and Correll and Correll (1982), the name misapplied to *C. albicans.*

Chromolaena DC. 1836. THOROUGHWORT

Herbs or subshrubs. Leaves opposite, simple, pinnate-veined, petiolate or epetiolate. Flowers in capitula in corymbiform arrays, discoid, bisexual; receptacle epaleate; petals 5, basally connate, actinomorphic; stamens 5, epipetalous, the filaments free, the anthers connate; ovary inferior, 2-carpellate, 1-loculate. Fruit a cypsela; pappus present.

A genus of about 165 species; North America, West Indies, Mexico, Central America, and South America. [From the Greek *chroma,* color, and *khlaina* (Latinized *laena*), cloak, in reference to the colored phyllaries of some species resembling a decorated cloak.]

Ooclinium DC., 1836; *Osmia* Sch.-Bip., 1866.

Selected reference: Nesom (2006f).

1. Leaves sessile or with a petiole 1(3) mm long; phyllaries with a loose, somewhat expanded, slightly spreading, truncate, hyaline, petaloid apex .... **C. ivifolia**
1. Leaves with a petiole 5–20 mm long; phyllaries appressed, the apex not as above.
    2. Leaves 1.5–4 cm long; involucre 5–8 mm high .... **C. frustrata**
    2. Leaves (3.5)5–10 cm long; involucre 8–11 mm high .... **C. odorata**

Chromolaena frustrata (B. L. Rob.) R. M. King & H. Rob. [Annoying, in reference to past difficulty in identification due to confusion with *C. heteroclinia.*] CAPE SABLE THOROUGHWORT.

Eupatorium frustratum B. L. Robinson, Proc. Amer. Acad. Arts 47: 193. 1911. *Osmia frustrata* (B. L. Robinson) Small, Fl. Florida keys 147. 155. 1913. *Chromolaena frustrata* (B. L. Robinson) R. M. King & H. Robinson, Phytologia 20: 201. 1970. TYPE: FLORIDA: Monroe Co.: Lignum Vitae Key, s.d., *Curtiss 1195** (holotype: GH).

Erect perennial herb, to 1.5 m; stem striate-angled, puberulent or hispidulous. Leaves with the blade elliptic-lanceolate, ovate-lanceolate, or deltate-ovate, 1.5–4 cm long, 0.7–2.3 cm wide, 3-nerved from the base, the apex acute to obtuse, the base truncate or broadly cuneate, the margin serrate-dentate, the upper and lower surfaces sparsely puberulent, the petiole 5–12 mm long. Flowers in capitula in a corymbiform array, the peduncle 3–10 mm long; involucre cylindric; phyllaries numerous in 4–6 series, elliptic, 5–8 mm long, the apex appressed, rounded; receptacle slightly convex; flowers numerous; corolla blue or lavender, ca. 4 mm long. Cypsela prismatic, 3–4 mm long, 3- to 5-ribbed, scabrellous; pappus of numerous barbellate bristles 3–4 mm long.

Pine rockland hammocks. Rare; Miami-Dade and Monroe Counties. Endemic. All year.

Chromolaena frustrata is listed as endangered in Florida (Florida Administrative Code, Chapter 5B-40) and in the United States (U.S. Fish and Wildlife Service, 50 CFR 23).

Chromolaena ivifolia (L.) R. M. King & H. Rob. [With leaves like ivy.] IVYLEAF THOROUGHWORT.

Eupatorium ivifolium Linnaeus, Syst. Nat., ed. 10. 1205. 1759. *Osmia ivifolia* (Linnaeus) Schultz Bipontinus, Jahresber. Pollichia 22–24: 250, 252. 1866. *Eupatorium ivifolium* Linnaeus var. *genuinum* 14: 280. 1916, nom. inadmiss. *Chromolaena ivifolia* (Linnaeus) R. M. King & H. Robinson, Phytologia 20: 202. 1970.

Erect perennial herb or subshrub, to 1.5 m; stem striate-angled, hispidulous or coarsely short-pilose. Leaves with the blade linear-lanceolate or elliptic-lanceolate, 1.5–7 cm long, 0.5–1.2 cm wide, 3-nerved from near the base, the apex acute or obtuse, the base cuneate, the margin denticulate or subentire, the upper and lower surfaces sparsely pubescent, sessile or the petiole to 3 mm long. Flowers in capitula in a corymbiform array, the peduncle 5–15 mm long; involucre cylindric or campanulate; phyllaries numerous in 3–5 series, oblanceolate, 5–6 mm long, the apex spreading, truncate, slightly expanded, hyaline, petaloid, pinkish purple; receptacle slightly convex; flowers numerous; corolla light blue, purplish, or reddish, ca. 4 mm long. Cypsela prismatic, 2–3 mm long, 3- to 5-ribbed, scabrellous; pappus of numerous barbellate bristles 2–3 mm long.

Disturbed sites. Rare; Polk, Hardee, and Escambia Counties. Florida west to Texas. North America; West Indies, Central America, and South America. Native to North America west of Florida and tropical America. Summer.

Chromolaena odorata (L.) R. M. King & H. Rob. [With an odor.] JACK-IN-THE-BUSH.

Eupatorium odoratum Linnaeus, Syst. Nat., ed. 10. 1205. 1759. *Eupatorium conyzoides* Vahl, Symb. Bot. 3: 96. 1794; non Miller, 1768. *Osmia odorata* (Linnaeus) Schultz Bipontinus, Jahresber. Pollichia 22–24: 252. 1866. *Osmia conyzoides* Small, Fl. S.E. U.S. 1164, 1338. 1903. *Chromolaena odorata* (Linnaeus) R. M. King & H. Robinson, Phytologia 20: 204. 1970.

Erect or subscandent perennial herb or subshrub, to 2.5 m; stem striate-angled, hispidulous or short-pilose. Leaves with the blade lanceolate, deltate-lanceolate, or ovate-lanceolate, 30–12 cm long, 1–4 cm wide, 3-nerved from the base, the apex acute, the base rounded or subtruncate, the margin coarsely dentate-serrate or subentire, the upper and lower surfaces sparsely pubescent, the petiole 0.5–2 cm long. Flowers in capitula in a corymbiform array, the peduncle 5–15 mm long; involucre cylindric; phyllaries numerous in 4–6(8) series, 4–8(10) mm long, the apex obtuse to truncate, appressed; receptacle slightly convex; flowers numerous; corolla pale lilac, pinkish, or nearly white, ca. 4 mm long. Cypsela prismatic, ca. 4 mm long, 3- to 5-ribbed, scabrellous; pappus of numerous barbellate bristles 3–4 mm long.

Hammocks, pinelands, and disturbed sites. Occasional; central and southern peninsula. Florida and Texas; West Indies and Mexico. All year.

EXCLUDED TAXA

Chromolaena heteroclinia (Grisebach) R. M. King & H. Robinson—Reported for Florida by Chapman (1897, as *Eupatorium heteroclinium* Grisebach) and Small (1903, as *Osmia heteroclinia* (Grisebach) Small), the name misapplied to Florida material of *C. frustrata*.

Chromolaena rigida (Swartz) R. M. King & H. Robinson—Reported for Florida by Chapman (1878, as *Ooclinium rigidum* (Swartz) de Candolle), the name misapplied to Florida material of *C. frustrata*.

Chrysogonum Nutt. 1753.

Herbs. Leaves opposite, simple, pinnate-veined, petiolate. Flowers in solitary or sometimes paired capitula, radiate; receptacle paleate; ray flowers carpellate; petals 5, basally connate, zygomorphic; disk flowers staminate; petals 5, basally connate, actinomorphic; stamens 5, epipetalous, the filaments free, the anthers connate; ovary inferior, 2-carpellate, 1-loculate. Fruit a cypsela; pappus present.

A monotypic genus; North America. [From the Greek *chrysos,* gold, and *gonos,* seed, in reference to the yellow capitula or the ray flower cypsela-complex.]

Selected references: Nesom (2001, 2006g); Steussy (1977).

Chrysogonum virginianum L. var. **australe** (Alexander ex Small) Ahles [Of Virginia; southern.] GREEN-AND-GOLD.

Chrysogonum australe Alexander ex Small, Man. S.E. Fl. 1415, 1509. 1933. *Chrysogonum virginianum* Linnaeus var. *australe* (Alexander ex small) Ahles, J. Elisha Mitchell Scientific Soc. 80: 173. 1964. TYPE: FLORIDA: Jackson Co.: ca. 1 mi. E of Marianna, 16 Mar 1925, *Harper s.n.* (holotype: NY; isotypes: GH, US).

Stoloniferous, mat-forming, perennial herb, to 1 dm; stem (stolon) striate-angled, the internodes to 6 dm long, villous-hirsute and stipitate-glandular. Leaves with the blade elliptic-ovate, 1–7 cm long, 0.8–4 cm wide, the apex obtuse, the base narrowly cuneate, the margin crenate-dentate, the upper and lower surfaces villous-hirsute and stipitate-glandular, the petiole 2–7 cm long. Flowers in capitula, solitary or paired at a node, the peduncle 2–10 cm long; involucre campanulate-hemispheric; phyllaries 8–10 in 2 series, the outer oblong-oblanceolate, 7–10 mm long, foliaceous, the distal half spreading, the apex acuminate or mucronate, the outer surface strigose distally, the margin ciliate, the inner elliptic, ca. ⅔ as long as the outer, somewhat scarious, the outer surface strigose; receptacle flat or shallowly convex; paleae oblanceolate, ca. 4 mm long, scarious; ray flowers 5(5); corolla yellow, the lamina 6–17 mm long; disk flowers numerous; corolla yellow, ca. 3 mm long. Cypsela oblanceolate or obovate, 3–5 mm long, laterally obflattened, 2-winged, the face scabrellous, dark brown; pappus coroniform.

The cypsela shed as a "cypsela-complex" consisting of a subtending inner phyllary, its palea, and two contiguous cypsela with their palea, all attached to a basal elaiosome.

Disturbed sites. Rare; Leon, Gadsden, Jackson, and Walton Counties. Georgia south to Florida, west to Louisiana. Spring–summer.

EXCLUDED TAXON

Chrysogonum virginianum Linnaeus—Because infraspecific categories were not recognized, the typical variety was reported for Florida by implication by Chapman (1860, 1883, 1897), Small (1903, 1913a), and Clewell (1985), and in error by Small (1933). Florida plants are all of var. *australe.*

Chrysoma Nutt. 1834.

Shrubs. Leaves alternate, simple, pinnate-nerved, epetiolate. Flowers in capitula in cymiform arrays, radiate; receptacle epaleate; ray flowers carpellate; petals 5, basally connate, zygomorphic; disk flowers bisexual; petals 5, basally connate, actinomorphic; stamens 5, epipetalous, the filaments free, the anthers connate; ovary inferior, 2-carpellate, 1-loculate. Fruit a cypsela; pappus present.

A monotypic genus; North America. [From the Greek *chrysos,* gold, and *ome,* having the condition of, in reference to the golden yellow capitula.]

Selected reference: Nesom (2006h).

Chrysoma pauciflosulosa (Michx.) Greene [From the Greek *pauci,* few, and *flos,* flowers, in reference to the few-flowered capitula.] BUSH GOLDENROD; WOODY GOLDENROD.

Solidago pauciflosulosa Michaux, Fl. Bor.-Amer. 2: 116. 1803. *Aplactis paniculata* Rafinesque, Fl. Tellur. 2: 42. 1837 ("1836"), nom. illegit. *Aster pauciflosulosus* (Michaux) Kuntze, Revis. Gen. Pl. 1: 318. 1891. *Chrysoma pauciflosulosa* (Michaux) Greene, Erythea 3: 8. 1895.

Chrysoma solidaginoides Nuttall, J. Acad. Nat. Sci. Philadelphia 7: 67. 1834. TYPE: FLORIDA.

Erect shrubs, to 1 m; stem striate-angled, glabrous. Leaves with the blade oblanceolate or narrowly elliptic, 2–6 cm long, 2–10 mm wide, 1-nerved, the apex acute, obtuse, or rounded, mucronate, the base cuneate, the margin entire, the upper and lower surfaces glabrous, gland-dotted, sessile. Flowers in capitula in a cymiform array, radiate, the peduncle 2–6 mm long; involucre cylindric; phyllaries 8–12 in 3–4(5) series, lanceolate, 4–6 mm long, erect, 1-nerved, orange-resinous; receptacle flat, shallowly pitted; ray flowers 1–2(3) or absent; corolla yellow, the lamina 4–6 mm long; disk flowers (2)3–4(5); corolla yellow, ca. 2 mm long. Cypsela turbinate-oblong, 3–4 mm long, nearly terete, 8- to 10-ribbed, whitish strigose-sericeous; pappus of numerous, unequal, apically attenuate barbellate bristles 6–8(10) mm long.

Scrub, dunes, and sandhills. Frequent; central and western panhandle. North Carolina south to Florida, west to Mississippi. Summer–fall.

Chrysopsis (Nutt.) Elliott, nom. cons. 1923 ("1924"). GOLDENASTER

Herbs. Leaves alternate, simple, pinnate-veined, epetiolate. Flowers in capitula in corymbiform, subumbelliform, or paniculiform arrays, radiate; receptacle epaleate; ray flowers carpellate; petals 5, basally connate, zygomorphic; disk flowers bisexual; petals 5, basally connate, actinomorphic; stamens 5, epipetalous, the filaments free, the anthers connate; ovary inferior, 2-carpellate, 1-loculate. Fruit a cypsela; pappus present.

A genus of 11 species; North America. [From the Greek *chrysos,* gold, and *opsis,* appearance, likeness, in reference to to the yellow corollas.]

Selected references: DeLaney et al. (2003); Semple (1981, 2006d).

1. Achene with 1–several yellow or red-brown gland-like ridges.
    2. Phyllaries with the apex long-attenuate, twisted .... **C. subulata**
    2. Phyllaries other than as above.
        3. Phyllaries stipitate-glandular, 1.5–2 mm wide, foliaceous .... **C. latisquamea**
        3. Phyllaries not stipitate-glandular, 1 mm wide or less, not foliaceous.
            4. Leaf margins distinctly serrate-pilose .... **C. gossypina**
            4. Leaf margins entire or obscurely serrate.
                5. Stem erect .... **C. linearifolia**
                5. Stem decurent .... **C. gossypina**

1. Achene lacking yellow or red-brown gland-like ridges.
   6. Upper leaves with long-silky-sericeous (arachnoid) trichomes .... **C. mariana**
   6. Upper leaves variously pubescent or glabrous, but not with long-silky-sericeous (arachnoid) trichomes.
      7. Phyllaries with the apex attenuate or subulate, spreading or reflexed.
         8. Stem erect; inland sites .... **C. lanuginosa**
         8. Stem procumbent or ascending; coastal dunes .... **C. godfreyi**
      7. Phyllaries with the apex acute to acuminate or obtuse, erect.
         9. Midstem leaves lanate or sericeous-tomentose; phyllaries yellow-green prior to anthesis.
            10. Midstem leaves short-spatulate, appressed sericeous-tomentose, the base cuneate, the margin usually flat .... **C. floridana**
            10. Midstem leaves oblong, oblong-elliptic or slightly oblanceolate, loosely lanate, the base truncate or rounded, the margin often undulate .... **C. highlandsensis**
         9. Midstem leaves glandular-scabrous or stipitate-glandular (rarely also lanate); phyllaries green prior to anthesis.
            11. Plant usually unbranched below the inflorescence; stem erect, the tip erect or ascending prior to inflorescence development; inflorescence open, corymbose; midstem leaves glandular scabrous or finely stipitate-glandular (trichomes mostly less than 0.4 mm long) .... **C. scabrella**
            11. Plant often branched near the base; stem ascending, the apex nodding prior to inflorescence development; inflorescence compact, obconic or paniculate-corymbose; midstem leaves densely stipitate-glandular (trichomes mostly 0.4 mm long or longer) .... **C. delaneyi**

Chrysopsis delaneyi Wunderlin & Semple [Commemorates Kris Robert DeLaney (1951–), Florida botanist.] DELANEY'S GOLDENASTER.

Chrysopsis delaneyi Wunderlin & Semple, in DeLaney et al. Bot. Explorer (Florida) 3: 2. 2003. FLORIDA: Highlands County: Avon Park, ca. 200 m N of W Pleasant St. and 100 m E of N Central Ave., T33S, R38E, Sec. 22, 9 Nov 1987, *DeLaney 1530* (holotype: USF; isotypes: FLAS, MO, NY, USF, WAT).

Erect or ascending perennial herb, to 1.5 m; stem densely stipitate-glandular, the tips usually nodding before capitula development. Basal leaves with the blade spatulate, oblanceolate, narrowly lanceolate, or linear, 10–18 cm long, (0.8)1.5–2.5(3.5) cm wide, the apex rounded or obtuse, the base narrowly cuneate, the margin distally serrate-denticulate, coarsely toothed, or serrate-denticulate and/or distally shallowly or conspicuously lobed, the upper and lower surfaces densely woolly and/or densely stipitate-glandular, the cauline linear-oblanceolate, linear, or elliptic-oblong, (3)4.5–11 cm long, (0.7)1–1.5(2.5) cm long, the apex acute, mucronate, the base slightly auriculate-clasping, truncate, the margin usually entire, sometimes distally coarsely toothed or serrate-denticulate, the upper and lower surfaces moderately hirsute, densely stipitate-glandular, all sessile. Flowers in capitula in a compact to moderately open array, the peduncle 0.5–11 cm long, stipitate-glandular; bracteoles foliose, linear or linear-lanceolate, stipitate-glandular; involucre campanulate or turbinate; phyllaries in 3–5 series, linear-lanceolate, 9–10 mm long, erect, unequal, the apex acute, acuminate, or sometimes aristate, the inner ones sometimes obtuse or rounded, the surface densely stipitate-glandular; receptacle slightly convex, shallowly pitted. Ray flowers 18–28; corolla yellow, the lamina (9)13–17(19) mm long; disk flowers numerous; corolla yellow, 5–6 mm long. Cypsela obconic-obovoid, 2–4 mm long, lacking gland-like ridges, shallowly ribbed or smooth, sparsely long-strigose; pappus in 2–3 series, the outer of linear scales to 1 mm long, the inner of numerous moderately clavate bristles 4–6 mm long.

Sandhills. Occasional; central peninsula, Broward County. Endemic. Fall.

Chrysopsis floridana Small [Of Florida.] FLORIDA GOLDENASTER.

Chrysopsis floridana Small, Fl. S.E. U.S. 1183, 1339, 1903. *Chrysopsis mariana* (Linnaeus) Elliott var. *floridana* (Small) Fernald, Rhodora 39: 455. 1937. *Heterotheca mariana* (Linnaeus) Shinners subsp. *floridana* (Small) V. L. Harms, Wrightia 4: 13. 1968. *Heterotheca floridana* (Small) R. W. Long, Rhodora 72: 44. 1970. FLORIDA: Manatee Co.: Bradenton, 28 Nov 1901, *Tracy 7344* (holotype: NY; isotypes: G, GH, MIN, NY, MO, US).

Erect or ascending perennial herb, to 7 dm; stem densely short-woolly. Basal leaves with the blade spatulate or oblanceolate, 4–10 cm long, 1.5–2.5 cm wide, the apex rounded, the base cuneate, the margin entire or distally dentate, the upper and lower surfaces densely short-woolly, the cauline with the blade obovate or oblanceolate, gradually or only slightly reduced distally, the apex rounded, mucronulate, the base cuneate or slightly auriculate clasping, the margin entire or sometimes undulate, with cilia rarely more than 1 mm long, the upper and lower surfaces densely appressed-tomentose, sparsely stipitate-glandular, all sessile. Flowers in capitula, solitary or in a subumbelliform or paniculiform array, the peduncle 1–4 cm long, densely stipitate-glandular; bracteoles foliose, linear or oblanceolate, proximally woolly, distally stipitate-glandular; involucre campanulate; phyllaries in 3–4 series, linear, erect, 5–8 mm long, unequal, yellow prior to anthesis, the apex acuminate or aristate, the outer surface stipitate-glandular; receptacle slightly convex, shallowly pitted. Ray flowers 15–20; corolla yellow, the lamina 6–8 mm long; disk flowers numerous; corolla 6–7 mm long. Cypsela obconic-obovoid, 2–3 mm long, lacking gland-like ridges, smooth or faintly ribbed, moderately strigose; pappus in 3 series, the outer of linear scales to 1 mm long, the inner of numerous weakly clavate bristles 5–6 mm long.

Scrub, rarely oak hammocks. Rare; Hillsborough, Pinellas, Manatee, Hardee, and Highlands Counties. Endemic. Fall.

Chrysopsis floridana is listed as endangered in Florida (Florida Administrative Code, Chapter 5B-40) and in the United States (U.S. Fish and Wildlife Service, 50 CFR 23).

Chrysopsis godfreyi Semple [Commemorates Robert Kenneth Godfrey (1911–2000), botanist at Florida State University.] GODFREY'S GOLDENASTER.

Chrysopsis godfreyi Semple, Canad. J. Bot. 56: 2092. 1978. FLORIDA: Okaloosa Co.: US 98 E of Destin, dunes N of Silver Beach, 5 Nov 1977, *Semple & Godfrey 3148* (holotype: WAT; isotypes: FSU, GH, MO, NY, US, USF).

Chrysopsis godfreyi Semple forma *viridis* Semple, Canad. J. Bot. 56: 2093. 1978. TYPE: FLORIDA: Okaloosa Co.: US 98 E of Destin, dunes N of Silver Beach, 5 Nov 1977, *Semple & Godfrey 3139* (holotype: WAT; isotypes: FSU, GH, MO, NY, US, USF).

Decumbent, ascending, or erect biennial or perennial herb, to 5 dm; stem proximally woolly, distally woolly or densely stipitate-glandular. Basal leaves with the blade oblanceolate, 1.7–10 cm long, 0.9–1.2 cm wide, the apex obtuse, the base narrowly cuneate, the upper and lower surfaces long-woolly, the cauline with the blade ovate or linear-lanceolate, 2–3.5 cm long, 0.6–1.7 cm wide, reduced upward, the apex acute, the base sometimes clasping, the margin entire, the upper and lower surfaces woolly or glabrous, densely stipitate-glandular, all sessile. Flowers in capitula in a corymbiform or paniculiform array, the peduncle 2–8 cm long, densely stipitate-glandular; bracteoles foliose, lanceolate or linear, glabrous and densely stipitate-glandular or densely woolly and stipitate-glandular; involucre campanulate; phyllaries in 4–5 series, linear-lanceolate, 9–12 mm long, unequal, the apex attenuate or long-attenuate, spreading or recurved, the outer surface densely stipitate-glandular; receptacle slightly convex, shallowly pitted; ray flowers 16–numerous; corolla yellow, the lamina 1–1.5 cm long; disk flowers numerous; corolla yellow, 6–8 mm long. Cypsela obconic-obovoid, 2–3 mm long, lacking gland-like ridges, shallowly ribbed, moderately strigose; pappus in 3 series, the outer of linear scales to 1 mm long, the inner of numerous weakly clavate bristles 6–7 mm long.

Dunes and scrub. Rare; Bay County, western panhandle. Florida and Alabama. Fall.

Chrysopsis godfreyi hybridizes with *C. gossypina* subsp. *cruiseana* in Florida.

Chrysopsis godfreyi is listed as endangered in Florida. (Florida Administrative Code, Chapter 5B-40).

Chrysopsis gossypina (Michx.) Elliott [*Gossypinus*, cottony, in reference to the leaves.] COTTONY GOLDENASTER.

Decumbent, ascending, or erect biennial or short-lived perennial herb, to 1 m; stem glabrous or densely woolly, proximally often sparsely glandular, distally sometimes densely so or eglandular. Basal leaves with the blade oblanceolate, 3–10 cm long, 1–3 cm wide, the apex rounded, the base cuneate, the margin entire or sometimes apically dentate, the upper and lower surfaces sparsely or densely woolly, the cauline with the blade linear, linear-lanceolate, oblanceolate, oblong, or ovate-elliptic, spreading or ascending, reduced upward, the apex acute or obtuse, sometimes mucronate, the base cuneate, the margin entire, pilose-ciliate, the upper and lower surfaces sparsely to densely woolly, sometimes glabrescent, all sessile. Flowers in capitula, solitary or in a compact or lax subumbelliform or corymbiform array, the peduncle 1–6 cm long, glabrous, woolly, or sparsely to densely stipitate-glandular; bracteoles foliose, oblong, lanceolate, or linear, the margin often ciliate, glabrous or woolly, eglandular or sparsely to densely stipitate-glandular; involucre campanulate or hemispheric; phyllaries in 4–5 series, linear or linear-lanceolate, 8–13 mm long, unequal, the apex acute or acuminate, appressed or sometimes spreading or recurved, the outer surface glabrate, sparsely or densely woolly, sparsely or densely stipitate-glandular, rarely eglandular; receptacle slightly convex, shallowly pitted; ray flowers (9)16–many; corolla yellow, the lamina 7–12 mm long; disk flowers numerous; corolla 6–8 mm long. Cypsela obconic-obovoid, 2–3 mm long, lacking gland-like ridges or with 1–6 yellow or red-brown, translucent, clavate ridges, shallowly ribbed, the surface sparsely to long strigose; pappus in 3 series, the outer of linear or linear-triangular scales to 1 mm long, the inner of numerous weakly clavate bristles 5–8 mm long.

Chrysopsis gossypina is highly variable with three nearly allopatric subspecies recognized. Intermediates are sometimes found where the ranges overlap.

1. Stem leaves sparsely or densely woolly .... subsp. **gossypina**
1. Stem leaves pilose or glabrate.
   2. Leaf margin distinctly serrate-pilose .... subsp. **hyssopifolia**
   2. Leaf margin entire or obscurely serrate ....subsp. **cruiseana**

Chrysopsis gossypina subsp. **gossypina**

Inula gossypina Michaux, Fl. Bor.-Amer. 2: 122. 1803. *Diplopappus lanatus* Cassini, in Cuvier, Dict. Sci. Nat. 13: 309. 1819, nom. illegit. *Chrysopsis gossypina* (Michaux) Elliott, Sketch Bot. S. Carolina 2: 337. 1823. *Heterotheca gossypina* (Michaux) Shinners, Field & Lab. 19: 71. 1951.

Inula trichophylla Nuttall, Gen. N. Amer. Pl. 2: 150. 1818. *Chrysopsis trichophylla* (Nuttall) Elliott, Sketch Bot. S. Carolina 2: 336. 1823. *Diplopappus trichophyllus* (Nuttall) Hooker, Companion Bot. Mag. 1: 97. 1836. *Diplogon trichophyllum* (Nuttall) Kuntze, Revis. Gen. Pl. 1: 334. 1891. *Heterotheca trichophylla* (Nuttall) Shinners, Field & Lab. 19: 1951. *Chrysopsis gossypina* (Michaux) Elliott forma *trichophylla* (Nuttall) Semple, Canad. J. Bot. 58: 148. 1980.

Chrysopsis decumbens Chapman, Fl. South. U.S. 217. 1860. *Chrysopsis gossypina* (Michaux) Elliott forma *decumbens* (Chapman) R. K. Godfrey, Rhodora 51: 113. 1949. TYPE: FLORIDA: Franklin Co.: St. Vincent Island.

Stem leaves sparsely or densely woolly.

Sandhills and scrub. Frequent; northern counties, central peninsula. West Virginia south to Florida, west to Alabama. Fall.

Chrysopsis gossypina subsp. **cruiseana** (Dress) Semple [Commemorates James Edwin Cruise (b. 1925), Canadian botanist who collected in Florida in 1953 with W. J. Dress.] CRUISE'S GOLDENASTER.

Chrysopsis cruiseana Dress, Gentes Herbar. 8: 409. 1954. *Chrysopsis gossypina* (Michaux) Elliott subsp. *cruiseana* (Dress) Semple, Canad. J. Bot. 58: 148. 1980. TYPE: FLORIDA: Escambia Co.: Santa Rosa Island, ca. 2 mi. E of Pensacola Beach, 27 Dec 1953, *Dress & Cruise 2822* (holotype: BH).

Stem leaves pilose or glabrate, the margin entire or obscurely serrate.

Dunes and scrub. Rare; western panhandle. South Carolina, Florida, and Alabama. Fall.

Chrysopsis gossypina subsp. *cruiseana* hybridizes with *C. godfreyi* in Florida.

Chrysopsis gossypina subsp. *cruiseana* (as *C. crusiana*) is listed as endangered in Florida (Florida Administrative Code, Chapter 5B-40).

Chrysopsis gossypina subsp. **hyssopifolia** (Nutt.) Semple [With leaves like hyssop (*Hyssopus officinalis*, Lamiaceae).]

Chrysopsis hyssopifolia Nuttall, J. Acad. Nat. Sci. Philadelphia 7: 67. 1834. *Chrysopsis trichophylla* (Nuttall) Torrey & A. Gray, Fl. N. Amer. 2: 254. 1842. *Diplogon hyssopifolium* (Nuttall) Kuntze, Revis. Gen. Pl. 1: 334. 1891. *Heterotheca hyssopifolia* (Nuttall) R. W. Long, Rhodora 72: 42. 1970.

Chrysopsis gossypina (Michaux) Elliott subsp. *hyssopifolia* (Nuttall) Semple, Canad. J. Bot. 58: 148. 1980. TYPE: FLORIDA: s.d., *Ware s.n.* (holotype: PH).

Chrysopsis gigantea Small, Fl. S.E. U.S. 1183, 1339. 1903. TYPE: FLORIDA: s.d., *Leavenworth s.n.* (holotype: NY; isotype; GH).

Chrysopsis mixta Dress, Gentes Herb. 8: 404. 1954. TYPE: FLORIDA: Walton Co.: 5.5 mi. E of Argyle on US 90, 2 Oct 1953, *Dress 2689* (holotype: BH).

Stem leaves pilose or glabrate, the margin distinctly serrate-pilose.

Scrub, sandhills, and dry flatwoods. Occasional; Alachua County, central and western panhandle. Florida, Alabama west to Louisiana. Fall.

Chrysopsis highlandsensis DeLaney & Wunderlin [Of Highlands County, Florida.] HIGHLANDS GOLDENASTER.

Chrysopsis highlandsensis DeLaney & Wunderlin, Bot. Explor. (Florida) 2: 2. 2002. *Chrysopsis floridana* Small var. *highlandsensis* (DeLaney & Wunderlin) D. B. Ward, Phytologia 94: 466. 2012. TYPE: FLORIDA: Polk Co.: NE of Avon Park along N side of CR-64 (Bombing Range Road), T35S, R29E, Sec. 35, 12 Nov 2001, *DeLaney 5113* (holotype: USF; isotypes: USF).

Erect or ascending, rarely decumbent, short-lived perennial herb, to 1.5 m; stem sparsely lanate, stipitate-glandular. Basal leaves with the blade spatulate or oblanceolate, rarely obovate, (3.5)5.5–12(15) cm long, 2–3 cm wide, the apex rounded or obtuse, the base narrowly cuneate, the margin entire or distally serrate-denticulate, the upper and lower surfaces long-lanate, obscurely stipitate-glandular, viscid, the cauline with the blade oblong, oblong-elliptic, ovate, or lanceolate, the apex obtuse or acute, mucronate or mucronulate, the base truncate or rounded, slightly auriculate-clasping, the margin undulate, ciliate, the upper and lower surfaces moderately woolly-lanulate, stipitate-glandular, all sessile. Flowers in capitula in a compact corymbiform or paniculiform array, the peduncle 2–6 cm long, stipitate-glandular; bracteoles foliose, linear or linear-lanceolate, stipitate-glandular; involucre cylindric-campanulate; phyllaries in 5–6 series, linear-lanceolate, 6–9 mm long, erect, unequal, yellow-green prior to anthesis, the apex acute, acuminate, sometimes aristate, the inner obtuse or rounded, the outer surface stipitate-glandular, viscid; receptacle slightly convex, shallowly pitted; ray flowers 16–22; corolla yellow, the lamina 8–10 mm long; disk flowers numerous; corolla yellow, 4–6 mm long. Cypsela obconic-obovoid, 2–3 mm long, lacking gland-like ridges, shallowly ribbed or smooth, moderate long-strigose; pappus in 3–4 series, the outer of linear scales ca. 1 mm long, the inner of numerous moderately clavate bristles.

Scrub and scrubby flatwoods. Rare; Polk, Highlands, and Glades Counties. Endemic. Fall.

Chrysopsis highlandsensis is listed as endangered in Florida. (Florida Administrative Code, Chapter 5B-40).

Chrysopsis lanuginosa Small [*Lanuginosus*, woolly, downy, in reference to leaves.] LYNN HAVEN GOLDENASTER.

Chrysopsis lanuginosa Small, Man. S.E. Fl. 1339, 1508, 1933. TYPE: FLORIDA: Bay Co.: near Lynn Haven, 3 Dec 1920, *Van Cleve 11* (holotype: NY: isotype: US).

Erect biennial or short-lived perennial herb, to 1 m; stem densely stipitate-glandular. Basal leaves with the blade oblanceolate, 3–8 cm long, 6–15 mm wide, the apex rounded or obtuse, the base cuneate, the margin entire, the upper and lower surfaces densely woolly, the cauline reduced upward, the blade linear-lanceolate or lanceolate, the apex acute or obtuse, the base cuneate, clasping, the margin entire, the upper and lower surfaces densely stipitate-glandular, all sessile. Flowers in capitula in a paniculiform array, the peduncle 2–12 cm long, densely stipitate-glandular; bracteoles foliose, linear or linear-lanceolate, the base clasping, stipitate-glandular; involucre campanulate; phyllaries in 3–4 series, linear, 8–12 mm long, ca. 1 mm wide, unequal, the apex long-attenuate, often twisted and reflexed, the surface densely stipitate-glandular; receptacle slightly convex, shallowly pitted; ray flowers 12–many; corolla yellow, the lamina 8–11 mm long; disk flowers numerous; corolla yellow, 5–6 mm long. Cypsela obconic-obovoid, 2–3 mm long, lacking gland-like ridges, shallowly ribbed or smooth, long-strigose; pappus in 3 series, the outer of linear scales ca. 1 mm long, the inner of numerous moderately clavate bristles 5–6 mm long.

Sandhills. Occasional; central and western panhandle. Endemic. Fall.

Chrysopsis lanuginosa hybridizes with *C. linearifolia* var. *linearifolia* in Florida.

Chrysopsis latisquamea Pollard [*Latus,* broad, wide, and *squama,* scale, in reference to the phyllaries.] PINELAND GOLDENASTER.

Chrysopsis latisquamea Pollard, Proc. Biol. Soc. Wash. 13: 131. 1900. *Heterotheca latisquamea* (Pollard) V. L. Harms, Castanea 39: 163. 1974. TYPE: FLORIDA: Orange Co.: Clarcona, 1900, *Meislahn 150* (holotype: US; isotype: NY).

Erect biennial or short-lived perennial herb, to 7 dm; stem proximally long-woolly, distally densely stipitate-glandular. Basal leaves with the blade oblanceolate, 4–8 cm long, 1–2 cm wide, the apex obtuse, the base cuneate, the margin entire, the upper and lower surfaces densely long-woolly, the cauline reduced upward, the blade elliptic or ovate, the apex obtuse, mucronate, the base cuneate, clasping or subclasping, the margin entire, the upper and lower surfaces densely long-woolly, stipitate-glandular, all sessile. Flowers in capitula in a lax to compact corymbiform array, the peduncle 1–6(10) cm long, densely stipitate-glandular; bracteoles foliose, ovate or lanceolate, stipitate-glandular; involucre campanulate; phyllaries in 2–3 series, oblanceolate, 8–11 mm long, unequal, erect, the apex acute, the surface stipitate-glandular; receptacle slightly convex, shallowly pitted; ray flowers 10–18; corolla yellow, the lamina 1–1.5 cm long; disk flowers numerous; corolla yellow, 6–9 mm long. Cypsela obconic-obovoid, 2–3 mm long, with 2–6 yellow or red-brown, translucent, clavate ridges, ribbed, sparsely strigose; pappus in 3 series, the outer of linear scales ca. 1 mm long, the inner of numerous weakly clavate bristles 7–9 mm long.

Sandhills. Occasional; northern and central peninsula, eastern panhandle. Endemic. Fall.

Chrysopsis linearifolia Semple [With narrow leaves.] NARROWLEAF GOLDENASTER.

Erect biennial or short-lived perennial herb, to 2 m; stem glabrous. Basal leaves with the blade oblanceolate or linear, 2–7 cm long, 1–7 mm wide, the apex rounded or obtuse, the base narrowly cuneate, the margin entire or undulate, the upper and lower surfaces densely woolly or glabrous, the cauline with the blade linear or linear-lanceolate, the apex acute, the base cuneate or sometimes rounded, the margin entire or undulate, the upper and lower surfaces glabrous or sparsely woolly proximally, all sessile. Flowers in capitula in a subumbelliform, compact paniculiform, corymbiform, or loose, open corymbiform array, the peduncle 2–8 cm long, glabrous; bracteoles foliose, linear, glabrous; involucre campanulate; phyllaries in 4–5 series, linear-lanceolate, 6–10(12) mm long, erect, unequal, the apex acute, acuminate, or long-acuminate, glabrous, the bases and the outer surface sparsely stipitate-glandular; receptacle slightly convex, shallowly pitted; ray flowers 10–many; corolla yellow, the lamina 9–12 mm long. Disk flowers numerous; corolla yellow, 5–6 mm long. Cypsela obconic-obovoid, 2–3 mm long, yellowish, with 2–6 golden yellow or red-brown, translucent clavate ridges, weakly ribbed, sparsely strigose; pappus in 3 series, the outer of linear or narrowly triangular scales ca. 1 mm long, the inner of numerous moderately clavate bristles 4–6 mm long.

1. Inflorescence compact, subumbellate; leaf margin entire or only slightly undulate or twisted; central and western panhandle .... subsp. **linearifolia**
1. Inflorescence loosely corymbose; leaf margin sometimes undulate, strongly twisted; peninsula .... subsp. **dressii**

Chrysopsis linearifolia subsp. **linearifolia**

Chrysopsis linearifolia Semple, Brittonia 30: 493. 1978. TYPE: FLORIDA: Franklin Co.: US 98 just S of the Ochlockonee River, 2 Nov 1976, *Godfrey 75745* (holotype: MO; isotypes: FSU, NY, USF, WAT).

Leaves with the margin entire or only slightly undulate or twisted. Inflorescence a compact, subumbelliform array.

Scrub and sandhills. Occasional; central and western panhandle. Endemic. Fall.

Chrysopsis linearifolia hybridizes with *C. lanuginosa* in Florida.

Chrysopsis linearifolia subsp. **dressii** Semple [Commemorates William John Dress (1918–2012), Cornell University botanist.] DRESS'S GOLDENASTER.

Chrysopsis linearifolia Semple subsp. *dressii* Semple, Brittonia 30: 494. 1978. *Chrysopsis linearifolia* Semple var. *dressii* (Semple) D. B. Ward, Phytologia 94: 466. 2012. TYPE: FLORIDA: Brevard Co.: Merritt Island, Cummit Cove, Rte. 3, 2 Oct 1976, *Semple et al. 2530* (holotype: MU; isotype: US, USF, WAT).

Leaves with the margin sometimes undulate, strongly twisted. Inflorescence a loosely corymbiform array.

Sandhills. Occasional; peninsula. Endemic. Fall.

Chrysopsis mariana (L.) Elliott [Of Maryland.] MARYLAND GOLDENASTER.

Inula mariana Linnaeus, Sp. Pl., ed. 2. 1240. 1763. *Chrysopsis mariana* (Linnaeus) Elliott, Sketch Bot. S. Carolina 2: 335. 1823. *Diplopappus marianus* (Linnaeus) Cassini ex Hooker, Companion Bot. Mag. 1: 97. 1836. *Inula glandulosa* Lamarck, Encycl. 3: 259. 1789, nom. illegit. *Diplogon marianum* (Linnaeus) Rafinesque ex Kuntze, Revis. Gen. Pl. 1: 334. 1891. *Heterotheca mariana* (Linnaeus) Shinners, Field & Lab. 19: 71. 1951.

Erect or ascending biennial or short-lived perennial herb, to 9 dm; stem silky-sericeous. Basal leaves with the blade spatulate or oblanceolate, 2–25 cm long, 0.5–4 cm wide, the apex acute or obtuse, the base narrow cuneate, the margin entire or obscurely dentate distally, the upper surface long-sericeous or glabrescent, the cauline with the blade lanceolate or elliptic-oblong, 1–3 cm long, 3–8 mm wide, reduced upward, the apex acute, the base cuneate, the margin entire or obscurely dentate, ciliate, the upper and lower surfaces long-silky-sericeous or glabrate, all sessile. Flowers in capitula, usually in a compact subumbelliform array, the peduncle 0.5–5 cm long, densely stipitate-glandular; bracteoles foliose, linear, stipitate-glandular; involucre campanulate; phyllaries in 4–5 series, linear, 7–10 mm long, erect, unequal, the apex acute, the outer surface densely stipitate-glandular; receptacle slightly convex, shallowly pitted; ray flowers 10–22; corolla yellow, the lamina 8–11 mm long; disk flowers numerous; corolla yellow, 5–7 mm long. Cypsela obconic to obovoid, 2–3 mm long, yellowish to purple, lacking gland-like ridges, shallowly ribbed or smooth, short-strigose; pappus in 2–3 series, the outer of bristles ca. 1 mm long, the inner of numerous weakly to moderately clavate bristles 4–6 mm long.

Dry flatwoods and sandhills. Frequent; nearly throughout. New York south to Florida, west to Texas. Fall.

Chrysopsis scabrella Torr. & A. Gray [Minutely scabrous, in reference to the leaves.] COASTALPLAIN GOLDENASTER.

Chrysopsis scabrella Torrey & A. Gray, Fl. N. Amer. 2: 255. 1842. *Diplogon scabrellum* (Torrey & A. Gray) Kuntze, Revis. Gen. Pl. 1: 334. 1891. *Heterotheca scabrella* (Torrey & A. Gray) R. W. Long, Sida 3: 449. 1969. TYPE: FLORIDA: s.d., *Leavenworth s.n.* (holotype: NY; isotype: GH).

Erect biennial herb, to 1 m; stem proximally woolly, mid to distally stipitate-glandular. Basal leaves with the blade oblanceolate or ovate, 4–10 cm long, 1–2 mm wide, the apex rounded or obtuse, the base narrowly cuneate, the margin entire, distally dentate-serrate, the upper and lower surfaces densely woolly, the cauline with the blade linear, linear-elliptic, elliptic, linear-lanceolate, or lanceolate, reduced upward, the apex acute, rarely obtuse, the base cuneate, the margin entire, the upper and lower surfaces sparsely srigillose, densely short-stipitate-glandular, or glabrate, the proximal and mid cauline sometimes woolly, all sessile. Flowers in capitula in a lax, corymbiform array, the peduncle 2–10 cm long, stipitate-glandular; bracteoles foliose, oblong or distally linear, short-stipitate-glandular; involucre campanulate; phyllaries in 4–5 series, linear-lanceolate, erect, unequal, 6–9 mm long, the apex acute or acuminate, rarely aristate, the surface short stipular-glandular; receptacle slightly convex, shallowly pitted; ray flowers 19–many; corolla yellow, 6–8 mm long; disk flowers numerous; corolla yellow, 5–7 mm long. Cypsela obconic-obovoid, 2–3 mm long, lacking gland-like ridges, shallowly ribbed, sparsely to moderately long-strigose; pappus in 3 series, the outer of linear scales ca. 1 mm long, the inner of numerous weakly to moderately clavate bristles 5–7 mm long.

Sandhills and scrub. Frequent; peninsula, eastern panhandle. North Carolina, South Carolina, Florida, Alabama, and Mississippi. Fall.

Chrysopsis subulata Small [*Subula,* a fine sharp point, in reference to the leaf apex.] SCRUBLAND GOLDENASTER.

Chrysopsis subulata Small, Man. S.E. Fl. 1338, 1508. 1933. *Heterotheca hyssopifolia* (Nuttall) R. W. Long var. *subulata* (Small) R. W. Long, Rhodora 72: 43. 1970. TYPE: FLORIDA: Highlands Co.: between Avon Park and Sebring, 17 Jul 1924, *Small et al. 11495* (holotype: NY; isotypes: GH, US).

Erect biennial herb, to 7 dm; stem glabrous or sparsely pilose, eglandular. Basal leaves with the blade oblanceolate or linear, 5–10 cm long, 4–10 mm wide, the apex obtuse, the base narrowly cuneate, the margin entire, sometimes obscurely dentate distally, the upper and lower surfaces densely woolly, the cauline with the blade linear-lanceolate or linear, reduced upward, the apex acute, the base cuneate, the margin entire or undulate, long-pilose-ciliate, the upper and lower surfaces sparely pilose, all sessile. Flowers in capitula in a lax corymbiform array, the peduncle 1–10 cm long, glabrous or glabrate; bracteoles foliose, linear-lanceolate, the margin ciliate, glabrous or sparsely pilose, stipitate-glandular; involucre campanulate; phyllaries in 4–5 series, linear, 8–10 mm long, unequal, the apex long-attenuate or acute, spreading or reflexed, twisted, proximally glabrous or sparsely stipitate-glandular; receptacle slightly convex, shallowly pitted; ray flowers 10–28; corolla yellow, the lamina 5–8(10) mm long; disk flowers numerous; corolla yellow, 5–6 mm long. Cypsela obconic-obovoid, 2–3 mm long, with 6–10 yellow or red-brown, translucent, clavate ridges, weakly ribbed, glabrous or sparsely strigose proximally; pappus in 3 series, the outer of linear or linear-triangular scales ca. 1 mm long, the inner of numerous moderately clavate bristles 6–7 mm long.

Flatwoods and disturbed sites. Frequent; peninsula. Endemic. Fall.

Cichorium L. 1754. CHICORY

Herbs. Leaves alternate, simple, pinnate-veined, petiolate or epetiolate. Flowers in capitula in solitary or glomerulate arrays, liguliflorous, bisexual; receptacle epaleate; petals 5, basally connate, zygomorphic; stamens 5, epipetalous, the filaments free, the anthers connate; ovary inferior, 2-carpellate, 1-loculate. Fruit a cypsela; pappus present.

A genus of 6 species; North America, South America, Europe, Africa, Asia, Australia, and Pacific Islands. [From the Arabic *chicouryeh,* ancient name for the plant.]

Selected reference: Strother (2006f).

Cichorium intybus L. [From the Arabic *tybi,* January, in reference to the month in which it was generally eaten.] CHICORY.

Cichorium intybus Linnaeus, Sp. Pl. 813. 1753. *Cichorium commune* Pallas, Reise Russ. Reich. 3: 655. 1776, nom. illegit. *Cichorium sylvestre* Lamarck, Fl. Franc. 2: 120. 1779 ("1778"), nom. illegit. *Cichorium rigidum* Salisbury, Prodr. Chap. Allerton 183. 1796, nom. illegit. *Cichorium perenne*

Stokes, Bot. Mat. Med. 4: 133. 1812, nom illegit. *Cichorium perenne* Stokes var. *sylvestre* Stokes, Bot. Mat. Med. 4: 133. 1812, nom. inadmiss. *Cichorium intybus* Linnaeus var. *sylvestre* Bischoff, Beitr. Deutschl. 25. 1851, nom. inadmiss. *Cichorium intybus* Linnaeus var. *caeruleum* Alefeld, Landw. Fl. 282. 1866, nom. inadmiss.

Erect perennial herb, latex present, to 1.5 m; stem striate-angled, setose, hispid, pilose, or glabrous. Basal leaves with the blade oblanceolate or linear-lanceolate, 5–35 cm long, 1–8(12) cm wide, the apex acute, the base cuneate, the margin runcinate-pinnate or dentate, rarely entire, short-petiolate or subsessile, the cauline similar, smaller, the margin dentate or entire, the distal mostly linear, sessile, sometimes clasping. Flowers in capitula in an axillary, subsessile glomerulate array, the peduncle 1–2 mm long, sometimes narrowly clavate, 1–4.5(8.5) mm long, or absent; involucre cylindric, 10–12 cm long; phyllaries in 2 series, the outer 5–6, lanceolate or lanceolate-ovate, 4–7 mm long, somewhat spreading, the apex acute or obtuse, proximally cartilaginous, the inner 8–12, linear or linear-lanceolate, 6–12 mm long, the apex acute or obtuse, usually with some glandular trichomes on the margin proximally or on the outer surface distally; receptacle flat, pitted, slightly hispid; flowers numerous; corolla blue, sometimes pink or white, the lamina 1–2 cm long. Cypsela prismatic, 3-to 5-angled, 2–3 mm long, glabrous; pappus of 2–3 series, coroniform, of numerous whitish, subequal, blunt, erose scales to 0.2 mm long.

Disturbed sites. Rare; Hillsborough and Escambia Counties. Nearly throughout North America; South America; Europe, Africa, Asia, Australia, and Pacific Islands. Native to Europe, Africa, and Asia. Summer–fall.

The leaves of *Cichorium intybus* have sometimes been used in salads or the ground roasted roots in coffee.

Cirsium Mill. 1754. THISTLE

Herbs. Leaves alternate, simple, pinnate-veined, petiolate or epetiolate. Flowers in capitula in solitary, corymbiform, subcapitate, or paniculiform arrays, discoid, bisexual; receptacle paleate; petals 5, basally connate, actinomorphic; stamens 5, epipetalous, the filaments free, the anthers connate; ovary inferior, 2-carpellate, 1-loculate. Fruit a cypsela; pappus present.

A genus of about 200 species; nearly cosmopolitan. [From the Greek *kirsion,* thistle.]

Selected reference: Keil (2006a).

1. Stems conspicuously winged with spiny, decurrent leaf bases .... **C. vulgare**
1. Stem inconspicuously or not at all winged with decurrent leaf bases.
    2. Capitula subtended by a secondary involucre of spinescent bracts .... **C. horridulum**
    2. Capitula not subtended by a secondary involucre of spinescent bracts.
        3. Phyllaries lacking an apical spine or with a vestigial spicule to ca. 0.5 mm long...... **C. muticum**
        3. Middle and outer phyllaries tipped with an evident spine at least 1 mm long.
            4. Lower leaf surface densely tomentose.
                5. Involucre 1.5–2.5 cm high; middle and outer phyllaries with a medial glutinous ridge .... **C. virginianum**

5. Involucre (2)2.5–3.5 cm high; phyllaries lacking a medial glutinous ridge .... **C. altissimum**

4. Lower leaf surface arachnoid-villous, thinly tomentose, or glabrate.

6. Plant usually much branching distally; involucre 1.5–2.5 cm long .... **C. nuttallii**

6. Plant with simple or with only a few branches; involucre 2.5–4 cm long..........**C. lecontei**

Cirsium altissimum (L.) Spreng. [Tallest.] TALL THISTLE.

Carduus altissimus Linnaeus, Sp. Pl. 824. 1753. *Cnicus altissimus* (Linnaeus) Willdenow, Sp. Pl. 3: 1671. 1803. *Cirsium altissimum* (Linnaeus) Sprengel, Syst. Veg. 3: 373. 1826.

Erect biennial or short-lived perennial herb, to 4 m; stem striate-angled, villous with septate trichomes, sometimes glabrate or distally thinly tomentose. Leaves with the blade oblanceolate or elliptic, 10–40 cm long, 1–13 cm wide, the apex acute, the base cuneate, the margin finely or coarsely spine-toothed or shallowly pinnatifid, the lobes triangular, the spines 1–5 mm long, the upper surface glabrate or villous with septate trichomes, the lower surface white-tomentose, the basal usually absent at flowering, the petiole winged, the principal cauline gradually reduced upward, the base sometimes weakly clasping. Flowers in capitula, solitary or in a corymbiform or paniculate array, the peduncle to 5 cm long, leafy-bracted, or absent; involucre ovoid, broadly cylindric, or campanulate, 2–4 cm long, 1.5–4 cm wide, thinly arachnoid; phyllaries numerous in 10–20 series, the surface greenish with the subapical central area darker, the outer surface with a narrow glutinous ridge, the apical spine 3–4 mm long, the inner lanceolate, the apex spreading, the margin suberose or finely serrate, the apical spine spreading, the middle and outer ovate, the margin entire, the spine abruptly spreading; flowers numerous; corolla pink to purple, rarely white, 2–3.5 cm long. Cypsela ovoid, compressed, tan to dark brown, 4–6 mm long, the apical collar yellowish; pappus in 3–5 series of numerous plumose bristles 1.2–2.4 cm long.

Open floodplain forests and calcareous hammocks. Rare; Jackson County. Massachusetts and New York south to Florida, west to North Dakota, South Dakota, Nebraska, Kansas, Oklahoma, and Texas. Fall.

Cirsium horridulum Michx. [*Horridulus,* somewhat rough, prickly, in reference to the numerous spines.] PURPLE THISTLE.

Carduus spinosissimus Walter, Fl. Carol. 194. 1788. *Cnicus spinosissimus* (Walter) Darlington, Fl. Cestr., ed. 2. 438. 1837; non Linnaeus, 1753; nec Lapeyrouse, 1813.

Cirsium horridulum Michaux, Fl. Bor.-Amer. 2: 90. 1803. *Carduus horridulus* (Michaux) Persoon, Syn. Pl. 2: 390. 1807. *Cnicus horridulus* (Michaux) Pursh, Fl. Amer. Sept. 507. 1814.

Cirsium megacanthum Nuttall, Trans. Amer. Philos. Soc., ser. 2. 7: 421. 1841. *Cirsium horridulum* Michaux var. *megacanthum* (Nuttall) D. J. Keil, Sida 21: 214. 2004.

Cirsium horridulum Michaux var. *elliottii* Torrey & A. Gray, Fl. N. Amer. 2: 460. 1843. *Carduus spinosissimus* Walter var. *elliottii* (Torrey & A. Gray) Porter, in Porter & Britton, Mem. Torrey Bot. Club 5: 345. 1894. *Cirsium horridulum* Michaux forma *elliottii* (Torrey & A. Gray) Fernald, Rhodora 45: 353. 1943.

Carduus pinetorum Small, Fl. S.E. U.S. 1308, 1341. 1903. *Cirsium pinetorum* (Small) Small, Fl. Miami 199, 200. 1913; non Greenman, 1905. *Cirsium smallii* Britton, in Britton & Millspaugh, Bahama Fl.

458. 1920. *Carduus smallii* (Britton) Ahles, J. Elisha Mitchell Sci. Soc. 80: 173. 1964, nom. illegit. TYPE: FLORIDA: Miami-Dade Co.: Coconut Grove, 2 & 5 Nov 1901, *Small & Nash s.n.* (holotype: NY).

Carduus vittatus Small, Bull. New York Bot. Gard. 3: 439. 1905. *Cirsium vittatum* (Small) Small, Fl. Miami 199, 200. 1913. *Cirsium horridulum* Michaux var. *vittatum* (Small) R. W. Long, Rhodora 72: 45. 1970. TYPE: FLORIDA: Miami-Dade Co.: W of Camp Jackson, May 1904, *Small & Wilson 1874* (holotype: NY).

Cirsium smallii Britton forma *purpureum* Murrill, Quart. J. Florida Acad. Sci. 12: 66. 1949. TYPE: FLORIDA: Alachua Co.

Erect biennial or perennial herb, to 2.5 m; stem striate-angled, glabrous or densely tomentose. Leaves with the blade linear, oblanceolate, or oblong-elliptic, 10–40 cm long, 2–10 cm wide, the apex acute, the base cuneate, the margin unlobed and spiny dentate or deeply pinnatifid and the lobes spiny-dentate or coarsely lobed, the principal spines 0.5–3 cm long, the upper surface glabrous or densely villous with septate trichomes, the lower surface subglabrate to loosely tomentose, the basal with the petiole spiny-winged, the cauline sessile, somewhat auriculate-clasping. Flowers in capitula, solitary or in a subcapitate or corymbiform array, subtended by involucre-like spine-margined bracts, the peduncle to 5 cm long or absent; involucre hemispheric to campanulate, 3–5 cm long, 3–8 cm wide; phyllaries in 5–9 series, linear or lanceolate, light green or yellowish, distally with a reddish margin, the outer surface usually sparsely tomentose, often scabridulous in submarginal bands, the outer and middle usually reddish tinged, the margin setulose-ciliolate, the apex acuminate, the spines 1–2 mm long; flowers numerous; corolla pink, purple, yellow, or white, 3–4.5 cm long. Cypsela ovoid, compressed, yellowish or tan, 4–6 mm long, the apical collar similar; pappus of 3–5 series of numerous plumose bristles 2.5–3.5 cm long.

Flatwoods and disturbed sites. Common; nearly throughout. Maine south to Florida, west to Oklahoma and Texas; Mexico. Spring–summer.

A polymorphic species with several variants segregated as species, varieties, and forms by various authors. Keil (2006a) recognized three varieties in Florida (var. *horridulum*, var. *megacanthum*, and var. *vittatum*) which are separated by the amount of pubescence on the stems and involucres, leaf shape, length of the leaf spines, and flower color. These features all vary greatly, sometimes within populations and sometimes on a populational or regional basis. Herbarium specimens are sometimes difficult to assign to a variety.

Cirsium lecontei Torr. & A. Gray [Commemorates John Eatton Le Conte (1784–1860), American botanist and zoologist.] LECONTE'S THISTLE.

Cirsium lecontei Torrey & A. Gray, Fl. N. Amer. 2: 458. 1843. *Cnicus lecontei* (Torrey & A. Gray) A. Gray, Proc. Amer. Acad. Arts 10: 39. 1874. *Carduus lecontei* (Torrey & A. Gray) Pollard, Bull. Torrey Bot. Club 24: 157. 1897.

Erect perennial herb, to 1 m; stem striated-angled, loosely arachnoid. Leaves with the blade linear, narrowly elliptic, or oblong, 15–25 cm long, 1–4 cm wide, reduced upward and becoming bract-like distally, the apex acute or acuminate, the base cuneate, the margin coarsely toothed or shallowly pinnatifid, the lobes coarsely few-toothed, the principal spines 3–6 mm long, the upper surface glabrous or sparsely villous with multicellular trichomes, the lower surface glabrate or arachnoid, the basal petiolate, the cauline sessile, the base clasping or somewhat decurrent. Flowers in capitula, solitary or in an open corymbiform array, the peduncle 5–30 cm long; involucre broadly cylindric or campanulate, 2.5–4 cm long, 1.5–4 cm wide; phyllaries in 6–10 series, the outer ovate or lanceolate, the inner linear-lanceolate, the outer surface with a prominent glutinous ridge, the outer and middle with the apex acute, the margin serrate-spinulose, the spines to 2 mm long, ascending; flowers numerous; corolla pink-purple, 2.2–4.5 cm long. Cypsela ovoid, compressed, light brown, 5–6 mm long, the apical collar paler; pappus of 3–5 series of numerous plumose bristles 2–4 cm long.

Wet flatwoods and bogs. Occasional; central and western panhandle. North Carolina south to Florida, west to Louisiana. Fall.

Cirsium muticum Michx. [*Muticus,* without a point, blunt, in reference to the phyllaries lacking an apical spine.] SWAMP THISTLE.

Cirsium muticum Michaux, Fl. Bor.-Amer. 2: 89. 1803. *Carduus muticus* (Michaux) Persoon, Syn. Pl. 2: 386. 1807. *Cnicus muticus* (Michaux) Pursh, Fl. Amer. Sept. 506. 1814.

Erect biennial herb, to 2.5 m; stem striate-angled, villous with septate trichomes or glabrate. Leaves with the blade ovate, broadly elliptic, or obovate, 15–55 cm long, 4–20 cm wide, the margin deeply pinnatifid, the apex acute to acuminate, the base cuneate, the lobes linear or lanceolate, irregularly few-toothed or -lobed, the spines 2–3 mm long, the upper surface thinly pilose, the lower surface thinly tomentose or glabrate, the veins villous with septate trichomes on the veins, the distal with narrow linear lobes and more spiny, the basal with the petiole spine-winged, the principal cauline sessile, reduced upward and becoming subclasping. Flowers in capitula, solitary or in an open corymbiform or paniculiform array, the peduncle to 2–15 cm long or absent; involucre ovoid to broadly cylindric or campanulate, 1.7–3 cm long, 1–2 cm wide; phyllaries in 8–12 series, green with a darker subapical patch, the outer surface with a narrow glutinous ridge, the outer ovate, the inner linear-lanceolate, the apex obtuse or acute, minutely spinulose; flowers numerous; corolla lavender or purple, rarely white, 1.6–3.2 cm long. Cypsela ovoid, compressed, dark brown, 4–6 mm long, the apical collar yellow; pappus of 3–5 series of numerous plumose bristles 1.2–2 cm long.

Floodplain forests and disturbed sites. Rare; Putnam County, central panhandle, and Escambia County. Labrador and Quebec south to Florida, west to Saskatchewan, North Dakota, Minnesota, Iowa, Missouri, Oklahoma, and Texas. Fall.

Cirsium nuttallii DC. [Commemorates Thomas Nuttall (1786–1859), English botanist and zoologist who worked in America 1808–1841.] NUTTALL'S THISTLE.

Carduus glaber Nuttall, Gen. N. Amer. Pl. 2: 129. 1818. *Cnicus glaber* (Nuttall) Elliott, Sketch Bot. S. Carolina 2: 270. 1824. *Cirsium nuttallii* de Candolle, Prodr. 6: 651. 1838. *Cnicus nuttallii* (de Candolle) A. Gray, Proc. Amer. Acad. Arts 10: 41. 1874, nom. illegit. *Carduus nuttallii* (de Candolle) Pollard, Bull. Torrey Bot. Club 24: 157, nom. illegit.

Erect biennial herb, to 3.5 m; stem striate-angled, glabrous or villous with septate trichomes, leaves with the blade oblong or elliptic, 7–30 cm long, 2–12 cm wide, gradually reduced distally, the uppermost reduced to linear bracts, the apex acute or acuminate, the base cuneate, the margin spinulose or irregularly dentate or shallowly to deeply pinnatifid, the lobes narrow, coarsely dentate or lobed, the primary spines 2–5 mm long, the upper surface glabrous or sparsely villous with septate trichomes, the lower surface sparsely tomentose, glabrate in age, the basal with the petiole winged, the principal cauline gradually reduced distally, becoming sessile, the base spine-lobed, sometimes decurrent. Flowers in capitula in an open corymbiform or paniculiform array, the peduncle 1–15 cm long; involucre hemispheric or campanulate, 1.5–2.5 cm long, 1–2.5 cm wide; phyllaries in 6–10 series, the outer ovate or elliptic, the inner linear-lanceolate, green or brownish, the outer surface with a narrow glutinous ridge, thinly arachnoid or glabrate, the outer and middle appressed, the margin entire, the apical spine abruptly spreading, 1–2(3) mm long, the inner flexuous, flat, the apex attenuate; flowers numerous; corolla white, pink, lavender, or purple, 1.7–2.5 mm long. Cypsela ovoid, compressed, 3–4 mm long, dark brown, the apical collar yellowish; pappus of 3–5 series of many plumose bristles 1.7–2.1 cm long, the longer bristles shorter than the corolla.

Wet to dry hammocks and disturbed sites. Frequent; nearly throughout. Virginia south to Florida, west to Texas. Spring–fall.

Cirsium virginianum (L.) Michx. [Of Virginia.] VIRGINIA THISTLE.

Carduus virginianus Linnaeus, Sp. Pl. 824. 1753. *Cirsium virginianum* (Linnaeus) Michaux, Fl. Bor.-Amer. 2: 90. 1803. *Cnicus virginianus* (Linnaeus) Pursh, Fl. Amer. Sept. 506. 1814.

Carduus revolutus Small, Fl. S.E. U.S. 1307, 1341. 1903. *Cirsium revolutum* (Small) Petrak, Beih. Bot. Centralbl. 35(2): 558. 1917. *Cirsium virginianum* (Linnaeus) Michaux forma *revolutum* (Small) Fernald, Rhodora 45: 509. 1943. TYPE: FLORIDA: s.d., *Chapman s.n.* (holotype: NY).

Erect biennial or perennial herb, to 1.5(2) m; stem striate-angled, thinly appressed-tomentose, glabrate in age. Leaves with the blade linear, linear-elliptic, or narrowly ovate, 3–15 cm long, 0.5–4 cm wide, the apex acute, the base cuneate, the margin spinulose or deeply lobed, the lobes remote, few-toothed or -lobed, the principal spines 3–5(9) mm long, the upper surface glabrous or thinly tomentose, the lower surface white-tomentose, the basal usually absent at flowering, the cauline reduced distally, linear, the margin entire or few-lobed, the basal and proximal leaves with the petiole winged, the middle and distal sessile. Flowers in capitula, solitary or in an open corymbiform or paniculiform array, the peduncle 10–15 mm long; involucre ovoid, cylindric, or narrowly campanulate, 1.7–2.4 cm long, 1–2 cm wide; phyllaries in 8–13 series, light green to brownish with a dark apex, the outer ovate, the inner linear-elliptic, the outer surface with a narrow glutinous ridge, the outer and middle with the margin entire, the apex erect or spreading, muricate or with a spine 1–2 mm long, the inner with the apex slightly flexuous, erose; flowers numerous; corolla purple, 2.1–2.6 cm long. Cypsela ovoid, compressed, 4–5 mm long, dark brown, the apical collar yellowish; pappus of 3–5 series of numerous plumose bristles 1.7–2.0 cm long.

Wet savannas and bogs. Rare; Duval, Clay, and Santa Rosa Counties. New Jersey, Delaware, and Virgina south to Florida, west to Texas. Summer–fall.

Cirsium vulgare (Savi) Ten. [*Vulgaris,* common, ordinary.] BULL THISTLE.

Carduus lanceolatus Linnaeus, Sp. Pl. 821. 1753. *Cirsium lanceolatum* (Linnaeus) Scopoli, Fl. Carniol., ed. 2. 2: 130. 1772; non Hill, 1769. *Cnicus lanceolatus* (Linnaeus) Willdenow, Fl. Berol. Prodr. 259. 1787. *Eriolepis lanceolata* (Linnaeus) Cassini, in Cuvier, Dict. Sci. Nat. 41: 331. 1825.

Carduus vulgaris Savi, Fl. Pis. 2: 241. 1798. *Cirsium vulgare* (Savi) Tenore, Fl. Napol. 5: 209. 1835–1836. *Cirsium lanceolatum* (Linnaeus) Scopoli var. *vulgare* Naegeli, in W.D.J. Koch, syn. Deut. Schweiz. Fl., ed. 2. 990. 1846, nom. inadmiss.

Erect or ascending biennial herb, to 2(3) m; stem striate-angled, villous with septate trichomes, conspicuously winged with spiny decurrent leaf bases. Leaves with the blade oblong-lanceolate or obovate, 15–40 cm long, 6–15 cm wide, the apex acute or acuminate, the base cuneate, the margin 1- to 2-pinnatifid, with divergent lobes, sometimes merely spinose-dentate, the lobes triangular or lanceolate, entire or spiny dentate, the edge flat or revolute, the spines 2–10 mm long, the upper surface with short appressed bristlelike spines, sometimes tomentose when young, reduced upward and the distal more deeply lobed than the proximal, the main lobes rigidly spiny, the margin spinulose or entire, the lower surface gray-tomentose, villous with septate trichomes along the veins, the basal and proximal cauline leaves with a winged petiole, decurrent on the stem, the middle and distal sessile, decurrent on the stem. Flowers in capitula in a corymbiform or paniculiform array, the peduncle 1–6 cm long; involucre hemispheric to campanulate, 3–4 cm long, 2–4 cm wide; phyllaries in 10–12 series, the outer linear-lanceolate, the inner linear, the outer and middle with the apex radiating, the apical spine 2–5 mm long, the apex of the inner flat, serrulate to minutely erose; flowers numerous; corolla purple or rarely white, 2.5–3.5 cm long. Cypsela ovoid, compressed, light brown with darker brown streaks, 3–5 mm long, the apical collar not differentiated; pappus of 3–5 series of numerous plumose bristles 2–3 cm long.

Disturbed sites. Rare; Wakulla County. Nearly throughout North America; South America; Europe, Africa, Asia, Australia, and Pacific Islands. Native to Europe, Africa, and Asia. Spring.

EXCLUDED TAXON

Cirsium repandum Michaux—Reported for Florida by Chapman (1860, 1883, 1897, as *Cnicus repandus* Elliott) and Small (1903, as *Carduus repandus* (Michaux) Persoon). No Florida specimens seen.

Cladanthus Cass. 1816.

Herbs. Leaves alternate, simple, petiolate or epetiolate. Flowers in capitula in solitary or corymbiform arrays, radiate; receptacle paleate; ray flowers carpellate or sterile; petals 5, basally connate, zygomorphic; disk flowers bisexual; petals 5, basally connate, actinomorphic; stamens 5, epipetalous, the filaments free, the anthers connate; ovary inferior, 2-carpellate, 1-loculate. Fruit a cypsela; pappus absent.

A genus of 5 species; North America, South America, Europe, Africa, Asia. [From the Greek, *klados,* branch, and *anthos,* flower, in reference to the branching stems at the base of the sessile capitulum in the original species.]

Selected reference: Watson (2006b).

Cladanthus mixtus (L.) Chevall. [Mingled.] WEEDY DOGFENNEL.

Anthemis mixta Linnaeus, Sp. Pl. 894. 1753. *Chamaemelum mixtum* (Linnaeus) Allioni, Fl. Pedem. 1: 185. 1785. *Ormenis bicolor* Cassini, in Cuvier, Dict. Sci. Nat. 36: 356. 1825, nom. illegit. *Ormenis mixta* (Linnaeus) Dumortier, Fl. Belg. 69. 1827. *Cladanthus mixtus* (Allioni) Chevallier, Fl. Gén. Env. Paris, ed. 2. 2: 576. 1836.

Erect annual herb, to 60 cm; stem striate-angled, puberulent. Leaves with the blade obovate, spatulate, oblong, or linear, 2–6(8) cm long, 0.5–2.5(3.5)cm wide, 1-to 2(3)-pinnately lobed, the lobes lanceolate, linear, or filiform, the apex apiculate, the base cuneate, the ultimate margins entire or dentate, the upper and lower surfaces villous, arachnoid, or glabrescent, the proximal petiolate, the distal epetiolate. Flowers numerous in a solitary capitulum or a lax corymbiform array; involucre hemispheric, 5–8(12) mm wide; phyllaries numerous in 2–3 series, linear-lanceolate, lanceolate, oblong, or obovate, subequal, the apex somewhat dilated, rounded, the margin hyaline, scarious, the outer surface subvillous, arachnoid, or glabrescent; receptacle hemispheric, narrowly columnar, or conic; palea carinate, with a central red-brown resin duct; ray flowers numerous; corolla orange, yellow, or white, the lamina 5–6(10) mm long, spreading or reflexed; disk flowers numerous; corolla orange or yellow, 2–3 mm long, the tube basally saccate, obliquely spurred, distally clasping the cypsela. Cypsela obovoid, ca. 1 cm long, slightly compressed, weakly ribbed, the 2 laterals and the inner surface finely striate, glabrous; pappus absent.

Disturbed sites. Rare; Escambia County. New Jersey, Pennsylvania, Ohio, North Carolina, Florida, and Oregon; South America; Europe, Africa, and Asia. Native to Europe, Africa, and Asia. Summer.

Conoclinium DC. 1836. THOROUGHWORT

Herbs. Leaves opposite, simple, pinnipalmate-veined, petiolate. Flowers in capitula in corymbiform arrays, discoid, bisexual; receptacle epaleate; petals 5, basally connate, actinomorphic; stamens 5, epipetalous, the filaments free, the anthers connate; ovary inferior, 2-carpellate, 1-loculate. Fruit a cypsela; pappus present.

A genus of 4 species; North America and Mexico. [From the Greek *konos,* cone, and *kline,* bed, in reference to the conic receptacle.]

Selected references: Patterson and Nesom (2006).

Conoclinium coelestinum (L.) DC. [*Coelestis,* heavenly. In reference to the beautiful flowers.] BLUE MISTFLOWER.

Eupatorium coelestinum Linnaeus, Sp. Pl. 838. 1753. *Conoclinium coelestinum* (Linnaeus) de Candolle, Prodr. 5: 135. 1836.

Conoclinium dichotomum Chapman, Bot. Gaz. 3: 5. 1878. TYPE: FLORIDA.
Conoclinium flaccidum Greene, Pittonia 4: 274. 1901. TYPE: FLORIDA: Gadsden Co.: River Junction, s.d., *Nash 2572* (holotype: US?).
Conoclinium nepetifolium Greene, Pittonia 4: 274. 1901. TYPE: FLORIDA: Indian River, Merritt, 31 Jul 1896, *Curtiss 5730* (holotype: US?).

Erect, decumbent, or procumbent perennial herb, to 1 m; stem rooting at the nodes, sparsely pilose. Leaves with the blade triangular, deltate, or ovate, 2–7(13) mm long, 2–5 mm wide, the apex acute or obtuse, the base cuneate, truncate, or rarely subcordate, 3- or 5-nerved from the base, the margin serrate or serrate-dentate, sometimes slightly lobed proximally, the upper and lower surfaces glabrous, the petiole 1–2 cm long. Flowers in capitula in a compact corymbiform array, discoid; involucre hemispheric, 3–6 mm wide; phyllaries ca. 25 in 2–3 series, lanceolate or linear, ca. 4 mm long, the apex acuminate, obscurely 2- to 3-nerved, subequal, glabrous; receptacle conic; flowers numerous; corolla blue, blue-violet, or rose-violet, 2–3 mm long. Cypsela prismatic, 1–2 mm long, 5-ribbed, glabrous; pappus in 1 series of numerous barbellate bristles 5–8 mm long.

Floodplain forests, wet hammocks, and pond margins. Common; nearly throughout. New York and Ontario south to Florida, west to Michigan, Nebraska, Kansas, Oklahoma, and Texas. Summer–fall.

Conyza Less., nom. Cons. 1832. HORSEWEED

Herbs. Leaves alternate, simple, pinnate-veined, petiolate or epetiolate. Flowers in capitula in paniculiform, corymbiform, or racemiform arrays, radiate (outer flowers with the lamina often absent or minute and flowers appearing disciform); receptacle epaleate; ray flowers carpellate; petals 5, basally connate, zygomorphic; disk flowers bisexual; petals 5, basally connate, actinomorphic; stamens 5, epipetalous, the filaments free, the anthers connate; ovary inferior, 2-carpellate, 1-loculate. Fruit a cypsela; pappus present.

A genus of about 30 species; North America, Mexico, Central America, South America, Europe, Africa, and Asia. [From the Greek *konops,* flea, or *konis,* dust, in reference to the powdered dry plant used to repel insects.]

As traditionally circumscribed, *Conyza,* along with *Aphanostephus* and three South American genera, is nested within *Erigeron* (Noyes, 2000). Until the taxonomy is resolved, we follow the traditional circumscription.

Selected reference: Strother (2006g).

1. Proximal leaves with the blade spatulate, to 6 cm wide .... C. **laevigata**
1. Proximal leaves with the blade narrowly oblanceolate or linear, to 25 cm wide.
    2. Involucre (4)5–6(8) mm long, conspicuously strigose or hispidulous; leaves densely strigose or hispidulous, the margin densely ciliate .... C. **bonariensis**
    2. Involucre 3–4(5) mm long, glabrous or sparsely strigillose; leaves sparsely pubescent or glabrous, the margin remotely ciliate .... C. **canadensis**

Conyza bonariensis (L.) Cronquist [Of the Santuario di Nostra Signora de Bonaria (shrine to the Virgin Mary) in Cagliari, Italy.] ASTHMAWEED.

Erigeron bonariensis Linnaeus, Sp. Pl. 863. 1753. *Leptilon bonariense* (Linnaeus) Small, Fl. S.E. U.S. 1231, 1340. 1903. *Conyza bonariensis* (Linnaeus) Cronquist, Bull. Torrey Bot. Club 70: 632. 1943. *Marsea bonariensis* (Linnaeus) V. M. Badillo, Bol. Soc. Venez. Ci. Nat. 10: 256. 1946.

Erigeron linifolius Willdenow, Sp. Pl. 3: 1955. 1903. *Conyzella linifolia* (Willdenow) Greene, Fl. Francisc. 386. 1897. *Leptilon linifolium* (Willdenow) Small, Fl. S.E. U.S. 1231, 1340. 1903.

Conyza floribunda Kunth, in Humboldt et al. Nov. Gen. Sp. 4: 73. 1820. *Erigeron floribundus* (Kunth) Schultz Bipontinus, Bull. Soc. Bot. France 12: 81. 1865. *Erigeron bonariensis* Linnaeus var. *floribundus* (Kunth) Cuatrecasas, Trab. Mus. Nac. Ci. Nat. Jard. Bot. Madric, Ser. Bot. 33: 132. 1936. *Conyza sumatrensis* (Retzius) E. Walker var. *floribunda* (Kunth) J. B. Marshall, Watsonia 10: 167. 1974.

Erect annual herb, to 1(1.5) m; stem striate-angled, strigose or hispidulous. Proximal leaves with the blade narrowly oblanceolate, 3–8(12) cm long, 1–2.5 cm wide, the apex acute or obtuse, usually mucronulate, the base narrowly cuneate onto a narrowly winged petiole, the margin lobed, coarsely toothed, or entire, the upper and lower surfaces strigose or hispidulous, the margin densely ciliate, the petiole to 1 cm long, the distal narrowly oblanceolate or linear, 1–5 cm long, 2–10 mm wide, the apex and base as the proximal, the margin obscurely toothed or entire, the surface as the proximal, the petiole to 5 mm long. Flowers in capitula in a paniculiform or racemiform array, the peduncle 0.5–2.5 cm long, strigose or hispidulous; involucre turbinate, (4)5–6(8) mm long; phyllaries numerous in 2–4 series, the outer lanceolate, greenish or purplish, the apex acute, the outer surface strigose or hispidulous, shorter than the inner, the inner linear, yellowish or purplish, the apex attenuate, more chartaceous, the outer surface sparsely strigose or hispidulous; receptacle slightly convex, pitted; ray flowers numerous; corolla white or purplish, the lamina minute or absent, the capitula appearing disciform; disk flowers 8–12, white or purplish. Cypsela oblong or elliptic, compressed, 4-angled, 1–2 mm long, pale tan, glabrous or sparsely strigillose; pappus of numerous, pinkish or tawny bristles 3–4 mm long.

Disturbed sites. Occasional; northern counties, central peninsula. Nearly cosmopolitan. Native to South America. Summer–fall.

Some authors treat *C. floribunda,* in synonymy here, as a distinct species (e.g., Strother, 2006g), as a hybrid with *C. bonariensis* or *C. canadensis,* or as a synonym of *C. sumatrensis* (Retzius) E. walker (Marshall, 1974; Pruski and Sancho, 2006).

Conyza canadensis (L.) Cronquist [Of Canada.] CANADIAN HORSEWEED.

Erigeron canadensis Linnaeus, Sp. Pl. 863. 1753. *Erigeron paniculatus* Lamarck, Fl. Franc. 2: 141. 1779 ("1778"), nom. illegit. *Erigeron ruderale* Salisbury, Prodr. Stirp. Chap. Allerton 195. 1796, nom. illegit. *Conyzella canadensis* (Linnaeus) Ruprecht, Mém. Acad. Imp. Sci. Saint Petersbourg, ser. 7. 14(4): 51. 1869. *Leptilon canadense* (Linnaeus) Britton, in Britton & A. Brown, Ill. Fl. N. U.S. 3: 391. 1898. *Aster canadensis* (Linnaeus) E.H.L. Krause, in Sturm, Deutschl. Fl., ed. 2. 13: 59. 1905; non (Linnaeus) Kuntze, 1891. *Conyza canadensis* (Linnaeus) Cronquist, Bull. Torrey Bot. Club 70: 632. 1943.

Erigeron pusillus Nuttall, Gen. N. Amer. Pl. 2: 148. 1818. *Tessenia canadensis* (Linnaeus) Bubani, Fl. Pyren. 2: 267. 1899. *Leptilon canadense* (Linnaeus) Britton var. *pusillum* (Nuttall) Daniels, Univ.

Missouri Stud., Sci. Ser. 2(2): 239. 1911. *Leptilon pusillum* (Nuttall) Britton, Torreya 14: 198. 1914. *Conyza parva* Cronquist, Bull. Torrey Bot. Club 70: 632. 1943. *Marsea canadensis* (Linnaeus) V. M. Badillo, Bol. Soc. Venez. Ci. Nat. 10: 256. 1946. *Conzya canadensis* (Linnaeus) Cronquist var. *pusilla* (Nuttall) Cronquist, Bull. Torrey Bot. Club 74: 150. 1947. *Erigeron canadensis* Linnaeus var. *pusillus* (Nuttall) B. Boivin, Phytologia 23: 55. 1972.

Erect annual herb, to 2(3.5) m; stem striate-angled, glabrate. Proximal leaves with the blade oblanceolate or linear, 2–5(10) cm long, 4–10(15) mm wide, the apex acute, the base cuneate, the margin toothed or entire, remotely ciliate, the upper and lower surfaces glabrate, the nerves usually hispid, the petiole to 1 cm long, the distal similar, smaller, the margin entire. Flowers in capitula in a paniculiform or corymbiform array, the peduncle 2–5 mm long, glabrate; involucre turbinate, 3–4(5) mm long; phyllaries numerous in 2–4 series, the outer lanceolate or linear, shorter than the inner, greenish or yellowish, the outer surface glabrous or sparsely strigose, the margin chartaceous or scarious, the inner similar, yellowish or reddish; receptacle slightly convex, pitted; ray flowers numerous; corolla white or purplish, the lamina to 1 mm long; disk flowers 8–many, corolla white or purplish. Cypsela oblong or elliptic, compressed, 4-angled, 1–2 mm long, pale tan to light gray-brown, sparsely strigillose; pappus of numerous white bristles 2–3 mm long.

Disturbed sites. Common; nearly throughout. Nearly cosmopolitan. Native to North America, Mexico, and Central America. All year.

Conyza laevigata (Rich.) Pruski [Smooth, in reference to leaves.] MANZANILLA HORSEWEED.

Erigeron laevigatus Richard, Actes Soc. Hist. Nat. Paris 1: 112 ("105"). 1792. *Conyza laevigata* (Richard) Pruski, Brittonia 50: 475. 1998.

Erect annual herb, to 1.5 m; stem striate-angled, pilose. Proximal leaves spatulate, with the blade suborbicular or elliptic-ovate, 4–10 cm long, 1–6 cm wide, the apex obtuse, the base cuneate, the margin crenate-serrate in the upper ½, lobed proximally, the upper and lower surfaces glabrous or sparsely pilose, especially on the veins, the petiole 1–2 cm long, narrowly winged to the base, the leaves reduced upward and becoming linear-oblanceolate, the margin entire, sessile. Flowers in capitula in an open paniculiform array, the peduncle to 2 cm long; involucre turbinate, 3–4 mm long; phyllaries numerous, sparsely pilose, the margin hyaline; receptacle slightly convex, muricate; ray flowers numerous; corolla white or purplish, the ligule to 1 mm long; disk flowers numerous; corolla white or purplish. Cypsela oblong or elliptic, compressed, 4-angled, densely pubescent; pappus of numerous white bristles 3–5 mm long.

Disturbed sites. Rare; Miami-Dade County. Florida; West Indies, Mexico, Central America, and South America. Native to tropical America. All year.

EXCLUDED TAXON

Conyza ramosissima Cronquist—Reported for Florida by Long and Lakela (1971), based on a misidentification of *C. canadensis* by Wilhelm (1984). No Florida material seen.

Coreopsis L. 1753. TICKSEED

Herbs. Leaves opposite and/or alternate, simple or compound, pinnate-veined, petiolate or epetiolate. Flowers in capitula in corymbiform arrays, radiate; calyculi present; receptacle paleate; ray flowers sterile or carpellate; petals 5, basally connate, zygomorphic; disk flowers bisexual; petals 4–5, basally connate, actinomorphic; stamens 4–5, epipetalous, the filaments free, the anthers connate; ovary inferior, 2-carpellate, 1-loculate. Fruit a cypsela; pappus present or absent.

A genus of about 35 species; North America, Mexico, South America, Europe, Africa, Asia, Australia, and Pacific Islands. [From the Greek *korios,* bedbug, and *opsis,* resembling, in reference to the fruit shape resembling a bedbug.]

Selected reference: Sherff (1936); Strother (2006h).

1. Ray flowers lavender; leaves subterete .... **C. nudata**
1. Ray flowers yellow or orange (sometimes marked with purple or brown at the base); leaves flat.
    2. Disk flowers apically 5-lobed; anthers 5.
        3. Leaves (at least some) palmately lobed or compound; ray flowers entire or slightly 2- to 3-toothed apically; receptacular bracts linear or linear-clavate.
            4. Leaves with evident petioles 1–5 cm long .... **C. tripteris**
            4. Leaves sessile or subsessile, the blade tapering to the base or the petiole less than 1 mm long .... **C. major**
        3. Leaves simple or pinnatifid; ray flowers sharp-toothed apically; receptacular bracts chaffy-flattened below, caudate-attenuate above.
            5. Disk flowers dark reddish purple apically; ray flowers commonly with a reddish brown spot basally; outer phyllaries linear and at least as long as the inner phyllaries .... **C. basalis**
            5. Disk and ray flowers solid yellow; outer phyllaries lanceolate to deltoid-oblong, shorter than or equaling the inner phyllaries.
                6. Stem leafy mainly toward the base, the peduncle about as long as or longer than the leafy part of the stem.
                    7. Plant annual; ray flowers with dark flecks at the base .... **C. nuecensis**
                    7. Plant perennial; ray flowers without basal flecks.
                        8. Leaves oblanceolate, the margin flat to slightly incurved, the larger often basally lobed, the surfaces short-pubescent at least on the veins .... **C. lanceolata**
                        8. Leaves linear or narrowly oblanceolate, the margin infolded, rarely lobed, the surfaces glabrous .... **C. bakeri**
                6. Stem leafy almost to the top, the peduncle less than ½ as long as the leafy part of the stem.
                    9. Leaves simple or with 1(2) pairs of broad pinnae at the base .... **C. pubescens**
                    9. Leaves pinnately compound with linear segments .... **C. grandiflora**
    2. Disk flowers apically 4-lobed; anthers 4.
        10. Leaves 1- or 2-pinnately compound; cypsela wingless or with broad or narrow, entire or subentire wings.
            11. Lamina of the ray flowers all yellow; cypsela with conspicuous awns .... **C. leavenworthii**
            11. Lamina of the ray flowers reddish brown or purple at the base or throughout (rarely all yellow); cypsela without awns or the awns minute .... **C. tinctoria**

10. Leaves simple, entire or bearing only a few basal auricles; cypsela with deeply laciniate wings.
    12. Leaves elliptic or ovate, not evidently basally disposed, pubescent, the margin ciliolate .... **C. integrifolia**
    12. Leaves linear or slightly oblanceolate, mainly basally disposed, glabrous, the margin eciliolate.
        13. Leaves with minute dark spots .... **C. linifolia**
        13. Leaves lacking minute dark spots.
            14. Outer phyllaries lanceolate, ½–⅔ as long as the inner .... **C. gladiata**
            14. Outer phyllaries deltoid, less than ½ as long as the inner .... **C. floridana**

Coreopsis bakeri E. E. Schill. [Commemorates Wilson Baker (b. 1940), Florida naturalist.] BAKER'S COREOPSIS.

Coreopsis bakeri E. E. Schilling, in E. E. Schilling et al., Phytotaxa 231(2): 179, f. 1. 2015. TYPE: FLORIDA: Jackson Co.: "Brooks 1" limestone glade, 30.81273°N, 85.25771°W, 18 Apr 2013, *Johnson & Baker 10623* (holotype: FSU; isotype: TENN).

Erect perennial herb, to 3.5 dm; stem striate, glabrate or pilose or glabrate. Leaves opposite, the blade simple or sometimes with 1–2 small lateral lobes, linear or narrowly oblanceolate, 11–29+ cm long, 3–14 mm wide, the apex acute or obtuse, the base cuneate, the margin infolded, with a few minute appressed cilia, the upper and lower surfaces glabrous, the petiole 2–15 cm long. Flowers in a solitary terminal capitulum, the peduncle 15–30 cm long; calyculus of linear-lanceolate or lanceolate bractlets 4–8 mm long; receptacle convex; palea linear-lanceolate; phyllaries narrowly ovate, 6–11 mm long; ray flowers 8; corolla yellow, 5-lobed, the lamina 2–2.5 cm long; disk flowers numerous; corolla yellow, 3–5 mm long. Cypsela oblong-elliptic, ca. 3 mm long, the lateral wings entire; pappus of awns to 1 mm long.

Limestone glades. Rare; Jackson County. Endemic. Spring.

Coreopsis basalis (A. Dietr.) S. F. Blake [Basal, in reference to the leaves.] GOLDENMANE TICKSEED.

Calliopsis basalis A. Dietrich, in Otto & A. Dietrich, Allg. Gartenzeitung 3: 329. 1835. *Coreopsis basalis* (A. Dietrich) S. F. Blake, Proc. Amer. Acad. Arts 51: 525. 1916. *Coreopsis basalis* (A. Dietrich) S. F. Blake var. *typica* Sherff, Brittonia 6: 341. 1948, nom. inadmiss.

Erect annual herb, to 5 dm; stem striate, sparsely pilose or glabrate. Leaves opposite, the blade simple or 1- to 3-pinnately lobed with 3–9+ lobes, the simple blade or terminal lobe elliptic, lanceolate, oblanceolate, or linear, 2.5–5.5 cm long, (1)2–9(20) mm wide, the apex obtuse, the base broadly cuneate, the upper and lower surfaces sparsely pilose or glabrous, the petiole 0.8–3.5(12) cm long. Flowers in a solitary terminal capitulum, the peduncle 6–15 cm long; calyculus of lanceolate-deltate to linear bractlets 6–9(12) mm long; receptacle convex; palea basally lanceolate or linear, distally attenuate; phyllaries lanceolate-ovate, 7–9 mm long; ray flowers 8; corolla yellow, the lamina with a proximal red-brown to purple spot or band, 1.5–2 cm long; disk flowers numerous; corolla yellow, red-brown to purple distally, 5-lobed, 3–4 mm long. Cypsela obovate to oblong, 1–2 mm long, without lateral wings; pappus of (1)2 cusps or of subulate or lanceolate scales less than 1 mm long or absent.

Dry, disturbed sites. Occasional; northern counties, central peninsula. North Carolina south to Florida, west to Texas, also Illinois. Native to the western United States. Spring–summer.

Coreopsis floridana E. B. Smith [Of Florida.] FLORIDA TICKSEED.

Coreopsis floridana E. B. Smith, Sida 6: 192, f. 40. 1976. TYPE: FLORIDA: Dixie Co.: US 27Alt, 11.4 mi NW of Shamrock, 31 Oct 1974, *Smith 1840* (holotype: US; isotypes: FSU, UARK).

Erect perennial herb, to 7 dm; stem striate, glabrous. Leaves alternate, the blade linear-lanceolate or linear, (2)4–8(15) cm long, (3)7–12 mm wide, the apex acute or obtuse, the base narrowly cuneate, the margin entire, the upper and lower surfaces glabrous, the petiole 3–10 cm long. Flowers in capitula in a corymbiform array, the peduncle 4–10 cm long; calyculus of linear-lanceolate bractlets 2–6 mm long; receptacle convex; palea linear; phyllaries ovate, 6–10 mm long; ray flowers 8+; corolla yellow, the lamina 1.5–2 cm long; disk flowers numerous; corolla purplish or yellow with purplish tips, 4-lobed, 3–5 mm long. Cypsela obovate or oblong, 3–4 mm long, the lateral wings narrow, deeply laciniate; pappus of 2 subulate scales.

Wet flatwoods. Frequent; nearly throughout. Endemic. Fall–winter.

Coreopsis gladiata Walter [*Gladius*, sword, in reference to the leaf-shape.] COASTALPLAIN TICKSEED.

Coreopsis gladiata Walter, Fl. Carol. 215. 1788.
Coreopsis angustifolia Aiton, Hort. Kew. 3: 253. 1789; non Linnaeus, 1753. TYPE: "Carolina and Florida," s.d. (holotype: BM).
Coreopsis longifolia Small, Bull. Torrey Bot. Club 22: 47. 1895. TYPE: FLORIDA: Duval Co.: near Jacksonville, 16 Oct 1893, *Curtiss 4489* (holotype: NY; isotype: NY).
Coreopsis helianthoides Beadle, Bot. Gaz. 25: 448. 1898; non C. Forster, 1786. TYPE: FLORIDA: Gadsden Co.: Aspalaga, s.d., *Chapman s.n.* (holotype: US?).
Coreopsis falcata F. E. Boynton, Biltmore Bot. Stud. 1: 141. 1902.
Coreopsis palustris Sorrie, in Weakly et al., J. Bot. Res. Inst. Texas 5: 439, f. 1. 2011.

Erect perennial herb, to 7 dm; stem striate, glabrous. Leaves alternate or sometimes alternate and opposite, the blade elliptic, lanceolate, or somewhat linear, (2)4–8(15) cm long, (3)7–12(40) mm wide, the apex acute or obtuse, the base cuneate, the margin entire, the upper and lower surfaces glabrous, the petiole 1–10(15) cm long. Flowers in capitula in a corymbiform array, the peduncle (1)4–8(10) cm long; calyculus of broadly deltoid-ovate or linear-lanceolate bractlets 2–6+ mm long; receptacle convex; paleae linear; phyllaries lanceolate or lanceolate-ovate, 6–9(12) mm long; ray flowers 8+; corolla yellow, the lamina (1.2)1.5–2+ cm long; disk flowers numerous; corolla purplish or yellow with purplish tips, 4-lobed, 3–5 mm long. Cypsela obovate or oblong, 3–4 mm long, the lateral wings narrow, deeply laciniate; pappus of 2 subulate scales.

Wet flatwoods. Occasional; northern and central peninsula, central and western panhandle. North Carolina south to Florida, west to Texas. Summer–fall.

Coreopsis grandiflora Hogg ex Sweet [Large-flowered.] LARGEFLOWER TICKSEED.

Coreopsis grandiflora Hogg ex Sweet, Brit. Fl. Gard. Pl. 175. 1826.

Erect perennial herb, to 6 dm; stem striate, glabrous. Leaves opposite, the blade usually 1(2) irregularly pinnate or pedately lobed with (3)5–9+ lobes, rarely simple, the simple blade or terminal lobe narrowly lanceolate or linear, 1.5–4.5(9) cm long, 1–8(12) mm wide, the apex acute, the base cuneate, the upper and lower surfaces glabrous, epetiolate or the petiole to 3.5 cm long. Flowers in a solitary terminal capitulum, the peduncle 8–15(25) cm long; calyculus of lanceolate or linear bractlets 3–9+ mm long; receptacle convex; palea basally lanceolate or linear; phyllaries lanceolate-ovate, 7–9 mm long; ray flowers 8; corolla yellow, the lamina 1.2–2.5 cm long; disk flowers numerous; corolla yellow, 5-lobed, 3–5 mm long. Cypsela obovate to oblong, 2–3 mm long, the lateral wings chartaceous, entire or irregularly toothed or laciniate; pappus absent or of (1)2 cusps or subulate or lanceolate scales less than 1 mm long.

Disturbed sites. Rare; Lafayette and Liberty Counties. Quebec and Ontario south to Florida, west to Wisconsin, Iowa, Missouri, Kansas, and New Mexico, also California. Spring.

Coreopsis integrifolia Poir. [With entire leaf margins.] FRINGELEAF TICKSEED; DYE-FLOWER.

Coreopsis integrifolia Poiret, in Lamarck, Encycl., Suppl. 2: 353. 1811.

Erect perennial herb, to 6 dm; stem striate, glabrous. Leaves opposite, the blade elliptic or lanceolate, 2–4.5 cm long, 1–1.8 cm wide, the apex acute or obtuse, the base broadly cuneate, the margin entire, the upper and lower surfaces glabrous, the petiole 5–15 mm long, ciliate. Flowers in capitula in a corymbiform array, the peduncle 3–8 cm long; calyculus of suboblong or linear bractlets 3–5 mm long; receptacle convex; paleae linear; phyllaries lance-ovate, 7–8+ mm long; ray flowers 8+; corolla yellow, 2–2.5 cm long; disk flowers numerous; corolla purplish or yellow with purplish tips, 4-lobed, 3–4 mm long. Cypsela narrowly obovate, 4–5 mm long, the lateral wings narrow, slightly laciniate; pappus of 2 minute cusps.

Floodplain forests and riverbanks. Rare; Nassau and St. Johns Counties, central panhandle. North Carolina south to Florida. Fall.

Coreopsis integrifolia is listed as endangered in Florida (Florida Administrative Code, Chapter 5B-40.)

Coreopsis lanceolata L. [Lance-shaped, in reference to the leaf blade.] LANCELEAF TICKSEED.

Coreopsis lanceolata Linnaeus, Sp. Pl. 908. 1753. *Coreopsoides lanceolata* (Linnaeus) Moench, Methodus 594. 1794. *Coreopsis lanceolata* Linnaeus var. *glabella* Michaux, Fl. Bor.-Amer. 2: 137. 1803. *Leachia lanceolata* (Linnaeus) Cassini, in Cuvier, Dict. Sci. Nat. 25: 388. 1822. *Coreopsis lanceolata* Linnaeus var. *succisifolia* de Candolle, Prodr. 5: 570. 1836, nom. inadmiss. *Coreopsis lanceolata* Linnaeus var. *angustifolia* Torrey & A. Gray, Fl. N. Amer. 2: 344. 1842, nom. inadmiss.

Erect perennial herb, to 3(6) dm; stem striate, glabrous. Leaves opposite, the blade simple, the margin with 1–2+ lateral lobes, the simple blade or terminal lobe lanceolate-ovate, lanceolate, oblanceolate, or linear-lanceolate, 5–12 cm long, 0.8–1.5 cm wide, the apex acute or obtuse, the base cuneate, the upper and lower surfaces glabrous, the petiole 1–5(8) cm long. Flowers in a solitary terminal capitulum, the peduncle (8)12–20(35) cm long; calyculus of lanceolate-ovate, lanceolate-linear, or linear bractlets 4–8(12) mm long; receptacle convex; palea basally lanceolate or linear, the apex attenuate; phyllaries deltate or lanceolate-deltate, 8–12 mm long; ray flowers 8; corolla yellow, the lamina 1.5–3 cm long; disk flowers numerous; corolla yellow, 5-lobed, 6–8 mm long. Cypsela obovate to oblong, 3–4 mm long, the lateral wings chartaceous, spreading, entire; pappus of (1)2 cusps or subulate or lanceolate scales less than 1 mm long or absent.

Sandhills and disturbed sites. Occasional; northern counties, central peninsula. Nearly throughout North America, South America; Europe, Africa, Asia, Australia, and Pacific Islands. Spring.

Coreopsis leavenworthii Torr. & A. Gray [Commemorates Melines Conkling Leavenworth (1796–1862), army physician who collected in Florida in 1838–1839.] LEAVENWORTH'S TICKSEED.

Coreopsis leavenworthii Torrey & A. Gray, Fl. N. Amer. 2: 346. 1842. *Coreopsis leavenworthii* Torrey & A. Gray var. *typica* Sherff, Brittonia 6: 341. 1948, nom. inadmiss. TYPE: FLORIDA: Hillsborough Co.: Tampa Bay, s.d., *Leavenworth s.n.* (holotype: GH).

Coreopsis leavenworthii Torrey & A. Gray var. *garberi* A. Gray, Syn. Fl. N. Amer. 1(2): 291. 1884. TYPE: FLORIDA: Hillsborough Co.: Tampa, s.d., *Garber s.n* (holotype: GH).

Coreopsis lewtonii Small, Bull. Torrey Bot. Club 25: 146. 1898. *Coreopsis leavenworthii* Torrey & A. Gray var. *lewtonii* (Small) Sherff, Bot. Gaz. 94: 592. 1933. TYPE: FLORIDA: Seminole Co.: Forest City, 1894, *Lewton s.n.* (holotype: NY).

Coreopsis angustata Greene, Pittonia 4: 239. 1901. TYPE: FLORIDA: Manatee Co.: Palma Sola, 16 May 1900, *Tracy 6921* (holotype: US?).

Coreopsis leavenworthii Torrey & A. Gray var. *curtissii* Sherff, Bot. Gaz. 94: 592. 1933. TYPE: FLORIDA: Suwannee Co.: Suwannee River, near Branford, 23 Oct 1900, *Curtiss 6734* (holotype: GH).

Erect annual herb, to 1(1.5) m; stem striate, glabrous. Leaves opposite, the blade of the proximal 1(2)-pinnate, the terminal lobe subelliptic, oblanceolate, lanceolate, or linear, 1.5–3 cm long, 4–8(12) mm wide, the cauline simple or 1(3)-pinnate, the simple blade or the terminal lobe narrowly oblanceolate or linear, 0.8–3.5(7) cm long, 1–3(5) mm wide, the apex acute, the base narrowly cuneate, the upper and lower surfaces glabrous, the petiole 3–6(15) cm long. Flowers in capitula in a corymbiform array, the peduncle 2–5(8) cm long; calyculus of deltoid-ovate, oblong, or linear bractlets 2–6 mm long; receptacle convex; palea linear; phyllaries lanceolate-oblong or lanceolate, (4)5–9 mm long; ray flowers (5)8; corolla yellow, rarely with a proximal red-brown spot, the lamina 8–15 mm long; disk flowers numerous; corolla red-brown or purple, 4-lobed, 2–4 mm long. Cypsela oblong, 2–3 mm long, the lateral wings entire; pappus of 2 subulate scales.

Wet flatwoods, ditches, and wet disturbed sites. Common; nearly throughout. Alabama and Florida. All year.

Coreopsis linifolia Nutt. [With leaves like *Linum* (Linaceae).] TEXAS TICKSEED.

Coreopsis linifolia Nuttall, J. Acad. Nat. Sci. Philadelphia 7: 75. 1834. *Coreopsis gladiata* Walter var. *linifolia* (Nuttall) Cronquist, Rhodora 47: 396. 1945.

Erect perennial herb, to 7 dm; stem striate, glabrous. Leaves alternate, opposite, or both, the blade simple, linear-lanceolate or linear, (2)4–8(15) cm long, (0.3)0.7–1.2(4) cm wide, the apex acute or obtuse, the base cuneate, the margin entire, the upper and lower surfaces glabrous, with dark or translucent dots, the petiole 1–10(15) cm long. Flowers in capitula in a corymbiform array, the peduncle (1)4–8(10) cm long; calyculus of linear-lanceolate bractlets 2–6 mm long; receptacle convex; palea linear; phyllaries lanceolate-ovate or linear-lanceolate, 6–10 mm long; ray flowers 8+; corolla yellow, the lamina (1)1.5–2 cm long; disk flowers numerous; corolla purplish or yellow and the tips purplish, 4-lobed, 3–5 mm long. Cypsela obovate or oblong, 3–4 mm long, the lateral wings narrow, slightly laciniate; pappus of 2 subulate scales.

Wet flatwoods and bogs. Occasional; northern peninsula, central and western panhandle. Virginia south to Florida, west to Texas. Fall.

Coreopsis major Walter [Greater, in reference to the leaf size.] GREATER TICKSEED.

Coreopsis major Walter, Fl. Carol. 214. 1788.

Coreopsis senifolia Michaux, Fl. Bor.-Amer. 2: 138. 1803.

Coreopsis senifolia Michaux var. *rigida* Nuttall, Gen. N. Amer. Pl. 2: 180. 1818. *Coreopsis rigida* (Nuttall) Nuttall, Trans. Amer. Philos. Soc., ser. 2. 7: 358. 1841. *Coreopsis delphinifolia* Lamarck var. *rigida* (Nuttall) Torrey & A. Gray, Fl. N. Amer. 2: 342. 1842. *Coreopsis major* Walter var. *rigida* (Nuttall) F. E. Boynton, in Small, Fl. S.E. U.S. 1276, 1340. 1903.

Coreopsis oemleri Elliott, Sketch Bot. S. Carolina 2: 435. 1823. *Coreopsis major* Walter var. *oemleri* (Elliott) Britton ex Small & Vail, Mem. Torrey Bot. Club 4: 131. 1894. *Coreopsis major* Walter forma *oemleri* (Elliott) Sherff, Bot. Leaf. 6: 4. 1952.

Coreopsis stellata Nuttall, J. Acad. Nat. Sci. Philadelphia 7: 76. 1834. *Coreopsis senifolia* Michaux var. *stellata* (Nuttall) Torrey & A. Gray, Fl. N. Amer. 2: 342. 1842. *Coreopsis major* Walter var. *stellata* (Nuttall) B. L. Robinson, Rhodora 10: 68. 1908.

Erect perennial herb, to 1 m; stem striate, glabrous. Leaves opposite, the blade simple or 3-foliolate, the simple blade or terminal leaflet lanceolate-elliptic to narrowly lanceolate, 2.5–5.5(10) cm long, 0.5–1.5(3) cm wide, the apex obtuse, the base broadly cuneate, the margin entire, the upper and lower surfaces glabrous, epetiolate or the petiole to 1 mm long. Flowers in capitula in an open corymbiform array, the peduncle 2–6 cm long; calyculus of 7–8 oblong to linear bractlets 3–6 mm long; receptacle slightly convex; palea linear to subulate, often dilated distally; phyllaries 8 in 1–2 series, lanceolate-oblong or lanceolate-ovate, 6–8 mm long; ray flowers 8; corolla yellow, the lamina 1.2–3 cm long; disk flowers numerous; corolla yellow, 5-lobed, 6–7 mm long. Cypsela obovate or oblong, 4–5 mm long, the lateral margins corky to chartaceous, not winged; pappus of (1)2 cusps or absent.

Sandhills. Occasional; central and western panhandle. New York and Massachusetts south to Florida, west to Illinois, Kentucky, Tennessee, Mississippi, and Louisiana. Summer–fall.

Coreopsis nudata Nutt. [Naked, in reference to the apparent lack of leaves.] GEORGIA TICKSEED.

Coreopsis nudata Nuttall, Gen. N. Amer. Pl. 2: 180. 1818. *Calliopsis nudata* (Nuttall) Sprengel, Syst. Veg. 3: 611. 1836.

Erect perennial herb, to 6(10) dm; stem striate, glabrous. Leaves alternate, simple, the blade subterete, apparently consisting of just the midrib, 5–30 cm long, 1–2 mm wide, the apex acuminate, the base sheathing, the margin entire, the upper and lower surfaces glabrous, epetiolate. Flowers in capitula in a corymbiform array, the peduncle 4–8 cm long; calyculus of lanceolate-deltate or lanceolate-ovate bractlets 2–5 mm long; receptacle convex; paleae linear; phyllaries lanceolate-ovate, 7–8 mm long; ray flowers 8+; corolla lavender, the lamina 1–2(3) cm long; disk flowers numerous, yellowish white or yellow, 4-lobed, 3–5 mm long. Cypsela narrowly oblong or linear, 2–3 mm long, the lateral wings to 1 mm wide, laciniate; pappus of 2 subulate scales 1–2 mm long.

Wet flatwoods, bogs, and cypress ponds. Frequent; northern counties. Georgia and Florida, west to Louisiana. Spring.

Coreopsis nuecensis A. Heller [Of along the Nueces River, Texas.] CROWN TICKSEED.

Coreopsis coronata Hooker, Bot. Mag. 63: t. 3460. 1836; non Linnaeus, 1763. *Coreopsis nuecensis* A. Heller, Bot. Expl. Texas 1: 106. 1895.

Erect annual herb, to 3(5) dm; stem striate, glabrous. Leaves opposite, simple, the blade entire or 1(2)-pinnately lobed, with 3–9+ lobes, the simple blade or terminal lobe elliptic or lanceolate, 1–3 cm long, 0.5–1.5 cm wide, the apex obtuse, the base broadly cuneate, the upper and lower surfaces glabrous, the petiole 1–8 cm long. Flowers in a solitary terminal capitulum, the peduncle (5)10–25 cm long; calyculus of subdeltate to lanceolate-oblong bractlets 4–6(8) mm long; receptacle convex; palea basally lanceolate or linear, the apex attenuate; phyllaries lanceolate-oblong, 6–8 mm long; ray flowers 8; corolla yellow, usually with (2)4–8+ proximal red-brown to purple flecks, the lamina 1–2 cm long; disk flowers numerous; corolla yellow, 5-lobed, ca. 3 mm long. Cypsela obovate or oblong, 3–4 mm long, the lateral wings chartaceous, entire; pappus of (1)2 cusps, subulate or lanceolate scales, or absent.

Disturbed sites. Rare; St. Johns County. Not recently collected. Escaped from cultivation. Florida, Louisiana, and Texas. Native to Texas. Spring.

Coreopsis pubescens Elliott [With trichomes, in reference to the foliage.] STAR TICKSEED.

Coreopsis pubescens Elliott, Sketch Bot. S. Carolina 2: 441. 1823. *Coreopsis pubescens* Elliott var. *typica* Sherff, Brittonia 6: 341. 1948, nom. inadmiss.

Coreopsis pubescens Elliott var. *robusta* A. Gray ex Eames, Rhodora 18: 239. 1916.

Coreopsis debilis Sherff, Bot. Gaz. 59: 366. 1930. *Coreopsis pubescens* Elliott var. *debilis* (Sherff) E. B. Smith, Sida 6: 151. 1976.

Erect perennial herb, to 7(9) dm; stem striate, sparsely pubescent. Leaves opposite, the blade usually simple, rarely with 1–2(3+) lateral lobes, the simple blade or terminal lobe lanceolate-elliptic, oblanceolate, or lanceolate, 2.5–6(8) cm long, 0.5–2.5(3.5) cm wide, the apex acute or obtuse, the base broadly cuneate, the upper and lower surfaces sparsely pubescent, the petiole to 1(2.5) cm long. Flowers in a solitary terminal capitulum, the peduncle (7)12–15 cm long; calyculus of lanceolate or linear-lanceolate bractlets 3–7 mm long; receptacle convex; palea basally lanceolate or linear, the apex attenuate; phyllaries lanceolate-ovate, 7–9 mm long; ray flowers 8; corolla yellow, the lamina 1.2–1.5 cm long; disk flowers numerous; corolla yellow, 5-lobed, 4–5 mm long. Cypsela obovate to oblong, 2–3 mm long, the lateral wings chartaceous, spreading, entire; pappus of (1)2 cusps, subulate or lanceolate scales, or absent.

Flatwoods. Rare; Jackson, Bay, Walton, and Escambia Counties. Massachusetts, Connecticut, and Virginia south to Florida, west to Kansas, Oklahoma, and Texas. Fall.

Coreopsis tinctoria Nutt. [*Tinctorius,* used in dyeing.] GOLDEN TICKSEED.

Coreopsis tinctoria Nuttall, J. Acad. Nat. Sci. Philadelphia 2: 114. 1821. *Diplosastera tinctoria* (Nuttall) Tausch, Hort. Canal. 1: xvi, t. 4. 1823. *Calliopsis tinctoria* (Nuttall) de Candolle, Prodr. 5: 568. 1836.

Erect annual herb, to 7(15) dm; stem striate, glabrous. Leaves opposite, the proximal 1(3)-pinnate, the terminal lobe lanceolate-ovate or oblanceolate, 1–6 cm long, 0.5–2.5 cm wide, the cauline 1–2(3)-pinnate, rarely simple, the simple blade or the terminal lobe lanceolate-linear or linear, 1–4.5 cm long, 1–2(5+) mm wide, the apex acute, the base narrowly cuneate, the upper and lower surfaces glabrous, the petiole 1–4 cm long. Flowers in capitula in a corymbiform array, the peduncle 1–5(15+) cm long; calyculus of deltoid-lanceolate bractlets 1–3 mm long; receptacle convex; palea linear; phyllaries lanceolate-oblong or lanceolate-ovate, 4–7(9) mm long; ray flowers (5)8; corolla yellow, red-brown proximally or with a proximal red-brown spot, rarely yellow throughout, the lamina 1.2–1.8 cm long; disk flowers numerous; corolla red-brown or purple, 4-lobed, 2–3 mm long. Cypsela oblong, 2–3(4) long, the lateral wings entire; pappus of 1–2 subulate scales.

Wet, disturbed sites. Rare; central and southern peninsula, panhandle. Escaped from cultivation. Nearly throughout North America; Mexico. Native to the eastern and western United States and Mexico. Summer–fall.

Coreopsis tripteris L. [*Tri,* three, and *pteris,* winged, in reference to the 3-foliolate leaves.] TALL TICKSEED.

Coreopsis tripteris Linnaeus, Sp. Pl. 908. 1753. *Anacis tripteris* (Linnaeus) Schrank, Denkschr. Königl. Akad. Wiss. München 5: 7. 1817. *Chrysostemma tripteris* (Linnaeus) Lessing, Syn. Gen. Compos. 227. 1832. *Bidens tripteris* (Linnaeus) E.H.L. Krause, in Sturm, Deutschl. Fl., ed. 2. 13: 158. 1905.

Erect perennial herb, to 1.8 dm; stem striate, glabrous. Leaves opposite, the blade 3-foliolate, the leaflet simple or sometimes somewhat pinnately lobed, lanceolate, 4–9(12) cm long, 1–3.5 cm wide, the apex acute, the base cuneate, the upper and lower surfaces glabrous, the petiole 0.5–4.5 cm long. Flowers in capitula in an open corymbiform array, the peduncle 2–5 cm long; calyculus of 5–6 oblong bractlets 2–5 mm long; receptacle slightly convex; palea linear to subulate, often dilated distally; phyllaries 8, oblong or lanceolate-oblong; ray flowers 8; corolla yellow, the lamina 1–2 cm long; disk flowers numerous; corolla red-brown to purplish, 5-lobed, 5–6 mm long. Cypsela obovate or oblong 4–5(6) mm long; the lateral margins narrow, corky to chartaceous, not winged; pappus of (1)2 cusps or absent.

Wet hammocks. Rare; central panhandle. Quebec and Ontario south to Florida and Texas. Summer.

EXCLUDED TAXA

Coreopsis auriculata Linnaeus—This southeastern U.S. species was reported for Florida by Chapman (1860), Small (1903, 1913a, 1933), and Strother (2006h). Excluded from Florida by Smith (1976). No Florida material seen.

Coreopsis verticillata Linnaeus—This southeastern U.S. species was reported for Florida by Sherff (1936). Excluded from Florida by Smith (1976). No Florida material seen.

Cosmos Cav. 1791.

Herbs. Leaves opposite, simple, pinnate-veined, petiolate or epetiolate. Flowers in capitula, solitary or in corymbiform arrays, radiate; calyculi present; receptacle paleate; ray flowers sterile; petals 5, basally connate, zygomorphic; disk flowers bisexual; petals 5, basally connate, actinomorphic; stamens 5, epipetalous, the filaments free, the anthers connate; ovary inferior, 2-carpellate, 1-loculate. Fruit a cypsela; pappus present or absent.

A genus of about 26 species; North America, tropical America, Africa, Asia, and Pacific Islands. [From the Greek *kosmos,* harmoniously ordered universe, or *kosmo,* ornament.]

Selected reference: Kiger (2006).

1. Ray flowers orange-red or orange-yellow .... **C. sulphureus**
1. Ray flowers pink, purple, or white.
   2. Ultimate leaf segments (2)3–10 mm wide; lamina of the ray flowers 0.5–1.5 cm long ... **C. caudatus**
   2. Ultimate leaf segments 1 mm wide or less; lamina of the ray flowers 1.5–4 cm long .... **C. bipinnatus**

Cosmos bipinnatus Cav. [Twice-pinnately divided, in reference to the leaves.] GARDEN COSMOS.

Cosmos bipinnatus Cavanilles, Icon. 1: 10, t. 14. 1791. *Georgina bipinnata* (Cavanilles) Sprengel, Syst. Veg. 3: 611. 1826. *Cosmos bipinnatus* Cavanilles var. *typicus* Sherff, Brittonia 6: 341. 1948, nom. inadmiss.

Erect or ascending annual herb, to 2 m; stem striate-angled, glabrous or sparsely puberulent. Leaves with the blade 6–11 cm long, the ultimate lobes to 1 mm wide or less, the apex acute, the margin entire, epetiolate or the petiole to 1 cm long. Flowers in capitula, solitary or in a corymbiform array, the peduncle 10–20 cm long; calyculus of 8 linear or lanceolate bractlets 6–13 mm long, spreading, the apex acuminate; involucre hemispheric, 7–15 mm wide; receptacle flat; phyllaries in 2 series; lanceolate or ovate-lanceolate, 7–13 mm long, subequal, erect, the apex rounded or obtuse; ray flowers 8; corolla pink, purple, or white, the lamina 1.5–4 cm long; disk flowers numerous; corolla yellow, 5–7 mm long. Cypsela fusiform, longitudinally 4- to 5-angled, 7–16 mm long, the apex tapering to a beak, glabrous; pappus of 1–3 ascending or erect retrorsely barbed awns 1–3 mm long or absent.

Disturbed sites. Rare; Alachua and Miami-Dade Counties. Escaped from cultivation. Quebec south to Florida, west to Ontario and California; West Indies, Mexico, Central America, and South America; Asia. Native to the southwestern United States and Mexico. Summer–fall.

Cosmos caudatus Kunth [Ending in a taillike appendage, in reference to beaked cypsela.] WILD COSMOS.

Cosmos caudatus Kunth, in Humboldt et al., Nov. Gen. Sp. 4: 240. 1820. *Bidens caudata* (Kunth) Schultz Bipontinus, in Seemann, Voy. Bot. Herald 308. 1856. *Bidens artemisiifolia* (Jacquin) Kuntze var. *caudata* (Kunth) Kuntze, Revis. Gen. Pl. 1: 321. 1891.

Erect or ascending annual herb, to 2.5 m; stem striate-angled, glabrous or sparsely hispid. Leaves with the blade 10–20 cm long, the ultimate lobes (2)3–10 mm wide, the apex acute, often mucronulate, the margin spinulose-ciliate, the petiole 1–7 cm long. Flowers in capitula, solitary or in a corymbiform array, the peduncle 10–30 cm long; calyculus of 8 linear-subulate bractlets 7–11 mm long, usually spreading, the apex acuminate; involucre hemispheric, 5–15 mm wide; receptacle flat; phyllaries 8 in 2 series, oblong-lanceolate, 7–11 mm long, subequal, erect, the apex acute to obtuse; ray flowers 8; corolla pink, purple, or white, the lamina, 0.5–1.5 cm long, the apex obtusely 3-lobed; disk flowers numerous; corolla yellow, 5–6 mm long. Cypsela fusiform, longitudinally 4-to 5-angled, 12–35 mm long, the apex tapering to a beak, glabrous or scabridulous proximally, setose distally; pappus of 2–3 widely divergent or reflexed awns 3–5 mm long.

Disturbed sites. Rare; Monroe County keys. Escaped from cultivation. Not recently collected. Florida; West Indies, Mexico, Central America, and South America; Asia and Pacific Islands. Native to tropical America. Summer–fall.

Cosmos sulphureus Cav. [*Sulphureus*, in reference to the color of the ray flowers.] SULPHUR COSMOS.

Cosmos sulphureus Cavanilles, Icon. 1: 56, t. 79. 1971. *Bidens sulphurea* (Cavanilles) Schultz Bipontinus, in Seemann, Bot. Voy. Herald 308. 1956. *Bidens artemisiifolia* (Jacquin) Kuntze var. *sulphurea* (Cavanilles) Kuntze, Revis. Gen. Pl. 1: 321. 1891. *Cosmos sulphureus* Cavanilles var. *typicus* Sherff, Brittonia 6: 341. 1948, nom. inadmiss.

Erect or ascending annual herb, to 2 m; stem striate-angled, glabrous or sparsely pilose or hispid. Leaves with the blade 5–12(25) cm long, the ultimate lobes 2–5 mm wide, the apex apiculate, the margin sparsely spinulose-ciliate, the petiole 1–7 cm long. Flowers in capitula, solitary or in a corymbiform array, the peduncle 10–20 cm long; calyculus of 8 linear-subulate bractlets 5–7(10) mm long, spreading-ascending, the apex acute; involucre hemispheric, 6–10 mm wide; receptacle flat; phyllaries 8 in 2 series, oblong-lanceolate, 9–13(18) mm long, subequal, erect, the apex acute, rounded, or obtuse; ray flowers 8; corolla orange-red or orange-yellow, the lamina 1.5–3(4) cm long, the apex subtruncate, denticulate; disk flowers numerous; corolla yellow, 6–7 mm long. Cypsela fusiform, longitudinally 4-to 5-angled, 15–30 mm long, the apex tapering to a beak, glabrous or hispidulous or rarely glabrous; pappus of 2–3 widely divergent awns 1–7 mm long or absent.

Disturbed sites. Rare; Madison, Putnam, Polk, and Miami-Dade Counties. Escaped from cultivation. New York and Massachusetts south to Florida, west to Michigan, Illinois, Missouri, Oklahoma, and Texas, also California; West Indies, Mexico, Central America, and South America; Africa, Asia, and Pacific Islands. Native to tropical America. All year.

Crassocephalum Moench 1794.

Herbs. Leaves alternate, simple, pinnate-veined, petiolate. Flowers in capitula in corymbiform arrays, discoid, bisexual; calyculi present; receptacle epaleate; petals 5, basally connate, actinomorphic; stamens 5, epipetalous, the filaments free, the anthers connate; ovary inferior, 2-carpellate, 1-loculate. Fruit a cypsela; pappus present.

A genus of about 24 species; North America, West Indies, Mexico, Central America, South America, Africa, Asia, Australia, and Pacific Islands. [From the Latin *Crassus,* thick, and the Greek *cephalus,* head, in reference to the thick capitula.]

Crassocephalum crepidioides (Benth.) S. Moore [*Crepid,* weak, feeble, in comparison to more substantial species of the genus.] REDFLOWER RAGLEAF.

Gynura crepidioides Bentham, in Hooker, Niger Fl. 438. 1849. *Senecio crepidioides* (Bentham) Ascherson, in Schweinfurth, Beitr. Fl. Aethiop. 155. 1867. *Crassocephalum diversifolium* Hiern var. *crepidioides* (Bentham) Hiern, Cat. Afr. Pl. 1(3): 595. 1898. *Crassocephalum crepidioides* (Bentham) S. Moore, J. Bot. 50: 211. 1912.

Erect annual herb, to 1 m; stem striate-angled, glabrous. Leaves with the blade elliptic or oblong-elliptic, 7–12 cm long, 4–5 cm wide, the apex acuminate, the base cuneate, the margin irregularly serrate or double-serrate, sometimes pinnately lobed in the lower half, the upper surface glabrous, the lower surface glabrous or sparsely hirsute on the main veins, the petiole 2–2.5 cm long. Flowers in capitula in a terminal corymbiform array, the peduncle to 20 cm long; calyculus of 6–21 linear bractlets 2–6 mm long; involucre cylindric, 1–1.2 cm long, the base truncate; receptacle flat; phyllaries numerous in 1 series, linear-lanceolate, 1–1.3 mm long, equal, erect at anthesis, later reflexed, the margin narrowly scarious, the apex pubescent; flowers numerous; corolla red-brownish or orange, 8–10 mm long. Cypsela narrowly oblong, ca. 2 mm long, ribbed, with a gray-white ring at the apex, pubescent; pappus of numerous white bristles 7–13 mm long.

Disturbed sites. Occasional; central and southern peninsula. Florida; West Indies, Mexico, Central America, South America, Africa, Asia, Australia, and Pacific Islands. Native to Africa. All year.

Crassocephalum crepidioides is listed as a Category II invasive species by the Florida Exotic Pest Plant Council (FLEPPC, 2017).

Crepis L. 1753. HAWKSBEARD

Herbs, latex present. Leaves alternate, simple, pinnate-veined, petiolate or epetiolate. Flowers in capitula in corymbiform arrays, ligulate, bisexual; calyculi present; receptacle epaleate; petals 5, basally connate, zygomorphic; stamens 5, epipetalous, the filaments free, the anthers connate; ovary inferior, 2-carpellate, 1-loculate. Fruit a cypsela; pappus present.

A genus of about 200 species; nearly cosmopolitan. [From the Greek *krepis,* slipper, sandal, in reference to the fruit shape.]

Selected reference: Babcock (1947); Bogler (2006a).

Crepis pulchra L. [*Pulcher,* beautiful.] SMALLFLOWER HAWKSBEARD.

Crepis pulchra Linnaeus, Sp. Pl. 806. 1753. *Chondrilla pulchra* (Linnaeus) Lamarck, Fl. Franc. 2: 106. 1779. ("1778"). *Lapsana pulchra* (Linnaeus) Villars, Hist. Pl. Dauphine 3: 163. 1788. *Prenanthes pulchra* (Linnaeus) de Candolle, in de Candolle & Lamarck, Fl. Franc. 4: 7. 1805. *Phaecasium lampsanoides* Cassini, in Cuvier, Dict. Sci. Nat. 39: 387. 1826, nom. illegit. *Idianthes pulchra* (Linnaeus) Desvaux, Fl. Anjou 199. 1827. *Sclerophyllum pulchrum* (Linnaeus) Gaudin, Fl. Helv. 5: 48. 1829. *Phaecasium pulchrum* (Linnaeus) Reichenbach f., in Reichenbach, Icon. Fl. Germ. Helv. 19: 39. 1858. *Crepis pulchra* Linnaeus subsp. *typica* Babcock, Univ. Calif. Publ. Bot. 19 402. 1941, nom. inadmiss.

Erect annual herb, to 1 m; stem striate-angled, proximally hispid and stipitate-glandular, distally glabrous. Basal and proximal cauline leaves with the blade oblanceolate, 5–20 cm long, 1–5 cm wide, the apex obtuse or acute, the base cuneate, the margin pinnately lobed or denticulate, the lobes triangular, the terminal the largest, the upper and lower surfaces stipitate-glandular, the petiole to 2 cm long, winged, the distal smaller, lanceolate, sessile. Flowers in capitula in a terminal corymbiform array, the peduncle 1–5 cm long; calyculus of 5–7 ovate or lanceolate bractlets 1–2 mm long, glabrous; involucre cylindric, 8–12 mm long; receptacle flat or convex, pitted; phyllaries 10–14 in 1–2 series, lanceolate, 8–10 mm long, equal, the base strongly keeled and thickened, the margin scarious, the outer surface glabrous; corolla light yellow, 5–12 mm long. Cypsela subcylindric, 10-to 12-ribbed, pale brown, 4–6 mm long, the apex attenuate; pappus of numerous white bristles 4–5 mm long.

Disturbed sites. Rare; Gadsden and Holmes Counties. Ontario south to Florida and Texas, also Oregon. North America; Europe, Africa, and Asia. Native to Europe and Asia. Spring.

EXCLUDED TAXON

Crepis foetida Linnaeus—Reported for Florida by Small (1933), misapplied to Florida material of *Hypochaeris ciliensis.* Small's report was apparently repeated by Bogler (2006a).

Croptilon Raf. 1837. SCRATCHDAISY

Herbs. Leaves alternate, simple, pinnate-veined, epetiolate. Flowers in capitula in paniculiform arrays; radiate; receptacle epaleate; ray flowers carpellate; petals 5, basally connate, zygomorphic; disk flowers bisexual; petals 5, basally connate, actinomorphic; stamens 5, epipetalous, the filaments free, the anthers connate; ovary inferior, 2-carpellate, 1-loculate. Fruit a cypsela; pappus present.

A genus of 3 species; North America and Mexico. [From the Greek *kropion,* scythe, and *ptilon,* wing, feather, in reference to the winglike or feather-like appearance of the curved, scythe-shaped leaves.]

Isopappus Torrey & A. Gray 1842.

Selected reference: Nesom (2006i).

Croptilon divaricatum (Nutt.) Raf. [Spreading at a wide angle, in reference to the inflorescence branches.] SLENDER SCRATCHDAISY.

Inula divaricata Nuttall, Gen. N. Amer. Pl. 2: 152. 1818. *Aster divaricatus* (Nuttall) Kuntze, Gen. Pl. 1: 318. 1891; non Linnaeus, 1753; nec Lamarck, 1783; nec Sprengel, 1826; nec Rafinesque ex de Candolle, 1836; nec (Nuttall) Torrey & A. Gray, 1842. *Chrysopsis divaricata* (Nuttall) Elliott, Sketch Bot. S. Carolina 2: 338. 1823. *Diplopappus divaricatus* (Nuttall) Hooker, Companion Bot. Mag. 1: 97. 1836. *Croptilon divaricatum* (Nuttall) Rafinesque, Fl. Tellur. 2: 47. 1837 ("1836"). *Isopappus divaricatus* (Nuttall) Torrey & A. Gray, Fl. N. Amer. 2: 239. 1842. *Haplopappus divaricatus* (Nuttall) A. Gray, in H. L. Abbot & R. S. Williamson, Pacif. Railr. Rep. 4: 99. 1856.

Erect annuals or perennials, to 1 m; stem striate-angled, sparsely hispid. Leaves with the blade lanceolate or linear, reduced and becoming bract-like distally, the apex acute, the base cuneate, the margin serrate in the upper ½, rarely entire, the upper surface glabrous, the lower surface sparsely arachnoid, sessile. Flowers in capitula in a paniculiform array, the peduncle 1–4 cm long, glabrate, sometimes hispid, stipitate-glandular, with minute bracteoles; involucre turbinate, (2)3–5 mm long; receptacle flat, pitted; phyllaries numerous in 3–5 series, lanceolate or linear-lanceolate, 3–5 mm long, unequal, the margin scarious, the outer surface glabrous, hirsute, or stipitate-glandular; ray flowers 5–11; corolla yellow, the lamina 4–6 mm long; disk flowers numerous; corolla yellow, 2–3 mm long. Cypsela turbinate, 6- to 10-nerved, tan, strigose; pappus of numerous tawny, barbellate bristles 2–3 mm long.

Disturbed sandhill, longleaf pineland, sand pine, and ruderal sites. Frequent; northern counties, central peninsula. Virginia south to Florida, west to Oklahoma and Texas. Summer–fall.

Cyanthillium Blume 1826. IRONWEED

Herbs. Leaves alternate, simple, pinnate-veined, petiolate. Flowers in capitula in corymbiform arrays, discoid, bisexual; receptacle epaleate; petals 5, basally connate, actinomorphic; stamens 5, epipetalous, the filaments free, the anthers connate; ovary inferior, 2-carpellate, 1-loculate. Fruit a cypsela; pappus present.

A genus of about 8 species; nearly cosmopolitan. [From the Greek *cyanos,* dark blue, and *anthyllion,* little flower, in reference to the small flowers.]

Selected reference: Strother (2006i).

Cyanthillium cinereum (L.) H. Rob. [ash-gray in color, in reference to its general appearance.] LITTLE IRONWEED.

Conyza cinerea Linnaeus, Sp. Pl. 862. 1753. *Serratula cinerea* (Linnaeus) Roxburgh, Hort. Bengal. 60. 1814. *Vernonia cinerea* (Linnaeus) Lessing, Linnaea 4: 291. 1829. *Cacalia cinerea* (Linnaeus) Kuntze, Revis. Gen. Pl. 1: 323. 1891. *Senecioides cinerea* (Linnaeus) Kuntze, in T. Post & Kuntze, Lex Gen. Phan. 515. 1903. *Cyanthillium cinereum* (Linnaeus) H. Robinson, Proc. Biol. Soc. Wash. 103: 252. 1990.

Vernonia fasciculata Blume, Bijdr. 893. 1826; non Michaux, 1903.

Erect annual herb, to 6(12) dm; stem striate-angled, pilose or strigillose, resin-dotted. Leaves with the blade deltate, oblanceolate, or spatulate, 2–4 cm long, 1–3 cm wide, often with diminutive leaves at the node, the apex rounded or acute, the base cuneate, the margin serrate, the upper surface scabrellous or glabrate, the lower surface hirtellous or pilose-strigillous, resin-dotted, the petiole 5–10 mm long. Flowers in capitula in a corymbiform array, the peduncle 0.5–1(2) cm long; involucre campanulate, 4–5 mm wide; phyllaries numerous in 3–4 series, 3–4 mm long, subulate or lanceolate, the apex acute or spinulose, the outer faces strigillose, resin-dotted; receptacle flat or convex; flowers numerous; corolla lavender, pink, or purple, 4–5 mm long. Cypsela narrowly elliptic, ca. 2 mm long, strigillose; pappus of numerous minute outer scales, the inner of numerous barbellate bristles 3–4 mm long.

Disturbed sites. Occasional; central and southern peninsula. Nearly cosmopolitan. Native to Africa and Asia. All year.

Dittrichia Greuter 1973. FALSE YELLOWHEAD

Herbs. Leaves alternate, simple, pinnate-veined, petiolate or epetiolate. Flowers in capitula in racemiform or paniculiform arrays; radiate; receptacle epaleate; ray flowers carpellate; petals 5, basally connate, zygomorphic; disk flowers bisexual; petals 5, basally connate, actinomorphic; stamens 5, epipetalous, the filaments free, the anthers connate; ovary inferior, 2-carpellate, 1-loculate. Fruit a cypsela; pappus present.

A genus of 2 species; North America, Europe, Africa, Asia, and Australia. [Commemorates Manfred Dittrich (b. 1934), German botanist.]

Dittrichia viscosa (L.) Greuter [Sticky, in reference to the viscous leaves.] FALSE YELLOWHEAD.

Erigeron viscosus Linnaeus, Sp. Pl. 863. 1753. *Solidago viscosa* (Linnaeus) Lamarck, Fl. Franç. 2: 144. 1779 ("1778"). *Inula viscosa* (Linnaeus) Aiton, Hort. Kew. 3: 223. 1789. *Pulicaria viscosa* (Linnaeus) W.D.J. Koch, Syn. Deut. Schweiz. Fl. 1: 361. 1837. *Cupularia viscosa* (Linnaeus) Godron & Grenier ex Godron, in Grenier & Godron, Fl. France 2: 181. 1850. *Dittrichia viscosa* (Linnaeus) Greuter, Exsicc. Genav. 4: 71. 1973.

Erect suffrutescent, perennial herb, to 1.3 m; stem striate-angled, viscid stipitate-glandular and stipitate-nonglandular. Leaves with the blade oblong-lanceolate or linear-triangular, 3–7 cm long, 0.5–3 cm wide, sometimes with 1 or more smaller ones in the axil of the larger, the apex obtuse or acute, the base rounded or broadly cuneate, the margin serrulate or denticulate, rarely entire, the upper and lower surfaces viscid stipitate-glandular and stipitate-nonglandular, short-petiolate, the upper sessile, sometimes semiamplexicaul. Flowers in capitula in a racemiform or paniculiform array, the peduncle 1–2 cm long, viscid stipitate-glandular and stipitate-nonglandular; involucre subconic, 5–6 mm wide; phyllaries in 3–4 series, unequal, the outer linear-triangular, 1–2 mm long, appressed, the outer surface pilose, glandular-stipitate, the inner linear-lanceolate, 4–8 mm long; receptacle flat, alveolate, the coronule minutely dentate; ray flowers numerous; corolla yellow, the lamina ca. 10–12 mm long; disk flowers numerous; corolla yellow, 6–8 mm long. Cypsela ellipsoid or terete, ca. 2 mm long, stipitate-glandular distally; pappus of numerous basally connate, pale brown, barbellate scales ca. 6 mm long.

Disturbed sites. Rare; Escambia County. Not recently collected. New Jersey, Pennsylvania, and Florida. North America, Europe, Africa, and Asia. Native to Europe, Africa, and Asia. Summer–fall.

Doellingeria Nees 1832. WHITETOP

Herbs. Leaves alternate, simple, pinnate-veined, petiolate or epetiolate. Flowers in capitula in corymbiform arrays, radiate; receptacle epaleate; ray flowers carpellate; petals 5, basally connate, zygomorphic; disk flowers bisexual; petals 5, basally connate, actinomorphic; stamens 5, epipetalous, the filaments free, the anthers connate; ovary inferior, 2-carpellate, 1-loculate. Fruit a cypsela; pappus present.

A genus of 3 species; North America. [Commemorates Ignatz Dollinger (1770–1841), German botanist.]

Selected reference: Semple and Chmielewski (2006).

1. Branches of the inflorescence leafless or nearly so; midvein of the phyllaries usually swollen from the base to the apex .... **D. infirma**
1. Branches of the inflorescence leafy; midvein of the phyllaries swollen only toward the apex .... **D. sericocarpoides**

Doellingeria infirma (Michx.) Greene [Disabled, in reference to the flexuous stems making the plant appear sickly disabled.] CORNEL-LEAF WHITETOP.

Aster infirmis Michaux, Fl. Bor.-Amer. 2: 109. 1803. *Doellingeria infirma* (Michaux) Greene, Pittonia 3: 52. 1896.

Aster humilis Willdenow, Sp. Pl. 2038. 1803. *Aster umbellatus* Miller var. *humilis* (Willdenow) Britton et al., Prelim. Cat. 27. 1888. *Doellingeria humilis* (Willdenow) Britton, in Britton & A. Brown, Ill. Fl. N. U.S. 3: 392. 1898. *Doellingeria umbellata* (Nees von Esenbeck subsp. *humilis* (Willdenow) W. Stone, Pl. S. New Jersey 763. 1912 ("1911").

Ascending or erect, suffrutescent, perennial herb, to 1.2 m; stem striate-angled, slightly to strongly flexuous, glabrous. Proximal leaves with the blade broadly lanceolate or oblanceolate, 5–13 cm long, 1.5–4.5 cm wide, reduced and narrower distally, the apex acuminate, the base cuneate, the margin finely ciliate, the upper and lower surfaces glabrous or sparsely pubescent, short-petiolate or sessile. Flowers in capitula in a corymbiform array, the peduncle 1–10 mm long, leafless or nearly so, sparsely to moderately canescent; bracts lanceolate or linear-lanceolate; involucre campanulate; phyllaries numerous in 4–5 series, lanceolate, 4–7 mm long, the apex narrowly rounded, the midvein usually swollen, translucent, the outer surface glabrate;

receptacle slightly convex, pitted; ray flowers (3)4–8(11); corolla white, the lamina 6–12(15) mm long; disk flowers few to many; corolla yellow, 4–7 mm long. Cypsela obconic, 2–4 mm long, 6- to 10-ribbed, glabrous or sparsely strigose, sometimes sparsely glandular; pappus of 4 series, the outer of whitish, linear or subulate, short scales to 1 mm long, the 3 inner of numerous, white or tan barbellate bristles 4–5 mm long.

Dry hammocks. Rare; Gadsden County. Massachusetts and New York south to Florida, west to Alabama. Summer–fall.

Doellingeria sericocarpoides Small [To resemble *Sericocarpus* (Asteraceae).] SOUTHERN WHITETOP.

Doellingeria sericocarpoides Small, Bull. Torrey Bot. Club 25: 620. 1898. *Aster sericocarpoides* (Small) K. Schumann, Just's Bot. Jahresber. 26: 375. 1900.
Aster umbellatus Miller var. *latifolius* A. Gray, Syn. Fl. N. Amer. 1(2): 197. 1884. *Doellingeria umbellata* (Miller) Nees von Esenbeck var. *latifolia* (A. Gray) House, Bull. New York State Mus. 254: 712. 1924.

Ascending or erect, suffrutescent, perennial herb, to 1.2 m; stem striate-angled, glabrous. Proximal leaves with the blade lanceolate, the distal ovate, 3–11 cm long, 1.5–4 cm wide, reduced upward, the apex acuminate, the base cuneate, the margin involute or weakly revolute, finely ciliate, the upper and lower surfaces glabrous or sparsely pubescent. Flowers in capitula in a corymbiform array, the peduncle 1–10 mm long, leafless or nearly so, sparsely to moderately canescent; bracts lanceolate or linear-lanceolate; involucre campanulate; phyllaries numerous in 3–4 series, lanceolate, 3–6 mm long, the apex broadly rounded, the midvein usually swollen and translucent only apically, the outer surface glabrate; receptacle slightly convex, pitted; ray flowers 2–7; corolla white, the lamina 6–12(15) mm long; disk flowers few to many; corolla yellow, 4–7 mm long. Cypsela obconic, 2–4 mm long, 6- to 8-ribbed, sparsely strigose, sometimes sparsely glandular; pappus of 4 series, the outer of whitish, linear or subulate, short scales to 1 mm long, the 3 inner of numerous, white or tan barbellate bristles 4–7 mm long.

Swamps. Rare; Duval and Holmes Counties, western panhandle. New York south to Florida, west to Oklahoma and Texas. Fall.

EXCLUDED TAXA

Doellingeria amygdalina (Lamarck) Nees von Esenbeck—This heterotypic synonym of the more northern *D. umbellata* (Miller) Nees von Esenbeck was reported for Florida by Chapman (1860, as *Diplopappus amydgalinus* (Lamarck) Hooker), the name misapplied to Florida material of *D. sericocarpoides*.
Doellingeria umbellata (Miller) Nees von Esenbeck—Reported for Florida by Radford et al. (1964, 1968, as *Aster umbellatus* Miller). No Florida material seen.

Dracopis (Cass.) Cass. 1825. CONEFLOWER

Herbs. Leaves alternate, simple, pinnate-veined, petiolate or epetiolate. Flowers in capitula, solitary or in corymbiform arrays; radiate; receptacle paleate; ray flowers sterile; petals 5, basally connate, zygomorphic; disk flowers bisexual; corolla basally connate, actinomorphic; stamens 5, epipetalous, the filaments free, the anthers connate; ovary inferior, 2-carpellate, 1-loculate. Fruit a cypsela; pappus absent.

A monotypic genus; North America. *Dracopis* is sometimes placed in *Rudbeckia* by various authors (e.g., Urbatsch and Cox, 2006b).

Dracopis amplexicaulis (Vahl) Cass. ex Less. [Stem-clasping, in reference to the leaves.] CLASPING CONEFLOWER.

Rudbeckia amplexicaulis Vahl, Skr. Naturhist.-Selsk. 2(2): 29, t. 4. 1793. *Dracopis amplexicaulis* (Vahl) Cassini ex Lessing, Syn. Gen. Compos. 226. 1832.

Annual herb, to 1 m; stem striate-angled, glabrous, glaucous. Leaves with the blade elliptic, lanceolate, oblong, or ovate, 3–15 cm long, 0.5–4 cm wide, the apex acute or acuminate, the base auriculate or amplexicaul, the margin crenate, entire, or serrate, the upper and lower surfaces glabrous, short-petiolate or sessile. Flowers in capitula, solitary or in a corymbiform array, the peduncle 1–2 cm long, glabrous; involucre hemispheric, 1–4 cm wide; phyllaries in 2 series, linear or lanceolate, the outer 2–5 times as long as the inner, spreading or reflexed; receptacle ovoid, conic, or columnar; paleae obovate, concave, each clasping a flower, the apex acute or obtuse, often mucronate, the margin ciliate, the surface subapically pubescent; ray flowers 6–10; corolla yellow or sometimes orange, the lamina often with a proximal maroon spot, the lamina 1–3 cm long, the lower surface hirsute, spreading or reflexed; disk flowers numerous; corolla proximally greenish yellow, distally purplish, 3–4 mm long. Cypsela obpyramidal, 2–3 mm long, 4- or 5-striate-angled, minutely cross-rugose, black, glabrous; pappus absent, but each with a ring of apical tan tissue ca. 1 mm long.

Dry, open sites; Alachua and Levy Counties where apparently introduced, Escambia County where probably native. South Carolina south to Florida, west to New Mexico, north to Illinois, also North Dakota. Spring–summer.

Echinacea Moench 1794. PURPLE CONEFLOWER

Herbs. Leaves alternate, simple, petiolate or epetiolate. Flowers in capitula, solitary, radiate; receptacle paleate; ray flowers sterile; petals 5, basally connate, zygomorphic; disk flowers bisexual; petals 5, basally connate, actinomorphic; stamens 5, the filaments free, the anthers connate; ovary inferior, 2-carpellate, 1-loculate. Fruit a cypsela; pappus present.

A genus of 9 species; North America. [From the Greek *Echinos,* sea urchin or hedgehog, in reference to the spiny tips of the receptacle paleae.]

Selected reference: Urbatsch et al. (2006).

Echinacea purpurea (L.) Moench [Purple, in reference to the flower color.] EASTERN PURPLE CONEFLOWER.

Rudbeckia purpurea Linnaeus, Sp. Pl. 707. 1753. *Echinacea purpurea* (Linnaeus) Moench, Methodus 591. 1794. *Brauneria purpurea* (Linnaeus) Britton, in Porter & Britton, Mem. Torrey Bot. Club 5: 334. 1894.

Erect perennial herb, to 1.2 m; stem striate-angled, pilose or glabrous. Basal leaves with the blade ovate or lanceolate, 5–50 cm long, (1)5–12 cm wide, the apex acute, the base rounded or cordate, 3- to 5-nerved from near the base, the margin serrate, dentate, or rarely entire, the upper and lower surfaces pilose or glabrous, sessile or the petiole to 15(20) cm long, the cauline leaves similar and reduced upward. Flowers in capitula, solitary terminally and on the branches, the peduncle 8–25 cm long; involucre hemispheric, 3–4 cm wide; phyllaries linear or lanceolate, 8–17 mm long, the apex attenuate, the outer surface sparsely pilose or glabrous, spreading or recurved; receptacle conic; paleae lanceolate, 9–15 mm long, keeled, the apex awn-like, straight or slightly curved, the base surrounding the cypsela, purplish distally; ray flowers 10–20; corolla pink or purple, the lamina 3–8 cm long, spreading or recurved; disk flowers numerous; corolla pink or purple, 5–6 mm long. Cypsela obpyramidal, 3- or 4-angled 4–5 mm long, light tan; pappus coroniform with 3–4 teeth ca. 1 mm long.

Openings in calcareous hammocks. Rare; Gadsden County. New York and Ontario south to Florida, west to Iowa, Colorado, Oklahoma, and Texas. Spring–fall.

Echinacea purpurea is listed as endangered in Florida (Florida Administrative Code, Chapter 5B-40).

Eclipta L., nom. cons. 1771.

Herbs. Leaves opposite, simple, pinnate-veined, petiolate or epetiolate. Flowers in capitula, solitary or in corymbiform arrays, radiate; receptacle paleate; ray flowers carpellate; petals 5, basally connate, zygomorphic; disk flowers bisexual; petals 5, basally connate, actinomorphic; stamens 5, epipetalous, the filaments free, the anthers connate; ovary inferior, 2-carpellate, 1-loculate. Fruit a cypsela; pappus present.

A genus of about 8 species; nearly cosmopolitan. [From the Greek *ekleipsis,* a failure to appear, apparently in reference to its small size.]

Selected reference: Strother (2006j).

Eclipta prostrata (L.) L. [Procumbent.] FALSE DAISY.

Verbesina prostrata Linnaeus, Sp. Pl. 902. 1753. *Eclipta prostrata* (Linnaeus) Linnaeus, Mant. Pl. 286. 1771. *Eclipta strumosa* Salisbury, Prodr. Stirp. Chap. Allerton 205. 1796, nom. illegit. *Eclipta alba* (Linnaeus) Hasskarl var. *prostrata* (Linnaeus) Kuntze, Revis. Gen. Pl. 1: 334. 1891. *Eclipta alba* (Linnaeus) Hasskarl forma *prostrata* (Linnaeus) Huber, Bull. Herb. Boissier, ser. 2. 1: 328. 1901.

Verbesina alba Linnaeus, Sp. Pl. 902. 1753. *Cotula alba* (Linnaeus) Linnaeus, Syst. Nat., ed. 12. 2: 564. 1767. *Eclipta erecta* Linnaeus, Mant. Pl. 2: 286. 1771, nom. illegit. *Eclipta adpressa* Moench, Suppl. Meth. 245. 1802, nom. illegit. *Eclipta dubia* Rafinesque, New Fl. 2: 40. 1837 ("1836"), nom. illegit. *Eclipta alba* (Linnaeus) Hasskarl, Pl. Jav. Rar. 528. 1848. *Eclipta alba* (Linnaeus) Hasskarl forma *erecta* Hasskarl, Pl. Jav. Rar. 528. 1848, nom. inadmiss. *Eclipta alba* (Linnaeus) Hasskarl var. *erecta* Kuntze, Revis. Gen. Pl. 1: 334. 1891, nom. inadmiss. *Eupatoriophalacron album* (Linnaeus) Hitchcock, Rep. (Annual) Missouri Bot. Gard. 4: 99. 1893.

Eclipta longifolia Schrader ex de Candolle, Prodr. 5: 490. 1836.

Eclipta longifolia Rafinesque New Fl. 2: 40. 1837 ("1836"); non Schrader ex de Candolle, 1836. TYPE: "Florida to New Jersey."

Paleista procumbens Rafinesque, New Fl. 2: 43. 1837 ("1836"). TYPE: "Carolina and Florida."

Erect or decumbent annual or perennial herb, to 5(7) dm; stem striate-angled, sparsely scabrellous. Leaves with the blade linear-lanceolate or lanceolate, 2–10 cm long, 0.5–3 cm wide, the apex acute, the base cuneate, the margin serrate or subentire, the upper and lower surfaces sparsely scabrellous, short-petiolate or sessile. Flowers in capitula, solitary or in a corymbiform array, the peduncle 1–3 cm long, sparsely scabrellous; involucre hemispheric, 3–5 mm wide; phyllaries 8–12 in 2–3 series, lanceolate, unequal, the outer surface sparsely scabrellous; receptacle flat or slightly convex; paleae linear or filiform, falling with the fruit; ray flowers numerous in 2–3 series; corolla white, the laminae ca. 2 mm long; disk flowers numerous; corolla white, 4- or 5-lobed, 1–2 mm long. Cypsela oblong, slightly compressed, 3- to 4-angled; pappus coroniform, sometimes with 2 teeth, 2–3 mm long.

Marshes, floodplains, and wet disturbed sites. Common; nearly throughout. Nearly cosmopolitan. Native to North America and tropical America. All year.

Elephantopus L. 1753. ELEPHANTSFOOT

Herbs. Leaves alternate, simple, pinnate-veined, petiolate or epetiolate. Flowers in capitula in bracteate clusters in corymbiform-paniculiform arrays, discoid, bisexual; petals 5, basally connate, actinomorphic; stamens 5, epipetalous, the filaments free, the anthers connate; ovary inferior, 2-carpellate, 1-loculate. Fruit a cypsela; pappus present.

A genus of about 20 species; North America, West Indies, Mexico, Central America, South America, Africa, Asia, Australia, and Pacific Islands. [From the Greek *elephantos,* elephant, and *pous,* foot (derived from a unit of measurement), in reference to the basal leaves of the original species resembling an elephant's foot.]

Selected reference: Strother (2006k).

1. Stem leafy, the basal leaves mostly absent at flowering .... **E. carolinianus**
1. Stem scapose or subscapose.
    2. Phyllaries 9–11 mm long; pappus 6–8 mm long .... **E. tomentosus**
    2. Phyllaries 6–8 mm long; pappus 3–5 mm long.
        3. Phyllaries densely villous with long, white trichomes; pappus bristles gradually dilated proximally into a narrowly triangular base .... **E. elatus**
        3. Phyllaries sparsely pubescent with short trichomes, sometimes with a few long, white ones; pappus bristles abruptly dilated proximally into a broadly triangular base .... **E. nudatus**

Elephantopus carolinianus Raeusch. [Of Carolina.] CAROLINA ELEPHANTSFOOT.

Elephantopus carolinianus Raeuschel, Nomencl. Bot. 256, 1797. *Elephantopus carolinianus* Willdenow, Sp. Pl. 3: 2390. 1803; non Raeuschel, 1797. *Elephantopus scaber* Linnaeus var. *carolinianus* Kuntz, Revis. Gen. Pl. 335. 1891.

Erect perennial herb, to 8(12) dm; stem striate-angled, sparsely pilose or hirsute. Leaves mostly cauline at flowering, the blade elliptic, ovate, or lanceolate, 6–12(18) cm long, 3–8 cm wide, the apex acute or obtuse, the base cuneate, the margin serrate-crenate or subentire, the upper and lower surfaces sparsely pilose or hirsute, short-petiolate or sessile. Flowers (1)4(5) in numerous sessile capitula in bracteate clusters in a corymbiform-paniculiform array; bracts 2–3, rounded- or lanceolate-deltate, (0.5)1–1.5(2.5) cm long, (4)6–12 mm wide, sparsely pilosulous; involucre cylindric, 1–3 mm wide; phyllaries 8 in 4 decussate pairs, the outer ovate, the inner lanceolate, 8–10 mm long, all chartaceous, the apex spinulose or mucronate, the outer surface of the inner 4 sparsely hispidulous or pilosulous; corolla white, pink, or purple, 5–6 mm long. Cypsela subclavate, 3–4 mm long, 10-ribbed, strigulose; pappus of 5 barbellate bristles 4–5 mm long.

Moist hammocks and floodplain forests. Frequent; nearly throughout. New Jersey and Pennsylvania south to Florida, west to Kansas, Oklahoma, and Texas. Summer–fall.

Elephantopus elatus Bertol. [*Elatus*, tall.] TALL ELEPHANTSFOOT.

Elephantopus elatus Bertoloni, Mem. Reale Accad. Sci. Inst. Bologna 2: 607. 1850.
Elephantopus elatus Bertoloni var. *intermedius* Gleason, Bull. Torrey Bot. Club 46: 252. 1919.

Erect perennial herb, to 7 dm; stem striate-angled, sparsely pilose or hirsute. Leaves mostly basal at flowering, the blade oblanceolate, 9–14(20) cm long, 2–3(4.5) cm wide, the apex obtuse or acute, the base cuneate, the margin serrate-crenate, the upper and lower surfaces sparsely pilose or hirsute, sessile or subsessile, often clasping the stem. Flowers (1)4(5) in numerous sessile capitula in bracteate clusters in a corymbiform-paniculiform array; bracts 2–3, rounded- or lanceolate-deltate, 0.8–1.2 cm long, 6–8 mm wide, densely pilose or hirsute; involucre cylindric, 1–3 mm wide; phyllaries 8 in 4 decussate pairs, the outer ovate, the inner lanceolate, 6–8 mm long, all chartaceous, the apex spinulose or mucronate, the outer faces of the inner 4 densely villous with long white trichomes; corolla white, pink, or purple, 4–5 mm long. Cypsela subclavate, 3–4 mm long, 10-ribbed, strigulose; pappus of 5 barbellate bristles 3–4 mm long gradually dilated proximally into a narrow triangular base.

Sandhills and dry flatwoods. Common; nearly throughout. South Carolina south to Florida, west to Louisiana. Summer–fall.

Elephantopus nudatus A. Gray [Naked, leaves mostly basal and the stem nearly devoid of leaves at flowering.] SMOOTH ELEPHANTSFOOT.

Elephantopus nudatus A. Gray, Proc. Amer. Acad. Arts 15: 47. 1879.

Erect perennial herb, to 11 dm; stem striate-angled, sparsely strigose, pilose, or hirsute. Leaves mostly basal at flowering, the blade oblanceolate, spatulate, or elliptic, 7–15(20) cm long, 2–3.5(4.5) cm wide, the apex acute or obtuse, the base cuneate, the margin irregularly shallow serrate-crenate, the upper and lower surfaces sparsely strigose, pilose, or hirsute, short-petiolate or subsessile. Flowers (1)4(5) in numerous sessile capitula in bracteate clusters in a corymbiform-paniculiform array; bracts 2–3, rounded- or lanceolate-deltate, 0.6–1.5 cm long, 4–9 mm wide, sparsely pilose or hirsute; involucre cylindric, 1–3 mm wide; phyllaries 8 in 4 decussate pairs, the outer ovate, the inner lanceolate, 6–8 mm long, all chartaceous, the apex spinulose or mucronate, the outer faces of the inner 4 sparsely pubescent with short trichomes, sometimes with a few long trichomes or hispidulous; corolla white, pink, or purple, 3–4 mm long. Cypsela subclavate, ca. 4 mm long, 10-ribbed, strigulose; pappus of 5 barbellate bristles 3–5 mm long abruptly dilated below into a broadly triangular base.

Wet flatwoods, bogs, and wet hammocks. Occasional; northern counties, Marion County. Virginia and Maryland south to Florida, west to Oklahoma and Texas. Summer–fall.

Elephantopus tomentosus L. [Thickly covered with trichomes, in reference to the leaves, stems, bracts, and phyllaries.] DEVIL'S GRANDMOTHER.

Elephantopus tomentosus Linnaeus, Sp. Pl. 814. 1753.

Erect perennial herb, to 6 dm; stem striate-angled, sparsely pilose or hirsute. Leaves mostly basal, the blade obovate, oblanceolate, or spatulate, 3–8(10) cm long, 0.5–10(2.5) cm wide, the apex acute or obtuse, the base cuneate, the margin serrate-crenate, the upper and lower surfaces pilose or hirsute, sessile, usually clasping the stem. Flowers (1)4(5) in numerous sessile capitula in bracteate clusters in a corymbiform-paniculiform array; bracts 2–3, rounded-cordate, 0.8–1.5(2) cm long, 7–1.5 mm wide, pilose or hirsute; involucre cylindric, 1–3 mm wide; phyllaries 8 in 4 decussate pairs, the outer ovate, the inner lanceolate, 9–11 mm long, all chartaceous, the apex spinulose or mucronate, the outer faces of the inner 4 pilosulous; corolla white, pink, or purple, 5–6 mm long. Cypsela subclavate, (3)4–5 mm long, 10-ribbed, strigulose; pappus of 5 barbellate bristles 6–8 mm long.

Sandhills and calcareous hammocks. Occasional; panhandle. Maryland and Virginia south to Florida, west to Oklahoma and Texas; Mexico. Summer–fall.

Emilia Cass. 1817. TASSELFLOWER

Herbs. Leaves alternate, simple, pinnate-veined, petiolate or epetiolate. Flowers in capitula in cymiform or corymbiform arrays, discoid, bisexual or sometimes the inner staminate; receptacle epaleate; petals 5, basally connate, actinomorphic; stamens 5, epipetalous, the filaments free, the anthers connate; ovary inferior, 2-carpellate, 1-loculate. Fruit a cypsela; pappus present.

A genus of about 50 species; nearly cosmopolitan. [Presumed to be named for someone named Emile.]

Selected reference: Barkley (2006a).

1. Flowers exserted 2–5 mm from the involucre; style appendages tapered; lower leaves mostly sessile .... **E. fosbergii**
1. Flowers exserted less than 2 mm from the involucre; style appendages truncate; lower leaves mostly petiolate .... **E. sonchifolia**

Emilia fosbergii Nicolson [Commemorates Francis Raymond Fosberg (1908–1993), American botanist.] FLORIDA TASSELFLOWER.

Emilia fosbergii Nicolson, Phytologia 32: 1975.

Erect annual herb, to 1 m; stem striate-angled, sparsely arachnoid villous proximally or glabrous. Leaves basal and cauline, the blade oblanceolate, 5–10 cm long, 3–5 cm wide, the distal smaller and bract-like, the apex acute, the base cuneate or auriculate, the margin irregularly dentate-serrate, lobed, or entire, winged-petiolate and auriculate-clasping or sessile and auriculate, the lower leaves mostly sessile. Flowers in capitula in a cymiform or corymbiform array; involucre campanulate or cylindric; phyllaries 8 or 13 in 1–2 series, linear-oblong, 9–14 mm long, equal, the margin scarious; receptacle flat or slightly convex; flowers numerous; corolla pink, purple, or reddish, surpassing the involucre by 2–5 mm; style appendages tapered. Cypsela fusiform-prismatic, 5-ribbed, with blunt trichomes on the ribs, yellowish brown; pappus of numerous white barbellate bristles ca. 5 mm long.

Disturbed sites. Frequent; peninsula, central and western panhandle. Florida, Louisiana, and Texas; West Indies, Mexico, Central America, and South America; Asia and Pacific Islands. Native to Asia. All year.

Emilia sonchifolia (L.) DC. [With leaves like *Sonchus.*] LILAC TASSELFLOWER.

Cacalia sonchifolia Linnaeus, Sp. Pl. 835. 1753. *Senecio sonchifolius* (Linnaeus) Moench, Suppl. Meth. 231. 1802. *Crassocephalum sonchifolium* (Linnaeus) Lessing, Linnaea 6: 252. 1831. *Emilia sonchifolia* (Linnaeus) de Candolle, in Wight, Contr. Bot. India 24. 1834. *Emilia sonchifolia* (Linnaeus) de Candolle var. *typica* Domin, Biblioth. Bot. 89: 686. 1930, nom. inadmiss.

Hieracium javanicum Burman f., Fl. Indica 174, t. 57(1). 1768. *Prenanthes javanica* (Burman f.) Willdenow, Sp. Pl. 3: 1534. 1803. *Sonchus javanicus* (Burman f.) Sprengel, Syst. Veg. 3: 648. 1826. *Emilia javanica* (Burman f.) C. B. Robinson, Philipp. J. Sci. 3: 217. 1907. *Emilia sonchifolia* (Linnaeus) de Candolle var. *javanica* (Burman f.) Mattfeld, Bot. Jahrb. Syst. 62: 445. 1929.

Cacalia sagittata Vahl, Symb. Bot. 3: 91. 1794, nom. illegit.

Erect annual herb, to 8 dm; stem striate-angled, villous or glabrous. Leaves mostly in the proximal ½, the blade ovate, obovate, or oblanceolate, 5–12 cm long, 1.5–4.5 cm wide, the distal smaller and bract-like, the apex acute or acuminate, the base of the proximal truncate or cordate, that of the mid-cauline and distal auriculate clasping, the margin usually deeply lobed, irregularly serrate, the upper and lower surfaces sparsely villous or glabrous, the lower leaves with the petiole 1–5 cm long, the mid-cauline and upper sessile or subsessile. Flowers in capitula in a cymiform or corymbiform array; involucre campanulate or cylindric; phyllaries usually 8 in 1 series, linear-oblong, 9–12 mm long, equal, the margin scarious; receptacle flat or slightly convex; disk flowers numerous; corolla lavender pink, or purple, rarely reddish, surpassing the involucre by less than 2 mm; style appendages truncate. Cypsela fusiform-prismatic, 5-ribbed, with blunt trichomes on the ribs, yellowish brown; pappus of numerous white barbellate bristles ca. 5 mm long.

Disturbed sites. Occasional; Alachua County, central and southern peninsula, Leon and Escambia Counties. South Carolina, Georgia, and Florida, also Ohio, Louisiana, and California; West Indies, Mexico, Central America, and South America; Africa, Asia, Australia, and Pacific Islands. Native to Asia. All year.

EXCLUDED TAXON

Emilia coccinea (Sims) G. Don—Reported for Florida by Small (1933) and Cronquist (1980), the name misapplied to Florida material of *E. fosbergii.*

Enydra Lour. 1790. BUFFALO SPINACH

Herbs. Leaves opposite, simple, pinnate-veined, epetiolate. Flowers in capitula, solitary, radiate; receptacle paleate; ray flowers carpellate; petals 5, basally connate, zygomorphic; disk flowers bisexual; petals 5, basally connate, actinomorphic; stamens 5, the filaments free, the anthers connate; ovary inferior, 2-carpellate, 1-loculate. Fruit a cypsela; pappus absent.

A genus of about 5 species; North America, South America, Africa, Asia, and Australia. [An ancient Greek coastal city, the name apparently used in reference to the plant's aquatic habitat.]

Enydra fluctuans Lour. [Fluctuating, in reference to its aquatic habitat.] BUFFALO SPINACH.

Enydra fluctuans Loureiro, Fl. Cochinch. 511. 1790. *Meyera fluctuans* (Loureiro) Sprengel, Syst. Veg. 3: 602. 1826.

Basally decumbent annual herb; stem to 3(6) dm long, rooting at the lower nodes, striate-angled, slightly succulent, slightly pubescent or glabrous. Leaves with the blade oblong or linear-oblong, 2–6 cm long, 4–14 mm wide, the apex acute or obtuse, the base truncate, the margin sparsely serrate, sessile, amplexicaul. Flowers in a solitary terminal or axillary capitula, subsessile; involucre cupulate, 8–10 mm wide, ca. 3 mm wide at the base; phyllaries 4, foliose, yellow-green, the outer pair ovate-oblong, 10–12 mm long, the inner pair somewhat smaller, the apex rounded, the outer surface glabrous; receptacle convex or conical; palea scalelike, ca. 5 mm long, the apex toothed, glandular, and sparsely pubescent; ray flowers numerous; corolla white or greenish white, sometimes slightly pinkish, the lamina ca. 2 mm long; disk flowers numerous; corolla white or greenish white, sometimes slightly pinkish. Cypsela obovoid-cylindric, 3–4 mm long, glabrous, enclosed by a rigid palea; pappus absent.

Ditches, canals, and other disturbed aquatic sites; Rare; Hillsborough County. Escaped from cultivation. Florida; Africa, Asia, and Australia. Native to Africa and Asia. Spring–fall.

Erechtites Raf. 1817. BURNWEED

Herbs. Leaves alternate, simple, pinnate-nerved, petiolate or epetiolate. Flowers in capitula in corymbiform arrays, discoid, bisexual; calyculi present; receptacle epaleate; petals 5, basally connate, actinomorphic; stamens 5, epipetalous, the filaments free, the anthers connate; ovary inferior, 2-carpellate, 1-loculate; pappus present.

A genus of about 12 species; North America, Mexico, West Indies, Central America, South America, Australia, and Pacific Islands. [A name used by Dioscoriedes for some plant in the tribe Senecioneae.]

Selected reference: Barkley (2006b).

Erechtites hieraciifolius (L.) Raf. ex DC. [With leaves like the genus *Hieracium*.] AMERICAN BURNWEED; FIREWEED.

Senecio hieraciifolius Linnaeus, Sp. Pl. 866. 1753. *Neoceis hieraciifolia* (Linnaeus) Cassini, Bull. Sci. Soc. Philom. Paris 1820: 91. 1820. *Erechtites hieraciifolius* (Linnaeus) Rafinesque ex de Candolle,

Prodr. 6: 294. 1838. *Erechtites hieraciifolius* (Linnaeus) Rafinesque ex de Candolle var. *glabrescens* Kuntze, Revis. Gen. Pl. 1: 335. 1891, nom. inadmiss. *Erechtites hieraciifolius* (Linnaeus) Rafinesque ex de Candolle var. *typicus* Fernald, Rhodora 19: 27. 1917, nom. inadmiss.

Erect, annual herb, to 2 m; stem striate-angled, coarsely pilose or glabrate. Leaves with the blade lanceolate or ovate-lanceolate, (3)6–20 cm long, (1)2–8 cm wide, the apex acute, the base cuneate, the margin subentire, serrate, or lobed, the upper and lower surfaces sparsely pilose or glabrous, winged petiolate or sessile. Flowers few to many capitula in a corymbiform array, the peduncle 1–3 cm long, with several linear bracteate leaves; calyculus of several linear bractlets; involucre urceolate or obconic; phyllaries usually ca. 20 in 1–2 series, linear-lanceolate, 10–17 mm long, the apex acute usually pinkish, glabrous; receptacle flat or nearly so; disk flowers numerous; corolla pale yellow, ca. 1 cm long. Cypsela prismatic, 10- to 12-ribbed, 2–3 mm long, puberulent between the nerves or glabrous; pappus of numerous white barbellate bristles ca. 1 cm long.

Wet to moist disturbed sites. Common; nearly throughout. Quebec south to Florida, west to Ontario, Minnesota, South Dakota, Nebraska, Kansas, Oklahoma, and Texas, also Washington, Oregon, and California; Mexico and West Indies.

Erigeron L. 1753. FLEABANE

Herbs. Leaves alternate, simple, pinnate-veined, petiolate or epetiolate. Flowers in capitula, solitary or in corymbiform or paniculiform arrays, radiate; receptacle epaleate; ray flowers carpellate; petals 5, basally connate, zygomorphic; disk flowers bisexual; petals 5, basally connate, actinomorphic; stamens 5, epipetalous, the filaments free, the anthers connate; ovary inferior, 2-carpellate, 1-loculate. Fruit a cypsela; pappus present.

A genus of about 390 species; nearly cosmopolitan. [From the Greek *erion,* woolly, and *geron,* old man, in reference to the pappus in some species resembling a white-haired old man.]

Molecular studies show that *Erigeron,* as currently recognized, is paraphyletic. Further study may result in a revised circumscription of the genus.

Selected reference: Nesom (2006j).

1. Pappus of the ray and disk flowers not alike, that of the disk flowers with inner capillary barbellate bristles, that of the ray flowers lacking the inner capillary bristles.
    2. Stem strigose; leaves reduced and sparse on the upper stem .... E. **strigosus**
    2. Stem villous; leaves ample and not much reduced on the upper stem .... E. **annuus**
1. Pappus of the ray and disk flowers alike, both with evident inner long barbellate bristles.
    3. Cypsela 4-nerved; plant subscapose; stem glabrous or sparsely appressed strigose .... E. **vernus**
    3. Cypsela 2-nerved (rarely obscurely 4-nerved); plant with moderately leafy stems; stem hirsute, the trichomes spreading.
        4. Flowering capitula ca. 5 mm wide, the phyllaries 3–4 mm long .... E. **bellioides**
        4. Flowering capitula 1 cm wide or wider, the phyllaries 5–15 mm long.
            5. Plants with rhizomes .... E. **pulchellus**
            5. Plants lacking rhizomes.

6. Pappus double, with some short outer setae in addition to the longer inner capillary bristles.......... **E. tenuis**
6. Pappus simple, the outer absent or a minute crown.
   7. Stems villous-hirsute, the trichomes up to ⅔ the width of the stem in length, sometimes appressed; phyllaries 3–4 mm long.......... **E. quercifolius**
   7. Stems with long-spreading trichomes longer than the width of the stem; phyllaries 4–6 mm long.......... **E. philadelphicus**

Erigeron annuus (L.) Pers. [Plant an annual.] EASTERN DAISY FLEABANE.

Aster annuus Linnaeus, Sp. Pl. 875. 1753. *Erigeron annuus* (Linnaeus) Persoon, Syn. Pl. 2: 431. 1807. *Diplopappus dubius* Cassini, Bull. Sci. Soc. Philom. Paris 1817: 368. 1817, nom. illegit. *Diplopappus annuus* (Linnaeus) Bluff & Fingerhuth, Comp. Fl. German. 2: 368. 1825. *Phalacroloma acutifolium* Cassini, in Cuvier, Dict. Sci. Nat. 39: 405. 1826, nom. illegit. *Phalacroloma annuum* (Linnaeus) Dumortier, Fl. Belg. 67. 1827. *Stenactis annua* (Linnaeus) Lessing, Syn. Gen. Compos. 189. 1832. *Aster stenactis* E.H.L. Krause, in Sturm, Deutschl. Fl., ed, 2. 13: 54. 1905, nom. illegit. *Erigeron annuus* (Linnaeus) Persoon var. *typicus* Cronquist, Brittonia 6: 266. 1947, nom. inadmiss.

Erect annual herb, to 1.5 m; stem striate-angled, sparsely villous. Basal leaves (usually withering by flowering) with the blade lanceolate, oblanceolate, or ovate, 1.5–8 cm long, 3–20 mm wide, the apex acute, the base cuneate, the margin irregularly serrate or subentire, the upper and lower surfaces sparsely strigose-hirsute, eglandular, epetiolate, the cauline lanceolate, oblong, similar and little reduced upward. Flowers in capitula in a paniculiform or corymbiform array, the peduncle to 1 cm long; involucre turbinate or hemispheric, 6–12 mm wide; phyllaries numerous, in 2–3(4) series, linear-elliptic, 3–5 mm long, the outer surface sparsely villous, minutely glandular; ray flowers numerous; corolla white, the lamina 4–10 mm long; disk flowers numerous; corolla yellow, 2–3 mm long. Cypsela oblong-ovoid, laterally compressed, ca. 1 mm long, 2-nerved, sparsely strigose; pappus of an outer minute crown of setae or narrow scales, the inner absent (ray flowers) or of 8–11 barbellate bristles (disk flowers).

Disturbed sites. Rare; Union County, panhandle. Nearly throughout North America. North America, Mexico, Central America, Europe, Africa, Asia, and Pacific Islands. Native to eastern North America. Spring.

Erigeron bellioides DC. [Resembling the genus *Bellis* (Asteraceae).] BELLORITA.

Erigeron bellioides de Candolle, Prodr. 5: 288. 1836.

Ascending or procumbent annual herb; stem with stolons to 15 cm long, the flowering stem erect, striate-angled, sparsely villous. Leaves mainly basal and rosulate, the blade suborbicular or broadly ovate, 1–3.5 cm long, 3–10 mm wide, those on the stolons and flowering stems reduced, the apex obtuse, the base broadly cuneate, the margin entire or crenulate, the upper and lower surfaces sparsely villous or glabrate, the cauline few, much reduced, the blade obovate or oblong, the petiole winged, 1–2.5 cm long. Flowers in capitula, solitary, terminal or sometimes also a few axillary, the peduncle to 1 cm long; involucre hemispheric, ca. 5 mm wide; phyllaries in 2 series, the inner ca. 2 mm long, lanceolate, 3–4 mm long, the outer shorter, linear, the apex acute, the outer sparsely villous; ray flowers numerous; corolla white, the lamina ca. 0.5

mm long; disk flowers numerous; corolla yellow, 1–2 mm long. Cypsela elliptic, flat, ca. 1 mm long, puberulent, pale straw-colored, stramineous; pappus of an outer minute crown of setae or narrow scales, the inner absent or of 8–11 barbellate bristles.

Moist disturbed sites. Rare; Miami-Dade County. Florida; West Indies and Pacific Islands. Native to West Indies. All year.

Erigeron philadelphicus L. [Of Philadelphia.] PHILADELPHIA FLEABANE.

Erigeron philadelphicus Linnaeus, Sp. Pl. 863. 1753. *Tessenia philadelphica* (Linnaeus) Lunell, Amer. Midl. Naturalist 5: 59. 1917.

Erect annual, biennial, or short-lived perennial, to 8 dm; stem striate-angled, loosely strigose to sparsely hirsute, hirsute-villous, or villous, minutely glandular. Basal leaves with the blade oblanceolate or obovate, 3–11(15) cm long, 1–2.5(4) cm wide, the margin crenate, serrate, or pinnate-lobed, the apex obtuse, the base auriculate-clasping, the upper and lower surfaces sparsely hirsute or villous, the cauline with the blade oblong-oblanceolate or lanceolate, gradually reduced upward. Flowers in capitula, solitary or few in a corymbiform array, the peduncle to 2 cm long; involucre turbinate, 6–15 mm wide; phyllaries in 2–3 series, linear-lanceolate, 4–6 mm long, the outer surface hirsute-villous or glabrous, sometimes minutely glandular; ray flowers numerous; corolla white or pinkish, the lamina 5–10 mm long; disk flowers numerous, corolla yellow, 2–3 mm long. Cypsela oblong-ovoid, laterally compressed, ca. 1 mm long, 2-nerved, sparsely strigose; pappus of an outer minute crown of setae or narrow scales, the inner of numerous barbellate bristles.

Floodplain forests, calcareous hammocks, and disturbed sites. Occasional; northern peninsula, central and western panhandle. Nearly throughout North America. North America; Europe, Africa, and Asia. Native to North America. Spring.

Erigeron pulchellus Michx. [Beautiful and little.] ROBIN'S PLANTAIN.

Erigeron pulchellus Michaux, Fl. Bor.-Amer. 2: 124. 1803. *Musteron pulchellum* (Michaux) Rafinesque, Fl. Tellur. 2: 50. 1837 (1836"). *Erigeron pulchellus* Michaux var. *typicus* Cronquist, Brittonia 6: 248. 1947, nom. illegit.

Erect or ascending perennial herb, to 6 dm; stem rhizomatous, hirsute or villous, eglandular. Basal leaves with the blade oblanceolate, obovate, suborbiculate, or subspatulate, 2–13(18) cm long, 0.6–3(5) cm wide, the base clasping and slightly auriculate, the margin dentate or rarely entire, ciliate, the upper and lower surfaces hirsute or villous, the cauline gradually reduced upward. Flowers in capitula, solitary or few in a corymbiform array, the peduncle to 6 cm long; involucre turbinate, 6–20 mm wide; phyllaries in 2–3(4) series, linear-lanceolate, 5–7 mm long, hirsute, minutely glandular or stipitate-glandular; ray flowers numerous; corolla blue or pinkish, the lamina 6–10 mm long; disk flowers numerous; corolla yellow, 5–6 mm long. Cypsela oblong-ovoid, laterally compressed, 1–2 mm long, 2(4)-nerved, glabrate or glabrous; pappus of an outer minute crown of setae or absent, the inner of numerous barbellate bristles.

Open hammocks. Rare; Jackson County. Quebec south to Florida, west to Ontario, Minnesota, Iowa, Kansas, Oklahoma, and Texas. Spring.

Erigeron quercifolius Poir. [With leaves like *Quercus* (Fagaceae).] OAKLEAF FLEABANE.

Erigeron quercifolius Poiret, in Lamarck, Tabl. Encycl. 3: 258. 1823. *Erigeron philadelphicus* Linnaeus var. *quercinus* de Candolle, Prodr. 5: 285. 1836.

Erect or ascending annual, biennial, or short-lived perennial herb, to 6 dm; stem striate-angled, villous or villous-hirsute proximally, loosely strigose or hirsute distally. Basal leaves with the blade oblanceolate, obovate, or spatulate, 1.5–11(15) cm long, 0.5–2.5(3) cm wide, the apex rounded, the margin serrate, crenate, or pinnate-lobed, the upper and lower surfaces hirsute or villous, the cauline with the blade oblong, reduced upward and becoming bract-like, clasping. Flowers in capitula, few to many in a corymbiform array, the peduncle to 2 cm long; involucre turbinate, 5–10 mm wide; phyllaries in (2)3–4 series, linear-lanceolate, 3–4 mm long, the outer surface hirsute-villous, eglandular or minutely glandular; ray flowers numerous; corolla blue, pink, or white, the lamina 4–6 mm long; disk flowers numerous; corolla yellow, 2–3 mm long. Cypsela oblong-ovoid, laterally compressed, ca. 1 mm long, 2-nerved, strigose; pappus of outer minute crown of setae or absent, the inner of numerous barbellate bristles.

Moist hammocks and disturbed sites. Common; nearly throughout. Virginia south to Florida. Spring–summer.

Erigeron strigosus Muhl. ex Willd. [With sharp, straight, appressed trichomes.] PRAIRIE FLEABANE.

Erigeron strigosus Muhlenberg ex Willdenow, Sp. Pl. 3: 1956. 1803. *Stenactis strigosa* (Muhlenberg ex Willdenow) de Candolle, Prodr. 5: 299. 1836. *Erigeron strigosus* Muhlenberg ex Willdenow var. *typicus* Cronquist, Brittonia 6: 267. 1947, nom. inadmiss. *Stenactis annua* (Linnaeus) Lessing subsp. *strigosa* (Muhlenberg ex Willdenow) Soo, in Soo & Javorka, Magyar Novenyek Kazik 2: 664. 1951. *Erigeron annuus* (Linnaeus) Persoon subsp. *strigosus* (Muhlenberg ex Willdenow) Wagenitz, in Hegi, Ill. Fl. Mitt. Eur., ed. 2. 6/3(2): 96. 1965.

Doronicum ramosum Walter, Fl. Carol. 205. 1788. *Erigeron ramosus* (Walter) Britton et al., Prelim. Cat. 27: 1888; non Rafinesque, 1817. *Tessenia ramosa* (Walter) Lunell, Amer. Midl. Naturalist 5: 59. 1917. *Stenactis ramosa* (Walter) Domin, Preslia 13–15: 226. 1935. *Erigeron annuus* (Linnaeus) Persoon var. *ramosus* (Walter) Hylander, Uppsala Univ. Arsskr. 1945(7): 309. 1945.

Stenactis beyrichii Fischer & C. A. Meyer, Index Sem. Hort. Petrop. 5: n. 2169. 1839. *Phalacroloma beyrichii* (Fischer & C. A. Meyer) Fischer & C. A. Meyer, Index Sem. Hort. Petrop. 6: 63. 1840. *Erigeron strigosus* Muhlenberg ex Willdenow var. *beyrichii* (Fischer & C. A. Meyer) A. Gray, Syn. Fl. N. Amer. 1(2): 219. 1884. *Erigeron ramosus* (Walter) Britton et al. var. *beyrichii* (Fischer & C. A. Meyer) Trelease ex Branner & Coville, Rep. (Annual) Arkansas Geol. Surv. 1888(4): 192. 1891.

Erect or ascending annual or biennial herb, to 7 dm; stem striate-angled, strigose, the trichomes appressed or ascending. Basal leaves with the blade spatulate or oblanceolate, (1)3–15(17) cm long, 0.3–1.5(2) cm wide, the cauline gradually reduced upward, the apex acute or obtuse, the base cuneate, the margin usually entire, sometimes crenate or serrate, the upper and lower surfaces sparsely strigose or hirsute, the petiole to 5 cm long on the basal blade, the leaves reduced upward and becoming sessile. Flowers in capitula, 10 to many in a corymbiform or paniculiform array, the peduncle to 2 cm long; involucre turbinate, 5–12 mm wide;

phyllaries in 2–4 series, linear-lanceolate, 2–4 mm long, the outer surface sparsely strigose; ray flowers numerous; corolla white, pinkish, or bluish, the lamina 4–6 mm long; disk flowers numerous; corolla yellow, 2–3 mm long. Cypsela oblong-ovoid, laterally compressed, ca. 1 mm long, 2-nerved, sparsely strigose; pappus of an outer minute crown of setae or scales, the inner absent (ray flowers) or of 8–15 barbellate bristles (disk flowers).

Disturbed sites. Frequent; northern counties, central peninsula. Nearly throughout North America. Spring–summer.

Erigeron tenuis Torr. & A. Gray [Slender.] SLENDER FLEABANE.

Erigeron tenuis Torrey & A. Gray, Fl. N. Amer. 2: 175. 1841.

Ascending biennial or short-lived perennial, to 4.5 dm; stem striate-angled, hirsute or pilose. Basal leaves with the blade oblanceolate, obovate, or spatulate, 2–10(13) cm long, 0.4–2 cm wide, the apex obtuse, the base cuneate, the margin serrate, crenate, or pinnate-lobed, the upper and lower surfaces sparsely hirsute or strigose-hirsute, the petiole 3–5 cm long, the cauline leaves becoming narrower and entire, abruptly reduced near midstem, sessile or subsessile. Flowers in capitula, few to many in a corymbiform array, the peduncle 1–4 cm long; involucre turbinate, 5–10 mm wide; phyllaries in 2–4 series, 3–4 mm long, the outer surface sparsely strigose; ray flowers numerous; corolla blue, pale lavender, or white, the lamina 3–5 mm long; disk flowers numerous; corolla yellow, 2–3 mm long. Cypsela oblong-ovoid, laterally compressed, ca. 1 mm long, 2-nerved, sparsely strigose; pappus of short outer bristles, the inner of numerous longer barbellate bristles.

Disturbed sites. Rare; Okaloosa County. Missouri and Kansas south to Florida and Texas. Spring.

Erigeron vernus (L.) Torr. & A. Gray [Vernal, pertaining to spring.] EARLY WHITETOP FLEABANE.

Aster vernus Linnaeus, Sp. Pl. 876. 1753. *Stenactis verna* (Linnaeus) Nees von Esenbeck, Gen. Sp. Aster. 275. 1832. *Erigeron vernus* (Linnaeus) Torrey & A. Gray, Fl. N. Amer. 2: 176. 1841.

Erect biennial or short-lived perennial herb, to 5 dm; stem rhizomatous, these sometimes producing rosulate offshoots, striate-angled, glabrous or sparsely appressed strigose. Leaves mostly basal, the blade oblanceolate or spatulate, 2–10(15) cm long, 0.5–2.5 cm wide, usually somewhat thickened, the apex rounded or obtuse, the base cuneate, the margin entire or denticulate, the upper and lower surfaces glabrate, the cauline bract-like. Flowers in capitula, few to many in a corymbiform array, the peduncle to 1.5 cm long; involucre turbinate, 5–11 mm wide; phyllaries in 2–3(4) series, linear-lanceolate, 3–4 mm long, the outer surface sparsely hirsute or strigose, sometimes glabrous, sometimes viscid; ray flowers numerous, corolla white, the lamina 4–8 mm long; disk flowers numerous; corolla yellow, 3–4 mm long. Cypsela oblong-ovoid, laterally compressed, ca. 1 mm long, 4-nerved, sparsely strigose; pappus of outer setae, the inner of numerous barbellate bristles.

Flatwoods, bogs, pond margins, and wet disturbed sites. Common; nearly throughout. Virginia south to Florida, west to Louisiana. Spring.

Eupatorium L. 1753. THOROUGHWORT

Herbs. Leaves opposite or alternate, simple, 3-veined or pinnate-veined, petiolate or epetiolate. Flowers in capitula in corymbiform or paniculiform arrays, discoid, bisexual; receptacle epaleate; petals 5, basally connate, actinomorphic; stamens 5, epipetalous, the filaments free, the anthers connate; ovary inferior, 2-carpellate, 1-loculate. Fruit a cypsela; pappus present.

A genus of about 45 species; North America, West Indies, Mexico, Europe, Asia. [Named for Mithridates Eupator, King of Pontus (132–63 BC).]

Eupatorium, as treated here, follows the work of King and Robinson (1987), who segregated many genera found in Florida that were formerly included in the broad circumscription of the genus: *Ageratina, Chromolaena, Conoclinium, Fleistmannia, Koanophyllon,* and *Eutrochium.* In addition, identification of taxa within *Eupatorium* s. str. is highly confusing as a result of frequent hybridization, polyploidy, and apomixis.

Selected references: Schilling (2011); Siripum and Schilling (2006).

1. Cauline leaves deeply lobed, pinnate, or bipinnately dissected.
    2. Capitula secund on the branches; stems and peduncles glabrous .... **E. leptophyllum**
    2. Capitula not secund on the branches; stems and peduncles puberulent.
        3. Upper cauline leaf divisions more than 1 mm wide; plant viscid to the touch when fresh .... **E. compositifolium**
        3. Upper cauline leaf divisions 0.5 mm wide or less; plant not viscid to the touch when fresh .... **E. capillifolium**
1. Cauline leaves entire, toothed, or shallowly lobed.
    4. Principal cauline leaves with the petiole 5 mm long or longer.
        5. Leaves succulent, the margin crenate or dentate .... **E. mikanioides**
        5. Leaves submembranaceous or coriaceous, the margin serrate .... **E. serotinum**
    4. Principal cauline leaves sessile or with the petiole less than 5 mm long.
        6. Principal cauline leaves connate-perfoliate .... **E. perfoliatum**
        6. Principal cauline leaves sessile or subsessile (petiole less than 5 mm long).
            7. Phyllaries with a prominent white filiform or acuminate apex.
                8. Leaves 0.4–1(1.5) cm wide .... **E. leucolepis**
                8. Leaves 1–4.5 cm wide.
                    9. Leaves little, if at all gland-dotted; phyllaries glabrous, not gland-dotted .... **E. petaloideum**
                    9. Leaves usually gland-dotted; phyllaries puberulent or villous (at least toward the bases and on the midviens), gland-dotted.
                        10. Leaf blades mostly 3–4 times their width, the base narrowly cuneate; phyllaries linear .... **E. album**
                        10. Leaf blades mostly 1–2 times their width, the base rounded or rounded-cuneate; phyllaries oblong or lanceolate-oblong .... **E. pilosum**
            7. Phyllaries with a round or short-acute white or green apex.
                11. Leaf base broadly cuneate, rounded, truncate, or subcordate.
                    12. Leaves lanceolate; inflorescence branches mostly alternate; resin dots on the phyllaries colorless or nearly so .... **E. pilosum**

12. Leaves ovate or broadly lanceolate; inflorescence branches mostly opposite; resin dots on the phyllaries yellow or golden.. **E. rotundifolium**

11. Leaf bases narrowly cuneate.

13. Leaf blades 6–40 times as long as wide...**E. hyssopifolium**

13. Leaf blades 2.5–7 times as long as wide.

14. Principal pair of lateral veins arising from the base of the leaf.........**E. altissimum**

14. Principal pair of lateral veins arising from the midrib, or the blade 1-nerved.

15. Plant branching at or near the base; branches arising above the midstem elongated ....**E. linearifolium**

15. Plant unbranched below the midstem; branches arising above the midstem short.

16. Leaves mostly 5–7 cm long, 15–20 mm wide, 3-nerved from near the base, usually sharply serrate, the blades spreading .... **E. semiserratum**

16. Leaves mostly 2–4 cm long, 3–10 mm wide, 1-nerved, usually entire or irregularly crenate-serrate, the blades recurved ....**E. mohrii**

Eupatorium album L. [White, in reference to the flower color.] WHITE THOROUGHWORT.

Eupatorium album Linnaeus, Mant. Pl. 1: 111. 1767. *Uncasia alba* (Linnaeus) Greene, Leafl. Bot. Observ. Crit. 1: 13. 1903. *Eupatorium album* Linnaeus var. *typicum* Fernald, Rhodora 39: 451. 1937, nom. inadmiss.

Erect perennial herb, to 1 m; stem sometimes short-rhizomatous, striate-angled, pubescent. Leaves opposite, the distal sometimes alternate, the blade elliptic or oblanceolate, 5–9 cm long, 1–2.5(4) cm wide, 3-veined from the base or pinnate-veined, the apex obtuse or rounded, the base cuneate, the margin coarsely serrate, the upper and lower surfaces pubescent, gland-dotted, sessile. Flowers in capitula in a corymbiform array; involucre obconic; phyllaries numerous in 2–4 series, linear, 2–9 mm long, the apex acuminate or attenuate, white, the outer surface pubescent, gland-dotted; flowers 5; corolla white, 4–5 mm long. Cypsela prismatic, 5-ribbed, 3–4 mm long, glabrous, gland-dotted; pappus of numerous white barbellate bristles 4–5 mm long.

Sandhills, flatwoods, and hammocks. Occasional; northern counties, central peninsula. New York and Connecticut south to Florida, west to Indiana, Kentucky, Tennessee, Arkansas, and Texas. Summer–fall.

Eupatorium altissimum L. [Tall.] TALL THOROUGHWORT.

Eupatorium altissimum Linnaeus, Sp. Pl. 837. 1753. *Uncasia altissima* (Linnaeus) Greene, Leafl. Bot. Obser. Crit. 1: 13. 1903.

Erect perennial herb, to 1.5 m; stem sometimes short-rhizomatous, striate-angled, pubescent. Leaves opposite, the nodes sometimes appearing leafy due to the lateral buds producing 2 or more pairs of leaves, the blade lanceolate-elliptic or oblanceolate, 5–12 cm long, 0.5–2 cm wide, 3-veined from the base, the apex acuminate, the base cuneate, the margin serrate distally, the upper and lower surfaces puberulent or villous, gland-dotted, sessile or subsessile. Flowers in capitula in a corymbiform array; involucre obconic; phyllaries numerous in 2–3 series, oblong,

1–4 mm long, the apex rounded or acute, the outer surface pubescent; flowers 5; corolla white, 3–4 mm long. Cypsela prismatic, 5-ribbed, 2–3 mm long, glabrous, gland-dotted; pappus of numerous white barbellate bristles 3–4 mm long.

Disturbed sites. Rare; Liberty County. Massachusetts and New York south to Florida, west to Minnesota, Nebraska, Kansas, Oklahoma, and Texas. Fall.

Eupatorium capillifolium (Lam.) Small ex Porter & Britton [*Capillus,* hair's width, in reference to the very narrow leaf lobes.] DOGFENNEL.

Artemisia capillifolia Lamarck, Encyl. 1: 267. 1783. *Artemisia tenuifolia* Willdenow, Sp. Pl. 3: 1819. 1803, nom. illegit. *Eupatorium foeniculaceum* Willdenow var. *traganthes* de Candolle, Prodr. 5: 176. 1836. *Eupatorium capillifolium* (Lamarck) Small ex Porter & Britton, Mem. Torrey Bot. Club 5: 311. 1894.
Eupatorium foeniculoides Walter, Fl. Carol. 199. 1788. *Eupatorium foeniculaceum* Willdenow, Sp. Pl. 3: 1750. 1803, nom. illegit.
Chrysocoma capillifolia Michaux, Fl. Bor.-Amer. 2: 101. 1803. TYPE: FLORIDA/SOUTH CAROLINA.
Eupatorium foeniculaceum Willdenow var. *lateriflorum* Torrey & A. Gray, Fl. N. Amer. 2: 83. 1841. TYPE: FLORIDA: s.d., *Chapman s.n.* (holotype: NY?).

Erect perennial herb, to 2 m; stem striate-angled, puberulent. Leaves opposite or alternate, the nodes often appearing leafy due to leaf development on lateral buds without axis elongation, the blade lanceolate, pinnate-lobed, the lobes linear, 0.5–10 cm long, to 1 mm wide, the apex rounded or acute, the base cuneate, the margin entire, strongly revolute, the upper and lower surfaces glabrate, gland-dotted, sessile. Flowers in capitula in a paniculiform array; involucre obconic; phyllaries numerous in 2–3 series, oblong, 1–3 mm long, the apex acuminate or mucronulate, the outer surface glabrous or glabrate, sometime sparsely gland-dotted; flowers 5; corolla white, 2–3 mm long. Cypsela prismatic, 5-ribbed, 1–2 mm long, glabrous, gland-dotted; pappus of numerous white barbellate bristles 2–3 mm long.

Wet flatwoods, marshes, and wet disturbed sites. Common; nearly throughout. Massachusetts south to Florida, west to Missouri, Oklahoma, and Texas; West Indies. Summer–fall.

Eupatorium compositifolium Walter [*Compositus,* united, in reference to the compound leaves.] YANKEEWEED.

Eupatorium compositifolium Walter, Fl. Carol. 199. 1788. *Traganthes compositifolia* (Walter) Greene, Leafl. Bot. Observ. Crit. 1: 13. 1903.
Chrysocoma coronopifolia Michaux, Fl. Bor.-Amer. 2: 102. 1803. *Eupatorium coronopifolium* (Michaux) Willdenow, Sp. Pl. 3: 1750. 1803.

Erect perennial herb, to 2 m; stem striate-angled, puberulent. Leaves opposite, the distal sometimes alternate, the blade lanceolate, pinnate-lobed, the lobes linear, 2–8 cm long, 1–3(4) mm wide, the apex acuminate, the base cuneate, the margin entire, the upper and lower surfaces puberulent, gland-dotted, sessile. Flowers in capitula in a paniculiform array; involucre obconic; phyllaries numerous in 2–3 series, elliptic or oblong, 1–3 mm long, the apex acuminate, mucronate, the outer surface puberulent, gland-dotted; flowers 5; corolla white, sometimes with a purple throat, 2–3 mm long. Cypsela prismatic, 5-ribbed, 1–2 mm long, glabrous, gland-dotted; pappus of numerous white barbellate bristles 3–4 mm long.

Wet flatwoods, cypress domes, marshes, and wet disturbed sites. Common; nearly throughout. Virginia south to Florida, west to Oklahoma and Texas. Summer–fall.

Eupatorium hyssopifolium L. [With leaves like hyssop (*Hyssopus officinalis*, Lamiaceae).] HYSSOPLEAF THOROUGHWORT.

Erect perennial herb, to 1 m; stem sometimes short-rhizomatous, striate-angled, pubescent. Leaves opposite, the distal sometimes alternate, often appearing whorled due to leaf development on lateral buds without axis elongation, the blade linear-lanceolate, lanceolate-oblong, or linear, 2–6 cm long, 2–15 mm wide, 3-veined from the base, the apex acute, the base cuneate, the margin entire, serrulate, or laciniate-toothed, the upper and lower surfaces scabrous (at least the lower), sessile. Flowers in capitula in a corymbiform array; involucre obconic; phyllaries numerous in 2–3 series, elliptic or oblong, 2–5 mm long, the apex obtuse or acute, the outer surface pubescent; flowers 5; corolla white, 3–4 mm long. Cypsela prismatic, 5-ribbed, 2–3 mm long, glabrous, gland-dotted; pappus of numerous white barbellate bristles 3–4 mm long.

1. Leaves mostly linear, 2–5 mm wide, the margin usually entire or obscurely serrulate .... var. **hyssopifolium**
1. Leaves mostly linear-lanceolate, 5–15 mm wide, the margin usually laciniate-toothed .... var. **laciniatum**

Eupatorium hyssopifolium var. **hyssopifolium**

Eupatorium hyssopifolium Linnaeus, Sp. Pl. 836. 1753. *Uncasia hyssopifolia* (Linnaeus) Greene, Leaflet. Bot. Observ. Crit. 1: 13. 1903.
Eupatorium lecheifolium Greene, Pittonia 3: 177. 1897. *Uncasia lecheifolia* (Greene) Leafl. Bot. Observ. Crit. 1: 13. 1903. TYPE: FLORIDA: Gadsden Co.: Quincy, 4 Sep 1895, *Nash 2566* (holotype: NY).

Leaves mostly linear, 2–5 mm wide, the margin usually entire or obscurely serrulate.

Flatwoods and hammocks. Occasional; northern and central peninsula, eastern and central panhandle. New York and Massachusetts south to Florida, west to Wisconsin, Illinois, Missouri, Arkansas, and Texas. Summer–fall.

Eupatorium hyssopifolium var. **laciniatum** A. Gray [*Laciniatus*, slashed in narrow divisions, in reference to the leaf margin.]

Eupatorium torreyanum Short & Peter, Transylvania J. Med. Assoc, Sci. 8: 575. 1836. *Uncasia torreyana* (Short & Peter) Greene, Leafl. Bot. Observ. Crit. 1: 13. 1903.
Eupatorium hyssopifolium Linnaeus var. *laciniatum* A. Gray, Syn. Fl. N. Amer. 1(2): 98. 1884.

Leaves mostly linear-lanceolate, 5–15 mm wide, the margin usually laciniate-toothed.

Flatwoods and hammocks. Rare; Taylor County. New York and Massachusetts south to Florida, west to Indiana, Kentucky, Tennessee, Arkansas, and Louisiana. Fall.

Eupatorium leptophyllum DC. [From the Greek *leptos,* slender, in reference to the leaves.] FALSEFENNEL.

Eupatorium leptophyllum de Candolle, Prodr. 5: 176. 1836. *Eupatorium foeniculaceum* Willdenow var. *glabrum* Torrey & A. Gray, Fl. N. Amer. 2: 83. 1841. *Eupatorium capillifolium* (Lamarck) Small var. *leptophyllum* (de Candolle) Ahles, J. Elisha Mitchell Sci. Soc. 80: 173. 1964, nom. illegit.

Erect perennial herb, to 3 m; stem striate-angled, glabrous, gland-dotted. Leaves opposite, the distal sometimes alternate, the nodes often appearing leafy due to the development of leaves on the lateral buds without axis elongation, the blade pinnate-lobed, the lobes 2–10 cm long, less than 1 mm wide, 1-veined or pinnate-veined, the apex rounded or acute, the base slightly expanded, the margin entire, the upper and lower surfaces puberulent or glabrate, gland-dotted, sessile or subsessile. Flowers in capitula in a paniculiform array, the branches recurved, secund; involucre obconic; phyllaries numerous in 1–2 series, lanceolate, 2–3 mm long, the apex acuminate, mucronate, the outer surface glabrous or glabrate; flowers 5; corolla white, 2–3 mm long. Cypsela prismatic, 5-ribbed, 1–2 mm long, glabrous, gland-dotted; pappus of numerous white barbellate bristles 2–3 mm long.

Moist to dry flatwoods. Frequent; nearly throughout. North Carolina south to Florida, west to Mississippi. Summer–fall.

Eupatorium leucolepis (DC.) Torr. & A. Gray [From the Greek *leucon,* white, and *lepis,* scale, in reference to the white scruffy appearance.] JUSTICEWEED.

Eupatorium glaucescens Elliott var. *leucolepis* de Candolle, Prodr. 5: 177. 1836. *Eupatorium leucolepis* (de Candolle) Torrey & a. Gray, Fl. N. Amer. 2: 84. 1841. *Uncasia leucolepis* (de Candolle) Greene, Leafl. Bot. Observ. Crit. 1: 13. 1903.

Erect perennial herb, to 1 m; stem sometimes short-rhizomatous, striate-angled, puberulent. Leaves opposite, the distal sometimes alternate, sometimes appearing in verticels due to the development of leaves on the lateral buds without axis elongation, the blade lanceolate- or linear-oblong, 2–6 cm long, 4–10 mm wide, pinnate-veined, strongly folded along the midrib, curved, the apex acute, the base rounded or cuneate, the margin entire or serrate, the upper surface scabrous, gland-dotted, the lower surface villous, gland-dotted, sessile. Flowers in capitula in a corymbiform array; involucre obconic; phyllaries numerous in 2–3 series, elliptic, 3–8 mm long, the apex acuminate or attenuate, mucronate, white, the outer surface puberulent, gland-dotted; flowers 5; corolla white, 3–4 mm long. Cypsela prismatic, 5-ribbed, 2–3 mm long, glabrous, gland-dotted; pappus of numerous white barbellate bristles 4–5 mm long.

Wet flatwoods and bogs. Occasional; northern peninsula, central and western panhandle. New York south to Florida, west to Texas. Summer–fall.

Eupatorium linearifolium Walter [With linear leaves.] WAXY THOROUGHWORT.

Eupatorium linearifolium Walter, Fl. Carol. 199. 1788. *Eupatorium hyssopifolium* Linnaeus var. *linearifolium* (Walter) Fernald, Rhodora 44: 460. 1942. *Uncasia linearifolia* (Walter) Greene, Leafl. Bot. Observ. Crit. 1: 13. 1903.

Eupatorium glaucescens Elliott, Sketch Bot. S. Carolina 2: 297. 1823.

Eupatorium tortifolium Chapman, Bot. Gaz. 3: 5. 1878. *Eupatorium hyssopifolium* Linnaeus var. *tortifolium* (Chapman) A. Gray, Syn. Fl. N. Amer. 1(2): 98. 1884. *Uncasia tortifolia* (Chapman) Greene, Leafl. Bot. observ. Crit. 1: 13. 1903.

Erect perennial herb, to 1 m; stem striate-angled, pubescent. Leaves opposite, the distal sometimes alternate, the blade oblong or lanceolate-oblong, 2–4.5 cm long, 5–10 mm wide, 3-veined from the base, the apex acute, the base cuneate, the margin entire or serrate, the teeth mostly proximal, the upper and lower surfaces finely puberulent, gland-dotted, sessile or subsessile. Flowers in capitula in a corymbiform array; involucre obconic; phyllaries numerous in 1–2 series, lanceolate, 2–5 mm long, the apex rounded or acute, the outer surface puberulent, gland-dotted; flowers 5; corolla white, 3–4 mm long. Cypsela prismatic, 5-ribbed, 2–3 mm long, glabrous, gland-dotted; pappus of numerous white barbellate bristles 3–5 mm long.

Sandhills and flatwoods. Occasional; northern and central peninsula, west to central panhandle. Virginia south to Florida, west to Texas. Summer–fall.

Eupatorium mikanioides Chapm. [Resembling *Mikania* (Asteraceae).] SEMAPHORE THOROUGHWORT.

Eupatorium mikanioides Chapman, Fl. South. U.S. 195. 1860. *Uncasia mikanioides* (Chapman) Greene, Leafl. Bot. Observ. Crit. 1: 13. 1903. TYPE: FLORIDA: Franklin Co.: St. Vincent Island.

Erect perennial herb, to 1 m; stem short-rhizomatous, striate-angled, puberulent. Leaves opposite, the blade deltate or rhombic, 2.5–8 cm long, 2–6 cm wide, 3-veined above the base, succulent, the apex acute, the base broadly cuneate, the margin irregularly serrate or crenate, the upper and lower surfaces sparsely puberulent or glabrate, gland-dotted, the petiole 1–3 cm long. Flowers in capitula in a corymbiform array; involucre obconic; phyllaries numerous in 2–3 series, elliptic or oblong, 2–5 mm long, the apex acute or acuminate, the outer surface puberulent, gland-dotted; flowers 5; corolla white, 3–4 mm long. Cypsela prismatic, 5-ribbed, 1–2 mm long, glabrous, gland-dotted; pappus of numerous white barbellate bristles 4–5 mm long.

Coastal flatwoods, salt marshes, and glades. Frequent; central and southern peninsula, eastern and central panhandle. Endemic. Summer–fall.

Eupatorium mohrii Greene [Commemorates Charles Theodor Mohr (1824–1901), German-born pharmacist and botanist who lived and worked in the United States.] MOHR'S THOROUGHWORT.

Eupatorium mohrii Greene, in C. Mohr, U.S. Natl. Herb. 6: 762, pl. 11. 1901. *Uncasia mohrii* (Greene) Greene, Leafl. Bot. Observ. Crit. 1: 13. 1903.
Eupatorium recurvans Small, Fl. S.E. U.S. 1167, 1338. 1903. TYPE: FLORIDA: Duval Co.: Jul, *Curtiss* [miscited as Nash] *1196* (holotype: NY).

Erect, perennial herb, to 1 m; stem tuberous-rhizomatous, striate-angled, puberulent. Leaves opposite, the distal sometimes alternate, the blade oblanceolate, 2–8 cm long, 0.5–1(2) cm wide, 3-veined above the base, the apex acute, the base cuneate, the margin serrate proximally, entire distally, the upper surface glabrous or glabrate, the lower surface puberulent, sessile or subsessile. Flowers in capitula in a corymbiform array; involucre obconic; phyllaries numerous in 1–2 series, oblanceolate, 1–3 mm long, the apex rounded, the outer surface puberulent, gland-dotted; flowers 5; corolla white, 2–4 mm long. Cypsela prismatic, 5-ribbed, 1–2 mm long, glabrous, gland-dotted; pappus of numerous white barbellate bristles ca. 3 mm long.

Flatwoods. Common; nearly throughout. Virginia south to Florida, west to Texas. Summer–fall.

Eupatorium perfoliatum L. [With the leaf base surrounding the stem so that the stem appears to pass through it.] COMMON BONESET.

Eupatorium perfoliatum Linnaeus, Sp. Pl. 838. 1753. *Eupatorium connatum* Michaux, Fl. Bor.-Amer. 2: 99. 1803, nom. illegit. *Uncasia perfoliata* (Linnaeus) Greene, Leafl. Bot. Observ. Crit. 1: 13. 1903. *Cunigunda perfoliata* (Linnaeus) Lunell, Amer. Midl. Naturalist 5: 35. 1917.

Eupatorium chapmanii Small, Fl. S. E. U.S. 1168, 1338. 1903. TYPE: FLORIDA: s.d., *Chapman* s.n. (holotype: NY).

Erect perennial herb, to 1 m; stem striate-angled, puberulent. Leaves opposite, the blade lanceolate, 5–15 cm long, 1.5–4 cm wide, the apex acute, the base connate-perfoliate, the margin serrate, the upper surface glabrate, the lower surface pilose, gland-dotted, sessile. Flowers in capitula in a corymbiform array; involucre obconic; phyllaries numerous in 1–2 series, oblong, 2–5 mm long, the apex acute or acuminate, the outer surface puberulent, gland-dotted; flowers 5; corolla white, ca. 3 mm long. Cypsela prismatic, 5-ribbed, ca. 2 mm long, glabrous, gland-dotted; pappus of numerous white barbellate bristles 3–4 mm long.

Marshes and wet hammocks. Occasional; northern counties, central peninsula. Quebec south to Florida, west to Manitoba, North Dakota, South Dakota, Nebraska, Kansas, Oklahoma, and Texas. Summer–fall.

Eupatorium petaloideum Britton ex Small [Petal-like, in reference to the phyllaries.] SNOWY WHITE THOROUGHWORT.

Eupatorium petaloideum Britton ex Small, Bull. Torrey Bot. Club 24: 492. 1897. *Uncasia petaloidea* (Britton ex Small) Greene, Leafl. Bot. Observ. Crit. 1: 13. 1903. *Eupatorium album* Linnaeus var. *petaloideum* (Britton ex Small) R. K. Godfrey ex D. B. Ward, Novon 14: 367. 2004. TYPE: FLORIDA: Duval Co.: s.n., *Curtiss 1190* (lectotype: NY). Lectotypified by Ward (2004: 367).

Erect perennial herb, to 7 dm; stem short-rhizomatous, striate-angled, puberulent. Leaves opposite, the distal sometimes alternate, the blade elliptic or ovate, 2–8 cm long, 1–3 cm wide, the apex rounded or acute, the base cuneate, the margin serrate or crenate-serrate, the upper surface glabrate, the lower surface villous, sessile. Flowers in capitula in a corymbiform array; involucre obconic; phyllaries numerous in 2–4 series, linear, 4–10 mm long, the apex acuminate or attenuate, strongly mucronate, white, the outer surface glabrous; flowers 5; corolla white, 3–4 mm long. Cypsela prismatic, 5-ribbed, 2–3 mm long, glabrous, gland-dotted; pappus of numerous white barbellate bristles 3–4 mm long.

Scrub oak hammocks and longleaf pinelands. Rare; northern peninsula, eastern and central panhandle. Georgia, Alabama, Florida, and Mississippi. Summer–fall.

Eupatorium pilosum Walter [With long, shaggy trichomes.] ROUGH BONESET.

Eupatorium pilosum Walter, Fl. Carol. 199. 1788.
Eupatorium verbenifolium Michaux var. *saundersii* Porter ex Britton, Man. Fl. N. States 923. 1901. *Eupatorium rotundifolium* Linnaeus var. *saundersii* (Porter ex Britton) Cronquist, Rhodora 50: 29. 1948.

Erect perennial herb, to 1 m; stem short-rhizomatous, striate-angled, puberulent or pilose. Leaves opposite, the distal sometimes alternate, the blade elliptic, lanceolate, or lanceolate-ovate, 3–9 cm long, 2–4.5 cm wide, 3-veined above the base, the apex acute or attenuate, white or green, the base rounded-cuneate, the margin irregularly serrate, the upper and lower surfaces puberulent, gland-dotted, sessile or subsessile. Flowers in capitula in a corymbiform array; involucre obconic; phyllaries numerous in 2–3 series, oblong or lanceolate-oblong, 2–7 mm long, the apex acute or acuminate, the outer surface puberulent, gland-dotted; flowers 5; corolla white, 3–4 mm long. Cypsela prismatic, 5-ribbed, 3–4 mm long, glabrous, gland-dotted; pappus of numerous white barbellate bristles 3–4 mm long.

Wet flatwoods and bogs. Occasional; northern counties, central peninsula. New York, Massachusetts, and Quebec south to Florida, west to Louisiana. Summer–fall.

Eupatorium rotundifolium L. [With the leaf blade rounded.] ROUNDLEAF THOROUGHWORT; FALSE HOARHOUND.

Eupatorium rotundifolium Linnaeus, Sp. Pl. 837. 1753. *Uncasia rotundifolia* (Linnaeus) Greene, Leafl. Bot. Observ. Crit. 1: 13. 1903. *Eupatorium rotundifolium* Linnaeus var. *typicum* Fernald & Griscom, Rhodora 37: 181. 1935, nom. inadmiss.
Eupatorium marrubium Walter, Fl. Carol. 199. 1788. *Eupatorium cuneifolium* Willdenow, Sp. Pl. 3: 1753. 1803, nom. illegit. *Uncasia cuneifolia* Greene, Leafl. Bot. Observ. Crit. 1: 13. 1903.
Eupatorium pubescens Muhlenberg ex Willdenow, Sp. Pl. 3: 1755. 1803. *Eupatorium rotundifolium* Linnaeus var. *pubescens* (Muhlenberg ex Willdenow) Britton et al., Prelim. Cat. 25. 1888. *Uncasia pubescens* (Muhlenberg ex Willdenow) Greene, Leafl. Bot. Observ. Crit. 1: 13. 1903.
Eupatorium teucrifolium Willdenow, Sp. Pl. 3: 1753. 1803.
Eupatorium verbenifolium Michaux, Fl. Bor.-Amer. 2: 98. 1803. *Eupatorium teucrifolium* Willdenow var. *verbenifolium* (Michaux) de Candolle, Prodr. 5: 178. 1836. *Uncasia verbenifolia* (Michaux) Greene, Leaf. Bot. Observ. Crit. 1: 13. 1903.
Eupatorium scabridum Elliott, Sketch Bot. S. Carolina 2: 299. 1823. *Eupatorium rotundifolium* Linnaeus var. *scabridum* (Elliott) A. Gray, Syn. Fl. N. Amer. 1(2): 99. 1884. *Uncasia scabrida* (Elliott) Greene, Leafl. Bot. Observ. Crit. 1: 13. 1903.
Eupatorium ovatum Bigelow, Fl. Boston, ed. 2. 296. 1824. *Eupatorium rotundifolium* Linnaeus var. *ovatum* (Bigelow) Torrey ex de Candolle, Prodr. 5: 178. 1836. *Eupatorium rotundifolium* Linnaeus subsp. *ovatum* (Bigelow) Montgomery & Fairbrothers, Brittonia 23: 149. 1970.

Erect perennial herb, to 1 m; stem short-rhizomatous, striate-angled, puberulent. Leaves opposite, sometimes alternate distally, the blade deltate, suborbicular, or ovate, 1.5–5(7) cm long, 1.5–4(6) cm wide, 3-veined at the base, the apex acute, the base rounded or truncate, the margin crenate or serrate, the upper and lower surfaces puberulent, gland-dotted, sessile or subsessile. Flowers in capitula in a corymbiform array; involucre obconic; phyllaries numerous in 2–3 series, elliptic or oblanceolate, 2–6 mm long, the apex acute, the outer surface puberulent, gland-dotted; flowers 5; corolla white, 3–4 mm long. Cypsela prismatic, 5-ribbed, 2–3 mm long, glabrous, gland-dotted; pappus of numerous white barbellate bristles 3–4 mm long.

Wet flatwoods and bogs. Frequent; nearly throughout. Maine south to Florida, west to Missouri, Oklahoma, and Texas. Summer–fall.

Eupatorium semiserratum DC. [*Semi*, half, *serratus*, saw-edged, in reference to the irregularly toothed leaf margin.] SMALLFLOWER THOROUGHWORT.

Eupatorium parviflorum Elliott, Sketch Bot. S. Carolina 2: 299. 1823; non Aublet, 1775; nec Swartz, 1788.

Eupatorium semiserratum de Candolle, Prodr. 5: 177. 1836. *Uncasia semiserrata* (de Candolle) Greene, Leaf. Bot. Observ. Crit. 1: 13. 1903. *Eupatorium cuneifolium* Willdenow var. *semiserratum* (de Candolle) Fernald & Griscom, Rhodora 37: 179. 1935.

Erect perennial herb, to 1 m; stem short short-rhizomatous, striate-angled, puberulent. Leaves opposite, sometimes alternate distally, the blade elliptic or elliptic-lanceolate, (3)5–7 cm long, 0.5–2.5 cm wide, 3-veined above the base, the apex acute, the base narrowly cuneate, the margin serrate, the upper and lower surfaces puberulent or villous, gland-dotted, sessile or subsessile. Flowers in capitula in a corymbiform array; involucre obconic; phyllaries numerous in 2–3 series, elliptic, 1–3 mm long, the apex rounded or acute, the outer surface puberulent, gland-dotted; flowers 5; corolla white, ca. 3 mm long. Cypsela prismatic, 5-ribbed, ca. 2 mm long, glabrous, gland-dotted; pappus of numerous white barbellate bristles ca. 3 mm long.

Wet flatwoods, bogs, and floodplain forests. Occasional; northern counties. North Carolina south to Florida, west to Tennessee, Arkansas, and Louisiana. Summer–fall.

Eupatorium serotinum Michx. [Late-coming, in reference to the plant flowering late in the season.] LATEFLOWERING THOROUGHWORT.

Eupatorium serotinum Michaux, Fl. Bor.-Amer. 2: 100. 1803. *Uncasia serotina* (Michaux) Greene, Leafl. Bot. Observ. Crit. 1: 13. 1903.

Erect perennial herb, to 1.5 m; stem striate-angled, pubescent. Leaves opposite, the blade lanceolate, 2–9 cm long, 0.5–4 cm wide, 3-veined, the apex acute, the base rounded, the margin entire or serrate, the upper and lower surfaces puberulent, gland-dotted, the petiole 1–2.5 cm long. Flowers in capitula in a corymbiform array; involucre obconic; phyllaries numerous in 1–2 series, elliptic or oblong, 1–3 mm long, the apex obtuse or acute, the outer surface puberulent, gland-dotted; flowers 5; corolla white, ca. 3 mm long. Cypsela prismatic, 5-ribbed, 1–2 mm long, glabrous, gland-dotted; pappus of numerous white barbellate bristles 2–3 mm long.

Wet hammocks, tidal marshes, and pond margins. Frequent; nearly throughout. New York, Massachusetts, and Ontario south to Florida, west to Minnesota, Iowa, Nebraska, Kansas, Oklahoma, and Texas; Mexico. Summer–fall.

HYBRIDS

Eupatorium ×anomalum Nash (*E. mohri* × *E. rotundifolium*) [Abnormal.] FLORIDA THOROUGHWORT.

Eupatorium anomalum Nash, Bull. Torrey Bot. Club 23: 106. 1896, pro sp. *Uncasia anomala* (Nash) Greene, Leafl. Bot. Observ. Crit. 1: 13. 1903. TYPE: FLORIDA: Jefferson Co.: Lloyds, 1 Sep 1895, *Nash 2515* (holotype: NY).

Wet flatwoods. Occasional; northern counties, central peninsula. Summer.

Eupatorium ×anomalum has been proposed as a hybrid of *E. mohrii* and *E. rotundifolium*. However, molecular data suggest that it is a hybrid between *E. mohrii* and *E. serotinum* (Siripum and Schilling, 2006).

Eupatorium anomalum is sometimes treated as a species by authors (e.g., Schilling and Grubbs, 2016; Siripum and Schilling, 2006).

Eupatorium ×pinnatifidum Elliott (*E. capillifolium* or *E. compositifolium* × *E. perfoliatum*) [Pinnately divided, in reference to the leaves.]

Eupatorium pinnatifidum Elliott, Sketch Bot. S. Carolina 2: 295. 1823, pro sp. *Traganthes pinnatifida* (Elliott) Greene, Leafl. Bot. Observ. Crit. 1: 13. 1903.

Eupatorium smithii Greene & C. Mohr, in C. Mohr, Contr. U.S. Natl. Herb. 6: 761, pl. 10. 1901; non B. L. Robinson, 1900. *Eupatorium eugenei* Small, Fl. S.E. U.S. 1165, 1338. 1903. *Traganthes eugenei* (Small) Greene, Leafl. Bot. Observ. Crit. 1: 13. 1903.

Eupatorium pectinatum Small, Fl. S.E. U.S. 1165, 1338. 1903; non Rafinesque, 1832. *Traganthes pectinata* Greene, Leafl. Bot. Observ. Crit. 1: 13. 1903. TYPE: FLORIDA: Leon Co.: 1836, *Chapman s.n.* (holotype: NY).

Marshes, wet hammocks, and swamps. Occasional; northern counties. Fall.

Molecular data show that *E. serotinum* rather than *E. perfoliatum* is sometimes the second parent (Siripum and Schilling, 2006).

Eurybia (Cass.) Cass. 1820. GRASSLEAF ASTER

Herbs. Leaves alternate, simple, pinnate-veined, petiolate or epetiolate. Flowers in capitula in corymbiform, racemiform, or spiciform arrays, radiate; receptacle epaleate; ray flowers carpellate; petals 5, basally connate, zygomorphic; disk flowers bisexual; petals 5, basally connate, actinomorphic; stamens 5, epipetalous, the filaments free, the anthers connate; ovary inferior, 2-carpellate, 1-loculate. Fruit a cypsela; pappus present.

A genus of about 22 species; North America, Europe, and Asia. [From the Greek *eurys,* wide and *baios,* few, in reference to the few ray flowers.]

Selected reference: Brouillet (2006a).

1. Leaves conspicuously spinulose-dentate; phyllaries spine-tipped; stem villous; ray flowers white or pinkish .... **E. eryngiifolia**
1. Leaves entire or only slightly spinulose-dentate; phyllaries acute or acuminate; stem glabrous or glabrescent proximally, villous or strigilose distally; ray flowers lavender or purple.

2. Capitula in an open corymbiform array (rarely solitary) .... **E. paludosa**
2. Capitula in a racemiform or spiciform array.
   3. Capitula relatively few-flowered (ray flowers usually 8–15(17); phyllary apices not spreading..... .... **E. spinulosa**
   3. Capitula many-flowered (ray flowers usually 15–30); phyllary apices usually squarrose or reflexed.. **E. hemispherica**

Eurybia eryngiifolia (Torr. & A. Gray) G. L. Nesom [With leaves resembling *Eryngium* (Apiaceae).] THISTLELEAF ASTER.

Prionopsis chapmanii Torrey & A. Gray, Fl. N. Amer. 2: 245. 1842. *Heleastrum chapmanii* (Torrey & A. Gray), Greene, Pittonia 3: 49. 1896. TYPE: FLORIDA: s.d., *Chapman s.n.* (holotype: NY?)
Aster eryngiifolius Torrey & A. Gray, Fl. N. Amer. 2: 502. 1843. *Eurybia eryngiifolia* (Torrey & A. Gray) G. L. Nesom, Phytologia 77: 259. 1995 ("1994"). TYPE: FLORIDA: s.d., *Chapman s.n.* (holotype: NY?).

Erect perennial herb, to 7 dm; stem with short rhizomes, striate-angled, villous. Basal leaves with the blade linear, 7–35 mm long, 3–8 mm wide, coriaceous, finely parallel-veined, the apex acute or acuminate, revolute, the base narrowly cuneate, the margin entire or remotely spinose-serrate, remotely and obscurely scabridulous, indurate, the upper and lower surfaces glabrous, the petiole barely distinct, the cauline with the blade linear or linear-lanceolate, 1.5–13 cm long, 2–7 mm wide, reduced upward, the margin more spiny distally, the upper surface becoming progressively villous distally, the lower surface glabrous, sessile. Flowers in capitula, solitary or 2–11 in a racemiform array, the peduncle villous, sometimes sparsely stipitate-glandular distally; bracts absent or 1–7, lanceolate, appressed, the margin spinulose, the upper surface villous, the lower surface glabrous; involucre campanulate-hemispheric; phyllaries numerous in 4–5 series, lanceolate or linear, 9–12 mm long, unequal, coriaceous, the base of the outer indurate and rounded, the margin slightly scarious or erose, sometimes villous-ciliate, stipitate-glandular, the apex long acuminate, squarrose, often purple-tinged, the outer surface villosulous, sometimes sparsely stipitate-glandular; ray flowers numerous; corolla white or pinkish, 1–2 cm long; disk flowers numerous; corolla yellow, 6–7 mm long; cypsela elliptic or obovoid, with numerous ribs, 2–4 mm long, glabrous; pappus of numerous tawny barbellate bristles ca. 8 mm long.

Wet flatwoods and bogs. Occasional; central and western panhandle. Georgia, Alabama, and Florida. Spring–summer.

Eurybia hemispherica (Alexander) G. L. Nesom [In reference to the involucre.] SOUTHERN PINE ASTER.

Aster hemisphericus Alexander, in Small, Man. S. E. U.S. Fl. 1391, 1509. 1933. *Aster paludosus* Aiton subsp. *hemisphericus* (Alexander) Cronquist, Bull. Torrey Bot. Club 74: 145. 1947. *Heleastrum hemisphericum* (Alexander) Shinners, Field & Lab. 17: 170. 1949. *Aster paludosus* Aiton var. *hemisphericus* (Alexander) Waterfall, Rhodora 62: 320. 1960. *Eurybia hemispherica* (Alexander) G. L. Nesom, Phytologia 77: 260. 1995 ("1994").
Aster pediomonus Alexander, in Small, Man. S.E. Fl. 1391, 1509. 1933.

Erect or ascending perennial herb, to 1 m; stem with creeping, tangled, scaly, often woody rhizomes, striate-angled, glabrous or glabrescent, often reddish. Basal leaves (withering by flowering) with the blade ovate-lanceolate, linear-lanceolate, or linear, 5–17 cm long, 3–12 mm wide, coriaceous, the midvein sometimes with 1–2 parallel pairs of secondary veins, the apex acute or obtuse, the base narrowly cuneate, the margin entire or sometimes remotely spinulose-toothed, scabrous, often revolute apically, the upper surface scabridulous, the lower surface glabrous or glabrate, the petiole winged, the base sheathing, the cauline reduced distally, 2.5–12 cm long, 2–8 mm wide, sessile or subsessile. Flowers in capitula, solitary or numerous in a racemiform or corymbiform array, the peduncle to 8 cm long; bracts absent or 3–4, linear-lanceolate, 1–3.5 cm long, grading into the phyllaries, the apex mucronulate; involucre hemispheric; phyllaries numerous in 4–6 series, 8–12(15) mm long, unequal, the outer oblong-lanceolate or lanceolate, coriaceous, the base rounded, the inner oblong or linear-lanceolate, membranaceous, the apex acute or acuminate, often strongly reflexed, the outer surface glabrous; ray flowers numerous, carpellate, corolla violet-purple or rarely white, the lamina (1)1.5(2) cm long; disk flowers numerous; corolla yellow, ca. 6 mm long. Cypsela cylindric-obovoid or fusiform, 3–4 mm long, with numerous ribs, strigillose; pappus of numerous tawny barbellate bristles 6–7 mm long.

Longleaf pine savannas. Rare; Washington and Escambia Counties. Kentucky and Tennessee south to Florida, west to Nebraska, Oklahoma, and Texas. Summer–fall.

Eurybia hemispherica is listed as endangered in Florida (Florida Administrative Code, Chapter 5B-40).

Eurybia paludosa (Aiton) G. L. Nesom [Growing in marshy places.] SOUTHERN SWAMP ASTER.

Aster paludosus Aiton, Hort. Kew. 3: 201. 1789. *Eurybia paludosa* (Aiton) G. L. Nesom, Phytologia 77: 261. 1995 ("1994").

Erect perennial herb, to 8 dm; stem with scaly rhizomes, striate-angled, glabrous or glabrescent, often reddish. Leaves with the blade elliptic, lanceolate, linear-lanceolate, 1.5–10 cm long, 5–9 mm wide, the apex acute or obtuse, mucronate, the base cuneate, the margin entire or remotely spinulose-serrate, scabrous, the upper surface sparsely hirtellous, the lower surface scabrous or glabrous, the petiole winged, the basal and proximalmost absent or withering at flowering, the mid and distal cauline reduced upward, sessile or subsessile. Flowers in capitula in a corymbiform array, rarely solitary, the peduncle 1–10 cm long, pubescent, villous, or hirsute; involucre campanulate, phyllaries numerous in 4–5 series, 9–11 mm long, unequal, linear or oblanceolate, the apex obtuse or acute, mucronate, spreading or squarrose; ray flowers numerous; corolla lavender or purple, the lamina (1)1.5–2 cm long; disk flowers numerous; corolla yellow, 1–2 mm long. Cypsela cylindric or obovoid, 3–4 mm long, with numerous ribs, strigillose; pappus of numerous tawny barbellate bristles 5–7 mm long.

Seepage slope bogs. Rare; Nassau County. North Carolina south to Florida. Summer–fall.

Eurybia spinulosa (Chapman) G. L. Nesom [In reference to the spine-tipped leaves, phyllaries, and bracts.] APALACHICOLA ASTER; PINEWOODS ASTER.

Aster spinulosus Chapman, Fl. South. U.S. 199. 1860. *Heleastrum spinulosum* (Chapman) Greene, Pittonia 3: 50. 1896. *Eurybia spinulosa* (Chapman) G. L. Nesom, Phytologia 77: 262. 1995 ("1994"). TYPE: FLORIDA.

Erect perennial herb, to 7 dm; stem with rhizomes, striate-angled, glabrous or glabrescent. Leaves with the blade linear, 7–35 cm long, 3–8 mm wide, the apex acute, the base cuneate, the margin entire or spinose-serrate, the upper and lower surfaces glabrous, petiolate, the cauline progressively reduced upward and becoming more spinose, the upper surface becoming villous, sessile. Flowers 3–16 in a spiciform or racemiform array, the peduncle absent or to 1.5 cm long; bracts absent or 1–2, lanceolate, glabrous; involucre campanulate; phyllaries numerous in 4–5 series, lanceolate, 5–10 mm long, unequal, the apex acute, apiculate, the outer surface glabrous or sparsely villous; ray flowers numerous; corolla pale purple, the lamina 1–1.5(2) cm long; disk flowers numerous; corolla yellow, 5–7 mm long. Cypsela fusiform, 2–3 mm long, with numerous ribs, strigillose; pappus of numerous tawny barbellate bristles 6–7 mm long.

Pine flatwoods. Rare; central panhandle. Endemic. Summer–fall.

Eurybia spinulosa is listed as endangered in Florida (Florida Admnistrative Code, Chapter 5B-40).

EXCLUDED TAXON

Eurybia spectabilis (Aiton) G. L. Nesom—This northern species was reported for Florida by Chapman (1860, as *Aster spectabilis* Aiton. No Florida material seen.

Euryops (Cass.) Cass. 1820.

Shrubs. Leaves alternate, simple, pinnate-veined, petiolate. Flowers in solitary capitula, radiate; receptacle epaleate; ray flowers carpellate; petals 5, basally connate, zygomorphic; disk flowers bisexual; petals 5, basally connate, actinomorphic; stamens 5, epipetalous, the filaments free, the anthers connate; ovary inferior, 2-carpellate, 1-loculate. Fruit a cypsela; pappus absent.

A genus of about 100 species; North America, Africa, Asia, and Australia. [From the Greek *eurys*, wide, and *opsis*, eye, apparently in reference to the large capitula.]

Euryops chrysanthemoides (DC.) B. Nord. [Resembling the genus *Chrysanthemum* (Asteraceae).] AFRICAN BUSHDAISY.

Gamolepis chrysanthemoides de Candolle, Prodr. 6: 40. 1838. *Euryops chrysanthemoides* (de Candolle) B. Nordenstam, Opera Bot. 20: 365. 1968.

Erect shrub, to 2 m; stem striate-angled, glabrous. Leaves with the blade elliptic-obovate, 3–8 cm long, 1.5–2.5 cm wide, reduced upward, deeply pinnatifid, the ultimate lobes linear-lanceolate, the apex acute, apiculate, the margin entire, the upper and lower surfaces glabrous, the petiole 1–2 cm long, narrowly winged to the base. Flowers in a solitary terminal capitulum, the peduncle 10–15 cm long; phyllaries numerous in 2 series, basally connate, 5–6 mm long, ovate or suborbicular, the apex acuminate; ray flowers numerous; corolla yellow, the lamina 1–1.5 cm long; disk flowers numerous; corolla yellow, 5–6 mm long. Cypsela narrowly obovoid, ca. 5 mm long, glabrous; pappus absent.

Disturbed sites. Rare; Miami-Dade County. Escaped from cultivation. Florida; Africa and Australia. Native to Africa. All year.

Euthamia (Nutt.) Cass. 1825. GOLDENTOP.

Herbs. Leaves alternate, simple, pinnate-veined, epetiolate. Flowers in capitula in corymbiform or paniculiform arrays, radiate; receptacle epaleate; ray flowers carpellate; petals 5, basally connate, zygomorphic; disk flowers bisexual; petals 5, basally connate, actinomorphic; stamens 5, epipetalous, the filaments free, the anthers connate; ovary inferior, 2-carpellate, 1-loculate. Fruit a cypsela; pappus present.

A genus of 5 species; North America, Mexico, Europe, and Asia. [From the Greek *eu*, good, well-developed, and *thama*, crowded, in reference to branching habit.]

Selected references: Haines (2006); Sieren (1981).

1. Leaves usually less than 3 mm wide, 1- to 3-veined, lax, and deflexed, the axillary fascicles usually numerous and well developed .... **E. caroliniana**
1. Leaves usually more than 3 mm wide, 3- to 5-veined, firm, divergent or ascending, the axillary fascicles absent or few .... **E. graminifolia**

Euthamia caroliniana (L.) Greene ex Porter & Britton [Of Carolina.] SLENDER FLATTOP GOLDENROD.

Erigeron carolinianus Linnaeus, Sp. Pl. 863. 1753. *Solidago caroliniana* (Linnaeus) Britton et al., Prelim. Cat. 26. 1888. *Euthamia caroliniana* (Linnaeus) Greene ex Porter & Britton, Mem. Torrey Bot. Club 5: 321. 1894.

Solidago lanceolata Linnaeus var. *minor* Michaux, Fl. Bor.-Amer. 2: 116. 1803. *Euthamia minor* (Michaux) Greene, Pittonia 5: 78. 1902. *Solidago minor* (Michaux) Fernald, Rhodora 10: 93. 1908; non Miller, 1768. *Solidago michauxii* House, Bull. New York State Mus. Nat. Hist. 254: 695. 1924. *Solidago tenuifolia* Pursh, Fl. Amer. Sept. 540. 1814. *Euthamia tenuifolia* (Pursh) Nuttall, Gen. N. Amer. Pl. 2: 162. 1818.

Euthamia microcephala Greene, Pittonia 5: 79. 1902. *Solidago microcephala* (Greene) Bush, Amer. Midl. Naturalist 5: 176. 1918.

Euthamia microphylla Greene, Pittonia 5: 79. 1902. *Solidago microphylla* (Greene) Bush, Amer. Midl. Naturalist 5: 177. 1918.

Euthamia tenuifolia Pursh var. *glutinosa* Nuttall, Trans. Amer. Philos. Soc., ser. 2. 7: 326. 1840. TYPE: FLORIDA.

Erect perennial herb, to 1 m; stem striate-angled, glabrate or glabrous. Leaves linear, 2.5–7 cm long, 1–3 mm wide, 1- to 3-veined, lax and deflexed, the apex obtuse or acuminate, the base cuneate, the upper and lower surfaces glabrous or glabrate, gland-dotted, sessile, the axillary fascicles usually numerous and well developed. Flowers in capitula in a corymbiform or paniculiform array, short-pedunculate or sessile and glomerulate; involucre campanulate or turbinate; phyllaries 3–5 mm long, the outer ovate, the inner linear-oblong, the apex acute or obtuse, the outer surface glabrous; ray flowers few to many; corolla yellow, the lamina 3–5 mm long; disk flowers few to many; corolla yellow, 3–5 mm long. Cypsela oblong to narrowly elliptic, 2- to 4-nerved, strigose; pappus of 1 series of numerous, white, subequal, antrorsely barbellate bristles.

Sandhills, flatwoods, and disturbed sites. Common; nearly throughout. Nova Scotia south to Florida, west to Michigan, Illinois, Mississippi, and Louisiana. Summer–fall.

Euthamia graminifolia (L.) Nutt. [*Gramini*, grass, in reference to the grasslike leaves.] FLATTOP GOLDENROD.

Chrysocoma graminifolia Linnaeus, Sp. Pl. 841. 1753. *Solidago graminifolia* (Linnaeus) Salisbury, Prodr. Stirp. Chap. Allerton 199. 1796. *Euthamia graminifolia* (Linnaeus) Nuttall, Trans. Amer. Philos. Soc., ser. 3. 7: 325. 1840. *Aster graminifolius* Kuntze, Revis. Gen. Pl. 1: 316. 1891. *Solidago graminifolia* (Linnaeus) Salisbury var. *typica* Rosendahl & Cronquist, Amer. Midl. Naturalist 33: 253. 1945, nom. inadmiss.

Euthamia fastigiata Bush, Amer. Midl. Naturalist 5: 164. 1918. *Solidago fastigiata* (Bush) Bush, Amer. Midl. Natualist 5: 164. 1918, nom. alt.

Solidago hirtipes Fernald, Rhodora 48: 65, pl. 1011. 1946, pro. Hybr. *Euthamia hirtipes* (Fernald) Sieren, Phytologia 23: 304. 1972. *Euthamia graminifolia* (Linnaeus) Nuttall var. *hirtipes* (Fernald) C.E.S. Taylor & R. J. Taylor, Sida 10: 176. 1983.

Erect perennial herb, to 1.5 m; stem striate-angled, glabrate or glabrous. Leaves linear or lanceolate, 3.7–13 cm long, 3–12 mm wide, 3- to 5-veined, firm, divergent or ascending, the apex obtuse or acuminate, the base cuneate, the upper and lower surfaces glabrous or spreading hirtellous, few and obscurely gland-dotted, sessile, the axillary fascicles absent or few. Flowers in capitula in a corymbiform or paniculiform array, short-pedunculate or sessile and glomerulate; involucre campanulate or turbinate; phyllaries 3–5 mm long, the outer ovate, the inner oblong, the apex acute or obtuse, the outer surface glabrous; ray flowers numerous; corolla yellow, the lamina 3–5 mm long; disk flowers few to many; corolla yellow, 3–4 mm long. Cypsela oblong to narrowly elliptic, 2- to 4-nerved, strigose; pappus of 1 series of numerous, white, subequal, antrosely barbellate bristles.

Wet flatwoods and cypress pond margins. Occasional; northern counties, central peninsula. Nearly throughout North America except for the southwestern states; Europe and Asia. Native to North America. Summer–fall.

EXCLUDED TAXA

Euthamia gymnospermoides Greene—Reported for Florida by Haines (2006). No Florida material seen.

Euthamia leptocephala (Torrey & A. Gray) Greene—Reported for Florida by Correll and Johnston (1970), Cronquist (1980), Wunderlin (1982), and Haines (2006), based on a misidentification of Florida material of *E. graminifolia*.

Eutrochium Raf. 1836 ("1838"). JOE PYE WEED

Herbs. Leaves whorled, simple, pinnate-veined, petiolate. Flowers in capitula in corymbiform arrays, discoid, bisexual; receptacle epaleate; petals 5, basally connate, actinomorphic; stamens 5; epipetalous, the stamens free, the anthers connate; ovary inferior, 2-carpellate, 1-loculate. Fruit a cypsela; pappus present.

A genus of 5 species; North America. [From the Greek *eu-*, truly, and *trochos,* wheel, in reference to the whorled leaves.]

Eupatoriadelphis R. M. King, 1970.

Selected reference: Lamont (2006a).

1. Stems purple at the nodes, solid.......... **E. purpureum**
1. Stems usually purple throughout, hollow.......... **E. fistulosum**

Eutrochium fistulosum (Barratt) E. E. Lamont [Hollow, in reference to the stem.] QUEEN-OF-THE-MEADOW; JOE PYE WEED.

Eupatorium laevigatum Torrey, Cat. Pl. New York 92. 1819; non Lamarck, 1788. *Eupatorium purpureum* Linnaeus var. *angustifolium* Torrey & A. Gray, Fl. N. Amer. 2: 82. 1841. *Eupatorium fistulosum* Barratt, *Eupatoria verticillata* (1). 1841. *Eupatorium purpureum* Linnaeus subsp. *angustifolium* (Torrey & A. Gray) Voss, Vilm. Blumengärtn. ed. 3. 1: 447. 1894. *Eupatoriadelphus fistulosus* (Barratt) R. M. King & H. Robinson, Phytologia 19: 432. 1970. *Eutrochium fistulosum* (Barratt) E. E. Lamont, Sida 21: 902. 2004.

Erect perennial herb, to 3.5 m; stem striate-angled, glabrous or glabrate, glandular-puberulent distally, purple throughout, sometimes greenish or purple-spotted, hollow. Leaves usually 4 or 6 (rarely 7) at a node, the blade narrow or broadly lanceolate, (8)12–25(28) cm long, (1.5)2–6(9) cm wide, the apex acute or acuminate, the base cuneate, the margin serrate-crenate, the upper surface glabrous or sparsely puberulent, the lower surface sparsely scabrellous or glabrate, the petiole (0.5)1–3(5) cm long, glabrous. Flowers in capitula in a corymbiform array, sessile or the peduncle to 8 mm long; involucre cylindric; phyllaries numerous in 5–6 series, lanceolate, 6–9 mm long, usually purplish, glabrous or the outer with a few trichomes on the midvein; flowers (4)5(7); corolla pink or purple, 5–6 mm long. Cypsela prismatic, 5-ribbed, 3–5 mm long, dark brown, gland-dotted; pappus of numerous cream-colored or purplish barbellate bristles in 1 series, 5–6 mm long.

Moist hammocks and flatwoods. Occasional; northern counties, central peninsula. Quebec south to Florida, west to Michigan, Illinois, Missouri, Oklahoma, and Texas. Summer–fall.

Eutrochium purpureum (L.) E. E. Lamont [In reference to the flower-color.] SWEETSCENTED JOE PYE WEED.

Eupatorium purpureum Linnaeus, Sp. Pl. 838. 1753. *Eupatorium purpureum* Linnaeus subsp. *typicum* Voss, Vilm. Blumengärtn., ed. 3. 1: 447. 1894, nom. inadmiss. *Cunigunda purpurea* (Linnaeus) Lunell, Amer. Midl. Naturalist 5: 35. 1917. *Eupatoriadelphis purpureus* (Linnaeus) R. M. King & H. Robinson, Phytologia 19: 432. 1970. *Eutrochium purpureum* (Linnaeus) E. E. Lamont, Sida 21: 902. 2004.

Erect perennial herb, to 2 m; stem striate-angled, glabrous proximally, slightly glandular-puberulent distally, purple at the nodes, solid, rarely hollow near the base. Leaves usually 3 or 4 at a node, the blade lanceolate-ovate, ovate, or deltate-ovate, (7)9–26(30) cm long, (2.5)3–15(18) cm wide, the apex acute or acuminate, the base broadly cuneate, the margin serrate-crenate, the upper surface sparsely puberulent or glabrous, the lower surface glabrous or sparsely to densely fine-pubescent along the major veins, the petiole 0.5–1.5(2) cm long, glabrous. Flowers in capitula in a corymbiform array, the peduncle 2–6 mm long; involucre cylindric; phyllaries numerous in 5–6 series, lanceolate, 6–9 mm long, usually purplish, glabrous or the outer sometimes sparsely puberulent; flowers (4)5–7(8); corolla pink or purple, 5–7 mm long. Cypsela prismatic, 5-ribbed, 3–5 mm long, dark brown, gland dotted; pappus of numerous cream-colored or purplish barbellate bristles in 1 series, 5–6 mm long.

Bluff forests. Occasional; central and western panhandle. Maine and Ontario south to Florida, west to Minnesota, Iowa, Nebraska, Kansas, Oklahoma, and Louisiana. Summer.

EXCLUDED TAXON

Eutrochium maculatum (Linnaeus) E. E. Lamont—Reported for Florida by Small (1933, as *Eupatorium maculatum* Linnaeus), misapplied to Florida material of *Eutrochium fistulosum*.

Facelis Cass. 1819. TRAMPWEED

Herbs. Leaves alternate, simple, pinnate-veined, epetiolate. Flowers in capitula in spicifom arrays, discoid, carpellate or bisexual; receptacle epaleate; petals 5, basally connate, actinomorphic; stamens 5, epipetalous, the filaments free, the anthers connate; ovary inferior, 2-carpellate, 1-loculate. Fruit a cypsela; pappus present.

A genus of 3 species; North America, South America, Africa, and Australia. [Etymology unkwnown, perhaps based on *faecalis* (Latin), in reference to the plant being weedy and of waste places.]

Selected reference: Nesom (2006k).

Facelis retusa (Lam.) Sch. Bip. [In reference to the retuse leaf apex.] ANNUAL TRAMPWEED.

Gnaphalium retusum Lamarck, Encycl. 2: 758. 1788. *Facelis apiculata* Cassini, Bull. Sci. Philom. Paris 1819: 94. 1819, nom. illegit. *Facelis retusa* (Lamarck) Schultz Bipontinus, Linnaea 34: 532. 1866. *Facelis retusa* (Lamarck) Schultz Bipontinus var. *typica* Beauverd, Bull. Soc. Genève, ser. 2. 5: 214. 1913, nom. inadmiss.

Erect, decumbent, or procumbent annual herb, to 3 dm; stem striate-angled, tomentose. Leaves with the blade spatulate, oblanceolate, or linear-lanceolate, 0.7–2(3) cm long, 2–4 mm wide, the apex truncate-apiculate or retuse, the base cuneate, the margin entire, often revolute, the upper surface glabrous, the lower surface gray-tomentose, sessile. Flowers in capitula in a subcapitate or loose spiciform array, rarely solitary in the leaf axil, subtended by bract-like leaves; involucre narrowly ovoid to cylindric; phyllaries in 3–5 series, 8–11 mm long, unequal, chartaceous distally; receptacle flat; peripheral flowers numerous, carpellate; corolla white or purplish; inner flowers 3–5, bisexual; corolla white or purplish. Cypsela obovoid, 2- to 3-ribbed, ca. 2 mm long, sericeous; pappus of numerous basally connate, slightly plumose bristles 10–11 mm long.

Disturbed sites. Occasional; northern counties. Virginia south to Florida, west to Kansas, Oklahoma, and Texas; South America; Africa and Australia. Native to South America. Spring.

Filago L., nom. cons. 1753. COTTONROSE

Herbs. Leaves alternate, simple, pinnate-veined, epetiolate. Flowers in capitula in glomerules in dichasiiform arrays, discoid, carpellate or bisexual; receptacle paleate; petals 4, basally connate, actinomorphic; stamens 5, epipetalous, the filaments free, the anthers connate; ovary inferior, 2-carpellate, 1-loculate. Fruit a cypsela; pappus present or absent.

A genus of about 12 species; North America, Europe, Africa, Asia, and Pacific Islands. [*Filum,* thread, *-ago,* to resemble, in reference to the cotton-like trichomes.]

Gifola Cass. (1819).

Selected reference: Morefield (2006).

Filago germanica (L.) Huds. [Of Germany.] COMMON COTTONROSE.

Gnaphalium germanicum Linnaeus, Sp. Pl. 857. 1753. *Filago germanica* (Linnaeus) Hudson, Fl. Angl. 328. 1762. *Filago vulgaris* Lamarck, Fl. Franç. 2: 61. 1779 ("1778"), nom. illegit. *Filago rotundata* Moench, Methodus 577. 1794, nom. illegit. *Gifola vulgaris* (Lamarck) Cassini, in Cuvier, Dict. Sci. Nat. 18: 531. 1820. *Impia germanica* Bluff & Fingerehuth, Comp. Fl. German. 1(2): 342. 1825, nom. illegit. *Gifola germanica* Dumortier, Fl. Belg. 68. 1827, nom. illegit.

Ascending annual herb, to 15 cm; stem white-woolly. Leaves oblong or lanceolate, 1.5–2.5(3) cm long, 2–3(4) mm wide, reduced upward, the apex acute, the base cuneate, the margin undulate or entire, the upper and lower surfaces white-woolly, sessile. Flowers in capitula in glomerules in a dichasiiform array; phyllaries 1–4, unequal, similar to the palea or absent; receptacle clavate; palea numerous in 2–3 series, surrounding numerous flowers, to 4 mm long, yellow with a reddish tinge, the apex acute; outer flowers numerous, carpellate, the inner (1)2–3(4), bisexual. Cypsela cylindric or subobovoid, subterete, papillate; pappus absent on the outer flowers, of numerous bristles on the inner carpellate and bisexual flowers.

Disturbed sites. Reported for Florida based on a specimen collected in 1829 (*Mettauer s.n.,* US) that only states "Florida." Massachusetts, New York, and Ontario south to Florida, also British Columbia, Washington, and Oregon. North America; Europe, Africa, Asia, and Pacific Islands. Native to Europe, Asia, and Africa. Spring–summer.

Flaveria Juss. 1789. YELLOWTOPS

Herbs. Leaves opposite, simple, pinnate-veined, petiolate or epetiolate. Flowers in cymiform or corymbiform arrays or glomerules, radiate or discoid; calyculi present or absent; receptacle epaleate; ray flowers (if present) carpellate; petals 5, basally connate, zygomorphic; disk flowers bisexual; petals 5, basally connate, actinomorphic; stamens 5, epipetalous, the filaments free, the anthers connate; ovary 2-carpellate, 1-loculate. Fruit a cypsela; pappus absent.

A genus of about 21 species; North America, West Indies, Mexico, Central America, South America, Europe, Africa, Asia, and Australia. [*Flavus*, yellow, in reference to the flower color.]

Selected reference: Yarborough and Powell (2006).

1. Capitula sessile, in sessile glomerules in the upper leaf axils .... **F. trinervia**
1. Capitula in corymbiform or cymiform arrays.
    2. Capitula in tight subglomerules in scorpioid cymiform arrays; leaves prominently 3-nerved, the margin serrate .... **F. bidentis**
    2. Capitula in dense but loose scorpioid cymiform or corymbiform arrays; leaves 1-nerved or with the lateral nerves indistinct, the margin entire or remotely minutely serrate.
        3. Capitula with 10–15 flowers; calyculi subtending the capitula often overtopping them .... **F. floridana**
        3. Capitula with 5–7(8) flowers; calyculi subtending the capitula not overtopping them .... **F. linearis**

Flaveria bidentis (L.) Kuntze [Two-toothed, apparently in reference to the often two calyculi.] COASTALPLAIN YELLOWTOPS.

Ethulia bidentis Linnaeus, Mant. Pl. 110. 1767. *Flaveria bidentis* (Linnaeus) Kuntze, Revis. Gen. Pl. 3(2): 148. 1898.

Erect annual herb, to 1 m; stem striate-angled, sparsely villous. Leaves with the blade lanceolate-elliptic, 5–12(18) cm long, 1–2.5(7) cm wide, the apex acute, the base cuneate, the margin serrate, the upper and lower surfaces sparsely pubescent, the petiole 3–15 mm long, winged, the bases clasping-connate, becoming sessile distally. Flowers in capitula in tight subglomerules in a scorpioid cymiform array; calyculi of 1–2 linear bracts 1–2 mm long; involucre oblong, angular; phyllaries 3–4 in 1 series, oblong, ca. 5 mm long; ray flowers 1 or absent; corolla pale yellow, the lamina to 1 mm long; disk flowers (2)3–8; corolla pale yellow, to 1 mm long. Cypsela oblanceolate or subclavate, 10-ribbed, 2–3 mm long; pappus absent.

Disturbed sites. Rare; Escambia and Miami-Dade Counties. Massachusetts, Georgia, Alabama, and Florida. North America; West Indies, Central America, and South America; Europe, Africa, and Asia. Native to Asia. Summer–fall.

Flaveria floridana J. R. Johnst. [Of Florida.] FLORIDA YELLOWTOPS.

Flaveria floridana J. R. Johnston, Proc. Amer. Acad. Arts 39: 291. 1903. FLORIDA: Lee Co.: Sanibel Island, 29 Jan 1896, *Webber 175* (lectotype: US; isolectotypes: F, MO). Lectotypified by Powell (1978: 714).

Flaveria pinetorum S. F. Blake, Bull. Torrey Bot. Club 50: 204. 1923. TYPE: FLORIDA: Lee Co.: vicinity of Fort Myers, 14 Dec 1919, *Standley 18909* (holotype: US).

Erect perennial or annual herb, to 1.2 m; stem striate-angled, glabrous or sometimes sparsely pubescent distally. Leaves with the blade linear, linear-lanceolate, or rarely elliptic, 5–14 cm long, 4–17 mm wide, the apex acute, the base cuneate, the margin entire, or remotely and minutely serrate, the upper and lower surfaces glabrous, sessile, the base clasping-connate. Flowers in capitula in a dense scorpioid cymiform array; calyculi of 1–3 linear-lanceolate bracts 4–6 mm long; involucre suburceolate, oblong, angular; phyllaries 5–6(9) in 1 series, oblong or ovate-orbicular, ca. 4 mm long; ray flowers 1 or absent; corolla yellow, the lamina 2–3 mm long; disk flowers 9–14; corolla yellow, ca. 1 mm long. Cypsela linear or oblong-lanceolate, 10-ribbed, 1–2 mm long; pappus absent.

Open coastal areas. Frequent; western central and southern peninsula. Endemic. Summer–fall.

Flaveria linearis Lag. [Linear, in reference to the leaf shape.] NARROWLEAF YELLOWTOPS.

Flaveria linearis Lagasca y Segura, Gen. Sp. Nov. 33. 1816.

Selloa nudata Nuttall, Amer. J. Sci. Arts 5: 300. 1822. *Gymnosperma nudatum* (Nuttall) de Candolle, Prodr. 5: 312. 1836. TYPE: FLORIDA: s.d., *Ware s.n.* (holotye: PH?).

Flaveria tenuifolia Nuttall, J. Acad. Nat. Sci. Philadelphia 7: 81. 1834. TYPE: FLORIDA: s.d., *Peale s.n.* (holotype: PH?).

Flaveria linearis Lagasca y Segura var. *latifolia* J. R. Johnston, Proc. Amer. Acad. Arts 39: 289. 1903. *Flaveria latifolia* (J. R. Johnston) Rydberg, in Britton, N. Amer. Fl. 34: 145. 1915. *Flaveria ×latifolia* (J. B. Johnston) Long & Rhamstine, Brittonia 20: 249. 1968. TYPE: FLORIDA: Palm Beach Co.: shore of Lake Worth, near Palm Beach, 31 Aug 1895, *Curtiss 5524* (lectotype: GH; isolectotypes: GA, KSC, MO, NY, UC, US). Lectotypified by Powell (1978: 615).

Erect perennial herb, to 8 dm; stem striate-angular, glabrous or sometimes sparsely pubescent distally. Leaves with the blade linear, 5–10(13) cm long, 1–4(15) cm wide, the apex acute, the base cuneate, the margin entire or remotely and minutely serrate, sessile, the bases clasping-connate. Flowers in capitula in clusters in a corymbifom array; calyculi of 1–3 linear bracts 1–3 mm long; involucre oblong, angular; phyllaries 5–6 in 1 series, linear or oblong, 3–5 mm long; ray flowers 1 or absent; corolla yellow, the lamina 2–3 mm long; disk flowers (2)5–7(8); corolla yellow, ca. 1 mm long. Cypsela linear, 10-ribbed, 1–2 mm long; pappus absent.

Coastal beaches, hammocks, and pinelands. Frequent; peninsula west to central panhandle. Florida; West Indies and Mexico. Summer–fall.

Flaveria trinervia (Spreng.) C. Mohr [Three-nerved, in reference to the leaves.] CLUSTERED YELLOWTOPS.

Oedera trinervia Sprengel, Bot. Gart. Halle 63. 1800. *Flaveria trinervia* (Sprengel) C. Mohr, Contr. U.S. Natl. Herb. 6: 810. 1901.

Flaveria repanda Lagasca y Segura, Gen. Sp. Nov. 33. 1816.

Erect annual herb, to 2 m; stem striate-angled, glabrous or glabrate. Leaves with the blade lanceolate, oblanceolate, elliptic, or subovate, 3–15 cm long, 1–4 cm wide, the apex acute or obtuse, the base cuneate, the margin serrate or serrate-dentate, petiolate proximally, sessile distally, the bases clasping-connate. Flowers in capitula in tight, axillary, sessile glomerules; calyculi absent; involucre oblong, cylindric or angular; phyllaries usually 2, oblong, 4–5 mm long; ray flowers 1 or absent; corolla yellow, the lamina ca. 1 mm long; disk flowers 1(2) or absent; corolla yellow, ca. 1 mm long. Cypsela oblanceolate or subclavate, 10-ribbed, 2–3 mm long; pappus absent.

Disturbed sites. Occasional; central and southern peninsula. Massachusetts, Virginia, Alabama, Florida, and Missouri, also Texas to California; West Indies, Mexico, central America, and South America; Africa, Asia, and Pacific Islands. Native to North America, West Indies, Central America, and South America. All year.

Fleischmannia Sch. Bip. 1850. THOROUGHWORT

Herbs. Leaves opposite, simple, pinnate-veined, petiolate. Flowers in capitula in corymbiform arrays, discoid, bisexual; receptacle epaleate; petals 5, basally connate, actinomorphic; stamens 5, epipetalous, the filaments free, the anthers connate; ovary inferior, 2-carpellate, 1-loculate. Fruit a cypsela; pappus present.

A genus of about 80 species; North America, Mexico, Central America, and South America. [Commemorates Gottfried Friedrich Fleischmann (1777–1850), professor of anatomy at Ehlangen, Germany, and teacher of Schultz Bipontinus.]

Selected reference: Nesom (2006l).

Fleischmannia incarnata (Walter) R. M. King & H. Rob. [*Incarnatus*, flesh-colored, in reference to the flower color.] PINK THOROUGHWORT.

Eupatorium incarnatum Walter, Fl. Carol. 200. 1788. *Kyrstenia incarnata* (Walter) Greene, Leafl. Bot. Observ. Crit. 1: 8. 1903. *Fleischmannia incarnata* (Walter) R. M. King & H. Robinson, Phytologia 19: 203. 1970.

Ascending or scandent annual or perennial herb, 1(2) m; stem striate-angled, puberulent. Leaves with the blade triangular-deltate, (1)2–5(7) cm long, (1)1.5–3.5(5) cm wide, the apex acute or acuminate, the base truncate, cordate, or sometimes broadly cuneate, the margin serrate or serrate-crenate, the upper and lower surfaces glabrous. Flowers in capitula in a loose corymbiform array; involucre obconic or hemispheric; phyllaries numerous in 2–4 series, 4–5 mm long, the outer linear-lanceolate, the inner linear-lanceolate or oblong-lanceolate, the apex acute, attenuate, or rounded, the outer surface sparsely pubescent or glabrous; receptacle flat or slightly convex; disk flowers numerous; corolla pink-purple or pink-lilac, 4–5 mm long. Cypsela prismatic, 5- to 8-ribbed, 2–3 mm long, sparsely strigose-hirtellous or glabrate; pappus of 1 series of numerous barbellate bristles 3–4 mm long.

Calcareous hammocks. Occasional; northern and central peninsula west to central panhandle. Virginia south to Florida, west to Illinois, Missouri, Oklahoma, and Texas. North America and Mexico. Summer–fall.

Gaillardia Foug. 1786. BLANKETFLOWER

Herbs. Leaves alternate, simple, pinnate-veined, petiolate. Flowers in solitary terminal capitula, radiate or discoid; ray flowers sterile; petals 5, basally connate, zygomorphic; disk flowers bisexual; petals 5, basally connate, actinomorphic, stamens 5, epipetalous, the filaments free, the anthers connate; ovary 2-carpellate, 10-loculate. Fruit a cypsela; pappus present.

A genus of about 25 species; North America, West Indies, Mexico, Central America, South America, Europe, and Africa. [Commemorates M. Gaillard de Merentonneau, eighteenth-century French magistrate and patron of botanists.]

Selected reference: Strother (2006l)

1. Receptacle with soft, toothlike seta smaller than the cypsela or obsolete; stem trichomes short-appressed .... **G. aestivalis**
1. Receptacle with firm, subulate setae equaling or longer than the cypsela; stem trichomes coarse, spreading .... **G. pulchella**

Gaillardia aestivalis (Walter) H. Rock [Of summer.] LANCELEAF BLANKETFLOWER.

Helenium aestivale Walter, Fl. Carol. 210. 1788. *Gaillardia aestivalis* (Walter) H. Rock, Rhodora 58: 315. 1956.
Gaillardia lanceolata Michaux, Fl. Bor.-Amer. 2: 142. 1803. TYPE: "Carolina ad Floridam."
Gaillardia lanceolata Michaux var. *flavovirens* C. Mohr, Contr. U.S. Natl. Herb. 6: 812. 1901. *Gaillardia aestivalis* (Walter) H. Rock var. *flavovirens* (C. Mohr) Cronquist, Brittonia 29: 221. 1977.
Gaillardia lutea Greene, Pittonia 5: 57. 1902.

Erect perennial herb, to 6 dm; stem sometimes rhizomatous, striate-angled, short-appressed. Leaves with the blade elliptic, linear, obovate, or spatulate, 1.5–6 cm long, 0.3–1.2(2) cm wide, the apex acute or obtuse, the base cuneate, the distal usually clasping, the margin remotely toothed or entire, the upper and lower surfaces scabrellous, the petiole to 3 cm long or sessile. Flowers in a solitary terminal capitulum, the peduncle 1–10(20) cm long; involucre hemispheric; phyllaries numerous in 2–3 series, lanceolate or lanceolate-ovate, 6–14 mm long, the apex acute, the outer surface scabrellous; receptacle convex or hemispheric, with soft, toothlike seta to ca. 3 mm long, smaller than the cypsela or these obsolete; ray flowers 8; corolla yellow, the lamina 1.5–2 cm long; disk flowers numerous; corolla yellow proximally, usually purplish distally, rarely all yellow, 1–2 mm long. Cypsela obpyramidal, ca. 2 mm long, 4-angled, pilose; pappus of 8–10 lanceolate, aristate scales 5–7 mm long.

Sandhills. Frequent; northern counties, central peninsula. North Carolina south to Florida, west to Kansas, Oklahoma, and Texas. Spring–summer.

Gaillardia pulchella Foug. [Beautiful and little.] FIREWHEEL.

Gaillardia pulchella Fougeroux de Bondaroy, Mem. Acad. Sci. (Paris) 1786: 5. 1788. *Gaillardia bicolor* Lamarck, Encycl. 2: 590. 1788, nom. illegit. *Gaillardia bicolor* Lamarck var. *vulgaris* Hooker, Bot. Mag. sub t. 3368. 1834, nom. inadmiss.
Gaillardia bicolor Lamarck var. *drummondii* Hooker, Bot. Mag. 61: t. 3368. 1834. *Gaillardia drummondii* (Hooker) de Candolle, Prodr. 5: 652. 1836. *Gaillardia pulchella* Fougeroux de Bondaroy var. *drummondii* (Hooker) B. L. Turner, in B. L. Turner et al., Sida, Bot. Misc. 24: 6. 2003.
Gaillardia picta D. Don, in Sweet, Brit. Fl. Gard., ser. 2. 3: t. 267. 1834. *Gaillardia pulchella* Fougeroux de Bondaroy var. *picta* (D. Don) A. Gray, Syn. Fl. N. Amer. 1(2): 352. 1884. *Gaillardia pulchella* Fougeroux de Bondaroy forma *picta* (D. Don) Voss, Vilm. Blumengärtn., ed. 3. 1: 497. 1894.

Erect annual or sometimes short-lived perennial herb, to 4(6) dm; stem striate-angled, strigillose, hirtellous, or slightly villous, the trichomes coarse, spreading. Leaves with the blade linear, oblong, or spatulate, 1–5(12) cm long, 0.4–1.2(3.5) cm wide, the apex acute or obtuse, the base cuneate, the distal usually clasping, the margin entire, toothed, or lobed, the upper and lower surfaces strigillose, hirtellous, or villous, the petiole to 3 cm long or sessile. Flowers in a solitary terminal capitulum, the peduncle 3–10(20) cm long; involucre hemispheric; phyllaries numerous in 2–3 series, narrowly triangular or linear, 6–14 mm long, the apex acute, the margin ciliate, the outer surface scabrellous; receptacle convex or hemispheric, with firm, subulate setae 2–3 mm long, equaling or longer than the cypsela; ray flowers 8–14 or rarely absent; corolla reddish or purplish proximally, yellow or orange distally, rarely yellow, reddish, or purplish throughout, the lamina 1.3–3 cm long; disk flowers numerous; corolla yellow proximally, usually purplish distally, rarely yellow, purple, or brown, often bicolored, ca. 1 mm long. Cypsela obpyramidal, ca. 2 mm long, 4-angled, pilose; pappus of 7–8 deltate or lanceolate, aristate scales 4–7 mm long.

Dry, open sites, mostly disturbed areas. Frequent; nearly throughout. North Carolina south to Florida, west to South Dakota, Nebraska, Kansas, Colorado, and Arizona. North America; West Indies, Mexico, Central America, and South America; Europe and Africa. Native to North America and Mexico. All year.

Galinsoga Ruiz & Pav. 1794. GALLANT-SOLDIER

Herbs. Leaves opposite, simple, pinnate-veined, petiolate. Flowers in capitula in corymbiform arrays, radiate; receptacle paleate; ray flowers carpellate; petals 5, basally connate, zygomorphic; disk flowers bisexual; petals 5, basally connate, actinomorphic; stamens 5, epipetalous, the filaments free, the anthers connate; ovary inferior, 2-carpellate, 1-loculate. Fruit a cypsela; pappus present or absent.

A genus of about 30 species; nearly cosmopolitan. [Commemorates Ignacio Mariano Martínez de Galinsoga (1756–1797), physician to the Spanish queen consort and director of the Real Jardín Botánico de Madrid.]

Selected reference: Canne-Hilliker (2006).

Galinsoga quadriradiata Ruiz & Pav. [In reference to the sometimes 4 ray flowers in the capitula.] PERUVIAN DAISY; SHAGGYSOLDIER.

Galinsoga quadriradiata Ruiz López & Pavón, Syst. Veg. Fl. Peruv. Chil. 198. 1798. *Galinsoga parviflora* Cavanilles var. *quadriradiata* (Ruiz López & Pavón) Poiret, in Lamarck, Encycl., Suppl. 2: 701. 1812. *Adventina ciliata* Rafinesque, New Fl. 1: 67. 1836. *Galinsoga ciliata* (Rafinesque) S. F. Blake, Rhodora 24: 35. 1922.

Erect annual herb, to 6 dm; stem striate-angled, pilose. Leaves with the blade lanceolate or ovate, 2–6 cm long, 1.5–4.5 cm wide, the apex acute, the base cuneate, the margin entire, serrate, the upper and lower surfaces pilose, 0.5–2 cm long. Flowers in capitula in a corymbiform array; involucre hemispheric or campanulate, 3–6 mm wide; phyllaries 6 to many in 2 series, elliptic, lanceolate, ovate, or oblong, 3–5 mm long, the apex acute or obtuse, the outer surface pilose; receptacle conic; palea of ray flowers elliptic or obovate, 2–3 mm long, in groups of 2–3 and shed with the adjacent phyllary and ray cypsela, that of the disk flowers lanceolate or obovate, shorter, convex or duplicate; ray flowers (4)5(8); corolla white or pink, the lamina 1–3 mm long; disk flowers numerous; corolla yellow, 2–3 mm long. Cypsela obpyramidal, ca. 2 mm long, strigose, shed with subtending phyllary and 2–3 adjacent palea; pappus of ray flowers of numerous fimbriate scales to 1 mm long, that of the disk flowers usually numerous, sometimes of 1–5 or absent, of white lanceolate or oblanceolate, fimbriate or aristate scales to 2 mm long.

Disturbed sites. Occasional; peninsula, central and western panhandle. Nearly throughout North America. Nearly cosmopolitan. Native to tropical America. All year.

Gamochaeta Wedd. 1856. EVERLASTING; CUDWEED

Herbs. Leaves alternate, simple, pinnate-veined, epetiolate. Flowers in capitula in glomerules in spiciform arrays, discoid, carpellate or bisexual; receptacle epaleate; petals 5, basally connate, actinomorphic; stamens 5, epipetalous, the filaments free, the anthers connate; ovary inferior, 2-carpellate, 1-loculate. Fruit a cypsela; pappus present.

A genus of about 50 species; nearly cosmopolitan. [From the Greek *gamos,* union, and *chaeta,* loose flowing hair, in reference to the basally connate pappus bristles.]

Selected reference: Nesom (2006m).

1. Leaves concolor or weakly bicolor (lower and upper surfaces nearly equally greenish to gray-green, the indumentum usually loosely tomentose or arachnose).
    2. Basal and lower leaves obovate-spatulate, the blades 4–16 mm wide; bracts among the heads spatulate or oblanceolate .... **G. pensylvanica**
    2. Basal and lower cauline leaves linear or oblanceolate, the blades 2–6 mm wide; bracts among the heads linear or narrowly oblanceolate .... **G. antillana**
1. Leaves bicolor (lower surface closely white pannose to white tomentose, the trichomes obscuring the surface, the upper surface glabrous, glabrate, or sparsely arachnose).
    3. Basal leaves not in a rosette, usually withering and deciduous at flowering; plant (3)5–8.5 dm tall; apex of the inner phyllaries acute-acuminate; flowering July–August .... **G. simplicicaulis**
    3. Basal leaves in a rosette, persistent at flowering; plant 1–5 dm tall; apex of the inner phyllaries acute, obtuse, rounded, or blunt; flowering April–June.
        4. Upper leaf surface glabrous or glabrate; involucre 2–3 mm long, the base glabrous; outer phyllaries elliptic-obovate to broadly ovate-elliptic, the apex rounded or obtuse .... **G. coarctata**
        4. Upper leaf surface sparsely arachnose (evident at 10x magnification); involucre 3–5 mm long, the base usually sparsely arachnose on the lower ⅓–½; outer phyllaries ovate, ovate-triangular, or ovate-lanceolate, the apex acute to acuminate.
            5. Stems pannose (trichomes closely appressed and not individually evident); apex of the inner phyllaries acute or acute-acuminate; fruit purple .... **G. chionesthes**
            5. Stems pannose-tomentose (trichomes individually evident); apex of the inner phyllaries acute, obtuse, or truncate-rounded or obtuse, sometimes apiculate; fruit tan or brownish.
                6. Blade of the inner phyllaries triangular, the apex acute, never apiculate .... **G. purpurea**
                6. Blade of the inner phyllaries elliptic-oblong, the apex truncate-rounded or obtuse, apiculate .... **G. argyrinea**

Gamochaeta antillana (Urb.) Anderb. [Of the Antilles.] CARIBBEAN PURPLE EVERLASTING; DELICATE EVERLASTING.

Gnaphalium antillanum Urban, Repert. Spec. Nov. Regni Veg. 13: 482. 1915. *Gamochaeta antillana* (Urban) Anderberg, Opera Bot. 104. 1991.
Gnaphalium subfalcatum Cabrera, Revista Mus. La. Plata Bot., ser. 2. 4: 174. 1941. *Gamochaeta subfalcata* (Cabrera) Cabrera, Bol. Soc. Argent. Bot. 9: 370. 1961.

Erect, ascending, or decumbent annual herb, to 4 dm, taprooted; stem striate-angled, loosely arachnose-tomentose. Leaves with the blade spatulate, oblanceolate, or linear, 2–3(4) cm long, 2–6 mm wide, the apex acute, the base narrowly cuneate, the margin entire, the upper and lower surfaces concolor, loosely tomentose, sessile. Flowers in capitula in a continuous, cylindric-spiciform array, becoming glomerulate-interrupted, with linear or narrowly oblanceolate leafy bracts throughout; involucre campanulate; phyllaries in 3–4(5) series, 3–4 mm long, the outer ovate-lanceolate, shorter than the inner, the apex acute, sometimes purple-tinged, the inner oblong, usually purple, the apex rounded or obtuse, whitish-tan; receptacle flat; carpellate flowers numerous, the bisexual flowers 3–5, all corollas yellow, purplish distally. Cypsela oblong, slightly flattened, ca. 0.5 mm long, with papilliform trichomes; pappus of 1 series of numerous basally connate, barbellate bristles.

Flatwoods and disturbed sites. Common; nearly throughout. Virginia south to Florida, west to Oklahoma and Texas. North America, West Indies, and South America; Europe and Pacific Islands. Native to North America, West Indies, and South America. Spring–fall.

Gamochaeta argyrinea G. L. Nesom [From the Greek *argyros*, silvery, in reference to the pannose stems and lower leaf surfaces.] SILVERY CUDWEED.

Gamochaeta argyrinea G. L. Nesom, Sida 21: 718. 2004.

Decumbent or ascending annual herb, to 4 dm, fibrous-rooted, rarely taprooted; stem striate-angled, white-pannose. Leaves with the blade oblanceolate, oblanceolate-oblong, or oblanceolate-obovate, 1.5–5(8) cm long, 5–10(15) mm wide, the apex obtuse to rounded, the base narrowly cuneate, the margin entire, the surfaces bicolor, the upper sparsely arachnose (evident at 10x), the lower closely white-pannose, sessile. Flowers in capitula in a continuous cylindric array, sometimes later becoming interrupted, producing axillary glomerules at the proximal nodes; involucre campanulate; phyllaries in 4–6 series, 3–4 mm long, the outer ovate-lanceolate, shorter than the inner, tawny, the apex acute or acuminate, the inner elliptic-oblong, purple-tinged, the apex truncate-rounded or obtuse, apiculate; receptacle flat; carpellate flowers numerous, the bisexual flowers 4–5(6), all corollas purple- or yellow-brown distally. Cypsela oblong, slightly flattened, ca. 0.5 mm long, tan or brownish, with papilliform trichomes; pappus of 1 series of numerous basally connate, barbellate bristles.

Disturbed sites. Occasional; northern counties, Pasco County. Pennsylvania south to Florida, west to Kansas, Oklahoma, and Texas. North America; West Indies. Native to North America. Spring.

Gamochaeta chionesthes G. L. Nesom [From the Greek *chioneos,* snow-white, and *esthes,* clothing, in reference to the white polished cloth-like indumentum of the stems and lower leaf surfaces.] WHITE-CLOAKED CUDWEED.

Gamochaeta chionesthes G. L. Nesom, Sida 21: 725. 2004.

Erect, ascending, or decumbent annual herb, to 4.5 dm, fibrous-rooted; stem striate-angled, with white, closely appressed, polished cloth-like indumentum (pannose). Leaves with the blade oblanceolate or oblanceolate-spatulate, 2–6(7) cm long, 5–13 mm wide, the apex obtuse or rounded, the base narrowly cuneate, the margin entire, the surfaces bicolor, the upper sparsely appressed-arachnose, the lower appressed white-pannose, sessile. Flowers in capitula in a continuous cylindric array, sometimes interrupted and producing axillary glomerules proximally; involucre cylindric-campanulate; phyllaries in 4–5 series, 3–4 mm long, the outer ovate, shorter than the inner, the apex acute or acuminate, the inner oblong-lanceolate, the apex acute or acuminate, purplish; receptacle flat; carpellate flowers numerous, the bisexual flowers 2–3, all corollas yellow, purplish distally. Cypsela oblong, slightly flattened, ca. 0.5 mm long, purple, with papilliform trichomes; pappus of 1 series of numerous basally connate, barbellate bristles.

Disturbed sites. Rare; Baker County, central and western panhandle. North Carolina south to Florida, west to Arkansas and Mississippi. Spring.

Gamochaeta coarctata (Willd.) Kerguélen [*Coarctatus,* pressed together, close-set, or narrowed, in reference to the narrow leaf blade.] AMERICAN EVERLASTING; ELEGANT CUDWEED.

Gnaphalium spicatum Lamarck, Encycl. 2: 757. 1788; non Miller, 1768. *Gnaphalium purpureum* Linnaeus var. *spicatum* Klatt, Linnaea 42: 140. 1878. *Gnaphalium coarctatum* Willdenow, Sp. Pl. 3: 1886. 1803. *Gamochaeta spicata* Cabrera, Bol. Soc. Argent. Bot. 9: 380. 1961, nom. illegit. *Gamochaeta coarctata* (Willdenow) Kerguélen, Lejeunia 120: 104, 1987.

Decumbent-ascending annual or biennial herb, to 3.5(5) dm, fibrous-rooted, white-pannose. Leaves with the blade spatulate or oblanceolate-obovate, (1.5)3–8(12) cm long, 6–15(20) mm wide, the apex rounded, sometimes slightly emarginate, the base narrowly cuneate, the margin entire or crenulate, the surfaces bicolor, the upper glabrous or glabrate, the lower appressed white-pannose, sessile. Flowers in cypsela in a continuous spiciform array, later branched, interrupted; involucre cylindric-campanulate; phyllaries in 4–5 series, 2–3 mm long, the outer elliptic-obovate to broadly ovate-elliptic, shorter than the inner, the apex rounded or obtuse, purple-tinged, the inner oblong, the apex rounded or obtuse, apiculate, tan; receptacle flat; carpellate flowers numerous, the bisexual flowers 2–3, all corollas yellow, purplish distally. Cypsela oblong, slightly flattened, ca. 0.5 mm long, with papilliform trichomes; pappus of 1 series of numerous basally connate, barbellate bristles.

Pond margins and wet disturbed sites. Occasional; Columbia and Miami-Dade Counties, panhandle. North Carolina south to Florida, west to Texas. North America; West Indies and Mexico; Europe, Asia, Australia, and Pacific Islands. Native to South America. Winter–spring.

Gamochaeta pensylvanica (Willd.) Cabrera [Of Pennsylvania.] PENNSYLVANIA EVERLASTING; PENNSYLVANIA CUDWEED.

Gnaphalium spatulatum Lamarck, Encycl. 2: 758. 1788; non Burman f., 1768. *Gnaphalium pensylvanicum* Willdenow, Enum. Pl. 867. 1809. *Gnaphalium purpureum* Linnaeus var. *spathulatum* Baker, in Martius, Fl. Bras. 6(3): 125. 1882. *Gamochaeta pensylvanica* (Willdenow) Cabrera, Bol. Soc. Argent. Bot. 9: 375. 1961.
Gnaphalium peregrinum Fernald, Rhodora 45: 479, t. 795. 1943.

Erect, decumbent, or procumbent annual herb, to 5 dm, tap-rooted; stem striate-angled, arachnose-tomentose. Leaves with the blade obovate-spatulate or oblanceolate, 2–7 cm long, 4–16 mm wide, the apex obtuse or rounded, sometimes apiculate, the base narrowly cuneate, the margin entire or sinuate, the surfaces concolor or weakly bicolor, the upper surface loosely tomentose, the lower surface more densely so, sessile. Flowers in glomerules in a continuous or interrupted spiciform array, with spatulate or oblanceolate leafy bracts throughout; involucre cupulate-campanulate; phyllaries in 3–4 series, 3–4 mm long, the outer ovate-triangular, shorter than the inner, the apex attenuate, apiculate, the inner oblong, usually purple-tinged, the apex acute or obtuse, sometimes yellow; receptacle flat; carpellate flowers numerous, the bisexual flowers 3–4, all corollas yellow (or at least the bisexual), purplish distally. Cypsela oblong, slightly flattened, ca. 0.5 mm long, with papilliform trichomes; pappus of 1 series of numerous basally connate, barbellate bristles.

Pond margins and wet disturbed sites. Common; nearly throughout. Pennsylvania south to Florida, west to Oklahoma and Texas. Nearly cosmopolitan. Native to West Indies, Central America, and South America. Spring–summer.

Gamochaeta purpurea (L.) Cabrera [Purple-colored, in reference to the purplish capitula.] SPOONLEAF PURPLE EVERLASTING; SPOONLEAF CUDWEED.

Gnaphalium purpureum Linnaeus, Sp. Pl. 854. 1753, nom. cons. *Gnaphalium purpureum* Linnaeus var. *normale* Kuntze, Revis. Gen. Pl. 1: 340. 1891, nom. inadmiss. *Gamochaeta purpurea* (Linnaeus) Cabrera, Bol. Soc. Argent. Bot. 9: 370. 1961.

Erect, ascending, or decumbent annual herb, to 4(5) dm, fibrous-rooted or taprooted; stem striate-angled, pannose-tomentose. Leaves with the blade oblanceolate or spatulate, 1–6 cm long, 5–14 mm wide, the apex obtuse or rounded, the base narrowly cuneate, the margin entire or sinuate, the surfaces bicolor, the upper surface sparsely arachnose, the lower surface appressed white-pannose, sessile. Flowers in capitula in a continuous spiciform array, later in interrupted bracteate glomerules, the proximal sometimes on long peduncles; involucre turbinate-campanulate; phyllaries in 4–5 series, 4–5 mm long, the outer ovate-triangular, shorter than the inner, the apex acute or acuminate, the inner triangular-lanceolate, the apex acute, purple-tinged; carpellate flowers numerous, the bisexual flowers 3–4, all corollas yellow, purplish distally. Cypsela oblong, slightly flattened, 0.5–1 mm long, tan or brownish, with papilliform trichomes; pappus of 1 series of numerous basally connate, barbellate bristles.

Dry, open sites, especially disturbed sites. Common; nearly throughout. North America,

West Indies, Mexico, Central America, and South America; Pacific Islands. Native to North America. Spring–summer.

Gamochaeta simplicicaulis (Willd. ex Spreng.) Cabrera [*Simplici,* simple, undivided, and *caulis,* stem, in reference to the plant usually having a single stem.] SINGLESTEM EVERLASTING; SIMPLE-STEM CUDWEED.

Gnaphalium simplicicaule Willdenow ex Sprengel, Syst. Veg. 3: 481. 1826. *Gnaphalium purpureum* Linnaeus var. *simplicicaule* (Willdenow ex Sprengel) Klatt, Linnaea 42: 140. 1878–1879. *Gamochaeta simplicicaulis* (Willdenow ex Sprengel) Cabrera, Bol. Soc. Argent. Bot. 9: 379. 1961.

Erect or ascending annual or biennial herb, to 5(8.5) dm, fibrous-rooted; stem striate-angled, white-pannose. Leaves with the blade oblanceolate or oblanceolate-spatulate, 5–9 cm long, 6–18 mm wide, the apex obtuse, the base narrowly cuneate, the margin undulate, the surfaces bicolor, the upper surface glabrous, the lower surface appressed white-pannose, the distal ones becoming linear-lanceolate, the apex acuminate, the mid and distal with sessile clusters of smaller leaves produced in the axil, sessile. Flowers in capitula in an interrupted spiciform array, sometimes with lateral branches, the glomerules often subtended by longer linear bracts; involucre cylindric-campanulate; phyllaries in 4–5 series, 3–4 mm long, the outer ovate or oblong, shorter than the inner, the apex acute or acuminate, brownish or tan, the inner narrowly oblong, the apex acute-acuminate, apiculate, brownish or tan; carpellate flowers numerous, the bisexual flowers (2)3, all yellow. Cypsela oblong, slightly flattened, ca. 0.5 mm long, with papilliform trichomes; pappus of 1 series of numerous basally connate, barbellate bristles.

Disturbed sites. Rare; Walton and Santa Rosa Counties. North Carolina south to Florida, west to Alabama. North America and South America; India, Australia, and Pacific Islands. Native to South America. Summer.

EXCLUDED TAXA

Gamochaeta americana (Miller) Weddell—Reported for Florida by Correll and Correll (1982, as *Gnaphalium americanum* Miller), Wunderlin (1998, as *Gnaphalium americanum* Miller), and Wunderlin and Hansen (2003), the name misapplied to Florida material of *Gamochaeta coarctata.*

Gamochaeta falcata (Lamarck) Cabrera—Reported for Florida by Small (1903, 1913a, 1933, all as *Gnaphalium falcatum* Lamarck), Radford et al. (1964, 1968, both as *Gnaphalium purpureum* var. *falcatum* (Lamarck) Torrey & A. Gray), Correll and Johnston (1970, as *Gnaphalium falcatum* Lamarck), Long and Lakela (1971, as *Gnaphalium purpureum* var. *falcatum* (Lamarck) Torrey & A. Gray), Cronquist (1980, as *Gnaphalium purpureum* var. *falcatum* (Lamarck) Torrey & A. Gray), Wunderlin (1982, 1998, both as *Gnaphalium falcatum* Lamarck), Wunderlin and Hansen (2003), Wilhelm (1984, as *Gnaphalium falcatum* Lamarck), and by Clewell (1985, as *Gnaphalium falcatum* Lamarck). All as misapplications of the name to Florida material of *Gamochaeta antillana.*

Garberia A. Gray 1880.

Shrubs. Leaves alternate or opposite, simple, pinnate-veined, petiolate. Flowers in capitula in corymbiform or paniculiform arrays, discoid, bisexual; receptacle epaleate; petals 5, basally connate, actinomorphic; stamens 5, epipetalous, the filaments free, the anthers connate; ovary inferior, 2-carpellate, 1-loculate. Fruit a cypsela; pappus present.

A monotypic genus; North America. [Commemorates Abram Paschal Garber (1838–1881), physician, botanist, and teacher, noted for his Florida collections.]

Selected reference: Lamont (2006b).

Garberia heterophylla (W. Bartram) Merr. & F. Harper [From the Greek *heteros,* different, and *phyllon,* leaf, in reference to the different-shaped leaves.] GARBER'S SCRUB STARS.

Cacalia heterophylla W. Bartram, Travels Carolina 164. 1791. *Garberia heterophylla* (W. Bartram) Merrill & F. Harper, in Merrill, Bartonia 23: 2. 1945. TYPE: FLORIDA: Marion Co.: near Salt Springs (holotype: PH?).

Liatris fruticosa Nuttall, Amer. J. Sci. Arts 5: 299. 1822. *Garberia fruticosa* (Nuttall) A. Gray, Proc. Acad. Nat. Sci. Philadelphia 31: 380. 1879. *Leptoclinium fruticosum* (Nuttall) A. Gray, Proc. Amer. Acad. Arts 15: 48. 1880. TYPE: FLORIDA: s.d., *Ware s.n.*

Erect shrub, to 2.5 m; stem terete or striate-angled, puberulent, glabrescent, often gland-dotted. Basal and proximal cauline leaves sometimes opposite or subopposite, the distal alternate, the blade spatulate, obovate, or orbicular-obovate, 1.5–3.5 cm long, 7–20 mm wide, the apex rounded or slightly retuse, the base cuneate, the margin entire, the upper and lower surfaces grayish green, gland-dotted, often viscid, farinaceous when young, short-petiolate or subsessile. Flowers in capitula in a corymbiform or paniculiform array, the peduncle ca. 1 cm long; involucre cylindric, 3–5(6) mm wide; phyllaries numerous in 3–5 series, lanceolate or linear-lanceolate, unequal, the outer ca. 3 mm long, the inner 5–6 mm long, the apex acute or acuminate, the outer surface farinaceous, usually gland-dotted; receptacle slightly convex; flowers usually 5; corolla pink or purplish, 8–10 mm long. Cypsela prismatic, ca. 10-ribbed, (6)7–8 mm long, scabrellous; pappus of numerous barbellate bristles in 2–3 series, 8–11(12) mm long, the outer shorter, often purple-tinged.

Sand pine-oak scrub. Frequent; northern and central peninsula. Endemic. Spring–fall.

Garberia heterophylla is listed as threatened in Florida (Florida Administrative Code, Chapter 5B-40).

Gerbera L., Nom. Cons. 1758. TRANSVAAL DAISY

Herbs. Leaves alternate, simple, pinnate-veined, petiolate. Flowers in solitary capitula, terminal, radiate; receptacle epaleate; ray flowers carpellate; petals 5, basally connate, zygomorphic; disk flowers bisexual or staminate; petals 5, basally connate, actinomorphic; stamens 5, epipetalous, the filaments free, the anthers connate; the ovary inferior, 2-carpellate, 1-loculate. Fruit a cypsela; pappus present.

A genus of about 30 species; North America, Africa, and Asia. [Commemorates Traugott Gerbera (1710–1743), German physician and botanist.]

Selected reference: Hansen (1985).

Gerbera jamesonii Adlam 1888. [Commemorates Robert Jameson (1774–1854), Scottish naturalist, mineralogist, and professor of natural history, University of Edinburgh.] TRANSVAAL DAISY; AFRICAN DAISY.

Gerbera jamesonii Adlam, Gard. Chron., ser. 3. 3: 775. 1888.
Gerbera jamesonii Bolus ex Hooker f., Bot. Mag. 115: t. 7087. 1889; non Adlam, 1888.

Erect perennial herb, to 6 dm; stem striate-angled, sparsely hirsute. Leaves with the blade oblong-spatulate or obovate, 20–30 cm long, 10–15 cm wide, the apex obtuse, the base cuneate, the margin irregularly pinnate-lobed or undulate, the upper and lower surfaces sparsely hirsute, the petiole 15–40 cm long. Flowers in a solitary capitulum, the peduncle 30–50 cm long; involucre campanulate; phyllaries numerous, in 2–3 series, lanceolate, 1–1.5 cm long, sparsely hirsute; ray flowers numerous; corolla yellow, pink, orange, red, or white, the lamina 3–4 cm long; disk flowers numerous; corolla yellow, pink, orange, red, or white, 5–6 mm long. Cypsela fusiform, laterally flattened, 10–15 mm long, ribbed; pappus of numerous barbellate bristles, 5–9 mm long, cream or gray.

Disturbed sites. Rare; Alachua County. Escaped from cultivation. North America; Africa. Native to Africa. Summer–fall.

Glebionis Cass. 1826. GARLAND CHRYSANTHEMUM

Herbs. Leaves alternate, simple, pinnate-veined, petiolate or epetiolate. Flowers in capitula, terminal, solitary, radiate; receptacle epaleate; ray flowers carpellate; petals 5, basally connate, zygomorphic; disk flowers bisexual; petals 5, basally connate, actinomorphic; anthers 5, epipetalous, the filaments free, the anthers connate; ovary inferior, 2-carpellate, 1-loculate. Fruit a cypsela; pappus absent.

A genus of 2 species; North America, Europe, Africa, and Asia. [*Gleba,* lump, mass, and *ionis,* state of, apparently in reference to the floral heads.]

Selected reference: Strother (2006m).

Glebionis coronaria (L.) Cass. ex Spach [*Coronarius,* wreath, crown, in reference to the floral heads used as garlands or crowns.] GARLAND CHRYSANTHEMUM.

Chrysanthemum coronarium Linnaeus, Sp. Pl. 890. 1753, nom. cons. *Matricaria coronaria* (Linnaeus) Desrousseaux, in Lamarck, Encycl. 3: 737. 1791. *Pinardia coronaria* (Linnaeus) Lessing, Syn. Gen. Compos. 255. 1832. *Glebionis coronaria* (Linnaeus) Cassini ex Spach, Hist. Nat. Vég. 10: 181. 1841.

Erect or ascending annual herb, to 5(8) dm; stem striate-angled, glabrous. Leaves with the blade oblong or obovate, 3–5.5 cm long, 1.5–3 cm wide, 2- to 3-pinnate-lobed, the ultimate margin dentate, the apex acute or acuminate, the base cuneate, the upper and lower surfaces glabrous, winged short-petiolate or sessile. Flowers in capitula, terminal, solitary, the peduncle 5–15 cm long; involucre hemispheric, 1.5–2.5 cm wide; phyllaries numerous in 3–4 series, ovate, obovate, or lanceolate, 5–10 mm long, the apex acute or obtuse, the margin and apex scarious, stramineous; receptacle convex; ray flowers numerous; corolla pale yellow, sometimes white-tipped, the lamina 1.5–2.5 cm long; disk flowers numerous; corolla pale yellow, ca. 5 mm long. Cypsela dimorphic, that of the ray flowers 3-angled, 2–3 mm long, the angles slightly winged, obscurely 10-ribbed, that of the disk flowers prismatic, slightly compressed, 2–3 mm long, obscurely 10-ribbed, sometimes the inner (rarely the outer) rib slightly winged.

Disturbed sites. Rare; Bay County. Escaped from cultivation. Scattered localities in North America; Europe, Africa, and Asia. Native to Europe, Africa, and Asia. Spring–fall.

EXCLUDED TAXON

Glebionis segeta (Linnaeus) Fourreau—Reported for Florida by Small (1933, as *Chrysanthemum segetum* Linnaeus) and Arriagada and Miller (1997, as *Chrysanthemum segetum*). No Florida material seen.

Gynura Cass., nom. cons. 1828.

Herbs. Leaves alternate, simple, pinnate-veined, petiolate or epetiolate. Flowers in capitula in corymbiform arrays or solitary, discoid, bisexual; calyculi present; receptacle epaleate; petals 5, basally connate, actinomorphic; stamens 5, epipetalous, the filaments free, the anthers connate; ovary inferior, 2-carpellate, 1-loculate. Fruit a cypsela; pappus present.

A genus of about 40 species; North America, Africa, Asia, Australia, and Pacific Islands. [From the Greek *gyne,* female, and *ura,* tail, apparently in reference to the conspicuous style branches.]

Selected reference: Barkley (2006c).

Gynura aurantiaca (Blume) DC. VELVETPLANT.

Cacalia aurantiaca Blume, Bijdr. 908. 1827. *Gynura aurantiaca* (Blume) de Candolle, Prodr. 6: 300. 1838.

Spreading or clambering perennial herb, to 1(3) m; stem somewhat fleshy, purplish villous. Leaves with the blade ovate, elliptic, or rhombic, 4–10(15) cm long, 2–5(8) cm wide, the apex acute, the base cuneate, the margin irregularly toothed, usually pinnate-lobed proximally, the upper and lower surfaces purplish villous, the petiole to 3 cm long or sessile. Flowers in capitula in a corymbiform array or solitary; calyculi of 3–8 bractlets 3–6 mm long; involucre cylindric or campanulate, 8–12 mm wide; phyllaries numerous in (1)2 series, linear, 10–12 mm long, subequal, the margin scarious; receptacle flat, foveolate; flowers numerous; corolla yellow, orange, or red, 8–12 mm long. Cypsela columnar or prismatic, 5-to 10-angled or -ribbed, glabrous; pappus of numerous white smooth or barbellate bristles 10–12 mm long.

Moist disturbed sites. Rare; Lee and Palm Beach Counties. Escaped from cultivation. Florida; Central America and South America; Africa, Asia, and Australia. Native to Africa and Asia. Spring–fall.

Haplocarpha Less. 1831. ONEFRUIT

Herbs. Leaves alternate, simple, pinnate-veined, epetiolate. Flowers in solitary capitula, radiate; receptacle epaleate; ray flowers carpellate; disk flowers bisexual; stamens 5, epipetalous, the filaments free, the anthers connate; ovary inferior, 2-carpellate, 1-loculate. Fruit a cypsela; pappus absent.

A genus of about 12 species; North America and Africa. [From the Greek *haplo,* single, and *carphos,* small dry body, in reference to the fruit.]

Haplocarpha lyrata Harv. [Resembling a lyre, a stringed instrument, in reference to the leaf shape.] HARP ONEFRUIT.

Haplocarpha lyrata Harvey, in Harvey & Sonder, Fl. Cap. 3: 465. 1865.

Perennial herb, to 2.5 dm; stem rhizomatous, rooting at the nodes, striate-angled, sparsely lanate. Leaves in basal rosettes, the blade oblanceolate, 8–15 cm long, 1–2.5 cm wide, the apex acute or obtuse, the base cuneate, the margin pinnate-lobed, the ultimate lobes irregularly dentate, the upper surface glabrous, the lower surface densely white-lanate, sessile. Flowers in a solitary capitulum, the peduncle 10–25 cm long, striate-angled, sparsely lanate; ray flowers numerous; corolla yellow, the lower surface green or red, the lamina ca. 2 cm long; disk flowers numerous; corolla yellow, 3–5 mm long. Cypsela turbinate, 3–4 mm long, 3- to 4-ribbed, silky and with a silky basal tuft; pappus of 2–several series of slender, bristle-tipped scales, ca. 1 mm long, tawny.

Disturbed sites. Rare; Hillsborough County. Escaped from cultivation. Florida; Africa. Native to Africa. Spring–summer.

Hartwrightia A. Gray ex S. Watson 1888.

Herbs. Leaves alternate, simple, pinnate-veined, petiolate or epetiolate. Flowers in capitula in corymbiform arrays, discoid, bisexual; receptacle paleate; petals 5, basally connate, actinomorphic; stamens 5, epipetalous, the filaments free, the anthers connate; ovary inferior, 2-carpellate, 1-loculate. Fruit a cypsela; pappus absent or present.

A monotypic genus; Georgia and Florida. [Commemorates Samuel Hart Wright (1825–1905), physician, farmer, astronomer, and botanist.]

Selected reference: Nesom (2006n).

Hartwrightia floridana A. Gray ex S. Watson. [Of Florida.]

Hartwrightia floridana A. Gray ex S. Watson, Proc. Amer. Acad. Arts 23: 265. 1888. TYPE: FLORIDA: Volusia Co.: Lake Helen.

Erect perennial herb, to 12 dm; stem striate-angled, glabrous. Leaves with the blade elliptic or linear, 5–25 cm long, 1–8 cm wide, the distal cauline bract-like, and the apex acute, obtuse, or rounded, the base cuneate, the margin entire or irregular and shallowly lobed near the base, the upper and lower surfaces glabrous, gland-dotted, the petiole of the basal to 2 dm long, that of the distal cauline sessile or subsessile. Flowers in capitula in a loose corymbiform array, short-pedunculate; involucre broadly obconic, 2–3 mm wide, the phyllaries numerous in 2–3 series, oblong-elliptic or lanceolate, 2–3 mm long, unequal; receptacle convex; palea of peripheral scales; flowers 7–10; corolla pink, bluish, or white, ca. 3 mm long. Cypsela obpyramidal, 5-angled, 3–4 mm long, gland-dotted; pappus of few setae or absent.

Open seepage areas. Occasional; northern and central peninsula. Summer–fall.

Hartwrightia floridana is listed as threatened in Florida (Florida Administrative Code, Chapter 5B-40).

Hasteola Raf. 1838. FALSE INDIAN PLANTAIN

Herbs. Leaves alternate, simple, pinnate-veined, petiolate or epetiolate. Flowers in capitula in corymbiform arrays, discoid, bisexual; calyculi present; receptacle epaleate; petals 5, basally connate, actinomorphic; stamens 5, epipetalous, the filaments free, the anthers connate; ovary inferior, 2-carpellate, 1-loculate. Fruit a cypsela; pappus present.

A genus of 2 species; North America. [*Hasta,* spear, and *ola,* diminutive, in reference to the leaf-shape of the type species.]

Selected references: Anderson (1994, 2006b); Anderson et al. (1995).

Hasteola robertiorum L. C. Anderson [Commemorates Robert Godfrey, Robert Kral, and Robert Simmons, who studied the species in Florida.] HAMMOCKHERB; GULF HAMMOCK INDIAN PLANTAIN.

Hasteola robertiorum L. C. Anderson, Syst. Bot. 19: 212. 1994. TYPE: FLORIDA: Levy Co.: ca. 1 mi (by air) of Otter Creek community, 15 Oct 1990, *Anderson & Simmons 13319* (holotype: NY; isotypes: FSU, MO, UNC, US, VDB).

Erect perennial herb, to 1(1.5) m; stem striate-angular, glabrous. Basal leaves 25–54 cm long, 8.5–15 cm wide, the blade hastate or deltoid, the apex acute or obtuse, the base cuneate, slightly decurrent on the petiole, the petiole 1–3 cm long, the proximal cauline 25–32 cm long, 8–10 cm wide, the blade ovate or narrowly elliptic-lanceolate, the apex acute, the base cuneate, slightly decurrent on the petiole, the middle cauline leaves pandurate or lanceolate, (10)12–20(30) cm long, 5–9 cm wide, the apex acute, the base auriculate, sessile. Flowers in capitula in a somewhat flat-topped corymbiform array; involucre broadly cylindric; calyculi 4–7 subulate or filiform bractlets 5–9 mm long; phyllaries 7–9; receptacle flat, foveolate; flowers numerous; corolla pale greenish white, 9–10 mm long. Cypsela cylindric-fusiform, 8- to 12-ribbed, 6–9 mm long, light golden brown with darker brown ribs; pappus of numerous barbellate bristles 4–5 mm long.

Wet hammocks. Rare; Levy and Lake Counties. Endemic. Fall.

Hasteola robertiorum is listed as endangered in Florida (Florida Administrative Code, Chapter 5B-40).

Helenium L. SNEEZEWEED

Herbs. Leaves alternate, simple, pinnate-veined, petiolate or epetiolate, estipulate. Flowers in capitula, solitary or in paniculiform or corymbiform arrays, radiate; phyllaries present; receptacle epaleate; ray flowers carpellate or sterile; petals 5, basally connate, zygomorphic; disk flowers bisexual; petals 5, basally connate, actinomorphic; stamens 5, epipetalous, the filaments free, the anthers connate; ovary inferior, 2-carpellate, 1-loculate. Fruit a cypsela; pappus present.

A genus of about 40 species; North America, West Indies, Mexico, Central America, and South America. [Named for Helen of Troy of Greek mythology.]

Selected reference: Bierner (2006).

1. Plant branched, the leaves not much reduced upward on the stem; capitula numerous in open corymbs.
    2. Leaves filiform (1–4 mm wide), not decurrent; plant annual .... **H. amarum**
    2. Leaves lanceolate (5 mm wide or wider), decurrent; plant perennial.
        3. Disk flowers reddish .... **H. flexuosum**
        3. Disk flowers yellow .... **H. autumnale**
1. Plant subscapose, the upper leaves reduced and relatively few, the lower leaves larger, tufted; capitula 1, rarely 2–5.
    4. Lobes of the disk flowers reddish .... **H. brevifolium**
    4. Lobes of the disk flowers yellow.
        5. Pappus scales deeply lacerate ½ or more their length .... **H. drummondii**
        5. Pappus scales entire or somewhat lacerate, but not deeply lacerate.
            6. Cypsela pubescent; midstem leaves decurrent to 0.5 cm on the stem .... **H. pinnatifidum**
            6. Cypsela glabrous; midstem leaves decurrent 2 cm or more on the stem .... **H. vernale**

Helenium amarum (Raf.) H. Rock [*Amarus,* bitter, in reference to the taste.] SPANISH DAISY; BITTERWEED.

Gaillardia amara Rafinesque, Fl. Ludov. 69. 1817. *Helenium amarum* (Rafinesque) H. Rock, Rhodora 59: 131. 1957.

Helenium tenuifolium Nuttall, J. Acad. Nat. Sci. Philadelphia 7: 66. 1834. *Heleniastrum tenuifolium* (Nuttall) Kuntze, Revis. Gen. Pl. 1: 342. 1891.

Annual herb, to 6(10) dm; stem branched distally, striate-angled, not winged, glabrous or sparsely pubescent. Leaves with the blade to 8 cm long, to 3 mm wide, slightly reduced upward, the upper and lower surfaces glabrous or sparsely pubescent, gland-dotted, the basal and proximal cauline usually withered by flowering, the basal with the blade linear or ovate, the margin entire or pinnately toothed, sometimes pinnatifid, the proximal cauline with the blade linear, the margin entire or pinnately toothed, the distal with the blade linear, entire, epetiolate. Flowers in capitula in a paniculiform array, the peduncle 3–11 cm long, sparsely pubescent; involucre hemispheric or subglobose, 6–10 mm wide; phyllaries numerous, in (1)–2 series, linear-lanceolate, 5–9 mm long, the outer surface moderately to densely pubescent, gland-dotted; ray flowers 8–10, carpellate; corolla yellow, the lamina 7–14 mm long; disk flowers numerous; corolla yellow proximally, yellow or yellow-brown distally, 2–3 mm long. Cypsela obpyramidal, ca. 1 mm long 4- to 5-angled, moderately to densely pubescent; pappus of 6–8 aristate scales 1–2 mm long.

Dry pine flatwoods, scrub, beach hammocks and dunes, and dry disturbed sites. Common; nearly throughout. Massachusetts, south to Florida, west to Wisconsin, Iowa, Kansas, and New Mexico, also California. Spring–fall.

Helenium autumnale L. [Of autumn, in reference to its flowering time.] COMMON SNEEZEWEED.

Helenium autumnale Linnaeus, Sp. Pl. 886. 1753. *Helenia autumnalis* (Linnaeus) Hill, Hort. Kew. 6. 1769. *Helenia decurrens* Moench, Methodus 489. 1794, nom. illegit. *Heleniastrum autumnale* (Linnaeus) Kuntze, Revis. Gen. Pl. 1: 342. 1891. *Helenium autumnale* Linnaeus var. *normale* Voss, Vilm. Blumengärtn., ed. 3. 1: 495. 1894, nom. inadmiss.

Helenium parviflorum Nuttall, Trans. Amer. Philos. Soc., ser. 2. 7: 384. 1841. *Heleniastrum parviflorum* (Nuttall) Kuntze, Revis. Gen. Pl. 1: 342. 1891. *Helenium autumnale* Linnaeus var. *parviflorum* (Nuttall) Fernald, Rhodora 45: 492. 1943.

Perennial herb, to 13 m; stem branched distally, striate-angled, strongly winged, sparsely to densely pubescent. Leaves with the blade to 25 cm long, to 5.5 cm wide, slightly reduced upward, the upper and lower surfaces glabrous or moderately to densely pubescent, the basal usually withered by flowering, the blade lanceolate, oblanceolate, or obovate, the margin entire or weakly lobed, the proximal and midcauline with the blade obovate or oblanceolate, the margin dentate or entire, the distal oblanceolate or lanceolate, the margin entire or dentate, petiolate or epetiolate. Flowers in capitula in a paniculiform array, the peduncle 3–10 cm long, moderately to densely pubescent; involucre hemispheric or subglobose, 8–23 mm wide; phyllaries numerous, in (1)–2 series, linear-lanceolate, 8–20 mm long, the outer surface moderately to densely pubescent, gland-dotted; ray flowers 8–20, carpellate; corolla yellow, the lamina 1–2 cm long; disk flowers numerous; corolla yellow 3–4 mm long. Cypsela obpyramidal, 1–2 mm long, 4- to 5-angled, sparsely to moderately pubescent; pappus of 5–7 aristate scales 1–2 mm long.

Wet flatwoods and floodplain forests. Occasional; eastern and central panhandle, Lake and Orange Counties. Nearly throughout North America. Summer–fall.

Helenium brevifolium (Nutt.) A. W. Wood [With short leaves.] SHORTLEAF SNEEZEWEED.

Leptopoda brevifolia Nuttall, Trans. Amer. Philos. Soc., ser. 2. 7: 373. 1841. *Helenium brevifolium* (Nuttall) A. W. Wood, Amer. Bot. Fl. 182. 1870. *Heleniastrum brevifolium* (Nuttall) Kuntze, Revis. Gen. Pl. 1: 342. 1891.

Perennial herb, to 1 m; stem unbranched or sparingly branched distally, striate-angled, weakly to moderately winged, glabrous proximally, sparsely to moderately pubescent distally. Leaves with the blade to 5 cm long, to 1.5 cm wide, reduced upward, the upper and lower surfaces glabrous or sparsely pubescent, the basal with the blade oblanceolate or spatulate, the margin entire, undulate, or undulate-serrate, the proximal and midcauline oblanceolate or lanceolate, the margin entire, undulate, or undulate-serrate, the distal linear-lanceolate, the margin entire, petiolate or epetiolate. Flowers in solitary or 2–10 capitula in a paniculiform or corymbiform array, the peduncle (3)6–11(15) cm long, sparsely to moderately pubescent; involucre hemispheric or subglobose, 1–2 mm wide; phyllaries numerous, in (1)–2 series, linear-lanceolate, 1–2 cm long, the outer surface sparsely to moderately gland-dotted; ray flowers 9–24, sterile; corolla yellow, the lamina 1–2 cm long; disk flowers numerous; corolla yellow proximally, reddish distally, 4–6 mm long. Cypsela obpyramidal, 1–2 mm long, 4- to 5-angled, sparsely to moderately pubescent; pappus of 6–8 nonaristate scales 1–2 mm long.

Seepage bogs. Rare; central and western panhandle. Virginia south to Florida, west to Louisiana. Spring.

Helenium drummondii H. Rock [Commemorates Thomas Drummond (1780–1835), Scottish botanist who collected in North America.] FRINGED SNEEZEWEED.

Helenium drummondii H. Rock, Rhodora 59: 173. 1957.
Leptopoda fimbriata Torrey & A. Gray, Fl. N. Amer. 2: 387. 1842; non Eaton, 1829.

Perennial herb, to 8 dm; stem unbranched or sparingly branched distally, striate-angled, moderately to strongly winged, glabrous or sparsely to moderately pubescent proximally, sparsely to moderately pubescent distally. Leaves with the blade to 6 cm long, to 1 cm wide, reduced upward, the upper and lower surfaces glabrous, the basal with the blade narrowly obovate or narrowly oblanceolate, the margin entire or sometimes undulate-serrate, the proximal and midcauline narrowly lanceolate, the margin entire or sometimes undulate-serrate, the distal linear-lanceolate, the margin entire or rarely undulate-serrate, petiolate or epetiolate. Flowers in solitary or in 2–3 capitula, the peduncle 10–30 cm long, sparsely pubescent; involucre hemispheric or subglobose, 1.5–2.5 cm wide; phyllaries numerous, in (1)–2 series, linear-lanceolate, 10–18 mm long, the outer surface sparsely to moderately pubescent, gland-dotted; ray flowers numerous, sterile; corolla yellow, the lamina 1.5–2.5 cm long; disk flowers numerous; corolla yellow, 4–5 mm long. Cypsela obpyramidal, 1–2 mm long, 4- to 5-angled, moderately pubescent; pappus of 5–12 deeply lacerate, non-aristate scales 2–4 mm long.

Wet pinelands and pond margins. Rare; locality unknown. Not recently collected. Spring.

Helenium drummondii was reported from Florida by Rock (1957) based on two collections: *Treat s.n.*, 1876 (PENN); *Leavenworth s.d.*, s.d (NY). It currently is known only from Arkansas, Louisiana, and Texas (Bierner, 2006).

Helenium flexuosum Raf. [Bent alternately in opposite directions, zigzag, in reference to the branches.] PURPLEHEAD SNEEZEWEED.

Helenium flexuosum Rafinesque, New Fl. 4: 81. 1838 ("1836").
Helenium anceps Rafinesque, New Fl. 4: 81. 1838 ("1836"). TYPE: FLORIDA.
Helenium nudiflorum Nuttall, Trans. Amer. Philos. Soc., ser. 2. 7: 384. 1841. *Heleniastrum nudiflorum* (Nuttall) Kuntze, Revis. Gen. Pl. 1: 342. 1891.
Leptopoda brachypoda Torrey & A. Gray, Fl. N. Amer. 2: 388. 1842. *Helenium brachypodum* (Torrey & A. Gray) A. W. Wood, Amer. Bot. Fl. 182. 1870.
Helenium floridanum Fernald, Rhodora 45: 494, pl. 799(3–4). 1943. TYPE: FLORIDA: Hernando Co.: Fitzgerald, s.d., *Curtiss 6663* (holotype: NY; isotypes: MIN, MO, NY, UC, US.

Perennial herb, to 1 m; stem branched distally, striate-angled, strongly winged, glabrous or sparsely pubescent proximally, sparsely to moderately pubescent distally. Leaves with the blade to 5 cm long, to 1.5 cm wide, slightly reduced upward, the upper and lower surfaces glabrous or sparsely to moderately pubescent, the basal with the blade oblanceolate, obovate, or spatulate, the margin entire or serrate, the proximal and midcauline oblanceolate or lanceolate, the margin entire or toothed, the distal lanceolate or linear-lanceolate, the margin entire, petiolate or epetiolate. Flowers in few to many capitula in a paniculiform array, the peduncle 3–10 cm long, sparsely to moderately pubescent; involucre subglobose, 1–1.5 cm wide; phyllaries numerous, in (1)–2 series, linear-lanceolate, 8–18 mm long, the outer surface moderately to densely pubescent, gland-dotted; ray flowers 8–12, sterile; corolla yellow, reddish brown, red, or purple, the lamina 1–2 cm long; disk flowers numerous; corolla yellow proximally and purple distally or entirely purple, 2–4 mm long. Cypsela obpyramidal, ca. 1 mm long, 4- to 5-angled, moderately pubescent; pappus of 5–6 aristate scales 1–2 mm long.

Wet flatwoods and floodplain forests. Occasional; nearly throughout. Quebec south to Florida, west to Ontario, Minnesota, Wisconsin, Illinois, Missouri, Kansas, Oklahoma, and Texas. Spring–fall.

Helenium pinnatifidum (Schwein. ex Nutt.) Rydb. [Pinnately cleft, in reference to the leaves.] SOUTHEASTERN SNEEZEWEED.

Leptopoda pinnatifida Schweinitz ex Nuttall, Trans. Amer. Philos. Soc., ser. 2. 7: 372. 1841. *Leptopoda puberula* Macbride ex Elliott var. *pinnatifida* (Schweinitz ex Nuttall) Torrey & A. Gray, Fl. N. Amer. 2: 387. 1942. *Helenium pinnatidum* (Schweinitz ex Nuttall) Rydberg, in Britton, N. Amer. Fl. 34: 130. 1915. TYPE: FLORIDA: s.d., *Baldwin s.n.* (holotype: PH). Lectotypified by Rock (1957: 149).

Leptopoda fimbriata Torrey & A. Gray, Fl. N. Amer. 2: 387. 1942; non (Michaux) Eaton, 1829. TYPE: FLORIDA/TEXAS.

Helenium puberulum A. W. Wood, Amer. Bot. Fl. 182. 1870; non de Candolle, 1836.

Perennial herb, to 8 dm; stem usually unbranched distally, striate-angled, weakly winged, glabrous or sparsely to moderately pubescent proximally, sparsely to moderately pubescent distally. Leaves with blade to 10 cm long, to 2 cm wide, reduced upward, the upper and lower surfaces glabrous or rarely sparsely pubescent, the basal with the blade obovate or oblanceolate, the margin usually pinnatifid, sometimes undulate or undulate-serrate, rarely entire, the proximal and midcauline linear-oblanceolate or linear-lanceolate, the margin entire or undulate-serrate, decurrent on the stem, the distal linear-lanceolate, the margin entire, petiolate or epetiolate. Flowers in solitary or 2–3 capitula, the peduncle 3–20 cm long, moderately to densely pubescent; involucre hemispheric, 1.5–3 cm wide; phyllaries numerous, in (1)2 series, linear-lanceolate, 12–20 mm long, the outer surface moderately pubescent, gland-dotted; ray flowers numerous, sterile; corolla yellow, the lamina 1.5–2 cm long; disk flowers numerous; corolla yellow, 4–6 mm long. Cypsela obpyramidal, ca. 1 mm long, 4- to 5-angled, moderately pubescent; pappus of 8–11 entire or slightly lacerate, non-aristate scales 1–2 mm long.

Wet flatwoods. Common; nearly throughout. North Carolina south to Florida, west to Alabama. Spring.

Helenium vernale Walter [*Vernalis,* pertaining to spring, in reference to the flowering time.] SAVANNAH SNEEZEWEED.

Helenium vernale Walter, Fl. Carol. 210. 1788. *Leptopoda puberula* Macbride ex Elliott, Sketch Bot. S. Carolina 2: 445. 1823, nom. illegit. *Heleniastrum vernale* (Walter) Kuntze, Revis. Gen. Pl. 1: 342. 1891.

Gaillardia fimbriata Michaux, Fl. Bor.-Amer. 2: 142. 1803. *Leptopoda fimbriata* (Michaux) Eaton, Man. Bot. 275. 1829. *Helenium fimbriatum* (Michaux) A. Gray, Proc. Amer. Acad. Arts 9: 204. 1874; non A. W. Wood, 1870. *Heleniastrum fimbriatum* (Michaux) Kuntze, Revis. Gen. Pl. 1: 342. 1891. TYPE: "Carolina ad Floridam."

Leptopoda helenium Nuttall, Gen. N. Amer. Pl. 2: 174. 1818. *Leptopoda decurrens* Macbride ex Elliott, Sketch Bot. S. Carolina 2: 446. 1823, nom. illegit. *Leptopoda helenioides* Cassini, in Cuvier, Dict. Sci. Nat. 28: 80. 1823, nom. illegit. *Helenium nuttallii* A. Gray, Proc. Amer. Acad. Arts 9: 204. 1874. *Heleniastrum helenium* (Nuttall) Kuntze, Revis. Gen. Pl. 1: 342. 1891. *Helenium helenium* (Nuttall) Small, Fl. S.E. U.S. 1292, 1341. 1903, nom. inadmiss. *Helenium decurrens* Moldenke, Bull. Torrey Bot. Club 62: 230. 1935, nom. illegit.

Leptopoda floridana Rafinesque, Atl. J. 147. 1832. TYPE: FLORIDA.

Helenium discovatum Rafinesque, New Fl. 4: 81. 1838 ("1836"). TYPE: FLORIDA.

Helenium fimbriatum A. W. Wood, Amer. Bot. Fl. 182. 1870. TYPE: FLORIDA.

Helenium leptopoda A. W. Wood, Amer. Bot. Fl. 182. 1870. TYPE: "S. Car. to Fla."

Perennial herb, to 8 dm; stem usually unbranched distally, striate-angled, weakly winged, glabrous proximally, glabrous or sparsely pubescent distally. Leaves usually with the blade to 7 cm long, to 1.5 cm wide, reduced upward, the upper and lower surfaces glabrous, rarely sparsely pubescent, the basal with the blade obovate or narrowly oblanceolate, the margin entire, undulate, or undulate-serrate, the proximal and midcauline narrowly lanceolate or narrowly oblanceolate, the margin entire or sometimes toothed, decurrent on the stem, the distal linear-lanceolate, the margin entire, petiolate or epetiolate. Flowers in solitary or rarely 2–3 capitula, the peduncle 2–23 cm long, glabrous or sparsely pubescent; involucre hemispheric, 1.5–2.5 cm wide; phyllaries numerous, in (1)–2 series, linear-lanceolate, 1–1.5 cm long, the outer surface glabrous or sparsely pubescent, gland-dotted; ray flowers numerous, sterile; corolla yellow, the lamina 1.5–2 cm long; disk flowers numerous; corolla yellow proximally, yellow, 5–6 mm long. Cypsela obpyramidal, 1–2 mm long, 4- to 5-angled, glabrous; pappus of 8 entire or lacerate, non-aristate scales ca. 2 mm long.

Bogs. Occasional; northern peninsula, central and western panhandle. North Carolina south to Florida, west to Louisiana. Spring.

EXCLUDED TAXON

Helenium quadridentatum Labillardiere—Reported for Florida by Small (1903, 1913a, 1933). No Florida specimens seen.

Helianthus L. SUNFLOWER

Herbs. Leaves opposite or alternate, simple, pinnate or pinnipalmate, petiolate or epetiolate. Flowers in capitula, solitary or in corymbiform arrays, radiate; receptacle paleate; ray flowers sterile; petals 5, basally connate, zygomorphic; disk flowers bisexual; petals 5, basally connate, actinomorphic; stamens 5, epipetalous, the filaments free, the anthers connate; ovary inferior, 2-carpellate, 1-loculate. Fruit a cypsela; pappus present.

A genus of about 70 species; nearly cosmopolitan. [From the Greek *Helios,* sun, and *anthos,* flower, in reference to the capitulum with yellow ray flowers resembling the sun.]

Selected reference: Heiser et al. (1969); Schilling (2006a).

1. Plant with the basal leaves well developed, the upper leaves few and reduced.
    2. Ray flowers absent or if present, these few and the laminae 1–2(10) mm long .... **H. radula**
    2. Ray flowers present, the lamina 1.5–3.5 cm long.
        3. Disk flowers reddish or purplish .... **H. heterophyllus**
        3. Disk flowers yellow.
            4. Leaves scabrous or hirsute, the larger blades 2–8 cm wide, 1.5 times as long as wide .... **H. occidentalis**
            4. Leaves glabrous, the larger blades less than 2 cm wide, more than 5 times as long as wide.... .... **H. carnosus**
1. Plant with many well-developed leaves, the basal leaves few or absent.
    5. Disk flowers reddish or purplish.
        6. Plant densely silvery white-pubescent .... **H. argophyllus**
        6. Plant not densely white-lanate.
            7. Phyllaries oblong or ovate.
                8. Leaves mostly on the lower part of the stem, usually opposite, the base broadly decurrent on the petiole; plant perennial .... **H. atrorubens**
                8. Leaves distributed evenly along the stem, usually alternate, the base scarcely or not decurrent on the petiole; plant annual .... **H. annuus**
            7. Phyllaries linear or lanceolate.
                9. Leaves distinctly long-petiolate, the petiole 1.5–7 cm long .... **H. debilis**
                9. Leaves sessile or short-petiolate, the petiole less than 0.5(1.2) cm long.
                    10. Leaves linear, the margin strongly revolute, only the midvein evident .... **H. angustifolius**
                    10. Leaves lanceolate, narrowly elliptic, or narrowly ovate, the margin not strongly revolute, with evident lateral veins.
                        11. Leaf margin undulate; leaf base and petiole not conspicuously hispid-ciliate; plant perennial .... **H. floridanus**
                        11. Leaf margin not undulate; leaf base and petiole conspicuously hispid-ciliate; plant annual .... **H. agrestis**
    5. Disk flowers yellow.
        12. Leaves linear, the margin strongly revolute, only the midvein evident .... **H. angustifolius**
        12. Leaves lanceolate, narrowly elliptic, or narrowly ovate, the margin not strongly revolute, with the lateral veins at least faintly evident.

13. Involucre conspicuously reflexed, at least in the upper half, conspicuously gland-dotted..... .... **H. resinosus**
13. Involucre merely loose or spreading, not conspicuously reflexed, not conspicuously gland-dotted.
 14. Leaf margin revolute........ **H. floridanus**
 14. Leaf margin flat.
  15. Petiole of the lower leaves 4–12 cm long; plant tuber-bearing........ **H. tuberosus**
  15. Petiole of the lower leaves to 3 cm long; plant not tuber bearing.
   16. Disk to 1.5 cm wide; ray flowers 5–8(10)........ **H. microcephalus**
   16. Disk 1.5–3.5 cm wide; ray flowers 10–25.
    17. Stem glabrous or rarely with a few long trichomes.
     18. Leaves sessile or subsessile........ **H. divaricatus**
     18. Leaves with a petiole 1–3 cm long........ **H. strumosus**
    17. Stem evidently hirsute.
     19. Leaves all or mostly opposite; petiole 5–20 mm long........ **H. hirsutus**
     19. Leaves all or mostly alternate; petiole less than 1 mm long....... **H. simulans**

Helianthus agrestis Pollard [Pertaining to fields or cultivated land.]
SOUTHEASTERN SUNFLOWER.

Helianthus agrestis Pollard, Proc. Biol. Soc. Wash. 13: 184. 1900. TYPE: FLORIDA: Volusia Co.: between Lake Beresford and the St. Johns River, 12 Jul 1900, *Curtiss s.n.* (holotype: US).
Helianthus tuberosus Linnaeus forma *oswaldiae* Oswald, Phytologia 44: 419. 1979. TYPE: FLORIDA: Lee Co.: 110 ft. W of Magnolia Drive and 300 ft. S of Bayshore Road, North Fort Myers, 4 Jul 1979, *Oswald & Oswald s.n.* (holotype: LL).
Helianthus agrestis Pollard forma *almae* Wolde, Phytologia 63: 310. 1987. TYPE: FLORIDA: Lee Co.: 460 ft. S of Bayshore Road, halfway between Crescent Lake and Magnolia Drives, North Fort Myers, 2 Nov 1986, *Wolde s.n.* (holotype: LL).
Helianthus agrestis Pollard forma *oswaldii* Wolde, Phytologia 63: 310. 1987. TYPE: Lee Co.: 440 ft. S of Bayshore Road, halfway between Crescent Lake and Magnolia Drives, North Fort Myers, 19 Oct 1986, *Oswalde s.n.* (holotype: LL).

Erect annual herb, to 1(2) m; stem striate-angled, glabrous or glabrate, glaucous. Leaves mostly opposite, the blade, lanceolate, 6–11 cm long, 0.7–1.9 cm wide, the apex acute, the base cuneate, the margin serrate, the upper and lower surfaces scabrous, the petiole 0.5–1.2 cm long. Flowers in a solitary or 2–15 capitula in a corymbiform array, the peduncle 2–8 cm long; involucre hemispheric, 1–1.5 cm wide; phyllaries numerous, lanceolate, 8–9 mm long, the apex acuminate, the outer surface glabrate, the margin ciliate; receptacle flat or slightly convex; paleae 5–7 mm long, subentire or rarely 3-toothed, the apex purplish; ray flowers ca. 12; corolla yellow, the lamina 1.5–2.5 cm long; disk flowers numerous; corolla yellow with the lobes reddish purple, ca. 4 mm long. Cypsela obpyramidal, slightly compressed, ca. 3 mm long, glabrous, slightly tuberculate; pappus of 2 aristate scales ca. 2 mm long.

Marshes and wet flatwoods. Frequent; peninsula. Florida and Georgia. Summer–fall.

Helianthus angustifolius L. [*Angusti*, narrow, with narrow leaves.] NARROWLEAF SUNFLOWER; SWAMP SUNFLOWER.

Helianthus angustifolius Linnaeus, Sp. Pl. 906. 1753. *Leighia bicolor* Cassini, in Cuvier, Dict. Sci. Nat. 25: 436. 1822, nom. illegit.
Coreopsis angustifolia Linnaeus, Sp. Pl. 908. 1753.

Erect perennial herb, to 1.5 m; stem striate-angled, usually strigose, hispid, or hirsute. Leaves opposite or alternate, the blade linear, 8–15 cm long, 2–5(10) mm long, the apex acute, the base cuneate, the margin entire, revolute, the upper and lower surfaces strigose, hispid, or hirsute, the lower surface sparsely glaucous, sometimes gland-dotted, sessile or subsessile. Flowers in 3–15 capitula in a corymbiform array, the peduncle 5–15 cm long; involucre hemispheric, 1–2 cm wide; phyllaries numerous, lanceolate, 4–9 mm long, the apex acute or acuminate, the outer surface scabrous, rarely glabrous, usually gland-dotted; receptacle flat or slightly convex; paleae oblanceolate, 5–7 mm long, entire or 3-toothed, the apex mucronate, purplish, gland-dotted; ray flowers numerous; corolla yellow, the lamina 1–2 cm long, the lower surface gland-dotted; disk flowers numerous; corolla yellow, 4–5 mm long. Cypsela obpyramidal, slightly compressed, 2–3 mm long, glabrate; pappus of 2 aristate scales ca. 2 mm long.

Marshes, wet flatwoods, and wet disturbed sites. Frequent; northern counties, central peninsula. New York south to Florida, west to Illinois, Missouri, Oklahoma, and Texas. Summer–fall.

Helianthus annuus L. [Yearly, in reference to the plant an annual.] COMMON SUNFLOWER.

Helianthus annuus Linnaeus, Sp. Pl. 904. 1753. *Helianthus platycephalus* Cassini, in Cuvier, Dict. Sci. Nat. 20: 352. 1821, nom. illegit.

Erect annual herb, to 3 m; stem striate-angled, usually hispid. Leaves mostly alternate, the blade lanceolate-ovate or ovate, 10–40 cm long, 5–40 cm wide, the apex acute, the base cuneate, subcordate, or cordate, the margin serrate, the upper surface glabrate, the lower surface somewhat hispid, sometimes gland-dotted, the petiole 2–20 cm long. Flowers in a solitary or 2–9 capitula in a corymbiform array, the peduncle 2–20 cm long; involucre hemispheric, 1.5–4(20) cm wide; phyllaries numerous, ovate or ovate-lanceolate, 1.3–2.5 cm long, the apex long-acuminate, the outer surface hirsute or hispid, rarely glabrate or glabrous, the margin usually ciliate; receptacle flat or slightly convex; paleae 9–11 mm long, 3-toothed, the middle tooth long-acuminate, glabrous or hispid; ray flowers numerous; corolla yellow, the lamina 2.5–5 cm long; disk flowers numerous; corolla yellow with the lobes reddish-yellow, 5–8 mm long. Cypsela obpyramidal, slightly compressed, (3)4–5(15) mm long, glabrate; pappus of 2 lanceolate scales 2–4 mm long and 0–4 obtuse scales ca. 1 mm long.

Disturbed sites. Occasional; peninsula, central and western panhandle. Escaped from cultivation. Nearly throughout North America. Nearly cosmopolitan. Native to the western United States and Mexico. Summer–fall.

Helianthus argophyllus Torr. & A. Gray [*Argo*, silver, and the Greek *phyllon*, leaf, in reference to silvery indumentum of the leaves.] SILVERLEAF SUNFLOWER.

Helianthus argophyllus Torrey & A. Gray, Fl. N. Amer. 2: 318. 1842. *Helianthus annuus* Linnaeus var. *argophyllus* (Torrey & A. Gray) Anashchenko, Bot. Zhurn. (Moscow & Leningrad) 59: 1476. 1974.

Erect annual herb, to 3 m; stem striate-angled, silvery white tomentose or floccose. Leaves mostly alternate, the blade ovate or lanceolate-ovate, 15–25 cm long, 10–20 cm wide, the apex acute, the base truncate or subcordate, the margin entire or serrulate, the upper surface glabrous, the lower surface silvery white tomentose, floccose, or sericeous, gland-dotted, the petiole 2–10 cm long. Flowers in a solitary or 4–5 capitula in a corymbiform array, the peduncle 2–8 cm long; involucre hemispheric, 2–3 cm wide; phyllaries numerous, ovate or lanceolate-ovate, 1.5–1.8 cm long, the apex long-attenuate, the outer surface silvery white villous, gland-dotted; receptacle flat or slightly convex; paleae 10–11 mm long, 3-toothed, the apex sparsely villous or glabrous; ray flowers numerous; corolla yellow, the lamina 2–3 cm long; disk flowers numerous; corolla yellow with the lobes reddish, 7–8 mm long. Cypsela obpyramidal, slightly compressed, 4–6 mm long, glabrate; pappus of 2 aristate scales 2–3 mm long.

Disturbed sites. Rare; Columbia, Volusia, Leon, and Franklin Counties. North Carolina, Florida, and Texas; South America, Africa, and Argentina. Native to Texas. Summer–fall.

Helianthus atrorubens L. [Dark reddish, in reference to disk flowers.] PURPLEDISK SUNFLOWER.

Helianthus atrorubens Linnaeus, Sp. Pl. 906. 1753. *Helianthus atrorubens* Linnaeus var. *normalis* Kuntze, Revis. Gen. Pl. 1: 343. 1891, nom. inadmiss.

Erect perennial herb, to 2 m; stem striate-angled, proximally villous or strigose-hispid, distally hispid or glabrate. Leaves opposite, the blade lanceolate or ovate, 7–26 cm long, 3–10 cm wide, the apex acute, the base cuneate or subtruncate, often decurrent to ½ of the petiole, the margin serrate or crenate, the upper and lower surfaces strigose-hispid, the lower surfaces sparsely glaucous, the petiole 4–25 cm long. Flowers in a solitary or 3–5 capitula in a corymbiform array, the peduncle 0.5–7 cm long; involucre broadly hemispheric, 1–1.5 cm wide; phyllaries numerous, ovate or oblong, 7–9 mm long, the apex obtuse or acute, sometimes mucronate, the outer surface usually glabrous, the margin ciliolate; receptacle flat or slightly convex; paleae 4–6 mm long, slightly 3-toothed or entire; ray flowers 8–13; corolla yellow, the lamina 1.5–2 cm long; disk flowers numerous; corolla yellow with the lobes reddish, 6–7 mm long. Cypsela obpyramidal, slightly compressed, ca. 3 mm long, glabrous or distally puberulent; pappus of 2 aristate scales ca. 2 mm long.

Hammocks. Rare; Jackson County. Virginia south to Florida, west to Kentucky, Tennessee, Mississippi, and Alabama. Fall.

Helianthus carnosus Small [Slightly fleshy, in reference to the leaves.] LAKESIDE SUNFLOWER; FLATWOODS SUNFLOWER.

Helianthus carnosus Small, Torreya 2: 75. 1902. TYPE: FLORIDA: Duval Co.: San Pablo, 27 Jul 1897, *Lighthipe 320* (holotype: NY).

Erect perennial herb, to 6 dm; stem striate-angled, glabrous. Leaves mostly basal, alternate or the proximal sometimes opposite, the blade lanceolate, linear-lanceolate, or linear, 10–25 cm long, 0.5–1.5 cm wide, the apex acute, the base cuneate, the margin entire, the upper and lower surfaces glabrous, the petiole intergrading with the blade. Flowers in a solitary or 2–3 capitula, the peduncle 5–10 mm long; involucre hemispheric, 2–2.5 cm wide; phyllaries numerous, ovate or ovate-lanceolate, 8–16 mm long, the apex acuminate, the outer surface glabrous or sometimes puberulent, with purple veins; receptacle flat or slightly convex; paleae 8–9 mm long, 3-toothed, the middle tooth long-acuminate; ray flowers numerous; corolla yellow, the lamina 2–4 cm long; disk flowers numerous; corolla yellow, 6–7 mm long. Cypsela obpyramidal, slightly compressed, ca. 3 mm long, glabrous; pappus of 2 aristate scales ca. 2 mm long and 2–4 erose scales less than 1 mm long.

Wet prairies and flatwoods. Occasional: northern peninsula, Volusia County. Endemic. Summer–fall.

Helianthus carnosus is listed as endangered in Florida (Florida Administrative Code, Chapter 5B-40).

Helianthus debilis Nutt. [Weak, in reference to its often reclining habit.] DUNE SUNFLOWER.

Erect or decumbent annual or perennial herb, to 2 m; stem glabrous, hirsute, hispid, or puberulent. Leaves mostly alternate, the blade deltate-ovate, lanceolate-ovate, or ovate, 2.5–14 cm long, 2–13 cm wide, the apex acute or subobtuse, the base cordate, truncate, or broadly cuneate, the margin subentire or serrate, the upper surface glabrous or glabrate, the lower surface glabrate or hispid, the petiole 1–7 cm long. Flowers solitary or 2–3 in capitula, the peduncle 9–50 cm long; involucre hemispheric, 1–2 cm wide; phyllaries numerous, linear-lanceolate, 8–17 mm long, the apex acute or long-acuminate, the outer surface glabrous or slightly hispid; receptacle flat or slightly convex; paleae ca. 8 cm long, 3-toothed, the middle tooth acuminate, hispid or glabrous; ray flowers numerous; corolla yellow, the lamina 1–2 cm long; disk flowers numerous; corolla yellow with the lobes usually reddish, ca. 5 mm long. Cypsela obpyramidal, slightly compressed, 3–4 mm long, glabrate; pappus of 2 lanceolate scales 2–3 mm long.

Distribution of the various subspecies of *H. debilis* is somewhat problematic as it is frequently in cultivation and can become locally established.

1. Stem erect, usually conspicuously red-brown mottled .... subsp. **cucumerifolius**
1. Stem decumbent with ascending or erect flowering branches, slightly or not at all red-brown mottled.
   2. Leaves entire or shallowly and fairly evenly toothed; stem glabrous or puberulent .... subsp. **debilis**
   2. Leaves coarsely and irregularly toothed; stem villous .... subsp. **vestitus**

Helianthus debilis subsp. **debilis** EAST COAST DUNE SUNFLOWER.

Helianthus debilis Nuttall, Trans. Amer. Philos. Soc., ser. 2. 7: 367. 1841. *Helianthus annuus* Linnaeus var. *debilis* (Nuttall) Anashchenko, Bot. Zhurn. (Moscow & Leningrad) 59: 1476. 1974. TYPE: FLORIDA: s.d., *Baldwin s.n.* (holotype: BM).

Stem decumbent with ascending or erect flowering branches, glabrous or puberulent, slightly or not at all red-brown mottled. Leaves entire or shallowly and fairly evenly toothed.

Coastal dunes. Frequent; eastern peninsula, Pinellas County. Endemic. All year.

Helianthus debilis subsp. **cucumerifolius** (Torr. & A. Gray) Heiser [With leaves resembling that of a cucumber (*Cucumis sativa,* Cucurbitaceae).] CUCUMBERLEAF DUNE SUNFLOWER.

Helianthus cucumerifolius Torrey & A. Gray, Fl. N. Amer. 2: 319. 1842. *Helianthus debilis* Nuttall var. *cucumerifolius* (Torrey & A. Gray) A. Gray, Syn. Fl. N. Amer. 1(2): 273. 1884. *Helianthus debilis* Nuttall forma *cucumerifolius* (Torrey & A. Gray) Voss, Vilm. Blumengärten., ed. 3. 1: 482. 1894. *Helianthus debilis* Nuttall subsp. *cucumerifolius* (Torrey & A. Gray) Heiser, Madroño 13: 160. 1956.

Helianthus debilis Nuttall subsp. *tardiflorus* Heiser, Madroño 13: 156. 1956. *Helianthus debilis* Nuttall var. *tardiflorus* (Heiser) Cronquist, Brittonia 29: 223. 1977. TYPE: FLORIDA: Sarsota Co.: Sarasota Key, s.d., *Curtiss 1455* (holotype: MO).

Stem erect, hispid, usually conspicuously red-brown mottled. Leaves coarsely and irregularly serrate.

Open sandy sites, often coastal dunes. Frequent; northern counties, central peninsula. Maine south to Florida, west to Texas, also Michigan. Summer.

Helianthus debilis subsp. **vestitus** (E. Watson) Heiser [Clothed, in reference to the leaf surface densely gland-dotted.] WEST COAST DUNE SUNFLOWER.

Helianthus vestitus E. Watson, Pap. Michigan Acad. Sci. 9: 347. 1929. *Helianthus debilis* Nuttall subsp. *vestitus* (E. Watson) Heiser, Madroño 13: 154. 1956. *Helianthus debilis* Nuttall var. *vestitus* (E. Watson) Cronquist, Brittonia 29: 223. 1977. TYPE: FLORIDA: Pinellas Co.: Hog (Caladesi) Island, 18 Apr 1900, *Tracy 6919* (holotype: MSC; isotype: NY).

Stem decumbent with ascending or erect flowering branches, hispid, slightly or not at all red-brown mottled. Leaves coarsely and irregularly toothed.

Coastal dunes. Occasional; western central peninsula. Endemic. All year.

Helianthus divaricatus L. [Spreading at a wide angle, in reference to the opposite leaves.] WOODLAND SUNFLOWER.

Helianthus divaricatus Linnaeus, Sp. Pl. 906. 1753.

Erect perennial herb, to 1.5 m; stem striate-angled, usually glabrous, rarely slightly hispid, often glaucous. Leaves opposite, the blade lanceolate or lanceolate-ovate, 6–15 cm long, 1–5 cm wide, the apex acute, the base rounded or cordate, the margin subentire or serrate, the upper surface glabrous, the lower surface hispid, sparsely glaucous, gland-dotted, sessile or subsessile. Flowers in a solitary or 2–10 capitula in a corymbiform array, the peduncle 0.5–9 cm long; involucre hermispheric, 1–1.5 cm wide; phyllaries numerous, lanceolate or linear-lanceolate, 6–12 mm long, the apex acuminate or attenuate, the outer surface hispidulous or glabrate, the margin ciliate; receptacle flat or slightly convex; paleae 5–8 mm long, 3-toothed, the teeth ciliate; ray flowers 8–12; corolla yellow, the lamina 1.5–3 cm long; disk flowers numerous; corolla yellow,

4–6 mm long. Cypsela obpyramidal, slightly compressed, 3–4 mm long, glabrate; pappus of 2 aristate scales ca. 2 mm long.

Flatwoods. Rare; Wakulla and Jackson Counties. Maine south to Florida, west to Ontario, Wisconsin, Iowa, Missouri, Oklahoma, and Louisiana. Summer.

Helianthus floridanus A. Gray ex Chapm. [Of Florida.] FLORIDA SUNFLOWER.

Helianthus floridanus A. Gray ex Chapman, Fl. South. U.S., ed. 2. 629. 1883. TYPE: FLORIDA: s.d., *Palmer 283* (lectotype: GH). Lectotypified by Heiser et al. (1969: 191).
Helianthus undulatus Chapman, Fl. South. U.S., ed. 3. 253. 1897. TYPE: FLORIDA: s.d., *Chapman s.n.* (lectotype: GH). Lectotypified by Heiser et al. (1969: 191).

Erect perennial herb, to 2 m; stem striate-angled, hispid or hispidulous. Leaves opposite or alternate (distally), rarely whorled, the blade lanceolate, elliptic, or lanceolate-ovate, 4–15 cm long, 0.5–6 cm wide, the apex acute, the base rounded or cuneate, the margin entire, undulate, or serrulate, often revolute, the upper surface somewhat scabrous, the lower surface scabrous or tomentulose, sparsely glaucous, gland-dotted, sessile or the petiole to 0.5(1) cm long. Flowers in a solitary or 2–6 capitula in a corymbiform array, the peduncle 2–15 cm long; involucre hemispheric, 1–2 cm wide; phyllaries numerous, lanceolate, 5–10 mm long, the apex obtuse, the outer surface hispid or hispidulous, gland-dotted; receptacle flat or slightly convex; paleae 5–6 mm long, entire or 3-toothed, the apex puberulent or glabrate, purplish, often gland-dotted; ray flowers numerous; corolla yellow, the lamina 2–2.5 cm long, the lower surface gland-dotted; disk flowers numerous; corolla yellow with the lobes reddish or yellow, 4–5 mm long. Cypsela obpyramidal, slightly compressed, ca. 3 mm long, glabrous; pappus of 2 aristate scales ca. 2 mm long.

Moist flatwoods. Occasional; northern and central peninsula. North Carolina south to Florida, west to Louisiana. Summer–fall.

Helianthus heterophyllus Nutt. [Having leaves of more than one form.] VARIABLELEAF SUNFLOWER.

Helianthus heterophyllus Nuttall, J. Acad. Nat. Sci. Philadelphia 7: 74. 1834. *Helianthus atrorubens* Linnaeus subsp. *heterophyllus* (Nuttall) Anashchenko, Bot. Zhurn. (Moscow & Leningrad) 59: 1477. 1974.
Helianthus elongatus Small, Fl. S.E. U.S. 1266, 1340. 1903. TYPE: FLORIDA: Wakulla Co.: St. Marks, 3 Sep 1895, *Nash 2541* (holotype: NY).

Erect perennial herb, to 1.2 m; stem striate-angled, usually hispid or subhirsute. Leaves mostly basal, mostly opposite, the blade ovate, lanceolate, or spatulate, 6–28 cm long, 1–4 cm wide, the cauline relatively few and narrowly lanceolate or linear, much reduced, the apex acute, the base cuneate, the margin entire, often revolute, the upper and lower surfaces hispid or subhirsute, the sessile or the petiole to 3 cm long. Flowers in a solitary or 2–3(5) capitula, the peduncle 10–15 cm long; involucre hemispheric, 1.5–2.5 cm wide; phyllaries numerous, lanceolate or ovate-lanceolate, 8–12 mm long, the apex acute or acuminate, the outer surface sparsely hispid or glabrate, the margin sometimes ciliate; receptacle flat or slightly convex; paleae 7–9 mm long, 3-toothed, the apex purplish; ray flowers numerous; corolla yellow, the lamina 1.5–3.5 cm long; disk flowers numerous; corolla yellow with the lobes reddish, 6–7 mm long. Cypsela obpyramidal, slightly compressed, 4–5 mm long, glabrate; pappus of 2 aristate scales ca. 2 mm long and 1–3 deltate scales ca. 1 mm long.

Wet prairies, wet flatwoods, and bogs. Occasional; Jefferson County, central and western panhandle. North Carolina south to Florida, west to Texas. Summer–fall.

Helianthus hirsutus Raf. [With coarse trichomes, in reference to the leaf and stem surfaces.] HAIRY SUNFLOWER.

Helianthus hirsutus Rafinesque, Ann. Nat. 14. 1820.

Erect perennial herb, to 2 m; stem striate-angled, hirsute. Leaves mostly opposite, the blade lanceolate or ovate, 6.5–18 cm long, 1–8 cm wide, the apex acute, the base truncate, broadly rounded, or cuneate, the margin subentire or serrate, the upper and lower surfaces slightly hirsute, the lower surface gland-dotted, the petiole 0.5–2 cm long. Flowers in a solitary or 2–7 capitula, the peduncle 1–5 cm long; involucre hemispheric, 1–1.5 cm wide; phyllaries numerous, lanceolate or linear-lanceolate, 6–12 mm long, the apex acuminate or attenuate, the outer surface hispidulous or glabrate, the margin ciliate; receptacle flat or slightly convex; paleae 7–10 mm long, 3-toothed, the apex yellowish, pubescent; ray flowers numerous; corolla yellow, the lamina 1.5–2 cm long; disk flowers numerous; corolla yellow, 6–7 mm long. Cypsela obpyramidal, slightly compressed, 4–5 mm long, glabrate or distally puberulent; pappus of 2 aristate scales 2–3 mm long.

Dry hammocks. Rare; Columbia County, panhandle. Connecticut and New York south to Florida, west to Ontario, Minnesota, Iowa, Nebraska, Kansas, Oklahoma, and Texas; Mexico. Fall.

Helianthus microcephalus Torr. & A. Gray [With small flower heads.] SMALL WOODLAND SUNFLOWER.

Helianthus microcephalus Torrey & A. Gray, Fl. N. Amer. 2: 329. 1842.
Helianthus parviflorus Bernhardi ex Sprengel, Syst. Veg. 3: 617. 1826; non Kunth, 1820.

Erect perennial herb, to 2 m; stem striate-angled, glabrous. Leaves opposite or alternate, the blade lanceolate, 7–15 cm long, 1–4 mm wide, the apex acute, the base cuneate, the margin entire or serrate, the upper surface glabrous, the lower surface tomentellous, gland-dotted, the petiole to 3 cm long. Flowers in 3–15 capitula in a corymbiform array, rarely solitary, the peduncle 1–3(8) cm long; involucre cylindric, 5–7 mm wide; phyllaries numerous, linear-lanceolate, 3–7 mm long, the apex acuminate, the outer surface glabrate, the margin ciliate; receptacle flat or slightly convex; paleae 5–7 mm long, 3-toothed, the apex pubescent; ray flowers 5–8; corolla yellow, the lamina 1–1.5 cm long; disk flowers numerous; corolla yellow, 4–5 mm long. Cypsela obpyramidal, slightly compressed, ca. 4 mm long, glabrous; pappus of 2 aristate scales ca. 2 mm long.

Hammocks. Rare; central and western panhandle. Connecticut and New York south to Florida, west to Ontario, Minnesota, Iowa, Nebraska, Kansas, Oklahoma, and Texas. Summer.

Helianthus occidentalis Riddell [Western.] FEWLEAF SUNFLOWER.

Helianthus occidentalis Riddell, W. J. Med. Phys. Sci. 9: 580. 1836.

Erect perennial herb, to 1.5 m; stem rhizomatous and sometimes also stoloniferous, striate-angled, often reddish, pilose to appressed pubescent proximally. Leaves mostly basal, opposite, the blade oblong-lanceolate, elliptic, or ovate, 5–20 cm long, 1.5–7 cm wide, usually much reduced upward, the apex acute, the base cuneate, the margin entire or nearly so, the upper and lower surfaces scabrous or hirsute, the lower surface gland-dotted, the petiole 2.5–10 cm long. Flowers in a solitary or 2–4(12) capitula in a corymbiform array, the peduncle 1–14 cm long; involucre cylindric, 1–1.5 cm wide; phyllaries numerous, lanceolate, 5–10 mm long, the apex acute or short-acuminate, the outer surface sparsely pilose or glabrate, the margin usually ciliate; receptacle flat or slightly convex; paleae 5–7 mm long; 3-toothed or subentire; ray flowers 8–many; corolla yellow, the lamina ca. 2 cm long, the lower surface gland-dotted; disk flowers numerous; corolla yellow, 5–6 mm long. Cypsela obpyramidal, slightly compressed, 3–4(5) mm long, sparsely villous or glabrate; pappus of 2 aristate scales 2–3 mm long and 0–4 deltate scales to 1 mm long.

Dry hammocks. Rare; Washington, Holmes, and Walton Counties. Massachusetts, Pennsylvania, south to Florida, west to Minnesota, Iowa, Missouri, Kansas, Arkansas, and Texas. Summer.

Helianthus radula (Pursh) Torr. & A. Gray [With the basal leaves scabrous like a tongue.] STIFF SUNFLOWER.

Rudbeckia radula, Pursh, Fl. Amer. Sept. 575. 1814. *Helianthus radula* (Pursh) Torrey & A. Gray, Fl. N. Amer. 2: 321. 1842. *Helianthus atrorubens* Linnaeus subsp. *radula* (Pursh) Anashchenko, Bot. Zhurn. (Moscow & Leningrad) 59: 1477. 1974.

Neactelis strigosa Rafinesque, New Fl. 4: 69. 1838. TYPE: FLORIDA.

Erect perennial herb, to 1 m; stem striate-angled, hispid distally. Leaves mostly basal, opposite, alternate distally, the blade obovate or suborbicular, the proximal cauline often broadly elliptic, 4.5–14.5 cm long, 2–12 cm wide, the distal much reduced, the apex acute or obtuse, the base broadly cuneate or rounded, the margin entire or serrulate, the upper and lower surfaces strigose-hispid, the petiole obscure, intergrading with the blade. Flowers usually in a solitary capitulum, the peduncle 10–20 cm long; involucre hemispheric, 5–8 mm wide; phyllaries numerous, lanceolate or ovate, 1–1.5 cm long, the apex acute or acuminate, the outer surface hispid or glabrous, often dark purple; receptacle flat or slightly convex; paleae 9–10 mm long, 3-toothed or subentire, the apex purplish; ray flowers absent or 2–8; corolla yellow or sometimes purplish, 1–2(10) mm long; disk flowers numerous; corolla yellow with the lobes reddish, 7–8 mm long. Cypsela obpyramidal, slightly compressed, 3–4 mm long, glabrate; pappus of 2 often unequal aristate scales ca. 1–3 mm long.

Wet flatwoods. Frequent; nearly throughout. South Carolina south to Florida, west to Louisiana. Summer.

Helianthus resinosus Small [In reference to the resinous glandular dots on the leaves.] RESINDOT SUNFLOWER.

Helianthus resinosus Small, Fl. S.E. U.S. 1269, 1340. 1903. TYPE: FLORIDA: Gadsden Co.: River Junction, 5 Sep 1895, *Nash 2581* (holotype: NY).

Erect perennial herb, to 3 m; stem striate-angled, often reddish or purplish, hirsute or villous. Leaves mostly alternate, the blade lanceolate or ovate, 6.5–20 cm long, 3–9 cm wide, the apex acute, the base gradually narrowed onto the petiole, the margin entire or serrate, the upper and lower surfaces hirsute, gland-dotted, the lower surface sparsely glaucous, the petiole 0.5–1 cm long. Flowers in 1 or 2–5 capitula, the peduncle 1–3 cm long; involucre hemispheric, 1.5–2.5 cm wide; phyllaries numerous, lanceolate, 1–2 cm long, the apex acute or attenuate, squarrose or reflexed, the outer surface hispid or villous, gland-dotted; receptacle flat or slightly convex; paleae 9–11 mm long, 3-toothed, gland-dotted; ray flowers numerous; corolla yellow (often light yellow), the lamina 2–3 cm long; disk flowers numerous; corolla yellow, 8–9 mm long. Cypsela obpyramidal, slightly compressed, 5–7 mm long, glabrate; pappus of 2 aristate scales 2–3 mm long.

Dry hammocks and bluff forests. Rare; Gadsden, Jackson, Calhoun, and Walton Counties. North Carolina south to Florida, west to Mississippi. Summer–fall.

Helianthus simulans E. Watson [In reference to its resemblance to other species of *Helianthus*, i.e., *H. angustifolius*.] MUCK SUNFLOWER.

Helianthus simulans E. Watson, Pap. Michigan Acad. Sci. 363. 1929.

Erect perennial herb, to 2.5 m; stem rhizomatous, striate-angled, strigose or hirsute. Leaves opposite or alternate, the blade linear-lanceolate or lanceolate, 9–22 cm long, 0.7–4 cm wide, the apex acute, the base cuneate, the margin entire or subentire, usually slightly revolute, the upper and lower surfaces hispidulous, the lower surface gland-dotted, the petiole 0.5–1 cm long. Flowers in a solitary or in 2–15 capitula in a corymbiform array, the peduncle 1–13 cm long; involucre hemispheric, 1.3–1.8 cm wide, linear-lanceolate or lanceolate, 6–17 mm long, the apex acuminate, the outer surface sparsely strigose or glabrate, gland-dotted; receptacle flat or slightly convex; paleae 6–7 mm long, 3- or 5-toothed, the apex strigose, gland-dotted; ray flowers numerous; corolla yellow, the lamina 1.5–4 cm long, the lower surface gland-dotted; disk flowers numerous; corolla yellow with the lobes sometimes reddish, 5–6 mm long. Cypsela obpyramidal, slightly compressed, ca. 3 mm long, glabrous; pappus of 2 usually unequal aristate scales ca. 2 mm long.

Moist or wet disturbed sites. Rare; northern counties, central peninsula. Escaped from cultivation. North Carolina south to Florida, west to Texas. Native to Alabama, Arkansas, Louisiana, and Texas. Fall.

Some Florida plants may possibly be of hybrid origin.

Helianthus strumosus L. [With a swelling, this at the stem base.] PALELEAF WOODLAND SUNFLOWER.

Helianthus strumosus Linnaeus, Sp. Pl. 905. 1753.
Helianthus australis Small, Fl. S.E. U.S. 1268. 1903; non Philippi, 1845. *Helianthus montanus* E. Watson, Pap. Michigan Acad. Sci. 9: 403. 1929.

Erect perennial herb, to 2 m; stem rhizomatous, striate-angled, glabrate. Leaves opposite or sometimes alternate distally, the blade lanceolate, lanceolate-ovate, or ovate, 7–18 cm long, 2–10 cm wide, the apex acute, the base subcordate or cuneate, the margin entire or slightly serrate, the upper surface glabrous, the lower surface glabrous or sparsely tomentulose, glaucous, usually densely gland-dotted, the petiole 1–3 cm long. Flowers in 3–15 capitula in a corymbiform array, the peduncle 1–9 cm long; involucre cylindric or hemispheric, 1–2 cm wide; phyllaries numerous, lanceolate, 5–10 mm long, the apex acute or acuminate, sometimes squarrose, the outer surface hispidulous or glabrous, the margin ciliate, rarely gland-dotted; receptacle flat or slightly convex; 5–7 mm long, 3-toothed; ray flowers numerous; corolla yellow, the lamina 1–2(3) cm long; disk flowers numerous; corolla yellow, 6–7 mm long. Cypsela obpyramidal, slightly compressed, 4–6 mm long, glabrate; pappus of 2 aristate scales 2–3 mm long.

Flatwoods and open hammocks. Occasional; northern peninsula west to central panhandle. Maine south to Florida, west to Ontario, North Dakota, Minnesota, Iowa, Nebraska, Kansas, Oklahoma, and Texas. Summer–fall.

Helianthus tuberosus L. [Producing tubers.] JERUSALEM ARTICHOKE.

Helianthus tuberosus Linnaeus, Sp. Pl. 905. 1753. *Helianthus tuberosus* Linnaeus var. *typicus* Cockerell, Amer. Naturalist 53: 188. 1919, nom. inadmiss.
Helianthus tomentosus Michaux, Fl. Bor.-Amer. 2: 141. 1803. *Helianthus strumosus* Linnaeus subsp. *tomentosus* (Michaux) Anashchenko, Bot. Zhurn. (Moscow & Leningrad) 59: 1480. 1974.

Erect perennial herb, to 2 m; stem rhizomatous, producing tubers late in the growing season, striate-angled, scabrous-hispid or hirsute, sometimes glaucous. Leaves opposite or alternate, the blade lanceolate or ovate, 10–23 cm long, 7–15 cm wide, the apex acute, the base cuneate, the margin entire or serrate, the upper surface scabrous, the lower surface puberulent, hirtellous, or tomentulose, gland-dotted, the petiole 2–8 cm long. Flowers in 3–15 capitula in a corymbiform array, the peduncle 1–15 cm long; involucre hemispheric, 1–1.5(2.5) cm wide; phyllaries numerous, lanceolate, 1–1.5 cm long, the apex acuminate; the outer surface hispidulous or puberulent, the margin ciliolate, gland-dotted; receptacle flat or slightly convex; paleae 8–9 mm long, 3-toothed, the apex pubescent; ray flowers numerous; corolla yellow, the lamina 2.5–4 cm long; disk flowers numerous; corolla yellow, 6–7 mm long. Cypsela obpyramidal, slightly compressed, 5–7 mm long, distally puberulent ot glabrous; pappus of 2 aristate scales 2–3 mm long and 0–1 deltate scales less than 1 mm long.

Disturbed sites. Rare; Jefferson, Leon, Jackson, and Washington Counties. Escaped from cultivation. Nearly throughout the United States and southeastern Canada; Europe. Native to the central United States and adjacent south-central Canada. Summer–fall.

HYBRID TAXON

Helianthus ×glaucus Small (*H. divaricatus* × *H. microcephalus*) [Glaucous, in reference to the stem].

Helianthus glaucus Small, Bull. Torrey Bot. Club 25: 480. 1898, pro sp.

Disturbed sites. Rare; Alachua County. Summer–fall.

EXCLUDED TAXA

Helianthus debilis Nuttall subsp. *praecox* (Engleman & A. Gray) Heiser—Reported for Florida by Chapman (1860, as *H. praecox* Engelman & A. Gray) and Small (1903, as *H. praecox* Engelman & A. Gray), the name misapplied to Florida material of *H. debilis* subsp. *vestitus*.
Helianthus giganteus Linnaeus—Reported for Florida by Small (1903). No Florida material seen.
Helianthus mollis Lamarck—Reported for Florida by Correll and Johnston (1970) as occurring in "most of e. U.S.," which would presumably include Florida. No Florida material seen.

Heliopsis Pers. 1807. OXEYE

Herbs. Leaves opposite, simple, pinnate-veined, petiolate. Flowers in capitula, solitary, radiate; receptacle paleate; ray flowers carpellate; petals 5, basally connate, zygomorphic; disk flowers bisexual; petals 5, basally connate, actinomorphic; stamens 5, epipetalous, the filaments free, the anthers connate; ovary inferior, 2-carpellate, 1-loculate. Fruit a cypsela; pappus absent.

A genus of about 15 species; North America, Mexico, Central America, and South America. [From the Greek *helios,* sun, and *opsis,* to resemble, in reference to the bright yellow flowers.]

Selected reference: Smith (2006a).

Heliopsis helianthoides (L.) Sweet var. **gracilis** (Nutt.) Gandhi & R. D. Thomas [Resembling the genus *Helianthus;* thin, slender.] SMOOTH OXEYE.

Heliopsis gracilis Nuttall, Trans. Amer. Philos. Soc., ser. 2. 7: 353. 1841. *Heliopsis laevis* (Linnaeus) Persoon var. *gracilis* (Nuttall) Torrey & A. Gray, Fl. N. Amer. 2: 303. 1842. *Heliopsis helianthoides* (Linnaeus) Sweet var. *gracilis* (Nuttall) Gandhi & R. D. Thomas, Sida, Bot. Misc. 4: 106. 1989.

Erect perennial herb, to 4(8) dm; stem rhizomatous, erect flowering stems striate-angled, stramineous or red-brown, glabrous. Leaves with the blade lanceolate or ovate-lanceolate, 4–8 cm long, 1.5–3 cm wide, the apex acuminate, the base broadly cuneate or rounded, the margin dentate or subentire, the upper surface sparsely scabrellous, the lower surface glabrous or sparsely short-pubescent, the petiole 0.5–2.5 cm long. Flowers in 1–3(5) capitula, the peduncle mostly 10–15 cm long; involucre hemispheric, 7–15 mm wide; phyllaries numerous in 2–3 series, oblong, 6–10 mm long, the outer surface minutely pubescent; receptacle convex; paleae lanceolate, yellowish, conduplicate, the apex obtuse, the surface glabrous; ray flowers 6–8(13); corolla yellow, the lamina 1–2 cm long; disk flowers numerous; corolla pale brown-yellow, the lobes dull yellow, 3–4 mm long. Cypsela obscurely 3- or 4-angled, 4–5 mm long, dark brown, glabrous or minutely pubescent on the angles; pappus a minute laciniate crown and with 1–3 toothlike scales.

Hammocks. Rare; Jefferson County, central panhandle, Walton County. Georgia south to Florida, west to Louisiana. Spring–summer.

EXCLUDED TAXA

Heliopsis helianthoides (Linnaeus) Sweet—The typical variety of this species was reported for Florida by Small (1903, 1913a, 1933), Radford et al. (1964, 1968), and Correll and Johnston (1970), the name misapplied to Florida material of *H. helianthoides* var. *gracilis.*

Heliopsis helianthoides var. *scabra* (Dunal) Fernald—Reported for Florida by Clewell (1985). All Florida material is *H. helianthoides* var. *gracilis.*

Heliopsis minor (Hooker) C. Mohr—This heterotypic synonym of *H. helianthoides* var. *scabra* (Dunal) Fernald was reported for Florida by Small (1903, 1913a, 1933), the name misapplied to Florida material of *H. helianthoides* var. *gracilis.*

Heterotheca Cass. 1817. FALSE GOLDENASTER

Herbs. Leaves alternate, simple, pinnate-veined, petiolate or epetiolate. Flowers in capitula in corymbiform or paniculiform arrays, radiate; receptacle epaleate; ray flowers carpellate; petals 5, basally connate, zygomorphic; disk flowers bisexual; petals 5, basally connate, actinomorphic; stamens 5, epipetalous, the filaments free, the anthers connate; ovary inferior, 2-carpellate, 1-loculate. Fruit a cypsela; pappus absent (ray flowers) or present (disk flowers).

A genus of about 28 species; North America, Mexico, and Pacific Islands. [From the Greek *heteros,* different, and *thece,* container, in reference to the dimorphic cypsela.]

Selected reference: Semple (2006e).

Heterotheca subaxillaris (Lam.) Britton & Rusby [In reference to the flower position.] CAMPHORWEED.

Inula subaxillaris Lamarck, Encycl. 3: 259. 1789. *Inula scabra* Pursh, Fl. Amer. Sept. 531. 1814, nom. illegit. *Heterotheca lamarckii* Cassini, in Cuvier, Dict. Sci. Nat. 21: 131. 1821, nom. illegit. *Chrysopsis scabra* Elliott, Sketch Bot. S. Carolina 2: 339. 1823, nom. illegit. *Heterotheca scabra* de Candolle, Prodr. 5: 317. 1836, nom. illegit. *Stelmanis scabra* Rafinesque, Fl. Tellur. 2: 47. 1837 ("1836"), nom. illegit. *Chrysopsis lamarckii* Nuttall, Trans. Amer. Philos. Soc., ser. 2. 7: 315. 1841, nom. illegit. *Heterotheca subaxillaris* (Lamarck) Britton & Rusby, Trans. New York Acad. Sci. 7: 10. 1887.

Heterotheca latifolia Buckley, Proc. Acad. Nat. Sci. Philadelphia 13: 459. 1862. *Heterotheca subaxillaris* (Lamarck) Britton & Rusby var. *latifolia* (Buckley) Gandhi & R. D. Thomas, Sida, Bot. Misc. 4: 110. 1989. *Heterotheca subaxillaris* (Lamarck) Britton & Rusby subsp. *latifolia* (Buckley) Semple, Sida 21: 759. 2004.

Heterotheca subaxillaris (Lamarck) Britton & Rusby var. *procumbens* B. Wagenknecht, Rhodora 62: 75. 1960.

Procumbent or erect annual or perennial herb, to 2 m; stem striate-angled, reddish brown, hispid-strigose, stipitate-glandular. Basal and proximal leaves usually not persisting at flowering, the blade ovate elliptic, or lanceolate, 1–9 cm long, 0.5–5 cm wide, reduced upward, the apex acute, the base cuneate, the margin coarsely serrate or entire, basally long-ciliate, the upper and lower surfaces moderately hispid-scabrous, the basal to midcauline with the petiole 1–4 cm long, the distal sessile. Flowers capitula in a corymbiform or paniculiform array, the peduncle 2 mm to 3.5 cm long, hispid-strigose, stipitate-glandular, usually with 1–4 leaflike bracts, these ovate or lancelate, reduced and becoming linear distally; involucre hemispheric or campanulate, 4–8(10) mm wide; phyllaries in 4–6 series, lanceolate, unequal, the apex acute, the margin scarious, ciliate, the outer surface stipitate-glandular, the apex with a tuft of 10–15 strigose trichomes; receptacle slightly convex, pitted; ray flowers numerous; corolla yellow, the lamina 3–7(9) mm long; disk flowers numerous; corolla yellow, 2–9 mm long. Cypsela obconic, 2- to 3-ribbed, dimorphic, that of the ray flowers triangular, 2–3 mm long, the surface glabrous or slightly strigose, that of the disk flowers laterally compressed, 2–4 mm long, moderately to densely strigose; pappus absent in the ray flowers, that of the disk flowers tan or rust-colored, the outer of linear to triangular short scales, the inner of numerous bristles 4–9 mm long, the longest weakly clavate.

Sandhills, coastal dunes, scrub, dry pine flatwoods, and dry disturbed sites. Common; nearly throughout. All year.

Hieracium L. 1753. HAWKWEED

Herbs. Leaves alternate, simple, pinnate-veined, petiolate or epetiolate. Flowers in capitula in thyrsiform or corymbiform arrays, liguliflorous, bisexual; calyculi present; receptacle epaleate; petals 5, basally connate, zygomorphic; stamens 5, epipetalous, the filaments free, the anthers connate; ovary inferior, 2-carpellate, 1-loculate. Fruit a cypsela; pappus present.

A genus of about 800 species; North America, West Indies, Mexico, Central America, South America, Europe, Africa, and Asia.

Selected reference: Strother (2006n).

1. Capitula in a thyrsiform array; inner phyllaries 6–7 mm long .... **H. gronovii**
1. Capitula in a corymbiform array; inner phyllaries 8–11 mm long .... **H. megacephalon**

Hieracium gronovii L. [Commemorates Johannes Fredericus Gronovius (1690–1762), Dutch botanist and patron of Linnaeus.] QUEEN-DEVIL.

Hieracium gronovii Linnaeus, Sp. Pl. 802. 1753.
Hieracium floridanum Britton, Torreya 1: 42. 1901. TYPE: FLORIDA: Leon Co.: near Tallahassee, 1895, *Berg s.n.* (holotype: NY).

Erect perennial herb, to 5(8) dm; stem striate-angled, pilose-hirsute proximally, sometimes also stellate-pubescent, stellate-pubescent distally, sometimes also pilose-hirsute. Leaves with the blade elliptic, obovate, or oblanceolate, 2–3.5(9) cm long, 1–4(5) cm wide, reduced upward, the apex rounded or acute, the base cuneate or rounded, the margin entire, the upper surface pilose-hirsute, the lower surface pilose-hirsute and stellate-pubescent, the petiole of the proximal to 1 cm long, the upper sessile or subsessile. Flowers in capitula in a usually narrow thyrsiform array, the peduncle 1–3 cm long, stellate-pubescent and stipitate-glandular; calyculi

8–12+, lanceolate; involucre cylindric or campanulate; phyllaries numerous in 2 series, the inner 6–7 mm long, the apex rounded or acuminate, the outer surface glabrous or stellate-pubescent and stipitate-glandular; receptacle flat, pitted; flowers numerous; corolla yellow, 8–9+ mm long. Cypsela urceolate, 4–5 mm long, 10-ribbed; pappus of numerous yellowish bristles in 2 series, ca. 5 mm long.

Sandhills and dry flatwoods. Occasional; northern counties, central peninsula. Maine south to Florida, west to Ontario, Minnesota, Illinois, Kansas, Oklahoma, and Texas; West Indies, Mexico, and Central America. Summer–fall.

Hieracium megacephalon Nash [From the Greek *megas,* large, and *cephale,* head, in reference to its large capitula.] COASTALPLAIN HAWKWEED.

Hieracium megacephalon Nash, Bull. Torrey Bot Club 22: 152. 1895. TYPE: FLORIDA: Lake Co.: vicinity of Eustis, 1–15 Apr 1894, Nash *390* (holotype: NY; isotype: NY).
Hieracium argyraeum Small, Bull. Torrey Bot. Club 25: 148. 1898. TYPE: FLORIDA: Volusia Co.: DeLand, Mar 1891, *Hulst s.n.* (holotype: NY).
Hieracium floridanum Gandoger, Bull. Soc. Bot. France 65: 48. 1918; non Britton, 1901. TYPE: FLORIDA: Manatee Co.: Bradenton, s.d., *Tracy 7705* (holotype: ?).

Erect perennial herb, to 4 dm; stem striate-angled, pilose-hirsute and stellate-pubescent proximally, pilose-hirsute and/or stellate-pubescent distally. Leaves with the blade elliptic or obovate, 3.5–8 cm long, 1.5–3(4) cm wide, reduced upward, the apex rounded or obtuse, the base cuneate or rounded, the margin entire or denticulate, the upper surface pilose-hirsute, the lower surface pilose-hirsute, sometimes also stellate-pubescent, the proximal with the petiole 1–4 cm long, the distal sessile. Flowers in capitula in an open corymbiform array, the peduncle 1–6 cm long, stellate-pubescent and stipitate-glandular; calyculi 5–12+, linear-lanceolate; involucre campanulate; phyllaries numerous in 2 series, the inner 8–11 mm long, the apex acute or acuminate, the outer surface stellate-pubescent and stipitate-glandular; receptacle flat, pitted; flowers numerous; corolla yellow, 10–12+ mm long. Cypsela urceolate, ca. 4 mm long, 10-ribbed; pappus of numerous yellowish bristles in 2 series, 5–7 mm long.

Sandhills and dry flatwoods. Common; peninsula west to central panhandle. North Carolina south to Florida. Summer.

EXCLUDED TAXA

Hieracium venosum Linnaeus—Reported for Florida by Radford et al. (1964, 1968) and Cronquist (1980), both apparently a misidentification of Florida material of *H. gronovii.* Also by Long and Lakela (1970), who misapplied the name to Florida material of *H. megacephalon.*
Hieracium ×*marianum* Willdenow (*H. gronovii* × *H. venosum*)—Reported for Florida by Small (1903, 1913a, 1933, all as *H. marianum*), who misapplied the name to Florida material of *H. gronovii.*

Hymenopappus L'Her. 1788.

Herbs. Leaves alternate, simple, pinnate-lobed, petiolate or epetiolate. Flowers in capitula in corymbiform arrays, discoid; receptacle epaleate; flowers bisexual; petals 5, basally connate, actinomorphic; stamens 5, epipetalous, the filaments free, the anthers connate; ovary inferior, 2-carpellate, 1-loculate. Fruit a cypsela; pappus present.

A genus of 11 species; North America and Mexico. [From the Greek *hymen,* membrane, and *pappos,* pappus, in reference to the membranous pappus scales.]

Selected reference: Strother (2006o).

Hymenopappus scabiosaeus L'Hér. [In reference to the use of the plant in folk medicine to treat scabies.] CAROLINA WOOLLYWHITE.

Hymenopappus scabiosaeus L'Héritier de Brutelle, Hymenopappus 1788.

Rothia caroliniensis Lamarck, J. Hist. Nat. 1: 17, pl. 1. 1792. *Hymenopappus caroliniensis* (Lamarck) Porter, in Porter & Britton, Mem. Torrey Bot. Club 5: 338. 1894.

Erect perennial herb, to 1.5 m; stem striate-angled, white tomentose. Leaves with the blade 1- to 2-pinnate-lobed, 5–25 cm long, reduced upward, the lobes filiform, 0.5–2 cm long, 2–8 mm wide, the lobe apex rounded or obtuse, the base cuneate, the ultimate margin entire, the upper surface glabrous, the lower surface white tomentose, the basal and proximal cauline with the petiole to 5 cm long, the mid- and lower cauline sessile or subsessile. Flowers in capitula in a corymbiform array, the peduncle 1–5 cm long, usually subtended by membranous bracts 5–14 mm long, 3–10 mm wide; involucre obconic or hemispheric; phyllaries in 2–3 series, ovate, 7–15 mm long, 4–8 mm wide, unequal, whitish, the margin membranous; receptacle slightly convex; flowers numerous; corolla whitish, 3–6 mm long. Cypsela obpyramidal, 3–5 mm long, 4- to 5-angled, each face usually 1- to 4-ribbed, slightly hirtellous; pappus of numerous scales to 1 mm long.

Dry open pine or oak hammocks and sandhills. Occasional; northern and central peninsula, west to central panhandle. South Carolina south to Florida, west to Oklahoma and Texas, also Indiana, Illinois, and Missouri. Summer.

Hypochaeris L. 1753. CAT'SEAR

Herbs. Leaves alternate, simple, pinnate-veined, epetiolate. Flowers in capitula, solitary or in paniculiform, corymbiform, or cymiform arrays, liguliflorous, bisexual; receptacle paleate; petals 5, basally connate, zygomorphic; stamens 5, epipetalous, the filaments free, the anthers connate; ovary inferior, 2-carpellate, 1-loculate. Fruit a cypsela; pappus present.

A genus of about 60 species; North America, South America, Europe, Africa, and Asia. [From the Greek *hypo,* beneath, and *choiras,* pig, in reference to pigs digging for the roots.]

Selected reference: Bogler (2006b).

1. Leaves basal and proximally cauline; pappus bristles in 1 series of plumose bristle.
   2. Corolla yellow; involucre broadly campanulate; phyllaries subhirsute, at least medially .... **H. chillensis**
   2. Corolla white; involucre cylindric or narrowly campanulate; phyllaries glabrous or sparsely tomentulose .... **H. microcephala**

1. Leaves all or mostly basal; pappus in 2 series, the outer of of barbellate bristles shorter than the inner plumose.
   3. Leaves glabrous or with trichomes only on the veins; flowers subequaling the phyllaries at flowering; cypselae dimorphic, the outer truncate, the inner slender beaked .... **H. glabra**
   3. Leaves evidently hirsute; flowers longer than the phyllaries at flowering; cypselae monomorphic, all uniformly long-beaked ....**H. radicata**

Hypochaeris chillensis (Kunth) Britton [Of Chillon, Ecuador.] SOUTH AMERICAN CAT'SEAR; TWEEDIE'S CAT'SEAR.

Apargia chillensis Kunth, in Humboldt et al., Nov. Gen. Sp. 4: 3. 1820. *Leontodon chillense* (Kunth) de Candolle, Prodr. 7: 105. 1838. *Achyrophorus chillensis* (Kunth) Schultz Bipontinus, Nov. Actorum Acad. Caes. Leop.-Carol. Nat. Cur. 21: 104. 1845. *Hypochaeris chillensis* (Kunth) Britton, Bull. Torrey Bot. Club 19: 371. 1892.

Seriola tweediei Hooker & Arnott, in Hooker, Companion Bot. Mag. 1: 31. 1835. *Hypochaeris brasiliensis* (Lessing) Bentham & Hooker f. ex Grisebach var. *tweediei* (Hooker & Arnott) Baker, in Martius, Fl. Bras. 6(3): 334. 1884. *Hypochaeris tweediei* (Hooker & Arnott) Cabrera, Notas Mus. La Plata, Bot. 2: 203. 1937.

Erect or ascending biennial or perennial herb, to 7 dm; stem striate-angled, glabrous or pilose proximally. Leaves basal and proximally cauline, the basal leaves with the blade elliptic to oblanceolate, 6–20 cm long, 1–5 cm wide, the apex acute or obtuse, the base narrowly cuneate, the margin coarsely and sharply or 2-dentate, ciliate, the upper and lower surfaces glabrous or coarsely hirsute, sessile, clasping, the cauline with the blade lanceolate, 5–10 cm long, 1–3 cm wide, the apex acute, the base cuneate, the margin sharply dentate or pinnatifid, the upper and lower surfaces glabrous or hirsute, sessile, clasping, the distal leaves reduced, entire, sessile, clasping. Flowers in solitary or 2–10 capitula in a loose paniculiform or corymbiform array, the peduncle 3–8 cm long, minutely bracteate; involucre broadly campanulate, 0.5–2 cm wide; phyllaries numerous, linear-lanceolate, 4–15 mm long, unequal, the outer surface slightly hirsute; receptacle flat, slightly pitted; paleae linear, scarious; flowers numerous; corolla yellow, 5–7 mm long, equaling the phyllaries at flowering. Cypsela monomorphic, all beaked, the beak 4–5 mm long, the body fusiform, 8–10 mm long, golden brown, muricate, 4- to 5-ribbed; pappus of white plumose bristles in 1 series, 6–8 mm long.

Disturbed sites. Occasional; northern counties, central peninsula. North Carolina south to Florida, west to Texas; South America; Africa and Asia. Native to South America. Spring.

Hypochaeris glabra L. [Glabrous.] SMOOTH CAT'SEAR.

Hypochaeris glabra Linnaeus, Sp. Pl. 811. 1753. *Hypochaeris stellata* Gaterau, Descr. Pl. Montauban 140. 1789, nom. Illegit.

Erect or ascending annual herb, to 5 dm; stem striate-angled, glabrous. Leaves usually all or mostly basal, the blade oblanceolate or oblong, 2–11 cm long, 0.5–3 cm wide, the apex acute or obtuse, the base narrowly cuneate, the margin subentire, dentate, or pinnatifid, the upper and lower surfaces glabrous or glabrate, hirsute on the veins, sessile, clasping. Flowers in solitary or 2–3 capitula in a loose cymiform array, the peduncle 2–6 cm long, minutely bracteate;

involucre narrowly campanulate, 0.5–2 cm wide; phyllaries numerous, lanceolate, 3–18 mm long, unequal, the margin scarious, the outer surface glabrous, the apex reddish or brownish, sometimes ciliate; receptacle flat, slightly pitted; paleae linear, scarious; flowers numerous, corolla white or yellowish, 5–8 mm long, subequaling the phyllaries at flowering. Cypsela dimorphic, the outer cylindric, stout, truncate, the inner fusiform, 8–10 mm long, slender, beaked, the beak 3–4 mm long, the body dark brown, 10-nerved; pappus of tawny bristles in 2 series, the outer barbellate, shorter than the inner plumose, the longest 9–10 mm long.

Disturbed sites. Rare; northern counties, central peninsula. Maine south to Florida, west to Ontario, Ohio, Tennessee, Oklahoma, and Texas, also Illinois and British Columbia south to California; Europe, Africa, Asia, and Australia. Native to Europe and Africa. Spring.

Hypochaeris microcephala (Sch. Bip.) Cabrera var. **albiflora** (Kuntze) Cabrera [From the Greek *micro,* small, and *cephale,* head, in reference to the small capitula; with white corollas.] SMALLHEAD CAT'SEAR.

Hypochaeris brasiliensis (Lessing) Hooker & Arnott var. *albiflora* Kuntze, Revis. Gen. Pl. 3(3): 159. 1898. *Hypochaeris microcephala* (Schultz Bipontinus) Cabrera var. *albiflora* (Kuntze) Cabrera, Notas Mus. La Plata, Bot. 2: 201. 1937. *Hypochaeris albiflora* (Kuntze) Azevedo-Goncalves & Matzenbacher, Compositae Newslett. 42: 3. 2005.

Erect perennial herb, to 6 dm; stem striate-angled, glabrous or sparsely pubescent distally. Leaves basal and proximally cauline, the basal leaves with the blade narrowly oblanceolate, 4–10(25) cm long, 1–5 cm wide, the apex acute, the base narrowly cuneate, the margin entire, sharply dentate, or deeply pinnatifid, the lobes long and narrow, ciliate, the upper and lower surfaces glabrous or glabrate, sessile, clasping, the cauline blades reduced upward, the margin shallowly dentate or coarsely pinnatifid, sessile, not clasping. Flowers in solitary or 2–13 capitula in a loose paniculiform or corymbiform array, the peduncle 3–8 cm long, minutely bracteate; involucre cylindric or narrowly campanulate, 0.5–1 cm wide; phyllaries numerous, narrowly lanceolate, 1–14 mm long, unequal, the margin scarious, the apex darkened, the outer surface glabrous or sparsely tomentulose; receptacle flat, slightly pitted; paleae linear, scarious; flowers numerous; corolla white, 5–7 mm long, shorter than the phyllaries at flowering. Cypsela monomorphic, all beaked, the beak 4–5 mm long, the body fusiform, 7–8 mm long, brown, muricate-roughened, 10-ribbed; pappus of white plumose bristles in 1 series, 7–8 mm long.

Disturbed sites. Rare; Escambia County. South Carolina south to Florida, west to Oklahoma and Texas; South America; Africa, Asia, and Australia. Native to South America. Spring.

Hypochaeris radicata L. [*Radicatus,* having roots, in reference to the scapose growth form, the flowering stalk arising from the roots.] HAIRY CAT'SEAR.

Hypochaeris radicata Linnaeus, Sp. Pl. 811. 1753. *Achyrophorus radicatus* (Linnaeus) Scopoli, Fl. Carniol., ed. 2. 2: 117. 1772. *Hypochaeris infesta* Salisbury, Prodr. Stirp. Chap. Allerton 182. 1796, nom. illegit. *Porcellites radicatus* (Linnaeus) Cassini, in Cuvier, Dict. Sci. Nat. 43: 43. 1826. *Hypochaeris radicata* Linnaeus var. *typica* Beck, Fl. Nieder-Oesterreich 1310. 1893, nom. inadmiss. *Hypochaeris radicata* Linnaeus var. *vulgaris* Regel, Linnaea 16: 60. 1842, nom. inadmiss.

Erect perennial herb, to 6 dm; stem striate-angled, glabrous or hirsute proximally. Leaves all basal, the blade oblanceolate, 5–35 cm long, 0.5–3 cm wide, the apex obtuse, the base narrowly cuneate, the margin coarsely dentate or pinnatifid, the upper and lower surfaces hirsute, sessile, clasping. Flowers in 2–7 capitula in a loose paniculiform or corymbiform array, the peduncle 4–15 cm long, minutely bracteate; involucre cylindric or campanulate, 1–2 cm wide; phyllaries numerous, narrowly lanceolate, 3–20 mm long, unequal, the margin scarious, sometimes darkened; receptacle flat, slightly pitted; paleae linear, scarious; flowers numerous; corolla bright yellow or grayish green, 1–1.5 cm long, longer than the phyllaries at flowering. Cypsela monomorphic, all beaked, the beak 3–5 mm long, the body fusiform, 6–10 mm long, golden brown, 10- to 12-ribbed, muricate; pappus of whitish bristles in 2 series, the outer barbellate, shorter than the inner plumose, the longest 10–12 mm long.

Disturbed sites. Occasional; panhandle. Nearly throughout North America; nearly cosmopolitan. Native to Europe. Spring.

Ionactis Greene 1897. ANKLE-ASTER

Herbs. Leaves alternate, simple, pinnate-veined, epetiolate. Flowers in capitula, solitary or in corymbiform arrays, radiate; receptacle epaleate; ray flowers carpellate; petals 5, basally connate, zygomorphic; disk flowers bisexual; petals 5, basally connate, actinomorphic; stamens 5, epipetalous, the filaments free, the anthers connate; ovary inferior, 2-carpellate, 1-loculate. Fruit a cypsela; pappus present.

A genus of 5 species; North America. [From the Greek *ion,* violet, *actis,* ray, in reference to violet-colored ray flowers.]

Selected reference: Nesom (2006o).

Ionactis linariifolia (L.) Greene [With linear leaves.] FLAXLEAF ANKLE-ASTER.

Aster linariifolius Linnaeus, Sp. Pl. 874. 1753. *Chrysopsis linariifolia* (Linnaeus) Nuttall, Gen. N. Amer. Pl. 2: 152. 1818. *Diplostephium linariifolium* (Linnaeus) Nees von Esenbeck, Gen. Sp. Aster. 199. 1832. *Diplopappus linariifolius* (Linnaeus) Hooker, Fl. Bor.-Amer. 2: 21. 1834. *Ionactis linariifolia* (Linnaeus) Greene, Pittonia 3: 245. 1897.

Perennial herb or slightly woody, to 5(7) dm; stems striate-angled, glabrous. Leaves with the blade linear, 1.5–4 cm long, 2–3 mm wide, the apex acute, the base cuneate, the margin entire, the upper and lower surfaces glabrous, sessile. Flowers in many or sometimes solitary capitula in a loose corymbiform array, sometimes solitary, the peduncle 1–6 cm long, bracteate; involucre campanulate; phyllaries numerous in 2–6 series, linear-lanceolate, 2–9 mm long, unequal, slightly keeled, the outer surface glabrous; receptacle flat, pitted; ray flowers numerous; corolla violet or rarely white, the lamina 8–15 mm long; disk flowers numerous; corolla yellow, 5–7 mm long. Cypsela narrowly obovoid, laterally flattened, 3–4 mm long, 3- to 4-ribbed, densely strigose; pappus of 1–2 series, the outer of short bristles or scales, the inner of numerous yellowish, barbellate, apically attenuate bristles ca. 4 mm long.

Sandhills. Occasional; Nassau County, central and western panhandle. Quebec south to Florida, west to Wisconsin, Iowa, Kansas, Oklahoma, and Texas. Fall.

Iva L. 1753. MARSHELDER

Herbs or shrubs. Leaves opposite, simple, pinnate-veined, petiolate or epetiolate. Flowers in capitula in spiciform or racemiform arrays, discoid, unisexual; receptacle paleate; petals 5, basally connate, actinomorphic; stamens 5, epipetalous, the filaments connate, the anthers free; ovary inferior, 2-carpellate, 1-loculate. Fruit a cypsela; pappus absent.

A genus of about 9 species; North America, West Indies, and Mexico.

Selected reference: Strother (2006p).

1. Suffrutescent herb or shrub.
   2. Leaves distinctly petiolate, coarsely serrate; phyllaries 4–5, subequal, 2–4 mm long .... **I. frutescens**
   2. Leaves sessile or subsessile, entire or slightly serrulate; phyllaries 6–9, unequal, 4–7 mm long .... **I. imbricata**
1. Annual or perennial herb.
   3. Phyllaries partly connate to form a toothed or lobed involucre .... **I. angustifolia**
   3. Phyllaries free.
      4. Principal leaves linear, less than 0.3 cm wide .... **I. microcephala**
      4. Principal leaves lanceolate or ovate, 2–7 cm wide .... **I. annua**

Iva angustifolia Nutt. Ex DC. [*Angustus,* narrow, with narrow leaves.] NARROWLEAF MARSHELDER.

Iva angustifolia Nuttall ex de Candolle, Prodr. 5: 529. 1836.

Erect annual herb, to 6(10) dm; stem striate-angled, glabrous or sparsely scabrellous. Leaves with the blade linear or linear-lanceolate, 1.5–2.5(4.5) cm long, 1–4(7) mm wide, the apex acute, the base cuneate, the margin entire, the upper and lower surfaces scabrellous or hispidulous, gland-dotted, sessile or the petiole to 2 mm long. Flowers in capitula in a spiciform array, the peduncle to 1 mm long; involucre turbinate; phyllaries ca. 3 mm long, the outer 3–5 connate; receptacle convex; paleae linear, ca. 2 mm long, membranous; carpellate flowers 1–2 or absent; corolla whitish, less than 1 mm long; staminate flowers 2–5; corolla whitish or pinkish, ca. 2 mm long. Cypsela obovoid, 2–3 mm long.

Wet disturbed sites. Rare; Wakulla and Gulf Counties. Native to the southwestern United States and Mexico. Summer–fall.

Iva annua L. [Plant an annual.] ANNUAL MARSHELDER.

Iva annua Linnaeus, Sp. Pl. 933. 1753.

Erect annual herb, to 1(1.5) m; stem striate-angled, sparsely scabrellous or glabrate. Leaves with the blade deltate, ovate, elliptic, or lanceolate, 3–10(15) cm long, 0.8–4.5(8) cm wide, the apex acute, the base cuneate, the margin serrate-crenate, the upper and lower surfaces scabrellous, especially on the veins, gland-dotted, the petiole 0.5–3 cm long. Flowers in capitula in a spiciform array, the peduncle to 1 mm long; involucre hemispheric; phyllaries 3–4(5) mm long, the outer series 3–5, free; receptacle convex; paleae linear, 2–3 mm long, membranous; carpellate flowers 3–5; corolla whitish, ca. 1 mm long; staminate flowers 8–12; corolla whitish or pinkish, 2–3 mm long. Cypsela obovoid, 2–3 mm long.

Floodplain forests. Rare; Bradford County, eastern and central panhandle. Summer–fall.

Iva frutescens L. [Becoming shrubby.] BIGLEAF SUMPWEED.

Iva frutescens Linnaeus, Sp. Pl. 989. 1753.

Erect shrub or subshrub, to 3.5 m; stems striate-angled, closely scabrellous or glabrate. Leaves with the blade ovate, elliptic, or lanceolate, 3–6(12) cm long, 0.5–3(4) cm wide, the apex acute, the base cuneate, the margin irregularly toothed, the upper and lower surfaces closely scabrellous, gland-dotted, the petiole 0.5–1 cm long. Flowers in capitula in a racemiform array, the peduncle 1–3 mm long; involucre hemispheric; phyllaries 2–4 mm long, the outer 5 free; receptacle convex; paleae linear or subulate, 2–3 mm long, membranous; carpellate flowers (2)5; corolla whitish, ca. 1 mm long; staminate flowers 3–8; corolla whitish or pinkish, 2–3 mm long. Cypsela obovate, 2–3 mm long.

Salt marshes and coastal strands. Frequent; nearly throughout. Summer–fall.

Iva imbricata Walter [Flattened and overlapping, in reference to the leaves in the inflorescence.] SEACOAST MARSHELDER.

Iva imbricata Walter, Fl. Carol. 232. 1788.

Decumbent or erect perennial herb or subshrub, to 0.5(1) m; stem striate-angled, glabrous. Leaves with the blade narrowly lanceolate-elliptic, lanceolate, or spatulate, 1–3(6) cm long, 0.5–1(2) cm wide, the apex acute, the base cuneate, the margin entire, rarely irregularly toothed, the upper and lower surfaces glabrous, rarely gland-dotted, sessile or the petiole to 2 mm long. Flowers in capitula in a racemiform array, the peduncle 2–7 mm long; involucre campanulate; phyllaries 5–7 mm long, the outer 6–10 free; receptacle convex; paleae linear or spatulate, 5–6 mm long, membranous; carpellate flowers 2–5; corolla whitish, 1–2 mm long; staminate flowers 8–12; corolla whitish or pinkish, 4–5 mm long. Cypsela obovoid, 4–5 mm long.

Coastal dunes. Frequent; peninsula, central and western panhandle. Summer–fall.

Iva microcephala Nutt. [From the Greek *micro,* small, and *cephale,* head, in reference to the small capitula.] PIEDMONT MARSHELDER.

Iva microcephala Nuttall, Trans. Amer. Philos. Soc., ser. 2. 7: 346. 1841.

Erect annual herb, to 0.7(1) m; stem striate-angled, glabrate. Leaves with the blade linear or filiform, 1–3.5(6) cm long, 1–2(3) mm wide, the apex acute, the base narrowly cuneate, the margin entire, the upper and lower surfaces scabrellous, gland-dotted, sessile or subsessile. Flowers in capitula in a racemiform array, the peduncle to 1 mm long; involucre hemispheric; phyllaries ca. 2 mm long, the outer 5 free; receptacle convex; paleae cuneiform or setiform, 1–2 mm long, membranous; carpellate flowers 2–3; corolla whitish, ca. 1 mm long; staminate flowers 3–5; corolla whitish or pinkish, ca. 2 mm long. Cypsela obovoid, 1–2 mm long.

Wet flatwoods, wet prairies, and pond margins. Frequent; peninsula west to central panhandle. Summer–fall.

EXCLUDED TAXA

Iva asperifolia Lessing—Reported for Florida by Jackson (1960), the name of this Mexican species misapplied to Florida material of *I. angustifolia.*
Iva cheiranthifolia Kunth—Reported for Florida by Jackson (1960), based on a Rugel specimen marked "Florida." It was most likely collected by Rugel in Cuba.

Koanophyllon Arruda 1816. THOROUGHWORT

Shrubs. Leaves opposite, simple, pinnipalmate-veined, petiolate. Flowers in capitula in corymbiform or paniculiform arrays, discoid, bisexual; receptacle epaleate; petals 5, basally connate, actinomorphic; stamens 5, epipetalous; the filaments free, the anthers connate; ovary inferior, 2-carpellate, 1-loculate. Fruit a cypsela; pappus present.

A genus of about 150 species; North America, Mexico, West Indies, Central America, and South America. [In reference to the leaves as a source of an indigo-like dye from Kano, Nigeria.]

Selected reference: Nesom (2006p).

Koanophyllon villosum (Sw.) R. M. King & H. Robinson [With shaggy trichomes.] FLORIDA SHRUB THOROUGHWORT.

Eupatorium villosum Swartz, Prodr. 111. 1788. *Koanophyllon villosum* (Swartz) R. M. King & H. Robinson, Phytologia 32: 265. 1975.

Shrub, to 2 m; stems striate-angled, densely puberulent or pilose, often gland-dotted. Leaves with the blade ovate, ovate-deltate, or ovate-lanceolate, 2–7(8) cm long, 1–4 cm wide, the apex rounded, obtuse, or acute, the base cordate or truncate, the margin entire or irregularly shallow serrate-crenate, the upper surface sparsely puberulent, the lower surface pilose, gland-dotted. Flowers in capitula in a corymbiform or paniculiform array, the peduncle 2–4 mm long; involucre hemispheric; phyllaries in 1–2 series, elliptic-lanceolate, ca. 3 mm long, unequal, 3-nerved, the outer surface hispidulous-puberulent; receptacle slightly convex; flowers numerous; corolla white or pinkish, 2–3 mm long, the lobes gland-dotted. Cypsela prismatic, 1–2 mm long, 5-ribbed, sparsely hispidulous-strigose, gland-dotted; pappus in 1 series, of numerous barbellate bristles 1–2 mm long.

Hammocks and pinelands. Rare; Miami-Dade County. Florida; West Indies. All year.

Koanophyllon villosum is listed as endangered in Florida (Florida Administrative Code, Chapter 5B-40).

Krigia Schreber, nom. cons. 1791. DWARFDANDELION

Herbs. Leaves alternate, simple, pinnate-veined, petiolate or epetiolate. Flowers in capitula, solitary, liguliflorous, bisexual; receptacle epaleate; petals 5, basally connate, zygomorphic; stamens 5, epipetalous, the filaments free, the anthers connate; ovary inferior, 2-carpellate, 1-loculate. Fruit a cypsela; pappus present or absent.

A genus of 9 species; North America and Mexico. [Commemorates David Krieg (1667–1713), German-born physician and botanist in Maryland.]

Cynthia D. Don, 1829; *Serinia* Raf., 1817.

Selected reference: Chambers and O'Kennon (2006).

1. Pappus absent .... **K. cespitosa**
1. Pappus present.
    2. Pappus of 5 outer scales to 1 mm long and 5 inner scabrous bristles 4–6 mm long .... **K. virginica**
    2. Pappus of ca. 10 outer scales to 1 mm long and numerous inner barbellate bristles 5–8 mm long.... .... **K. dandelion**

Krigia cespitosa (Raf.) K. L. Chambers [*Caespitosus*, growing in tufts or patches.] WEEDY DWARFDANDELION.

Serinia cespitosa Rafinesque, Fl. Ludov. 57, 149. 1817. *Krigia cespitosa* (Rafinesque) K. L. Chambers, in B. B. Simpson, J. Arnold Arbor. 54: 52. 1973.
Apogon gracilis Elliott, Sketch Bot S. Carolina 2: 267. 1823.

Erect or ascending annual herb, to 4 dm; stem striate-angled, eglandular or stipitate-glandular, especially distally. Leaves with the blade oblanceolate, 2–15 cm long, 0.5–2 cm wide, the apex acute or obtuse, the base cuneate, the margin entire or remotely toothed or lobed, the upper and lower surfaces glabrous or sparsely glandular-villous, the proximal with the petiole to 1 cm long, the distal sessile. Flowers in a solitary capitulum, axillary or terminal on the stem, the peduncle 2–15 cm long; involucre campanulate; phyllaries 5–10, 2–5 mm long, lanceolate, the midvein evident, the apex acute, erect in fruit; receptacle low-convex, pitted; flowers numerous; corolla yellow, the lamina 2–6 mm long. Cypsela fusiform, broadest at or above the middle, 1–2 mm long, 10- to 20-ribbed; pappus absent.

Floodplain forests and wet disturbed sites. Occasional; northern counties, Hillsborough County. Spring.

Krigia dandelion (L.) Nutt. [From the French *dent-de-lion,* lion's tooth, in reference to the often remotely toothed or pinnately lobed leaf margin like that of *Taraxacum* (dandelion) which the plant resembles.] POTATO DWARFDANDELION.

Leontodon dandelion Linnaeus, Sp. Pl. 798. 1753. *Tragopogon dandelion* (Linnaeus) Linnaeus, Sp. Pl., ed. 2. 1111. 1763. *Troximon dandelion* (Linnaeus) F. W. Schmidt, Samml. Phys.-Oekon Aufs. 268. 1795. *Krigia dandelion* (Linnaeus) Nuttall, Gen. N. Amer. Pl. 2: 127. 1818. *Cynthia dandelion* (Linnaeus) de Candolle, Prodr. 7: 89. 1838. *Adopogon dandelion* (Linnaeus) Kuntze, Revis. Gen. Pl. 1: 304. 1891.

Erect perennial herb, to 5 dm; stem striate-angled, glabrous or minutely glandular-villous. Leaves with the blade linear, lanceolate, or oblanceolate, 6–24 cm long, 0.5–2 cm wide, the apex acute or obtuse, the base cuneate or winged petioloid, the margin entire, remotely toothed, or pinnately lobed, the lobes entire, the upper and lower surfaces glabrous or sparsely glandular or eglandular-villous, sessile. Flowers in a solitary capitulum from a basal rosette, the peduncle 20–30 cm long; involucre campanulate; phyllaries numerous, linear-lanceolate, 10–15 mm long, the midvein obscure, the apex acute, reflexed in fruit; receptacle low-convex, pitted; flowers numerous; corolla yellow or orangish, the lower surface often purple-tinged, the lamina 1.5–2.5 cm long. Cypsela columnar, 2–3 mm long, 10- to 15-ribbed, reddish brown; pappus of ca. 10 outer scales to 1 mm long and numerous barbellate inner bristles 5–8 mm long.

Moist disturbed sites. Rare; Gadsden and Jackson Counties. Spring.

Krigia virginica (L.) Willd. [Of Virginia.] VIRGINIA DWARFDANDELION.

Hyoseris virginica Linnaeus, Sp. Pl. 809. 1753. *Krigia virginica* (Linnaeus) Willdenow, Sp. Pl. 3: 1613. 1803.

Hyoseris caroliniana Walter, Fl. Carol. 194. 1788. *Adopogon carolinianus* (Walter) Britton, in Porter & Britton, Mem. Torrey Bot. Club 5: 346. 1894.

Krigia caroliniana Nuttall, Gen. N. Amer. Pl. 2: 126. 1818. TYPE: FLORIDA: Nassau Co.: St. Marys, s.d., *Baldwin s.n.* (lectotype: PH). Lectotypified by Shinners (1947: 197).

Krigia leptophylla de Candolle, Prodr. 7: 88. 1838. *Krigia caroliniana* Nuttall var. *leptophylla* (de Candolle) Torrey & A. Gray, Fl. N. Amer. 2: 468. 1843.

Erect annual herb, to 3 dm; stem striate-angled, eglandular or stipitate-glandular. Leaves initially in basal rosettes, later cauline, the blade oblanceolate or spatulate, 1.5–18 cm long, 5–8 mm wide, the apex acute or obtuse, the base cuneate, the margin irregularly pinnate-lobed, the terminal lobes usually denticulate, the upper and lower surfaces glabrous and eglandular or sparsely glandular-villous, the cauline produced late-season, narrowly oblanceolate or linear, usually entire, winged petiolate or sessile. Flowers in a solitary capitulum, from rosettes early or later from axillary or terminal on the stem, the peduncle 10–35 cm long; involucre campanulate; phyllaries 9–15, lanceolate, 5–8 mm long, the midvein obscure, the apex acute, reflexed in fruit; receptacle low-convex, pitted; flowers numerous; corolla yellow, the lower surface often purplish, the lamina 5–12 mm long. Cypsela narrowly obconic, ca. 2 mm long, 15- to 20-ribbed, dark reddish brown; pappus of 5 rounded outer scales to 1 mm long and 5 inner scabrous bristles 4–6 mm long.

Open hammocks, flatwoods, and disturbed sites. Frequent; northern counties, central peninsula. Spring.

EXCLUDED TAXON

Krigia oppositifolia "Rafinesque," nom. inval.—According to Chambers (1973), this name was not accepted by the author, thus it is not validly published. Small (1903, 1913a, 1933, all as *Serinia oppositifolia* Kuntze), Correll and Johnston (1970), and Cronquist (1980) all reported this taxon from Florida, misapplying the name to *K. cespitosa*.

Lactuca L. 1753. LETTUCE

Herbs, latex present. Leaves alternate, simple, pinnate-veined, petiolate or epetiolate. Flowers in capitula in corymbiform or paniculiform arrays, liguliflorous, bisexual; calyculi present; receptacle epaleate; petals 5, basally connate, zygomorphic; stamens 5, epipetalous, the filaments free, the anthers connate; ovary inferior, 2-carpellate, 1-loculate. Fruit a cypsela; pappus present.

A genus of about 75 species; North America, Mexico, Central America, Europe, Africa, and Asia. [*Lactis*, milky juice, in reference to the milky latex.]

Selected reference: Strother (2006q).

1. Cypsela with a short, stout beak to ⅓ as long as the body or beakless .... **L. floridana**
1. Cypsela with a distinct filiform beak more than ½ to nearly as long as the body.
   2. Cypsela prominently several-nerved on each face .... **L. serriola**
   2. Cypsela with a single median nerve on each face.
      3. Corolla lamina orange-yellow; leaves fairly well distributed along the stem .... **L. canadensis**
      3. Corolla lamina blue; leaves mostly basal .... **L. graminifolia**

Lactuca canadensis L. [Of Canada.] CANADA LETTUCE.

Lactuca canadensis Linnaeus, Syst. Nat., ed. 10. 1193. 1759. *Sonchus pallidus* Willdenow, Sp. Pl. 3: 1521. 1803. *Cicerbita canadensis* (Linnaeus) Wallroth, Sched. Crit. 434. 1822. *Mulgedium integrifolium* Cassini, in Cuvier, Dict. Sci. Nat. 33: 299. 1824, nom. illegit. *Lactuca canadensis* Linnaeus var. *typica* Wiegand, Rhodora 22: 10. 1920, nom. inadmiss.

Lactuca elongata Muhlenberg ex Willdenow, Sp. Pl. 1525. 1803. *Cicerbita elongata* (Muhlenberg ex Willenow) Wallroth, Sched. Crit. 434. 1822. *Lactuca canadensis* Linnaeus var. *elongata* (Muhlenberg ex Willdenow) Farwell, Pap. Michigan Acad. Sci. 2: 46. 1923.

Lactuca elongata Muhlenberg ex Willdenow var. *integrifolia* Chapman, Fl. South. U.S. 252. 1860; non Torrey & A. Gray, 1843. TYPE: "Florida and northward."

Erect biennial herb, to 2(4.5) m; stem striate-angled, glabrous, sometimes glaucous. Leaves with the blade oblong, lanceolate, linear, or obovate, to 15 cm long, to 8.5 cm wide, reduced upward, the apex acuminate, the base cuneate, the margin entire, denticulate, or pinnatifid, the upper surface glabrous, lower surface glabrous or sometimes sparsely pilose on the midrib, the petiole winged or sessile and sagittate-auriculate. Flowers in numerous capitula in a corymbiform or paniculiform array, the peduncle to 1 cm long, bracteate; calyculi 3–10, lanceolate; involucre campanulate; phyllaries numerous in 2 series, 10–12 mm long, subequal, usually reflexed in fruit; receptacle convex, pitted; flowers numerous; corolla orange-yellow. Cypsela elliptic, 3–4 mm long, slightly compressed, the faces 1(3)-nerved, brown, the beak filiform, 2–3 mm long; pappus of numerous white barbellate bristles 5–6 mm long.

Open floodplain forests, pond margins, and wet disturbed sites. Occasional; northern and central peninsula. Nearly throughout North America. North America, Mexico, Central America, and South America; Europe and Asia. Summer–fall.

Lactuca floridana (L.) Gaertn. [*Floridus*, abundant in flowers.] WOODLAND LETTUCE.

Sonchus floridanus Linnaeus, Sp. Pl. 794. 1753. *Lactuca floridana* (Linnaeus) Gaertner, Fruct. Sem. Pl. 2: 362. 1791. *Cicerbita floridana* (Linnaeus) Wallroth, Sched. Crit. 434. 1822. *Mulgedium lyratum* Cassini, in Cuvier, Dict. Sci. Nat. 33: 297. 1824, nom. illegit. *Mulgedium floridanum* (Linnaeus) de Candolle, Prodr. 7: 249. 1838.

Lactuca villosa Jacquin, Pl. Hort. Schoenbr. 3: t. 367. 1798. *Cicerbita villosa* (Jacquin) Beauverd, Bull. Soc. Bot. Genève, ser. 2. 2: 124. 1910. *Mulgedium villosum* (Jacquin) Small, Fl. Lancaster Co. 316. 1913. *Lactuca floridana* (Linnaeus) Gaertner var. *villosa* (Jacquin) Cronquist, Rhodora 50: 31. 1948.
Sonchus acuminatus Willdenow, Sp. Pl. 3: 1521. 1803. *Cicerbita acuminata* (Willdenow) Wallroth, Sched. Crit. 434. 1822. *Mulgedium acuminatum* (Willdenow) de Candolle, Prodr. 7: 250. 1838. *Lactuca acuminata* (Willdenow) A. Gray, Syn. Fl. N. Amer. 1(2): 443. 1884.

Erect annual or biennial herb, to 1.5(3) m; stem striate-angled, glabrous. Leaves with the blade oblong, ovate, or elliptic, to 30 cm long, to 15 cm wide, reduced upward, the apex acuminate, the base cuneate, the margin entire, denticulate, or pinnatifid, the upper surface glabrous, the lower surface glabrous or sparsely pilose on the midrib, the petiole winged, the base clasping. Flowers in capitula in a subpyramidal paniculiform array, the peduncle to 3 cm long, bracteate; calyculi 3–10, lanceolate; involucre campanulate; phyllaries numerous in 2 series, (8)10–12 mm long, subequal, usually reflexed in fruit; receptacle convex, pitted; flowers numerous; blue or rarely white. Cypsela fusiform, 4–5 mm long, slightly compressed, the faces 5- to 6-ribbed, brown, the beak stout, to 1 mm long; pappus of numerous white barbellate bristles 4–5 mm long.

Wet flatwoods and open hammocks. Occasional; nearly throughout. New York and Massachusetts south to Florida, west to Manitoba, Minnesota, South Dakota, Nebraska, Kansas, Oklahoma, and Texas; West Indies. Native to North America. Summer–fall.

Lactuca graminifolia Michx. [*Gramineus*, grasslike, with grasslike leaves.] GRASSLEAF LETTUCE.

Lactuca graminifolia Michaux, Fl. Bor.-Amer. 2: 85. 1803. *Lactuca elongata* Muhlenberg ex Willdenow var. *graminifolia* Chapman, Fl. South. U.S. 252. 1860.

Erect biennial herb, to 1(1.5) m; stem striate-angled, glabrous. Leaves with the blade linear or lanceolate, to 35 cm long, to 5 cm wide, the apex acute or acuminate, the blade cuneate, the margin entire, denticulate, or pinnatifid, the upper surface glabrous, the lower surface glabrous or setose on the veins, with a winged petiole or sessile, the base clasping. Flowers capitula in a paniculiform array, the peduncle to 3 cm long, bracteate; calyculi 3–10, lanceolate; involucre campanulate; phyllaries numerous in 2 series, 12–20 mm long, subequal, usually reflexed in fruit; receptacle convex, pitted; flowers numerous; corolla blue. Cypsela elliptic, 5–6 mm long, slightly flattened, the faces 1(3)-ribbed, brown, the beak filiform, 2–4 mm long; pappus of numerous white barbellate bristles 5–9 mm long.

Disturbed sites. Common; nearly throughout. Virginia south to Florida, west to Arizona, also New Jersey, Illinois, and Colorado; Mexico and Central America. Summer–fall.

Lactuca serriola L. [With serrate teeth, in reference to the leaf margin.] PRICKLY LETTUCE.

Lactuca serriola Linnaeus, Cent. Pl. 2: 29. 1756.

Erect annual or short-lived perennial, to 1(2) m; stem striate-angled, glabrous, sometimes glaucous, often prickly. Leaves with the blade oblong, ovate, or lanceolate, to 25 cm long, to 7.5

cm wide, the apex acute, the base cuneate, the margin entire, denticulate, or pinnatifid, often prickly, the upper surface glabrous, the lower surface glabrous, usually prickly on the midvein, sessile or auriculate-clasping. Flowers in capitula in a paniculiform array, the peduncle to 1 cm long, bracteate; calyculi 3–10, lanceolate; involucre campanulate; phyllaries numerous in 2 series, 9–10(12) mm long, subequal, usually reflexed in fruit; receptacle convex, pitted; flowers numerous; corolla yellow. Cypsela oblanceolate, 3–4 mm long, slightly flattened, the faces (3)5- to 9-ribbed, grayish or tan, the beak filiform, 3–4 mm long; pappus of numerous white barbellate bristles (3)4–5 mm long.

Disturbed sites. Rare; Columbia, Marion, Leon, and Escambia Counties. Nearly throughout North America. North America; Mexico and South America; Europe, Africa, Asia, Australia, and Pacific Islands. Native to Europe, Africa, and Asia. Summer–fall.

EXCLUDED TAXON

Lactuca hirsuta Muhlenberg ex Nuttall—Reported for Florida by Cronquist (1980), the basis unknown. No Florida material seen.

Lagascea Cav. 1803. SILKLEAF

Herbs. Leaves opposite, simple, pinnate-veined, petiolate. Flowers in capitula in head-like terminal glomerules (synflorescence), discoid, bisexual; receptacle paleate; disk flowers 1; petals 5, basally connate, actinomorphic; stamens 5, epipetalous, the filaments free, the anthers connate; ovary inferior, 2-carpellate, 1-loculate. Fruit a cypsela; pappus present.

A genus of 8 species; North America, West Indies, Mexico, Central America, South America, Asia, Africa, and Pacific Islands. [Commemorates Mariano la Gasca y Segura (1776–1839), Spanish botanist, physician, and director of Real Botánico de Madrid (Royal Botanical Garden of Madrid).]

Selected reference: Steussy (1978).

Lagascea mollis Cav. [Soft, in reference to the leaf surface.] SILKLEAF.

Lagascea mollis Cavanilles, Anales Ci. Nat. 6: 332, t. 44. 1803. *Nocco mollis* (Cavanilles) Jacquin, Fragm. Bot. 58, t. 85(1). 1805.

Erect annual or short-lived perennial herb, to 1 m; stem striate-angled, hirtellous and stipitate-glandular. Leaves with the blade ovate, the apex acute, the base broadly cuneate, the margin subentire or serrate, the upper and lower surfaces strigose, the petiole 0.5–2.5 cm long, strigose. Flowers solitary in numerous capitula in a terminal synflorescence ca. 1 cm high, 1–3 cm wide, subtended by small leaves the synflorescence peduncle 3.5–5(14) cm long, retrorsely strigose and stipitate-glandular; phyllaries 4–5 in 1 series, linear, 4–5 mm long, proximally connate, the outer surface pilose, gland-dotted in 2–3 rows; corolla white or pink-purple, 5–6 mm long, the outer surface sparsely pubescent. Cypsela obovoid, ca. 3 mm long, pubescent distally; pappus of a minute, erose, pubescent corona.

Disturbed sites. Rare; Franklin County. Not recently collected. Florida; West Indies, Mexico, Central America, and South America; Africa and Asia. Native to Mexico and Central America. Summer–fall.

Lagascea mollis was collected in Florida only at Apalachicola by Chapman in the 1850s, probably from ballast.

Launaea Cass. 1822.

Herbs. Leaves alternate, simple, pinnate-veined, epetiolate. Flowers in capitulae in spiciform, racemiform, or paniculiform arrays, liguliflorous, bisexual; receptacle epaleate; petals 5, basally connate, zygomorphic; stamens 5, epipetalous, the filaments free, the anthers connate; ovary inferior, 2-carpellate, 1-loculate. Fruit a cypsela; pappus present.

A genus of about 60 species; North America, West Indies, Mexico, Central America, South America, Europe, Africa, Asia, and Australia. [Commemorates J.-C.-M. Mordant de Launay (1750–1816), French naturalist and librarian at Muséum National d'Histoire Naturelle, Paris.]

Brachyrhamphus DC., 1838.

Selected references: Kilian (1997); Whetstone and Brodeur (2006).

Launaea intybacea (Jacq.) Beauverd [Latin for endive/chicory, from the Egyptian *tybi*, January, the month it was eaten.] ACHICORIA AZUL.

Lactuca intybacea Jacquin, Icon. Pl. Rar. 1: 16, t. 162. 1784. *Cicerbita intybacea* (Jacquin) Wallroth, Sched. Crit. 434. 1822. *Brachyrhamphus intybaceus* (Jacquin) de Candolle, Prodr. 7: 177. 1838. *Launaea intybacea* (Jacquin) Beauverd, Bull. Soc. Bot. Genève, ser. 2. 2: 114. 1910.

Erect annual herb, to 1.5 m; stem striate-angled, glabrous. Leaves basal and proximally cauline, the basal with the blade obovate or spatulate, (5)10–25 cm long, 2–6(12) cm wide, the apex acute or obtuse, the base subauriculate, the margin sinuate-lobed, the ultimate margin spinulose-denticulate, reduced upward, sessile. Flowers in a capitula in a spiciform, racemiform, or paniculiform array, the peduncle to 2 cm long, bracteate; involucre cylindric; phyllaries in 3–5 series, ovate, lanceolate, or linear, 1–1.5 cm long, unequal, the apex obtuse or acuminate, the margin scarious, the outer surface glabrous, reflexed in fruit; receptacle convex; flowers numerous; corolla yellow; lamina 5–6 mm long. Cypsela fusiform, often with a short beak, 4–5 mm long, 4-angular, each with 4 ribs, each with 2 secondary ribs, grayish; pappus in 2–3 series of white bristles 6–9 mm long, outer series barbellate, longer than the inner, setaceous.

Disturbed sites. Rare; Miami-Dade and Monroe Counties. Florida and Texas; West Indies, Mexico, Central America, and South America; Africa and Asia. Native to Africa. All year.

Leucanthemum Mill. 1754. CHRYSANTHEMUM; OXEYEDAISY

Herbs. Leaves alternate, simple, pinnate-veined, petiolate or epetiolate. Flowers in capitulae, radiate; receptacle epaleate; ray flowers carpellate; petals 5, basally connate, zygomorphic; disk flowers bisexual; petals 5, basally connate, actinomorphic; stamens 5, epipetalous, the filaments free, the anthers connate; ovary inferior, 2-carpellate, 1-loculate. Fruit a cypsela; pappus absent.

A genus of about 40 species; North America, Europe, and Asia. [From the Greek *leukos,* white, and *anthemon,* flower.]

Selected reference: Strother (2006r).

Leucanthemum vulgare Lam. [Common.] OXEYEDAISY.

Chrysanthemum leucanthemum Linnaeus, Sp. Pl. 888. 1753. *Matricaria leucanthemum* (Linnaeus) Scopoli, Fl. Carniol., ed. 2. 2: 147. 1772. *Leucanthemum vulgare* Lamarck, Fl. Franc. 2: 137. 1779 ("1778"). *Chrysanthemum vulgare* (Lamarck) Gaterau, Descr. Pl. Montauban 149. 1789, nom. illegit. *Chrysanthemum pratense* Salisbury, Prodr. Stirp. Chap. Allerton 203. 1796, nom. illegit. *Tanacetum leucanthemum* (Linnaeus) Schultz Bipontinus, Tanaceteen 35. 1844. *Chrysanthemum leucanthemum* Linnaeus var. *vulgare* Petermann, Deutschl. Fl. 301. 1849, nom. inadmiss. *Tanacetum leucanthemum* (Linnaeus) Schultz Bipontinus var. *pratense* Fenzl, Verh. Zool.-Bot. Vereins Wien 3: 335. 1853, nom. inadmiss. *Leucanthemum vulgare* Lamarck var. *baumgartnerianum* Schur, Enum. Pl. Transsilv. 338. 1866, nom. inadmiss. *Pyrethrum leucanthemum* (Linnaeus) Franchet, Fl. Loir-et-Cher 307. 1885. *Chrysanthemum leucanthemum* Linnaeus var. *typicum* Beck, Fl. Nieder-Oesterreich 1203. 1893, nom. inadmss. *Chamaemelum leucanthemum* (Linnaeus) E.H.L. Krause, in Sturm, Deutschl. Fl., ed. 2. 13: 210. 1905. *Leucanthemum leucanthemum* (Linnaeus) Rydberg, in Britton, N. Amer. Fl. 34: 235. 1916, nom. inadmiss.

Erect perennial herb, to 3(10) dm; stem striate-angled, glabrous or glabrate. Leaves with the blade obovate, oblong-lanceolate, or spatulate, 1–4(12) cm long, 0.8–2(3) cm wide, reduced upward, the apex rounded or obtuse, the base cuneate, the margin pinnately lobed, the ultimate margin irregularly dentate or entire, the upper and lower surfaces glabrous, gland-dotted, the petiole 1–3 cm long, winged, the upper leaves sessile or subsessile. Flowers in capitula, solitary or 2–3, the peduncle 3–15 cm long, glabrous or glabrate; involucre hemispheric, 1–2 cm wide; phyllaries in 3–4 series, ovate or lanceolate, 7–9 mm long, unequal, glabrous; receptacle slightly convex; ray flowers numerous; corolla white, the lamina 1–2(3.5) cm long; disk flowers numerous; corolla yellow, ca. 2 mm long. Cypsela columnar, 2–3 mm long, 10-ribbed, the apex of the ray cypsela sometimes coronate or auriculate.

Disturbed sites. Rare; Leon, Gadsden, Jackson, and Santa Rosa Counties. Escaped from cultivation. Nearly throughout North America; Europe and Asia. Native to Europe. Spring–fall.

Liatris Gaertn. ex Schreb., nom. cons. 1791. GAYFEATHER

Herbs. Leaves alternate, simple, pinnipalmate-veined, epetiolate. Flowers in capitula in corymbiform, racemiform, or spiciform arrays, discoid, bisexual; phyllaries present; receptacle epaleate; petals 5, basally connate, actinomorphic; stamens 5, epipetalous, the filaments free, the anthers connate; ovary inferior, 2-carpellate, 1-loculate. Fruit a cypsela; pappus present.

A genus of about 40 species; North America, West Indies, and Mexico. [From the Gaelic *liatrus,* spoon-shaped, in reference to the tuberous roots of some species.]

Ammopursus Small, 1924; *Lacinaria* Hill, 1762.

Selected reference: Nesom (2006q).

1. Pappus bristles plumose.
   2. Phyllaries with an elongated, erect, scarious, petaloid apex; capitula 4- to 5-flowered..... **L. elegans**
   2. Phyllaries with a spreading or recurved green apex; capitula 20- to 45-flowered..........**L. squarrosa**
1. Pappus bristles barbellate (sometimes strongly in *L. ohlingerae*).
   3. Capitula in a corymbiform to subcorymbiform array .... **L. ohlingerae**
   3. Capitula in a racemiform or spiciform array.
      4. Leaves (1)3- to 5-nerved.
         5. Corolla tube pilose at the base within .... **L. savannensis**
         5. Corolla tube glabrous as the base within .... **L. spicata**
      4. Leaves 1(3)-nerved.
         6. Phyllary apices (at least the mid and inner) usually acuminate, acute, rounded, or rounded-retuse and minutely involute-cuspidate, acuminate, or apiculate.
            7. Stem glabrous or sparsely pilose.
               8. Phyllary apex rounded-retuse and minutely involute-cuspidate or apiculate; corolla tube glabrous at the base within .... **L. tenuifolia**
               8. Phyllary apex acute, sometimes mucronulate; corolla tube pilose at the base within ... . **L. pauciflora**
            7. Stem various, but not glabrous or sparsely pilose.
               9. Stem hirsute, puberulent, pilose-puberulent, or strigose-puberulent.
                  10. Corm irregularly shaped and elongated into rhizomes; phyllaries 8–10 mm long.. .... **L. garberi**
                  10. Corm globose or elongate; phyllaries 4–6(7) mm long.
                     11. Proximal cauline leaves usually lanceolate, linear, linear-oblanceolate, or oblanceolate, sometimes narrowly spatulate, 2–8(10) mm wide; phyllary apices usually rounded or obtuse-truncate, rarely acute................................ **L. gracilis**
                     11. Proximal cauline leaves spatulate, (8)11–22(27) mm wide; phyllary apices usually acuminate or acute, rarely obtuse... **L. gholsonii**
               9. Stem hirtellous.
                  12. Capitula in a relatively loose, racemiform or spiciform (often secund) array......... .... **L. pauciflora**
                  12. Capitula in a dense, spiciform, cylindric array.
                     13. Capitula rigidly ascending, appressed to the rachis and each other, overlapping; phyllary margin usually not ciliolate, the apex erect .... **L. chapmanii**
                     13. Capitula spreading to ascending, not strongly overlapping; phyllary margin ciliate, the apex slightly spreading.. **L. provincialis**
         6. Phyllary apices (at least the mid and inner) usually obtuse, rounded, or truncate, rarely acute.
            14. Stem glabrous.
               15. Capitula in a dense spiciform array; involucre (8)9–11(12) mm wide .... **L. savannensis**
               15. Capitula in a loose racemiform array; involucre 5–7 mm wide............... **L. elegantula**
            14. Stem variously pubescent.
               16. Involucre (6)8–20 mm wide; flowers 11–24(30) in a capitula.
                  17. Phyllaries bullate, the hyaline margin relatively wide, often erose, lacerate, or irregular.. **L. aspera**
                  17. Phyllaries not bullate, the hyaline margin absent or relatively narrow, entire......... .... **L. squarrulosa**
               16. Involucre 2.5–7 mm wide; flowers 3–12 in a capitula.

18. Peduncle 10–15(30) mm long; involucre 5–7 mm wide .... **L. patens**
18. Peduncle 2–10(12) mm long; involucre 2.5–4(8) mm wide.
    19. Proximal floral bracts abruptly differentiated from the distal cauline leaves; phyllary apices usually rounded or obtuse-truncate, rarely acute .... **L. gracilis**
    19. Proximal floral bracts gradually differentiated from the distal cauline leaves; phyllary apices usually acuminate or acute, rarely obtuse .... **L. gholsonii**

Liatris aspera Michx. [Asper, rough, in reference to the puberulent-hispid leaves.] TALL GAYFEATHER.

Liatris aspera Michaux, Fl. Bor.-Amer. 2: 92. 1803. *Lacinaria aspera* (Michaux) Greene, Pittonia 4: 318. 1901. *Lacinaria scariosa* (Linnaeus) Hill var. *aspera* (Michaux) Farwell, Rep. (Annual) Michigan Acad. Sci. 15: 188. 1913. *Liatris aspera* Michaux var. *typica* Gaiser, Rhodora 48: 302. 1946, nom. inadmiss.

Lacinaria scariosa (Linnaeus) Hill var. *intermedia* Lunell, Amer. Midl. Naturalist 2: 173, 177. 1912. *Liatris aspera* Michaux var. *intermedia* (Lunell) Gaiser, Rhodora 48: 305. 1946. *Lacinaria scariosa* (Linnaeus) Hill var. *media* Lunell, Amer. Midl. Naturalist 2: 264. 1912, nom. illegit.

Erect perennial herb, to 1.8 m; roots forming a globose corm; stem striate-angled, hispidulous-puberulent. Basal and proximal cauline leaves with the blade broadly oblanceolate, elliptic, elliptic-spatulate, lanceolate-spatulate, or linear-lanceolate, 8–25 cm long, (4)6–25 mm wide, reduced upward, 1-nerved, the apex acute, the base cuneate or long-tapering, clasping, the margin entire, the upper and lower surfaces glabrous or sparsely to densely hispidulous-puberulent, gland-dotted, sessile. Flowers in capitula in a loose, spiciform array, the peduncle absent or 1–5(10) mm long, ascending, spreading, or deflexed; involucre campanulate, hemispheric, or turbinate, 1–2 cm wide; phyllaries in 4–5 series, obovate or oblong-spatulate, 9–16 mm long, unequal, the margin with broad white-hyaline, irregular, erose, or lacerate margin, the apex rounded to truncate, usually reflexed, the outer surface glabrate; receptacle flat; flowers numerous; corolla lavender or pinkish purple, rarely white, 5–6 mm long, the inner surface pilose. Cypsela prismatic, 4–6 mm long, 8- to 11-ribbed, hirtellous, gland-dotted; pappus of 1–2 series of numerous barbellate bristles subequaling the corolla.

Dry hammocks. Rare; northern peninsula, central panhandle. New York south to Florida, west to Ontario, North Dakota, South Dakota, Nebraska, Kansas, Oklahoma, and Texas. Fall.

Liatris chapmanii Torr. & A. Gray [Commemorates Alvan Wentworth Chapman (1809–1899), American physician and botanist who studied the flora of the southeastern United States.] CHAPMAN'S GAYFEATHER.

Liatris chapmanii Torrey & A. Gray, Fl. N. Amer. 2: 502. 1843. *Lacinaria chapmanii* (Torrey & A. Gray) Kuntze, Revis. Gen. Pl. 1: 349. 1891. TYPE: FLORIDA: s.d., *Chapman s.n.* (holotype: NY).

Liatris chapmanii (Torrey & A. Gray) Kuntze var. *longifolia* Nash, Bull. Torrey Bot. Club 23: 106. 1896. TYPE: FLORIDA: Hillsborough Co.: ca. 5 mi. from the Tampa along the electric RR line toward Ballast Point, 24 Aug 1895, *Nash* 2473 (holotype: NY; isotypes: F, GH, ND, US).

Lacinaria deamiae Lunell, Amer. Midl. Naturalist 2: 163. 1912. TYPE: FLORIDA: Pinellas Co.: near Veteran City, 8 mi. W of St. Petersburg, 29 Sep 1907, *Deam 2804* (holotype: ?; isotypes: GH, IND, NY, US).

Erect perennial herb, to 7.5(15) dm; roots forming a globose or elongate corm; stem striate-angled, hirtellous. Basal and proximal cauline leaves with the blade spatulate-oblanceolate, or narrowly oblanceolate, 4–15(18) cm long, 4–8(11) mm wide, reduced upward, 1-nerved, the apex acute, the base cuneate or long-tapering, clasping, the margin entire, the upper and lower surfaces glabrous or hirtellous, gland-dotted, sessile. Flowers in capitula in a dense, spiciform array, sessile; involucre cylindric, 3–5 mm wide; phyllaries in 3(4) series, oblong or oblong-lanceolate, 8–12 mm long, unequal, the margin hyaline, the apex acute or acuminate, the outer surface glabrous or rarely minutely puberulent; receptacle flat; flowers 3–4; corolla lavender or pinkish purple, rarely white, 5–6 mm long, the inner surface glabrous. Cypsela prismatic, 4–6 mm long, 8- to 11-ribbed, hirtellous, gland-dotted; pappus of 1–2 series of numerous barbellate bristles subequaling the corolla.

Scrub, sandhills, and coastal dunes. Frequent; peninsula, central and western panhandle. Georgia, Alabama, and Florida. Fall.

Liatris elegans (Walter) Michx. [Elegant, in reference to its attractive flowers.] PINKSCALE GAYFEATHER.

Staehelina elegans Walter, Fl. Carol. 202. 1788. *Serratula speciosa* Aiton, Hort. Kew. 3: 138. 1789, nom. illegit. *Liatris elegans* (Walter) Michaux, Fl. Bor.-Amer. 2: 91. *Eupatorium speciosum* Ventenat, Descr. Pl. Nouv. t. 79. 1802, nom. illegit. 1803. *Calostelma elegans* (Walter) D. Don, in Sweet, Brit. Fl. Gard., ser. 2. pl. 184. 1833. *Lacinaria elegans* (Walter) Kuntze, Revis. Gen. Pl. 1: 349. 1891. *Liatris elegans* (Walter) Michaux var. *typica* Gaiser, Rhodora 48: 341. 1946, nom. inadmiss.
Liatris elegans (Walter) Michaux var. *kralii* Mayfield, Sida 20: 601. 2002.

Erect perennial herb, to 3(12) dm; roots forming a depressed globose corm; stem striate-angled, puberulent or hirsute-pubescent. Basal and proximal cauline leaves with the blade narrowly oblanceolate, 6–20(30) cm long, 3–8 mm wide, reduced upward, 1-nerved, the apex acute, the base cuneate or long-tapering, clasping, the margin entire, the upper and lower surfaces glabrous or sparsely puberulent, gland-dotted, sessile. Flowers in capitula in a dense, spiciform array, sessile or the peduncle sometimes 1–5(10) mm long, ascending; involucre turbinate or cylindric, 4–6 mm wide; phyllaries in 3–4 series, lanceolate-triangular, 12–20 mm long, unequal, the apex prolonged, spreading, somewhat dilated, petaloid, pink, purplish, white, or yellow, the outer surface strigose or strigose-hispid; receptacle flat; flowers 4–5; corolla lavender or pinkish purple, rarely white, 5–7 mm long, the inner surface glabrous. Cypsela prismatic, 4–6 mm long, 8- to 11-ribbed, hirtellous, gland-dotted; pappus of 1–2 series of numerous plumose bristles subequaling the corolla.

Sandhills. Occasional; northern and central counties. South Carolina south to Florida, west to Oklahoma and Texas. Fall.

Liatris elegantula (Greene) K. Schum. [Elegantly, in reference to the attractive inflorescence.] GRASSLEAF GAYFEATHER.

Lacinaria elegantula Greene, Pittonia 4: 316. 1901. *Liatris elegantula* (Greene) K. Schumann, Bot. Jahresber. (Just) 29(1): 569. 1903. *Liatris graminifolia* Willdenow var. *elegantula* (Greene) Gaiser, Rhodora 48: 254. 1946.

Erect perennial herb, to 1 m; roots forming a subglobose corm; stem striate-angled, glabrous. Basal and proximal cauline leaves with the blade linear-lanceolate or narrowly oblanceolate, 8–21 cm long, 2–5(10) mm wide, reduced upward, 1-nerved, the apex acute, the base cuneate or long-tapering, clasping, the margin entire, the upper and lower surfaces glabrous, gland-dotted, sparsely pilose-ciliate along the proximal margin, sessile. Flowers in a capitula in a loose, racemiform array, sessile or the peduncle 2(7) mm long, ascending; involucre turbinate, 5–7 mm wide; phyllaries in 3–4(5) series, 6–8 mm long, unequal, the margin hyaline, sometimes sparsely ciliolate, the apex rounded, the outer surface glabrous or glabrate; receptacle flat; flowers numerous; corolla lavender or pinkish purple, rarely white, 4–6 mm long, the inner surface pilose. Cypsela prismatic, 3–4 mm long, 8- to 11-ribbed, hirtellous, gland-dotted; pappus of 1–2 series of numerous barbellate bristles subequaling the corolla.

Flatwoods and open hammocks. Occasional; northern counties, central peninsula. Georgia south to Florida, west to Mississippi. Summer.

Liatris garberi A. Gray [Commemorates Abram Paschal Garber (1838–1881), American physician and botanist who collected primarily in Pennsylvania and Florida.] GARBER'S GAYFEATHER

Liatris garberi A. Gray, Proc. Amer. Acad. Arts 15: 48. 1880. *Lacinaria garberi* (A. Gray) Kuntze, Revis. Gen. Pl. 349. 1891. TYPE: FLORIDA: Hillsborough Co.: Tampa, Sep 1877, *Garber s.n.* (holotype: GH; isotype; NY).

Lacinaria nashii Small, Fl. S.E. U.S. 1175, 1338. 1903. TYPE: FLORIDA: Manatee Co.: Palmetto, 21–23 Aug 1895, *Nash 2430* (holotype: NY; isotypes: F, GH, ND, PH).

Lacinaria chlorolepis Small ex Alexander, Man. S.E. Fl. 1333. 1933. TYPE: FLORIDA: Hillsborough Co.: Tampa, Sep 1877, *Garber s.n.* (lectotype: NY; isotype: GH). Lectotypified by Gaiser (1946: 236).

Erect perennial herb, to 8 dm; roots forming an elongated, irregularly shaped, rhizomatous corm; stem striate-angled, pilose-pubescent or hirsute. Basal and proximal leaves with the blade linear or linear-oblanceolate, 12–30 cm long, 3–8 mm wide, reduced upward to midstem, abruptly reduced distally to bracts, 1-nerved, the apex acute or sometimes the basal obtuse, the base cuneate or long-tapering, clasping, the margin entire, the upper and lower surfaces glabrous or glabrate, the margin sometimes sparsely ciliate, gland-dotted, sessile. Flowers in capitula in a dense, spiciform array, sessile or the peduncle 1–5(10) mm long, ascending; involucre cylindric or campanulate, 4–5 mm wide; phyllaries in 3–4(5) series, ovate, oblong, or lanceolate, 8–10 mm long, unequal, the margin hyaline, ciliolate, the apex acute, the outer surface glabrous or sometimes sparsely strigose or pilosulous; receptacle flat; flowers 6–10; corolla lavender or pinkish purple, rarely white, 4–5 mm long, the inner surface glabrous. Cypsela prismatic, 3–4 mm long, 8- to 11-ribbed, hirtellous, gland-dotted; pappus of 1–2 series of numerous barbellate bristles subequaling the corolla.

Flatwoods. Occasional, central and southern peninsula. Florida; West Indies. Summer–fall

Liatris gholsonii L. C. Anderson [Commemorates Angus Kemp Gholson (1921–2014), Florida botanist and conservationist.] GHOLSON'S GAYFEATHER.

Liatris gholsonii L. C. Anderson, Sida 20: 98. 2002. *Liatris gracilis* Pursh var. *gholsonii* (L. C. Anderson) D. B. Ward, Phytologia 94: 141. 2012. TYPE: FLORIDA: Liberty Co.: No Name Ravine on the Nature Conservancy's Apalachicola Bluffs and Ravines Preserve, T1N, R7W, Sec. 7, SW 1/4, 13 Sep 2001, *Anderson 19932* (holotype: BRIT; isotypes: FSU, MO, NY).

Erect perennial herb, to 1(1.8) m; roots forming a globose corm; stem striate-angled, sparsely puberulent. Basal and proximal leaves with the blade elliptic, lanceolate-elliptic, or spatulate, (10)15–25 cm long, (8)11–22(27) mm wide, 1-nerved, gradually reduced upward, the apex acute, the base cuneate or long-tapering, clasping, the margin entire, the upper and lower surfaces glabrous, gland-dotted, sessile. Flowers in a capitula in a loose, racemiform array, the peduncle (2)5–12(26) mm long, ascending or spreading, sometimes sparsely bracteate; involucre cylindric or turbinate, ca. 3 mm wide; phyllaries in 3(4) series, ovate or oblong, 4–5 mm long, unequal, the margin hyaline, ciliolate, the apex acute or acuminate, the outer surface glabrate; receptacle flat; flowers 3–5(6); corolla lavender or pinkish purple, 5–6 mm long, the inner surface pilose. Cypsela prismatic, 4–5 mm long, 8- to 11-ribbed, hirtellous, gland-dotted; pappus of 1–2 series of numerous barbellate bristles subequaling the corolla.

Open, xeric woods and ravine bluff tops. Rare; Gadsden and Liberty Counties. Endemic. Fall.

Liatris gholsonii is listed as endangered in Florida (Florida Administrative Code, Chapter 5B-40).

Liatris gracilis Pursh [Thin, slender, in reference to its habit.] SLENDER GAYFEATHER.

Liatris gracilis Pursh, Fl. Amer. Sept. 508. 1814. *Liatris pilosa* (Aiton) Willdenow var. *gracilis* (Pursh) Nuttall, Gen. N. Amer. Pl. 2: 131. 1818. *Lacinaria gracilis* (Pursh) Kuntze, Revis. Gen. Pl. 1: 349. 1891.
Liatris botrys Rafinesque, New Fl. 4: 76. 1838 ("1836"). TYPE: "Florida and Georgia."
Lacinaria laxa Small, Bull. Torrey Bot. Club 25: 472. 1898. *Liatris laxa* (Small) K. Schumann, Bot. Jahresber. (Just) 26(1): 378. 1900. TYPE: FLORIDA: Miami-Dade Co.: Cape Florida, 14 Mar 1892, *Simpson 537* (holotype: NY; isotype: GH).

Erect perennial herb, to 1 m; roots forming a globose or subglobose corm; stem striate-angled, puberulent, pilose-puberulent, or strigose-puberulent. Basal and proximal cauline leaves with the blade linear, linear-oblanceolate, oblanceolate, or narrowly spatulate, 5–15 cm long, 2–8(10) mm wide, reduced upward, 1-nerved, the apex acute, the base cuneate or long-tapering, clasping, the margin entire, the upper and lower surfaces glabrous, the margin of the proximal ciliate, sessile. Flowers in a capitula in a loose racemiform array, sessile or the peduncle 2–10(12) mm long, spreading or ascending; involucre cylindric or campanulate, 3–4(5) mm wide; phyllaries in 3(4) series, ovate or oblong, 4–6(7) mm long, unequal, the margin hyaline, ciliolate, the apex rounded or truncate, the outer surface sparsely puberulent, sometimes purple; receptacle flat; flowers 3–6(9); corolla lavender or pinkish purple, rarely white, 4–5 mm long, the inner surface pilose. Cypsela prismatic, 3–4 mm long, 8- to 11-ribbed, hirtellous, gland-dotted; pappus of 1–2 series of numerous barbellate bristles subequaling the corolla.

Sandhills and flatwoods. Common; nearly throughout. South Carolina south to Florida, west to Mississippi. Summer–fall.

Liatris ohlingerae (S. F. Blake) B. L. Rob. [Commemorates Sophronia Carson Ohlinger (1872–1940), Florida botanist.] FLORIDA GAYFEATHER; SCRUB BLAZING-STAR.

Lacinaria ohlingerae S. F. Blake, Bull. Torrey Bot. Club 50: 203, t. 9. 1923. *Ammopursus ohlingerae* (S. F. Blake) Small, Bull. Torrey Bot. Club 51: 393. 1924. *Liatris ohlingerae* (S. F. Blake) B. L. Robinson, Contr. Gray Herb. 104: 49. 1934. TYPE: FLORIDA: Polk Co.: 9 mi. SE of Frostproof, 1 Nov 1922, *Ohlinger s.n.* (holotype: US).

Erect perennial herb, to 1 cm; roots forming a subcylindric corm; stem striate-angled, minutely villous-puberulent or glabrescent. Basal and proximal cauline leaves with the blade linear, 8–15 cm long, 1–2(3) mm wide, reduced upward, 1-nerved, the apex acute, the base cuneate or long-tapering, clasping, the margin entire, the upper and lower surfaces sparsely puberulent or glabrous, gland-dotted, sessile. Flowers in a capitula in an open corymbiform or subcorymbiform array, rarely solitary, the peduncle 2–7 cm long, ascending or spreading, bracteate; involucre hemispheric, campanulate, or turbinate, 1.5–2 cm wide; phyllaries in 6–7 series, oblong, 1.5–2.3 cm long, unequal, the margin hyaline, faintly purplish, ciliolate, the apex rounded or obtuse, the outer surface glabrate; receptacle flat; flowers numerous; corolla lavender or pinkish purple, 8–10 mm long, the inner surface glabrous. Cypsela prismatic, 7–10 mm long, 8- to 11-ribbed, hirtellous, gland-dotted; pappus of 1–2 series of numerous barbellate or subplumose bristles subequaling the corolla.

Scrub. Rare; Polk and Highlands Counties. Endemic. Summer.

This species was reported in error for Desoto County by Nesom (2006q).

Liatris ohlingerae is listed as endangered in Florida (Florida Administrative Code, Chapter 5B-40) and in the United States (U.S. Fish and Wildlife Service, 50 CFR 23).

Liatris patens G. L. Nesom & Kral [Spreading, in reference to the long-pedunculate capitula spreading on the floral axis.] GEORGIA BLAZING-STAR.

Liatris patens G. L. Nesom & Kral, Sida 20: 1579, f. 2–3. 2003.

Erect perennial herb, to 1 m; roots forming a depressed-globose or globose corm; stem striate-angled, strigose-hirtellous. Basal and proximal cauline leaves with the blade linear-oblanceolate, 9–18 cm long, 2–4(7) mm wide, reduced upward, 1-nerved, the apex acute, the base cuneate, clasping, the margin entire, the upper and lower surfaces glabrous, the margin of the proximal leaves ciliate, sessile. Flowers in capitula in a loose racemiform array, the peduncle 1–2.5(3) cm long, spreading; involucre turbinate or campanulate, 5–7 mm wide; phyllaries in (2)3–4(5) series, obovate or oblong-obovate, 5–8 mm long, unequal, the margin with a pinkish or purplish hyaline, sometimes slightly erose, ciliate, the apex rounded or subtruncate, the outer surface sparsely strigose-villous or glabrate; receptacle flat; flowers numerous; corolla lavender or pinkish purple, rarely white, 4–5 mm long, the inner surface pilose. Cypsela prismatic, ca. 3 mm long, 8- to 11-ribbed, hirtellous, gland-dotted; pappus of 1–2 series of numerous barbellate bristles subequaling or longer than the corolla.

Flatwoods. Rare; Madison County. South Carolina, Georgia, and Florida. Fall.

Liatris pauciflora Pursh [Few-flowered.]

Erect perennial herb, to 9 dm; roots forming a globose, depressed-globose, or elongate corm; stem striate-angular, puberulent-hirtellous or glabrous. Basal and proximal cauline leaves with the blade linear-oblanceolate, 4–12 cm long, 2–7 mm wide, 1-nerved, reduced upward, the apex acute, the base cuneate or long-tapering, clasping, the margin entire, the upper and lower surfaces hirtellous or glabrous, weakly gland-dotted, the margin of the proximal leaves sometimes ciliate, sessile. Flowers in capitula in a dense, racemiform or spiciform array, sessile or the peduncle 1–8 mm long, ascending; involucre cylindric, 3–7 mm wide; phyllaries in 3–4 series, oblong or oblong-oblanceolate, (9)10–13 mm long, unequal, the margin hyaline, the apex acute, sometimes mucronulate, the outer surface glabrous, gland-dotted or eglandular; receptacle flat; flowers 3–6; corolla lavender or pinkish purple, rarely white, 4–5 mm long, the inner surface pilose. Cypsela prismatic, 3–4 mm long, 8- to 11-ribbed, hirtellous, gland-dotted; pappus of 1–2 series of numerous barbellate bristles subequaling the corolla.

1. Phyllaries eglandular; stems glabrous or rarely sparsely hirtellous .... var. **pauciflora**
1. Phyllaries glandular-dotted; stems puberulent-hirtellous .... var. **secunda**

Liatris pauciflora var. **pauciflora** FEWFLOWER GAYFEATHER.

Liatris pauciflora Pursh, Fl. Amer. Sept. 510. 1814. *Lacinaria pauciflora* (Pursh) Kuntze, Revis. Gen. Pl. 1: 349. 1891.

Stem glabrous, rarely sparsely hirtellous. Phyllaries eglandular.

Sandhills. Occasional; northern and central peninsula, eastern panhandle. Georgia and Florida. Summer–fall.

Liatris pauciflora var. **secunda** (Elliott) D. B. Ward [*Secundus*, turned toward the same side, in reference to the capitula.] PIEDMONT GAYFEATHER.

Liatris secunda Elliott, Sketch Bot. S. Carolina 2: 278. 1823. *Lacinaria secunda* (Elliott) Alexander, in Small, Man. S.E. Fl. 1333. 1933. *Liatris pauciflora* Pursh var. *secunda* (Elliott) D. B. Ward, Novon 14: 367. 2004.

Stem puberulent-hirtellous. Phyllaries gland-dotted.

Sandhills, coastal dunes, and scrub. Occasional; central and western panhandle. North Carolina south to Florida, west to Alabama. Summer–fall.

Liatris provincialis R. K. Godfrey [From one area, in reference to its localized distribution.] GODFREY'S GAYFEATHER; GODFREY'S BLAZING-STAR.

Liatris provincialis R. K. Godfrey, Amer. Midl. Naturalist 66: 466, f. 1. 1961. TYPE: FLORIDA: Franklin Co.: Alligator Point, 4 Oct 1959, *Godfrey 58985* (holotype: GH).

Erect perennial herb, to 9 dm; roots forming a globose or elongate corm; stem striate-angled, hirtellous. Basal and proximal cauline linear-oblanceolate, 6–15 cm long, 2–6 mm wide, 1-nerved, reduced upward and becoming linear, the apex acute, the base cuneate or long-tapering, clasping, the margin entire, the upper and lower surfaces glabrous or hirtellous (sometimes only on the midvein of the lower surface), sessile. Flowers in capitula in a dense, spiciform array; sessile; involucre cylindric, 3–5 mm wide; phyllaries in 3–4(5) series, oblong, obovate, or lanceolate, 9–11 mm long, unequal, the margin hyaline, ciliolate, the apex acute or acuminate, sometimes micronulate, the outer surface glabrous, sparsely puberulent, or hirtellous; receptacle flat; flowers 3–4; corolla lavender or pinkish purple, 5–6 mm long, the inner surface glabrous. Cypsela prismatic, 4–6 mm long, 8- to 11-ribbed, hirtellous, gland-dotted; pappus of 1–2 series of numerous barbellate bristles subequaling the corolla.

Sandhills and coastal dunes. Rare; Wakulla and Franklin Counties. Endemic. Summer–fall.

Liatris provincialis is listed as endangered in Florida (Florida Administrative Code, Chapter 5B-40).

Liatris savannensis Kral & G. L. Nesom [Growing in savannas.] SAVANNA GAYFEATHER.

Liatris savannensis Kral & G. L. Nesom, Sida 20: 1574. 2003. *Liatris spicata* (Linnaeus) Willdenow var. *savannensis* (Kral & G. L. Nesom) D. B. Ward, Phytologia 94: 141. 2012. TYPE: FLORIDA: Charlotte Co.: ca. 3 mi. S of Punta Gorda, 7 Oct 1979, *Kral 64559* (holotype: US; isotypes: FSU, NY, VDB, VSC).

Erect perennial herb, to 1.3 m; roots forming a globose or irregularly cylindric, narrowly ovoid, or fusiform corm; stem striate-angled, glabrous, sometimes sparsely gland-dotted. Leaves with the blade linear-elliptic, linear-oblanceolate, or linear-spatulate, (17)20–40 cm long, 3–10 mm wide, 1- or 3- to 5-veined, reduced upward, the apex acute, the base cuneate or long-tapering, clasping, the margin entire, the upper and lower surfaces glabrous or the proximal leaves with the margin ciliate, gland-dotted, sessile. Flowers in capitula in a dense, spiciform array, sessile or the peduncle 1–2 mm long, spreading or ascending; involucre turbinate or campanulate, (8)9–11(12) mm wide; phyllaries in (2)3–4(5) series, obovate or oblong, 5–8 mm long, unequal; the margin pinkish purple hyaline, sometimes erose, ciliate, the apex rounded or subtruncate, the outer surface sparsely strigose-villous or glabrate; receptacle flat; flowers numerous; corolla lavender or pinkish purple, rarely white, 5–6 mm long, the inner surface pilose. Cypsela prismatic, ca. 4 mm long, 8- to 11-ribbed, hirtellous, gland-dotted; pappus of 1–2 series of numerous barbellate bristles subequaling the corolla.

Flatwoods and open hammocks. Occasional; central peninsula. Endemic. Fall.

Liatris spicata (L.) Willd. [In reference to the capitula in a spiciform array.] DENSE GAYFEATHER.

Serratula spicata Linnaeus, Sp. Pl. 819. 1753. *Liatris spicata* (Linnaeus) Willdenow, Sp. Pl. 3: 1636. 1803. *Liatris macrostachya* Michaux, Fl. Bor.-Amer. 2: 91. 1803, nom. illegit. *Liatris spicata* (Linnaeus) Willdenow var. *macrostachya* de Candolle, Prodr. 5: 130. 1836, nom. inadmiss. *Lacinaria spicata*

(Linnaeus) Kuntze, Revis. Gen. Pl. 1: 349. 1891. *Liatris spicata* (Linnaeus) Willdenow var. *typica* Gaiser, Rhodora 48: 178. 1946, nom. inadmiss.

Liatris resinosa Nuttall, Gen. N. Amer. Pl. 2: 131. 1818. *Lacinaria spicata* (Linnaeus) Kuntze, var. *resinosa* (Nuttall) Voss, Vilm. Blumengärtn., ed. 3. 1: 450. 1894. *Liatris spicata* (Linnaeus) Willdenow var. *resinosa* (Nuttall) Gaiser, Rhodora 48: 216. 1946.

Erect perennial herb, to 1(1.8) m; roots forming a globose or slightly elongate corm; stem striate-angled, glabrous. Basal and proximal cauline leaves with the blade oblong-lanceolate or linear-lanceolate, 12–35 cm long, (2)2–10(20) mm wide, 3- to 5-veined, reduced distally, the apex acute, the base cuneate or long-tapering, clasping, the margin entire, the upper and lower surfaces glabrous or sparsely villous, gland-dotted, sessile. Flowers in capitula in a dense or loose spiciform array, sessile or the peduncle rarely 1–2 mm long, ascending; involucre cylindric or campanulate, 4–6 mm wide; phyllaries in (3)4–5 series, ovate or oblong, 7–11 mm long, unequal, the margin hyaline, sometimes ciliolate, the apex rounded or obtuse, the outer surface glabrate; receptacle flat; flowers (4)5–8(14); corolla lavender or pinkish purple, rarely white, 5–6 mm long, the inner surface glabrous. Cypsela prismatic, 4–6 mm long, 8- to 11-ribbed, hirtellous, gland-dotted; pappus of 1–2 series of numerous barbellate bristles subequaling the corolla.

Sandhills and flatwoods. Occasional; nearly throughout. Quebec south to Florida, west to Ontario, Michigan, Wisconsin, Illinois, Missouri, Arkansas, and Louisiana. Summer–fall.

Liatris squarrosa (L.) Michx. [Projecting outward, in reference to the sometimes spreading or reflexed phyllary apex.] SCALY GAYFEATHER.

Serratula squarrosa Linnaeus, Sp. Pl. 818. 1753. *Lacinaria squarrosa* (Linnaeus) Hill, Veg. Syst. 4: 49. 1762. *Liatris squarrosa* (Linnaeus) Michaux, Fl. Bor.-Amer. 2: 92. 1803. *Liatris squarrosa* (Linnaeus) Michaux var. *typica* Gaiser, Rhodora 48: 394. 1946.

Liatris squarrosa (Linnaeus) Michaux var. *gracilenta* Gaiser, Rhodora 48: 397. 1946.

Erect perennial herb, to 8 dm; roots forming a globose or slightly elongate corm; stem striate-angled, glabrous or villous-puberulent. Basal and proximal cauline leaves with the blade linear or linear-lanceolate, 7–22 cm long, 2–12 mm wide, 3- to 5-veined, reduced upward, the apex acute, the base cuneate or long-tapering, clasping, the margin entire, the upper and lower surfaces sparsely pilose or glabrescent, sessile. Flowers solitary or few to many in a loose, spiciform array, sessile or the peduncle 1–8(20) mm long, spreading or ascending; involucre cylindric or campanulate, 7–10 mm wide; phyllaries in 5–7 series, ovate, ovate-triangular, or oblong-triangular, 1–2 cm long, unequal, sometimes spreading-reflexed, the margin not hyaline, ciliate, the apex acute or acuminate, the outer surface glabrous or pubescent; receptacle flat; flowers numerous; corolla lavender or pinkish purple, 5–6 mm long, the inner surface glabrous. Cypsela prismatic, 4–6 mm long, 8- to 11-ribbed, hirtellous, gland-dotted; pappus of 1–2 series of numerous plumose bristles subequaling the corolla.

Sandhills. Rare; Gadsden, Washington, Santa Rosa, and Escambia Counties. Delaware and Maryland south to Florida, west to Michigan, Illinois, Missouri, Arkansas, and Louisiana. Fall.

Liatris squarrulosa Michx. [Spreading, in reference to the distally spreading or reflexed phyllary apices.] APPALACHIAN GAYFEATHER.

Liatris squarrulosa Michaux, Fl. Bor.-Amer. 2: 92. 1803. *Liatris scariosa* (Linnaeus) Willdenow var. *squarrulosa* (Michaux) A. Gray, Syn. Fl. N. Amer. 1(2): 110. 1884. *Lacinaria scariosa* (Linnaeus) Hill var. *squarrulosa* (Michaux) Small & Vail, Mem. Torrey Bot. Club 4: 28: 1894.
Lacinaria earlei Greene, Pittonia 4: 316. 1901. *Liatris earlei* (Greene) K. Schumann, Bot. Jahresber. (Just) 29(1): 569. 1903.

Erect perennial herb, to 8(13) dm; roots forming a globose corm; stem striate-angled, puberulent. Basal and proximal cauline leaves with the blade oblanceolate, 8–29 cm long, 1–2.5(4.5) cm wide, 1-nerved, reduced upward, the apex acute, the base cuneate or long-tapering, clasping, the margin entire, the upper and lower surfaces puberulent, hirtellous, or glabrate, sessile. Flowers in capitula in a loose or dense racemiform or spiciform array, sessile or the peduncle 2–8(30) mm long, ascending, spreading, or deflexed; involucre cylindric, campanulate, or turbinate, 0.8–1.5 cm wide; phyllaries in 4–5(6) series, obovate or oblong-spatulate, (6)8–10 mm long, unequal, the margin weakly hyaline, the apex acute, obtuse, or rounded, the outer surface puberulent, puberulent-hirtellous, or glabrate; receptacle flat; flowers numerous; corolla lavender or pinkish purple, 6–8 mm long, the inner surface pilose. Cypsela prismatic, (3)4–6 mm long, 8- to 11-ribbed, hirtellous, gland-dotted; pappus of 1–2 series of numerous barbellate bristles subequaling the corolla.

Dry open hammocks. Rare; Gadsden and Okaloosa Counties. West Virginia south to Florida, west to Illinois, Missouri, Oklahoma, and Texas. Fall.

Liatris tenuifolia Nutt. [With narrow leaves.] SHORTLEAF GAYFEATHER.

Erect perennial herb, to 1.8 m; roots forming a globose corm; stem glabrous or sparsely pilose. Basal and proximal cauline leaves with the blade linear or linear-lanceolate, 10–38 cm long, (1)2–6(9) mm wide, 1-nerved, reduced upward, the apex acute, the base cuneate or long-tapering, clasping, the margin entire, the upper and lower surfaces glabrous or the proximal margins sometimes ciliate, gland-dotted or eglandular, sessile. Flowers in capitula in a dense racemiform array, the peduncle 1–7 mm long, ascending; involucre turbinate or campanulate, 4–5 mm wide; phyllaries in 2–3(4) series, lanceolate- or elliptic-oblong, 5–7 mm long, unequal, the margin hyaline, sometimes pinkish purple, the apex rounded-retuse, minutely involute-cuspidate or apiculate, the outer surface glabrate; receptacle flat; flowers 3–6; corolla lavender or pinkish purple, rarely white, 4–5 mm long, the inner surface glabrous. Cypsela prismatic, 3–4 mm long, 8- to 11-ribbed, hirtellous, gland-dotted; pappus of 1–2 series of numerous barbellate bristles subequaling the corolla.

1. Leaves 1–2 mm wide, gland-dotted, the base sparsely ciliate .... var. **tenuifolia**
1. Leaves 2–6(9) mm wide, eglandular, the base eciliate .... var. **quadriflora**

Liatris tenuifolia var. **tenuifolia**

Liatris tenuifolia Nuttall, Gen. N. Amer. Pl. 2: 131. 1818. *Lacinaria tenuifolia* (Nuttall) Kuntze, Revis. Gen. Pl. 1: 349. 1891.

Leaves 1–2 mm wide, gland-dotted, the base sparsely ciliate.

Sandhills and open hammocks. Frequent; nearly throughout. South Carolina south to Florida, west to Mississippi. Summer–fall.

Liatris tenuifolia var. **quadriflora** Chapman [With four flowers commonly occurring in the capitulum.]

Liatris laevigata Nuttall, Trans. Amer. Philos. Soc., ser. 2. 7: 285. 1841. *Lacinaria laevigata* (Nuttall) Small, Fl. S.E. U.S. 1175, 1339. 1903. *Liatris tenuifolia* Nuttall var. *laevigata* (Nuttall) B. L. Robinson, Proc. Amer. Acad. Arts 47: 201. 1911. TYPE: FLORIDA: 26 Nov, *Read s.n.* (lectotype: BM; isolectotype: PH). Lectotypified by Gaiser (1946: 290–91).

Liatris tenuifolia Nuttall var. *quadriflora* Chapman, Fl. South. U.S., ed. 2. 626. 1883. *Liatris quadriflora* (Chapman) Bridges & Orzell, J. Bot. Res. Inst. Texas. 11: 298. 2017. TYPE: FLORIDA: Lee Co.: along the Caloosa River.

Leaves 2–6(9) mm wide, eglandular, the base eciliate.

Dry, open hammocks and dry flatwoods. Common; nearly throughout. Georgia and Florida. Summer–fall.

EXCLUDED TAXA

Liatris graminifolia Willdenow—Reported for Florida by many authors, including Chapman (1860), Small (1903, 1913a, 1933, all as *Lacinaria graminifolia* (Willdenow) Kuntze, Radford et al. (1964, 1968), Cronquist (1980), Wunderlin (1982, 1998), Wunderlin and Hansen (2003), and Clewell (1985), all who misapplied the name to Florida material of *L. elegantula.* According to Nesom and Stucky (2004), *L. graminifolia* is a heterotypic synonym of the northeastern United States taxon *L. pilosa* (Aiton) Willdenow.

Liatris scariosa (Linnaeus) Willdenow—Reported for Florida by Chapman (1860) and Small (1903, 1913a, both as *Lacinaria scariosa* (Linnaeus) Hill), the name misapplied to Florida material of *L. aspera,* and by Alexander (in Small, 1933, as *Lacinaria scariosa*), the name misapplied to Florida material of L. *squarrulosa.*

Lygodesmia D. Don 1829. ROSE-RUSH

Herbs. Leaves alternate, simple, pinnate-veined, epetiolate. Flowers in capitula, solitary or in corymbiform arrays, liguliform, bisexual; calyculi present; receptacle epaleate; petals 5, basally connate, zygomorphic; stamens 5, epipetalous, the filaments free, the anthers connate; ovary inferior, 2-carpellate, 1-loculate. Fruit a cypsela; pappus present.

A genus of 5 species; North America and Mexico. [From the Greek *lygos,* forked stick, and *desme,* bundle, or handful, in reference to the clumped stems with reduced leaves.]

Selected reference: Bogler (2006c); Tomb (1980).

Lygodesmia aphylla (Nutt.) DC. [Lacking leaves.] ROSE-RUSH.

Prenanthes aphylla Nuttall, Gen. N. Amer. Pl. 2: 123. 1818. *Lygodesmia aphylla* (Nuttall) de Candolle, Prodr. 7: 198. 1838. *Erythremia aphylla* (Nuttall) Nuttall, Trans. Amer. Philos. Soc., ser. 2. 7: 446. 1841.

Erect perennial herb, to 8 dm; roots or rhizomes fleshy; stem striate-angled, glabrous. Leaves basal in rosettes, sometimes withered at flowering, the blade linear, 10–35 cm long, 2–3 mm wide, the apex acute, the base cuneate, the margin entire, the upper and lower surfaces glabrous, the cauline similar or reduced to subulate scales to 3 mm long. Flowers in capitula, solitary or 2–5 in a corymbiform array; involucre cylindric, 5–7 mm wide; calyculi of 4–14 in 2 series, deltate bractlets, 1–5 mm long, the margin scarious, erose ciliate, the upper and lower faces tomentulose; phyllaries 8 in 1 series, linear, 1.4–2.2 cm long, the margin scarious, the apex acute, purplish, often with a keel-shaped appendage, the outer surface scabrate near the base; receptacle flat; flowers 8–10; corolla pink, lavender, or rarely white, the lamina ca. 2 cm long. Cypsela subcylindric, 11–14 mm long, the inner surface sulcate, the outer surface striate, the apex attenuate, glabrous; pappus of numerous tawny or white, smooth bristles 1–1.5 mm long.

Dry flatwoods, scrub, and dry disturbed sites. Common; peninsula west to central panhandle. Spring–summer.

Marshallia Schreb., nom. cons. 1791. BARBARA'S BUTTONS

Herbs. Leaves alternate, simple, pinnipalmate-veined, petiolate or epetiolate. Flowers in capitula, solitary or 2 to many in corymbiform arrays, discoid, bisexual; receptacle paleate; petals 5, basally connate, actinomorphic; stamens 5, epipetalous, the filaments free, the anthers connate; the ovary inferior, 2-carpellate, 1-loculate. Fruit a cypsela; pappus present.

A genus of 7 species; North America. [Commemorates Moses Marshall (1758–1813), American botanist, physician, and magistrate in Pennsylvania.]

Selected references: Channell (1957); Watson (2006c).

1. Cauline leaves absent or narrowly oblanceolate .... **M. obovata**
1. Cauline leaves always present, linear.
    2. Phyllaries strongly acuminate or subulate-tipped; peduncle with the trichomes purple or with purple cross-walls .... **M. graminifolia**
    2. Phyllaries acute or obtuse, often mucronulate; peduncle with the trichomes white .... **M. ramosa**

Marshallia graminifolia (Walter) Small [With grasslike leaves.] GRASSLEAF BARBARA'S BUTTONS.

Athanasia graminifolia Walter, Fl. Carol. 200. 1788. *Marshallia graminifolia* (Walter) Small, Bull. Torrey Bot. Club 25: 482. 1898.

Persoonia angustifolia Michaux, Fl. Bor.-Amer. 2: 106. 1803. *Trattenikia angustifolia* (Michaux) Persoon, Syn. Pl. 2: 403. 1807. *Marshallia angustifolia* (Michaux) Pursh, Fl. Amer. Sept. 520. 1814. *Phyteumopsis angustifolia* (Michaux) Poiret, in Lamarck, Encycl., Suppl. 4: 406. 1816.

Marshallia angustifolia (Michaux) Pursh var. *cyananthera* Elliott, Sketch Bot. S. Carolina 2: 317. 1823. *Marshallia graminifolia* (Walter) Small var. *cyananthera* (Elliott) Beadle & F. E. Boynton, Biltmore Bot. Stud. 1: 4. 1901.

Marshallia tenuifolia Rafinesque, New Fl. 4: 77. 1838 ("1836"). *Marshallia graminifolia* (Walter) Small subsp. *tenuifolia* (Rafinesque) L. E. Watson, in Watson & Estes, Syst. Bot. 15: 412. TYPE: "Florida and Georgia" s.d., *Baldwin s.n.* (lectotype: DWC). Lectotypified by Channell (1957: 120).

Erect perennial herb, to 8(12) dm; stem striate-angled, glabrous. Basal leaves with the blade oblanceolate, usually 3-nerved, (2)4–25 cm long, 2–13 mm wide, 3-veined, the apex acute or obtuse, the margin entire, the upper and lower surfaces glabrous, short-petiolate, the cauline linear, reduced upward, sessile. Flowers in capitula, solitary or to 30 in a corymbiform array, the peduncle 5–50 cm long, with trichomes purple or with purple cross-walls; involucre 2–3 cm wide; phyllaries numerous in 2 series, lanceolate, 4–10 mm long, the margin slightly winged proximally, the apex strongly acuminate or subulate-tipped, the outer surface glabrous; receptacle convex; paleae linear-attenuate, the apex subulate, often gland-dotted; flowers numerous; corolla pale lavender or purple, rarely white, 5–6 mm long, the outer surface pubescent. Cypsela turbinate, ca. 3 mm long, 5-angled, 10-ribbed, pubescent; pappus of 5(6) membranous short scales, the margin entire or denticulate.

Wet flatwoods and bogs. Occasional; northern and central peninsula, central and western panhandle. North Carolina south to Florida, west to Texas. Summer–fall.

Marshallia obovata (Walter) Beadle & F. E. Boynton [Reversed egg-shaped, in reference to the basal leaf blade.] SPOONSHAPE BARBARA'S BUTTONS.

Athanasia obovata Walter, Fl. Carol. 201. 1788. *Marshallia obovata* (Walter) Beadle & F. E. Boynton, Biltmore Bot. Stud. 1: 5. 1901.

Persoonia lanceolata Michaux, Fl. Bor.-Amer. 2: 105. 1803. *Trattenikia lanceolata* (Michaux) Persoon, Syn. Pl. 2: 403. 1807. *Marshallia lanceolata* (Michaux) Pursh, Fl. Amer. Sept. 519. 1814. *Phyteumopsis lanceolata* (Michaux) Poiret, in Lamarck, Encycl., Suppl. 4: 405. 1816.

Marshallia lanceolata (Michaux) Pursh var. *platyphylla* M. A. Curtis ex Chapman, Fl. South. U.S. 241. 1860. *Marshallia obovata* (Walter) Beadle & F. E. Boynton var. *platyphylla* (M. A. Curtis ex Chapman) Beadle & F. E. Boynton, Biltmore Bot. Stud. 1: 6. 1901.

Marshallia obovata (Walter) Beadle & F. E. Boynton var. *scaposa* Channell, Contr. Gray Herb. 181: 90, pl. 7. 1957.

Erect perennial herb, to 6 dm; stem striate-angled, glabrous. Basal and proximal cauline leaves narrowly oblanceolate, the apex acute or obtuse, 5–10 cm long, 5–15 mm wide, slightly reduced distally, 3-veined, the margin entire, the upper and lower surfaces glabrous, short-petiolate. Flowers in capitula, solitary or 2, the peduncle 20–40 cm long, the trichomes white; involucre 2–3 cm wide; phyllaries numerous in 2 series, linear-lanceolate, 5–10 mm long, the margin slightly winged proximally, the apex obtuse, the outer surface glabrous; receptacle convex; paleae linear-spatulate, the apex obtuse; flowers numerous; corolla white, 5–6 mm long, the outer surface of the tube pubescent. Cypsela turbinate, 3–4 mm long, 5-angled, 10-ribbed, pubescent; pappus of 5(6) membranous scales less than ½ as long as the cypsela, the margin entire or denticulate.

Open, dry hammocks and sandhills. Rare; Jackson County. Virginia south to Florida, west to Tennessee and Alabama. Spring.

Marshallia obovata is listed as endangered in Florida (Florida Administrative Code, Chapter 5B-40).

Marshallia ramosa Beadle & F. E. Boynton [*Ramosus*, branched, in reference to the stems.] SOUTHERN BARBARA'S BUTTONS.

Marshallia ramosa Beadle & F. E. Boynton, Biltmore Bot. Stud. 1: 8, pl. 2. 1901.

Erect perennial herb, to 6 dm; stem striate-angled, glabrous. Basal and cauline leaves with the blade linear, 6–18 cm long, 2–7 mm wide, 3-veined, the apex acute, the margin entire, the upper and lower surfaces glabrous, the basal leaves short-petiolate, the upper cauline reduced, sessile. Flowers in capitula, (2)4–10(20) in a corymbiform array, the peduncle 6–12 cm long, the trichomes white; involucre 1–2.5 cm wide; phyllaries numerous in 2 series, lanceolate, 4–10 mm long, the margin slightly winged proximally, the apex acute, the outer surface glabrous; phyllaries lanceolate, 5–8 mm long, the margin usually slightly winged proximally, the apex acute or obtuse, often mucronulate, the outer surface glabrous; receptacle convex; paleae linear, the apex obtuse, mucronate, or subulate; flowers numerous; corolla usually white, sometimes pale lavender, 4–6 mm long, the outer surface of the tube pubescent. Cypsela turbinate, ca. 4 mm long, 5-angled, 10-ribbed, pubescent; pappus of 5–6 membranous scales about ½ as long as the cypsela, the margin entire or denticulate.

Sandhills and pine savannas. Rare; Clay and Washington Counties. Georgia and Florida. Spring–summer.

Marshallia ramosa is listed as endangered in Florida (Florida Administrative Code, Chapter 5B-40).

Melampodium L. 1753. BLACKFOOT

Herbs. Leaves opposite, simple, pinnipalmate-veined, epetiolate. Flowers in capitula, solitary, radiate; receptacle paleate; ray flowers carpellate; petals 5, basally connate, zygomorphic; disk flowers staminate; petals 5; basally connate, actinomorphic; stamens 5, epipetalous, the filaments free, the anthers connate; ovary inferior, 2-carpellate, 1-loculate. Fruit a cypsela; pappus absent.

A genus of about 40 species; North America, Central America; Asia. [From the Greek *melas*, black, and *podion*, foot, in reference to the black stem base and roots.]

Selected reference: Strother (2006s).

Melampodium divaricatum (Rich.) DC. [Spreading at an angle, in reference to the stems.] BOTON DE ORO.

Dysodium divaricatum Richard, in Persoon, Syn. Pl. 2: 489. 1807. *Melampodium ovatifolium* Reichenbach, Inconogr. Bot. Exot. 30. 1824, nom. illegit. *Melampodium divaricatum* (Richard) de Candolle, Prodr. 5: 520. 1836. *Eleutheranthera divaricata* (Richard) Millspaugh, Publ. Field Columb. Mus., Bot.Ser. 1: 53. 1895.

Erect annual herb, to 1 m; stem striate-angled, pubescent in a line from between the leaves or glabrate. Flowers in a solitary capitulum on an axillary branch, the peduncle 1–12 cm long, sparsely pubescent; involucre hemispheric; phyllaries 8 to many in 2 series, the outer series of 5, ovate or suborbicular, connate ¼–⅓ their length, the outer surface pubescent, the inner each inclosing a ray flower ovary and shed with the cypsela; receptacle flat or convex; paleae of lanceolate scales, conduplicate; ray flowers 8–13; corolla yellow orange, the lamina 4–7 mm long; disk flowers numerous; corolla yellow-orange, ca. 2 mm long. Cypsela laterally compressed and D-shaped, shed enclosed with the ray flower phyllary and palea as a unit; pappus absent.

Disturbed sites. Rare; Leon County. Florida; Mexico, West Indies, Central America, and South America; Asia. Native to tropical America. Summer.

Melanthera Rohr 1792. SQUARESTEM

Herbs. Leaves opposite, simple, pinnate- or pinnipalmate-veined, petiolate. Flowers in capitula, solitary or in corymbiform arrays, discoid, bisexual; receptacle paleate; petals 5, basally connate, actinomorphic; stamens 5, epipetalous, the filaments free, the anthers connate; ovary inferior, 2-carpellate, 1-loculate. Fruit a cypsela; pappus present.

A genus of about 35 species; North America, West Indies, Mexico, Central America, South America, Africa, Asia, and Pacific Islands. [From the Greek *melas,* black, and *anthera,* anther.]

Considerable confusion exists in the number of *Melanthera* species in Florida. Alexander, in Small (1933) recognized six species (*M. angustifolia, M. deltoidea, M. hastata, M. ligulata, M. parvifolia,* and *M. radiata*). Parks (1973) recognized five (*M. angustifolia, M. aspera, M. ligulata, M. nivea,* and *M. parvifolia*). Later, Parks (2006) recognized three species: *M. angustifolia* with leaf blades usually linear and 4–5 × 2–3 mm, *M. nivea* with leaf blades usually ovate and 5–12 × 3–8 cm, and *M. parvifolia* with leaf blades usually ovate and 1.5–4 × 1–1.5 cm. We recognize only a single species, *M. nivea.* However, among Florida specimens, *M. nivea* appears to intergrade morphologically with *M. angustifolia* and *M. parvifolia* as delimited by Parks (2006), including leaf blade shape and size. Parks (2006) considered *M. parvifolia* to be endemic to south Florida and the other two widespread in the Americas. Due to taxonomic uncertainty, one species, *M. nivea,* is here recognized, although further taxonomic revision may be warranted.

Selected reference: Parks (2006).

Melanthera nivea (L.) Small [*Niveus,* snow-white, in reference to the disk corolla.] SNOW SQUARESTEM.

Bidens nivea Linnaeus, Sp. Pl. 833. 1753. *Melananthera hastata* Michaux, Fl. Bor.-Amer. 2: 107. 1803, nom. illegit. *Melananthera hastata* Michaux var. *lobata* Pursh, Fl. Amer. Sept. 519. 1814, nom. inadmiss. *Melanthera trilobata* Cassini, in Cuvier, Dict. Sci. Nat. 29: 485. 1823, nom. illegit. *Melanthera hastata* de Candolle, Prodr. 5: 545. 1836, nom. illegit. *Wulffia hastata* M. Gómez de la Maza y Jiménez, Anales Soc. Esp. Hist. Nat. 19: 274. 1890, nom. illegit. *Amellus niveus* (Linnaeus) Kuntze, Revis. Gen. Pl. 1: 305. 1891. *Melanthera nivea* (Linnaeus) Small, Fl. S.E. U.S. 1251, 1340. 1903. *Melanthera hastata* de Candolle subsp. *lobata* Borhidi, Bot. Közlem. 58: 177. 1971, nom. inadmiss.

Calea aspera Jacquin, Collectanea 2: 290. 1788. *Melananthera deltoidea* Michaux, Fl. Bor.-Amer. 2: 107. 1803, nom. illegit. *Melananthera linnaei* Kunth, in Humboldt et al., Nov. Gen. Sp. 4: 199. 1820, nom. illegit. *Melanthera urticifolia* Cassini, in Cuvier, Dict. Sci. Nat. 29: 485. 1823, nom. illegit. *Wulffia deltoidea* M. Gómez de la Maza y Jiménez, Dicc. Bot. Nombres Vulg. 26. 1889. *Amellus asper* (Jacquin) Kuntze, Revis. Gen. Pl. 1: 305. 1891. *Amellus asper* (Jacquin) Kuntze var. *normalis* Kuntze, Revis. Gen. Pl. 1: 305. 1891, nom. inadmiss. *Melanthera deltoidea* Small, Fl. S.E. U.S. 1251. 1903, nom. illegit. *Melanthera aspera* (Jacquin) Steudel ex Small, Bull. Torrey Bot. Club 36: 174. 1909.

Melanthera angustifolia A. Richard, in Sagra, Hist. Fis. Cuba, Bot. 11: 54. 1850. *Wulffia angustifolia* (A. Richard) M. Gómez de la Maza y Jiménez, Anales Soc. Esp. Hist. Nat. 19: 274. 1890.
Melanthera lanceolata Bentham ex Oersted, Vidensk. Meddel. Dansk Naturhist. Foren. Kjoebenhavn 1852: 88. 1852. *Amellus lanceolatus* (Bentham ex Oersted) Kuntze, Revis. Gen. Pl. 1: 306. 1891.
Amellus asper (Jacquin) Kuntze var. *glabriusculus* Kuntze, Revis. Gen. Pl. 1: 306. 1891. *Melanthera aspera* (Jacquin) Steudel ex Small var. *glabriuscula* (Kuntze) J. C. Parks, Rhodora 75: 194. 1973.
Melanthera lobata Small, Fl. S.E. U.S. 1251, 1340. 1903. TYPE: FLORIDA: Lake Co.: vicinity of Eustis, 1–15 Jul 1894, *Nash 1411* (holotype: NY; isotypes: GH, MICH, MO, UC, US).
Melanthera parvifolia Small, Fl. S.E. U.S. 1251, 1340. 1903. *Melanthera hastata* de Candolle var. *parvifolia* (Small) O. E. Schulz, in Urban, Symb. Antill. 7: 126. 1911. TYPE: Florida: Monroe Co.: Big Pine Key, s.d., *Blodgett s.n.* (holotype: NY; isotype: GH).
Melanthera ligulata Small, Bull. New York Bot. Gard. 3: 439. 1905. TYPE: FLORIDA: Broward Co.: below Fort Lauderdale, May 1904, *Small & Wilson 1775* (holotype: NY).
Melanthera radiata Small, Bull. Torrey Bot. Club 36: 173. 1909. TYPE: FLORIDA: Miami-Dade Co.: near Camp Longview, 13–16 May 1904, *Small & Wilson 1575* (holotype: NY; isotype: US).
Melanthera brevifolia O. E. Schulz, in Urban, Symb. Antill. 7: 123. 1911. Lectotype: Miami-Dade Co.: Elliott Key, s.d. (lectotypes: F, MISSA, NY, PH). Lectotypified by Parks (1973: 194).
Melanthera angustifolia A. Richard var. *subhastata* O. E. Schultz, Repert. Spec. Nov. Regni Veg. 26: 109. 1929. *Melanthera amellus* (Linnaeus) D'Arcy var. *subhastata* (O. E. Schulz) D'Arcy, Phytologia 30: 6. 1975. *Melanthera aspera* (Jacquin) Steudel ex Small var. *subhastata* (O. E. Schulz) D'Arcy, Ann. Missouri Bot. Gard. 62: 1119. 1976.

Erect or sometimes sprawling or scandent, perennial herb, sometimes proximally somewhat woody, to 2 m tall; stem 4-angled and striate, scabrid-hispid, strigose, hirsute, or glabrate. Leaves with the blade oblanceolate, ovate, deltate, 3-lobed, lanceolate, elliptic, or linear, 3–12 cm long, 0.5–8 cm wide, with 1 or 3 main veins, the apex acute, the base truncate, hastate, or cuneate, the margin undulate, crenate or serrate, often irregularly, often 3-lobed, the upper and lower surfaces strigose, hispid, or sometimes glabrescent, the petiole to 6 cm long. Flowers in capitula in the axil of reduced distal leaves, solitary or sometimes forming a loose corymbiform array, the peduncle 2–20 cm long, hispid, strigose, or glabrescent; involucre 6–20 mm wide, phyllaries ovate or lanceolate, 3–14 mm long; receptacle flat, convex, or hemispheric; paleae oblanceolate, 4–7 mm long, conduplicate, the apex mucronate, straight or recurved; flowers numerous; corolla white, 5–10 mm long; anthers black. Cypsela obpyramidal, 2–3 mm long, sometimes slightly compressed, (3)4-angled, striate or sometimes verrucose-tuberculate, slightly brown; pappus of 2–12 barbellate bristles or awns.

Moist to dry hammocks, pine flatwoods, and beaches. Common; nearly throughout. Illinois, Kentucky, south to Florida, west to Louisiana; West Indies, Mexico, Central America, and South America. All Year.

Mikania Willd. HEMPVINE

Vines. Leaves opposite, simple, pinnipalmate-veined, petiolate. Flowers in capitula in corymbiform arrays, discoid, bisexual; receptacle epaleate; petals 5, basally connate, actinomorphic; stamens 5, epipetalous, the filaments free, the anthers connate; ovary inferior, 2-carpellate, 1-loculate. Fruit a cypsela; pappus present.

A genus of about 450 species; North America, West Indies, Mexico, Central America, South America, Africa, Asia, Australia, and Pacific Islands. [Commemmorates Josef Gottfried Mikan (1743–1824), Czech botanist.]

Selected reference: Holmes (2006).

1. Phyllaries 6–8 mm long; corolla lobes linear; cypsela 3–4 mm long .... **M. cordifolia**
1. Phyllaries 3–6 mm long; corolla lobes triangular or deltate; cypsela ca. 2 mm long.
   2. Phyllaries linear-lanceolate 5–6 mm long; corolla usually pinkish or purplish, ca. 4 mm long.. **M. scandens**
   2. Phyllaries lanceolate or narrowly ovate; corolla white, ca. 3 mm long .... **M. micrantha**

Mikania cordifolia (L. f.) Willd. [*Cordatus,* heart-shaped, in reference to the heart-shaped leaves.] FLORIDA KEYS HEMPVINE.

Cacalia cordifolia Linnaeus f., Suppl. Pl. 351. 1782 ("1781"). *Mikania cordifolia* (Linnaeus) Willdenow, Sp. Pl. 3: 1746. 1803. *Willoughbya cordifolia* (Linnaeus f.) Kuntze, Revis. Gen. Pl. 1: 371. 1891.

Climbing perennial vine, to 30 m; stem 6-angled, gray-tomentulose or tomentose. Leaves with the blade ovate or deltate, 5–10 cm long, 3–8 cm wide, the apex acute or acuminate, the base cordate, the margin subentire or undulate-dentate, the upper and lower surfaces pilose or tomentose, the petiole 2–6 cm long. Flowers in capitula in a terminal or lateral corymbiform array; involucre cylindric, 2–3 mm wide; phyllaries 4 in 2 series, elliptic or narrowly ovate, 6–8 mm long, subequal, the apex acute to slightly rounded; receptacle flat; flowers 4; corolla white, 3–5 mm long, the lobes linear. Cypsela prismatic, 3–4 mm long, 5-ribbed, glabrous or sparsely gland-dotted; pappus of numerous white barbellate bristles 4–5 mm long.

Hammocks. Common; peninsula, central panhandle. Georgia and Florida west to Texas, also in New Jersey; West Indies, Mexico, Central America, and South America. All year.

Mikania micrantha Kunth [*Micro,* small, the Greek *anthos,* flower, in reference to the small flowers.] MILE-A-MINUTE.

Mikania micrantha Kunth, in Humboldt et al., Nov. Gen. Sp. 4: 134. 1818. *Willoughbya micrantha* (Kunth) Rusby, Mem. Torrey Bot. Club 4: 211. 1895. *Mikania micrantha* Kunth forma *typica* B. L. Robinson, Contr. Gray Herb. 64: 42. 1922, nom. inadmiss.

Climbing perennial vine, to 30 m; stem subterete or obscurely 6-angled, glabrous or sparsely pubescent. Leaves with the blade ovate or deltate, 1.5–6 cm long, 1–5 cm wide, the apex acute or acuminate, the base cordate, the margin dentate or small-lobed, the upper and lower surfaces glabrous, the petiole 2–6 cm long. Flowers in capitula in a terminal or lateral corymbiform array; involucre cylindric, 2–3 mm wide; phyllaries 4 in 2 series, lanceolate or narrowly ovate, ca. 3 mm long, subequal, the apex acuminate; receptacle flat; corolla white, ca. 4 mm long, the lobes triangular or deltate. Cypsela prismatic, ca. 2 mm long, 5-ribbed, densely gland-dotted; pappus of numerous white barbellate bristles ca. 3 mm long.

Disturbed sites. Rare; Miami-Dade County. Florida; West Indies, Mexico, Central America, South America; Asia, Australia, and Pacific Islands. Native to West Indies, Mexico, Central America, and South America. All year.

Mikania micrantha is listed as a category II invasive species in Florida by the Florida Exotic Pest Plant Council (FLEPPC, 2017).

Mikania scandens (L.) Willd. [Climbing.] CLIMBING HEMPVINE.

Eupatorium scandens Linnaeus, Sp. Pl. 836. 1753. *Mikania scandens* (Linnaeus) Willdenow, Sp. Pl. 3: 1743. 1803. *Willoughbya scandens* (Linnaeus) Kuntze, Revis. Gen. Pl. 1: 371. 1891. *Willoughbya scandens* (Linnaeus) Kuntze var. *normalis* Kuntze, Revis. Gen. Pl. 1: 372. 1891, nom. inadmiss.
Mikania batatifolia de Candolle, Prodr. 5: 197. 1836.
Willoughbya heterophylla Small, Fl. S.E. U.S. 1170, 1338. 1903. TYPE: FLORIDA: Monroe Co.: No Name Key, Feb, *Curtiss 1213** (holotype: NY; isotypes: F, GH, MISSA, NY, PH, US).

Climbing perennial vine, to 30 m; stem subterete or obscurely 6-angled, glabrous or pilose. Leaves with the blade triangular or triangular-ovate, 3–15 cm long, 2–11 cm wide, the apex acute or acuminate, the base cordate or hastate, the margin subentire, undulate, crenate, or dentate, the upper and lower surfaces puberulent, the petiole 3–6 cm long. Flowers in capitula in a terminal or lateral corymbiform array; involucre cylindric, 2–3 mm wide; phyllaries 4 in 2 series, linear or lanceolate, 5–6 mm long, subequal, the apex acuminate; receptacle flat; corolla usually pink or purplish, sometimes white, 3–5 mm long, the lobes linear. Cypsela prismatic, ca. 2 mm long, 5-ribbed, densely gland-dotted; pappus of numerous white, pinkish, or purplish slightly barbellate bristles 4–5 mm long.

Floodplain forests, riverbanks, pond margins, and coastal swales. Common; nearly throughout. Maine south to Florida, west to Michigan, Illinois, Missouri, Oklahoma, and Texas; West Indies, and Mexico. All year.

Nabalus Cass. 1825. RATTLESNAKEROOT

Herbs. Leaves alternate, simple, pinnate-veined, epetiolate. Flowers in capitula in racemiform, corymbiform, or paniculiform arrays, liguliform, bisexual; calyculi present; receptacle epaleate; petals 5, basally connate, zygomorphic; stamens 5, epipetalous, the filaments free, the anthers connate; ovary inferior, 2-carpellate, 1-loculate. Fruit a cypsela; pappus present.

A genus of 26 species; North America and Asia. [*Nabal,* wilt, in reference to the drooping flower heads.]

Nabalus is sometimes included in *Prenanthes* by various workers (such as Bogler, 2006d), although molecular study shows the two not to be closely related (Kim et al., 1996; Zang et al., 2011).

Selected reference: Bogler (2006d).

1. Inflorescence racemiform; flowers pinkish, pale lavender, or white .... **N. autumnalis**
1. Inflorescence corymbiform or paniculiform; flowers pale yellow or greenish yellow..... **N. serpentaria**

Nabalus autumnalis (Walter) Weakley [Of the fall season.] SLENDER RATTLESNAKEROOT.

Prenanthes autumnalis Walter, Fl. Carol. 193. 1788. *Prenanthes virgata* Michaux, Fl. Bor.-Amer. 2: 83. 1803, nom. illegit. *Harpalyce virgata* D. Don, Edinburgh, New Philos. J. 6: 308. 1829, nom. illegit.

Nabalus virgatus de Candolle, Prodr. 7: 242. 1838, nom. illegit. *Nabalus autumnalis* (Walter) Weakley, in Weakly et al., J. Bot. Res. Inst. Texas 5: 439. 2011.

Erect perennial herb, to 1.4 m; roots thickened, tuberous; stem striate-angled, glabrous. Leaves with the blade oblong or linear, 7–18 cm long, 0.5–6 cm wide, reduced upward, the apex acute or acuminate, the base attenuate, the margin deeply and narrowly pinnately lobed or divided, the upper and lower surfaces glabrous, sometimes with few trichomes on the veins on the lower surface, sessile and clasping at the base. Flowers in capitula in a racemiform array, nodding at flowering; calyculi of 6–8 triangular or linear bractlets 1–3 mm long, green or purple; involucre cylindric-campanulate, 1–4 mm wide; phyllaries 6–7, linear or elliptic, 8–13 mm long, equal, pink or purple, the outer surface glabrous; receptacle slightly convex; corolla pinkish, pale lavender, or white, the lamina 11–17 mm long. Cypsela linear or elliptic, 4–6 mm long, subterete or angular, indistinctly 8- to 10-ribbed, dark green; pappus of 1 series of numerous pale yellow or tan, subequal, barbellate bristles ca. 5 mm long.

Moist flatwoods. Rare; Duval County. New Jersey south to Florida, also Mississippi. Fall.

Nabalus serpentaria (Pursh) Hook. [As the genus *Serpentaria,* used for the treatment of snake bite.] CANKERWEED; LIONSFOOT.

Prenanthes serpentaria Pursh, Fl. Amer. Sept. 599, t. 24. 1814. *Nabalus serpentaria* (Pursh) Hooker, Fl. Bor.-Amer. 1: 294. 1833. *Nabalus albus* (Linnaeus) Hooker var. *serpentaria* (Pursh) Torrey & A. Gray, Fl. N. Amer. 2: 480. 1843,

Nabalus fraseri de Candolle, Prodr. 7: 241. 1838. *Prenanthes fraseri* (de Candolle) D. Dietrich, Syn. Pl. 4: 1309. 1847.

Erect perennial herb, to 2 m; roots short and thick, with lateral storage roots; stem striate-angled, proximally glabrous, distally sparsely tomentose. Leaves with the blade deltate, ovate, or elliptic, 5–20 cm long, 4–10 cm wide, reduced upward, the apex acute or obtuse, the base cordate or rounded, the margin deeply pinnately lobed or divided, the ultimate margin entire or dentate, the upper and lower surfaces glabrous or finely tomentose on the veins, sessile and clasping at the base. Flowers in capitula in a corymbiform or paniculiform array, nodding at flowering; calyculi of 5–7 triangular bractlets 1–3 mm long, green or purple; involucre cylindric-campanulate, 4–5 mm wide; phyllaries 7–8(10), linear-lanceolate or elliptic, 10–13 mm long, equal, green or purple, the outer surface sparsely hispid or appressed-setose; receptacle slightly convex; Flowers 10–14; corolla pale yellow or greenish yellow, the lamina 9–15 mm long. Cypsela subcylindric, 5–8 mm long, subterete or angled, 8- to 10-ribbed, tan or brown; pappus of 1 series of numerous pale yellow or tan, subequal, barbellate bristles 7–9 mm long.

Hammocks. Occasional; northern counties. New Hampshire and New York south to Florida, west to Ohio, Kentucky, Tennessee, and Mississippi. Fall.

Oclemena Greene 1903.

Herbs. Leaves alternate, simple, pinnate-veined, petiolate or epetiolate. Flowers in capitula in corymbiform arrays, radiate; receptacle epaleate; ray flowers carpellate; petals 5, basally connate, zygomorphic; disk flowers bisexual; petals 5, basally connate, actinomorphic; stamens 5, epipetalous, the filaments free, the anthers connate; ovary inferior, 2-carpellate, 1-loculate. Fruit a cypsela; pappus present.

A genus of 3 species; North America. [Etymology unknown.]

Selected reference: Brouillet (2006b).

Oclemena reticulata (Pursh) G. L. Nesom [Like a network, in reference to the leaf veins.] WHITETOP ASTER; PINEBARREN ASTER.

Aster reticulata Pursh, Fl. Amer. Sept. 548. 1814. *Doellingeria reticulata* (Pursh) Greene, Pittonia 3: 53. 1896. *Oclemena reticulata* (Pursh) G. L. Nesom, Phytologia 77: 264. 1995 ("1994"). TYPE: FLORIDA: Brevard Co.: Indian River, 1874, *Palmer 250* (neotype: GH). Neotypified by Semple et al. (1991: 270).

Chrysopsis obovata Nuttall, Gen. Amer. Pl. 2: 152. 1818. *Aster obovatus* (Nuttall) Elliott, Sketch Bot. S. Carolina 2: 368. 1823. *Doellingeria obovata* (Nuttall) Nees von Esenbeck, Gen. Sp. Aster. 182. 1832. *Diplostephium obovatum* (Nuttall) de Candolle, Prodr. 5: 273. 1836. *Diplopappus obovatus* (Nuttall) Torrey & A. Gray, Fl. N. Amer. 2: 184. 1841.

Erect perennial herb, to 9 dm; stem rhizomatous, striate-angled, densely villosulous, gland-dotted distally. Leaves with the blade obovate-elliptic, ovate-elliptic, or elliptic-oblong, 2.5–11 cm long, 1–4 cm wide, the acute or obtuse, mucronulate, the base cuneate or rounded, the margin entire, undulate, or serrate distally, revolute, the upper and lower surfaces densely villous-pubescent, stipitate-glandular, the petiole 1–2 mm long or sessile. Flowers in capitula in a corymbiform array, the peduncle 2–5 cm long, villosulous, stipitate-glandular; involucre cylindric-campanulate, 6–12 mm wide; phyllaries numerous in 3(4) series, ovate-lanceolate or linear-lanceolate, 5–10 mm long, slightly pilose, gland-dotted, often with 1(2) linear bractlets; receptacle slightly convex; ray flowers (5)7–11(14); corolla white or pinkish, the lamina 10–15 mm long; disk flowers numerous; corolla pinkish yellow, 5–8 mm long. Cypsela fusiform, slightly compressed, 2–4 m long, 6- to 8-ribbed, strigillose, gland-dotted, tan; pappus of numerous whitish bristles in 3 series, subequaling the disk corolla, the innermost subclavate.

Bogs and wet flatwoods. Frequent; northern counties, central peninsula. South Carolina south to Florida, west to Alabama. Spring.

Onopordum L. 1753. COTTONTHISTLE

Herbs. Leaves alternate, simple, pinnate-veined, epetiolate. Flowers in capitula, solitary or few in terminal corymbiform arrays, discoid, bisexual; receptacle epaleate; petals 5, basally connate, actinomorphic; stamens 5, epipetalous, the filaments free, the anthers connate; ovary inferior, 2-carpellate, 1-loculate. Fruit a cypsela; pappus present.

A genus of about 47 species; North America, Europe, Africa, Asia, and Australia. [From the Greek *onos*, donkey, and *perdo*, fart, "donkey fart food" the vernacular name for the plant.]

Selected reference: Keil (2006b).

Onopordum acanthium L. [From the Greek *acantha*, spine, in reference to the spiny stems, leaves, and phyllaries.] SCOTCH COTTONTHISTLE.

Onopordum acanthium Linnaeus, Sp. Pl. 827. 1753. *Acanos spina* Scopoli, Fl. Carniol., ed. 2. 2: 132. 1772, nom. illegit.

Erect biennial herb, to 4 m; stem striate-angled, white-tomentose, with spinose wing leaf extensions to 1.5 cm wide. Leaves with the blade elliptic, lanceolate, or triangular-ovate, 10–60 cm long and wide, usually longer than wide, reduced upward, the apex acute, spine-tipped, the base tapering to the stem and spiny winged down the stem, the margin dentate or shallowly pinnate-lobed, spiny, the upper and lower surfaces tomentose. Flowers solitary or in a terminal cluster of 2–3 capitula; involucre hemispheric or subspheric, ca. 2 cm wide; phyllaries numerous in 8–10+ series, linear, 1–1.5 cm long, the apex tapering to a stiff spine to 6 mm long, the outer surface loose-tomentose; receptacle flat or convex, alveolate with fringed pits; disk flowers numerous; corolla purple or white, 2–2.5 cm long, glabrous. Cypsela subcylindric, 4–5 mm long, 4- to 5-angled, transversely roughed; pappus of numerous pink or reddish, basally connate, minutely barbed bristles 7–9 mm long.

Disturbed sites. Rare; St. Johns County. Nearly throughout North America; Europe, Asia, and Australia. Native to Europe and Asia. Summer–fall.

Packera Á. Löve & D. Löve 1976.

Herbs. Leaves alternate, simple, pinnate-veined, petiolate or epetiolate. Flowers in solitary or corymbiform or umbelliform arrays, radiate; calyculi present; receptacle epaleate; ray flowers carpellate; petals 5, basally connate, zygomorphic; disk flowers bisexual; petals 5, basally connate, actinomorphic; stamens 5, epipetalous, the filaments free, the anthers connate; ovary inferior, 2-carpellate, 1-loculate. Fruit a cypsela; pappus present.

A genus of about 64 species; North America, Mexico, and Asia. [Commemorates John G. Packer (1929–), Canadian botanist.]

Selected reference: Trock (2006b).

1. Cauline leaves gradually reduced upward; plant an annual .... **P. glabella**
1. Cauline leaves much reduced upward, most leaves basal; plant a perennial.
    2. Basal leaves cordate at the base .... **P. aurea**
    2. Basal leaves cuneate or nearly truncate at the base.
        3. Basal leaves widest near the apex; plant producing filiform aerial stolons .... **P. obovata**
        3. Basal leaves widest at or below the middle; plant not producing aerial stolons.
            4. Phyllaries with parallel sides, tapering only from above the middle; cypsela hirsute on the ribs .... **P. anonyma**
            4. Phyllaries linear, tapering from near the base; cypsela glabrous .... **P. paupercula**

Packera anonyma (A. W. Wood) W. A. Weber & Á. Löve [Nameless, in reference to its going unnamed so long.] SMALL'S RAGWORT.

Senecio anonymus A. W. Wood, Class-Book Bot., ed. 1861. 464. 1861. *Packera anonyma* (A. W. Wood) W. A. Weber & Á. Löve, Phytologia 49: 45. 1981.

Senecio aureus Linnaeus var. *angustifolia* Britton, Mem. Torrey Bot. Club 2: 39. 1890. *Senecio smallii* Britton, Mem. Torrey Bot. Club 4: 132. 1894.

Erect or ascending perennial herb, to 7 dm; stem striate-angled, tomentose proximally and in the leaf axils distally. Basal leaves with the blade elliptic or oblanceolate, 3–9 cm long, 1–2 cm wide, the apex rounded or acute, the base cuneate, the margin serrate or crenate-serrate, the upper and lower surfaces glabrous, the petiole 5–8 cm long, the base clasping, the cauline reduced upward, the blade elliptic or lanceolate, the margin often pinnatifid, short-petiolate or sessile, the base clasping. Flowers in capitula in a corymbiform array; peduncle bracteate, distally tomentose; calyculi of 1–5 bractlets, inconspicuous; phyllaries numerous in 1–2 series, linear, 3–5 mm long, equal, the margins scarious, the outer surface glabrous; receptacle flat, foveolate; ray flowers ca. 8; corolla yellow, the lamina 5–7 mm long; disk flowers numerous; corolla yellow, 3–4 mm long. Cypsela cylindric, ca. 1 mm long, 5- to 10-ribbed, hirsute on the ribs; pappus of numerous white barbellate bristles ca. 3 mm long.

Hammocks and disturbed sites. Occasional; Jefferson County, central and western panhandle. New York, south to Florida, west to Indiana, Kentucky, Tennessee, Arkansas, and Louisiana. Spring.

Packera aurea (L.) Á. Löve & D. Löve [Golden, in reference to the flower color.] GOLDEN RAGWORT.

Senecio aureus Linnaeus, Sp. Pl. 870. 1753. *Senecio rotundifolius* Stokes, Bot. Mat. Med. 4: 215. 1812, nom. illegit. *Packera aurea* (Linnaeus) Á. Löve & D. Löve, Bot. Not. 128: 250. 1976.

Erect perennial herb, to 6 dm; stem rhizomatous, striate-angled, glabrous or tomentose in the leaf axils. Basal leaves with the blade cordate or reniform, 2–6 cm long and wide, the apex rounded, the base cordate, the margin crenate or crenate-serrate, the upper and lower surfaces glabrous, the petiole 3–5 cm long, the base clasping, the cauline reduced upward, the blade oblong, the margin pinnatifid, short petiolate or sessile, the base clasping. Flowers in capitula in a corymbiform or subumbelliform array, the peduncle bracteate, glabrous or sparsely tomentose; calyculi of 1–5 bractlets, inconspicuous; phyllaries numerous in 1–2 series, linear, 6–8 mm long, equal, the apex purple or black, the margin scarious, the outer surface glabrous or proximally sparsely tomentose; receptacle flat, foveolate; ray flowers (8)10–13; corolla yellow, the lamina 8–10 mm long; disk flowers numerous; corolla yellow, 5–6 mm long. Cypsela cylindric, 1–2 mm long, 5- to 10-ribbed, glabrous; pappus of numerous white barbellate bristles ca. 4 mm long.

Floodplains. Occasional; Taylor County, central panhandle. Newfoundland and Labrador south to Florida, west to Manitoba, North Dakota, South Dakota, Minnesota, Iowa, Missouri, Oklahoma, and Texas. Winter–spring.

Packera glabella (Poir.) C. Jeffrey [Almost smooth, in reference to the stems and leaves.] BUTTERWEED.

Senecio lyratus Michaux, Fl. Bor.-Amer. 2: 120. 1803; non Forskal, 1775; nec Linnaeus f., 1782. *Senecio glabellus* Poiret, in Lamarck, Encycl. 7: 102. 1806. *Senecio lobatus* Persoon, Syn. Pl. 2: 436. 1807, nom. illegit. *Senecio carolinianus* Sprengel, Syst. Veg. 3: 559. 1826, nom. illegit. *Packera glabella* (Poiret) C. Jeffrey, Kew Bull. 47: 101. 1992.

Erect annual or perennial herb, to 7 dm; stem striate-angled, glabrous or sparsely tomentose in the leaf axils. Basal proximal cauline leaves with the blade obovate or oblanceolate, 5–15 cm long, 1–3 cm wide, the apex rounded, the base tapering, the margin pinnatifid, the ultimate margin crenate or undulate, the upper and lower surfaces glabrous, sessile, the base clasping, the cauline reduced upward and similar to the basal. Flowers in an umbelliform or corymbiform array, the peduncle bracteate, glabrous or basally tomentose; calyculi of 1–5 bractlets, conspicuous; phyllaries numerous in 1–2 series, linear, 5–7 mm long, equal, the margin scarious, the outer surface glabrous; receptacle flat, foveolate; ray flowers ca. 13; corolla yellow, the lamina 7–9 mm long; disk flowers numerous; corolla yellow, 4–6 mm long. Cypsela cylindric, 1–2 mm long, 5- to 10-ribbed, sparsely hirtellous on the ribs; pappus of numerous white barbellate bristles 3–4 mm long.

Open wet sites. Common; nearly throughout. Ontario and Michigan, south to Florida, west to Nebraska, Kansas, Oklahoma, and Texas. Spring.

Packera obovata (Muhl. ex Willd.) W. A. Weber & Á. Löve [Egg-shaped, in reference to the shape of the proximal leaf blades.] ROUNDLEAF RAGWORT.

Senecio obovatus Muhlenberg ex Willdenow, Sp. Pl. 3: 1999. 1803. *Senecio aureus* Linnaeus var. *obovatus* (Muhlenberg ex Willdenow) Torrey & A. Gray, Fl. N. Amer. 2: 442. 1843. *Packera obovata* (Muhlenberg ex Willdenow) W. A. Weber & Á. Löve, Phytologia 59: 47. 1981.
Senecio elliottii Torrey & A. Gray, Fl. N. Amer. 2: 443. 1843. *Senecio obovatus* Muhlenberg ex Willdenow var. *elliottii* (Torrey & A. Gray) Fernald, Rhodora 45: 507. 1943. SYNTYPE: FLORIDA.
Senecio aureus Linnaeus var. *fastigiatus* Chapman, Fl. South. U.S. 245. 1860. TYPE: FLORIDA.

Erect perennial herb, to 5 dm; stem rhizomatous, striate-angled, glabrous or sometimes tomentose proximally and in the leaf axils. Basal and proximal cauline leaves with the blade suborbiculate, ovate, or obovate, 4–10 cm long, 2–8 cm wide, the apex rounded, the base tapering, the margin crenate, dentate, or serrate, the upper and lower surfaces glabrous, sessile, the base clasping, the cauline reduced upward, pinnatisect, sessile, the base clasping. Flowers in a corymbiform array, the peduncle bracteate, glabrous or proximally tomentose; calyculi of 1–5 bractlets, conspicuous; phyllaries numerous in 1–2 series, linear, 3–6 mm long, equal, the apex sometimes reddish, the outer surface glabrous or proximally tomentose; receptacle flat, foveolate; ray flowers ca. 13; corolla yellow, the lamina 7–10 mm long; disk flowers numerous; corolla yellow, 4–6 mm long. Cypsela cylindric, 1–2 mm long, 5- to 10-ribbed, glabrous or hirsute on the ribs; pappus of numerous white barbellate bristles 3–6 mm long.

Hammocks. Occasional; central panhandle. Quebec and Ontario south to Florida, west to Michigan, Illinois, Missouri, Kansas, and New Mexico; Mexico. Winter–spring.

Packera paupercula (Michx.) Á. Löve & D. Löve [Poor, in reference its small size.] BALSAM GROUNDSEL.

Senecio pauperculus Michaux, Fl. Bor.-Amer. 2: 120. 1803. *Senecio aureus* Linnaeus var. *pauperculus* (Michaux) MacMillan, Metasp. Minnesota Valley 556. 1892. *Senecio balsamitae* Muhlenberg ex Willdenow var. *pauperculus* (Michaux) Fernald ex Greenman, Rhodora 3: 6. 1901. *Senecio pauperculus* Michaux var. *typicus* B. Boivin, Naturaliste Canad. 75: 214. 1948, nom. inadmiss. *Packera paupercaula* (Michaux) Á. Löve & D. Löve, Bot. Not. 128: 520. 1976.

Senecio crawfordii Britton, Torreya 1: 21. 1901. *Senecio balsamitae* Muhlenberg ex Willdenow var. *crawfordii* (Britton) Greenman, Rhodora 10: 69. 1908. *Senecio pauperculus* Michaux var. *crawfordii* (Britton) T. M. Barkley, Trans. Kansas Acad. Sci. 65: 357. 1962. *Senecio pauperculus* Michaux subsp. *crawfordii* (Britton) G. W. Douglas & Ruyle-Douglas, Canad. J. Bot. 56: 1710. 1978.

Erect or ascending perennial herb, to 5 dm; stem striate-angled, glabrous or sparsely tomentose proximally. Basal leaves with the blade lanceolate, elliptic, or oblanceolate, 3–6 cm long, 1–2 cm wide, the apex obtuse or rounded, the base tapering, the margin subentire, dentate, or serrate, the upper and lower surfaces glabrous, the cauline reduced upward, pinnatisect, sessile, the base clasping. Flowers in a corymbiform array, the peduncle bracteate or ebracteate, glabrous; calyculi of 1–5 bractlets, inconspicuous; phyllaries numerous in 1–2 series, linear, 5–8 mm long, equal, the margin scarious, the outer surface glabrous; receptacle flat, foveolate; ray flowers ca. 8; corolla pale yellow, the lamina 5–10 mm long; disk flowers numerous; corolla yellow, 4–6 mm long. Cypsela cylindric, 1–2 mm long, 5- to 10-ribbed, glabrous; pappus of numerous white barbellate bristles 4–5 mm long.

Flatwoods. Rare; Bay County. Nearly throughout North America.

EXCLUDED TAXON

Packera tomentosa C. Jeffrey—Reported as *Senecio tomentosus* Michaux by Chapman (1860, 1883, 1897), Small (1903, 1913a, 1933), Radford et al. (1964, 1968), Correll and Johnston (1970), Cronquist (1980), Godfrey and Wooten (1982), Clewell (1985), and Wunderlin (1998). Also reported for Florida by Trock (2006b) as *Packera tomentosa* C. Jeffrey. The names were misapplied to Florida material of *Packera anonyma*. *Senecio tomentosus* Michaux (1803; non Salisbury, 1796) is a later homonym and thus illegitimate.

Palafoxia Lag. 1816. PALAFOX

Herbs or shrubs. Leaves opposite or alternate, simple, pinnate-veined, petiolate or epetiolate. Flowers in corymbiform arrays, discoid, bisexual; receptacle epaleate; petals 5, basally connate, actinomorphic; stamens 5, epipetalous, the filaments free, the anthers connate; ovary inferior, 2-carpellate, 1-loculate. Fruit a cypsela; pappus present.

A genus of 12 species, North America and Mexico. [Commemorates Don Juan de Palafox y Mendoza (1600–1659) Spanish bishop and founder of the University of Mexico.]

Polypteris Nutt., 1818.

Selected reference: Strother (2006t).

1. Pappus consisting of linear scales 6–7 mm long .... **P. integrifolia**
1. Pappus consisting of short scales 1–3(4) mm long.
    2. Phyllaries stipitate-glandular; corolla throat funnelform, shorter than the lobes .... **P. texana**
    2. Phyllaries not stipitate-glandular; corolla throat subcylindric, longer than the lobes .... **P. feayi**

Palafoxia feayi A. Gray [Commemorates William T. Feay (1803–1879), Georgia physician, teacher, and botanist who collected in Georgia and Florida.] FEAY'S PALAFOX.

Palafoxia feayi A. Gray, Proc. Amer. Acad. Arts 12: 59. 1877. TYPE: FLORIDA: s.d., *Feay s.n.* (lectotype: GH). Lectotypified by Turner and Morris (1976: 592).

Erect perennial herb or subshrub, to 3 m; stem strigillose or glabrate. Leaves opposite proximally, alternate distally, the blade elliptic, oblong, or ovate, 2–6 cm long, 0.5–2.5 cm wide, reduced upward, the apex obtuse or rounded, the base cuneate or rounded, the margin entire, the upper and lower surfaces sparsely strigillose or glabrate, the petiole 1–6 mm long. Flowers in capitula in a corymbiform array; involucre campanulate or turbinate, 1–2 cm wide; phyllaries 9–12 in 2–3 series, linear or oblong, 5–9 mm long, subequal, purplish tinged, the apex acute, the outer surface slightly strigillose; receptacle flat, pitted; flowers numerous; corolla pinkish, purplish or whitish, 10–14 mm long, the throat subcylindric, longer than the lobes. Cypsela obpyramidal, 6–8 mm long, 4-angled, pubescent; pappus of 8 short scales ca. 2 mm long.

Sandhills and scrub. Frequent; central and southern peninsula. Endemic. Fall.

Palafoxia integrifolia (Nutt.) Torr. & A. Gray [With entire leaves.] COASTALPLAIN PALAFOX.

Polypteris integrifolia Nuttall, Gen. N. Amer. Pl. 2: 139. 1818. *Palafoxia integrifolia* (Nuttall) Torrey & A. Gray, Fl. N. Amer. 2: 368. 1842. *Hymenopappus integrifolius* (Nuttall) Sprengel, Syst. Veg. 3: 449. 1826. *Paleolaria fastigiata* Lessing, Syn. Gen. Compos. 156. 1832, nom. illegit. *Palafoxia fastigiata* de Candolle, Prodr. 5: 125. 1836, nom. illegit. *Lomaxeta verrucosa* Rafinesque, New Fl. 4: 72. 1838 ("1836"), nom. illegit.

Erect subshrubs or shrubs, to 1.5 m; stem strigillose or glabrate. Leaves opposite proximally, alternate distally, the blade linear or lanceolate, 2.5–9 cm long, 2–13 mm wide, reduced upward, the apex acute or obtuse, the base cuneate, the margin entire, the upper and lower surfaces scabrous, the petiole 5–10 mm long. Flowers in capitula in a corymbiform array; involucre turbinate, 7–10 cm wide; phyllaries 12–15 in 2–3 series, linear or oblong, 8–11 mm long, subequal, the apex rounded, the outer surface glabrous or glabrate; receptacle flat, pitted; flowers numerous; corolla pinkish, purplish or whitish, 7–13 mm long, the throat funnelform, shorter than the lobes. Cypsela obpyramidal, 5–6 mm long, 4-angled, pubescent; pappus of 10 linear scales 6–7 mm long.

Sandhills. Frequent; peninsula west to central panhandle. Georgia and Florida. Fall.

Palafoxia texana DC. [Of Texas.] TEXAS PALAFOX.

Palafoxia texana de Candolle, Prodr. 5: 125. 1836. *Polypteris texana* (de Candolle) A. Gray, Proc. Amer. Acad. Arts 19: 30. 1883. *Othake texana* (de Candolle) Bush, Trans. Acad. Sci. St. Louis 14: 176. 1904.

Palafoxia rosea (Bush) Cory var. *ambigua* Shinners, Field & Lab. 20: 95. 1952. *Palafoxia texana* de Candolle var. *ambigua* (Shinners) B. L. Turner & M. I. Morris, Rhodora 78: 623. 1976.

Erect annual or perennial herb, to 8 dm; stem scabrous. Leaves opposite proximally, alternate distally, the blade ovate-lanceolate or linear-lanceolate, 3–8 cm long, (0.5)1–2 cm wide, reduced upward, the apex acute or obtuse, the base cuneate, the margin entire, the upper and lower surfaces scabrous, the petiole 1–3 cm long. Flowers in capitula in a corymbiform array; involucre turbinate, 6–15 mm wide; phyllaries 12–15 in 2–3 series, linear or oblanceolate, 5–8 mm long, subequal, the apex acute or obtuse, the outer surface scabrellous, stipitate-glandular; receptacle flat, pitted; flowers numerous; corolla pinkish, purplish, or whitish, 7–10 mm long, the throat funnelform, shorter than the lobes. Cypsela obpyramidal, 4–6 mm long, 4-angled, pubescent; pappus of 8 short scales 1–3(4) mm long.

Dry, disturbed sites. Rare; Bay County. Florida, Louisiana, and Texas; Mexico. Native to Texas and Mexico. Fall.

Parthenium L. 1753. FEVERFEW

Herbs. Leaves alternate, simple, pinnate-veined, petiolate or epetiolate. Flowers in capitula in paniculiform arrays, radiate; receptacle paleate; ray flowers carpellate; petals 5, basally connate, zygomorphic; disk flowers staminate; petals 5, basally connate, actinomorphic; stamens 5, epipetalous, the filaments free, the anthers connate; ovary inferior, 2-carpellate, 1-loculate. Fruit a cypsela; pappus absent.

A genus of about 16 species; North America, West Indies, Mexico, Central America, South America, Africa, Asia, Australia, and Pacific Islands. [From the Greek *parthenos,* virgin, in reference to only the pistillate ray flowers being fertile.]

Selected reference: Strother (2006u).

Parthenium hysterophorus L. [From the Greek *hystero* womb, uterus, and *phero,* to bear, in reference to carpellate ray flowers being fertile.] SANTA MARIA FEVERFEW.

Parthenium hysterophorus Linnaeus, Sp. Pl. 988. 1753. *Parthenium pinnatifidum* Stokes, Bot. Mat. Med. 4: 278. 1812, nom. illegit.

Erect annual herb, to 1 m; stem striate-angled, scabrellous, gland-dotted. Leaves with the blade ovate or elliptic, 3–18 cm long, 1–5(9) cm wide, reduced upward, (1)2-pinnately lobed, the ultimate lobes lanceolate or linear, 3–5 cm long, 2–15 mm wide, the upper and lower surfaces scabrellous, gland-dotted, tapering often to a winged petiole, the base clasping. Flowers in capitula in a paniculiform array, the peduncle 1–8(15) mm long; involucre hemispheric; phyllaries 10(12) in 2 series, the outer 5(6) elliptic-lanceolate, 2–3 mm long, the inner 5(6) ovate or suborbicular, 3–4 mm long, the outer surface of the outer series gland-dotted; receptacle flat; paleae flabelliform; ray flowers 5(6); corolla white, the lamina 1–2 mm long; disk flowers numerous; corolla white, to 1 mm long. Cypsela obovoid, ca. 2 mm long, slightly laterally flattened, shed with the subtending phyllary and the 2 adjacent disk flowers with the 3 enclosing paleae.

Disturbed sites. Occasional; peninsula, central and western panhandle. New York and Massachusetts south to Florida, west to Michigan, Illinois, Missouri, Kansas, Oklahoma, and Texas; West Indies, Mexico, Central America, and South America; Africa, Asia, Australia, and Pacific Islands. Native to tropical America. All year.

Pascalia Ortega 1797.

Herbs. Leaves opposite, simple, pinnate-veined, petiolate or epetiolate. Flowers in solitary capitula, radiate; receptacle paleate; ray flowers carpellate; petals 5, basally connate, zygomorphic; disk flowers bisexual; petals 5, basally connate, actinomorphic; stamens 5, epipetalous, the filaments free, the anthers connate; ovary inferior, 2-carpellate, 1-loculate. Fruit a cypsela; pappus present.

A monotypic genus; North America, South America, Europe, Africa, Asia, Australia and perhaps also other localities. [Commemorates Diego Baldassara Pascal (1768–1802), Italian physician and botanist.]

Selected reference: Strother (2006v).

Pascalia glauca Ortega [*Glaucus*, in reference to the waxy, whitish or bluish coating of the leaf surface.] BEACH CREEPING OXEYE.

Pascalia glauca Ortega, Nov. Pl. Descr. Dec. 39: t. 4. 1797. *Seruneum glaucum* (Ortega) Kuntze, Revis. Gen. Pl. 3(2): 179. 1898. *Wedelia glauca* (Ortega) O. Hoffmann ex Hicken, Apuntes Hist. Nat. 2: 254. 1910.

Erect perennial herb, to 1 m; stem striate-angled, glabrous or minutely scabrous. Leaves lanceolate or linear-lanceolate, 4–5 cm long, 0.8–1.5 cm wide, slightly reduced upward, the apex acute or obtuse, the base cuneate, the margin entire or irregularly few-toothed, the upper and lower surfaces sparsely scabrous, often glaucous, short petiolate or sessile, the base clasping. Flowers in capitula solitary, terminal capitula, the peduncle 1.5–2 cm long, strigillose; involucre campanulate or hemispheric, 1–1.5 cm wide; phyllaries numerous in 2–3 series, linear or linear-lanceolate, 1–1.5 cm long, unequal, the outer longer than the inner; receptacle hemispheric; paleae lanceolate or ovate, 5–6 mm long, the apex acuminate, membranous, conduplicate; ray flowers numerous, carpellate; corolla yellow, the lamina 10–15 mm long; disk flowers numerous, bisexual; corolla yellow, 5–6 mm long. Cypsela obovoid, 5–6 mm long, 3-angled (ray flowers) or 4-angled (disk flowers), slightly compressed, glabrous; pappus of short connate scales often with 1–2 awns, 2–3 mm long.

Coastal dunes. Rare; Escambia County. Georgia, Florida, and Alabama; South America; Europe, Africa, Asia, and Australia; perhaps also other localities. Native to South America. Summer.

Pectis L. 1759. CINCHWEED

Herbs. Leaves opposite, simple, pinnate-veined, epetiolate. Flowers in capitula, solitary or in cymbiform arrays; receptacle epaleate; ray flowers carpellate; petals 5, basally connate, zygomorphic; disk flowers bisexual; petals 5, basally connate, actinomorphic; stamens 5, epipetalous, the filaments free, the anthers connate; ovary inferior, 2-carpellate, 1-loculate. Fruit a cypsela; pappus present.

A genus of about 93 species; North America, West Indies, Mexico, Central America, South America, and Pacific Islands. [From the Greek *pecten,* comb, in reference to the ciliate leaf margins resembling the teeth of a comb.]

Selected reference: Keil (2006c).

1. Capitula distinctly pedunculate, the peduncle 5–15 mm long .... **P. glaucescens**
1. Capitula sessile or subsessile, the peduncle to 2(7) mm long.
    2. Lower leaf surface with 2 rows of large glands (rarely scattered.................................. **P. linearifolia**
    2. Lower leaf surface with scattered minute glands.
        3. Involucre cylindric or fusiform; phyllaries coherent at the base and falling together .... **P. prostrata**
        3. Involucre campanulate; phyllaries separate and falling individually......................... **P. humifusa**

Pectis glaucescens (Cass.) D. J. Keil [*Glaucedo,* in reference to the waxy, whitish or bluish coating of the leaf surface.] SANDDUNE CINCHWEED.

Chthonia glaucescens Cassini, in Cuvier, Dict. Sci. Nat. 9: 173. 1817. *Pectis glaucescens* (Cassini) D. J. Keil, Sida 11: 386. 1986.

Chthonia leptocephala Cassini, in Cuvier, Dict. Sci. Nat. 27: 206. 1823. *Pectis leptocephala* (Cassini) Urban, Symb. Antill. 5: 280. 1907.

Pectis lessingii Fernald, Proc. Amer. Acad. Arts 33: 78: 1897. TYPE: FLORIDA: Miami-Dade Co.: between the Everglades and Biscayne Bay, Jun, *Curtiss 1162* (lectotype: GH; isolectotypes: BM, CM, F, K, MICH, NY, P, US). Lectotypified by Keil (1986: 388).

Prostrate or erect annual or perennial herb, to 5 dm; stem pubescent, sometimes in decurrent lines, or glabrate. Leaves with the blade linear, 1–3.5 cm long, 1–2 mm wide, the apex acute, the base cuneate, the margin proximally with 1–5 pairs of setae 1–2 mm long, the upper and lower surfaces glabrous, the lower surface submarginally dotted with elliptic or circular oil-glands, sometimes with a few additional scattered oil-glands, sessile. Flowers in capitula, solitary or in a diffuse cymbiform array, the peduncle 1–3.5(4.5) cm long; involucre cylindric; phyllaries linear-oblanceolate, 4–5 mm long, with subapical oil-glands, sometimes with smaller submarginal or scattered oil-glands; receptacle convex; ray flowers 5; corolla yellow, the lamina 4–5 mm long; disk flowers 3–7; corolla yellow, 2–3 mm long, 2-lipped. Cypsela cylindric, ca. 3 mm long, ribbed or angled, strigillose, dark brown; pappus of 0–5 antrorsely scabrous bristles or slender scales 1–2 mm long and 0–5 entire or irregularly laciniate scales less than 1 mm long.

Pinelands, flatwoods, open hammocks, and disturbed sites. Frequent; central and southern peninsula. Florida; West Indies. Summer–fall.

Pectis humifusa Sw. [Spreading over the ground, procumbent.] YERBA DE SAN JUAN.

Pectis humifusa Swartz, Prodr. 114. 1788. *Lorentea humifusa* (Swartz) Lessing, Linnaea 6: 719. 1831.

Prostrate and mat-forming annual or perennial herb; stem puberulent in decurrent lines. Leaves with the blade oblong-lanceolate or obovate, 3–17 mm long, 2–4 mm wide, the apex obtuse or rounded, the base cuneate, the margin proximally with 2–6 pairs of setae 1–2 mm long, the upper and lower surfaces glabrous, dotted with small round oil-glands, sessile. Flowers in capitula, solitary or in a congested, leafy, cymbiform array, the peduncle 1–12 mm long; involucre campanulate; phyllaries coherent at the base and falling together, obovate, 5–6 mm long, with scattered small round oil-glands; receptacle convex; ray flowers 5; corolla yellow, the lamina 4–5 mm long; disk flowers 12–20; corolla yellow, 3–4 mm long, 2-lipped. Cypsela cylindric, 3–4 mm long, ribbed or angled, puberulent, dark brown, the ray pappus of 2–3 slender aristate scales ca. 2 mm long and 2–10 laciniate scales or bristles, the disk pappus of 4–15 antrorsely scabrid scales or bristles.

Dry disturbed sites. Rare; Collier County. Florida; West Indies and South America. Native to West Indies and South America. Summer–fall.

Pectis linearifolia Urb. [With linear leaves.] FLORIDA CINCHWEED.

Pectis linearifolia Urban, Symb. Antill. 5: 276. 1907. TYPE: FLORIDA: Hillsborough Co.: Tampa, 24 Aug 1895, *Nash 2479* (lectotype: US; isolectotype: E, GH, K, LE, MO, MSC, NYm PR, UC, WU). Lectotypified by Keil (1986: 388).

Decumbent or erect annual herb, to 4 dm; stem glabrous or puberulent in decurrent lines. Leaves with the blade linear, 1–5 cm long, 1–3 mm wide, the apex acute, the base cuneate, the margin proximally with 2–6 pairs of setae 1–3 mm long, the upper and lower surfaces glabrous, the lower surface submarginally dotted with 2 rows of large round oil-glands, these rarely scattered, sessile. Flowers in capitula, solitary or in a congested cymbiform array, sessile or the peduncle to 1 mm long; involucre cylindric or narrowly campanulate; phyllaries linear or linear-oblanceolate, 5–6 mm long, with scattered elliptic oil-glands; receptacle convex; ray flowers 5; corolla yellow, the lamina 5–6 mm long; disk flowers 4–10; corolla yellow, ca. 3 mm long, 2-lipped. Cypsela cylindric, 2–3 mm long, ribbed or angled, puberulent, dark brown; pappus of 2–5 antrorsely barbed bristles or awns ca. 2 mm long and shorter barbellate scales.

Dry disturbed sites. Frequent; Alachua County, central and southern peninsula. Endemic. Pine flatwoods, open hammocks, sand flats, shell middens, coastal dunes, and disturbed sites. Summer–fall.

Pectis prostrata Cav. [Lying flat on the ground, procumbent.] SPREADING CINCHWEED.

Pectis prostrata Cavanilles, Icon. 4: 12, t. 324. 1797.

Prostrate or ascending annual herb, often mat-forming; stem puberulent in decurrent lines. Leaves linear or narrowly oblanceolate, 1–4 cm long, 2–7 mm wide, the apex acute, the base cuneate, the margin proximally with 4–12 pairs of setae 1–3 mm long, the upper and lower surfaces glabrous, the lower surface dotted with small oil-glands. Flowers in capitula, solitary or in a congested, leafy, cymbiform array, the peduncle 1–2 mm long; involucre campanulate or cylindric; phyllaries oblong or obovate, 5–8 mm long, usually with elliptic oil-glands in submarginal rows and sometimes along the midrib; receptacle convex; ray flowers 5; corolla yellow, the lamina 3–4 mm long; disk flowers 3–17; corolla yellow, ca. 2 mm long, 2-lipped. Cypsela cylindric, 3–5 mm long, ribbed or angled, strigillose, dark brown; pappus of 2–5 lanceolate scales ca. 2 mm long.

Oak scrub, pinelands, flatwoods, and disturbed sites. Frequent; nearly throughout. North Carolina, Florida, and Alabama, west to Arizona; West Indies, Mexico, and Central America. Summer–fall.

HYBRID

Pectis ×floridana D. J. Keil (*P. glaucescens* × *P. prostrata*) [Of Florida.]

Pectis floridana G. J. Keil, Sida 11: 389. 1986. TYPE: FLORIDA: Collier Co.: 6 mi. SE of Royal Palm Hammock along US 41, 18 Nov 1983, *Keil 16488* (holotype: FTG).

Disturbed sites. Rare; Charlotte and Martin Counties, southern peninsula. Summer–fall.

EXCLUDED TAXA

Pectis ciliaris Linnaeus—Reported for Florida by Small (1903) and Fernald (1897), the name misapplied to Florida material of *P. linearifolia.*

Pectis linifolia Linnaeus—Reported for Florida by Chapman (1860), the name misapplied to Florida material of *P. linearifolia.* Also reported by Keil (2006c). No Florida specimens known.

Phoebanthus S. F. Blake 1916. FALSE SUNFLOWER

Herbs. Leaves alternate, simple, pinnate-veined, epetiolate. Flowers in capitula, solitary or in corymbiform arrays, radiate; receptacle paleate; ray flowers sterile; petals 5, basally connate, zygomorphic; disk flowers bisexual; petals 5, basally connate, actinomorphic; stamens 5, epipetalous, the filaments free, the anthers connate; ovary inferior, 2-carpellate, 1-loculate. Fruit a cypsela; pappus absent to present.

A genus of 2 species; North America. [From the Greek *phoebus,* also known as Apollo, the sun god of Greek and Roman mythology, and *anthos,* flower, in reference to the large yellow *Helianthus*-like flowers.]

Selected reference: Schilling (2006b).

1. Leaves linear or linear-lanceolate, 3–7 mm wide; phyllaries appressed............................**P. grandiflorus**
1. Leaves linear-filiform, 1–2 mm wide; phyllaries spreading..**P. tenuifolius**

Phoebanthus grandiflorus (Torr. & A. Gray) S. F. Blake [With large flowers.] FLORIDA FALSE SUNFLOWER.

Helianthella grandiflora Torrey & A. Gray, Fl. N. Amer. 2: 333. 1842. *Phoebanthus grandiflorus* (Torrey & A. Gray) S. F. Blake, Proc. Amer. Acad. Arts 51: 520. 1916. TYPE: FLORIDA.

Erect perennial herb, to 1 m; roots tuberous; stem slightly striate-angled, scabrous. Leaves with the blade linear or linear-lanceolate, 2–7 cm long, 3–7 mm wide, reduced upward, the apex acute, the base cuneate or rounded, the margin entire, revolute, the upper surface scabrous, the lower surface sparsely hispid-scabrous, gland-dotted, sessile. Flowers in capitula, solitary or in a corymbiform array; involucre hemispheric, 1–2 cm wide; phyllaries numerous in 2–3 series, linear-lanceolate, ca. 1 cm long, appressed; receptacle convex; paleae ca. 10 mm long, conduplicate, the central lobe subequaling the lateral lobes; ray flowers 10–20; corolla yellow, the lamina 2–4 cm long; disk flowers numerous; corolla yellow, 5–7 mm long. Cypsela obpyramidal, slightly compressed, 3–5 mm long, purplish black; pappus absent or of 1(2)laciniate scales 1–2 mm long and 0–2 minute scales.

Sandhills. Frequent; northern and central peninsula, also Jackson County, where probably accidentally introduced. Endemic. Spring–summer.

Phoebanthus tenuifolius (Torr. & A. Gray) S. F. Blake [With narrow leaves.] PINELAND FALSE SUNFLOWER.

Helianthella tenuifolia Torrey & A. Gray, Fl. N. Amer. 2: 333. 1842. *Phoebanthus tenuifolius* (Torrey & A. Gray) S. F. Blake, Proc. Amer. Acad. Arts 51: 520. 1916. TYPE: FLORIDA.

Erect perennial herb, to 1 m; roots tuberous; stem slightly striate-angled, scabrous. Leaves with the blade linear-filiform, 3–7 cm long, 1–2 mm wide, the apex acute, the base cuneate or rounded, the margin entire, revolute, the upper surface scabrous, the lower surface sparsely hispid-scabrous, gland-dotted, sessile. Flowers in capitula, solitary or in a corymbiform array; involucre campanulate or hemispheric, 10–15 cm wide; receptacle convex; paleae 10 mm long, the central lobe usually longer than the lateral lobes; phyllaries numerous in 2–3 series, linear-lanceolate, 7–8 mm long, spreading; receptacle convex; paleae 7–8 mm long, conduplicate, the central lobe usually longer than the lateral lobes; ray flowers 10–20; corolla yellow, the lamina 2–4 cm long; disk flowers numerous; corolla yellow, 4–5 mm long. Cypsela obpyramidal, slightly compressed, 4–5 mm long, purplish black; pappus of 1–2 laciniate scales 1–2 mm long and 0–4 minute scales.

Sandhills and flatwoods. Rare; central panhandle. Florida and Alabama. Spring–summer.

Phoebanthus tenuifolius is listed as threatened in Florida (Florida Administrative Code, Chapter 5B-40).

Pilosella Hill 1756.

Herbs. Leaves alternate, simple, pinnate-veined, epetiolate. Flowers in capitula in umbelliform arrays, liguliflorous, bisexual; calyculi present; receptacle epaleate; petals 5, basally connate, zygomorphic; stamens 5, epipetalous, the filaments free, the anthers connate; ovary inferior, 2-carpellate, 1-loculate. Fruit a cypsela; pappus present.

A genus of about 150 species; North America, Mexico, South America, Europe, Africa, Asia, Australia, and Pacific Islands. [*Pilose*, with long hairs, and *ellus*, diminutive, in reference to the pilose-hirsute trichomes.]

Some workers have maintained the traditional generic description of *Hieracium* s.l. with *Pilosella* as a subgenus. Most modern works now treat *Pilosella* as a separate genus based on morphological, cytological, and biochemical differences (Fehrer et al., 2007, 2009; Hand et al., 2015; Krak et al., 2013; Majeský et al., 2017).

Pilosella aurantiaca (L.) F. W. Schultz & Sch. Bip. [*Aurantiacus*, orange, in reference to the orange corolla.] KING-DEVIL; DEVIL'S PAINTBRUSH; ORANGE HAWKWEED.

Hieracium aurantiacum Linnaeus, Sp. Pl. 801. 1753. *Pilosella aurantiaca* (Linnaeus) F. W. Schultz & Schultz Bipontinus, in Schultz Bipontinus, Flora 45: 426. 1862.

Erect perennial herb, to 3.5(6) dm; stem striate-angled, pilose-hirsute, stipitate-glandular. Leaves mostly basal, 3–8+, the cauline 1–3, the blade spatulate or oblanceolate, 4.5–7(16) cm long, 1–3.5 cm wide, reduced upward and becoming bract-like, the apex acute, the base cuneate, the margin entire or denticulate, the upper and lower surfaces pilose-hirsute, sessile, clasping the stem. Flowers in capitula, 3–7(12) in an umbelliform array, the peduncle 1–3 cm long, stellate-pubescent, stipitate-glandular; calyculi of 5–10 linear bractlets; involucre campanulate, 6–8 mm wide; phyllaries numerous, linear, 12–14 mm long, subequal, the outer surface pilose-hirsute, stellate-pubescent, stipitate-glandular; receptacle flat, pitted; flowers numerous; corolla orange, the lamina 10–14 mm long. Cypsela columnar, 1–2 mm long, 10-ribbed, glabrous; pappus in 1 series, of numerous white barbellate bristles 3–4 mm long.

Disturbed sites. Rare; DeSoto County. Escaped from cultivation. Nearly throughout North America; South America; Europe, Asia, Australia, and Pacific Islands. Native to Europe. Spring–summer.

Pityopsis Nutt. 1840. SILKGRASS

Herbs. Leaves alternate, simple, parallel-veined, epetiolate. Flowers in capitula in corymbiform or paniculiform arrays, radiate; receptacle epaleate; ray flowers carpellate; petals 5, basally connate, zygomorphic; disk flowers bisexual; petals 5, basally connate, actinomorphic; stamens 5, epipetalous, the filaments free, the anthers connate; ovary inferior, 2-carpellate, 1-loculate. Fruit a cypsela; pappus present.

A genus of 7 species; North America, West Indies, Mexico, and Central America. [From the Greek *pitys*, pine, and *opsis*, resemblance, in reference to the leaves of *P. pinifolia* resembling those of *Pinus* (Pinaceae).]

Three recent treatments of the *Pityopsis graminifolia* complex (Semple, 2006f; Bridges and

Orzell, 2018; Nesom, 2019) differ with respect to taxonomic boundaries and ranks. Due to a lack of consensus, this treatment follows Semple.

Selected reference: Semple (2006f).

1. Basal leaves longer than the cauline.
   2. Stem and peduncles eglandular .... **P. graminifolia**
   2. Stem and peduncles stipitate-glandular.
      3. Stem leaves fewer than 10 .... **P. oligantha**
      3. Stem leaves more than 10 .... **P. aspera**
1. Basal leaves shorter than the cauline.
   4. Stem conspicuously flexuous; involucre equaling the pappus .... **P. flexuosa**
   4. Stem not conspicuously flexuous; involucre shorter than the pappus.
      5. Stem leaves linear, usually falcate .... **P. falcata**
      5. Stem leaves ovate-lanceolate, not falcate .... **P. graminifolia**

Pityopsis aspera (Shuttlew. ex Small) Small [Rough, in reference to the sparsely stipitate-glandular stem and leaf surfaces.] PINELAND SILKGRASS.

Chrysopsis graminifolia (Michaux) Elliott var. *aspera* A. Gray, Syn. Fl. N. Amer. 1(2): 121. 1884. TYPE: FLORIDA: Wakulla Co.: St. Marks, Aug 1843, *Rugel s.n.* (lectotype: GH; isolectotype: NY). Lectotypified by Semple and Bowers (1985: 17).

Chrysopsis aspera Shuttleworth ex Small, Fl. S.E. U.S. 1182, 1903. *Pityopsis aspera* (Shuttleworth ex Small) Small, Man. S.E. Fl. 1341, 1508. 1933. *Heterotheca aspera* (Shuttleworth ex Small) Shinners, Sida 3: 348. 1969. TYPE: FLORIDA: Wakulla Co.: St. Marks, Aug 1843, *Rugel s.n.* (lectotype: NY; isolectotype: GH). Lectotypified by Semple and Bowers, 1985: 17).

Chrysopsis adenolepis Fernald, Rhodora 44: 471, t. 742(1–2). 1942. *Heterotheca adenolepis* (Fernald) Ahles, J. Elisha Mitchell Sci. Soc. 80: 173. 1964. *Pityopsis adenolepis* (Fernald) Semple, Canad. J. Bot. 58: 148. 1980. *Pityopsis aspera* (Shuttleworth ex Small var. *adenolepis* (Fernald) Semple & Bowers, Univ. Waterloo Biol. Ser. 29: 18. 1985.

Erect perennial herb, to 5 dm; stem with basal rhizomes, striate-angled, silvery-sericeous, stipitate-glandular. Basal leaves with the blade linear-oblanceolate, 5–25(35) cm long, 5–10 mm wide, the apex acute, the base tapering, clasping, the margin entire, sometimes stipitate-glandular, the upper and lower surfaces silvery-sericeous, sessile, the cauline linear, greatly reduced upward. Flowers in capitula in a corymbiform array, the peduncle 1–6 cm long, stipitate-glandular, bracteolate; involucre turbinate, shorter than the pappus; phyllaries in 5–7 series, lanceolate, 4–8 mm long, unequal, the margin scarious, the outer surface stipitate-glandular, the apex sometimes long pilose-ciliate; receptacle slightly convex, pitted; ray flowers 6–10; corolla yellow, the lamina 4–6 mm long; disk flowers numerous; corolla yellow, 4–5 mm long. Cypsela fusiform, 2–3 mm long, 8- to 10-ribbed, strigose; pappus in 3–4 series, the outer of linear or linear-triangular scales to 1 mm long, the inner of numerous barbellate bristles 5–6 mm long.

Scrub, sandhills, dry flatwoods, and dry hammocks. Occasional; northern counties. Virginia south to Florida, west to Tennessee and Louisiana. Fall.

Pityopsis aspera is a variable species that is sometimes divided into var. *aspera* and var.

adenolepis in Florida (Semple, 2006f). However, the features of the stems and leaves used to separate them intergrade.

Pityopsis falcata (Pursh) Nutt. [*Falcatus,* curved like a sickle, in reference to the leaves.] SICKLELEAF SILKGRASS.

Inula falcata Pursh, Fl. Amer. Sept. 532. 1814. *Inula mariana* Linnaeus var. *falcata* (Pursh) Nuttall, Gen. N. Amer. Pl. 2: 151. 1818. *Chrysopsis falcata* (Pursh) Elliott, Sketch Bot. S. Carolina 2: 336. 1823. *Pityopsis falcata* (Pursh) Nuttall, Trans. Amer. Philos. Soc., ser. 2. 7: 318. 1841. *Heterotheca falcata* (Pursh) V. L. Harms, Castanea 34: 408. 1969.

Erect perennial herb, to 3(4) dm; stem with basal rhizomes, striate-angled, silvery-sericeous. Leaves with the blade 5–9 cm long, 2–7 mm wide, the basal shorter than the cauline, usually withering by flowering, the cauline reduced upward, linear, falcate, often conduplicate, the apex acuminate, the base tapering, clasping, the margin entire, the upper and lower surfaces sparsely sericeous or glabrate, sessile. Flowers in capitula in a corymbiform array, the peduncle 1–4 cm long, white-villous, bracteolate; involucre turbinate-campanulate, shorter than the pappus; phyllaries in 5–6 series, lanceolate, 5–8 mm long, unequal, the margin scarious, the outer surface strigose, the apex with tufts of trichomes; receptacle slightly convex, pitted; ray flowers 9–15; corolla yellow, the lamina 5–8 mm long; disk flowers numerous; corolla yellow, 5–6 mm long. Cypsela fusiform, 3–4 mm long, 8- to 10-ribbed, strigose; pappus in 3–4 series, the outer of linear-setiform scales to 1 mm long, the inner of numerous barbellate bristles 4–6 mm long.

Coastal beaches. Rare; Pinellas County. Not recently collected. Ontario, New York, Massachusetts, Connecticut, Rhode Island, New Jersey, and Florida. Native to the northeastern United States. Summer–fall.

Pityopsis flexuosa (Nash) Small [Zigzag, bent alternately in opposite directions, in reference to the stem.] ZIGZAG SILKGRASS.

Chrysopsis flexuosa Nash, Bull. Torrey Bot. Club 23: 107. 1896. *Pityopsis flexuosa* (Nash) Small, Man. S.E. Fl. 1341, 1508. 1933. *Heterotheca flexuosa* (Nash) V. L. Harms, Castanea 34: 408. 1969. TYPE: FLORIDA: Leon Co.: ca. 4 mi. S of Tallahassee, 3 Sep 1895, *Nash 2545* (holotype: NY; isotypes: F, GH, K, MO, US).

Erect perennial herb, to 5 dm; stem with basal rhizomes, striate-angled, flexuous (zigzag), silvery-sericeous. Leaves with the blade linear-lanceolate, 3–7 cm long, 3–7 mm wide, slightly reduced upward, the basal equaling or shorter than the cauline, withering by flowering, the apex acute, the base tapering, clasping, the upper and lower surfaces silvery-sericeous or glabrescent, sessile. Flowers in capitula in a corymbiform array, the peduncle 1–11 cm long, tomentose, minutely glandular, bracteolate; involucre narrowly campanulate, equaling the pappus; phyllaries in 4–6 series, lanceolate, 7–11 mm long, unequal, the margin scarious, fimbriate, the outer surface sparsely pilose, minutely stipitate-glandular or eglandular; receptacle slightly convex, pitted; ray flowers 9–13; corolla yellow, the lamina 5–8 mm long; disk flowers numerous; corolla yellow, 6–7 mm long. Cypsela fusiform, 3–4 mm long, 8- to 10-ribbed, strigose;

pappus in 3–4 series, the outer of linear scales to 1 mm long, the inner of numerous barbellate bristles 5–7 mm long.

Sandhills. Rare; Jefferson County, central panhandle. Endemic. Fall.

Pityopsis flexuosa is listed as endangered in Florida (Florida Administrative Code, Chapter 5B-40).

Pityopsis graminifolia (Michx.) Nutt. [With grasslike leaves.] NARROWLEAF SILKGRASS.

Inula graminifolia Michaux, Fl. Bor.-Amer. 2: 122. 1803. *Chrysopsis graminifolia* (Michaux) Elliott, Sketch Bot. S. Carolina 2: 334. 1823. *Diplopappus graminifolius* (Michaux) Lessing, Linnaea 5: 144. 1830. *Pityopsis graminifolia* (Michaux) Nuttall, Trans. Amer. Philos. Soc., ser. 2. 7: 317. 1841. *Diplogon graminifolium* (Michaux) Kuntze, Revis. Gen. Pl. 1: 334. 1891. *Heterothea graminifolia* (Michaux) Shinners, Field & Lab. 19: 71. 1951. TYPE: "Carolina ad Floridam," s.d., *Michaux s.n.* (holotype: P).

Erigeron nervosus Willdenow, Sp. Pl. 3: 1953. 1803. *Chrysopsis nervosa* (Willdenow) Fernald, Rhodora 44: 470. 1942. *Heterotheca nervosa* (Willdenow) Shinners, Field & Lab. 19: 68. 1951. *Pityopsis nervosa* (Willdenow) Dress, Baileya 19: 166. 1975.

Inula argentea Persoon, Syn. Pl. 2: 452. 1807. *Chrysopsis argentea* (Persoon) Elliott, Sketch Bot. S. Carolina 2: 334. 1823. *Pityopsis argentea* (Persoon) Nuttall, Trans. Amer. Philos. Soc., ser. 2. 7: 318. 1841.

Inula graminifolia Michaux var. *tenuifolia* Torrey, Ann. Lyceum Nat. Hist. New York 2: 212. 1828. *Pityopsis graminifolia* (Michaux) Nuttall var. *tenuifolia* (Torrey) Semple & Bowers, Univ. Waterloo Biol. Ser. 29: 24. 1985. *Pityopsis tenuifolia* (Torrey) Nesom, Phytoneuron 219–1: 15. 2019.

Chrysopsis graminifolia (Michaux) Elliott var. *latifolia* Fernald, Bot. Gaz. 24: 434. 1897. *Chrysopsis latifolia* (Fernald) Small, Fl. S.E. U.S. 1182, 1339. 1903. *Pityopsis graminifolia* (Michaux) Nuttall forma *latifolia* (Fernald) Fernald ex Small, Man. S.E. Fl. 1341. 1933. *Pityopsis graminifolia* (Michaux) Nuttall var. *latifolia* (Fernald) Semple & Bowers, Univ. Waterloo Biol. Ser. 29: 28. 1985. *Pityopsis latifolia* (Fernald) Bridges & Orzell, J. Bot. Res. Inst. Texas 12: 31. 2018. TYPE: FLORIDA: Martin Co.: Jensen, 25 Mar 1897, *Curtiss 5819* (lectotype: GH). Lectotypified by Semple and Bowers (1985: 28).

Chrysopsis microcephala Small, Fl. S.E. U.S. 1182, 1339. 1903. *Pityopsis microcephala* (Small) Small, Man. S.E. Fl. 1341, 1508. 1933. *Heterotheca microcephala* (Small) Shinners, Field & Lab. 19: 71. 1951. *Chrysopsis graminifolia* (Michaux) Elliott var. *microcephala* (Small) Cronquist, Brittonia 29: 219. 1977. *Pityopsis graminifolia* (Michaux) Nuttall var. *microcephala* (Small) Semple, Canad. J. Bot. 58: 148. 1980. TYPE: FLORIDA: Duval Co.: near Jacksonville, 13 Oct 1894, *Curtiss 5319* (holotype: NY; isotypes: GH, US).

Chrysopsis tracyi Small, Fl. S.E. U.S. 1182, 1339. 1903. *Pityopsis tracyi* (Small) Small, Man. S.E. Fl. 1341, 1508. 1933. *Heterotheca graminifolia* (Michaux) Shinners var. *tracyi* (Small) R. W. Long, Rhodora 72: 43. 1970. *Pityopsis graminifolia* (Michaux) Nuttall var. *tracyi* (Small) Semple & Bowers, Phytologia 58: 430. 1985. *Pityopsis nervosa* (Willdenow) Dress var. *tracyi* (Small) D. B. Ward, Novon 15: 368. 2004. TYPE: FLORIDA: Manatee Co.: Palma Sola, 3 Dec 1901, *Tracy 7713* (holotype: NY; isotypes, GH, US).

Pityopsis graminifolia (Michaux) Nuttall var. *aequilifolia* Semple & Bowers, Phytologia 58: 430. 1985. *Pityopsis aequilifolia* (Bowers & Semple) Bridges & Orzell, J. Bot. Res. Inst. Texas 12: 31. 2018. TYPE: FLORIDA: Lake Co.: Tavares, FL 19 S of old 441, 20 Sep 1971, *Wofford & Bowers 71–558* (holotype: TENN).

Erect perennial herb, to 8 dm; stem with basal rhizomes, striate-angled, silvery-sericeous. Leaves with the blade linear or lanceolate, 8–25(40) cm long, 2–20 mm wide, the basal longer or shorter than the cauline, the cauline reduced upward, the apex acute or acuminate, the base tapering, sessile, clasping, the margin entire, upper and lower surfaces silvery-sericeous. Flowers in capitula in a corymbiform or paniculiform array, the peduncle 1–10 cm long, sericeous, bracteolate; involucre turbinate-campanulate, usually shorter than the pappus; phyllaries in 4–6 series, lanceolate, 5–13 mm long, unequal, the margin scarious, fimbriate, ciliate, the outer surface pilose, sometimes stipitate-glandular; receptacle slightly convex, pitted; ray flowers 9–13; corolla yellow, the lamina 4–14 mm long; disk flowers numerous; corolla yellow, 4–9 mm long. Cypsela fusiform, 3–5 mm long, 8- to 10-ribbed, strigose; pappus in 3–4 series, the outer of linear or linear-triangular scales to 1 mm long, the inner of numerous barbellate bristles 5–9 mm long.

Scrub and sandhills. Common; nearly throughout. Delaware and Maryland south to Florida, west to Ohio, Kentucky, Oklahoma, and Texas; West Indies, Mexico, and Central America. Summer–fall.

Pityopsis graminifolia is a polymorphic species that is sometimes divided into as many as five varieties in Florida (Semple, 2006f). However, the features used to separate these slightly intergrade and it seems best to recognize a single taxon as we do here.

Pityopsis oligantha (Chapm. ex Torr. & A. Gray) Small. [From the Greek *oligos*, small, and *anthos*, flower, in reference to having few capitula.] GRASSLEAF GOLDENASTER.

Chrysopsis oligantha Chapman ex Torrey & A. Gray, Fl. N. Amer. 2: 253. 1842. *Pityopsis oligantha* (Chapman ex Torrey & A. Gray) Small, Man. S.E. Fl. 1341, 1508. 1933. *Heterotheca oligantha* (Chapman ex Torrey & A. Gray) V. L. Harms, Wrightia 4: 11. 1968.

Erect perennial herb, to 5 dm; stem with basal rhizomes, striate-angled, silvery-sericeous, distally dark stipitate-glandular. Leaves with the blade linear or oblanceolate, 8–30 cm long, 3–14 mm wide, reduced upward, the basal persistent at flowering, the cauline fewer than 10, the apex acute, the base tapering, sessile, clasping, the upper and lower surfaces silvery-sericeous. Flowers in capitula in a corymbiform array, the peduncle 2–14 cm long, stipitate-glandular, sometimes with a few long trichomes, bracteolate; involucre campanulate, shorter than the pappus; phyllaries in 4–5 series, lanceolate, 9–11 mm long, unequal, the margin scarious, fimbriate, the outer surface stipitate-glandular, the apex with twisted trichomes; receptacle slightly convex, pitted; ray flowers 11–16; corolla yellow, the lamina 9–13 mm long; disk flowers numerous; corolla yellow, 6–8 mm long. Cypsela fusiform, 4–5 mm long, 8- to 10-ribbed, strigose; pappus in 3–4 series, the outer of linear or linear-triangular scales to 1 mm long, the inner of numerous barbellate bristles 6–7 mm long.

Wet flatwoods, bogs, and cypress pond margins. Occasional; Jefferson County, central and western panhandle. Georgia and Florida west to Texas. Spring.

Pluchea Cass. 1817. CAMPHORWEED

Subshrubs and herbs. Leaves alternate, simple, pinnate-veined, petiolate or epetiolate. Flowers in capitula in corymbiform or paniculiform arrays, discoid, carpellate or staminate; receptacle epaleate; petals 5, basally connate, actinomorphic; stamens 5, epipetalous, the filaments free, the anthers connate; ovary inferior, 2-carpellate, 1-loculate. Fruit a cypsela; pappus present.

A genus of about 56 species; North America, West Indies, Mexico, Central America, South America, Africa, Asia, Australia, and Pacific Islands. [Commemorates Noël Antoine Pluche (1688–1761), French priest and naturalist.]

Selected reference: Nesom (2006r).

1. Shrubs .... **P. carolinensis**
1. Herbs.
   2. Stem evidently winged with decurrent leaf bases .... **P. sagittalis**
   2. Stem not winged.
      3. Leaves distinctly petiolate.
         4. Phyllaries sessile-glandular, the outer surface sometimes puberulent; lateral flowering branches shorter than the terminal ones .... **P. camphorata**
         4. Phyllaries glandular-pubescent; lateral flowering branches longer than the terminal ones .... **P. odorata**
      3. Leaves sessile or subsessile.
         5. Corolla rose-pink or purplish .... **P. baccharis**
         5. Corolla creamy-white or yellow, rarely pale pink.
            6. Phyllaries 9–12 mm long, the middle ones 2–3 mm wide; upper cauline leaves only slightly shorter than the lower; pappus of basally connate barbellate bristles .... **P. longifolia**
            6. Phyllaries 5–7 mm long, the middle ones less than 2 mm wide; upper cauline leaves much shorter than the lower; pappus of free barbellate bristles .... **P. foetida**

Pluchea baccharis (Mill.) Pruski [In reference to the inflorescence resembling that of the genus *Baccharis* (Asteraceae).] ROSY CAMPHORWEED.

Conyza baccharis Miller, Gard. Dict., ed. 8. Conyza no. 16. 1768. *Pluchea baccharis* (Miller) Pruski, Sida 21: 2035. 2035.

Pluchea rosea R. K. Godfrey, J. Elisha Mitchell Sci. Soc. 68: 266, pl. 21(5–6). 1952. TYPE: FLORIDA: Lake Co.: vicinity of Eustis, 16–31 May 1894, *Nash 758* (holotype: GH; isotypes: F, NY, UC, US).

Erect perennial herb, to 6 dm; stem angled, puberulent or villous, stipitate- or sessile-glandular. Leaves with the blade ovate, ovate-oblong, or elliptic-oblong, 2–7 cm long, 0.5–3 cm wide, the apex acute or obtuse, often apiculate, the base clasping, the margin shallowly serrate-dentate, apiculate, the upper and lower surfaces puberulent or sparsely villous, stipitate- or sessile-glandular, sometimes viscid, sessile. Flowers in capitula in a corymbiform array; involucre campanulate or turbinate; phyllaries in 3–6 series, ovate, ovate-lanceolate, or lanceolate, 4–6 mm long, unequal, rose-pink or purplish, the outer surface appressed-villous, puberulent, or arachnose, usually with glandular trichomes; receptacle flat; peripheral flowers carpellate, the inner staminate; corolla rose-pink or purplish. Cypsela oblong cylindric, 4- to 8-ribbed; pappus of 1 series of free barbellate bristles.

Marshes and wet flatwoods. Common; nearly throughout. North Carolina south to Florida, west to Texas; West Indies, Mexico, and Central America. Spring–summer.

Pluchea camphorata (L.) DC. [With a camphor-like odor.] CAMPHORWEED.

Erigeron camphoratus Linnaeus, Sp. Pl. 864. 1753. *Pluchea camphorata* (Linnaeus) de Candolle, Prodr. 5: 452. 1836.

Erect annual or perennial herb, to 2 m; stem angled, minutely puberulent or sessile-glandular. Leaves with the blade elliptic or oblong-elliptic, the apex acute, the base cuneate, the margin dentate-serrate or entire, the upper and lower surfaces puberulent, glandular-puberulent, or sessile-glandular, the petiole 1–2 cm long. Flowers in capitula in a paniculiform corymbiform array; involucre campanulate; phyllaries in 3–6 series, ovate or lanceolate, 4–6 mm long, unequal, creamy white, sometimes purplish, the outer surface sparsely puberulent or glabrate, minutely sessile-glandular; receptacle flat; peripheral flowers carpellate, the inner staminate; corolla rose-pink or purplish. Cypsela oblong cylindric, 4- to 8-ribbed; pappus of 1 series of free barbellate bristles.

Floodplains and swamps. Occasional; northern counties, central peninsula. New Jersey and Pennsylvania south to Florida, west to Wisconsin, Illinois, Missouri, Kansas, Oklahoma, and Texas. Summer–fall.

Pluchea carolinensis (Jacq.) G. Don [Of carolina.] CURE-FOR-ALL.

Conyza carolinensis Jacquin, Collectanea 2: 271. 1789. *Pluchea carolinensis* (Jacquin) G. Don, in Sweet, Hort. Brit., ed. 3. 350. 1839.

Erect shrubs, to 4 m; stem angled, matted-villous, proximally glabrescent. Leaves with the blade elliptic, oblong-obovate, or ovate, 5–6(20) cm long, 2–6(8) cm wide, the apex acute, obtuse, or rounded, the base cuneate, the margin entire or denticulate, the upper surface glabrate, the lower surface villous or puberulent, the petiole 1–4 cm long. Flowers in capitula in a corymbiform array; involucre campanulate or cupulate; phyllaries in 3–6 series, ovate or lanceolate, 5–6 mm long, unequal, greenish or creamy-white to tan, sometimes purplish, the outer surface glandular tomentose; receptacle flat; peripheral flowers carpellate, the inner staminate; corolla whitish or pink-lavender. Cypsela oblong cylindric, 4- to 8-ribbed; pappus of 1 series of free barbellate bristles.

Coastal hammocks, rarely inland. Occasional; central and southern peninsula. Florida and Texas; West Indies, Mexico, Central America, and South America; Pacific Islands. Native to West Indies, Central America, and South America. Winter–spring.

The specific epithet references the Carolinas, apparently in error, as the species is known in the United States only from Florida and Texas.

Pluchea foetida (L.) DC. [Foul-smelling.] STINKING CAMPHORWEED.

Baccharis foetida Linnaeus, Sp. Pl. 861. 1753. *Pluchea foetida* (Linnaeus) de Candolle, Prodr. 5: 452. 1836. *Conyza camphorata* Pursh, Fl. Amer. Sept. 323. 1814, nom. illegit.

Pluchea foetida (Linnaeus) de Candolle var. *imbricata* Kearney, Bull. Torrey Bot. Club 21: 483. 1894. *Pluchea imbricata* (Kearney) Nash, Bull. Torrey Bot. Club 23: 108. 1896. TYPE: FLORIDA: Lake Co.: vicinity of Eustis, 16–31 Jul 1894, *Nash 1434* (lectotype: NY; isolectotypes: GH, PH, UC, US). Lectotypified by Godfrey (1952: 265).

Pluchea tenuifolia Small, Man. S.E. Fl. 1399, 1509. 1933. TYPE: FLORIDA: Hardee Co.: 5.5 mi. N of Bowling Green on the road to Fort Meade, s.d., *Hand 360* (holotype: NY).

Erect annual or perennial herb, to 1 m; stem angled, arachnose, sessile-glandular. Leaves with the blade oblong, elliptic, ovate-lanceolate, or ovate, 3–10(13) cm long, 1–4 cm wide, the apex acute, obtuse, or rounded, the base clasping, the upper and lower surfaces sessile-glandular, sessile. Flowers in capitula in a corymbiform array; involucre cupulate, campanulate, or turbinate-campanulate; phyllaries in 3–6 series, ovate or ovate-lanceolate, 5–7 mm long, unequal, creamy-white, pinkish, purplish, or yellowish, the outer surface arachnoid-pubescent, sessile-glandular; receptacle flat; peripheral flowers carpellate, the inner staminate; corolla creamy-white or yellowish, rarely pale pink. Cypsela oblong cylindric, 4- to 8-ribbed; pappus of 1 series of free barbellate bristles.

Swamps and marshes. Frequent; nearly throughout. New Jersey south to Florida, west to Oklahoma and Texas, also Missouri; West Indies and Mexico. Summer–fall.

Pluchea longifolia Nash [With long leaves.] SWEETSCENT.

Pluchea longifolia Nash, Bull. Torrey Bot. Club 23: 108. 1896. TYPE: FLORIDA: Brevard Co.: Titusville, 30 Jul–1 Aug 1895, *Nash 2293* (holotype: NY; isotypes: F, GH, US).

Erect perennial herb, to 2 m; stem angled, sparsely arachnose. Leaves with the blade oblong, elliptic, ovate-lanceolate, or ovate, the apex acute or obtuse, the base clasping or subclasping, the margin coarsely and irregularly serrate-dentate, the upper surface hirtellous and sessile-glandular, the lower surface villous and sessile-glandular, sessile. Flowers in capitula in a corymbiform array; involucre cylindric-campanulate; phyllaries ovate or lanceolate, 9–12 mm long, unequal, creamy-white, the outer surface puberulent, sometimes sparsely glandular; receptacle flat; peripheral flowers carpellate, the inner staminate; corolla creamy-white. Cypsela oblong cylindric, 4- to 8-ribbed; pappus of 1 series of basally connate barbellate bristles.

Swamps. Frequent; northern and central peninsula, west to central panhandle. Endemic. Summer.

Pluchea odorata (L.) Cass [With an odor.] SWEETSCENT.

Conyza odorata Linnaeus, Syst. Nat., ed. 10. 1213. 1759. *Conyza purpurascens* Swartz, Prodr. 112. 1788, nom. illegit. *Pluchea odorata* (Linnaeus) Cassini, in Cuvier, Dict. Sci. Nat. 42: 3. 1826. *Pluchea purpurascens* de Candolle, Prodr. 5: 452. 1836. *Placus odoratus* (Linnaeus) Baillon ex M. Gómez de la Maza y Jiménez, Anales Soc. Esp. Hist. Nat. 19: 273. 1890. *Placus purpurascens* M. Gómez de la Maza y Jiménez, Anales Soc. Esp. Hist. Nat. 19: 273. 1890, nom. illegit. *Pluchea odorata* (Linnaeus) de Candolle var. *normalis* Kuntze, Revis. Gen. Pl. 1: 357. 1891, nom. inadmiss.

Pluchea purpurascens de Candolle var. *succulenta* Fernald, Rhodora 44: 227. 1942. *Pluchea odorata* (Linnaeus) Cassini var. *succulenta* (Fernald) Cronquist, in Radford et al. (eds.), Vasc. Fl. S.E. U.S. 1: 175. 1980.

Erect annual or perennial herb, to 2 m; stem angled, stipitate- or sessile-glandular, often also with eglandular trichomes. Leaves with the blade ovate, ovate-lanceolate, or ovate-elliptic, 4–15 cm long, 1–7 cm wide, the apex acute or obtuse, the base cuneate or clasping, the margin irregularly serrate, the upper and lower surfaces pubescent or glabrate, sessile or the petiole to 1 cm long. Flowers in capitula in a corymbiform array; involucre cylindric-campanulate; phyllaries in 3–6 series, ovate or lanceolate, 5–6 mm long, unequal, creamy-white, sometimes purplish, the outer surface minutely puberulent, minutely sessile-glandular; receptacle flat; peripheral flowers carpellate, the inner staminate; corolla pink, rosy, or purple. Cypsela oblong cylindric, 4- to 8-ribbed; pappus of 1 series of free barbellate bristles.

Swamps. Frequent; nearly throughout. Maine south to Florida, west to Ontario, Michigan, Kansas, and California. Summer–fall.

Pluchea sagittalis (Lam.) Cabrera [With two equal sharp basal lobes directed downward.] WINGSTEM CAMPHORWEED.

Conyza sagittalis Lamarck, Encycl. 2: 94. 1786. *Pluchea sagittalis* (Lamarck) Cabrera, Bol. Soc. Argent. Bot. 3: 36. 1949.

Gnaphalium suaveolens Vellozo, Fl. Flumin., Icon. 8: pl. 100. 1831 ("1827"). *Pluchea quitoc* de Candolle, Prodr. 5: 450. 1836, nom. illegit. *Pluchea suaveolens* (Vellozo) Kuntze, Revis. Gen. Pl. 3(2): 168. 1898.

Erect perennial herb, to 2 m; stem angled, minutely hirtellous or strigillose, sessile-glandular, winged with decurrent leaf bases. Leaves with the blade lanceolate or elliptic-lanceolate, the proximal sometimes spatulate or oblanceolate, 5–15 cm long, 1–3(4) cm wide, the apex acute or obtuse, the base cuneate, decurrent on the stem, the margin shallowly serrate, the upper and lower surfaces minutely hirtellous or strigillose, sessile-glandular, sessile. Flowers in capitula in a corymbiform array; involucre hemispheric or cupulate; phyllaries in 3–6 series ovate or lanceolate, 4–7 mm long, unequal, greenish or creamy-white, the outer surface stipitate-glandular; receptacle flat; peripheral flowers carpellate, the inner staminate; corolla whitish or rose-purple. Cypsela oblong cylindric, 4- to 8-ribbed; pappus of 1 series of free barbellate bristles.

Wet disturbed sites. Rare; Escambia County. Not recently collected. Florida and Alabama; West Indies and South America. Native to West Indies and South America. Summer.

EXCLUDED TAXA

Pluchea bifons (Linnaeus) de Candolle—This European plant was reported for Florida by Chapman (1860, 1883, 1897), the name misapplied to Florida material of *P. foetida*.

Pluchea symphytifolia (Miller) Gillis—This name, a synonym of *Neurolaena lobata* (Linnaeus) Cassini was reported for Florida by Cronquist (1980), Wunderlin (1982), and Correll and Correll (1982), all misapplied to Florida material of *P. carolinensis*.

Polymnia L. 1753. LEAFCUP

Herbs. Leaves opposite, simple, pinnate-veined, petiolate. Flowers in capitula in corymbiform arrays, radiate; receptacle paleate; ray flowers carpellate; petals 5, basally connate, zygomorphic; disk flowers staminate; petals 5, basally connate, actinomorphic; stamens 5, epipetalous, the filaments free, the anthers connate; ovary inferior, 2-carpellate, 1-loculate. Fruit a cypsela; pappus absent.

A genus of 3 species; North America. [From the Greek *poly*, many, and *hymnos*, praise, the name of the Greek muse *Polyhymnia* (*Polymnia*), the muse of sacred music, song, and dance, the plant apparently so named in reference to its delicate nature.]

Selected reference: Strother (2006w).

Polymnia laevigata Beadle [*Laevigatus*, smooth, in reference to the stems and leaves.] TENNESSEE LEAFCUP.

Polymnia laevigata Beadle, Bot. Gaz. 25: 378. 1898.

Erect perennial herb, to 2 m; stem striate-angled, glabrous or glabrate. Leaves with the blade deltate, cordate-lanceolate, or ovate, 4–20(40) cm long, 2–12(25) cm wide, reduced upward, 5- to 11-pinnately lobed, the apex acute, the base cordate or truncate, the ultimate margin irregularly and coarsely serrate or dentate, the upper and lower surfaces glabrous or glabrate, the lower surface lighter, the petiole 3–10 mm long. Flowers in capitula in a loose corymbiform array; involucres hemispheric; phyllaries in 2 series, 2–3 mm long, the outer lanceolate, the inner ovate, the margin with minute bristles; receptacle slightly convex; paleae oblanceolate; ray flowers 4–6, pistillate; corolla yellowish white, the lamina ca. 3 mm long; disk flowers numerous, staminate; corolla yellow, ca. 3 mm long. Cypsela pyriform, ca. 3 mm long, slightly compressed, 5-ribbed, finely striate between the ribs, brown; pappus absent.

Calcareous hammocks. Rare; Jackson County. Spring–fall.

Polymnia laevigata is listed as endangered in Florida (Florida Administrative Code, Chapter 5B-40).

EXCLUDED TAXON

Polymnia canadensis Linnaeus—Reported for Florida by Radford et al. (1968). Excluded from Florida by Wells (1965). No Florida specimens seen.

Praxelis Cass. 1826.

Herbs or subshrubs. Leaves opposite, simple, pinnipalmate-veined, petiolate. Flowers in capitula in corymbiform arrays, discoid, bisexual; receptacle epaleate; petals 5, basally connate, actinomorphic; stamens 5, epipetalous, the filaments free, the anthers connate; ovary inferior, 2-carpellate, 1-loculate. Fruit a cypsela; pappus present.

A genus of about 16 species; North America, South America, Asia, and Australia. [Etymology unknown.]

Selected reference: Abbott et al. (2008).

Praxelis clematidea (Kuntze) R. M. King & H. Rob. [Resembling *Clematis* (Ranunculaceae).]

Eupatorium clematideum Grisebach, Abh. Königl. Ges. Wiss. Göttingen 19: 172. 1879; non (Wallich ex de Candolle) Schultz Bipontinus, 1866. *Eupatorium urticifolium* Linnaeus f. var. *clematideum* Kuntze, Revis. Gen. Pl. 3(2): 148. 1898. *Praxelis clematidea* (Kuntze) R. M. King & H. Robinson, Phytologia 20: 194. 1970. *Eupatorium catarium* Veldkamp, Gard. Bull. Singapore 51(1): 121. 1999.

Erect or ascending annual or perennial herb or subshrub, to 1.3 m; stem striate-angled, pubescent, eglandular. Leaves with the blade ovate, 2–5 cm long, 1–3 cm wide, reduced upward, the apex acute, the base cuneate or rounded, the margin irregularly dentate or serrate, the upper surface glabrate, the lower surface pubescent, the trichomes eglandular on the veins, stipitate-glandular between the veins, the petiole 3–10 mm long. Flowers in capitula in a corymbiform array, the peduncle 4–7 mm long, pubescent, eglandular; involucre cylindric-campanulate, 4–5 mm wide; phyllaries in 2–3 series, linear-lanceolate, 6–8 mm long, the apex long-attenuate, the outer surface with short eglandular trichomes at the base, the margin ciliate; receptacle conic; flowers numerous; corolla bright blue, 3–5 mm long. Cypsela oblong-cylindric, 2–3 mm long, slightly compressed, 3- to 5-ribbed, with scattered setae, the body black, the ribs pale; pappus of numerous whitish, barbellate bristles 4–5 mm long.

Sandhill, dry flatwoods, and disturbed sites. Frequent; central peninsula. Florida; South America; Asia and Australia. Native to south America.

Praxelis clematidea is listed as a category II invasive species in Florida by the Florida Exotic Pest Plant Council (FLEPPC, 2017).

Pseudelephantopus Rohr 1792. DOG'S-TONGUE

Herbs. Leaves alternate, simple, pinnate-veined, epetiolate. Flowers in capitula in spiciform or paniculate-spiciform arrays, disciform, bisexual; receptacle epaleate; petals 5, basally connate, zygomorphic; stamens 5, epipetalous, the stamens free, the anthers connate; ovary inferior, 2-carpellate, 1-loculate. Fruit a cypsela; pappus present.

A monotypic genus; North America, West Indies, Mexico, Central America, South America, Africa, Asia, Australia, and Pacific Islands. [From the Greek *pseudo,* false, and the generic name *Elephantopus,* in reference to its resemblance to that genus.]

Selected reference: Strother (2006y).

Pseudelephantopus spicatus (B. Juss. ex Aubl.) C. F. Baker [In reference to the spiciform inflorescences.] DOG'S-TONGUE.

Elephantopus spicatus B. Jussieu ex Aublet, Hist. Pl. Guiane 808. 1775. *Distreptus spicatus* (B. Jussieu ex Aublet) Cassini, in Cuvier, Dict. Sci. Nat. 13: 367. 1819. *Pseudelephantopus spicatus* (B. Jussieu ex Aublet) C. F. Baker, Trans. Acad. Sci. St. Louis 12: 15. 1902.

Erect perennial herb, to 1 m; stem often rhizomatous, striate-angled, sparsely hirsute or glabrescent. Leaves mostly cauline at flowering, the blade obovate, lanceolate, or linear, 3–15(20) cm long, 1–13(4.5) cm wide, the apex acute, obtuse, or rounded, the base tapering and winged, the margin serrate or entire, the upper and lower surfaces sparsely pilose or hirsute, the lower surface resinous gland-dotted, sessile, clasping. Flowers in capitula in a spiciform or paniculate-spiciform array, each capitular cluster subtended by 1–2 lanceolate or linear leaflike bracts 1.5–4.5 cm long; involucre subcylindric or fusiform, 2–3 mm wide; phyllaries 8 in 4 decussate pairs, the outer ovate, the inner lanceolate, 9–12 mm long, chartaceous, the apex acuminate, the outer surface sparsely hispidulous, pilosulous, or glabrescent, gland-dotted; flowers (2)4(5); corolla white, pink, or purple, the lobes linear-lanceolate, 8–10 mm long, unequal, the outer sinus the deepest. Cypsela subclavate, 7–8 mm long, slightly flattened, 8- to 10-ribbed, appressed strigillose; pappus of 6–10 basally connate, laciniate or aristate scales, 1–9 mm long, 2(3) of the aristate scales with a distal awn-like plicate (2-folded) arista.

Open, sandy disturbed sites. Occasional; Alachua County, central and southern peninsula. Florida; West Indies, Mexico, Central America, and South America; Africa, Asia, Australia, and Pacific Islands. Native to tropical America. All year.

Pseudognaphalium Kirp. 1950. CUDWEED

Herbs. Leaves alternate, simple, pinnate-veined, epetiolate. Flowers in capitula in glomerules in corymbiform or paniculiform arrays, sometimes in terminal clusters, disciform, carpellate or bisexual; receptacle epaleate; petals 5, basally connate, actinomorphic; stamens 5, the filaments free, the anthers connate; ovary inferior, 2-carpellate, 1-loculate. Fruit a cypsela; pappus present.

A genus of about 85 species; nearly cosmopolitan. [From the Greek *pseudo,* false, and the genus name *Gnaphalium,* in reference to its resemblance to that genus.]

Selected reference: Nesom (2006s).

1. Upper and lower leaf surfaces with about equal amounts of indumentum .... **P. luteoalbum**
1. Lower leaf surface with the indumentum denser than on the upper leaf surface.
    2. Stem stipitate-glandular, scarcely woolly except in the inflorescence; upper surface of the leaves stipitate-glandular .... **P. helleri**
    2. Stem woolly; upper surface of the leaves sparsely woolly, glabrate in age.
        3. Leaves to 2 cm wide, the base somewhat dilated and partly clasping the stem .... **P. domingense**
        3. Leaves to 1 cm wide, the base not dilated or clasping the stem .... **P. obtusifolium**

Pseudognaphalium domingense (Lam.) Anderb. [Of the Dominican Republic.] DOMINICAN CUDWEED.

Gnaphalium domingense Lamarck, Encycl. 2: 743. 1788. *Pseudognaphalium domingense* (Lamarck) Anderberg, Opera Bot. 104: 147. 1991.

Erect annual herb, to 6 dm; stem striate-angled, white-woolly, glabrate in age. Leaves with the blade linear-lanceolate, 2–6 cm long, 4–20 mm wide, slightly reduced upward, the apex acute, the base cuneate, the margin entire, slightly revolute, the upper and lower surfaces bicolor, the upper surface sparsely woolly, glabrate in age, the lower surface white- or gray-tomentose, sessile, clasping. Flowers in capitula in a corymbiform array; involucre campanulate; phyllaries in 4–6 series, elliptic or oblong, 6–7 mm long, white, hyaline, glabrous; receptacle flat; peripheral flowers numerous, carpellate; corolla yellow; inner flowers 9–15, bisexual; corolla yellow. Cypsela cylindric-oblong, 4- to 6-ribbed, smooth; pappus of 10–12 basally connate barbellate bristles.

Disturbed sites. Rare; Brevard County. Florida; West Indies and Central America. Native to West Indies and Central America. Winter–spring.

Pseudognaphalium helleri (Britton) Anderb. [Commemorates Amos Arthur Heller (1867–1944), American botanist.] HELLER'S CUDWEED.

Gnaphalium helleri Britton, Bull. Torrey Bot. Club 20: 280. 1893. *Gnaphalium polycephalum* Michaux var. *helleri* (Britton) Fernald, Rhodora 10: 94. 1908. *Gnaphalium obtusifolium* Linnaeus var. *helleri* (Britton) S. F. Blake, Rhodora 20: 72. 1918. *Pseudognaphalium helleri* (Britton) Anderberg, Opera Bot. 104: 147. 1991.

Erect annual herb, to 1 m; stem striate-angled, villous, stipitate-glandular. Leaves with the blade oblong-lanceolate, 2.5–7 cm long, 4–20 mm wide, slightly reduced upward, the apex acute, the base cuneate, the margin entire, flat, the upper and lower surfaces bicolor, the upper surface green, minutely stipitate-glandular, the lower surface white- or gray-tomentose, minutely stipitate-glandular, sessile. Flowers in capitula in a corymbiform array; involucre campanulate; phyllaries in 4–6 series, ovate, ovate-oblong, or oblong, 6–7 mm long, white, hyaline, glabrous or slightly tomentose; receptacle flat; peripheral flowers numerous, carpellate; corolla yellow; inner flowers 9–15, bisexual; corolla yellow. Cypsela cylindric-oblong, 4- to 6-ribbed, smooth; pappus of 10–12 basally connate barbellate bristles.

Dry hammocks. Rare; Gadsden County. Virginia south to Florida, west to Minnesota, Missouri, Oklahoma, and Texas. Summer–fall.

Pseudognaphalium luteoalbum (L.) Hilliard & B. L. Burtt [Yellow-white, in reference to the flower color.] JERSEY CUDWEED.

Gnaphalium luteoalbum Linnaeus, Sp. Pl. 851. 1753. *Gnaphalium luteoalbum* Linnaeus var. *normalis* Kuntze, Revis. Gen. Pl. 1: 340. 1891, nom. inadmiss. *Dasyanthus conglobatus* Bubani, Fl. Pyren. 2: 199. 1900, nom. illegit. *Pseudognaphalium luteoalbum* (Linnaeus) Hilliard & B. L. Burtt, Bot. J. Linn. Soc. 82. 206. 1981.

Erect annual herb, to 4 dm; stem striate-angled, loosely white-tomentose. Leaves with the blade narrowly obovate or spatulate, 1–3(6) cm long, 2–10 mm wide, the distal smaller, narrowly oblanceolate, -oblong, or linear, the apex acute, obtuse, or rounded, the margin entire, weakly revolute, the upper and lower surfaces concolored or weakly bicolored, gray-tomentose, sessile, clasping and somtimes decurrent 1–2 mm. Flowers in capitula in terminal glomerules 1–2 cm wide, sometimes in a corymbiform or paniculiform array; involucre campanulate; phyllaries in 3–4 series, ovate or ovate-oblong, 3–4 mm long, silvery-gray or yellowish, hyaline; receptacle flat; peripheral flowers numerous, carpellate; corolla yellowish; inner flowers 5–10, bisexual;

corolla yellowish, the apex red. Cypsela cylindric-oblong, with whitish papilliform trichomes; pappus of 10–12 basally connate barbellate bristles.

Disturbed sites. Rare; Miami-Dade County. Massachusetts, Vermont, New York, and Pennsylvania, Alabama and Florida, Arkansas and Louisiana west to California, north to Washington; Mexico and South America; Europe, Africa, Asia, Australia, and Pacific Islands. Native to Europe, Africa, Asia, Australia, and Pacific Islands. Winter–spring.

Pseudognaphalium obtusifolium (L.) Hilliard & B. L. Burtt [With obtuse leaves.] SWEET EVERLASTING; RABBIT TOBACCO.

Gnaphalium obtusifolium Linnaeus, Sp. Pl. 851. 1753. *Pseudognaphalium obtusifolium* (Linnaeus) Hilliard & B. L. Burtt, Bot. J. Linn. Soc. 82: 205. 1981.

Erect annual or winter perennial, to 1 m; stem striate-angled, white-tomentose. Leaves with the blade linear-lanceolate, elliptic, or oblanceolate, 2.5–10 cm long, 2–10 mm wide, the apex acute, the base cuneate, the margin entire, flat, the upper and lower surfaces bicolor, the upper green, glabrous or sometimes with a very light tomentum, slightly gland-dotted, the lower white-tomentose, sessile. Flowers in capitula in a corymbiform array; involucre campanulate; phyllaries in 4–6 series, ovate or ovate-oblong, 5–7 mm long, white, hyaline, glabrous or basally tomentose; receptacle flat; peripheral flowers numerous, carpellate; corolla yellowish; inner flowers 4–8(11), bisexual; corolla yellow. Cypsela cylindric-oblong, 4- to 6-ribbed, smooth; pappus of 10–12 basally connate barbellate bristles.

Flatwoods, sandhills, and pond margins. Common; nearly throughout. Prince Edward Island, Nova Scotia, and New Brunswick south to Florida, west to Ontario, Minnesota, Iowa, Nebraska, Kansas, Oklahoma, and Texas. Summer–fall.

Pseudogynoxys Greenm. Cabrera 1950.

Vines. Leaves alternate, simple, pinnate-veined, petiolate. Flowers in capitula in corymbiform arrays or solitary, radiate; calyculi present; receptacle epaleate; ray flowers carpellate; petals 5, basally connate, zygomorphic; disk flowers bisexual; petals 5, basally connate, actinomorphic; stamens 5, epipetalous, the filaments free, the anthers connate; ovary inferior, 2-carpellate, 1-loculate. Fruit a cypsela; pappus present.

A genus of 14 species; North America, West Indies, Mexico, Central America, and South America. [From the Greek *pseudo,* false, and the genus name *gynoxys,* in reference to its resemblance to that genus.]

Selected reference: Barkley (2006d).

Pseudogynoxys chenopodioides (Kunth) Cabrera [*Chenopodium* (Amaranthaceae), and from the Greek *ioides,* to resemble, in reference to the leaves.] MEXICAN FLAMEVINE.

Senecio chenopodioides Kunth, in Humboldt et al. Nov. Gen. Sp. 4: 179. 1820. *Pseudogynoxys chenopodioides* (Kunth) Cabrera, Brittonia 7: 56. 1950.

Gynoxys berlandieri de Candolle, Prodr. 6: 326. 1838. *Gynoxys berlandieri* de Candolle var. *cordifolia* de Candolle, Prodr. 6: 326. 1838, nom. inadmiss. *Senecio berlandieri* (de Candolle) Hemsley, Biol. Cent.-Amer., Bot. 2: 236. 1881; non Schultz Bipontinus, 1845. *Senecio confusus* Britten, J. Bot. 36: 260. 1898. *Pseudogynoxys berlandieri* (de Candolle) Cabrera, Brittonia 7: 56. 1950.

Twining or climbing herbaceous or suffrutescent vine, to 50 m; stem striate-angled, sparsely puberulent or glabrate. Leaves with the blade ovate or lanceolate, 2–8(12) cm long, 1–5(7) cm wide, the apex acute or acuminate, the base truncate or rounded, the margin dentate, the upper and lower surfaces sparsely puberulent or glabrate, the petiole 1–3 cm long. Flowers in capitula in a corymbiform array or solitary; calyculi of 10–20 lanceolate-deltate or filiform bractlets 2–10 mm long; involucre hemispheric or campanulate, 1–2 cm wide; phyllaries numerous in 1–2 series, linear-lanceolate or linear, 8–12 mm long, membranous, puberulent or glabrescent; receptacle flat, foveolate; ray flowers 6–15, carpellate; corolla orange or brick-red, the lamina 12–20 mm long; disk flowers numerous, bisexual; corolla orange, 8–12 mm long. Cypsela subcylindric, (2)4 mm long, weakly ribbed, puberulent; pappus of numerous white barbellate bristles 5–8 mm long.

Disturbed sites. Occasional; central and southern peninsula. Escaped from cultivation. Florida and Texas; West Indies, Mexico, Central America, and South America. Native to Mexico, Central America, and South America. All year.

Pterocaulon Elliott 1823. BLACKROOT

Herbs. Leaves alternate, simple, pinnate-veined, epetiolate. Flowers in capitula in spiciform arrays, discoid, carpellate or staminate; petals 5, basally connate, actinomorphic; stamens 5, epipetalous, the filaments free, the anthers connate; ovary inferior, 2-carpellate, 1-loculate. Fruit a cypsela; pappus present.

A genus of about 20 species; North America, West Indies, Mexico, Central America, South America, Asia, and Australia. [From the Greek *pteron,* wing, and *kaulos,* stem, in reference to the stems winged by decurrent leaf bases.]

Selected reference: Nesom (2006t).

Pterocaulon pycnostachyum (Michx.) Elliott [From the Greek *pycno,* dense, compact, and *stachy,* spike, in reference to the dense flowering spiciform inflorescence.] BLACKROOT.

Conyza pycnostachya Michaux, Fl. Bor.-Amer. 2: 126. 1803. *Pterocaulon pycnostachyum* (Michaux) Elliott, Sketch. Bot. S. Carolina 2: 324. 1823. *Chlaenobolus pycnostachyus* (Michaux) Cassini, in Cuvier, Dict. Sci. Nat. 49: 338. 1827. *Pterocaulon undulatum* C. Mohr, Contr. U.S. Nat. Herb. 6: 790, nom. illegit. *Chlaenobolus undulatus* Small, Fl. S.E. U.S. 1236, 1340. 1903, nom. illegit. TYPE: "Carolina ad Floridam."

Erect perennial herb, to 8 dm; stem often rhizomatous, striate-angled, whitish-tomentose. Leaves with the blade lanceolate, obovate-lanceolate, oblong, or elliptic, 3–11 cm long, 1–3(3.5) cm wide, the apex acute or obtuse, the base winged and decurrent on the stem, the margin dentate or denticulate, sometimes nearly entire, slightly repand, the upper surface glabrous or glabrescent, the lower surface densely whitish-tomentose, sessile. Flowers in a dense, usually continuous, narrow, somewhat ovoid, spiciform array, rarely interrupted, then so near the base, (2)3–8(10) cm long, usually solitary, sometimes with 1–2 basal branches; involucre campanulate; phyllaries in 4–6 series, narrowly lanceolate, 4–5 mm long, unequal, whitish-tomentose; receptacle flat; peripheral flowers numerous, carpellate; corolla yellowish; inner flowers 6–10(15), staminate; corolla yellowish. Cypsela fusiform, ca. 1 mm long, 6- to 9-ribbed, strigose, minutely sessile-glandular between the ribs; pappus of 1–2 series of numerous barbellate bristles 2–3 mm long.

Pine flatwoods, sandhills, and disturbed sites. Common; nearly throughout. North Carolina south to Florida, west to Mississippi. Spring–fall.

EXCLUDED TAXON

Pterocaulon virgatum (Linnaeus) de Candolle—Reported for Florida by Wunderlin (1982), the name misapplied to Florida material of *P. pycnostachyum.*

Pulicaria Gaertn. 1791. FALSE FLEABANE

Herbs. Leaves alternate, simple, pinnate-veined, epetiolate. Flowers in capitulae in corymbiform arrays, radiate; receptacle epaleate; ray flowers carpellate; petals 5, basally connate, zygomorphic; disk flowers bisexual; petals 5, basally connate, actinomorphic; stamens 5, epipetalous, the filaments free, the anthers connate; ovary inferior, 2-carpellate, 1-loculate. Fruit a cypsela; pappus present.

A genus of about 67 species; North America, Europe, Africa, and Asia. [*Pulex,* flea, and *aris,* belonging to, in reference to the use of the plants as a flea repellent.]

Pulicaria arabica (L.) Cass. [Of Arabia (Arabian peninsula).] LADIES' FALSE FLEABANE.

Inula arabica Linnaeus, Mant. Pl. 114. 1767. *Pulicaria arabica* (Linnaeus) Cassini, in Cuvier, Dict. Sci. Nat. 44: 94. 1826.

Erect or ascending annual herb, to 6 dm long; stem striate-angled, pubescent. Leaves oblong, lanceolate, or oblanceolate, 2–4 cm long, 5–15 mm wide, the apex acute or obtuse, the margin irregularly dentate, the upper and lower surfaces pubescent, the lower leaves tapering to the base and subpetiolate, the upper sessile, somewhat half-clasping the stem. Flowers in capitulae in a corymbiform array, the peduncle 5–20 mm long; involucre hemispheric or campanulate, 5–10 mm wide; phyllaries in 3–4 series, lanceolate or linear, ca. 5 mm long, unequal; receptacle flat or slightly convex, alveolate; ray flowers numerous, carpellate; corolla yellow, the lamina 3–4 mm long; disk flowers numerous, bisexual; corolla yellow, 2–3 mm long. Cypsela ellipsoid, 1–2 mm long, weakly compressed, brown, sparsely appressed pubescent; pappus of outer short, connate, suberose scales and 7–12 free barbellate bristles 3–4 mm long.

Disturbed sites. Rare; Escambia County. Not recently collected. Florida, Alabama, and

California; North America; Europe, Africa, and Asia. Native to Europe, Africa, and Asia. Summer.

Pyrrhopappus DC. 1838. DESERTCHICORY

Herbs. Leaves alternate, simple, pinnate-veined, petiolate or epetiolate. Flowers in capitula in corymbiform arrays, liguliform, bisexual; calyculi present; receptacle epaleate; petals 5, basally connate, zygomorphic; stamens 5, epipetalous, the filaments free, the anthers connate; ovary inferior, 2-carpellate, 1-loculate. Fruit a cypsela; pappus present.

A genus of 5 species; North America and Mexico. [From the Greek *pyrrhos,* fire-like, yellowish red in color, and *pappos,* pappus, in reference to the pappus color.]

Sitilias Raf., 1838.

Selected reference: Strother (2006z).

Pyrrhopappus carolinianus (Walter) DC. [Of Carolina.] CAROLINA DESERTCHICORY.

Leontodon carolinianus Walter, Fl. Carol. 192. 1788. *Borkhausia caroliniana* (Walter) Nuttall, Gen. N. Amer. Pl. 2: 126. 1818. *Sitilias caroliniana* (Walter) Rafinesque, New Fl. 4: 85. 1838 ("1836"). *Pyrrhopappus carolinianus* (Walter) de Candolle, Prodr. 7: 144. 1838.

Pyrrhopappus georgianus Shinners, Field & Lab. 21: 93. 1953. *Pyrrhopappus carolinianus* (Walter) de Candolle var. *georgianus* (Shinners) Ahles, J. Elisha Mitchell Sci. Soc. 80: 173. 1964.

Erect annual herb, to 5(10) dm; stem striate-angled, glabrous proximally or sometimes pilosulous. Proximal leaves with the blade lanceolate, 8–15 cm long, 1–7 cm wide, reduced upward, the apex acuminate, the base cuneate, the margin dentate or pinnately lobed, the upper and lower surfaces glabrous, the lower surface sometimes sparsely pilose on the main veins, the distal with the blade narrowly lanceolate, the margin entire or 1- to 2-lobed near the base, sessile and clasping. Flowers in capitula in a loose corymbiform array; calyculi of 13–16 bractlets in 2–3 series, subulate or filiform, 8–12 mm long; involucre subcylindric or campanulate; phyllaries in 2 series, numerous, linear, 1.5–2.5 cm long, equal, the margin scarious, the apex acute, thickened and darkened, with a keel-like flap near the tip, the outer surface glabrous; receptacle slightly convex, pitted; flowers numerous; corolla yellow, the lamina 1–2 cm long. Cypsela cylindric, the body 4–6 mm long, the beak 8–10 mm long, 5-ribbed, transversely rugose, reddish brown, glabrous; pappus in 2 series, the outer coroniform with short whitish bristles, the inner of 2–3 series of numerous, reddish or yellowish, subequal, barbellate bristles 7–10 mm long.

Dry hammocks and disturbed sites. Frequent; northern counties, central peninsula. New Jersey and Pennsylvania south to Florida, west to Iowa, Nebraska, Kansas, Oklahoma, and Texas. Spring–summer.

EXCLUDED TAXA

Pyrrhopappus multicaulis de Candolle—Reported for Florida by Small (1903, 1913a, 1933, all as *Sitilias multicaulis* (de Candolle) Greene), Wilhelm (1984), and Clewell (1985), the name misapplied to Florida material of *P. carolinianus.*

Pyrrhopappus pauciflorus (D. Don) de Candolle—Reported for Florida by Wunderlin (1998) and Wunderlin and Hansen (2003), the name misapplied to Florida material of *P. carolinianus.*

Ratibida Raf. 1817. PRAIRIE CONEFLOWER

Herbs. Leaves alternate, simple, pinnate-veined, petiolate. Flowers in capitula in corymbiform arrays, radiate; receptacle paleate; ray flowers sterile; petals 5, basally connate, zygomorphic; disk flowers bisexual; petals 5, basally connate, actinomorphic; stamens 5, epipetalous, the filaments free, the anthers connate; ovary inferior, 2-carpellate, 1-loculate. Fruit a cypsela; pappus present or absent.

A genus of 7 species; North America and Mexico. [Etymology unknown.]

Selected reference: Urbatsch and Cox (2006a).

1. Receptacle columnar, 2–4.5 times as long as wide.. **R. columnifera**
1. Receptacle ellipsoid, globular, or ovoid, 1–1.5 times as long as wide.. **R. pinnata**

Ratibida columnifera (Nutt.) Wooton and Standl. [Column-bearing, in reference to the receptacle.] UPRIGHT PRAIRIE CONEFLOWER.

Rudbeckia columnifera Nuttall, Cat. Pl. Upper Louisiana 1813. *Rudbeckia columnaris* Sims, Bot. Mag. 39: pl. 1601. 1813. *Ratibida columnaris* Pursh, Fl. Amer. Sept. 575. 1814; nom. illegit; non Sims, 1813. *Ratibida sulcata* Rafinesque, J. Phys. Chim. Hist. Nat. Arts 89: 100. 1819, nom. illegit. *Obeliscaria columnaris* (Sims) de Candolle, Prodr. 5: 559. 1836. *Ratibida columnaris* (Sims) D. Don, in Sweet, Brit. Fl. Gard. 7: pl. 361. 1837. *Lepachys columnaris* Torrey & A. Gray, Fl. N. Amer. 2: 315. 1842, nom. illegit. *Ratibida columnifera* (Nuttall) Wooton & Standley, Contr. U.S. Natl. Herb. 19: 706. 1915. *Lepachys columnifera* (Nuttall) J. F. Macbride, Contr. Gray Herb. 65: 45. 1922.

Erect perennial herb, to 1 m; stem striate-angled, sparsely hirsute. Leaves with the blade elliptic-oblong or -obovate, 2–15 cm long, 0.8–6 cm wide, pinnatifid, the lobes 3–14, linear-lanceolate or oblong-ovate, 1–16 mm wide, the apex acute, the base cuneate, winged, the lobe margins entire, the upper and lower surfaces sparsely hirsute, gland-dotted. Flowers in 1–15 capitula, solitary or 2–15 in a corymbiform array, the peduncle (1.5)6.5–48 cm long; involucre subrotate, 8–16 mm wide; phyllaries 5–15 in 2 series, the outer linear, 4–14 mm long, the inner ovate-lanceolate, 1–3 mm long; receptacle subhemispheric or columnar, 1–5 cm long, 7–12 mm wide, 2–4.5 times as long as wide; palea linear or oblong-oblique, 2–4 mm long, chartaceous, conduplicate, the apex truncate, the surfaces densely strigose and gland-dotted, each margin with an oval-oblanceolate resin gland ca. 1 mm long on the inner surface; ray flowers 4–12; corolla yellow, purplish-yellow, or maroon and yellow, the lamina 7–35 mm long; disk flowers numerous; corolla greenish yellow, often purplish distally, 1–2 mm long. Cypsela oblong, 1–3 mm long, compressed, the outer margin glabrous, the inner margin glabrous or ciliate; pappus of 1–2 toothlike projections or absent.

Dry disturbed sites. Occasional; northern counties. Nearly throughout North America; Mexico. Native to the midwestern United States and Mexico. Summer.

Ratibida pinnata (Vent.) Barnhart [In reference to the pinnatifid leaf blades.] PINNATE PRAIRIE CONEFLOWER.

Rudbeckia pinnata Ventenat, Descr. Pl. Nouv. pl. 71. 1802. *Lepachys pinnatifida* Rafinesque, J. Phys. Chim. Hist. Nat. Arts 89: 100. 1819, nom. illegit. *Obeliscaria pinnata* (Ventenat) Cassini, in Cuvier, Dict. Sci. Nat. 35: 273. 1825. *Lepachys pinnata* (Ventenat) Torrey & A. Gray, Fl. N. Amer. 2: 314. 1842. *Ratibida pinnata* (Ventenat) Barnhart, Bull. Torrey Bot. Club 24: 410. 1897. *Helianthus pinnatus* (Ventenat) E.H.L. Krause, in Sturm, Deutchl. Fl., ed. 2. 13: 166. 1905.

Erect perennial herb, to 1.25 m; stem striate-angled, strigose. Leaves with the blade oblong-lanceolate, -obovate, or -elliptic, 5–40 cm long, 3–15 cm wide, pinnatifid or pinnate, the lobes 3–9, narrowly lanceolate or ovate, 1–15 cm long, 0.2–3.5 cm wide, the apex acute, the base cuneate, winged, the lobe margins entire or irregularly serrate, the upper and lower surfaces strigose, gland-dotted. Flowers in capitula, solitary or 2–12 in a corymbiform array, the peduncle 3–27 cm long; involucre 8–16 mm wide; phyllaries 10–15 in 2 series, the outer linear, 3–15 mm long, the inner ovate-lanceolate, 3–6 mm long; receptacle ellipsoid, globular or ovoid, 1–2.5 cm long, 10–18 mm wide, 1–1.5 times as long as wide; palea oblong, 2–4 mm long, chartaceous, conduplicate, the apex truncate, the surfaces densely strigose and gland-dotted, each margin with a linear or oblanceolate resin gland 1–2 mm long on the inner surface; ray flowers 6–15; corolla yellow, the lamina 16–60 mm long; disk flowers numerous; corolla yellow, often purplish distally, 3–4 mm long. Cypsela linear-oblanceolate, 2–4 mm long, compressed, the outer margin glabrous, the inner margin glabrous or ciliate; pappus of 1–2 toothlike projections or absent.

Calcareous hammocks. Rare; Madison, Gadsden, and Jackson Counties. Massachusetts and New York south to Florida, west to Ontario, Minnesota, North Dakota, Nebraska, Kansas, Oklahoma, and Texas. Spring.

Rayjacksonia R. I. Hartman & M. A. Lane 1996. TANSYASTER

Herbs. Leaves alternate, simple, pinnate-veined, epetiolate. Flowers in capitula in subcorymbiform arrays, discoid; receptacle epaleate; ray flowers carpellate; petals 5, basally connate, zygomorphic; disk flowers bisexual; petals 5, basally connate, actinomorphic; stamens 5, epipetalous; the filaments free, the anthers connate; ovary inferior, 2-carpellate, 1-loculate. Fruit a cypsela; pappus present.

A genus of 3 species; North America and Mexico. [Commemorates Raymond Carl Jackson (1928–2008), American botanist.]

Selected reference: Nesom (2006u).

Rayjacksonia phyllocephala (DC.) R. L. Hartman & M. A. Lane [From the Greek *phyllon*, leaf, and *cephale*, head, in reference to the leaflike phyllaries.] CAMPHOR DAISY.

Haplopappus phyllocephalus de Candolle, Prodr. 5: 347. 1836. *Haplopappus rubiginosus* Torrey & A. Gray var. *phyllocephalus* (de Candolle) A. Gray, Syn. Fl. N. Amer. 1(2): 130. 1884. *Aster phyllocephalus* (de Candolle) Kuntze, Revis. Gen. Pl. 1: 316. 1891. *Eriocarpum phyllocephalum* (de Candolle) Greene, Erythea 3: 15. 1895. *Eriocarpum rubiginosum* (Torrey & A. Gray) Britton var. *phyllocephalum* (de Candolle) A. Heller, Contr. Herb. Franklin Marshall Coll. 1: 101. 1895. *Sideranthus phyllocephalus* (de Candolle) Small, Fl. S.E. U.S. 1186, 1339. 1903. *Haplopappus phyllocephalus* de Candolle var. *genuinus* S. F. Blake, Contr. Gray Herb. 52: 22. 1917, nom. inadmiss. *Haplopappus phyllocephalus* de Candolle subsp. *typicus* H. M. Hall, Publ. Carnegie Inst. Wash. 389. 58. 1928, nom. inadmiss. *Machaeranthera phyllocephala* (de Candolle) Shinners, Field & Lab. 18: 40. 1950. *Rayjacksonia phyllocephala* (de Candolle) R. L. Hartman & M. A. Lane, Amer. J. Bot. 83: 369. 1996.

Eriocarpum megacephalum Nash, Bull. Torrey Bot. Club 23: 107. 1896. *Haplopappus megacephalus* (Nash) Hitchcock, Trans. Kansas Acad. Sci. 16: 131. 1899. *Sideranthus megacephalus* (Nash) Small, Fl. S.E. U.S. 1185, 1339. 1903. *Haplopappus phyllocephalus* de Candolle subsp. *megacephalus* (Nash) H. M. Hall, Publ. Carnegie Inst. Wash. 389: 59. 1928. *Machaeranthera phyllocephala* (Nash) Shinners var. *megacephala* (Nash) Shinners, Field & Lab. 18: 40. 1950. *Haplopappus phyllocephalus* de Candolle var. *megacephalus* (Nash) Waterfall, Rhodora 62: 321. 1960. *Rayjacksonia phyllocephala* (de Candolle) R. L. Hartman & M. A. Lane var. *megacephala* (Nash) D. B. Ward, Phytologia 94: 467. 2012. TYPE: FLORIDA: Manatee Co.: Palmetto, S Shore of Sneeds Island, near mouth of Manatee River, 21–23 Aug 1895, *Nash 2432* (holotype: NH; isotypes: G, GH, MICH, ND-G, PH, US).

Eriocarpum floridanum Gandoger, Bull. Soc. Bot. France 65: 41. 1918. TYPE: FLORIDA: Escambia County: Pensacola, s.d., *Tracy 8515* (holotype: LY; isotypes: G, GH, US).

Eriocarpum tracyi Gandoger, Bull. Soc. Bot. France 65: 41. 1918. TYPE: FLORIDA: Manatee Co.: Palmetto, Sneeds Island, 9 Sep 1899, *Tracy 6354* (holotype: LY; isotypes: GH, US).

Erect annual or perennial herbs, to 7(10) dm; stem slightly striate-angled, pubescent. Leaves with the blade oblong or oblanceolate, 2–8 cm long, 5–15 mm wide, the apex acute or obtuse, the base cuneate, the margin coarsely serrate, the teeth bristle-tipped, the upper and lower surfaces pubescent, sessile and clasping the stem. Flowers in capitula in a subcorymbiform array, the heads sessile, the distal leaves continuing to the capitulum base and often surpassing it; involucre hemispheric, 1.2–3 cm wide; phyllaries numerous in 3–4 series, 1–1.5 cm long, subequal, linear-lanceolate, the apex erect or spreading, subherbaceous, the outer surface glabrous; receptacle slightly convex, shallowly pitted; ray flowers numerous, carpellate; corolla yellow, the lamina 8–15 mm long; disk flowers numerous, bisexual; corolla yellow, 2–3 mm long. Cypsela dimorphic, the ray flowers obovoid, 3-angled, the disk flowers ellipsoid, compressed, 5- to 9-ribbed.

Dunes and open coastal sites. Frequent; central and southern peninsula, also Escambia County. Florida, Mississippi west to Texas, also Colorado; Mexico. Summer–fall.

Rudbeckia L. CONEFLOWER

Herbs. Leaves alternate, simple, pinnate-veined, petiolate or epetiolate. Flowers in capitula in solitary or corymbiform or paniculiform arrays, radiate; receptacle paleate; ray flowers sterile;

petals 5, basally connate, zygomorphic; disk flowers bisexual; petals 5, basally connate, actinomorphic; stamens 5, epipetalous, the filaments free, the anthers connate; the ovary inferior, 2-carpellate, 1-loculate. Fruit a cypsela; pappus present or absent.

A genus of about 25 species; North America and Europe. [Commemorates Olaf Johannes Rudbeck (1630–1702) and Olaf Olai Rudbeck (1660–1740), father and son botanists and professors at Uppsala University.]

Selected reference: Urbatsch and Cox (2006b).

1. Leaves linear or linear-lanceolate, grasslike.
   2. Capitula solitary; ray flowers orange-red or maroon; leaves pubescent .... **R. graminifolia**
   2. Capitula several; ray flowers yellow; leaves glabrous .... **R. mohrii**
1. Leaves lanceolate or ovate.
   3. Disk flowers yellow or yellow-green .... **R. laciniata**
   3. Disk flowers brown-purple.
      4. Lower stem leaves deeply lobed .... **R. triloba**
      4. Lower stem leaves entire or merely toothed.
         5. Stem and leaves glabrous or glabrate.
            6. Midstem and lower cauline leaves distinctly auriculate clasping .... **R. auriculata**
            6. Midstem and lower cauline leaves petiolate or sessile .... **R. nitida**
         5. Stem and leaves conspicuously pubescent, strigose, or hirsute.
            7. Leaves and the upper stem appressed pubescent .... **R. mollis**
            7. Leaves and stems hirsute or strigose.
               8. Leaves and stems hirsute; paleae hispid toward the apex; pappus absent .... **R. hirta**
               8. Leaves and stems strigose (sometimes short hirsute near the base); paleae smooth or ciliolate-margined; pappus an inconspicuous low crown .... **R. fulgida**

Rudbeckia auriculata (Perdue) Kral [*Auriculatus,* with ear-like appendages, in reference to the leaf bases.] EARED CONEFLOWER.

Rudbeckia fulgida Aiton var. *auriculata* Perdue, Rhodora 63: 119. 1961. *Rudbeckia auriculata* (Perdue) Kral, Rhodora 77: 49. 1975.

Erect perennial herb, to 3 m; stem rhizomatous, striate-angled, glabrate. Leaves with the blade lanceolate, elliptic, or ovate, 15–65 cm long, 4–20 cm wide, reduced upward, the apex acute, the margin crenate, serrate, dentate, or entire, the upper and lower surfaces scabrous, the lower sometimes glabrescent, the basal petiolate, the base attenuate, the cauline sessile, the base auriculate or truncate. Flowers in capitula in a loose paniculiform array; involucre hemispheric; phyllaries in 2–3 series, lanceolate, to 1 cm long, unequal; receptacle conic or ovoid; paleae obovate, 4–6 mm long, concave, each clasping a flower, the apex obtuse or acute, the outer surface of the tip pubescent; ray flowers 8–14; corolla yellow, the lamina 1.8–3 cm long; disk flowers numerous; corolla brown-purple, ca. 4 mm long. Cypsela obpyramidal, 4-angled, 3–5 mm long; pappus of 4–7 unequal scales, to 2 mm long.

Bogs. Rare; Walton County. Georgia, Florida, and Alabama. Fall.

Rudbeckia auriculata is listed as endangered in Florida (Florida Administrative Code, Chapter 5B-40).

Rudbeckia fulgida Aiton [*Fulgens,* brightly colored, in reference to the ray flowers.] ORANGE CONEFLOWER.

Rudbeckia fulgida Aiton, Hort. Kew. 3: 251. 1789. *Helianthus fulgidus* (Aiton) E.H.L. Krause, in Sturm, Deutschl. Fl., ed. 2. 13: 165. 1905.
Rudbeckia spathulata Michaux, Fl. Bor.-Amer. 2: 144. 1803. *Rudbeckia fulgida* Aiton var. *spathulata* (Michaux) Perdue, Rhodora 59: 298. 1957.
Rudbeckia foliosa C. L. Boynton & Beadle, in Small, Fl. S.E. U.S. 1256, 1340. 1903.

Erect perennial herb, to 1.2 m; stem stoloniferous, striate-angled, glabrate. Leaves with the blade lanceolate, ovate, elliptic, or spatulate, 2–30 cm long, 0.5–8 cm wide, the apex acute or obtuse, the base attenuate or cordate, the margin entire or irregularly serrate, the upper and lower surfaces hirsute, strigose, or glabrate, the proximal petiolate, the distal sessile, cordate or auriculate. Flowers in capitula, solitary or 2–7 in a loose corymbiform array; involucre hemispheric; phyllaries in 2–3 series, lanceolate, to 2 cm long, unequal; receptacle hemispheric or ovoid; paleae obovate, 3–4 mm long, concave, each clasping a flower, the apex obtuse or acute, the margin sometimes ciliate, the outer surface of the tip usually glabrous; ray flowers 10–15; corolla yellow, the lamina 1.5–2.5 cm long; disk flowers numerous; corolla proximally yellow-green, distally brown-purple, 3–4 mm long. Cypsela obpyramidal, 4-angled, 2–4 mm long; pappus coroniform, less than 1 mm long.

Flatwoods and rocky hammocks. Occasional; eastern and central panhandle. Quebec south to Florida, west to Ontario, Wisconsin, Iowa, Missouri, Oklahoma, and Texas. Fall.

Rudbeckia graminifolia (Torr. & A. Gray) C. L. Boynton & Beadle [With grasslike leaves.] GRASSLEAF CONEFLOWER.

Echinacea atrorubens (Nuttall) var. *graminifolia* Torrey & A. Gray, Fl. N. Amer. 2: 306. 1842. *Rudbeckia graminifolia* (Torrey & A. Gray) C. L. Boynton & Beadle, Biltmore Bot. Stud. 1: 12. 1901. TYPE: FLORIDA: s.d., *Chapman s.n.* (holotype: ?; isotypes: GH, NY).

Erect perennial herb, to 8 dm; stem ascending strigose. Leaves with the blade linear, linear-lanceolate, or narrowly elliptic, grasslike, the length 10 times the width, 1–25 cm long, 5–10 mm wide, reduced upward, the apex acute, the base attenuate, the margin entire, the upper and lower surfaces sparsely hirsute or glabrous, the basal petiolate or sessile, the upper sessile. Flowers in a solitary capitulum; involucre hemispheric; phyllaries in 2–3 series, lanceolate, to 1 cm long, unequal, spreading or reflexed; receptacle hemispheric or ovoid; paleae obovate, 4–5 mm long, concave, each clasping a flower, the apex acuminate-cuspidate, the outer surface of the tip sparsely strigose; ray flowers 8–16; corolla orangish red or maroon, the lamina 1–2.5 cm long; disk flowers numerous; corolla proximally maroon, distally brown-purple, ca. 4 mm long. Cypsela obpyramidal, 4-angled, 2–3 mm long; pappus coroniform, less than 1 mm long.

Bogs and cypress swamps. Occasional; central panhandle. Endemic. Spring–fall.

Rudbeckia hirta L. [*Hirtus*, hairy, in reference to the stems and leaves.] BLACKEYED SUSAN.

Rudbeckia hirta Linnaeus, Sp. Pl. 907. 1753. *Helianthus hirtus* (Linnaeus) E.H.L. Krause, in Sturm, Deutschl. Fl., ed. 2. 13: 165. 1905.
Rudbeckia divergens T. V. Moore, Pittonia 4: 177. 1900. TYPE: Duval Co.: near Jacksonville, 16 May 1894, *Curtiss 4759* (holotype: ?; isotype: NY).
Rudbeckia floridana T. V. Moore, Pittonia 4: 176. 1900. *Rudbeckia hirta* Linnaeus var. *floridana* (T. V. Moore) Perdue, Rhodora 59: 296. 1957. TYPE: FLORIDA: Seminole Co.: Sanford, 29 Jul 1895, *Nash 2272* (holotype: ?; isotypes: GH, NY).
Rudbeckia floridana T. V. Moore var. *angustifolia* T. V. Moore, Pittonia 4: 176. 1900. *Rudbeckia hirta* Linnaeus var. *angustifolia* (T. V. Moore) Perdue, Rhodora 59: 296. 1957.
Rudbeckia hirta Linnaeus var. *pulcherrima* Farwell, Rep. (Annual) Michigan Acad. Sci. 6: 209. 1904. *Rudbeckia serotina* Nuttall forma *pulcherrima* (Farwell) Fernald & B. G. Schubert, Rhodora 50: 175. 1948.

Annual, biennial, or perennial herb, to 1 m; stem striate-angled, hispid or hirsute. Leaves with the blade elliptic, lanceolate, or ovate, 3–30 cm long, 0.4–7 cm wide, reduced upward, the apex acute, the base attenuate or cuneate, the margin entire or irregularly serrate, the upper and lower surfaces hispid or hirsute, the basal petiolate, the cauline petiolate or sessile. Flowers in capitula, solitary or 2–5 in a loose corymbiform array; involucre hemispheric; phyllaries in 2–3 series, lanceolate to 1 cm long; phyllaries in 2–3 series, lanceolate, to 3 cm long, unequal, the outer surface hispid or hirsute; receptacle hemispheric or ovoid; paleae obovate, 4–6 mm long, the apex acute or cuspidate, concave, each clasping a flower, the outer surface of the tip hispid or hirsute; ray flowers 8–16; corolla yellow or orange-yellow, usually with a basal maroon spot, 1.5–4.5 cm long; disk flowers numerous; corolla proximally yellow-green, distally brown-purple, 3–4 mm long. Cypsela obpyramidal, 4-angled, 2–3 mm long; pappus absent.

Sandhills, flatwoods, and disturbed sites. Common; nearly throughout. Nearly throughout North America. Summer–fall.

Rudbeckia laciniata L. [Cut into narrow divisions, in reference to the pinnately or pinnatifid leaves.] CUTLEAF CONEFLOWER.

Rudbeckia laciniata Linnaeus, Sp. Pl. 906. 1753. *Helianthus laciniatus* (Linnaeus) E.H.L. Krause, in Sturm, Deutschl. Fl., ed. 2. 13: 164. 1905; non A. Gray, 1849.
Rudbeckia heterophylla Torrey & A. Gray, Fl. N. Amer. 2: 312. 1842. *Rudbeckia laciniata* Linnaeus var. *heterophylla* (Torrey & A. Gray) Fernald & B. G. Schubert, Rhodora 50: 172. 1948. TYPE: FLORIDA: s.d., *Chapman s.n.* (holotype: NY?; isotype: GH).

Erect perennial herb, to 3 m; stem rhizomatous, striate-angled, glabrate. Leaves with the blade ovate or lanceolate, 8–50 cm long, 3–25 cm wide, reduced upward, the apex acute or acuminate, the base cuneate or cordate, the margin 1- to 2-pinnatifid or pinnately lobed, the lobes 3–11, sometimes entire, the margin entire or irregularly dentate, the upper and lower surfaces glabrate, slightly caucous, the basal petiolate, the cauline petiolate or sessile. Flowers in capitula in a loose corymbiform array; involucre hemispheric; phyllaries in 2–3 series, ovate or lanceolate, to 2 cm long, unequal, the margin ciliate or glabrous; receptacle hemispheric, ovoid, or globose; paleae obovate, 3–7 mm long, concave, each clasping a flower, the apex truncate or rounded, the outer surface of the tip pubescent; ray flowers 8–12; corolla yellow, the lamina 1.5–5 cm long; disk flowers numerous; corolla yellow or yellow-green, 4–5 mm long. Cypsela obpyramidal, 4-angled, 3–5 mm long; pappus coroniform or of 4 scales, to ca. 1 mm long.

Moist open hammocks and floodplains. Rare; Levy County, central panhandle, Walton County. Nearly throughout North America except for Nevada, Washington, and California. Summer–fall.

Rudbeckia mohrii A. Gray [Commemorates Charles Theodore Mohr (1824–1901), German-born pharmacist and botanist who worked in the United States and Mexico.] MOHR'S CONEFLOWER.

Rudbeckia mohrii A. Gray, Proc. Amer. Acad. Arts 17: 217. 1882. *Rudbeckia bupleuroides* Shuttleworth ex Chapman, Fl. South. U.S., ed. 2. 629. 1883, nom. illegit. TYPE: FLORIDA: Wakulla Co.: Dead Lakes, 22 Jun 1880, *Mohr s.n.* (holotype: GH).

Erect perennial herb, to 1 m; stem rhizomatous, striate-angled, glabrous. Leaves with the blade linear or linear-lanceolate, grasslike, 2–40 cm long, 2–15 mm wide, reduced upward, the apex acute, the base attenuate, the margin entire, sometimes ciliate, the upper and lower surfaces glabrous, the basal petiolate, the cauline sessile or petiolate. Flowers in capitula in a loose corymbiform array; involucre hemispheric; phyllaries in 2–3 series, lanceolate, to 1 cm long; receptacle ovoid or hemispheric; paleae obovate, 4–5 mm long, concave, each clasping a flower, the apex obtuse or acuminate, the outer surface of the tip glabrous or sparsely pubescent; ray flowers 7–14; corolla yellow, the lamina 1.2–3 cm long; disk flowers numerous; corolla proximally yellow-green, distally maroon, 3–4 mm long. Cypsela obpyramidal, 4-angled, 3–4 mm long; pappus coroniform or of connate or free scales ca. 1 mm long.

Wet flatwoods and cypress swamps. Occasional; eastern and central panhandle. Georgia and Florida. Summer–fall.

Rudbeckia mollis Elliott [*Mollis,* soft.] SOFTHAIR CONEFLOWER.

Rudbeckia mollis Elliott, Sketch Bot. S. Carolina 2: 453. 1823.

Erect annual, biennial, or perennial herb, to 1 m; stem striate-angled, pilose. Leaves with the blade oblong, oblong-lanceolate, or elliptic, 1.5–12 cm long, 0.5–4 cm wide, reduced upward, the apex acute or obtuse, the base cuneate or the upper ones auriculate, the margin entire or irregularly serrate, the upper and lower surfaces softly pilose or woolly, gland-dotted, the basal petiolate, the cauline sessile. Flowers in capitula, solitary or 2–12 in a loose corymbiform array; involucre hemispheric; phyllaries in 2–3 series, lanceolate, to 1.5 cm long, unequal, the outer surface pubescent, gland-dotted; receptacle conic or hemispheric; paleae obovate, 5–7 mm long, concave, each clasping a flower, the apex acute, the outer surface of the tip hirsute, gland-dotted; ray flowers 10–16; corolla yellow, the lamina 2–4 cm long; disk flowers numerous;

corolla proximally yellow-green, distally brown-purple, 3–4 mm long. Cypsela obpyramidal, 4-angled, 2–4 mm long; pappus absent or coroniform, less than 1 mm long.

Sandhills and open hammocks. Occasional; northern counties, Marion County. South Carolina, Georgia, Florida, and Alabama. Summer.

Rudbeckia nitida Nutt. [*Nitidus*, shining, in reference to the leaf surface.] SHINY CONEFLOWER; ST. JOHN'S SUSAN.

Rudbeckia nitida Nuttall, J. Acad. Nat. Sci. Philadelphia 7: 78. 1834.
Rudbeckia glabra de Candolle, Prodr. 5: 556. 1836.
Rudbeckia nitida Nuttall var. *longifolia* A. Gray, Syn. Fl. N. Amer. 1(2): 62. 1884. TYPE: FLORIDA: Manatee Co.: Manatee River, Jun 1878, *Garber 13* (lectotype: GH; isolectotypes: F. US). Lectotypified by P. B. Cox and Urbatsch, Castanea 59: 313. 1994.

Erect perennial herb, to 1.4 m; stem rhizomatous, striate-angled, glabrous. Leaves with the blade elliptic or lanceolate, 5–60 cm long, 2–9 cm wide, reduced upward, the apex acute, the base cuneate, the margin irregularly crenate, serrate, or entire, the upper and lower surfaces glabrous or sparsely pilose, the basal petiolate, the cauline petiolate or sessile. Flowers in capitula, solitary or several in loose corymbiform array; involucre hemispheric; phyllaries in 2–3 series, lanceolate, to 2 cm long, unequal; receptacle ovate or columnar; paleae obovate, 4–5 mm long, concave, each clasping a flower, the apex obtuse or acuminate, the outer surface of the tip glabrous or sparsely pubescent; ray flowers 8–15; corolla yellow, the lamina 1.2–6 cm long; disk flowers numerous; corolla proximally yellow-green, distally brown-purple or maroon, 3–4 mm long. Cypsela obpyramidal, 4-angled, 3–6 mm long; pappus coroniform, to 2 mm long.

Wet flatwoods. Occasional; northern and central peninsula, central panhandle. Georgia, Alabama, and Florida, also Louisiana. Summer–fall.

Rudbeckia nitida is listed as endangered in Florida (Florida Administrative Code, Chapter 5B-40).

Rudbeckia triloba L. [Three-lobed, in reference to some of the cauline leaves.] BROWNEYED SUSAN.

Rudbeckia triloba Linnaeus, Sp. Pl. 907. 1753.
Rudbeckia triloba Linnaeus var. *pinnatiloba* Torrey & A. Gray, Fl. N. Amer. 2: 309. 1842. *Rudbeckia pinnatiloba* (Torrey & A. Gray) Beadle, Bot. Gaz. 25: 277. 1898. TYPE: FLORIDA: s.d., *Chapman s.n.* (holotype: ?; isotypes: GH, NY).

Erect perennial herb, to 1.5 m; stem rhizomatous, pilose or glabrate. Leaves with the blade ovate or elliptic, 2–30 cm long, 1.5–8 cm wide, reduced upward, the apex acute or acuminate, the base truncate, rounded, or cuneate, the margin sometimes proximally 3- to 5-lobed, the ultimate margin irregularly serrate, the upper and lower surfaces hirsute or strigose, the basal petiolate, the cauline petiolate or sessile, sometimes clasping. Flowers in capitula in a loose paniculiform array; involucre hemispheric; phyllaries in 2–3 series, lanceolate to 1.5 cm long, unequal, the outer surface hirsute; receptacle conic or subhemispheric; paleae obovate, 5–7 mm long, concave, each clasping a flower, the apex cuspidate, the outer surface of the tip glabrous; ray flowers 8–15; corolla yellow or yellow-orange with a basal maroon spot, the lamina (0.8)1–3 cm long; disk flowers numerous; corolla proximally yellow-green, distally brown-purple, 3–4 mm long. Cypsela obpyramidal, 4-angled, 2–3 mm long; pappus coroniform, less than 1 mm long.

Disturbed sites. Rare; Levy County, central panhandle. Quebec south to Florida, west to Ontario, Minnesota, Iowa, Nebraska, Utah, Oklahoma, and Texas. Spring.

Rudbeckia triloba is listed as endangered in Florida (Florida Administrative Code, Chapter 5B-40).

EXCLUDED TAXA

Rudbeckia atrorubens Nuttall—Reported for Florida by Chapman (1860, 1883, 1897), the name misapplied to material of *R. graminifolia*. *Rudbeckia atrorubens* is the basionym of *Echinacea atrorubens* (Nuttall) Nuttall, a lower midwestern species, not found in Florida.

Rudbeckia maxima Nuttall—Reported for Florida by Chapman (1860, 1883, 1897). Excluded from Florida by Cox and Urbatsch (1994). No Florida material seen.

Sachsia Griseb. 1866.

Herbs. Leaves alternate, simple, pinnate-veined, epetiolate. Flowers in capitula in corymbiform or paniculiform arrays, radiate or disciform; receptacle epaleate; phyllaries present; ray or peripheral flowers, lacking a lamina, carpellate; petals 5, basally connate, actinomorphic or zygomorphic; disk flowers staminate or bisexual; petals 5, basally connate, actinomorphic; stamens 5, epipetalous, the filaments free, the anthers connate; ovary inferior, 2-carpellate, 1-loculate. Fruit a cypsela; pappus present.

A genus of 2 species; North America and West Indies. [Commemorates Ferdinand Gustav Julius von Sachs (1832–1897), German botanist.]

Selected reference: Nesom (2006v).

Sachsia polycephala Griseb. [From the Greek *poly,* many, and *cephale,* head, in reference to the many capitula in the inflorescence.] BAHAMA SACHSIA.

Sachsia polycephala Grisebach, Cat. Pl. Cub. 151. 1866. *Placus polycephalus* (Grisebach) M. Gómez de la Maza y Jiménez, Anales Soc. Esp. Hist. Nat. 19: 273. 1890.

Sachia bahamensis Urban, Symb. Antill. 3: 408. 1903.

Erect perennial herb, to 6 dm; stem sometimes rhizomatous, puberulent or glabrate, finely glandular, proximally sericeous-woolly. Leaves with the blade mostly proximal, obovate, oblanceolate, or spatulate, 2–12 cm long, 1–4 cm wide, much reduced and becoming bract-like upward, the apex rounded, obtuse, or acute, the base cuneate, the margin denticulate or coarsely and irregularly dentate with callus-tipped teeth, the upper surface glabrous, the lower surface slightly sericeous and glandular, sessile. Flowers in capitula in a loose corymbiform or paniculiform array; involucre cylindric-ovoid or urceolate, 2–3 mm wide; phyllaries in 5–6 series, ovate- or linear-lanceolate, 5–8 mm long, unequal, the outer surface slightly strigose at the apex, glandular; receptacle flat; ray or peripheral flowers in 1(3) series corolla whitish, the lamina less than 1 mm long or absent; disk flowers 6–18; corolla whitish, 3–4 mm long. Cypsela cylindric, ca. 2–3 mm long, 8- to 12-ribbed, strigillose; pappus of whitish, barbellate bristles, 3–4 mm long.

Pine rocklands. Rare; Miami-Dade and Monroe Counties. Florida; West Indies. Fall–spring.

Sachsia polycephala is listed as threatened in Florida (Florida Administrative Code, Chapter 5B-40).

Sclerolepis Cass. 1816. BOGBUTTON

Herbs. Leaves whorled, simple, pinnate-veined, epetiolate. Flowers in capitula, solitary, discoid, bisexual; receptacle epaleate; petals 5, basally connate, actinomorphic; stamens 5, epipetalous, the filaments free, the anthers connate; ovary inferior, 2-carpellate, 1-loculate. Fruit a cypsela; pappus present.

A monotypic genus; North America. [From the Greek *scleros*, hard, and *lepis*, scale, in reference to the hard pappus scales.]

Selected reference: Lamont (2006c).

Sclerolepis uniflora (Walter) Britton et al. PINK BOGBUTTON.

Ethulia uniflora Walter, Fl. Carol. 195. 1788. *Sclerolepis uniflora* (Walter) Britton et al., Prelim. Cat. 25. 1888.

Sparganophorus verticillatus Michaux, Fl. Bor.-Amer. 2: 95, t. 42. 1803. *Sclerolepis verticillata* (Michaux) Cassini, in Cuvier, Dict. Sci. Nat. 48: 155. 1827. TYPE: "Carolina ad Floridam."

Decumbent, erect, or sometimes floating perennial herb, to 3 dm (submerged to 6 dm); stem rhizomatous, usually creeping and mat-forming, striate-angled, glabrous. Leaves 4–6 per node, the blade linear, 7–23 mm long, to 2 mm wide, the apex acute, callus-tipped, the base cuneate, the margin entire, the upper and lower surfaces glabrous, sometimes sparsely gland-dotted, sessile. Flowers in a solitary terminal capitulum, the peduncle 2–6 cm long; involucre campanulate or subhemispheric, to 1 cm wide; phyllaries numerous in 2–3 series, linear or lanceolate, 2–4 mm long, unequal, the margin minutely serrulate, the outer surface glabrous; receptacle conic or hemispheric; flowers numerous; corolla pink-lavender, 2–3 mm long, the outer surface sparsely stipitate-glandular, the linear-clavate style branches much exserted. Cypsela prismatic, 2–3 mm long, 5-ribbed, gland-dotted; pappus of 5 broadly oblong scales, ca. 1 mm long, the margin crenulate, the apex obtuse, erose.

Bogs, cypress swamps, and wet flatwoods. Occasional; northern counties, central peninsula. New Hampshire, Massachusetts, and Rhode Island, New Jersey south to Florida, west to Louisiana. Summer–fall.

Senecio L. 1753. RAGWORT

Herbs. Leaves alternate, simple, pinnate-veined, petiolate or epetiolate. Flowers in capitula in corymbiform arrays, radiate or discoid; calyculi present; ray flowers (if present) carpellate;

petals 5, basally connate; zygomorphic; disk flowers bisexual; petals 5, basally connate, actinomorphic; stamens 5, epipetalous, the filaments free, the anthers connate; ovary inferior, 2-carpellate, 1-loculate. Fruit a cypsela; pappus present.

A genus of about 1,250 species; nearly cosmopolitan. [*Senex*, old person, in reference to the white pappus bristles resembling the white hair of an elderly person.]

Studies in the past several decades have shown the concept of *Senecio* in the broad sense to be an assemblage of several genera that have subsequently been removed (e.g., *Packera*). Work is ongoing and further revision will likely occur.

Selected reference: Barkley (1978, 2006e).

1. Capitula radiate .... **S. brasiliensis**
1. Capitula discoid .... **S. vulgaris**

Senecio brasiliensis (Spreng.) Less. var. **tripartitus** (DC.) Baker [Of Brazil; divided into three parts, in reference to the leaves.] HEMPLEAF RAGWORT.

Senecio tripartitus de Candolle, Prodr. 6: 418. 1837. *Senecio brasiliensis* (Sprengel) Lessing var. *tripartitus* (de Candolle) Baker, in Martius, Fl. Bras. 6(3): 322. 1884.
Senecio cannabinifolius Hooker & Arnott, J. Bot. (Hooker) 3: 341. 1841.

Erect annual herb, to 8 dm; stem striate-angled, glabrous. Leaves with the blade narrowly ovate, 4.5–8 cm long, 0.4–7 cm wide, the apex acute, mucronate, the base cuneate, the margin pinnatisect, irregularly dentate toward the apex, the upper surface glabrescent, the lower surface gray-woolly pubescent, sessile, clasping. Flowers in capitula in a loose corymbiform array, the peduncle 1–2 cm long; calyculi of 2–4(6) bractlets, 3–4 mm long, the apex brownish, the margin ciliate; involucre cylindric-campanulate; phyllaries ca. 21 in 2 series, lanceolate, 9–10 mm long, the apex brownish; receptacle flat, foveolate; ray flowers 10–12; corolla yellow, the lamina 14–15 mm long; disk flowers 10–12; corolla yellow, 7–8 mm long. Cypsela cylindric, ca. 3 mm long, 8-ribbed, pubescent; pappus of numerous white barbellate bristles, 3–4 mm long.

Disturbed sites. Rare; Escambia County. Not recently collected. Florida and Alabama; South America. Native to South America. Summer.

Senecio vulgaris L. [Common.] COMMON GROUNDSEL.

Senecio vulgaris Linnaeus, Sp. Pl. 867. 1753. *Senecio semperflorens* Stokes, Bot. Mat. Med. 4: 210. 1812, nom. illegit.

Erect annual herb, to 6 dm; stem striate-angled, glabrous or sparsely and irregularly tomentose when young. Leaves with the blade ovate or oblanceolate, 2–10 cm long, 0.5–2(4) cm wide, reduced upward, the apex rounded, the base tapered, the margin irregularly lobed, the ultimate margin secondarily, irregularly dentate or denticulate, the upper and lower surfaces glabrous, slightly glaucous, the proximal winged-petiolate, clasping, the distal sessile, auriculate-clasping. Flowers in capitula in a loose corymbiform array; calyculi of 2–4(6) bractlets, 1–2 mm long, the apex black; involucre cylindric-campanulate; phyllaries ca. 21 in 2 series, linear, 4–6 mm long, the apex black; receptacle flat, foveolate; ray flowers absent; disk flowers numerous;

corolla yellow, 5–7 mm long. Cypsela prismatic, ca. 2 mm long, 5-ribbed, sparsely pubescent; pappus of numerous white barbellate bristles, 3–4 mm long.

Disturbed sites. Rare; Flagler, Lake, Gadsden, Okaloosa, and Escambia Counties. Nearly throughout North America. Nearly cosmopolitan. Native to Europe, Africa, and Asia. Spring–fall.

EXCLUDED TAXON

Senecio tomentosus Michaux—Reported for Florida by Chapman (1860, 1883, 1897), Small (1903, 1913a, 1933), Radford et al. (1964, 1968), Correll and Johnston (1970), Cronquist (1980), Godfrey and Wooten (1982), Clewell (1985), Wunderlin (1998), and Trock (2006b). All misapplied to Florida material of *Packera anonyma. Senecio tomentosus* Michaux (1803; non Salisbury, 1796) is a later homonym and thus illegitimate.

Sericocarpus Nees 1832. WHITETOP ASTER

Herbs. Leaves alternate, simple, pinnate-veined, petiolate or epetiolate. Flowers in capitula in corymbiform arrays, radiate; receptacle epaleate; ray flowers carpellate; petals 5, basally connate, zygomorphic; disk flowers bisexual; petals 5, basally connate, actinomorphic; stamens 5, epipetalous, the filaments free, the anthers connate; ovary inferior, 2-carpellate, 1-loculate. Fruit a cypsela; pappus present.

A genus of 5 species; North America. [From the Greek *sericos,* silky, and *carpos,* fruit, in reference to the pubescent cypselae.]

Selected reference: Semple and Leonard (2006).

1. Phyllaries spreading at the apex; leaves generally toothed, the basal and proximal cauline persistent at flowering .... **S. asteroides**
1. Phyllaries erect; leaves essentially entire, the basal and proximal cauline withering at flowering .... **S. tortifolius**

Sericocarpus asteroides (L.) Britton et al. [Resembling the genus *Aster.*] TOOTHED WHITETOP ASTER.

Conyza asteroides Linnaeus, Sp. Pl. 861. 1753. *Aster conyzoides* Willdenow, Sp. Pl. 3: 2043. 1803, nom. illegit. *Aster marilandicus* Michaux, Fl. Bor.-Amer. 2: 108. 1803, nom. illegit. *Sericocarpus conyzoides* Nees von Esenbeck, Gen. Sp. Aster. 150. 1832, nom. illegit. *Sericocarpus asteroides* (Linnaeus) Britton et al., Prelim. Cat. 26. 1888. *Aster asteroides* (Linnaeus) MacMillan, Metasp. Minnesota Valley 524. 1891. *Aster paternus* Cronquist, Bull. Torrey Bot. Club 74: 149. 1947.

Erect perennial herb, to 6.5 dm; stem striate-angled, puberulent. Leaves with the blade oblanceolate, obovate, or spatulate, 1–15 cm long, 0.5–3 cm wide, reduced upward, the apex acuminate or acute, the base cuneate, the margin serrate, the distal becoming entire, the upper and lower surfaces puberulent, the basal and proximal cauline winged-petiolate, the mid and distal cauline sessile. Flowers in capitula in sessile clusters in a corymbiform array; involucre cylindric, 4–7 mm wide; phyllaries numerous in 3–4 series, ovate, 2–5 mm long, unequal, the apex spreading, the outer surface glabrate; receptacle slightly convex, pitted; ray flowers 3–7; corolla white, the lamina 2–6 mm long; disk flowers 9–20; corolla white, 4–6 mm long. Cypsela fusiform-obconic, to 2 mm long, strigose; pappus of (2)3(4) series, of whitish or tan barbellate bristles, 4–5 mm long.

Hammocks. Occasional; western panhandle. Maine south to Florida, west to Michigan, Ohio, Kentucky, Tennessee, and Mississippi. Summer–fall.

Sericocarpus tortifolius (Michx.) Nees [With twisted leaves.] WHITETOP ASTER; DIXIE ASTER.

Aster tortifolius Michaux, Fl. Bor.-Amer. 2: 109. 1803. *Sericocarpus tortifolius* (Michaux) Nees von Esenbeck, Gen. Sp. Aster. 151. 1832.

Aster collinsii Nuttall, J. Acad. Nat. Sci. Philadelphia 7: 82. 1834. *Sericocarpus collinsii* (Nuttall) Nuttall, Trans. Amer. Philos. Soc., ser. 2. 7: 302. 1841. *Sericocarpus tortifolius* (Michaux) Nees von Esenbeck var. *collinsii* (Nuttall) Torrey & A. Gray, Fl. N. Amer. 2: 103. 1841. *Sericocarpus bifoliatus* Porter var. *collinsii* (Nuttall) S. F. Blake, Proc. Amer. Acad. Arts 51: 515. 1916. TYPE: FLORIDA: s.d., *Ware s.n.* (Lectotype: BM). Lectotypified by Blake (1916: 515).

Sericocarpus bifoliatus Porter, in Porter & Britton, Mem. Torrey Bot. Club 5: 322. 1894. *Aster bifoliatus* (Porter) Ahles, J. Elisha Mitchell Sci. Soc. 80: 173. 1964.

Sericocarpus acutisquamosus Small, Fl. S.E. U.S. 1206, 1339. 1903. TYPE: FLORIDA: Columbia Co.: Lake City, 29–31 Aug 1895, *Nash 2495* (holotype: NY).

Erect perennial herb, to 1.2 m; stem striate-angled, puberulent. Leaves with the blade obovate, 1–4 cm long, 3–10 mm wide, the basal and proximalmost cauline withering at flowering, the apex acuminate, slightly cuspidate, the base cuneate, the margin entire, the upper and lower surfaces puberulent, resinous, sessile. Flowers in capitula in a corymbiform array, the peduncle 5–10 mm long, the bracts lanceolate or ovate; involucre cylindric, 5–8 mm wide; phyllaries numerous in 4–5 series, ovate, 2–6 mm long, unequal, the apex erect, the outer surface puberulent; receptacle slightly convex, pitted; ray flowers 2–5; corolla white, the lamina 3–6 mm long; disk flowers 6–11; corolla white, 5–8 mm long. Cypsela fusiform-obconic, 1–3 mm long, strigose; pappus of (2)3(4) series, of whitish or tan barbellate bristles, 6–8 mm long.

Sandhills, flatwoods, scrub, oak hammocks, and disturbed sites. Frequent; nearly throughout. North Carolina south to Florida, west to Louisiana. Summer–fall.

Silphium L. 1753. ROSINWEED

Herbs. Leaves alternate, opposite, or whorled, simple, pinnate-veined, petiolate or epetiolate. Flowers in capitula in paniculiform or racemiform arrays, radiate; receptacle paleate; ray flowers carpellate; petals 5, basally connate, zygomorphic; disk flowers staminate; petals 5, basally connate, actinomorphic; stamens 5, epipetalous, the filaments free, the anthers connate; ovary inferior, 2-carpellate, 1-loculate. Fruit a cypsela; pappus present or absent.

A genus of 13 species; North America. [From the Greek *silphion*, an unknown resin-producing plant from the Middle East used as a seasoning and medicine, perhaps a *Ferula* (Apiaceae) species.]

Selected reference: Clevinger (2006).

1. Cauline leaves well developed .... **S. asteriscus**
1. Cauline leaves much reduced .... **S. compositum**

Silphium asteriscus L. [Starlike, in reference to the flowering capitula, an old generic name.] STARRY ROSINWEED.

Silphium asteriscus Linnaeus, Sp. Pl. 920. 1753. *Silphium reticulatum* Moench, Methodus 607. 1794, nom. illegit.
Silphium dentatum Elliott, Sketch Bot. S. Carolina 2: 468. 1823. *Silphium asteriscus* Linnaeus var. *dentatum* (Elliott) Chapman, Fl. South. U.S. 221. 1860.
Silphium scaberrimum Elliott, Sketch Bot. S. Carolina 2: 466. 1823.
Silphium gracile A. Gray, Proc. Amer. Acad. Arts 8: 653. 1873.
Silphium asteriscus Linnaeus var. *angustatum* A. Gray, Syn. Fl. N. Amer. 1(2, suppl.): 449. 1886. *Silphium angustum* Small, Fl. S.E. U.S. 1244, 1340. 1903. *Silphium dentatum* Elliott var. *angustatum* (A. Gray) L. M. Perry, Rhodora 39: 293. 1937. TYPE: FLORIDA: Gadsden Co.: Chattahoochee, 3 Sep 1884, *Curtiss s.n.* (holotype: GH).
Silphium simpsonii Greene, Pittonia 4: 44. 1899. *Silphium asteriscus* Linnaeus var. *simpsonii* (Greene) Clevinger, Novon 14: 275. 2004. TYPE: FLORIDA: Manatee Co.: Palma Sola, 8 Jul 1890, *Simpson 81* (holotype: US).

Erect perennial herb, to 2 m; stem terete, striate, hirsute, hispid, scabrous, or glabrous. Leaves alternate, opposite, or whorled, the blade lanceolate or ovate, 1.5–25 cm long, 0.5–5 cm wide, slightly reduced upward, the apex acuminate, acute, or obtuse, the base cuneate or rounded, the margin irregularly dentate, serrate, or entire, the upper and lower surfaces hirsute, scabrous, or glabrous. Flowers in capitula in a paniculiform or racemiform array; involucre campanulate or hemispheric, 1–2 cm wide; phyllaries 12–26 in 2–3 series, ovate, the outer appressed or reflexed, the apex acute or obtuse, the outer surface hispid or scabrous, slightly glaucous; receptacle slightly convex; paleae oblong or lanceolate; ray flowers 8–21; corolla yellow, the lamina 1.5–2 cm long; disk flowers numerous; corolla yellow, ca. 5 mm long. Cypsela obovate, 6–15 mm long, laterally flattened, slightly winged, brown; pappus absent or of 2 awns to 5 cm long.

Flatwoods. Occasional; northern counties, central panhandle. New York south to Florida, west to Illinois, Missouri, Oklahoma, and Alabama. Summer–fall.

Silphium compositum Michx. [Compound, in reference to the deeply divided or pinnatifid leaves.] KIDNEYLEAF ROSINWEED.

Silphium compositum Michaux, Fl. Bor.-Amer. 2: 145. 1803. TYPE: "Carolina ad Floridam," s.d., *Michaux s.n.* (holotype: P).
Silphium compositum Michaux var. *ovatifolium* Torrey & A. Gray, Fl. N. Amer. 2: 277. 1842. *Silphium compositum* Michaux var. *michauxii* Torrey & A. Gray, Fl. N. Amer. 2: 276. 1842, nom. inadmiss. *Silphium ovatifolium* (Torrey & A. Gray) Small, Fl. S.E. U.S. 1242, 1340. 1903. *Silphium compositum*

Michaux subsp. *ovatifolium* (Torrey & A. Gray) C. R. Sweeney & T. R. Fisher, in C. R. Sweeney, Ohio J. Sci. 70: 232. 1970. TYPE: FLORIDA: s.d., *Chapman s.n.* (holotype: NY).
Silphium venosum Small, Bull. Torrey Bot. Club 25: 478. 1898. *Silphium compositum* Michaux subsp. *venosum* (Small) C. R. Sweeney & T. R. Fisher, in C. R. Sweeney, Ohio J. Sci. 70: 232. 1970. *Silphium compositum* Michaux var. *venosum* (Small) Kartez & Gandhi, Phytologia 71: 272. 1991.

Erect perennial herb, to 2.5 m; stem terete, striate, glabrous, sometimes glaucous. Leaves alternate, the blade cordate, elliptic, hastate, ovate, reniform, or sagittate, 2–35 cm long, 0.5–52 cm wide, the cauline much reduced, the proximal sometimes pinnately lobed, the apex acute, the base oblique, truncate, cuneate, cordate, sagittate, or hastate, the ultimate margin irregularly toothed, the upper and lower surfaces glabrous, scabrous, or hispid, the basal petiolate, the cauline petiolate or sessile. Flowers in capitula in a paniculiform or racemiform array; involucre campanulate or hemispheric, 1–2 cm wide; phyllaries 11–18 in 2–3 series, ovate, the outer appressed or reflexed, the apex obtuse or cuspidate, the outer surface sparsely scabrous, slightly glaucous; receptacle slightly convex; paleae oblong or lanceolate; ray flowers 6–12; corolla yellow, the lamina 1.5–2 cm long; disk flowers numerous; corolla yellow, ca. 5 mm long. Cypsela obovate, 6–12 mm long, laterally flattened, slightly winged, brown; pappus of 2 awns 1–4 cm long.

Sandhills. Occasional; northern counties, central peninsula. Virginia and West Virginia south to Florida, west to Tennessee and Alabama. Summer–fall.

Smallanthus Mack. ex Small 1933.

Herbs. Leaves opposite or alternate, simple, pinnipalmate-veined, petiolate. Flowers in capitula, solitary or in corymbiform arrays, radiate; receptacle paleate; ray flowers carpellate; petals 5, basally connate, zygomorphic; disk flowers staminate; petals 5, basally connate, actinomorphic; stamens 5, epipetalous, the filaments free, the anthers connate; ovary inferior, 2-carpellate, 1-loculate. Fruit a cypsela; pappus absent.

A genus of about 24 species; North America, West Indies, Mexico, Central America, and South America. [Commemorates John Kunkel Small (1869–1938), American botanist, and from the Greek, *anthos,* flower.]

Polymniastrum Lam., 1823; Small, 1913, nom. illegit.

Selected reference: Strother (2006aa).

Smallanthus uvedalia (L.) Mack ex Small [Commemorates Robert Uvedale (1642–1722), English teacher and horticulturist, who grew the plant in his garden.] HAIRY LEAFCUP.

Osteospermum uvedalia Linnaeus, Sp. Pl. 923. 1753. *Polymnia uvedalia* (Linnaeus) Linnaeus, Sp. Pl., ed. 2. 1303. 1763. *Polymnia uvedalia* (Linnaeus) Linnaeus var. *genuina* S. F. Blake, Rhodora 19: 47. 1917, nom. inadmiss. *Polymniastrum uvedalia* (Linnaeus) Small, in Small & J. J. Carter, Fl. Lancaster Co. 302, 319. 1913. *Smallanthus uvedalia* (Linnaeus) Mackenzie ex Small, Man. S.E. Fl. 1406, 1509. 1933.
Polymnia uvedalia (Linnaeus) Linnaeus var. *densipilis* S.F. Blake, Rhodora 19: 48. 1917.

Polymnia uvedalia (Linnaeus) Linnaeus var. *floridana* S. F. Blake, Rhodora 19: 48. 1917. TYPE: FLORIDA: Brevard Co.: Indian River region, 2 Dec 1902, *Fredholm 5626* (holotype: GH; isotype: US).

Erect or ascending perennial herb, to 1–3 m; stem striate-angled, sparsely hirtellous or puberulent with gland-tipped trichomes distally, glabrate or glabrous proximally. Leaves opposite proximally, alternate distally, the blade deltate, obovate, or ovate, 10–35(60) cm long, 10–35 cm wide, the apex acute or acuminate, the base cuneate or long-tapering, the margin 3–5 pinnipalmately lobed, the ultimate margin irregularly dentate or denticulate, the upper surface sparsely scabrous, the lower surface hirtellous, pilosulous, or puberulent, especially on the veins, gland-dotted, the petiole 3–12 cm long, often winged, the base clasping the stem. Flowers in capitula, solitary or 2–5 in a corymbiform array; involucre hemispheric, 8–15 mm wide; phyllaries 4–6 in 1 series, lanceolate, elliptic, or ovate, 1–2 cm long, the apex acute or obtuse; receptacle flat or slightly convex; paleae obovate or spatulate, 6–14 mm long, those of the ray flowers longer than the disk flowers, concave and surrounding the flower, scarious; ray flowers 7–13; corolla yellow, the lamina 1–3 cm long; disk flowers numerous; corolla yellow, 3–4 mm long. Cypsela obovoid, 5–6 mm long, slightly compressed, finely ribbed or striate, dark brown; pappus absent.

Moist hammocks. Frequent; northern and central peninsula, central and western panhandle. New York south to Florida, west to Michigan, Illinois, Missouri, Kansas, Oklahoma, and Texas; West Indies, Mexico, and Central America. Native to North America, Mexico, and Central America. Summer–fall.

Solidago L. 1753. GOLDENROD

Herbs. Leaves alternate, simple, pinnate- or pinnipalmate-nerved, petiolate or epetiolate. Flowers in capitula in racemiform, paniculiform, or corymbiform arrays; radiate; receptacle epaleate; ray flowers (if present) carpellate; petals 5, basally connate, zygomorphic; disk flowers bisexual; petals 5, basally connate, actinomorphic; stamens 5, epipetalous, the filaments free, the anthers connate; ovary inferior, 2-carpellate, 1-loculate. Fruit a cypsela; pappus present.

A genus of about 100 species; nearly cosmopolitan. [*Solidus*, whole, and *-ago*, resembling or becoming, in reference to its healing properties.]

Solidago has long been a confusing complex of species because of the similarity and seeming intergrading of taxa. We follow Semple and Cook (2006) with some slight differences.

Selected reference: Semple and Cook (2006).

1. Capitula in small clusters in the axil of the distal cauline leaves .... **S. caesia**
1. Capitula in corymbiform, paniculiform, or racemiform arrays.
    2. Leaf bases auriculate and clasping completely around the stem .... **S. auriculata**
    2. Leaf bases not auriculate, if clasping, then not completely surrounding the stem.
        3. Leaves strongly rugose-veined.. **S. rugosa**
        3. Leaves not strongly rugose-veined.
            4. Cauline leaves closely appressed to the stem, entire, linear, and evidently smaller than the basal leaves; plant glabrous.

5. Cauline leaves abruptly reduced upward; inflorescence elongate, narrow, naked; stoloniform rhizomes present .... **S. stricta**
5. Cauline leaves gradually reduced upward; inflorescence paniculiform, often leafy at the base; stoloniform rhizomes absent .... **S. sempervirens**

4. Cauline leaves divergent, or if appressed, then the stem pubescent or the leaves not linear and evidently small.
6. Panicle branches not recurved or secund.
7. Middle cauline leaves reduced, the leaves mostly basal .... **S. puberula**
7. Middle cauline leaves not reduced; leaves mostly cauline .... **S. petiolaris**
6. Panicle branches recurved, secund.
8. Stem below the inflorescence glabrous or nearly so.
9. Basal leaves much larger than the middle and upper cauline leaves.
10. Leaves scabrous on the upper surface .... **S. patula**
10. Leaves glabrous on the upper surface .... **S. gracillima**
9. Basal leaves not evidently larger than the middle and upper cauline leaves.
11. Leaves lanceolate or linear-lanceolate.
12. Stem below the inflorescence glabrous and usually glaucous .... **S. gigantea**
12. Stem below the inflorescence sparsely pubescent, never glaucous .... **S. leavenworthii**
11. Leaves ovate, elliptic, or elliptic-lanceolate.
13. Leaves scabrous on the lower surface .... **S. ulmifolia**
13. Leaves glabrous or sparsely pilose on the lower surface.
14. Basal leaves present at flowering, long-pedunculate; cauline leaves rather abruptly contracted to the petiole .... **S. arguta**
14. Basal leaves usually absent at flowering, or if present, then short-petiolate; cauline leaves cuneate at the base .... **S. latissimifolia**
8. Stem below the inflorescence distinctly pubescent (sometimes with trichomes only in vertical lines).
15. Leaves with the margin entire .... **S. odora**
15. Leaves with the margin at least with a few teeth.
16. Leaves much reduced upward, the larger leaves basal.
17. Upper cauline leaves elliptic-ovate; ray flowers absent or 1–2 .... **S. brachyphylla**
17. Upper cauline leaves linear-spatulate; ray flowers 3–9 .... **S. nemoralis**
16. Leaves not much reduced upward, the cauline leaves well developed.
18. Leaves evidently 3-nerved .... **S. canadensis**
18. Leaves not evidently 3-nerved.
19. Stem (and often the leaves, especially on the veins of the lower surface) short-hirsute; leaves ovate-lanceolate or elliptic-oblong .... **S. fistulosa**
19. Stem and leaves short-pubescent; leaves linear or narrowly elliptic.
20. Leaves twisted, the apex usually obtuse or rounded, the larger leaves with low marginal serrations; ray flowers 2–8 .... **S. tortifolia**
20. Leaves not twisted, the apex acute or acuminate, the larger leaves with sharp marginal serrations; ray flowers 8–15 .... **S. leavenworthii**

Solidago arguta Aiton var. **caroliniana** A. Gray [*Argutus,* sharp-toothed, in reference to the leaf margin; of Carolina.] CAROLINA GOLDENROD.

Solidago arguta Aiton var. *caroliniana* A. Gray, Syn. Fl. N. Amer. 1(2): 155. 1884. *Solidago boottii* Hooker var. *caroliniana* (A. Gray) Cronquist, Rhodora 49: 79. 1947. *Solidago arguta* Aiton subsp. *caroliniana* (A. Gray) G. H. Morton, Phytologia 28: 1. 1974.

Solidago boottii Hooker var. *yadkinensis* Porter, Mem. Torrey Bot. Club 3: 27. 1892; Bull. Torrey Bot. Club 19: 129. 1892. *Solidago yadkinensis* (Porter) Small, Bull. Torrey Bot. Club 22: 368. 1895.

Solidago tarda Mackenzie, in Small, Man. S.E. Fl. 1355, 1509. 1933.

Erect perennial herb, to 1.2 m; stem striate-angled, proximally glabrous, distally strigose in the floral array. Basal and proximal leaves with the blade broadly ovate, 10–30 cm long, 3–10 cm wide, the apex acute or acuminate, the base tapering abruptly to a winged petiole, the margin serrate, the upper and lower surfaces glabrous or slightly scabrous, the petiole clasping, the mid and distal cauline with the blade lanceolate, 5–7 cm long, 1–1.5 cm wide, reduced upward, the margin distally becoming entire, sessile, clasping. Flowers in capitula in an open paniculiform array with recurved branches, secund, the peduncle 2–3 mm long, the bracteoles oblong-lanceolate; involucre campanulate; phyllaries numerous in 3–4 series, linear-oblong, 3–5 mm long, unequal, the margin ciliate, the apex obtuse, the outer surface glabrous; receptacle slightly convex, pitted; ray flowers 2–8; corolla yellow, the lamina 4–5 mm long; disk flowers 8–20; corolla yellow, ca. 4 mm long. Cypsela obconic or cylindric, ca. 2 mm long, somewhat compressed, the ribs 8–10, distinctly ridged, distally strigillose; pappus of an outer series of short, setiform scales and 2 inner series of numerous barbellate bristles ca. 3 mm long.

Wet to moderately dry hammocks. Occasional; northern counties, central peninsula. Delaware, Maryland, and Virginia south to Florida, west to Missouri, Arkansas, and Louisiana. Fall.

Solidago auriculata Shuttlew. ex S. F. Blake [*Auriculatus,* with ear-like appendages, in reference to the leaf base.] EARED GOLDENROD.

Solidago amplexicaulis Torrey & A. Gray ex A. Gray, Syn. Fl. N. Amer. 1(2): 153. 1884; non M. Martens, 1841. *Solidago auriculata* Shuttleworth ex S. F. Blake, J. Wash. Acad. Sci. 21: 326. 1931. *Solidago notabilis* Mackenzie, in Small, Man. S.E. Fl. 1353, 1509. 1933, nom. illegit.

Erect or ascending perennial herb, to 1.5 m; stem with short rhizomes, striate-angled, velutinous or loose hirsutulous. Basal and proximal cauline leaves with the blade ovate, 3.5–12 cm long, 3–7 cm wide, the apex acute or acuminate, the base cordate, the margin serrate, the upper surface sparsely scabrous or strigose, the lower surface pilose on the nerves, the petiole winged, to 10 cm long, clasping, the midcauline and distal similar, the blade ovate, 2–4 cm long, 1–2 cm wide, the petiole shorter, broadly winged, auriculate-clasping, the distal reduced to 1 cm long, the base auriculate, sessile clasping. Flowers in capitula in an open paniculiform array with relatively few arching branches, secund, the peduncle 1–3 mm, the bracteoles linear-lanceolate; involucre campanulate; phyllaries numerous in 2–3 series, lanceolate, 3–4(5) mm long, unequal, the apex acute or acuminate, the outer surface sparsely short-strigose; receptacle slightly convex, pitted; ray flowers 1–3; corolla yellow, the lamina 1–2 mm long; disk flowers 4–8; corolla yellow, 3–4 mm long. Cypsela obconic or cylindric, 2–3 mm long, somewhat compressed, the ribs 8–10, distinctly ridged, short strigose; pappus of an outer series of short, setiform scales and 2 inner series of numerous barbellate bristles ca. 2 mm long.

Bluff forests and calcareous hammocks. Rare; Gadsden, Franklin, and Jackson Counties. South Carolina and Tennessee south to Florida, west to Oklahoma and Texas. Summer–fall.

Solidago brachyphylla Chapman ex Torr. & A. Gray [From the Greek, *brachy*, short, and *phyllon*, leaf in reference to the short leaves.] DIXIE GOLDENROD.

Solidago brachyphylla Chapman ex Torrey & A. Gray, Fl. N. Amer. 2: 218. 1842. *Solidago boottii* Hooker var. *brachyphylla* (Chapman ex Torrey & A. Gray), Proc. Amer. Acad. Arts 17: 195. 1882. TYPE: FLORIDA.

Erect perennial herb, to 1.3 m; stem with short, caudex-like rhizomes, striate-angled, strigose-puberulent. Basal leaves with the blade oblanceolate, spatulate, ovate, or rounded, 2–4 cm long, the apex acute or obtuse, the margin irregularly serrate, the upper and lower surfaces strigose-puberulent, the petiole 3–5 cm long, winged, clasping, the cauline with the blade elliptic, lanceolate, or ovate, 2.5–5(6.5) cm long, 1–2.5 cm wide, the petiole ca. 1 mm long, clasping, the distal (array branches) much reduced to ca. 5 mm long, sessile. Flowers in capitula in an open paniculiform array, the branches divaricate, ascending to recurved, weakly or strongly secund, the peduncle 2–3 mm long, the bracteoles linear; involucre narrowly campanulate; phyllaries numerous in 3–4 series, lanceolate, 3–5 mm long, unequal, the margin ciliate, the apex acute, the outer surface glabrous; receptacle slightly convex, pitted; ray flowers absent or rarely 1–2; corolla yellow, 2–3 mm long; disk flowers 4–8; corolla yellow, 3 mm long. Cypsela obconic or cylindric, ca. 3 mm long, somewhat compressed, the ribs 8–10, short-strigose; pappus of an outer series of short, setiform scales and 2 inner series of numerous barbellate bristles ca. 2 mm long.

Bluff forests. Occasional; northern peninsula, central and western panhandle. South Carolina south to Florida, west to Mississippi. Summer–fall.

Solidago caesia L. [*Caesius*, lavender-blue, in reference to the glaucous stem.] BLUESTEM GOLDENROD; WREATH GOLDENROD.

Solidago caesia Linnaeus, Sp. Pl. 879. 1753. *Leioligo caesia* (Linnaeus) Rafinesque, Fl. Tellur. 2: 42. 1837. *Aster caesius* (Linnaeus) Kuntze, Revis. Gen. Pl. 1: 317. 1891.
Solidago caesia Linnaeus var. *zedia* R. E. Cook & Semple, Sida 21: 221. 2004.

Erect perennial herb, to 8(10) dm; stem with woody, caudex-like rhizomes, glabrous, blue or purple glaucous, glabrous or slightly strigose in the floral array. Basal leaves withering by flowering, the proximal to midcauline with the blade lanceolate, 5–10(15) cm long, 0.8–2(3) cm wide, the apex acuminate, the base cuneate, the margin irregularly serrate, the upper surface sparsely pubescent, the lower surface sparsely pubescent or glabrous, subsessile or sessile, the distal narrowly lanceolate, (2)4–7(10) cm long, 5–11(15) cm wide, the margin irregularly serrate or entire, the upper and lower surfaces sparsely pubescent or glabrous, sessile. Flowers in capitula in terminal and axillary racemiform-paniculiform clusters, secund, the peduncle 2–10 mm long, the bracteoles linear; involucre narrowly campanulate; phyllaries numerous in 3 series, linear-oblong or lanceolate, 2–6 mm long, unequal, the apex acute or obtuse, the outer surface glabrous; receptacle slightly convex, pitted; ray flowers 1–6; corolla yellow, the lamina 2–3(4) mm long; disk flowers 3–6(9); corolla yellow, 2–3(4) mm long. Cypsela obconic or cylindric, 1–2 mm long, somewhat compressed, the ribs 8–10, strigose; pappus of an outer series of short, setiform scales and 2 inner series of numerous barbellate bristles 2–3 mm long.

Bluff forests. Occasional; Suwannee County, central and western panhandle. Quebec south to Florida, west to Ontario, Wisconsin, Iowa, Missouri, Oklahoma, and Texas. Summer–fall.

Solidago canadensis L. var. **scabra** (Muhl. ex Willd.) Torr. & A. Gray [Of Canada; *scaber,* rough to the touch, in reference to the leaf surface.] CANADA GOLDENROD.

Solidago altissima Linnaeus, Sp. Pl. 878. 1753. *Doria altissima* (Linnaeus) Lunell, Amer. Midl. Naturalist 5: 43. 1917. *Solidago canadensis* Linnaeus subsp. *altissima* (Linnaeus) Bolos & Vigo, Collect. Bot (Barcelona) 14: 102. 1983.
Solidago hirsutissima Miller, Gard. Dict., ed. 8. 1768.
Solidago scabra Muhlenberg ex Willdenow, Sp. Pl. 3: 2059. 1803. *Solidago canadensis* Linnaeus var. *scabra* (Muhlenberg ex Willdenow) Torrey & A. Gray, Fl. N. Amer. 2: 224. 1842.

Erect perennial herb, to 2 m; stem with short, creeping rhizomes, striate-angled, pubescent throughout or sometimes proximally glabrescent. Basal leaves withering at flowering, the proximal and midcauline with the blade lanceolate, 9.5–15 cm long, 1.5–2 cm wide, the apex acute or acuminate, the base cuneate, the margin entire or irregularly serrate, the upper and lower surfaces sparsely strigose, short-petiolate or sessile, clasping, the mid to distal smaller and reduced upward, the margin irregularly serrate, serrulate, or entire, the upper surface scabrous, the lower surface strigillose, villous-strigillose along the veins, sessile, clasping. Flowers in capitula in a pyramidal paniculiform array, the branches divergent and recurved or ascending, secund, the peduncle 1–4 mm long, bracteoles linear; involucre narrowly campanulate; phyllaries numerous in 3 series, lanceolate or linear-lanceolate, 3–4 mm long, unequal, the margin minutely stipitate-glandular, the apex acute or obtuse, the outer surface glabrous; receptacle slightly convex, pitted; ray flowers (5)8–13(17); corolla yellow, the lamina 1–2 mm long; disk flowers (2)3–6(9); corolla yellow, 2–4 mm long. Cypsela obconic, ca. 1 mm long, somewhat compressed, the ribs 8–10, strigillose; pappus of an outer series of short, setiform scales and 2 inner series of numerous barbellate bristles 2–3 mm long.

Hammocks and disturbed sites. Frequent; northern counties, Polk and Lee Counties. Nearly throughout North America; Europe. Native to North America. Fall.

Solidago fistulosa Mill. [*Fistular,* hollow throughout, but closed at the ends, in reference to the stem.] PINEBARREN GOLDENROD.

Solidago fistulosa Miller, Gard. Dict., ed. 8. 1768. *Aster fistulosus* (Miller) Kuntze, Revis. Gen. Pl. 1: 316. 1891.
Solidago pilosa Walter, Fl. Carol. 207. 1788; non Miller, 1768.
Solidago aspericaulis A. H. Moore, Rhodora 16: 54. 1914. TYPE: FLORIDA: Volusia Co.: Orange City, 12 Oct 1913, *Tidestrom* 7051 (holotype: NA; isotypes: GH, L, US).

Erect perennial herb, to 1.5 m; stem with elongate, creeping rhizomes, striate-angled, short-hirsute, at least distally. Basal and proximal cauline usually withering by flowering, the midcauline with the blade with the blade ovate-lanceolate or elliptic-oblong, 3.5–12 cm long, 1–5 cm wide, reduced upward, the apex acute or obtuse, the base broad, the margin serrulate or entire, the upper surface sparsely strigose or glabrous, the lower surface hirsute-villous on the midnerve, sessile, clasping. Flowers in capitula in an open paniculiform array with recurved branches, secund, the peduncle 2–8 mm long, the bracteoles linear or linear-lanceolate; involucre narrowly campanulate; phyllaries numerous in 4–5 series, ovate-lanceolate or linear-lanceolate, 3–6 mm long, unequal, the apex acute, the outer surface glabrous; receptacle slightly convex, pitted; ray flowers (2)4–10; corolla yellow, the lamina 1–3 mm long; disk flowers (2)4–7; corolla yellow, 4–5 mm long. Cypsela obconic, ca. 2 mm long, somewhat compressed, the ribs 8–10, strigillose distally; pappus of an outer series of short, setiform scales and 2 inner series of numerous barbellate bristles 3–4 mm long.

Wet flatwoods, bogs, and cypress pond margins. Common; nearly throughout. New Jersey south to Florida, west to Louisiana. Summer–fall.

Solidago gigantea Aiton [*Giganteus,* very large, in reference to the size.] GIANT GOLDENROD.

Solidago gigantea Aiton, Hort. Kew. 3: 211. 1789. *Solidago serotina* Aiton var. *gigantea* (Aiton) A. Gray, Proc. Amer. Acad. Arts 17: 196. 1882.

Solidago serotina Aiton, Hort. Kew. 3: 211. 1789; non Retzius, 1781. *Aster latissimifolius* (Miller) Kuntze var. *serotinus* Kuntze, Revis. Gen. Pl. 1: 314. 1891. *Solidago gigantea* Aiton var. *leiophylla* Fernald, Rhodora 41: 457. 1939, nom. illegit. *Solidago gigantea* Aiton var. *serotina* (Kuntze) Cronquist, in C. L. Hitchcock et al., Vasc. Pl. Pacific Northw. 5: 306. 1955. *Solidago gigantea* Aiton subsp. *serotina* (Kuntze) McNeill, Bot. J. Linn. Soc. 67: 280. 1973.

Erect perennial herb, to 2 m; stem with rhizomes, striate-angled, glabrous or sparsely strigose in the floral array, sometimes glaucous. Basal and proximal cauline leaves withering at flowering, the blade lanceolate, 9–10 cm long, 1–1.5 cm wide, the apex acuminate, the base cuneate, the margin serrate, the upper surface glabrous, the lower surface pilose on the nerves or glabrous, sessile, the midcauline similar, the largest near midstem and reduced upward, sessile. Flowers in capitula in a pyramidal paniculiform array with divergent, recurved branches, secund, the peduncle 2–3 mm long, the bracteoles linear-lanceolate; involucre campanulate; phyllaries numerous in 3–4 series, lanceolate or linear-lanceolate, 2–5 mm long, unequal, the apex acute, the outer surface glabrous; receptacle slightly convex, pitted; ray flowers (7)9–15(24); corolla yellow, the lamina 1–3 mm long; disk flowers (4)7–12(17); corolla yellow, 3–4 mm long. Cypsela obconic or cylindric, 1–2 mm long, somewhat compressed, the ribs 8–10, sparsely strigose; pappus of an outer series of short, setiform scales and 2 inner series of numerous barbellate bristles 2–3 mm long.

Open wet sites. Rare; Liberty, Calhoun, and Holmes Counties. Nearly throughout North America; Mexico; Europe, Africa, and Asia. Native to North America. Summer–fall.

Solidago gracillima Torr. & A. Gray [*Gracillimus,* very slender, in reference to the elongate proximal inflorescence branches.] VIRGINIA GOLDENROD.

Solidago gracillima Torrey & A. Gray, Fl. N. Amer. 2: 215. 1842. *Aster gracillimus* (Torrey & A. Gray) Kuntze, Revis. Gen. Pl. 1: 318. 1891. *Solidago stricta* Aiton subsp. *gracillima* (Torrey & A. Gray) Semple, Sida 20: 1615. 2003. TYPE: FLORIDA: s.d., *Chapman s.n.* (lectotype: NY; isolectotype: GA, NY.) Lectotypified by Semple (2003: 1615).

Erect or ascending perennial herb, to 2 m; stem with long, stoloniform rhizomes, striate-angled, glabrous. Basal leaves with the blade oblanceolate or elliptic-lanceolate, 6–60 cm long, 3–20 mm wide, the apex acute, the base tapering, the margin irregularly serrulate, the upper and lower surfaces glabrous, subsessile, the petiole narrowly winged, the cauline oblong-lanceolate or linear, 1–3 cm long, 2–4 mm wide, abruptly reduced proximally, then gradually distally, the apex acute, the base cuneate, the margin entire, the upper and lower surfaces glabrous, sessile. Flowers in capitula in a terminal and often also with axillary branches, sometimes in a pyramidal paniculiform array with recurved branches, secund, the peduncle 2–10 mm long, the bracteoles linear; involucre narrowly campanulate; phyllaries numerous in 3–4 series, oblong, 4–6 mm long, unequal, the apex acute or rounded, the outer surface glabrous; receptacle slightly convex, pitted; ray flowers 3–7; corolla yellow, the lamina ca. 2 mm long; disk flowers 3–7; corolla yellow, 3–5 mm long. Cypsela obconic or cylindric, ca. 2 mm long, somewhat compressed, the ribs 8–10, strigose; pappus of an outer series of short, setiform scales and 2 inner series of numerous barbellate bristles ca. 3 mm long.

Sandhills and hammocks. Rare; central panhandle. Virginia south to Florida, west to Tennessee and Alabama. Fall.

Solidago latissimifolia Mill. [*Latissimus,* the widest, with the widest leaves.] ELLIOTT'S GOLDENROD.

Solidago latissimifolia Miller, Gard. Dict., ed. 8. 1768. *Solidago elliptica* Aiton, Hort. Kew. 3: 214. 1789, nom. illegit. *Aster latissimifolius* (Miller) Kuntze, Revis. Gen. Pl. 1: 314. 1891.

Solidago elliottii Torrey & A. Gray, Fl. N. Amer. 2: 218. 1842. *Aster sublitoralis* Kuntze, Revis. Gen. Pl. 317. 1891. *Solidago elliottii* Torrey & A. Gray var. *typica* Fernald, Rhodora 38: 215. 1936, nom. illegit.

Solidago edisoniana Mackenzie, in Small, Man. S.E. Fl. 1358, 1509. 1933. *Solidago elliottii* Torrey & A. Gray var. *edisoniana* (Mackenzie) Fernald, Rhodora 38: 216. 1936. TYPE: FLORIDA: Polk Co.: Fort Meade, 1 Jan 1930, *Moore 502* (holotype: NY).

Solidago mirabilis Small ex Mackenzie, in Small, Man. S.E. Fl. 1359, 1509. 1933. TYPE: FLORIDA: Brevard Co.: Turnbull Hammock, N of Titusville, 17 Dec 1927, *Small & Mosier s.n.* (holotype: NY).

Erect perennial herb, to 3(4) m; stem with creeping, elongate rhizomes, striate-angled, glabrous or puberulent in the floral array. Basal and proximal cauline leaves withering at flowering, the mid and distal with the blade elliptic or lanceolate-elliptic, 5–15 cm long, 1.5–3.5 cm wide, slightly reduced distally, the apex acute, the base cuneate, the margin serrate or entire, the upper and lower surfaces glabrous, the lower surface glaucous, sessile or subsessile. Flowers in capitula in a paniculiform array with recurved branches, secund, the peduncle 4–10 mm long, the bracteoles linear-lanceolate; involucre campanulate; phyllaries numerous in 4–5 series, ovate-lanceolate or linear-lanceolate, 4–6 mm long, the margin distally ciliate, the apex obtuse or rounded, the outer surface glabrous; receptacle slightly convex, pitted; ray flowers 6–10; corolla yellow, the lamina 2–3 mm long; disk flowers 4–7; corolla yellow, 3–4 mm long. Cypsela obconic, ca. 1 mm long, somewhat compressed, the ribs 8–10, sparsely strigose; pappus of an outer series of short, setiform scales and 2 inner series of numerous barbellate bristles 3–5 mm long.

Hammocks. Occasional; Alachua County, central peninsula, Liberty County. New Hampshire and Vermont south to Florida, west to Alabama, also Nova Scotia. Fall.

Solidago leavenworthii Torr. & A. Gray [Commemorates Melines Conkling Leavenworth (1796–1862), army physician who made botanical collections in Florida.] LEAVENWORTH'S GOLDENROD.

Solidago leavenworthii Torrey & A. Gray, Fl. N. Amer. 2: 223. 1842. *Aster leavenworthii* (Torrey & A. Gray) Kuntze, Revis. Gen. Pl. 1: 318. 1891. TYPE: FLORIDA: Alachua Co.: near Micanopy, s.d., *Leavenworth s.n.* (holotype: GH).
Solidago nashii Small, Bull. Torrey Bot. Club 25: 475. 1898. TYPE: FLORIDA.

Erect perennial herb, to 2 m; stem with elongate, creeping rhizomes, striate-angled, scabrous-puberulent proximal to the floral array, in strips below the leaves. Basal and proximal cauline absent at flowering, mid and distal cauline with the blade linear-lanceolate or lanceolate, 2.5–15 cm long, 2–15 mm wide, the apex acute or acuminate, the base cuneate, the margin serrate or entire, short-ciliate, the upper and lower surfaces glabrous or with short trichomes in lines on the main nerves, sessile. Flowers in capitula in an elongate, pyramidal, paniculiform array with recurved branches, secund, the peduncle 1–5 mm long, the bracteoles linear-lanceolate; involucre narrowly campanulate; phyllaries numerous in 3–4 series, lanceolate or oblong-lanceolate, 3–5 mm long, unequal, the margin ciliate-fimbriate distally, the apex obtuse or rounded, the outer surface glabrous; receptacle slightly convex, pitted; ray flowers 8–15; corolla yellow, the lamina 2–3 mm long; disk flowers 6–10; corolla yellow, 3–4 mm long. Cypsela obconic, 1 mm long, somewhat compressed, the ribs 8–10, strigillose; pappus of an outer series of short, setiform scales and 2 inner series of numerous barbellate bristles ca. 3 mm long.

Pond margins, wet flatwoods, and wet disturbed sites. Frequent; peninsula west to central panhandle. North Carolina south to Florida, west to Mississippi. Summer–fall.

Solidago nemoralis Aiton [Pertaining to woods and groves.] DYERSWEED GOLDENROD.

Solidago nemoralis Aiton, Hort. Kew. 3: 213. 1789. *Solidago nemoralis* Aiton var. *typica* Rosendahl & Cronquist, Amer. Midl. Naturalist 33: 249. 1945, nom. inadmiss.
Solidago nemoralis Aiton var. *haleana* Fernald, Rhodora 38: 227, pl. 431. 1936. *Solidago nemoralis* Aiton subsp. *haleana* (Fernald) G. W. Douglas, Canad. J. Bot. 64: 2726. 1986.

Erect perennial herb, to 1 m; stem striate-angled, short-canescent. Basal leaves oblanceolate or obovate, 2–9.5 cm long, 7–15 mm wide, the apex obtuse or rounded, the base tapering to a long, winged petiole, sometimes withering at flowering, the margin crenate, the upper and lower surfaces puberulent, the mid and distal cauline linear-oblanceolate, 1.5–4.5 cm long,

3–7 mm wide, reduced upward, the margin irregularly crenate or entire, sessile. Flowers in capitula in a paniculiform array with recurved branches, secund, the peduncle 2–4 mm long, the bracteoles linear; involucre narrowly campanulate; phyllaries numerous in 3 series, ovate or linear-lanceolate, 3–6 mm long, unequal, the apex acute or obtuse, the outer surface glabrous; receptacle slightly convex, pitted; ray flowers 5–11; corolla yellow, the lamina 3–6 mm long; disk flowers 3–10; corolla yellow, 3–5 mm long. Cypsela obconic, 1–2 mm long, somewhat compressed, the ribs 8–10, strigose; pappus of an outer series of short, setiform scales and 2 inner series of numerous barbellate bristles 2–4 mm long.

Sandhills and hammocks. Rare; Jefferson, Leon, Gadsden, and Jackson Counties. Quebec south to Florida, west to British Columbia, Montana, Wyoming, Colorado, and New Mexico; Europe. Native to North America. Summer–fall.

Solidago odora Aiton [With an odor.]

Erect or arching perennial herb, to 1.2 m; stem evenly puberulent or nearly so, sometimes with the pubescence in distinct vertical lines from the leaf bases. Basal and proximal cauline leaves usually withering by flowering, the mid and distal cauline lanceolate or linear-lanceolate, 3–11 cm long, 8–20 mm wide, reduced upward, the apex acute, the base rounded, the margin entire, the upper and lower surfaces glabrous, gland-dotted, sessile. Flowers in capitula in a paniculiform array with recurved branches, secund, the peduncle 2–8 mm long, the bracteoles linear-lanceolate; involucre narrowly campanulate; phyllaries numerous in 3–4 series, ovate, lanceolate, or linear-lanceolate, 3–5 mm long, unequal, the apex acute, the outer surface glabrous; receptacle slightly convex, pitted; ray flowers 3–4(6); corolla yellow, the lamina 2–3 mm long; disk flowers 3–5; corolla yellow, 3–4 mm long. Cypsela obconic, ca. 2 mm long, somewhat compressed, the ribs 8–10, strigose or glabrate; pappus of an outer series of short, setiform scales and 2 inner series of numerous barbellate bristles 2–3 mm long.

1. Leaves linear-lanceolate; stem with the pubescence in distinct vertical lines from the leaf bases (at least the upper ones) .... var. **odora**
1. Leaves lanceolate or ovate-lanceolate; stem with the pubescence fairly evenly distributed (sometimes with a glabrous or subglabrous strip below each leaf base) .... var. **chapmanii**

Solidago odora var. **odora** ANISESCENTED GOLDENROD; SWEET GOLDENROD.

Solidago odora Aiton, Hort. Kew. 3: 214. 1789. *Aster odorus* (Aiton) Kuntze, Revis. Gen. Pl. 1: 318. 1891. *Solidago odora* Aiton var. *inodora* A. Gray, Manual, ed. 5. 244. 1867. *Solidago odora* Aiton forma *inodora* (A. Gray) Britton, Bull. Torrey Bot. Club 17: 124. 1890.

Stem with the pubescence in distinct vertical lines from the leaf bases (at least the upper ones). Leaves linear-lanceolate.

Sandhills. Frequent; northern counties, central peninsula. New Hampshire, Vermont, and New York south to Florida, west to Ohio, Missouri, Oklahoma, and Texas; Mexico. Summer–fall.

Solidago odora var. **chapmanii** (A. Gray) Cronquist [Commemorates Alvan Wentworth Chapman (1809–1899), American physician and botanist.] CHAPMAN'S GOLDENROD.

Solidago chapmanii A. Gray, Proc. Amer. Acad. Arts 16: 80. 1880. *Aster commutatus* Kuntze, Revis. Gen. Pl. 1: 315. 1891; non A. Gray, 1884. *Solidago odora* Aiton var. *chapmanii* (A. Gray) Cronquist, Brittonia 29: 224. 1977. *Solidago odora* Aiton subsp. *chapmanii* (A. Gray) Semple, Sida 20: 1611. 2003. TYPE: FLORIDA: Levy Co.: Nov 1877, *Garber s.n.* (lectotype: GH; isolectotype: GH). Lectotypified by Semple (2003: 1612).

Stem with the pubescence fairly evenly distributed (sometimes with a glabrous or subglabrous strip below each leaf base). Leaves lanceolate or ovate-lanceolate.

Sandhills and dry open hammocks. Common; peninsula west to central panhandle. Georgia and Florida. Summer–fall.

Solidago patula Muhl. ex Willd. var. **strictula** Torr. & A. Gray [*Patula,* spreading outward, in reference to the leaves; small and close together, in reference to the leaves.] ROUNDLEAF GOLDENROD.

Solidago patula Muhlenberg ex Willdenow var. *strictula* Torrey & A. Gray, Fl. N. Amer. 2: 213. 1842. *Solidago patula* Muhlenberg ex Willdenow subsp. *strictula* (Torrey & A. Gray) Semple, Sida 20: 1612. 2003.

Solidago salicina Elliott, Sketch Bot. S. Carolina 2: 389. 1823.

Erect perennial herb, to 1.5 m; stem striate-angled, sometimes winged on the angles, glabrous or sparsely pubescent in the floral array. Basal and proximal cauline leaves with the blade ovate, 15–20 cm long, 5–6 cm wide, the apex acute, the base abruptly narrowed to long, winged petioles, the margin serrate, the upper surface scabrous, the lower surface glabrous, the winged petiole clasping, the distal cauline reduced to bracts in the floral array, usually finely serrate or entire. Flowers in capitula in a pyramidal paniculiform array with recurved branches, secund, the peduncle 1–4 mm long, the bracteoles lanceolate; involucre campanulate; phyllaries numerous in 3–4 series, ovate or linear-ovate, 3–5 mm long, unequal, the apex obtuse, the outer surface glabrous; receptacle slightly convex, pitted; ray flowers 5–12; corolla yellow, the lamina ca. 2 mm long; disk flowers 5–15; corolla yellow, ca. 3 mm long. Cypsela obconic or cylindric, ca. 2 mm long, somewhat compressed, the ribs 8–10, strigillose; pappus of an outer series of short, setiform scales and 2 inner series of numerous barbellate bristles 2–3 mm long.

Bogs. Rare; central and western panhandle. Virginia south to Florida, west to Missouri, Oklahoma, and Texas. Fall.

Solidago petiolaris Aiton [Petiolate.] DOWNY RAGGED GOLDENROD.

Solidago petiolaris Aiton, Hort. Kew. 3: 216. 1789. *Aster petiolaris* (Aiton) Kuntze, Revis. Gen. Pl. 1: 318. 1891.

Solidago milleriana Mackenzie, in Small, Man. S.E. Fl. 1350, 1509. 1933. TYPE: "Fla. to Ala. and N. C."

Erect perennial herb, to 1.5 m; stem striate-angled, finely puberulent or scabrous-puberulent at least distally. Basal leaves absent at flowering, the cauline with the blade lanceolate-elliptic or ovate, 3–15 cm long, 0.5–3 cm wide, reduced upward, the apex acute, the base rounded or cuneate, the margin entire or irregularly serrate, the upper surface glabrous or scabrous, the lower surface strigillose or glabrous, sometimes resinous, short-petiolate or sessile. Flowers in capitula in a paniculiform or racemiform array, stiffly ascending, not secund, the peduncle 2–15 mm long, the bracteoles lanceolate; involucre campanulate; phyllaries numerous in 3–4 series, linear-lanceolate, 4–8 mm long, unequal, the apex acute or acuminate, subsquarrose-tipped, the outer surface strigose or glabrous, stipitate-glandular; receptacle slightly convex, pitted; ray flowers (5)7–9; corolla yellow, the lamina 3–7 mm long; disk flowers (8)10–16; corolla yellow, 4–5 mm long. Cypsela obconic or cylindric, 3–4 mm long, somewhat compressed, the ribs 8–10, glabrous or glabrate; pappus of an outer series of short, setiform scales and 2 inner series of numerous barbellate bristles ca. 4 mm long.

Sandhills and hammocks. Occasional; northern counties. North Carolina south to Florida, west to Mississippi, also Illinois, Missouri, Nebraska, and Colorado south to Louisiana, Texas, and New Mexico; Mexico. Summer–fall.

Solidago puberula Nutt. var. **pulverulenta** (Nutt.) Chapman [Puberulent, in reference to the stems and leaves; powdery or dusty colored, in reference to the stems and leaves.] DOWNY GOLDENROD.

Solidago pulverulenta Nuttall, Gen. N. Amer. Pl. 2: 161. 1818. *Leioligo pulverulenta* (Nuttall) Rafinesque, Fl. Tellur. 2: 42. 1837 ("1836"). *Solidago puberula* Nuttall var. *pulverulenta* (Nuttall) Chapman, Fl. South. U.S. 210. 1860. *Aster pulverulentus* (Nuttall) Kuntze, Revis. Gen. Pl. 1: 316, 318. 1891. *Solidago puberula* Nuttall subsp. *pulverulenta* (Nuttall) Semple, Sida 20: 1612. 2003. TYPE: "Georgia and Florida."

Erect perennial herb, to 1 m; stem striate-angled, puberulent. Basal and proximal leaves with the blade oblanceolate or obovate, (2)5–11 cm long, 1–5 cm wide, the apex acute, obtuse, or rounded, the base cuneate, the margin serrate, the upper and lower surfaces puberulent, the petiole long, winged, clasping, the mid to distal cauline lanceolate or linear-elliptic, 1–5 cm long 3–10 mm wide, reduced upward, the apex acute, the base cuneate, the margin becoming entire, sessile. Flowers in capitula in an elongate paniculiform array, the lateral racemiform clusters short and ascending, not secund, sometimes lateral branches elongated and sharply ascending, the peduncle 2–3 mm long, the bracteoles lanceolate; involucre campanulate; phyllaries numerous in 3–4 series, linear, 3–5 mm long, unequal, the apex acuminate, the outer surface puberulent; receptacle slightly convex, pitted; ray flowers 9–16; corolla yellow, the lamina 2–4 mm long; disk flowers (6)10–15; corolla yellow, ca. 3 mm long. Cypsela obconic or cylindric, ca. 1 mm long, somewhat compressed, the ribs 8–10, sparsely strigose or glabrate; pappus of an outer series of short setiform scales and 2 inner series of numerous barbellate bristles ca. 2 mm long.

Wet flatwoods, bogs, and swamps. Occasional; panhandle. New Jersey south to Florida, west to Louisiana. Fall.

Solidago rugosa Mill. subsp. **aspera** (Aiton) Cronquist [Wrinkled, in reference to the leaf surface; rough, in reference to the leaf surface.] WRINKLELEAF GOLDENROD.

Solidgo aspera Aiton, Hort. Kew. 3: 212. 1789. *Solidago rugosa* Miller var. *aspera* (Aiton) Fernald, Rhodora 17: 7. 1915. *Solidago rugosa* Miller subsp. *aspera* (Aiton) Cronquist, Rhodora 49: 78. 1947.
Solidago celtidifolia Small, Fl. S.E. U.S. 1198, 1339. 1915. *Solidago rugosa* Miller var. *celtidifolia* (Small) Fernald, Rhodora 38: 223. 1936.

Erect perennial herb, to 2 m; stem with long-creeping rhizomes, striate-angled, densely hispid or strigose. Basal and proximal cauline leaves withering by flowering, the blade elliptic or lanceolate, 7–10 cm long, 2–2.5 cm wide, the apex acute or attenuate, the base cuneate, the margin serrate, the upper and lower surfaces strongly rugose-veined, the upper surface glabrate, the lower usually hispid-strigose, at least on the main veins, sessile, the mid to distal cauline leaves lanceolate or elliptic, 4–8(9) mm long, 1–2(3) cm wide, the largest at midstem, not much reduced distally, similar to the proximal, sessile. Flowers in capitula in a pyramidal paniculiform array with divergent and recurved branches, secund, the peduncle 1–2 mm long, the bracteoles linear-lanceolate or ovate; involucre narrowly campanulate; phyllaries numerous in 3–4 series, lanceolate or linear-lanceolate, 2–5 mm long, unequal, the apex acute or obtuse, the outer surface glabrous; receptacle slightly convex, pitted; ray flowers 5–9; corolla yellow, the lamina 1–2 mm long; disk flowers (2)4–6(8); corolla yellow, 2–4 mm long. Cypsela obconic, ca. 1 mm long, somewhat compressed, the ribs 8–10, strigillose; pappus of an outer series of short, setiform scales and 2 inner series of numerous barbellate bristles 2–3 mm long.

Swamps and wet hammocks. Occasional; Jefferson County, central and western panhandle. Maine south to Florida, west to Ontario, Michigan, Illinois, Missouri, Oklahoma, and Texas. Fall.

Solidago sempervirens L. [Evergreen.] SEASIDE GOLDENROD.

Solidago sempervirens Linnaeus, Sp. Pl. 878. 1753. *Aster sempervirens* (Linnaeus) Kuntze, Revis. Gen. Pl. 1: 318. 1891. *Solidago sempervirens* Linnaeus var. *typica* Fernald, Rhodora 37: 448. 1935, nom. inadmiss.
Solidago mexicana Linnaeus, Sp. Pl. 879. 1753. *Aster mexicanus* (Linnaeus) Kuntze, Revis. Gen. Pl. 1: 318. 1891. *Solidago sempervirens* Linnaeus var. *mexicana* (Linnaeus) Fernald, Rhodora 37: 447. 1935. *Solidago sempervirens* Linnaeus subsp. *mexicana* (Linnaeus) Semple, Sida 20: 1615. 2003.

Erect perennial herb, to 2 m; stem striate-angled, glabrous throughout or pubescent in the floral array. Basal and proximal leaves with the blade narrowly ovate or oblanceolate, 10–40 cm long, 1–6 cm wide, the apex acute, the base tapering to a long, winged petiole, the margin entire, the upper and lower surfaces glabrous, the mid to distal cauline with the blade lanceolate, 4–6 cm long, 5–10 mm wide, reduced upward, closely appressed to the stem, the apex acute, the base rounded, sometimes somewhat clasping, the margin entire, the upper and lower surfaces glabrous, sessile. Flowers in capitula in a paniculiform array with recurved branches, ascending or slightly secund, the peduncle 2–3 mm long, the bracteoles linear-lanceolate; involucre campanulate; phyllaries numerous in 3–4 series, lanceolate, 3–7 mm long, unequal, the margin ciliate, the apex acute, the outer surface glabrous; receptacle slightly convex, pitted; ray flowers 8–17; corolla yellow, the lamina 5–6 mm long; disk flowers 10–22; corolla yellow, ca. 3 mm long. Cypsela obconic, ca. 1 mm long, somewhat compressed, the ribs 8–10, strigose; pappus of an outer series of short, setiform scales and 2 inner series of numerous barbellate bristles ca. 4 mm long.

Coastal dunes and tidal marshes, rarely inland in wet hammocks, and cypress swamp, river, and lake margins. Common; peninsula, central and western panhandle. Labrador, Newfoundland, and Quebec south to Florida, west to Ontario, Wisconsin, Illinois, Mississippi, and Texas; West Indies, Mexico, and Central America. Summer–fall.

Solidago stricta Aiton [Upright, very straight.] WAND GOLDENROD.

Solidago stricta Aiton, Hort. Kew. 3: 216. 1789. *Aster strictus* (Aiton) Kuntze, Revis. Gen. Pl. 1: 318. 1891.
Solidago virgata Michaux, Fl. Bor.-Amer. 2: 117. 1803. *Lepiactis virgata* (Michaux) Rafinesque, Fl. Tellur. 2: 43. 1837 "1836."
Solidago angustifolia Elliott, Sketch Bot. S. Carolina 2: 388. 1823; non Miller, 1768. *Solidago stricta* Aiton var. *angustifolia* A. Gray, Proc. Amer. Acad. Arts 17: 192. 1882. *Solidago salaria* House, Amer. Midl. Naturalist 7: 131. 1921.
Solidago flavovirens Chapman, Fl. South. U.S. 211. 1860. *Aster flavovirens* (Chapman) Kuntze, Revis. Gen. Pl. 1: 318. 1891. TYPE: FLORIDA: Franklin Co.: near Appalachicola, s.d., *Chapman* s.*n.* (holotype: GH; isotype: US).
Solidago chrysopsis Small, Bull. Torrey Bot. Club 25: 477. 1898. TYPE: FLORIDA: Monroe Co.: Big Pine Key, s.d., *Blodgett s.n.* (holotype: NY).

Ascending or erect perennial herb, to 2 m; stem with long, stoloniferous rhizomes, striate-angled, glabrous. Basal leaves with the blade oblanceolate or elliptic-lanceolate, 6–60 cm long, 3–20 mm wide, the apex obtuse or rounded, the base tapering, the margin entire, the upper and lower surfaces glabrous, the petiole winged, nearly completely sheathing the stem, the cauline oblong-lanceolate or linear, 1–3 cm long, 2–4 mm wide, abruptly reduced proximally, then gradually reduced distally, the apex acute, the base cuneate, the margin entire, the upper and lower surfaces glabrous, sessile. Flowers in capitula in a narrow paniculiform array, not secund, the peduncle 2–10 mm long, the bracteoles linear; involucre narrowly campanulate; phyllaries numerous in 3–4 series, oblong, 4–6 mm long, unequal, the apex acute or rounded, the outer surface glabrous; receptacle slightly convex, pitted; ray flowers 3–7; corolla yellow, the lamina ca. 2 mm long; disk flowers 8–12; corolla yellow, 3–5 mm long. Cypsela obconic or cylindric, ca. 2 mm long, somewhat compressed, the ribs 8–10, strigose; pappus of an outer series of short, setiform scales and 2 inner series of numerous barbellate bristles ca. 3 mm long.

Flatwoods, sandhills, bogs, and tidal marsh margins. Common; nearly throughout. New Jersey south to Florida, west to Texas; Mexico and Central America. Summer–fall.

Solidago tortifolia Elliott [With twisted leaves.] TWISTEDLEAF GOLDENROD.

Solidago tortifolia Elliott, Sketch Bot. S. Carolina 2: 377. 1823. *Aster purshianus* Kuntze, Revis, Gen. Pl. 1: 316.

Ascending or erect perennial herb, to 1.3 m; stem with small, woody, elongate, creeping rhizomes, striate-angled, strigillose-villous mid to distal. Basal and cauline leaves withering by flowering, midcauline to distal with the blade linear or linear-oblanceolate, 2–7 cm long, 2–7(10) mm wide, somewhat twisted, the apex acute, obtuse, or rounded, the base rounded, the margin irregularly serrulate or entire, the upper and lower surfaces glabrous or finely strigillose, sessile. Flowers in capitula in a pyramidal paniculiform array with recurved branches, secund, the peduncle 1–4 mm long, the bracteoles linear; involucre narrowly campanulate; phyllaries numerous in 3–4 series, ovate or oblong, 2–4 mm long, unequal, the apex acute, obtuse, or rounded, the outer surface glabrous; receptacle slightly convex, pitted; ray flowers 2–8; corolla yellow, the lamina 1–2 mm long; disk flowers 2–4(6); corolla yellow, 3–4 mm long. Cypsela obconic, ca. 1 mm long, somewhat compressed, the ribs 8–10, strigillose; pappus of an outer series of short, setiform scales and 2 inner series of numerous barbellate bristles 2–3 mm long.

Sandhills and flatwoods. Frequent; peninsula west to central panhandle. Maryland south to Florida, west to Texas. Summer–fall.

Solidago ulmifolia Muhl. ex Willd. [With leaves like *Ulmus* (Ulmaceae).] ELMLEAF GOLDENROD.

Solidago ulmifolia Muhlenberg ex Willdenow, Sp. Pl. 3: 2060. 1803. *Aster ulmifolius* (Muhlenberg ex Willdenow) Kuntze, Revis. Gen. Pl. 1: 318. 1891.

Erect perennial herb, to 1.2 m; stem glabrous, sparsely pubescent in the floral array. Basal and proximal cauline leaves often withering by flowering, the blade ovate-lanceolate, 6–10(15) cm long, 2–3(5) cm wide, the apex acute, the base tapering, the margin coarsely serrate, the upper surface sparsely hirsute or scabrous, the lower surface hirsute on the main veins, the petiole short, winged, mid to distal cauline with the blade lanceolate or elliptic, 2–5 cm long, 5–20 mm wide, gradually reduced upward, the apex acute, the base cuneate, the margin serrate, becoming entire distally, the upper and lower surfaces similar to the lower leaves, subsessile or sessile. Flowers in capitula in an open paniculiform array with recurved branches, secund, the peduncle ca. 2 mm long, the bracteoles lanceolate; involucre narrowly campanulate; phyllaries numerous in 2–3 series, ovate or linear-lanceolate, 3–4 mm long, unequal, the apex acute or obtuse, the outer surface glabrous; receptacle slightly convex, pitted; ray flowers 3–6; corolla yellow, the lamina 1–2 mm long; disk flowers 4–7; corolla yellow, ca. 3 mm long. Cypsela obconic or cylindric, ca. 1 mm long, somewhat compressed, the ribs 8–10, puberulent; pappus of an outer series of short, setiform scales and 2 inner series of numerous barbellate bristles 2–3 mm long.

Moist hammocks. Rare; Jackson and Calhoun Counties. Nova Scotia and Maine south to Florida, west to Ontario, Minnesota, Nebraska, Kansas, Oklahoma, and Texas. Fall.

EXCLUDED TAXA

Solidago arguta Aiton—Because no infraspecific categories were recognized, the typical variety was reported for Florida by implication by Radford et al. (1968). All Florida material is of variety *caroliniana*.

Solidago arguta Aiton var. *boottii* (Hooker) E. J. Palmer & Steyermark—This more northern taxon was reported for Florida by Chapman (1860, 1883, 1897), Small (1903, 1913a, 1933), Radford et al. (1964), and Correll and Johnston (1970), all as *S. boottii* Hooker. All Florida material is of *S. arguta* var. *caroliniana.*

Solidago canadensis Linnaeus—Because no infraspecific categories were recognized, the typical variety was reported for Florida by implication by Chapman (1860, 1883, 1897) and Small (1903). All Florida material is of var. *scabra.*

Solidago juncea Aiton—Reported for Florida by Kral (1966), who misapplied the name to Florida material of *S. gracillima.*

Solidago patula Muhlenberg ex Willdenow—Because infraspecific categories were not recognized, the typical variety was reported for Florida by implication by Godfrey and Wooten (1981). All Florida material is of var. *strictula.*

Solidago petiolata Miller—Reported for Florida by Mackenzie, in Small (1933), the name misapplied to Florida material of *S. stricta.*

Solidago puberula Nuttall—The typical variety of this species was reported by Small (1903, 1913a, 1933) and Clewell (1985). All Florida material is of var. *pulverulenta.*

Solidago rugosa Miller—The typical variety of this species was reported for Florida by Small (1903), the name apparently misapplied to Florida material of *S. ulmifolia.* Also reported for Florida by implication by Chapman (1897) and Godfrey and Wooten (1981), who did not distinguish infraspecific taxa, and by Semple and Cook (2006), who listed the taxon in error. All Florida material is of subsp. *aspera.*

Solidago serotina Aiton—This is a synonym of the more northern species *S. gigantea,* but was reported for Florida by Chapman (1860, 1883, 1897), Small (1903, 1913a, 1933), and Cronquist (1980, as *S. gigantea* var. *serotina*). These reports are misapplications of the name to our material of *S. leavenworthii.*)

Solidago speciosa Nuttall—Reported for Florida by Chapman (1860). According to Correll and Johnston (1970), "the species as a whole from Mass., N.Y., and s. Minn. s. to the Gulf States," which would presumably include Florida. No Florida material is known. It was excluded from Florida by Cronquist (1980).

Solidago speciosa Nuttall var. *rigidiuscula* Torrey & A. Gray—Reported for Florida by Buswell (1942, as *S. rigidiuscula* Torrey & A. Gray), apparently a misapplication of the name to Florida material of *S. arguta.*

Soliva Ruiz & Pav. 1794. BURRWEED

Herbs. Leaves alternate, simple, pinnipalmately veined, petiolate. Flowers in capitula solitary in the leaf axils, disciform; peripheral flowers carpellate; petals absent; inner flowers staminate; petals 5, basally connate, actinomorphic; stamens 5, epipetalous, the filaments free, the anthers connate; ovary inferior, 2-carpellate, 1-loculate. Fruit a cypsela; pappus absent.

A genus of about 7 species; North America, Mexico, South America, Europe, Africa, Asia, Australia, and Pacific Islands. [Commemorates Salvador Soliva, eighteenth-century physician to the Spanish court.]

Gymnostyles Juss.

Selected reference: Watson (2006d).

1. Cypsela with broad lateral wings prolonged into a spinelike projection at the apex on either side of the larger terminal spine, glabrous distally .... **S. sessilis**
1. Cypsela with thickened lateral appendages that are scarcely winglike and not prolonged into a spinelike projection at the apex, pilose distally.

2. Cypsela wings smooth distally .... **S. anthemifolia**
2. Cypsela wings transversely rugose or cross-ribbed to the apex .... **S. stolonifera**

Soliva anthemifolia (Juss.) Sweet [With leaves like *Anthemis* (Asteraceae).] BUTTONWEED.

Gymnostyles anthemifolia Jussieu, Ann. Mus. Hist. Nat. 4: 262, t. 61(1). 1804. *Soliva anthemifolia* (Jussieu) Sweet, Hort. Brit. 243. 1826.

Procumbent or erect annual herb, to 15(30) cm; stem sometimes stoloniferous and mat-forming, striate-angled, villous or glabrescent. Leaves mostly basal, the blade obovate or spatulate, 3–8(15) cm long, 2–4 cm wide, 2- to 3-pinnipalmately lobed, the ultimate lobes oblong, lanceolate, or oblanceolate, the apex acute, the ultimate margin entire, the upper and lower surfaces villous or glabrescent, petiolate, slightly clasping. Flowers in capitula mostly clustered in the leaf axils at ground level, rarely scattered along the stem; involucre hemispheric, 4–8 mm wide; phyllaries 5–8 in 1–2 series, lanceolate or ovate, subequal, the margin scarious; receptacle slightly convex, epaleate; peripheral flowers numerous in 1–8 series; corolla absent; inner flowers 2–4+; corolla yellowish white, ca. 2 mm long. Cypsela body oblanceolate, ca. 2 mm long, flattened, the thickened lateral wings transversely rugose or ribbed proximally, the shoulders not spinelike, the surface villose or pilose, sometimes glabrescent, the persistent stylar sheath indurate, spinelike, 2–3 mm long, usually inflexed; pappus absent.

Disturbed sites. Rare; Alachua, Franklin, and Escambia Counties, central peninsula. Florida west to Arkansas and Texas; Mexico and South America; Asia and Australia. Native to South America. Spring.

Soliva sessilis Ruiz & Pav. [Stalkless, in reference to the axillary flower heads.] FIELD BURRWEED.

Soliva sessilis Ruiz López & Pavón, Syst. Veg. Fl. Peruv. Chil. 215. 1798. *Gymnostyles chilensis* Sprengel, Syst. Veg. 3: 500. 1826.

Gymnostyles pterosperma Jussieu, Ann. Mus. Natl. Hist. Nat. 4: 262, t. 16(3). 1804. *Soliva pterosperma* (Jussieu) Lessing, Syn. Gen. Compos. 368. 1832.

Prostrate or ascending annual herb, to 5(25) cm; stem sometimes rooting at the nodes, striate-angled, villous or glabrescent. Leaves basal and cauline, the blade oblanceolate, 1–2(3) cm long, 5–10 mm wide, 2(3)-pinnipalmately lobed, the ultimate lobes elliptic, the apex acute, the ultimate margin entire, the upper and lower surfaces villous or glabrescent, petiolate, clasping. Flowers in capitula mostly in the leaf axils scattered along the stem; involucre hemispheric, 2–4(5) mm wide; phyllaries 5–8 in 1–2 series, lanceolate or ovate, subequal, the margin scarious; receptacle slightly convex, epaleate; peripheral flowers 5–8(17) in 1–2 series; corolla absent; inner flowers 4–8+; corolla yellowish white, ca. 2 mm long. Cypsela body obovate or lanceolate, 2–3 mm long, flattened, the lateral wing entire or incised, each shoulder with a distally projecting spinelike tooth, the surface glabrous, scabrellous or hirtellous proximally, the persistent stylar sheath indurate, spinelike, 1–2 mm long, erect or slightly inflexed; pappus absent.

Disturbed sites. Occasional; northern counties, central peninsula. Virginia south to Florida, west to Oklahoma and Texas, also British Columbia south to California and Arizona; Mexico and South America; Europe, Africa, Asia, and Pacific Islands. Native to South America. Spring.

Soliva stolonifera (Brot.) Sweet [Stolon-bearing.] CARPET BURRWEED.

Hippia stolonifera Brotero, Phytogr. Lusitan. Select. 14: 1800. *Gymnostyles lusitanica* Sprengel, Syst. Veg. 3: 500. 1826, nom. illegit. *Soliva stolonifera* (Brotero) Sweet, Hort. Brit. 243. 1826. *Soliva lusitanica* Lessing, Syn. Gen. Compos. 268. 1832, nom. illegit. *Gymnostyles stolonifer*a (Brotero) Tutin, Bot. J. Linn.Soc. 70: 18. 1975.

Procumbent annual herb, to 5(15) cm; stem rooting at the nodes, striate-angled, strigillose or villous. Leaves basal and cauline, the blade spatulate, 1(2) cm long, 3–5 mm wide, 1(2)-pinnipalmately lobed, the ultimate lobes oblong, lanceolate, or oblanceolate, the apex acute or obtuse, the ultimate margin entire, the upper and lower surfaces strigillose or villous, petiolate, slightly clasping. Flowers in capitula mostly in the leaf axils scattered along the stem; involucre hemispheric, 3–6(8) mm wide; phyllaries 5–8 in 1–2 series, lanceolate or ovate, subequal, the margin scarious; receptacle slightly convex, epaleate; peripheral flowers numerous in 2–4 series; corolla absent; inner flowers 4–6+; corolla yellowish white, ca. 2 mm long. Cypsela body oblanceolate, ca. 2 mm long, compressed, the thickened lateral wings transversely rugose or ribbed, the shoulders not spinelike, the surface distally villous, pilose, or arachnose, the persistent stylar sheath indurate, spinelike, 1–2 mm long, usually inflexed; pappus absent.

Disturbed sites. Rare; Leon and Gadsden Counties. South Carolina south to Florida, west to Arkansas and Texas; South America; Europe. Native to South America. Spring.

EXCLUDED TAXON

Soliva nasturtiifolia de Candolle—This illegitimate renaming of *Hippia minuta* Linnaeus f. (=*Cotula mexicana* (de Candolle) Cabrera was reported for Florida as *S. nasturtiifolia* by Small (1903, 1913a, 1933), who misapplied the name to Florida material of *S. stolonifera.*

Sonchus L. 1753. SOWTHISTLE

Herbs. Leaves alternate, simple, pinnate-veined, petiolate or epetiolate. Flowers in capitula in corymbiform arrays, liguliflorous, bisexual; receptacle epalete; petals 5, basally connate, zygomorphic; stamens 5, epipetalous, the filaments free, the anthers connate; ovary inferior, 2-carpellate, 1-loculate. Fruit a cypsela; pappus present.

A genus of about 90 species; nearly cosmopolitan. [From the Greek *sonchos,* an ancient name for a kind of thistle.]

Selected reference: Hyatt (2006).

1. Cypsela longitudinally ribbed, otherwise smooth .... **S. asper**
1. Cypsela longitudinally ribbed and finely transverse rugulose-tuberulate .... **S. oleraceus**

Sonchus asper (L.) Hill [Rough, uneven, in reference to the spinulose leaf margin.] SPINY SOWTHISTLE.

Sonchus oleraceus Linnaeus var. *asper* Linnaeus, Sp. Pl. 794. 1753. *Sonchus asper* (Linnaeus) Hill, Herb. Brit. 1: 47. 1769. *Hieracium asperum* (Linnaeus) Bernhardi, Syst. Verz. 136. 1800. *Sonchus fallax* Wallroth var. *asper* (Linnaeus) Wallroth, Sched. Crit. 432. 1822. *Sonchus asper* (Linnaeus) Hill var. *pungens* Bischoff, Beitr. Fl. Deutschl. 222. 1851, nom. inadmiss. *Sonchus asper* (Linnaeus) Hill forma *typicus* Beck, Fl. Nieder-Oesterreich 1322. 1893, nom. inadmiss. *Hieracium sonchus* E.H.L. Krause, in Sturm, Deutschl. Fl., ed. 2. 14: 113. 1906, nom. illegit.

Erect annual or biennial herb, to 1(2) m; stem striate-angled, glabrous or reddish brown stipitate-glandular distally, usually somewhat glaucous. Mid cauline leaves with the blade spatulate, oblong, obovate, or lanceolate, 6–30 cm long, 1–15 cm wide, reduced upward, the apex acute, the base auriculate, the auricles usually recurved or curled, rounded, the margin usually irregularly pinnately lobed, the lobes somewhat deltate, the ultimate margin prickly-dentate, the upper and lower surfaces glabrous, the lower surface somewhat glaucous, often with spines along the main veins, sessile, clasping the stem. Flowers in capitula in a corymbiform array, the peduncle usually sparsely reddish brown stipitate-glandular, sometimes glabrous; involucre urceolate-campanulate, 9–13 mm wide; phyllaries numerous in 3–5 series, linear-lanceolate, 9–13 mm long, unequal, the apex acute, the outer surface usually stipitate-glandular; receptacle slightly convex, slightly pitted; flowers numerous; corolla yellow, the lamina 8–15 mm long. Cypsela ellipsoid, 2–3 mm long, compressed, longitudinally 3- to 5-ribbed, otherwise smooth, yellowish brown or reddish brown; pappus of 3–4 series of numerous white, barbellate bristles 6–9 mm long.

Disturbed sites. Frequent; peninsula west to central panhandle. Nearly throughout North America. Nearly cosmopolitan. Native to Europe. Spring–summer.

Sonchus oleraceus L. [Pertaining to a kitchen garden.] COMMON SOWTHISTLE.

Sonchus oleraceus Linnaeus, Sp. Pl. 794. 1753. *Hieracium oleraceum* (Linnaeus) Bernhardi, Syst. Verz. 136. 1800. *Lactuca oleracea* (Linnaeus) M. Gómez de la Maza y Jiménez, Anales Soc. Esp. Nat. 19: 268. 1890.

Erect annual or biennial herb, to 1.5(2) m; stem striate-angled, glabrous or reddish brown stipitate-glandular distally, usually somewhat glaucous. Midcauline leaves with the blade spatulate, oblong, obovate, or lanceolate, 6–35 cm long, 1–15 cm wide, reduced upward, the apex acute, the base auriculate, the auricles deltate or lanceolate, more or less straight, the margin usually irregularly pinnately lobed, the lobes somewhat deltate or lanceolate, the ultimate margin entire or prickly-dentate, the upper and lower surfaces glabrous, the lower surface somewhat glaucous, often with spines along the main veins, sessile, clasping the stem. Flowers in capitula in a corymbiform array, the peduncle usually glabrous, sometimes stipitate-glandular; involucre urceolate-campanulate, 9–13 mm wide; phyllaries numerous in 3–5 series, linear-lanceolate, 9–13 mm long, the apex acute, the outer surface usually glabrous, sometimes tomentose and/or stipitate-glandular; receptacle slightly convex, slightly pitted; flowers numerous; corolla yellow, the lamina 8–15 mm long. Cypsela oblanceolate, compressed, longitudinally 2- to 4-ribbed, finely transverse rugulose-tuberculate across and between the ribs, dark brown; pappus of 3–4 series of numerous white, barbellate bristles 5–8 mm long.

Disturbed sites. Frequent; nearly throughout. Nearly throughout north America. Nearly cosmopolitan. Native to Europe. Spring–summer.

Sphagneticola O. Hoffm. 1900. CREEPING OXEYE

Herbs. Leaves opposite, simple, pinnate-veined, petiolate or epetiolate. Flowers in capitula, solitary, axillary, radiate; receptacle paleate; ray flowers carpellate; petals 5, basally connate, zygomorphic; disk flowers bisexual; petals 5, basally connate, actinomorphic; stamens 5, epipetalous, the filaments free, the anthers connate; ovary inferior, 2-carpellate, 1-loculate. Fruit a cypsela; pappus present.

A genus of 5 species; West Indies, Mexico, Central America, South America, Asia, Australia, and Pacific Islands. [*Sphagnum,* a moss, and *cola,* dwelling, in reference to its growing in bog-like habitats.]

Selected reference: Strother (2006bb).

Sphagneticola trilobata (L.) Pruski [Three-lobed, in reference to the leaves.] CREEPING OXEYE.

Silphium trilobatum Linnaeus, Syst. Nat., ed. 10. 1233. 1759. *Buphthalmum repens* Lamarck, Encycl. 1: 515. 1785, nom. illegit. *Wedelia carnosa* Richard, in Persoon, Syn. Pl. 2: 490. 1807, nom. illegit. *Polymnia carnosa* Poiret, in Lamarck, Encycl., Suppl. 4: 482. 1816, nom. illegit. *Verbesina carnosa* M. Gómez de la Maza y Jiménez, Anales Soc. Esp. Hist. Nat. 19: 274. 1890, nom. illegit. *Seruneum trilobatum* (Linnaeus) Kuntze, Revis. Gen. Pl. 1: 365. 1891. *Wedelia trilobata* (Linnaeus) Hitchcock, Rep. (Annual) Missouri Bot. Gard. 4: 99. 1893. *Stemmodontia trilobata* (Linnaeus) Small, Fl. S.E. U.S. 1262, 1340. 1903. *Complaya trilobata* (Linnaeus) Strother, Syst. Bot. Monogr. 33: 14. 1991. *Thelechitonia trilobata* (Linnaeus) H. Robinson & Cuatrecasas, Phytologia 72: 142. 1992. *Sphagneticola trilobata* (Linnaeus) Pruski, in Acevedo-Rodriguez, Mem. New York Bot. Gard. 78: 114. 1996.

Procumbent perennial herb, to 2 dm; stem rooting at the nodes, striate-angled, scabrous. Leaves with the blade elliptic or lanceolate, 5–18 cm long, 1–5 cm wide, the apex acute, obtuse, or rounded, mucronate, the base cuneate, the margin often 3-lobed, irregularly serrate, the upper and lower surfaces scabrous, sessile or the petiole to 5 mm long. Flowers in capitula, solitary in the leaf axils, radiate, the peduncle 5–10 cm long; involucre obconic, 6–12 mm wide; phyllaries 12–15 in 2–3 series, lanceolate, 10–15 mm long, foliaceous; receptacle convex; paleae conduplicate, scarious, the apex acute; ray flowers 4–10; corolla yellow, the lamina 5–15 mm long; disk flowers numerous; corolla yellow, 4–6 mm long. Cypsela broadly clavate, 3–5 mm long, laterally flattened, 3- to 4-angled, the surface smooth or tuberculate; pappus coroniform.

Moist disturbed sites. Occasional; peninsula, central and western panhandle. Escaped from cultivation. Florida west to Texas; West Indies, Mexico, Central America, and South America; Asia, Australia, and Pacific Islands. Native to tropical America and Asia. All year.

Sphagneticola trilobata is listed as a Category II invasive species in Florida by the Florida Exotic Pest Plant Council (FLEPPC, 2017).

Stokesia L'Her. 1789. STOKES' ASTER

Herbs. Leaves alternate, simple, pinnate-veined, petiolate or epetiolate. Flowers in capitula, solitary or in corymbiform arrays, pseudo-radiate, bisexual; receptacle epaleate; petals 5, basally connate, the peripheral flowers zygomorphic, the inner flowers actinomorphic; stamens 5, epipetalous, the filaments free, the anthers connate; ovary inferior, 2-carpellate, 1-loculate. Fruit a cypsela; pappus present.

A monotypic genus; North America. [Commemorates Jonathan S. Stokes (1755–1831), Scottish physician, botanist, and paleontologist.]

Selected reference: Strother (2006cc).

Stokesia laevis (Hill) Greene [Smooth, in reference to the upper leaf surface.] STOKES' ASTER.

Carthamus laevis Hill, Hort. Kew. 57, t. 5. 1768. *Stokesia cyanea* L'Heritier de Brutelle, Sert. Angl. 28. 1789 ("1788"), nom. illegit. *Stokesia laevis* (Hill) Greene, Erythea 1: 3. 1893.

Erect perennial herb, to 5 dm; stem rhizomatous, striate-angled, tomentulose or glabrescent. Leaves ovate, lanceolate, or linear-lanceolate, 8–15 cm long, 1–5 cm wide, reduced upward, the apex acute or obtuse, mucronate, the base cuneate, the entire or spinose-toothed, the upper surface glabrous, the lower surface tomentulose or glabrate, gland-dotted, the basal with the petiole 3–12 cm long, winged, the base clasping, the cauline sessile, the base clasping. Flowers in capitula, solitary or in a corymbiform array, pseudo-radiant, bisexual; involucre hemispheric, 2.5–4.5 cm wide; phyllaries numerous in 5–7 series, 2.5–4.5 cm long, the outer elliptic or spatulate, with an appressed subchartaceous base, distally foliaceous, the margin pectinate spine-toothed, at least at the base, the inner oblong or linear, subchartaceous, usually entire, the apex spiny, the outer surface tomentulose, gland-dotted; receptacle convex; flowers numerous; corolla blue or purplish blue, the peripheral with the inner sinus much deeper than the others, the corolla zygomorphic and somewhat ray-like, the inner flowers actinomorphic. Cypsela columnar, 5–8 mm long, 3- to 4-angled, glabrous; pappus of 4–5 scales, 8–12 mm long.

Wet flatwoods and savannas, bogs, and seepage areas. Occasional; northern peninsula, central and western panhandle. North Carolina south to Florida, west to Louisiana. Summer.

Symphyotrichum Nees 1832. ASTER

Herbs or woody vines. Leaves alternate, simple, pinnate-veined, petiolate or epetiolate. Flowers in capitula in racemiform, paniculiform, or corymbiform arrays, radiate; receptacle epaleate; ray flowers carpellate; petals 5, basally connate, zygomorphic; disk flowers bisexual; petals 5, basally connate, actinomorphic; stamens 5, epipetalous, the filaments free, the anthers connate; ovary inferior, 2-carpellate, 1-loculate. Fruit a cypsela; pappus present.

About 90 species; North America, West Indies, Mexico, Central America, South America, Europe, and Asia. [From the Greek *symphysis*, growing together, and *trichos*, hair, in reference to the basally connate bristles as seen in a European cultivar of the type species.]

Ampelaster G. L. Nesom, 1995.

Selected references: Brouillet et al. (2006); Semple (2006g).

1. Basal lowermost cauline leaves both cordate and petiolate.
    2. Cauline leaves sessile and cordate-clasping or with a conspicuously auriculate-clasping petiole .... **S. undulatum**
    2. Cauline leaves not sessile and cordate-clasping or with an auriculate-clasping petiole.
        3. Leaf margins shallowly toothed; phyllaries glabrous or sometimes with ciliate margins, with an elongate, narrow green apex; pappus white .... **S. urophyllum**
        3. Leaf margins entire or occasionally toothed; phyllaries minutely pubescent, with a diamond-shaped green apex; pappus tawny, reddish brown, or rose-tinged.
            4. Basal leaf bases shallowly cordate, rounded, or cuneate, the proximal cauline leaf bases rounded or cuneate .... **S. oolentangiense**
            4. Basal leaf bases mostly deeply cordate, the proximal cauline leaf bases cordate, truncate, or rounded .... **S. shortii**
1. Basal and lowermost leaves not cordate and petiolate.
    5. Leaves silvery-silky on both sides.
        6. Phyllaries appressed .... **S. concolor**
        6. Phyllaries loosely spreading or recurved .... **S. plumosum**
    5. Leaves not silvery-silky on both sides.
        7. Principal stem leaves to 3 cm long (usually ca. 1.5 cm long).
            8. Phyllaries little imbricate (more or less 2-ranked), mostly green .... **S. sericeum**
            8. Phyllaries regularly imbricate (3-ranked or more), only the apex green.
                9. Stem leaves squarrose spreading or reflexed; stems, leaves, and phyllaries glabrous or with only a few scattered trichomes .... **S. walteri**
                9. Stem leaves ascending; stems, leaves, and phyllaries pubescent .... **S. adnatum**
        7. Principal stem leaves 3 cm long or longer.
            10. Leaves auriculate- or cordate-clasping.
                11. Plant scrambling or climbing, often several meters long .... **S. carolinianum**
                11. Plant erect, usually less than 1 m tall.
                    12. Phyllaries not evidently glandular .... **S. laeve**
                    12. Phyllaries evidently glandular.
                        13. Involucre 5–9 mm long .... **S. patens**
                        13. Involucre 9–12 mm long .... **S. georgianum**
            10. Leaves not auriculate- or cordate-clasping.
                14. Plant somewhat succulent; phyllaries scarcely herbaceous.
                    15. Leaves primarily basally disposed, linear and grasslike; lamina of the ray flowers 1–2 cm long .... **S. chapmanii**
                    15. Basal leaves absent at flowering, or if present, then not linear and grasslike; lamina of the ray flowers less than 1 cm long.
                        16. Plant a perennial from a fibrous-rooted rhizome or crown; lamina of the ray flowers 5–9 mm long; stem usually conspicuously flexuous .... **S. tenuifolium**
                        16. Plant a taprooted annual; lamina of the ray flowers 1–4 mm long; stem essentially straight.

17. Capitula usually dense in an elongated, pyramidal, paniculiform array; inner phyllaries 6–7 mm long; phyllary apices linear-acuminate, the green zone of the phyllaries narrowly lanceolate; pappus accrescent, 4–6 mm long at maturity .... **S. subulatum**
17. Capitula in a corymbiform to thyrsiform, diffusely paniculate, or secund to subsecund and paniculiform array or at the tips of elongate branches; inner phyllaries 4–7 mm long; phyllary apices acute to acuminate, the green zone of the phyllaries lanceolate to elliptic, the chartaceous base usually conspicuous; pappus not accressent, 3–4(5) mm long.
    18. Phyllary apices appressed, acute, flat, the inner phyllaries with a broad lanceolate, distinctly demarcated, apical green zone .... **S. squamatum**
    18. Phyllary apices loose, linear-acuminate, the distal margin often inrolled-involute, the inner phyllaries with a narrowly lanceolate, often weakly demarcated, apical green zone.
        19. Capitula usually in a corymbiform or racemiform array; phyllary apices acute to abruptly short-acuminate; lamina of the ray flowers white, light pink, or slightly blue; disk flowers (6)8–15 .... **S. expansum**
        19. Capitula in a diffuse paniculiform, pyramidal-paniculiform, corymbiform, or secund to subsecund array; phyllary apices long-acuminate; lamina of the ray flowers blue or purple; disk flowers 11–23 .... **S. bahamense**

14. Plant not succulent; phyllaries evidently herbaceous.
    20. Phyllaries with a subulate, marginally inrolled green apex .... **S. pilosum**
    20. Phyllaries flat, not with a subulate, marginally inrolled green apex.
        21. Leaves scabrous, conspicuously rough to the touch.
            22. Lower leaf surface glabrous .... **S. praealtum**
            22. Lower leaf surface scabrous .... **S. oolentangiense**
        21. Leaves glabrous or scabrellous, smooth or nearly so to the touch.
            23. Capitula 8–11 mm long; phyllaries subfoliaceous; principal stem leaves 2–5 cm wide .... **S. elliottii**
            23. Capitula 3–6(8) mm high; phyllaries other than subfoliaceous; principal stem leaves to 2 cm wide.
                24. Corolla of the disk flowers deeply lobed ½ or more the length of the limb .... **S. lateriflorum**
                24. Corolla of the disk flowers lobed less than 1/2 the length of the limb.
                    25. Peduncles either all short (less than 2 mm long) or sparsely bracteolate or the bracts leaflike.
                        26. Involucre 3–4 mm long; lamina of the ray flowers 3–6 mm long; branches of the floral array often recurved and unilaterally racemiform .... **S. racemosum**
                        26. Involucre 4–7 mm long; lamina of the ray flowers usually 6–15 mm long; branches of the floral array usually not recurved and unilaterally racemiform.

27. Lower surface of the leaves with the veinlets forming an obscure reticulum, or if the reticulum is evident, then the areoles clearly longer than wide .... **S. lanceolatum**
27. Lower surface of the leaf with the veinlets forming a conspicuous reticulum, the areoles nearly isodiametric .... **S. praealtum**

25. Peduncles conspicuously long, copiously bracteolate.
28. Leaves and stems puberulent.. **S. fontinale**
28. Leaves and stems glabrous or glabrate.
29. Capitula 1 cm wide or wider; midstem leaves usually narrowly elliptic, at least some 1 cm wide or wider; plant usually of wet habitats.. **S. simmondsii**
29. Capitula less than 1 cm wide; midstem leaves linear, usually less than 0.5 cm wide; plant of dry habitats .... **S. dumosum**

Symphyotrichum adnatum (Nutt.) G. L. Nesom [Broadly attached, in reference to the cauline leaf bases.] SCALELEAF ASTER.

Aster adnatus Nuttall, J. Acad. Nat. Sci. Philadelphia 7: 82. 1834. *Lasallea adnata* (Nuttall) Semple & Brouillet, Amer. J. Bot. 67: 1023. 1980. *Virgulus adnatus* (Nuttall) Reveal & Keener, Taxon 30: 650. 1981. *Symphyotrichum adnatum* (Nuttall) G. L. Nesom, Phytologia 77: 275. 1995 ("1994"). FLORIDA: s.d., *Nuttall s.n.* (lectotype: BM; isolectotypes: GH, P). Lectotypified by Jones and Lowry (1986: 395).

Erect or scandent-sprawling perennial herb, to 1 m; stem with thick, woody, sometimes cormoid caudex, striate-angled, scabrous. Basal leaves early deciduous, the blade oblanceolate or obovate, 1–4.3 cm long, 7–15 mm wide, the apex acute or obtuse, the base cuneate, the margin entire, the upper and lower surfaces finely scabrous, sessile or subpetiolate, the proximal cauline with the blade oblanceolate, 2–3.5 cm long, 5–13 mm wide, the apex acute or obtuse, the base auriculate-clasping, the margin entire, the upper and lower surfaces finely scabrous, shiny, sessile, the distal with the blade lanceolate or linear, 3–10 mm long, 1–4 mm wide, gradually reduced upward to bracts, the apex acuminate, spinulose, the base subclasping and adnate to the stem for ½ or more its length, the upper and lower surfaces finely scabrous, viscid, minutely stipitate-glandular, sessile. Flowers in capitula in an open paniculiform array, the branches at first spreading, then ascending and divaricate, the peduncle to 5 cm long, densely short-pilose, glandular, the bracts linear, adnate, short-strigose; involucre campanulate; phyllaries numerous in 4–5 series, oblong or lanceolate, 4–7 mm long, unequal, the subapical green zone lenticular, the apex obtuse, mucronate or subspinulose, the margin hyaline, the outer surface scabrellous-puberulent, stipitate-glandular; receptacle flat or slightly convex, pitted; ray flowers 10–20 in 1 series; corolla lavender, the lamina 5–8 mm long; disk flowers numerous; corolla yellow, 4–7 mm long. Cypsela obovoid, not compressed, 2–3 mm long, faintly 6- to 10-nerved, sparsely strigose, tan or brown; pappus of numerous tan, subequal, barbellate bristles 4–8 mm long.

Sandhills and dry flatwoods. Frequent; peninsula, central and western panhandle. Georgia south to Florida, west to Louisiana. Fall.

Symphyotrichum bahamense (Britton) G. L. Nesom [Of Bahama.] BAHAMAN ASTER.

Aster bahamensis Britton, Bull. Torrey Bot. Club 41: 14. 1914. *Aster subulatus* Michaux var. *bahamensis* (Britton) Bosserdet, Taxon 19: 249. 1970. *Symphyotrichum bahamense* (Britton) G. L. Nesom, Phytologia 77: 276. 1995 ("1994").

Aster subulatus Michaux var. *elongatus* Bosserdet ex A. G. Jones & Lowry, Bull. Mus. Natl. Hist. Nat., B, Adansonia 8: 406. 1986. *Symphyotrichum subulatum* (Michaux) G. L. Nesom var. *elongatum* (Bosserdet ex A. G. Jones & Lowry) S. D. Sundberg, Sida 21: 907. 2004. TYPE: FLORIDA: Hillsborough Co.: Tampa, 20 Aug 1895, *Nash 2416* (holotype: P; isotype: NY).

Erect annual herb, to 2 m; stem striate-angled, glabrous or glabrate. Basal and proximal cauline leaves withering by flowering, the blade ovate or oblanceolate, 1–9 cm long, 6–14 mm wide, the apex rounded, obtuse, or acute, the base cuneate, rounded, the margin entire, serrulate, or crenulate, often strigillose-ciliolate, the upper and lower surfaces glabrous, the petiole long, the base sheathing, sparsely ciliate, the proximal cauline with the blade narrowly lanceolate or subulate, 2–10(20) cm long, 2–10(20) mm wide, the apex acute or acuminate, the base narrowly cuneate, the margin subentire, entire, or serrulate, the upper and lower surfaces glabrous, subpetiolate or sessile, the distal with the blade narrowly lanceolate or subulate, 0.5–11 cm long, 1–6 mm wide, the apex acuminate, the base cuneate, the margin entire or subentire, the upper and lower surfaces glabrous, sessile. Flowers in capitula in an open, diffuse array on the primary branches, the peduncle to 4 cm long, the bracts linear; involucre cylindric-campanulate; phyllaries numerous in 3–5 series, 6–8 mm long, unequal, linear, lanceolate, the subapical green zone narrowly lanceolate, the apex acute, the margin hyaline, the outer surface glabrous; receptacle flat or slightly convex, pitted; ray flowers numerous in (2)3 series; corolla pink or lavender, the lamina (2)3–4 mm long; disk flowers numerous; corolla yellow, 4–5 mm long. Cypsela obovoid or fusiform, sometimes slightly compressed, 1–3 mm long, 5-nerved, sparsely pilose, light brown or purple; pappus of numerous white, subequal, barbellate bristles (3)4–5 mm long.

Wet flatwoods and marshes. Frequent; peninsula west to Wakulla County. Georgia, Florida, and Alabama, also Louisiana and California; West Indies. Summer–fall.

Symphyotrichum carolinianum (Walter) Wunderlin & B. F. Hansen [Of Carolina.] CLIMBING ASTER.

Aster carolinianus Walter, Fl. Carol. 208. 1788. *Aster scandens* J. Jacquin ex Sprengel, Syst. Veg. 3: 524. 1826, nom. illegit. *Lasallea caroliniana* (Walter) Semple & Brouillet, Amer. J. Bot. 67: 1023. 1980. *Virgulus carolinianus* (Walter) Reveal & Keener, Taxon 30: 650. 1981. *Ampelaster carolinianus* (Walter) G. L. Nesom, Phytologia 77: 250. 1995 ("1994"). *Symphyotrichum carolinianum* (Walter) Wunderlin & B. F. Hansen, Novon 11: 366. 2001.

Clambering, sprawling, or climbing perennial woody vine, to 4 m; stem striate-angled, short-pilose or hirsute. Leaves with the blade elliptic, lanceolate, oblanceolate, or ovate, 3–7 cm long, 1–1.5(2) cm wide, reduced upward, the apex acuminate, the base auriculate-clasping, the margin entire, the upper and lower surfaces pilose, sessile. Flowers in capitula in a paniculiform array, terminal on the branches, the peduncle 1–4 cm long, pilose, the bracts lanceolate or ovate, to 10 mm long, foliaceous; involucre turbinate-campanulate; phyllaries numerous in 5–6 series, 8–12 mm long, unequal, spreading or reflexed, the outer linear-lanceolate or spatulate, the inner linear, proximally whitish, herbaceous distally, the green zone restricted to the apical ¼ or less, the apex acute, the margin hyaline proximally, the outer surface pilose; receptacle flat or slightly convex, pitted; ray flowers numerous in 1 series; corolla pale rose-purple or pink, the lamina 9–15(20) mm long; disk flowers numerous; corolla yellow or rose-purple, 6–8 mm long. Cypsela narrowly cylindric or fusiform, not compressed, ca. 4 mm long, 9- to 12-ribbed, glabrous, tan or brown, sometimes mottled purple or black between the light-colored ribs; pappus of numerous light stramineous or faintly rust-colored, unequal, barbellate bristles shorter than the disk flower corolla.

Swamps and riverbanks. Common; peninsula west to central panhandle. North Carolina south to Florida, also Mississippi. Fall.

Some authors (e.g., Semple, 2006g) place this species in the genus *Ampelaster,* although the recognition of *Almutaster, Ampelaster,* and *Psilactis* may render the current sense of *Symphyotrichum* nonmonophyletic (Vaezi and Brouillet, 2009; Morgan and Holland, 2012).

Symphyotrichum chapmanii (Torr. & A. Gray) Semple & Brouillet [Commemorates Alvan Wentworth Chapman (1809–1899), American physician and botanist.] SAVANNAH ASTER.

Aster chapmanii Torrey & A. Gray, Fl. N. Amer. 2: 161. 1842. *Heleastrum chapmanii* (Torrey & A. Gray) Shinners, Sida 3: 348. 1969; non Greene, 1896. *Eurybia chapmanii* (Torrey & A. Gray) G. L. Nesom, Phytologia 77: 259. 1995 ("1994"). *Symphyotrichum chapmanii* (Torrey & A. Gray) Semple & Brouillet, in Semple et al., Univ. Waterloo Biol. Ser. 41: 133. 2002. TYPE: FLORIDA: s.d., *Chapman s.n.* (lectotype: NY; isolectotypes: K, NY). Lectotypified by Semple and Brouillet, in Semple et al. (2002: 134).

Erect perennial herb, to 8 dm; stem rhizomatous, striate-angled, glabrous. Basal leaves with the blade linear or linear-lanceolate, 1–3 cm long, 2–7 mm wide, the apex acuminate, the base cuneate, the margin sparsely denticulate or entire, ciliate, indurate-translucent, the upper and lower surfaces strigillose, the petiole to 15 cm long, the base sheathing, the proximal cauline with the blade linear, 4–10 cm long, 1–3 mm wide, the apex acute, the base clasping, the margin entire, often revolute, the upper and lower surfaces sparsely strigillose, sessile or subsessile, the distal with the blade linear, sometimes awl-shaped, 5–18 cm long, 5–10 mm wide, much reduced upward, the apex acute, the base subclasping or adnate for ⅙ or more its length, the margin entire, often revolute, the upper and lower surfaces sparsely strigulose, sessile. Flowers in capitula in an open, usually corymbiform or sometimes paniculiform array, the peduncle 1–20 cm long, glabrous, the bracts linear; involucre cylindric-campanulate; phyllaries numerous in 4–5(6) series, 6–9 mm long, the outer oblong-lanceolate, the inner linear-lanceolate, the subapical green zone narrowly lanceolate, the apex acute or acuminate, sometimes mucronate, purplish, the margin scarious, hyaline, purplish distally, the outer surface villous; receptacle flat or slightly convex, pitted; ray flowers 8–numerous in 1 series; corolla purple or blue-lavender, the lamina (10)14–15(20) mm long; disk flowers numerous; corolla yellow, 5–6 mm long. Cypsela obovoid, compressed, 3–5 mm long, 8–10(14)-nerved, glabrous, tan or gray, the nerves yellowish; pappus of numerous yellowish, subequal, barbellate bristles 4–5 mm long.

Wet flatwoods, bogs, acid swamps. Occasional; Alachua and St. Lucie Counties, central and western panhandle. Florida and Alabama. Fall.

Symphyotrichum concolor (L.) G. L. Nesom [Of the same color, in reference to the upper and lower leaf surfaces.] EASTERN SILVER ASTER.

Aster concolor Linnaeus, Sp. Pl. 874. 1753. *Lasallea concolor* (Linnaeus) Semple & Brouillet, Amer. J. Bot. 67: 1022. 1980. *Virgulus concolor* (Linnaeus) Reveal & Keener, Taxon 30: 649. 1981. *Symphyotrichum concolor* (Linnaeus) G. L. Nesom, Phytologia 77: 278. 1995 ("1994").

Aster simulatus Small, Bull. Torrey Bot. Club 51: 388. 1924. *Aster concolor* Linnaeus var. *simulatus* (Small) R. W. Long, Rhodora 72: 39. 1970. TYPE: FLORIDA: Miami-Dade Co.: Silver Palm Schoolhouse, 26 Nov 1913, *Small & Small s.n.* (holotype: NY).

Aster concolor Linnaeus var. *devestitus* S. F. Blake, Rhodora 32: 145. 1930. *Symphyotrichum concolor* (Linnaeus) G. L. Nesom var. *devestitum* (S. F. Blake) Semple, Sida 21: 762. 2004. TYPE: FLORIDA: Bay Co.: Lynn Haven, 13 Oct 1921, *Billington 80* (holotype: US).

Ascending or erect perennial herb, to 8 dm; stem long-rhizomatous, with a subcormoid woody caudex, striate-angled, glabrous or canescent distally. Basal leaves withering by flowering, the blade elliptic-lanceolate, 1–4 cm long, 5–15 mm wide, the apex acute or obtuse, the base cuneate, the margin entire, rarely remotely serrate, ciliate, the upper and lower surfaces silvery silky-pilose, sessile, the proximal cauline withering by flowering, the blade oblanceolate, 2–3.5 cm long, 5–15 mm wide, the apex acute or obtuse, cuspidate, the base rounded, subclasping, the margin entire, the upper and lower surfaces silvery silky-pilose, sessile, the distal with the blade oblong or lanceolate, rarely ovate, 9–15 mm long, 2–5 mm wide, reduced upward, the apex acute, mucronate, the base cuneate, the margin entire, the upper and lower surfaces usually silvery silky-pilose, sessile. Flowers in capitula, solitary or in a narrow erect paniculiform array of 2–5 per branch, the peduncle pubescent, the bracts linear; involucre campanulate; phyllaries numerous in 3–5 series, linear-lanceolate, 5–7 mm long, unequal, oblong-lanceolate, the green zone restricted to the apex, the apex acute or acuminate, usually mucronate or subspinulose, the margin scarious, the outer surface densely silky; receptacle flat or slightly convex, pitted; ray flowers 8–12 in 1 series; corolla rose-purple, rarely white, the lamina 4–9 mm long; disk flowers (9)11–numerous; corolla pink or purple, 5–6 mm long. Cypsela obovoid, not compressed, 3–4 mm long, 7- to 10-nerved, strigose, tan; pappus of numerous tan, subequal, barbellate bristles 4–6 mm long.

Sandhills and dry flatwoods. Frequent; nearly throughout. Massachusetts and New York south to Florida, west to Louisiana. Summer–fall.

Symphyotrichum dumosum (L.) G. L. Nesom [Of bushy habit.] RICE BUTTON ASTER.

Aster dumosus Linnaeus, Sp. Pl. 873. 1753. *Aster dumosus* Linnaeus var. *verus* Torrey & A. Gray, Fl. N. Amer. 2: 128. 1841, nom. inadmiss. *Symphyotrichum dumosum* (Linnaeus) G. L. Nesom, Phytologia 77: 280. 1995 ("1994").

Aster coridifolius Michaux, Fl. Bor.-Amer. 2: 112. 1803. *Aster foliolosus* Aiton var. *coridifolius* (Michaux) Nuttall, Gen. N. Amer. Pl. 2: 155. 1818. *Aster dumosus* Linnaeus var. *coridifolius* (Michaux) Torrey & A. Gray, Fl. N. Amer. 2: 128. 1841.

Aster dumosus Linnaeus var. *subulifolius* Torrey & A. Gray, Fl. N. Amer. 2: 128. 1841. *Symphyotrichum dumosum* (Linnaeus) G. L. Nesom var. *subulifolium* (Torrey & A. Gray) G. L. Nesom, Phytologia 77: 280. 1995 ("1994").

Aster dumosus Linnaeus var. *gracilentus* Torrey & A. Gray, Fl. N. Amer. 2: 128. 1841.

Aster dumosus Linnaeus var. *gracilipes* Wiegand, Rhodora 30: 166. 1928. *Aster gracilipes* (Wiegand) Alexander, in Small, Man. S.E. Fl. 1389, 1509. 1933. TYPE: FLORIDA: Duval Co.: near Jacksonville, 1894, *Curtiss 5345* (holotype: CU; isotypes: FSU, NY).

Aster dumosus Linnaeus var. *pergracilis* Wiegand, Rhodora 30: 166. 1928. *Symphyotrichum dumosum* (Linnaeus) G. L. Nesom var. *pergracilis* (Wiegand) G. L. Nesom, Phytologia 77: 280. 1995 ("1994").

Erect perennial herb, to 1 m; stem usually long-rhizomatous, often with a thick, woody, stout caudex, striate-angled, strigose, glabrescent, or glabrous. Basal leaves withering by flowering, the blade oblanceolate or spatulate, 1–5 cm long, 3–15 mm wide, the apex obtuse, rounded or subacute, the base cuneate, the margin crenate-serrate, the upper surface glabrous or submarginally short-strigose, the lower surface glabrous or sparsely strigose, with a short, winged, basally clasping petiole or sessile and usually only slightly or not clasping, the proximal cauline with the blade linear-oblanceolate, 2.5–12 cm long, 2–7 mm wide, otherwise as the basal, the distal with the blade oblong, linear-lanceolate, or linear, 0.5–4.5 cm long, 1–4 mm wide, reduced upward, the bases cuneate or rounded, not clasping, the margin entire or serrate, otherwise as the basal and proximal cauline. Flowers in capitula in a diffuse, open paniculiform array, the branches ascending, the secondary ones stiffly racemiform, the peduncle to 5 cm long, sparsely strigose or glabrous, the bracts linear, linear-oblong, or linear-elliptic; involucre cylindric-campanulate; phyllaries numerous in 4–6 series, (3)5–6 mm long, unequal, slightly spreading, the outer linear-oblanceolate, the inner linear-oblanceolate or linear, the subapical green zone oblanceolate or elliptic, the apex acute or obtuse, mucronate, the margin distally scarious, the outer surface glabrous or glabrate; receptacle flat or slightly convex, pitted; ray flowers numerous in 1 series; corolla blue, pink, lavender, or white, the lamina (4)5–7(8) mm long; disk flowers numerous; corolla yellow or pink, 4–5 mm long. Cypsela oblong-obovoid, sometimes slightly compressed, 2–3 mm long, 3- to 4-nerved, strigillose, pink or yellowish with pink streaks or gray with yellowish nerves; pappus of numerous white or whitish, subequal, barbellate bristles ca. 4 mm long.

Sandhills, flatwoods, and hammocks. Common; nearly throughout. New Brunswick south to Florida, west to Ontario, Wisconsin, Iowa, Missouri, Oklahoma, and Texas. Summer–fall.

Symphyotrichum elliottii (Torr. & A. Gray) G. L. Nesom [Commemorates Stephen Elliott (1771–1830), American legislator, banker, educator, and botanist.] ELLIOTT'S ASTER.

Aster elliottii Torrey & A. Gray, Fl. N. Amer. 2: 140. 1841. *Aster puniceus* Linnaeus subsp. *elliottii* (Torrey & A. Gray) A. G. Jones, Phytologia 55: 384. 1984. *Symphyotrichum elliottii* (Torrey & A. Gray) G. L. Nesom, Phytologia 77: 280. 1995 ("1994").

Erect perennial herb, to 1.5(2) m; stem long-rhizomatous, striate-angled, glabrous or hirtellous distally. Basal leaves withering by flowering, the blade elliptic, 5–25 cm long, 1–5 cm wide, the apex acute or short-acuminate, apiculate or mucronate, the base narrowly cuneate, the margin serrate or serrulate, the upper surface scabridulous, the lower surface glabrous, the petiole long, winged, the base sheathing, the proximal cauline withering by flowering, the blade elliptic, lanceolate, or oblanceolate, 7–11 cm long, 1.5–3.5 cm wide, otherwise like the basal, the distal with the blade elliptic, 1–7 cm long, to 3 cm wide, reduced upward, the margin serrulate or entire, otherwise like the lower leaves, subpetiolate or sessile. Flowers in capitula in a paniculiform array, the branches strongly ascending, the peduncle to 2.5 cm, hirsute, the bracts linear; involucre campanulate; phyllaries numerous in 5–6 series, 8–11 mm long, slightly unequal, linear-lanceolate, oblanceolate, or linear, the outer distally foliaceous, the subapical green zone lanceolate or linear, the apex acuminate, apiculate, spreading or squarrose, the margin narrowly scarious, hyaline, sparsely and remotely ciliolate, the outer surface glabrous; receptacle flat or slightly convex, pitted; ray flowers numerous in 1 series; corolla pink or lavender, the lamina 7–14 mm long; disk flowers numerous; corolla yellow, pink, or brown, ca. 6 mm long. Cypsela oblanceolate, compressed, 2–3 mm long, 3- to 4-nerved, glabrous or sparsely pubescent, tan or pale brown; pappus of numerous whitish, subequal, barbellate bristles 5–6 mm long.

Swamps and marshes. Frequent; peninsula west to central panhandle. Virginia south to Florida, west to Louisiana. Summer–fall.

Symphyotrichum expansum (Poeppig ex Spreng.) G. L. Nesom [Spread out, in reference to the leaves and inflorescence.] SOUTHWESTERN ANNUAL SALTMARSH ASTER.

Erigeron expansus Poeppig ex Sprengel, Syst. Veg. 3: 518. 1826. *Aster inconspicuus* Lessing, in Schlechtendal & Chamisso, Linnaea 5: 143. 1830. *Tripolium subulatum* (Michaux) Nees von Esenbeck var. *cubensis* de Candolle, Prodr. 5: 254. 1836. *Aster exilis* Elliott var. *inconspicuus* (Lessing) Hieronymus, Bot. Jahrb. Syst. 29: 19. 1900, nom. illegit. *Symphyotrichum expansum* (Poeppig & Sprengel) G. L. Nesom, Phytologia 77: 281. 1995 ("1994").

Tripolium subulatum (Michaux) Nees von Esenbeck var. *parviflorum* Nees von Esenbeck, Gen. Sp. Aster. 157. 1832. *Symphyotrichum subulatum* (Michaux) G. L. Nesom var. *parviflorum* (Nees von Esenbeck) S. D. Sundberg, Sida 21: 907. 2004.

Aster divaricatus (Nuttall) Torrey & A. Gray var. *sandwicensis* A. Gray, in H. Mann, Proc. Amer. Acad. Arts 7: 173. 1867. *Aster exilis* Elliott var. *australis* A. Gray, Syn. Fl. N. Amer. 1(2): 203. 1884, nom. illegit. *Aster sandwicensis* (A. Gray) Hieronymus, Bot. Jahrb. Syst. 29: 20. 1901. *Aster subulatus* Michaux var. *australis* Shinners, Field & Lab. 21: 158. 1953, nom. illegit. *Aster subulatus* Michaux var. *sandwicensis* (A. Gray) A. G. Jones, Brittonia 32: 465. 1984.

Erect annual herb, to 1.5 m; stem striate-angled, glabrous or glabrate. Basal leaves withering by flowering, the blade ovate, or oblanceolate, 1–9 cm long, 6–14 mm wide, the apex rounded, obtuse, or acute, the base cuneate, the margin entire, serrulate, or crenulate, often ciliolate, the upper and lower surfaces glabrous, long-petiolate, the bases sheathing, sparsely ciliate, the proximal cauline withering by flowering, the blade narrowly lanceolate or subulate, 2–10 cm long, 2–10(20) mm wide, the apex acute or acuminate, the base narrowly cuneate, the margin entire or serrulate, the upper and lower surfaces glabrous, petiolate, subpetiolate or sessile, the distal with the blade narrowly lanceolate or subulate, 0.5–11 cm long, 1–6 mm wide, similar to the proximal cauline, sessile. Flowers in capitula in an open, diffuse array, the peduncle to 4 cm long, the bracts subulate; involucre cylindric-campanulate; phyllaries numerous in 3–5 series, 5–7(8) mm long, unequal, narrowly lanceolate, the subapical green zone narrowly lanceolate, the apex acute, the margin hyaline, often purple tinged, the outer surface glabrous; receptacle flat or slightly convex, pitted; ray flowers numerous in (1)2 series; corolla white, sometimes pink, the lamina 2–3 mm long; disk flowers (6)8–15; corolla yellow, 3–4(5) mm long. Cypsela narrowly obovoid or fusiform, sometimes slightly compressed, 2–3 mm long, 5-nerved, sparsely strigillose, light brown or purple; pappus of numerous white, subequal, barbellate bristles 3–4 mm long.

Marshes and wet disturbed sites. Rare; Okaloosa, Lake, and Orange Counties, southern peninsula. Florida, Oklahoma, and Texas, west to California; West Indies, Mexico, Central America, and South America; Asia and Pacific Islands. Native to North America and tropical America. Summer–fall.

Symphyotrichum fontinale (Alexander) G. L. Nesom [Growing in or near springs.] FLORIDA WATER ASTER.

Aster fontinalis Alexander, in Small, Man. S.E. Fl. 1382, 1509. 1933. *Symphyotrichum fontinale* (Alexander) G. L. Nesom, Phytologia 77: 282. 1995 ("1994"). TYPE: FLORIDA: Collier Co.: S of Deep Lake, 7 Dec 1925, *Small & Buswell s.n.* (holotype: NY).

Aster patens Aiton var. *floridanus* R. W. Long, Rhodora 72: 40. 1970. TYPE: FLORIDA: Collier Co.: Big Cypress, Brown Lake 15 mi. S of Immokalee, 12 Dec 1945, *Brass 15780* (holotype: GH).

Erect perennial herb, to 9 dm; stem long-rhizomatous, striate-angled, hispidulous-strigillose. Basal leaves absent at flowering, the blade oblanceolate or ovate, 2.5–4 cm long, 12–15 mm wide, the apex obtuse, mucronate, the base cuneate, the margin entire or remotely serrulate, the upper and lower surfaces scabrous, sessile, the proximal cauline usually withering by flowering, the blade obovate or oblanceolate, 2.5–8 cm long, 5–18 mm wide, the apex spinulose-mucronate, the base rounded or auriculate-clasping, the margin distally shallow serrate or subentire, the upper and lower surfaces scabrellous, sessile, the distal with the blade lanceolate, oblanceolate, or oblong-lanceolate, 0.5–5 cm long, 2–10 mm wide, reduced upward and becoming bract-like, the apex acute, mucronate, or subspinulose, the base slightly clasping or rounded, the margin entire or subentire, the upper and lower surfaces scabrellous, sometimes minutely gland-dotted, sessile. Flowers in capitula in a paniculiform array, the branches widely spreading, the peduncle to 4.5 cm, strigose-hirsute, sometimes sparsely stipitate-glandular, the bracts linear-oblong or -lanceolate, ca. 3 mm long, spinulose, spreading or reflexed, strigulose, sometimes minutely stipitate-glandular; involucre cylindric-campanulate; phyllaries numerous in 4–5 series, 6–8 mm long, unequal, the outer oblong or oblanceolate, the inner linear-oblanceolate, the subapical green zone elliptic or lanceolate, the apex acuminate, mucronulate or apiculate, often reddish tinged, the outer surface glabrous or glabrate; receptacle flat or slightly convex, pitted; ray flowers numerous in 1 series; corolla purplish-blue or lavender, the lamina 7–13 mm long; disk flowers 9–numerous; corolla yellow or reddish purple, sometimes brownish, 5–6 mm long. Cypsela obovoid-fusiform, slightly compressed, ca. 2 mm long, 5-nerved, sparsely strigillose or glabrescent, tan; pappus of numerous whitish, subequal, barbellate bristles 5–6 mm long.

Wet pinelands and marshes. Occasional; Pasco, Lee, and Collier Counties, eastern and central panhandle. Georgia and Florida. Summer–fall.

Symphyotrichum georgianum (Alexander) G. L. Nesom [Of Georgia.] GEORGIA ASTER.

Aster georgianus Alexander, in Small, Man. S.E. Fl. 1381, 1509. 1933. *Aster patens* Aiton var. *georgianus* (Alexander) Cronquist, Brittonia 29: 218. 1977. *Virgulus patens* (Aiton) Reveal & Keener var. *georgianus* (Alexander) Reveal & Keener, Taxon 30: 650. 1981. *Virgulus georgianus* (Alexander) Semple, Amer. J. Bot. 71: 523. 1984. *Symphyotrichum georgianum* (Alexander) G. L. Nesom, Phytologia 77: 282. 1995 ("1994").

Erect perennial herb, to 1 m; stem woody-rhizomatous, striate-angled, short-strigose, stipitate-glandular. Basal leaves early deciduous, the blade spatulate or obovate, 2–7 cm long, 1–2 cm wide, the apex acute, the base narrowly cuneate, the margin entire or irregularly serrate, the upper and lower surfaces finely scabrous, sessile, the proximal cauline withering by flowering, the blade oblong, lanceolate, or oblanceolate, 2–7 cm long, 1–2 cm wide, the apex acute, the base auriculate-clasping, the margin entire, the upper and lower surfaces scabrellous, stipitate-glandular, sessile, the distal with the blade lanceolate or oblanceolate, 1.5–5 cm long, 6–11 mm wide, reduced upward and becoming bract-like, the apex mucronate or spinulose, the base subauriculate-clasping, the margin entire, the upper and lower surfaces scabrous, sometimes stipitate-glandular, sessile. Flowers in capitula in a narrow, racemiform or broad, paniculiform array, sometimes solitary on the branches, the branches spreading-ascending, the peduncle strigose, stipitate-glandular, the bracts linear or oblong-lanceolate, spreading or reflexed; involucre campanulate or hemispheric; phyllaries numerous in 4–7 series, 9–12 mm long, unequal, oblong-spatulate, lanceolate, or linear-lanceolate, unequal, the subapical green zone lanceolate, the apex foliaceous, spreading or reflexed, the margin hyaline, narrowly scarious, hispid-ciliate, sometimes also stipitate-glandular, the outer surface strigillose, stipitate-glandular; receptacle flat or slightly convex, pitted; ray flowers 12–numerous in 1 series; corolla lavender-violet or reddish purple, the lamina 14–24 mm long; disk flowers numerous; corolla white with purplish lobes, 7–9 mm long. Cypsela obovoid, compressed, 3–4 mm long, 7- to 10-nerved, sparsely strigose, gray-brown; pappus of numerous tan or tawny, subequal, barbellate bristles 9–11 mm long.

Oak-hickory woods. Rare; Leon County. North Carolina south to Florida, west to Alabama. Fall.

Symphyotrichum laeve (L.) Á. Löve & D. Löve var. **concinnum** (Willd.) G. L. Nesom [Smooth, in reference to the stems and leaves; elegant.] SMOOTH BLUE ASTER.

Aster concinnus Willdenow, Enum. Pl. 884. 1809. *Aster laevis* Linnaeus var. *concinnus* (Willdenow) House, Bull. New York State Mus. 243/244: 15. 1923. *Symphyotrichum laeve* (Linnaeus) Á. Löve & D. Löve var. *concinnum* (Willdenow) G. L. Nesom, Phytologia 77: 283. 1995 ("1994"). *Symphyotrichum laeve* (Linnaeus) Á. Löve & D. Löve subsp. *concinnum* (Willdenow) Semple & Brouillet, in Semple et al., Univ. Waterloo Biol. Ser. 41: 134. 2002.

Erect perennial herb, to 1 m; stem short-rhizomatous, with a thick woody caudex or a few long rhizomes, glabrous, glaucous. Basal leaves withering by flowering, the blade linear-lanceolate or linear, 3–20 cm long, 1–2.5(3) cm wide, the apex acute or obtuse, the base cuneate or rounded, slightly auriculate-clasping, the margin crenate-serrate, the upper and lower surfaces glabrous, the petiole slightly winged, the base sheathing, the proximal cauline often withering by flowering, the blade spatulate, oblong, ovate, or ovate-lanceolate, (4)8–15 cm long, (1)2–4.5 cm wide, the apex acute or obtuse, or rounded, the base broadly cuneate, rounded, or auriculate-clasping, the margin entire or crenate-serrate, the upper and lower surfaces scabrous, with a winged petiole, subsessile, or sessile, the distal with the blade ovate-lanceolate, lanceolate, or linear, 1–4.5 cm long, 1–14 mm wide, reduced upward, the apex acute or obtuse, the base auriculate-clasping or rounded, the margin entire, the upper and lower surfaces glabrous, sessile. Flowers in capitula in a paniculiform array, the peduncle to 6 cm long, glabrous or puberulent in lines, the bracts linear or linear-lanceolate, subclasping; involucre cylindric or campanulate; phyllaries numerous in 4–6 series, (4)5–7(8) mm long, the outer subulate or lanceolate, the inner oblong-lanceolate, linear-lanceolate, or linear-oblanceolate, unequal, the subapical green zone diamond-shaped, the apex acute, acuminate, or obtuse, red-mucronulate or apiculate, the margin scarious, erose, hyaline, distally ciliate, the outer surface glabrous; receptacle flat or slightly convex, pitted; ray flowers (11)13–numerous in 1 series; corolla blue or purple, rarely white, the lamina (6)7–11(15) mm long; disk flowers numerous; corolla yellow or purplish red, 4–6 mm long. Cypsela oblong-obovoid, compressed, 2–4 mm long, 4- to 5-nerved, glabrous or glabrate, purple or brown; pappus of numerous tawny or red or rose-tinged, subequal, barbellate bristles 5–7 mm long.

Dry hammocks. Rare; Jackson County. New York south to Florida, west to Illinois, Kentucky, Tennessee, and Mississippi. Summer–fall.

Symphyotrichum lanceolatum (Willd.) G. L. Nesom var. **latifolium** (Semple & Chmiel.) G. L. Nesom [With lanceolate leaves; with broad leaves.] WHITE PANICLE ASTER.

Aster lanceolatus Willdenow var. *latifolius* Semple & Chmielewski, Canad. J. Bot. 65: 1060. 1987. *Symphyotrichum lanceolatum* (Willdenow) G. L. Nesom var. *latifolium* (Semple & Chmielewski) G. L. Nesom, Phytologia 77: 285. 1995 ("1994").

Erect perennial herb, to 1.5(2) m; stem striate-angled, glabrous or pubescent in lines. Basal leaves withering by flowering, the blades linear, lanceolate, or oblanceolate, 1–8 cm long, 5–20 mm wide, the apex acute or obtuse, mucronate, the base cuneate, the margin entire, the upper and lower surfaces glabrous, the petiole winged, the margin ciliate, the proximal cauline withering by flowering, the blade lanceolate-ovate, linear-lanceolate, or oblanceolate, (4)5–15 cm long, (0.5)1–2(3.5) cm wide, the apex acute or acuminate, mucronate, the base cuneate, somewhat decurrent, the margin entire, the upper and lower surfaces glabrous, sessile or subsessile, the distal with the blade oblanceolate or linear, 3–10(14) cm long, to 3–10(15) mm wide, the apex acute or acuminate, the base cuneate, the margin entire, sessile. Flowers in capitula in a paniculiform array, sometimes congested distally on the lateral branches, the peduncle to 1 cm long, the bracts foliaceous; involucre cylindric or campanulate; phyllaries numerous in (3)4–6 series, 2–5 mm long, unequal, linear or linear-lanceolate, the subapical green zone lanceolate or linear-lanceolate, the apex acute or acuminate, usually mucronulate, the margin scarious, erose, hyaline, sparsely ciliolate, the outer surface glabrous; receptacle flat or slightly convex, pitted; ray flowers numerous in 1 series; corolla white, the lamina 4–9 mm long; disk flowers numerous; corolla yellow or purple, 3–6 mm long. Cypsela obovoid, slightly compressed, 1–2 mm long, 4- to 5-nerved, strigillose, gray or tan; pappus of numerous white or tawny, subequal, barbellate bristles 3–5 mm long.

Moist hammocks. Rare; Jefferson, Leon, Franklin, and Jackson Counties. Ontario south to Florida, west to Manitoba, North Dakota, South Dakota, Nebraska, Kansas, Oklahoma, and Texas. Fall.

Symphyotrichum lateriflorum (L.) Á. Löve & D. Löve [With the flowers along the sides, in reference to the axillary capitula.] CALICO ASTER.

Solidago lateriflora Linnaeus, Sp. Pl. 879. 1753. *Aster lateriflorus* (Linnaeus) Britton, Trans. New York Acad. Sci. 9: 11. 1884. *Symphyotrichum lateriflorum* (Linnaeus) Á. Löve & D. Löve, Taxon 31: 359. 1982.

Aster tradescantii Linnaeus, Sp. Pl. 876. 1753. *Symphyotrichum tradescantii* (Linnaeus) G. L. Nesom, Phytologia 77: 293. 1995 ("1994").

Aster vimineus Lamarck, Encycl. 1: 306. 1783. *Aster miser* Linnaeus var. *vimineus* (Lamarck) Farwell, Rep. (Annual) Michigan Acad. Sci. 6: 214. 1904.

Aster pendulus Aiton, Hort. Kew. 3: 204. 1789. *Aster divergens* Aiton var. *pendulus* (Aiton) Nuttall, Gen. N. Amer. Pl. 2: 159. 1818. *Aster miser* Linnaeus var. *pendulus* (Aiton) L. C. Beck, Bot. North. Middle States 186. 1833. *Aster lateriflorus* (Linnaeus) Britton var. *pendulus* (Aiton) E. S. Burgess, in Britton & A. Brown, Ill. Fl. N. U.S. 3: 380. 1898.

Aster horizontalis Desfontaines, Tabl. Ecole Bot., ed. 3. 402. 1829. *Aster diffusus* Aiton var. *horizontalis* (Desfontaines) A. Gray, Syn. Fl. N. Amer. 1(2): 187. 1884. *Aster lateriflorus* (Linnaeus) Britton var. *horizontalis* (Desfontaines) Farwell, Asa Gray Bull. 3: 21. 1895. *Symphyotrichum lateriflorum* (Linnaeus) Á. Löve & D. Löve var. *horizontale* (Desfontaines) G. L. Nesom, Phytologia 77: 285. 1995 ("1994").

Aster agrostifolius E. S. Burgess, in Small, Fl. S.E. U.S. 1226, 1340. 1903.

Aster spatelliformis E. S. Burgess, in Small, Fl. S.E. U.S. 1225, 1340. 1903. *Aster lateriflorus* (Linnaeus) Britton var. *spatelliformis* (E. S. Burgess) A. G. Jones, Phytologia 55: 379. 1984. *Symphyotrichum lateriflorum* (Linnaeus) Á. Löve & D. Löve var. *spatelliforme* (E. S. Burgess) G. L. Nesom, Phytologia 77: 285. 1995 ("1994"). TYPE: FLORIDA: Duval Co.: near Jacksonville, 1895, *Curtiss s.n.* (holotype: NY).

Ascending or erect perennial herb, to 1(1.5) m; stem short-rhizomatous or with a short, woody caudex, pilose, villous, glabrate, or glabrous. Basal leaves usually withering by flowering, the blade oblanceolate, ovate-lanceolate, spatulate, or suborbicular, 0.5–3.5 cm long, 1–2.5 cm wide, the apex acute, obtuse, or rounded, the base cuneate, the margin crenate-serrate, the upper surface scabrous or glabrate, the lower surface with the midvein usually pilose or glabrate, rarely glabrous, the petiole usually short, winged, the proximal cauline mostly withering by flowering, the blade ovate, elliptic, elliptic-oblanceolate, or lanceolate, rarely linear-lanceolate, (3)4–10 cm long, (0.5)1–2(3.5) cm wide, reduced upward, the apex acute or obtuse, the base cuneate, the margin crenate-serrate, the surfaces as in the basal, sessile or with a short winged petiole, the distal with the blade ovate, ovate-lanceolate, elliptic, lanceolate, linear-lanceolate, or linear, 1–15 cm long, 1–30 mm wide, reduced upward, the apex acute or obtuse, the base cuneate, the margin sometimes entire, the surfaces as in the proximal cauline, sessile. Flowers in capitula in a paniculiform array, the branches divaricate, long-arching or ascending, slender, secund, the peduncle to 1 cm long, the bracts foliaceous; involucre cylindric-campanulate; phyllaries numerous in 3–4(6) series, sometimes slightly spreading, 3–6(7) mm long, unequal, the outer oblong-lanceolate or -oblanceolate, the inner linear, the subapical green zone lanceolate or slightly diamond-shaped, the apex acute, acuminate, or obtuse, the margin scarious, erose, hyaline, somewhat ciliolate, the outer surface glabrous or sparsely puberulent; receptacle flat or slightly convex, pitted; ray flowers numerous in 1 series; corolla white, rarely pinkish or purplish, the lamina (3)4–5(8) mm long; disk flowers numerous; corolla yellow, pink, or reddish purple, 3–5 mm long. Cypsela oblong-obovoid, sometimes slightly compressed, 1–2 mm long, 3- to 5-nerved, strigillose, gray or tan; pappus of numerous white or pinkish, subequal, barbellate bristles 3–4 mm long.

Bluff forests, floodplain forests, and creek swamps. Occasional; northern counties, central peninsula. Nova Scotia, Prince Edward Island, and New Brunswick south to Florida, west to Manitoba, Minnesota, South Dakota, Nebraska, Kansas, Oklahoma, and Texas, also British Columbia. Fall.

Symphyotrichum oolentangiense (Riddell) G. L. Nesom [Along the Oolentangy River in Ohio.] SKYBLUE ASTER.

Aster oolentangiense Riddell, W. J. Med. Phys. Sci. 8: 495. 1835. *Symphyotrichum oolentangiense* (Riddell) G. L. Nesom, Phytologia 77: 288. 1995 ("1994").

Erect perennial herb, to 1.5 m; stem short-rhizomatous or with a branched woody caudex, striate-angled, hispidulous, pilosulous, strigillose, or glabrate. Basal leaves withering by flowering, the blade ovate or ovate-lanceolate, 4–15(18) cm long, 1–4(6) cm wide, the apex rounded, obtuse, or acute, the base subcordate, rounded, or cuneate, the margin crenate-serrate or subentire, the upper surface strigose, the lower surface strigose-pilose, the petiole slightly winged, the base sheathing, ciliate, the proximal cauline with the blade ovate, ovate-lanceolate, or lanceolate, 3–13 cm long, 1–7 cm wide, reduced upward, the apex acute or acuminate, callus-tipped, the base rounded or cuneate, the margin entire, the upper and lower surfaces as in the basal, the petiole narrowly winged, clasping, the distal with the blade lanceolate or linear-lanceolate,

0.5–8 cm long, 1–10 mm wide, much reduced upward, the apex acute, the base cuneate or slightly rounded, the margin entire, the upper and lower surfaces as in the proximal cauline, sessile or with a short, winged petiole. Flowers in capitula in an open paniculiform array, the branches spreading-ascending, the peduncle to 6 cm long, the bracts linear; involucre campanulate or cylindric-campanulate; phyllaries numerous in (3)4–6 series, the outer slightly spreading, lanceolate or oblanceolate, unequal, the apical green zone diamond-shaped, the apex acute, acuminate, or caudate with the tip involute, mucronulate or apiculate, the margin scarious, erose, hyaline, ciliolate, the outer surface glabrous; receptacle flat or slightly convex, pitted; ray flowers numerous in 1 series; corolla blue or violet-purple, rarely rose, white, or bluish, the lamina (5)8–12(14) mm long; disk flowers numerous; corolla yellow or purple, 4–5 mm long. Cypsela oblong-obovoid, slightly compressed, ca. 2 mm long, 4- to 5-nerved, glabrate or sparsely strigose, dull purple or yellowish with purple streaks; pappus of numerous tawny or rose-tinged, subequal, barbellate bristles 3–4 mm long.

Open disturbed woods. Rare; Jefferson, Leon, Gadsden, and Jackson Counties. Ontario and New York south to Florida, west to Minnesota, South Dakota, Nebraska, Kansas, Oklahoma and Texas; Mexico. Native to the north and west of Florida and Mexico. Fall.

Symphyotrichum patens (Aiton) G. L. Nesom [Spreading, in reference to the stem habit.] LATE PURPLE ASTER.

Aster patens Aiton, Hort. Kew. 3: 201. 1789. *Aster phlogifolius* Muhlenberg ex Willdenow var. *patens* (Aiton) A. W. Wood, Class-Book Bot. 192. 1845. *Lasallea patens* (Aiton) Semple & Brouillet, Amer. J. Bot. 67: 1023. 1980. *Virgulus patens* (Aiton) Reveal & Keener, Taxon 30: 650. 1981. *Symphyotrichum patens* (Aiton) G. L. Nesom, Phytologia 77: 288. 1995 ("1994").

Aster patens Aiton var. *gracilis* Hooker, Companion Bot. Mag. 1: 97. 1835. *Aster patens* Aiton var. *tenuicaulis* C. Mohr, Contr. U.S. Natl. Herb. 6: 730. 1901, nom. illegit. *Aster tenuicaulis* E. S. Burgess, in Small, Fl. S.E. U.S. 1220, 1340. 1903. *Virgulus patens* (Aiton) Reveal & Keener var. *gracilis* (Hooker) Reveal & Keener, Taxon 30: 650. 1981. *Symphyotrichum patens* (Aiton) G. L. Nesom var. *gracile* (Hooker) G. L. Nesom, Phytologia 77: 288. 1995 ("1994").

Ascending or erect perennial herb, to 1.2 m; stem with a short, woody, sometimes tangled or cormoid caudex, sometimes with long rhizomes, striate-angled, scabrous-hirsute or puberulent distally. Basal leaves early deciduous, the blade spatulate or obovate, 3–7 cm long, 1–3 cm wide, the apex acute or rounded, the base cuneate, the margin entire or irregularly serrulate, the upper and lower surfaces scabrous-hirsute, the petiole short, winged, the base sheathing, the proximal cauline with the blade ovate or lanceolate, rarely spatulate, (2)3–7(10) cm long, 1–3(4) cm wide, reduced upward, the apex acute, the base cordate-clasping or auriculate-amplexicaul, the margin entire or irregularly serrulate, the upper and lower surfaces as in the basal, sessile, the distal with the blade lanceolate or ovate, 1.5–3.5 cm long, 4–13 mm wide, the apex acute or sometimes obtuse, mucronate or spinulose, the base cordate-clasping or auriculate-amplexicaul, the margin entire or irregularly serrulate, the upper and lower surfaces scabrous, sometimes stipitate-glandular, sessile. Flowers in capitula in a paniculiform array, the branches divaricate, the peduncle to 10(15) cm long, scabrous-hirsute or puberulent, the bracts linear, foliaceous; involucre campanulate; phyllaries numerous in 4–5 series, 5–9 mm long, unequal, ovate-lanceolate or linear, squarrose, the subapical green zone diamond-shaped, the apex acute or acuminate, the margin hyaline, erose, ciliolate, the outer surface strigillose, stipitate-glandular; receptacle flat or slightly convex, pitted; ray flowers numerous in 1 series; corolla light lavender-violet or mauve, the lamina 10–18(20) mm long; disk flowers numerous; corolla yellow, cream, white, or purple, 5–8 mm long. Cypsela obovoid or oblong-obovoid, not compressed, 2–4 mm long, faintly 7- to 10-nerved, sericeous or strigillose, dull purple or brown; pappus of numerous tawny or sometimes rose-tinged, subequal, barbellate bristles 5–7 mm long.

Sandhills and open hammocks. Rare; northern counties. Maine south to Florida, west to Illinois, Missouri, Kansas, Oklahoma, and Texas. Fall.

Symphyotrichum pilosum (Willd.) G. L. Nesom [With pilose trichomes.] WHITE OLDFIELD ASTER.

Aster villosus Michaux, Fl. Bor.-Amer. 2: 113. 1803; non Thunberg, 1800. *Aster pilosus* Willdenow, Sp. Pl. 3: 2025. 1803. *Aster ericoides* Linnaeus var. *villosus* Torrey & A. Gray, Fl. N. Amer. 2: 124, 503. 1841. *Aster ericoides* Linnaeus var. *pilosus* (Willdenow) Porter, in Porter & Britton, Mem. Torrey Bot. Club 5: 323. 1894, nom. illegit. *Aster ericoides* Linnaeus forma *villosus* (Torrey & A. Gray) Voss, Vilm. Blumengärtn., ed. 3. 1: 466. 1894. *Symphyotrichum pilosum* (Willdenow) G. L. Nesom, Phytologia 77: 289. 1995 (“1994”).

Ascending or erect perennial herb, to 1(1.5) m, stem with a stout, branched caudex, sometimes long-rhizomatous, striate-angled, hirsute-hispid or pilose, sometimes glabrate. Basal leaves withering by flowering, new rosettes developing at flowering, the blade oblanceolate or spatulate, 1–6 cm long, 5–15 mm wide, the apex obtuse or rounded, the base cuneate, the margin crenate-serrate, mostly distally, the upper and lower surfaces sparsely hirsute or glabrous, the petiole winged, the base sheathing, the proximal cauline usually deciduous at flowering, often with axillary clusters of small leaves, the blade elliptic-oblanceolate, elliptic-oblong, linear-lanceolate, or linear-oblanceolate, 4–10 cm long, 0.5–2.5 cm wide, reduced upward, the apex acute, spinulose, the base cuneate, slightly clasping, the margin entire or serrate, the upper and lower surfaces as in the basal, subsessile or the petiole short, winged, the base clasping, the distal with the blade lanceolate-oblong, linear-lanceolate, or linear, 1–10 cm long, 1–8 mm wide, reduced upward, otherwise as in the proximal cauline, sessile. Flowers in capitula in a paniculiform array, the branches divaricate or arched, ascending and secund, sometimes racemiform, the peduncle to 3(5) cm long, the bracts linear, appressed or spreading; involucre campanulate or cylindric-campanulate; phyllaries numerous in 4–5 series, (3)4–5(6) mm long, unequal, sometimes slightly spreading, the outer oblong-lanceolate, the inner linear, the subapical green zone lanceolate or lanceolate-rhombic, the apex acute or acuminate, involute, spinulose, the margin hyaline, scarious, erose, distally ciliolate, the outer surface glabrous or sparsely hirsutulous; receptacle flat or slightly convex, pitted; ray flowers numerous in 1 series; corolla white, rarely pinkish or bluish, the lamina (4)5–8(11) mm long; disk flowers numerous; corolla yellow, reddish purple, or brown, 3–4(6) mm long. Cypsela oblong-obovoid, sometimes slightly compressed, 1–2 mm long, 4- to 6-nerved, strigillose, whitish or gray; pappus of numerous white, subequal, barbellate bristles ca. 4 mm long.

Sandhills and open hammocks. Occasional; Marion and Jefferson Counties, central and western panhandle. Nova Scotia, New Brunswick, and Quebec south to Florida, west to Ontario, Minnesota, South Dakota, Nebraska, Kansas, Oklahoma, and Texas, also British Columbia and Washington; Europe. Native to North America. Spring–fall.

Symphyotrichum plumosum (Small) Semple [Feathery, in reference to inflorescence.]

Aster plumosus Small, Bull. Torrey Bot. Club 51: 387. 1924. *Symphyotrichum concolor* (Linnaeus) G. L. Nesom var. *plumosum* (Small) Wunderlin & B. F. Hansen, Novon 11: 367. 2001. *Symphyotrichum plumosum* (Small) Semple, in Semple et al., Univ. Waterloo Biol. Ser. 41: 134. 2002. TYPE: FLORIDA: Gadsden Co.: Aspalaga, Oct 1897, *Chapman s.n.* (holotype: NY).

Ascending or erect perennial herb, to 6 dm; stem striate-angled, glabrous or sparsely strigose distally. Basal leaves withering by flowering, the blade elliptic-lanceolate, 2–4 cm long, 1–2 cm wide, the apex acute or obtuse, the base cuneate, the margin entire, rarely irregularly serrate, the upper and lower surfaces silvery silky, sessile, the proximal cauline withering by flowering, the blade oblanceolate, 2–3 cm long, 4–8 mm wide, the apex acute or obtuse, cuspidate-mucronate, the base rounded, subclasping, the margin entire, the upper and lower surfaces silvery silky, sessile, the distal lanceolate to narrowly ovate, 8–15 mm long, 3–6 mm wide, reduced upward, the apex acute, mucronate, the base cuneate, the margin entire, the upper and lower surfaces silvery silky, sessile. Flowers in capitula in a narrow paniculiform, wand-shaped array or sometimes in a compact, racemiform array, the peduncle to 3 cm long, the bracts linear; involucre campanulate; phyllaries numerous in 3–4 series, 7–9 mm long, unequal, spreading or reflexed, linear, distally foliaceous, the apex acute, the margin scarious proximally, green distally, the outer surface woolly-strigose; receptacle flat or slightly convex, pitted; ray flowers 7–12 in 1 series; corolla rose-purple, the lamina 6–9 mm long; disk flowers numerous; corolla pink or purple, 6–9 mm long. Cypsela fusiform-obovoid, not compressed, 2–3 mm long, 6- to 8-nerved, strigose, tan; pappus of numerous tan or yellowish tan, subequal, barbellate bristles 6–8 mm long.

Flatwoods. Rare; central panhandle. Endemic. Fall.

Symphyotrichum praealtum (Poir.) G. L. Nesom [Very high, in reference to the plant height.] WILLOWLEAF ASTER.

Aster salicifolius Aiton, Hort. Kew. 3: 203. 1789; non Lamarck, 1783. *Aster praealtus* Poiret, in Lamarck, Encycl., Suppl. 1: 493. 1811. *Symphyotrichum praealtum* (Poiret) G. L. Nesom, Phytologia 77: 289. 1995 ("1994").

Erect or ascending perennial herb, to 1.5(2) m; stem striate-angled, glabrate or moderately hispidulous-hirsute (mostly in lines) distally. Basal leaves withering by flowering, the blade spatulate, 4–7 cm long, 2–2.5 cm wide, the apex rounded or obtuse, mucronate, the base cuneate, the margin entire or serrulate, often revolute, the upper surface glabrate, strigose, or scabrous, the lower surface glabrous, the petiole short, winged, the base sheathing, ciliate, the proximal cauline withering by flowering, the blade elliptic, lanceolate, oblanceolate, or linear-lanceolate, 4–10(15) cm long, 3–18 mm wide, reduced upward, the apex acute or acuminate, callus-tipped, the base cuneate, sometimes rounded, sessile, the margin entire or serrulate, often revolute, the upper and lower surfaces as in the basal, sessile, the distal with the blade ovate, elliptic-lanceolate, lanceolate, or linear, 1–8.5 cm long, 2–17 mm wide, the apex acute or obtuse, callus-tipped, the base cuneate, the margin entire, the upper and lower surfaces as in the basal and proximal cauline, sessile. Flowers in capitula in a dense paniculiform array, the branches ascending and somewhat racemiform, the peduncle to 1 cm long, pilose, the bracts linear or oblong-lanceolate, foliaceous; involucre campanulate; phyllaries numerous in 4–6 series, (4)5–7(8) mm long, unequal, oblong-lanceolate or linear, the subapical green zone lanceolate or lanceolate-rhombic, the apex acute or acuminate, mucronate, often reddish purple, sometimes spreading, the margin scarious, erose, hyaline, sparsely ciliolate, the outer surface glabrous; receptacle flat or slightly convex, pitted; ray flower numerous in 1 series; corolla blue-violet, lavender, or rose-purple, rarely white, the lamina 5–10(12) mm long; disk flowers numerous; corolla pale blue-violet, lavender, or rose-purple, rarely white, 5–10(12) mm long. Cypsela obovoid, sometimes slightly compressed, ca. 2 mm long, 4- to 5-nerved, strigillose, purple or yellowish with purple; pappus of numerous white, subequal, barbellate bristles 4–7 mm long.

Disturbed sites. Rare; Franklin and Okaloosa Counties. New Brunswick, Maine, and Ontario south to Florida, west to North Dakota, South Dakota, Nebraska, Colorado, and New Mexico; Mexico; Europe. Native to North America and Mexico. Fall.

Symphyotrichum racemosum (Elliott) G. L. Nesom [In reference to the racemiform inflorescence.] SMOOTH WHITE OLDFIELD ASTER.

Aster racemosus Elliott, Sketch Bot. S. Carolina 2: 248. 1823. *Symphyotrichum racemosum* (Elliott) G. L. Nesom, Phytologia 77: 290. 1995 ("1994").

Aster brachypholis Small ex Alexander, in Small, Man. S.E. Fl. 1389, 1509. 1933. *Aster fragilis* Willdenow var. *brachypholis* (Small ex Alexander) A. G. Jones, Phytologia 55: 377. 1984. TYPE: FLORIDA: Gadsden Co.: Aspalaga Bluff, along Apalachicola River, 28 Nov 1923, *Small et al. 11027* (holotype: NY).

Erect perennial herb, to 9(10) dm; stem long-rhizomatous or with a woody caudex, striate-angled, glabrous or glabrate, the leaf axils usually with clusters of leaves. Basal leaves withering by flowering, new rosettes often present, the blade spatulate or oblanceolate, 0.5–4 cm long, 5–15 mm wide, the apex obtuse, acute, or acuminate, mucronate, the base cuneate or rounded, the margin crenate-serrate, often somewhat revolute, the upper surface glabrous, the lower surface pilosulous, the petiole narrowly winged, the base sheathing, strigose-ciliate, the proximal cauline withering by flowering, the blade elliptic, elliptic-lanceolate, or linear-lanceolate, 2–7 cm long, 3–20 mm wide, reduced upward, the apex obtuse, acute, or acuminate, mucronate, the base cuneate or rounded, the margin crenate-serrate, short-ciliate distally, often revolute, the upper surface glabrous, the lower surface pilosulous, sessile or with a winged, strigose-ciliate petiole, the base clasping, the distal with the blade linear-lanceolate or linear, 0.5–6 cm long, 1–8 mm wide, reduced distally, unequal in size, the apex obtuse, acute, or acuminate, mucronate, the base cuneate, the margin serrulate or entire, the upper surface glabrous, the lower surface pilosulous, sessile or subpetiolate. Flowers in capitula in a diffuse, paniculiform array, the branches lax, horizontally spreading or arching, racemiform, the peduncle to 3 cm long, pubescent in lines, the bracts linear-elliptic, glabrous; involucre cylindric; phyllaries numerous in 4–6 series, (3)4–5(6) mm long, unequal, oblong-lanceolate or linear, the subapical green zone oblanceolate or linear-oblanceolate, the apex acute or acuminate, the margin narrowly scarious, hyaline, ciliolate, the outer surface glabrous; receptacle flat or slightly convex, pitted; ray flowers numerous in 1 series; corolla white, rarely pink, the lamina 5–8 mm long; disk flowers numerous; corolla cream, pale yellow, pink, or red, 3–5 mm long. Cypsela obovoid, slightly compressed, ca. 1 mm long, faintly 4- to 5-nerved, strigillose or sericeus, gray or tan; pappus of numerous white, subequal, barbellate bristles 3–4 mm long.

Bluff forests. Rare; northern peninsula, central and western panhandle. New Brunswick and Quebec south to Florida, west to Wisconsin, Iowa, Missouri, Oklahoma, and Texas. Fall.

Symphyotrichum sericeum (Vent.) G. L. Nesom var. **microphyllum** (DC.) Wunderlin and B. F. Hansen [*Sericeus,* silky, in reference to the silky trichomes; with small leaves.] WESTERN SILVER ASTER.

Aster pratensis Rafinesque, Fl. Ludov. 67. 1817. *Lasallea sericea* (Ventenat) Greene subsp. *pratensis* (Rafinesque) Semple & Brouillet, Amer. J. Bot. 67: 1022. 1980. *Virgulus pratensis* (Rafinesque) Reveal & Keener, Taxon 30: 649. 1981. *Symphyotrichum pratense* (Rafinesque) G. L. Nesom, Phytologia 77: 290. 1995 ("1994").

Aster sericeus Ventenat var. *microphyllus* de Candolle, Prodr. 5: 233. 1836. *Symphyotrichum sericeum* (Ventenat) G. L. Nesom var. *microphyllum* (de Candolle) Wunderlin & B. F. Hansen, Novon 11: 367. 2001.

Aster ciliatus Nuttall, Trans. Amer. Philos. Soc., ser. 2. 7: 295. 1841; non Walter, 1788; nec Muhlenberg ex Willdenow, 1803. *Aster phyllolepis* Torrey & A. Gray, Fl. N. Amer. 2: 113. 1841. *Lasallea phyllolepis* (Torrey & A. Gray) Greene, Leafl. Bot. Observ. Crit. 1: 5. 1903.

Ascending or erect perennial herb, to 6 dm; stem with a cormoid, woody caudex, striate-angled, glabrous or sparsely strigose distally. Basal leaves withering by flowering, the blade elliptic-lanceolate, 2–3 cm long, 1–2 cm wide, the apex acute or obtuse, the base cuneate, the margin entire, rarely remotely serrate, the upper and lower surfaces pilose-scabrous, sessile, the proximal cauline withering by flowering, the blade oblanceolate, 2–3 cm long, 5–13 mm wide, reduced upward, the apex acute or obtuse, cuspidate-muronate, the base rounded, subclasping, the margin entire, the upper and lower surfaces glabrous, strigose, or somewhat scabrous, sessile, the distal with the blade oblong-lanceolate, 1.5–2.5 cm long, 4–6 mm wide, the apex acute, spinulose, the base cuneate, the margin entire, the upper and lower surfaces strigose, somewhat scabrous, or glabrous, sessile. Flowers in capitula, in an open paniculiform array, 1–5 terminal on a branch, the peduncle glabrous or sparsely strigose, the bracts oblong-lanceolate; involucre cylindric-campanulate; phyllaries numerous in 3–4 series, 6–11 mm long, subequal, the outer lanceolate, the inner linear-lanceolate, distally foliaceous, the apex acute or acuminate, the margin long-ciliate, the outer surface glabrous or sparsely pilose; receptacle flat or slightly convex, pitted; ray flowers numerous in 1 series; corolla rose-purple, the lamina 8–13 mm long; disk flowers numerous; corolla pink or purple, 5–8 mm long. Cypsela obovoid, not compressed, ca. 3 mm long, 5- to 8-nerved, glabrous, tan or light brown; pappus of numerous tan, subequal, barbellate bristles 4–6 mm long.

Open calcareous hammocks. Rare; Gadsden County. Virginia south to Florida, west to Oklahoma and Texas. Fall.

Symphyotrichum shortii (Lindl.) G. L. Nesom [Commemorates Charles Wilkins Short (1794–1863), American physician, professor of medical botany, and botanist.] SHORT'S ASTER.

Aster shortii Lindley, in Hooker, Fl. Bor.-Amer. 2: 9. 1834. *Symphyotrichum shortii* (Lindley) G. L. Nesom, Phytologia 77: 291. 1995 ("1994").

Aster camptosorus Small, Bull. Torrey Bot. Club 24: 339. 1897. *Aster shortii* Lindley var. *camptosorus* (Small) D. B. Ward, Phytologia 94: 465. 2012.

Ascending or erect perennial herb, to 1.5 m; stem short-rhizomatous or a subwoody caudex, striate-angled, proximally glabrous, distally hirtellous. Basal leaves withering by flowering, new basal rosettes sometimes developing at flowering, the blade ovate or ovate-lanceolate, 1–6 cm long, 1–3.5 cm wide, the apex acute or obtuse, mucronate, the base cordate or rounded, the margin crenate or crenate-serrate, the upper surface glabrous or strigose-hispid, the lower surface hispid, especially along the veins, the petiole 1–2 times as long as the blade, sometimes narrowly winged, the base sheathing, pilose or hirsute, the proximal cauline usually persistent, the blade ovate or lanceolate, 5–10 cm long, 2–6(7) cm wide, reduced upward, the apex acute or acuminate, mucronate, the base shallowly cordate, truncate, or rounded, sometimes oblique, the margin entire, shallowly crenate-serrate, the upper surface glabrous or strigose-hispid, the lower surface hispid, especially along the veins, the petiole short, slightly winged distally, slightly clasping, reduced upward, the distal with the blade ovate, lanceolate, or linear, 0.8–8 cm long, 1 mm to 4 cm wide, reduced upward, the apex acute or acuminate, mucronate, the base rounded to cuneate, the margin entire or subentire, the upper surface glabrous or strigose-hispid, the lower surface hispid, especially along the veins, the petiole short, sometimes narrowly winged, sometimes slightly clasping, or sessile. Flowers in capitula, usually in an open, diffuse, paniculiform or sometimes racemiform array, the branches usually divaricate or arching, sometimes secund, the peduncle to 3(5) cm long, the bracts ovate or lanceolate; involucre cylindric-campanulate; phyllaries numerous in 4–5(6) series, 4–6 mm long, unequal, the outer lanceolate or oblong-lanceolate, the inner linear-lanceolate or linear, the subapical green zone lanceolate or diamond-shaped, the apex acute or acuminate, often involute, mucronulate, the margin scarious, erose, hyaline, the outer surface strigillose-hirsutulous; receptacle flat or slightly convex, pitted; ray flowers numerous in 1 series; corolla blue or purple-blue, rarely pinkish or white, the lamina (10)11–16 mm long; disk flowers numerous; corolla yellow or reddish purple, 5–7 mm long. Cypsela oblong-obovoid, compressed, 3–4 mm long, 4- to 7-nerved, glabrous, dull purple or brown; pappus of numerous reddish brown, tawny, or rose-tinged, subequal, barbellate bristles 4–6 mm long.

Calcareous hammocks. Rare; Gadsden and Jackson Counties. Ontario south to Florida and Mississippi. Fall.

Symphyotrichum simmondsii (Small) G. L. Nesom [Commemorates Edward Simmonds (1866–1931), director of the Plant Introduction Field Station, Miami, Florida.] SIMMONDS' ASTER.

Aster simmondsii Small, Fl. Miami 190, 200. 1913. *Symphyotrichum simmondsii* (Small) G. L. Nesom, Phytologia 77: 291. 1995 ("1994"). TYPE: FLORIDA: Miami-Dade Co.: Everglades N of Long Pine Key, 18–26 Jan 1909, *Small & Carter 2892* (holotype: NY).

Aster sulznerae Small, Fl. Miami 190, 200. 1913. TYPE: FLORIDA: Miami-Dade Co.: Miami River, 3 Apr 1904, *Britton 436* (holotype: NY).

Aster pinifolius Alexander, in Small, Man. S.E. Fl. 1387, 1509. 1933; non Nees von Esenbeck, 1818; nec F. Mueller, 1866. *Symphyotrichum kralii* G. L. Nesom, Phytologia 82: 284. 1998 ("1997"). TYPE: FLORIDA: Miami-Dade Co.: Everglades W of Coconut Grove, 22 Nov 1916, *Small 7958* (holotype: NY).

Erect perennial herb, to 1.2 m; stem long-rhizomatous or with a woody caudex, striate-angled, sparsely scabrellous, glabrescent, or glabrous. Basal leaves withering by flowering, the blade oblanceolate, 2.5–8 cm long, 6–16 mm wide, the apex acute or obtuse, the base cuneate, the margin serrate, recurved, the upper surface scabridulous, the lower surface pilosulous, glabrate, or glabrous, the petiole short, winged, the base sheathing, the proximal cauline mostly withering by flowering, the blade elliptic, oblanceolate, lanceolate, or linear, 4.5–7 cm long, 8–15 mm wide, the apex acute or obtuse, the base cuneate, the margin serrate, crenate-serrate, or entire, upper surface scabridulous, the lower surface pilosulous, the petiole winged, the base sheathing or clasping, the distal with the blade lanceolate, oblanceolate, or linear, 1–12 cm long, 5–8 mm wide, reduced upward, the apex acute or obtuse, the base cuneate, the margin serrulate or entire, upper surface scabridulous, the lower surface pilosulous, sessile. Flowers in capitula in a diffuse, paniculiform array, the branches usually lax, spreading and often arching, racemiform, the peduncle to 6 cm, glabrous or sparsely scabrous, the bracts linear-lanceolate; involucre campanulate or campanulate-turbinate; phyllaries numerous in 5–6 series, (5)6–8 mm long, unequal, oblong-lanceolate, linear-lanceolate, or lanceolate, the subapical green zone, oblanceolate, the apex acute or acuminate, mucronate, often purplish tipped, the margin narrowly scarious, distally erose, hyaline, ciliolate, the outer surface glabrous; receptacle flat or slightly convex, pitted; ray flowers numerous in 1 series; corolla pale lilac, pale lavender, or pale purple, the lamina 5–8 mm long; disk flowers numerous; corolla yellow or reddish, 5–7 mm long. Cypsela oblong-obovoid or fusiform-obovoid, compressed, 2–3 mm long, 4- to 6-nerved, slightly strigillose, tan; pappus of numerous cream, subequal, barbellate bristles 5–7 mm long.

Wet flatwoods, ditches, and pond and lake margins. Frequent; nearly throughout. North Carolina, South Carolina, and Florida. Summer–fall.

Symphyotrichum squamatum (Spreng.) G. L. Nesom [*Squamatus,* with scales, in reference to the peduncle bracteoles.] SOUTHEASTERN ANNUAL SALTMARSH ASTER.

Conyza squamata Sprengel, Syst. Veg. 3: 515, 1836. *Aster squamatus* (Sprengel) Hieronymus, Bot. Jahrb. Syst. 29: 19. 1901. *Conyzanthus squamatus* (Sprengel) Tamamschjan, in Komarov, Fl. USSR 25: 186. 1959. *Symphyotrichum squamatum* (Sprengel) G. L. Nesom, Phytologia 77: 292. 1995

("1994"). *Symphyotrichum subulatum* (Michaux) G. L. Nesom var. *squamatum* (Sprengel) S. D. Sundberg, Sida 21: 908. 2004.

Erect perennial herb, to 1.5 m; stem striate-angled, glabrous or glabrate. Basal leaves withering by flowering, the blade ovate, or oblanceolate, 1–9 cm long, 6–14 mm wide, the apex rounded, obtuse, or acute, the base cuneate, the margin entire, serrulate, or crenulate, often ciliolate, the upper and lower surfaces glabrous, long-petiolate, the bases sheathing, sparsely ciliate, the proximal cauline withering by flowering, the blade narrowly lanceolate or subulate, 2–10 cm long, 2–10(20) mm long, the apex acute or acuminate, the base narrowly cuneate, the margin entire or serrulate, the upper and lower surfaces glabrous, petiolate, subpetiolate or sessile, the distal with the blade narrowly lanceolate or subulate, 0.5–11 cm long, 1–6 mm wide, similar to the proximal cauline, sessile. Flowers in capitula in an open, diffuse array, secund on the primary branches, the peduncle to 4 cm long, the bracts linear or lanceolate; involucre cylindric or campanulate; phyllaries numerous in 3–5 series, 5–7(8) mm long, unequal, subulate or lanceolate, the subapical green zone broadly lanceolate, the apex acute, the margin entire, hyaline, often purple-tinged, the outer surface glabrous; receptacle flat or slightly convex, pitted; ray flowers numerous in (2)3 series; corolla white, the lamina 1–2 mm long; disk flowers (3)7–14; corolla yellow, sometimes purple-tinged, 4–5 mm long. Cypsela narrowly obovoid or fusiform, sometimes slightly compressed, 2–3 mm long, 5-nerved, strigillose, light brown or purple; pappus of numerous white, subequal barbellate bristles 4–5 mm long.

Brackish marshes. Rare; Escambia County. North Carolina, Florida, Alabama, Louisiana, and Texas; West Indies and South America; Europe, Asia, Australia. Native to south America. Summer–fall.

Symphyotrichum subulatum (Michaux) G. L. Nesom [With a fine sharp point, in reference to the leaf apex.] ANNUAL SALTMARSH ASTER.

Aster subulatus Michaux, Fl. Bor.-Amer. 2: 111. 1803. *Tripolium subulatum* (Michaux) Nees von Esenbeck, Gen. Sp. Aster. 156. 1832. *Tripolium subulatum* (Michaux) Nees von Esenbeck var. *boreale* de Candolle, Prodr. 5: 254. 1836, nom. inadmiss. *Mesoligus subulatus* (Michaux) Rafinesque, Fl. Tellur. 2: 44. 1837. *Symphyotrichum subulatum* (Michaux) G. L. Nesom, Phytologia 77: 293. 1995 ("1994").

Aster subulatus Michaux var. *euroauster* Fernald & Griscom, Rhodora 37: 183, t. 351 (1, 3). 1935.

Erect annual herb, to 1.2(1.5) m; stem striate-angled, glabrous or glabrate. Basal leaves withering by flowering, the blade ovate, or oblanceolate, 1–9 cm long, 6–14 mm wide, the apex rounded, obtuse, or acute, the base cuneate, the margin entire, serrulate, or crenulate, often ciliolate, the upper and lower surfaces glabrous, long-petiolate, the bases sheathing, sparsely ciliate, the proximal cauline withering by flowering, the blade narrowly lanceolate or subulate, 2–10 cm long, 2–10(20) mm long, the apex acute or acuminate, the base narrowly cuneate, the margin entire or serrulate, the upper and lower surfaces glabrous, petiolate, subpetiolate or sessile, the distal with the blade narrowly lanceolate or subulate, 0.5–11 cm long, 1–6 mm wide, similar to the proximal cauline, sessile. Flowers in capitula in an open, diffuse, paniculiform array, the peduncle to 4 cm long, the bracts linear or lanceolate; involucre cylindric or campanulate; phyllaries numerous in 3–5 series, 5–7(8) mm long, unequal, subulate or lanceolate, the green zone lanceolate, extending the phyllary length, the apex acute, the margin hyaline, entire or erose, often ciliate, the outer surface glabrous; receptacle flat or slightly convex, pitted; ray flowers numerous in 2 series; corolla white, the lamina 2–3 mm long; disk flowers 4–10(13); corolla yellow, sometimes purple-tinged, 3–5 mm long. Cypsela narrowly obovoid or fusiform, sometimes slightly compressed, (1)2–3 mm long, 5-nerved, strigillose, light brown or purple; pappus of numerous white, barbellate bristles 4–6 mm long.

Wet flatwoods and marshes. Occasional; northern peninsula, central and western panhandle. New Brunswick south to Florida, west to Ontario, Michigan, Illinois, Arkansas, and Texas, also Nebraska. Summer–fall.

Symphyotrichum tenuifolium (L.) G. L. Nesom [With slender leaves.] PERENNIAL SALTMARSH ASTER.

Aster tenuifolius Linnaeus, Sp. Pl. 873. 1753. *Symphyotrichum tenuifolium* (Linnaeus) G. L. Nesom, Phytologia 77: 293. 1995 ("1994").

Aster bracei Britton ex Small, Fl. Miami 190, 200. 1913. *Symphyotrichum bracei* (Britton ex Small) G. L. Nesom, Phytologia 77: 276. 1995 ("1994").

Aster tenuifolius Linnaeus var. *aphyllus* R. W. Long, Rhodora 72: 40. 1970. *Symphyotrichum tenuifolium* (Linnaeus) G. L. Nesom var. *aphyllum* (R. W. Long) S. D. Sundberg, Sida 21: 905. 2004.

Ascending or erect perennial herb, to 6(10) dm; stem rhizomatous, striate-angled, glabrous or with trichomes in lines distally. Basal leaves withering by flowering, new basal rosettes appearing by flowering, the blade ovate or oblanceolate, 1.5–2 cm long, 5–15 mm wide, the apex rounded, the base cuneate, the margin entire, the upper and lower surfaces glabrous, the petiole short, the base sheathing, the proximal cauline withering by flowering, the blade lanceolate, linear-lanceolate, oblanceolate, or linear, 1–8(15) cm long, 1–6(12) mm wide, the apex acute or acuminate, the base narrowly cuneate, the margin entire, the upper and lower surfaces glabrous, sessile, the distal with the blade narrowly lanceolate or subulate, 1–11 cm long, 1–5 mm wide, the apex acuminate, the base narrowly cuneate, the margin entire, the upper and lower surfaces glabrous, sessile. Flowers in capitula in an open, diffuse, paniculiform array, the peduncle to 4(6) cm long, the bracts linear; involucre narrowly turbinate; phyllaries numerous in 4–5 series, 4–10(11) mm long, unequal, lanceolate or subulate, the subapical green zone spatulate or oblanceolate-rhombic, the apex acute, the margin hyaline, entire, often purple-tinged, the outer surface glabrous or minutely pubescent distally; receptacle flat or slightly convex, pitted; ray flowers numerous in 1 series; corolla white or pink, the lamina 5–9 mm long; disk flowers numerous; corolla yellow or purplish, 3–6(7) mm long. Cypsela narrowly obovoid or fusiform, sometimes slightly compressed, 2–4(5) mm long, 5- to 6-nerved, strigillose, light brown; pappus of numerous tawny or white, subequal barbellate bristles 3–6 mm long.

Brackish marshes. Frequent; nearly throughout. New Hampshire and New York south to Florida, west to Texas. Summer–fall.

Symphyotrichum undulatum (L.) G. L. Nesom [Wavy, in reference to the leaf margin.] WAVYLEAF ASTER.

Aster undulatus Linnaeus, Sp. Pl. 875. 1753. *Symphyotrichum undulatum* (Linnaeus) G. L. Nesom, Phytologia 77: 293. 1995 ("1994").
Aster diversifolius Michaux, Fl. Bor.-Amer. 2: 113. 1803. *Aster undulatus* var. *diversifolius* (Michaux) A. Gray, Syn. Fl. N. Amer. 1(2): 181. 1884.
Aster asperulus Torrey & A. Gray, Fl. N. Amer. 2: 120. 1841.
Aster baldwinii Torrey & A. Gray, Fl. N. Amer. 2: 127. 1841.
Aster undulatus Linnaeus var. *asperulus* A. W. Wood, Amer. Bot. Fl. 162. 1870.
Aster undulatus Linnaeus var. *loriformis* E. S. Burgess, in Britton & A. Brown, Ill. Fl. N. U.S. 3: 365. 1898. *Aster loriformis* (E. S. Burgess) E. S. Burgess, in Small, Fl. S.E. U.S. 1217, 1340. 1903.
Aster asperifolius E. S. Burgess, in Small, Fl. S.E. U.S. 1216, 1339. 1903.
Aster liguliformis E. S. Burgess, in Small, Fl. S.E. U.S. 1217, 1339. 1903. TYPE: FLORIDA: Duval Co.: Jacksonville, 27 Oct 1894, *Curtiss s.n.* (lectotype: US; isolectotype: K). Lectotypified by Jones and Lowry (1986: 401).
Aster mohrii E. S. Burgess, in Small, Fl. S.E. U.S. 1217, 1340. 1903.
Aster proteus E. S. Burgess, in Small, Fl. S.E. U.S. 1216, 1339. 1903. TYPE: FLORIDA: Duval Co.: Oct, *Curtiss 1278** (Lectotype: US; isolectotypes: BM, M, NCU, ND, NEB, NY, P). Lectotypified by Jones and Lowry (1986: 403).

Erect perennial herb, to 1.5 m; stem striate-angled, hirtellous. Basal leaves withering by flowering, sometimes developing new basal rosettes at flowering, the blade oblong-ovate or ovate, (1)3.5–6 cm long, (1)1.5–7 cm wide, the apex obtuse or rounded, mucronulate, the base cordate or rounded, sometimes cuneate, the margin shallowly crenate-serrate or subentire, the upper surface scabrous, sometimes strigose or hirsute, rarely glabrate, the petiole narrowly winged, the base dilated and sheathing, hirsute, the proximalmost cauline withering by flowering, the proximal persistent, the blade ovate or lanceolate, 3–12(14) cm long, (1)2–5 cm wide, reduced upward, the apex acute or acuminate, mucronate, the base cordate or auriculate-clasping, the margin shallowly crenate-serrate or subentire, the upper surface scabrous, sometimes strigose or hirsute, rarely glabrate, the petiole winged, the wings widening abruptly at the base and auriculate-clasping, the distal with the blade lanceolate-ovate, lanceolate, or linear-lanceolate, 0.5–10 cm long, 1–2.5 cm wide, reduced upward, the branch leaves ovate or lanceolate, smaller, the apex acute or acuminate, mucronate, the base clasping, sheathing, the margin serrulate or entire, sessile. Flowers in capitula in an open, paniculiform array, the peduncle to (3)5 cm long, the bracts linear, appressed; involucre campanulate or cylindric-campanulate; phyllaries numerous in 4–5(6) series, 4–6 mm long, unequal, the outer oblong-lanceolate, the inner linear-oblanceolate, the subapical green zone diamond-shaped or lanceolate, the apex acute, acuminate, or obtuse, often mucronulate, sometimes purplish tipped, the margin scarious, erose, hyaline, ciliolate, the outer surface hirsutulous; receptacle flat or slightly convex, pitted; ray flowers numerous in 1 series; corolla blue or purple, sometimes lilac, the lamina 6–12 mm long; disk flowers numerous; corolla light yellow or purple, 4–6 mm long. Cypsela oblong-obovoid, compressed, ca. 2 mm long, 3- to 4-nerved, strigillose, dull purple, light brown, or tan; pappus of numerous tawny or rose-tinged, subequal, barbellate bristles 4–5 mm long.

Dry hammocks and sandhills. Occasional; northern counties, central peninsula. Nova Scotia and Maine south to Florida, west to Ontario, Illinois, Kentucky, Tennessee, Arkansas, and Louisiana. Summer–fall.

Symphyotrichum urophyllum (Lindl. ex DC.) G. L. Nesom [From the Greek *urus,* tailed, and *phyllus,* leaf, in reference to the leaves with an elongated taillike apex.] WHITE ARROWLEAF ASTER.

Aster urophyllus Lindley ex de Candolle, Prodr. 233. 1836. *Aster sagittifolius* Wedemeyer ex Willdenow var. *urophyllus* (Lindley ex de Candolle) E. S. Burgess, in Britton & A. Brown, Ill. Fl. N. U.S. 365. 1898. *Symphyotrichum urophyllum* (Lindley ex de Candolle) G. L. Nesom, Phytologia 77: 294. 1995 ("1994").

Aster sagittifolius Wedemeyer ex Willenow var. *dissitifolius* E. S. Burgess, in Britton & A. Brown, Ill. Fl. N. U.S. 3: 365. 1898.

Erect perennial herb, to 1(2) m; stem short-rhizomatous or with a stout, branched caudex, striate-angled, glabrous or glabrate, distally sparsely pilose. Basal leaves usually withering by flowering, sometimes persistent, the blade ovate, ovate-lanceolate, or lanceolate, 2.5–12 cm long, 1.5–5 cm wide, the apex acute, sometimes rounded, the base cordate, truncate, or rounded, the margin serrate, ciliate, the upper surface sparsely pilose, scabrous, or glabrate, the lower surface pilose on the midvein, the petiole 5–15 cm long, sometimes narrowly winged, the base sheathing, the proximal cauline sometimes deciduous at flowering, the blade ovate or lanceolate, 5–12 cm long, 2–5 cm wide, reduced upward, the apex acute or acuminate, the base truncate, rounded, or cuneate, the petiole winged, the base clasping, the margin serrate, the upper surface sparsely pilose, scabrous, or glabrate, the lower surface pilose on the midvein, the distal with the blade lanceolate-ovate, lanceolate, or linear-lanceolate, 1–10 cm long, to 2.5 cm wide, reduced upward, the apex acuminate or subcaudate, the base cuneate, the margin entire or subentire, the upper surface sparsely pilose, scabrous, or glabrate, the lower surface pilose on the midvein, short, winged petiolate or sessile. Flowers in capitula in a dense, elongate, paniculiform array, the branches ascending, sometimes arched when long, the peduncle to 2 cm long, bracts linear, ascending; involucre cylindric; phyllaries numerous in 4–6 series, (4)5–6(7) mm long, unequal, sometimes recurved-spreading, linear-lanceolate, the subapical green zone, linear-lanceolate or linear, the apex long-acuminate or caudate, hyaline mucronate or aristate, the margin scarious, hyaline, erose, sparsely ciliolate, the outer surface glabrous; receptacle flat or slightly convex, pitted; ray flowers numerous in 1 series; corolla white or pale pink, lilac, or light blue, the lamina 4–8(10) mm long; disk flowers numerous; corolla white, cream, or pink, 4–5 mm long. Cypsela obovoid, slightly compressed, 2–3 mm long, 4- to 6-nerved, glabrous or glabrate, tan, brown, or dull purple; pappus of numerous white or rose-tinged, subequal, barbellate bristles 3–4 mm long.

Hammocks. Occasional; eastern and central panhandle. Vermont and New York south to Florida, west to Ontario, North Dakota, Minnesota, Nebraska, Kansas, Oklahoma, Arkansas, and Mississippi. Fall.

Symphyotrichum walteri (Alexander) G. L. Nesom [Commemorates Thomas Walter (1740–1789), British-born, American botanist.] WALTER'S ASTER.

Aster squarrosus Walter, Fl. Carol. 209. 1788; non Allioni, 1785. *Aster walteri* Alexander, in Small, Man. S.E. Fl. 1382, 1509. 1933. *Lasallea walteri* (Alexander) Semple & Brouillet, Amer. J. Bot. 67: 1023. 1980. *Virgulus walteri* (Alexander) Reveal & Keener, Taxon 30: 650. 1981. *Symphyotrichum walteri* (Alexander) G. L. Nesom, Phytologia 77: 294. 1995 ("1994").

Erect or scandent-sprawling perennial herb, to 1 m; stems long-rhizomatous, with a woody caudex, striate-angled, glabrous or sparsely fine-strigose, sometimes minutely stipitate-glandular. Basal leaves early deciduous, the blade oblanceolate or obovate, 1–4 cm long, 7–15 mm wide, the apex obtuse, mucronate, the base cuneate, the margin entire, the upper and lower surfaces finely scabrous, sessile, the proximal cauline with the blade ovate or lanceolate, 1–3 cm long, 2–10 mm wide, spreading or reflexed, the apex acute, sometimes spinulose, the base cordate-clasping, the margin entire, the upper and lower surfaces finely scabrous, shiny, sessile, the distal with the blade lanceolate or triangular, 5–15 mm long, 3–6 mm wide, spreading or reflexed, reduced to bracts distally, the apex acute, spinulose, the base clasping, the margin entire, the upper and lower surfaces finely scabrous, sessile. Flowers in capitula in a diffuse, paniculiform array, the peduncle to 15 cm long, foliaceous; involucre campanulate; phyllaries numerous in 4–5 series, 5–7 mm long, unequal, oblanceolate, the subapical green zone diamond-shaped, the apex acute or obtuse, mucronate or subspinlose, the margin finely scabrous, hyaline, the outer surface sparsely strigillose; receptacle flat or slightly convex, pitted; ray flowers numerous in 1 series; corolla bluish-purple, the lamina 5–9(11) mm long; disk flowers numerous; corolla yellow, 4–7 mm long. Cypsela obovoid, not compressed, 2–3 mm long, very faintly nerved, strigose, tan or brown; pappus of numerous tan, subequal, barbellate bristles 4–6 mm long.

Flatwoods. Occasional; northern and central peninsula, west to central panhandle. North Carolina south to Florida. Fall.

EXCLUDED TAXA

Symphyotrichum divaricatum (Nuttall) G. L. Nesom—Reported for Florida by Chapman (1860, as *Aster divaricatus* (Nuttall) Torrey & A. Gray; non Linnaeus, 1753; nec Lamarck, 1783; nec Sprengel, 1826; nec Rafinesque ex de Candolle, 1836). Considered a synonym of *S. subulatum* (Michaux) G. L. Nesom var. *ligulatum* S. D. Sunberg by Brouillet et al. (2006), who do not report it from Florida. No Florida specimens seen.

Symphyotrichum ericoides (Linnaeus) G. L. Nesom—Reported for Florida by Chapman (1860, 1883, 1897, all as *Aster ericoides* Linnaeus) and Small (1903, as *Aster ericoides* Linnaeus), the name misapplied to material of *S. pilosum*).

Symphyotrichum grandiflorum (Linnaeus) G. L. Nesom—This northern taxon was reported for Florida by Small (1903, as *Aster grandiflorus* Linnaeus) and Alexander (in Small, 1933, as *Aster grandiflorus* Linnaeus). No Florida specimens seen.

Symphyotrichum lanceolatum (Willdenow) G. L. Nesom—Because infraspecific categories were not recognized, the typical variety of this taxon was reported from Florida by implication by Wunderlin (1998, as *Aster lanceolatus* Willdenow). All Florida material are of var. *latifolium*.

Symphyotrichum prenanthoides (Muhlenberg ex Willdenow) G. L. Nesom—Reported for Florida by Wilhelm (1984, as *Aster prenanthoides* Muhlenberg ex Willdenow). No Florida specimens seen.

Symphyotrichum sagittifolium (Wedemeyer ex Willdenow) G. L. Nesom—Misapplied to Florida material of *S. urophyllum*. Among the many authors who have misapplied the name to Florida material of *S. urophyllum* are Chapman (1860, 1883, 1897, all as *Aster sagittifolius* Wedemeyer ex Willdenow), Small (1903, 1913a, 1933, all as *Aster sagittifolius* Wedemeyer ex Willdenow var. *dissitifolius* E. S. Burgess), Cronquist (1980, as *Aster sagittifolius* Wedemeyer ex Willdenow), Wilhelm (1984, as *Aster sagittifolius* Wedemeyer ex Willdenow), Clewell (1985, as *Aster sagittifolius* Wedemeyer ex Willdenow, and Wunderlin (1998, as *Aster sagittifolius* Wedemeyer ex Willdenow). The type of this taxon is conspecific with the extralimital *S. cordifolium* (Linnaeus) G. L. Nesom (Brouillet et al., 2006).

Symphyotrichum simplex (Willdenow) Á. Löve & D. Löve–This taxon, a synonym of the typical variety of *S. lanceolatum* (Willdenow) G. L. Nesom, was reported for Florida by Chapman (1860, as *Aster simplex* Willdenow). All Florida material is of *S. lanceolatum* var. *latifolium*.

Symphyotrichum subulatum (Michaux) G. L. Nesom var. *ligulatum* (Shinners) S. D. Sundberg—This is a synonym of the extralimital *S. divaricatum*, but has been misapplied to Florida material of *S. bahamense* (Long and Lakela, 1971; Correll and Correll, 1984, all as *Aster subulatus* Michaux var. *ligulatus* Shinners) and to *S. subulatum* (Cronquist, 1980; Wilhelm, 1984; Clewell, 1985, all as *Aster subulatus* Michaux).

Synedrella Gaertn. 1791.

Herbs. Leaves opposite, simple, pinnate-veined, petiolate. Flowers in axillary glomerules or solitary, radiate; receptacle paleate; ray flowers carpellate; petals 5, basally connate, zygomorphic; disk flowers bisexual; petals 5, basally connate, actinomorphic; stamens 5, epipetalous, the filaments free, the anthers connate; ovary inferior, 2-carpellate, 1-loculate. Fruit a cypsela; pappus present.

A monotypic genus; North America, West Indies, Mexico, Central America, South America, Africa, Asia, Australia, and Pacific Islands. [From the Greek, *syndrome*, occurring together, and *-ellus*, diminutive, in reference to the capitula occurring in small clusters.]

Selected reference: Strother (2006dd).

Synedrella nodiflora (L.) Gaertn. [With flowers at the nodes.] NODEWEED.

Verbesina nodiflora Linnaeus, Cent. Pl. 1: 28. 1755. *Synedrella nodiflora* (Linnaeus) Gaertner, Fruct. Sem. Pl. 2: 456. 1791. *Ucacou nodiflorum* (Linnaeus) Hitchcock, Rep. (Annual) Missouri Bot. Gard. 4: 100. 1893.

Erect or ascending annual herb, to 8 dm; stem striate-angled, slightly scabrous. Leaves with the blade ovate or elliptic, 3–10 cm long, 2–5 cm wide, the apex acute or obtuse, the base cuneate or rounded, the margin serrate-crenate, the upper and lower surfaces slightly scabrous, the petiole winged, the base clasping. Flowers in capitula in an axillary glomerule or solitary, radiate; involucre cylindric or campanulate, 3–6 mm wide; phyllaries 2–5+ in 1(2) series, lanceolate, ca. 5 mm long, the apex aristate; receptacle convex; paleae linear-lanceolate, scarious, flat or cupped at the base; ray flowers 2–9 in 1–2 series; corolla yellowish, the lamina 2–4 mm long; disk flowers 4–12+; corolla yellowish, ca. 3 mm long. Cypselae dimorphic, ca. 4 cm long, those of the ray flowers strongly obcompressed or flat, narrowly ovate or winged, the wings laciniate, those of the disk flowers linear and not winged; pappus of the ray flowers of 2 triangular scales, that of the disk flowers of 2–3 subulate scales or awns.

Disturbed sites. Rare; Franklin and Lee Counties, southern peninsula. Florida; West Indies, Mexico, Central America, and South America; Africa, Asia, Australia, and Pacific Islands. Native to tropical America. All year.

Tagetes L. 1753. MARIGOLD

Annual herbs or perennial subshrubs. Leaves opposite or sometimes distally alternate, simple, petiolate or epetiolate. Flowers in capitula in corymbiform arrays or solitary, radiate; receptacle epaleate; ray flowers carpellate; petals 5, basally connate, zygomorphic; disk flowers bisexual; petals 5, basally connate, actinomorphic; stamens 5, epipetalous, the filaments free, the anthers connate; ovary inferior, 2-carpellate, 1-loculate. Fruit a cypsela; pappus present.

A genus of about 50 species; nearly cosmopolitan. [*Tages,* an Etruscan god who is said to have sprung from the earth while it was being ploughed, in reference to the plants being a weed of arable land.]

Selected reference: Strother (2006ee).

1. Leaf blades not pinnate-lobed; plant a subwoody perennial .... **T. lucida**
1. Leaf blades 1- to 3-pinnate-lobed; plant annual.
    2. Ray flower laminae 10 mm long or longer .... **T. erecta**
    2. Ray flower laminae 1–2 mm long .... **T. minuta**

Tagetes erecta L. [In reference to the habit.] AZTEC MARIGOLD.

Tagetes erecta Linnaeus, Sp. Pl. 887. 1753. *Tagetes major* Gaertner, Fruct. Sem. Pl. 2. 437. 1791, nom. illegit.

Erect annual herb, to 1.2 m; stem striate-angled, glabrous. Leaves with the blade lanceolate or linear-lanceolate, 3–12(25) cm long overall, the apex acute, the base cuneate, the margin pinnately lobed, the lobes lanceolate or linear-lanceolate, 15–25 mm long, 3–8 mm wide, the apex acute or acuminate, the ultimate margin irregularly serrate or entire, the teeth usually bristle-tipped, the upper and lower surfaces glabrous, with scattered oil glands, the petiole slightly winged, the base clasping. Flowers in capitula, solitary or in a corymbiform array, the peduncle 3–10 cm long; involucre turbinate or broadly campanulate; phyllaries 2–20 in 1–2 series, broadly lanceolate, 1–2 cm long, basally connate, dotted with oil glands; receptacle convex or conic, pitted; ray flowers (3)5–8; corolla yellow, orange, or red-brown, the lamina 1–2 cm long; disk flowers numerous; corolla yellow, orange, or red-brown, 7–12 mm long. Cypsela narrowly obpyramidal or fusiform, 6–11 mm long, longitudinally ribbed, glabrous; pappus of 2 subulate-aristate scales 6–12 mm long and 2–4 free or connate linear-oblong, suberose scales 2–6 mm long.

Disturbed sites. Rare; Escambia, Wakulla, and Sumter Counties, southern peninsula. Escaped from cultivation. Nearly cosmopolitan. Native to Mexico. Summer–fall.

Tagetes lucida Cav. [*Lucidus,* shining, in reference to the glossy leaves.] SWEETSCENTED MARIGOLD.

Tagetes lucida Cavanilles, Icon. 3: 33, t. 264. 1796–1796 ("1794").

Subwoody perennial, to 8 dm; stem striate-angled, glabrous. Leaves linear or oblong, to 8 cm long, to 1 cm wide, the apex acute or obtuse, the base cuneate, the margin serrate or entire, the upper and lower surfaces glabrous with scattered oil glands, sessile, clasping. Flowers in capitula, solitary or in a corymbiform array, the peduncle 3–6 cm long; involucre cylindric or fusiform; phyllaries 5–10 in 1–2 series, lanceolate, ca. 1 cm long; receptacle convex or conic; ray flowers 3–5; corolla yellow, the lamina 8–10 mm long; disk flowers numerous; corolla golden yellow, 4–5 mm long. Cypsela narrowly obpyramidal or fusiform, 6–8 mm long; pappus of 1–2 subulate scales 2–3 mm long and 3–5 ovate or lanceolate scales to 1 mm long.

Disturbed sites. Rare; Leon County. Escaped from cultivation. Alabama and Florida; Mexico, Central America, and South America. Native to Mexico and Central America. Summer–fall.

Tagetes minuta L. [Very small, in reference to the ray flowers with small lamina.] MUSTER JOHN HENRY.

Tagetes minuta Linnaeus, Sp. Pl. 887. 1753.

Erect annual herb, to 1(1.8) m; stem striate-angled, glabrous. Leaves with the blade narrowly lanceolate, 8–15 cm long overall, the apex acute, the base cuneate, the margin pinnately lobed, the lobes lanceolate or linear-lanceolate, 12–25 mm long, (2)4–7 mm wide, the apex acute or acuminate, the ultimate margin serrate, the teeth usually bristle-tipped, the petiole slightly winged, the base clasping, the upper and lower surfaces glabrous, with scattered oil glands. Flowers in capitula in a corymbiform array, the peduncle 1–5 mm long; involucre cylindric; phyllaries 2–10 in 1–2 series, lanceolate, 7–10 mm long, basally connate, dotted with oil glands; receptacle convex or conic, pitted; ray flowers 1–3; corolla yellow, the lamina 1–2 mm long; disk flowers 3–5; corolla yellow, orange, or red-brown, 3–4 mm long. Cypsela narrowly obpyramidal or fusiform, (5)6–7 mm long, longitudinally ribbed, glabrous; pappus of 1–2 subulate scales 2–3 mm long and 3–5 free ovate or lanceolate scales to 1 mm long.

Disturbed sites. Rare; Levy, Madison, Leon, and Gadsden Counties. Escaped from cultivation. Nearly cosmopolitan. Native to South America. Fall.

Taraxacum F. H. Wigg., nom. cons. 1780. DANDELION

Herbs. Leaves alternate, simple, pinnate-veined, petiolate. Flowers in capitula, solitary, liguliform, bisexual; calyculi present; receptacle epaleate; petals 5, basally connate, zygomorphic; stamens 5, epipetalous, the filaments free, the anthers connate; ovary inferior, 2-carpellate, 1-loculate. Fruit a cypsela; pappus present.

A genus of about 60 species; nearly cosmopolitan. [From the Greek *taraxos,* disorder, and *akos,* remedy, in reference to its use to treat various medical disorders.]

Selected reference: Brouillet (2006c).

Taraxacum officinale Weber ex F. H. Wigg. [Used in medicine.] COMMON DANDELION.

Leontodon taraxacum Linnaeus, Sp. Pl. 798. 1753. *Hedypnois taraxacum* (Linnaeus) Scopoli, Fl. Carniol., ed. 2. 2: 99. 1772. *Leontodon vulgaris* Lamarck, Fl. Franc. 2: 113. 1779 ("1778"), nom. illegit. *Taraxacum officinale* Weber ex F. H. Wiggers, Prim. Fl. Holsat. 56. 1780. *Taraxacum vulgare* Schrank, Baier. Reise 11. 1786, nom. illegit. *Leontodon officinalis* (Webber ex F. H. Wiggers) Withering, Arr. Brit. Pl., ed. 3. 3: 679. 1796, nom. illegit.; non J. F. Gmelin, 1791. *Taraxacum dens-leonis* Desfontaines, Fl. Atlant. 2: 228. 1799, nom. illegit. *Leontodon taraxacum* Linnaeus var. *vulgare* Bentham, Cat. Pl. Pyrenees 94. 1826, nom. inadmiss. *Taraxacum leontodon* Dumortier, Fl. Belg. 61. 1827, nom. illegit. *Leontodon taraxacum* Linnaeus subsp. *officinalis* (Weber ex F. H. Wiggers) Gaudin, Fl. Helv. 5: 61. 1829, nom. inadmiss. *Leontodon taraxacum* Linnaeus var. *genuinus* Wimmer & Grabowski, Fl. Siles. 2(2): 225. 1831, nom. inadmiss. *Taraxacum officinale* Weber ex F. H. Wiggers var. *genuinum* W.D.J. Koch, Syn. Deut. Schweiz. Fl. 428. 1837, nom. inadmiss. *Taraxacum dens-leonis* Desfontaines var. *officinale* (Weber ex F. H. Wiggers) Cosson & Germain de Saint-Pierre, Fl. Env. Paris 432. 1845, nom. inadmiss. *Taraxacum officinale* Weber ex F. H. Wiggers var. *pratense* Neilreich, Fl. Nied.-Oesterr. 412. 1858, nom. inadmiss. *Taraxacum officinarum* Ruprecht, Fl. Ingr. 619. 1860, nom. illegit. *Taraxacum vulgare* Schrank var. *genuinum* Ascherson, Fl. Brandenburg 2: 375. 1864, nom. inadmiss. *Taraxacum taraxacum* (Linnaeus) H. Karsten, Deut. Fl. 1138. 1883, nom. inadmiss. *Taraxacum officinale* Weber ex F. H. Wiggers var. *typicum* Beck, Fl. Nieder-Oesterreich 1315. 1893, nom. inadmiss. *Taraxacum taraxacum* (Linnaeus) H. Kirsten subsp. *genuinum* Ascherson & Graebner, Fl. Nordostdeut. Flachl. 765. 1899, nom. inadmiss. *Taraxacum dens-leonis* Desfontaines var. *commune* Rouy, Fl. France 9: 188. 1905, nom. inadmiss. *Taraxacum officinale* Weber ex F. H. Wiggers subsp. *vulgare* Schinz & Thellung, in Schinz & R. Keller, Fl. Schweiz., ed. 2. 1: 542. 1905, nom. inadmiss. *Taraxacum palustre* (Lyons) de Candolle var. *vulgare* Fernald, Rhodora 35: 380. 1933.

Leontodon laevigatus Willdenow, Sp. Pl. 3: 1546. 1803. *Taraxacum laevigatum* (Willdenow) de Candolle, Cat. Pl. Horti. Monsp. 149. 1813. *Leontodon taraxacum* Linnaeus var. *laevigatus* (Willdenow) Bentham, Cat. Pl. Pyrenees 94. 1826. *Leontodon taraxacum* Linnaeus subsp. *laevigatus* (Willdenow) Gaudin, Fl. Helv. 5: 61. 1829. *Taraxacum officinale* Weber ex F. H. Wiggers var. *laevigatum* (Willdenow) Babington, Man. Brit. Bot. 179. 1843. *Taraxacum dens-leonis* Desfontaines var. *laevigatum* (Willdenow) Cosson & German de Saint-Pierre, Fl. Env. Paris 432. 1845. *Taraxacum taraxacoides* (Hoppe) Willkomm var. *laevigatum* (Willdenow) Willkomm, in Willkomm & Lange, Prodr. Fl. Hispan. 2: 231. 1865. *Taraxacum taraxacum* (Linnaeus) H. Karsten var. *laevigatum* (Willdenow) Kuntze, Revis. Gen. Pl. 3(2): 181. 1898. *Taraxacum taraxacum* (Linnaeus) H. Karsten subsp. *laevigatum* (Willdenow) Ascherson & Graebner, Fl. Nordostdeut. Flachl. 765. 1899. *Taraxacum laevigatum* (Willdenow) de Candolle var. *normale* Rouy, Fl. France 9: 189. 1905.

Taraxacum erythrospermum Andrzejowski ex Besser, Enum. Pl. 75, 106. 1822. *Leontodon erythrospermus* (Andrzejowski ex Besser) Eichwald, Naturhist. Skizze Lith. Volhyn. 150. 1830. *Taraxacum officinale* Weber ex F. H. Wiggers var. *erythrospermum* (Andrzejowski ex Besser) Babington, Man. Brit. Bot. 179. 1843. *Taraxacum caucasicum* de Candolle var. *erythrospermum* (Andrzejowski ex Besser) Ledebour, Fl. Ross. 2: 814. 1846. *Taraxacum vulgare* Schrank subsp. *erythrospermum* (Andrzejowski ex Besser) Acangeli, Comp. Fl. Ital. 428. 1882. *Taraxacum taraxacum* (Linnaeus) H. Karsten var. *erythrospermum* (Andrzejowski ex Besser) Aescherson & Graebner, Fl. Nordostdeut. Flachl. 765. 1899. *Taraxacum laevigatum* (Willdenow) de Candolle var. *erythrospermum* (Andrzejowski ex Besser) J. Weiss, in W.D.J. Koch, Syn. Deut. Schweiz. Fl., ed. 3. 1656. 1900.

Erect or ascending perennial herb, to 5(6) dm; stem subequaling or surpassing the leaves, glabrous or sparsely villous. Leaves horizontal or erect, the blade oblanceolate, oblong, or obovate,

(4)5–45 cm long, (0.7)1–10 cm wide, the apex acute, acuminate, or obtuse, the base cuneate, the margin shallowly to deeply lobed, laciniate, or toothed, the lobes retrorse, broadly to narrowly triangular or sublanceolate, the terminal about as large as the distal, the ultimate margin toothed or entire, the upper and lower surfaces glabrous or sparsely villous, the petiole narrowly winged, sometimes only slightly distally. Flowers in a solitary, axillary capitulum, the peduncle 5–40 cm long; calyculi of 12–18, lanceolate, reflexed or recurved, sometimes purplish or glaucous, bractlets in 2 series, 4–12 mm long, 1–4 mm wide, the apex acute or long-acuminate, erose, the margin white or purplish, narrowly scarious, sometimes villous ciliate; involucre campanulate; phyllaries 13–19 in 2 series, lanceolate or linear-lanceolate, 1–2.5 cm long, the apex acuminate, the margin scarious, erose distally, sometimes with a horned appendage; flowers numerous; corolla yellow, the lamina 1.5–2 cm long. Cypsela body oblanceoloid, 2–4 mm long, slightly compressed, 4- to 15-ribbed, the cone ca. 1 mm long, terete, the beak (5)7–9 mm long, terete, the body surface smooth or tuberulate, olivaceous, yellow, or grayish, sometimes red, reddish brown, or reddish purple; pappus of numerous white or sordid, equal, barbellate bristles 4–8 mm long.

Disturbed sites. Common; nearly throughout. Nearly cosmopolitan. Native to Europe and Asia. Spring–fall.

Tetragonotheca L. 1753. NERVERAY

Herbs. Leaves opposite, simple, pinnate-veined, petiolate or epetiolate. Flowers in capitula, solitary or in corymbiform arrays, radiate; receptacle paleate; ray flowers carpellate; petals 5, basally connate, zygomorphic; disk flowers bisexual; petals 5, basally connate, actinomorphic; stamens 5, the filaments free, the anthers connate; ovary inferior, 2-carpellate, 1-loculate. Fruit a cypsela; pappus absent.

A genus of 5 species; North America, West Indies, and Mexico. [From the Greek *tetra,* four, *gonio,* angle, and *theca,* container, in reference to the quadrangular involucre.]

Selected reference: Strother (2006ff).

Tetragonotheca helianthoides L. [Resembling *Helianthus.*] PINELAND NERVERAY; SQUAREHEAD.

Tetragonotheca helianthoides Linnaeus, Sp. Pl. 903. 1753. *Polymnia tetragonotheca* Linnaeus, Syst. Nat., ed. 12. 576. 1767, nom. illegit. *Gonotheca helianthoides* (Linnaeus) Rafinesque, Med. Repos., ser. 2. 5: 352. 1808.

Erect perennial herb, to 1 m; stem striate-angled, spreading or retrorsely subvillous. Leaves with the blade ovate, rhombic, or lanceolate, 7–20 cm long, 3–13 cm wide, the apex acute, the base broadly cuneate, the margin coarsely toothed, the upper and lower surfaces subvillous, sessile or subpetiolate, the base clasping. Flowers in capitula, solitary or in a loose corymbiform array; involucre hemispheric, 4-angled, 1–2 cm wide; phyllaries 10–25, the outer 4 broadly lanceolate, foliaceous, 2–2.5 cm long, the outer surface subvillous, the inner ovate or lanceolate, smaller, each subtending a ray flower; receptacle conic; paleae lanceolate or ovate-lanceolate, flat or weakly conduplicate, the apex acute; ray flowers 6–14; corolla yellow, the lamina 1.5–4 cm long; disk flowers numerous; corolla yellow, 8–10 mm long. Cypsela ovoid, 4–6 mm long, 4- or 5-angled, finely 30- to 40-ribbed, sparsely strigose; pappus absent.

Sandhills and open hammocks. Frequent; northern counties, central peninsula. Virginia south to Florida, west to Tennessee and Mississippi. Spring–summer.

Thymophylla Lag. 1816. PRICKLYLEAF

Herbs. Leaves opposite or alternate, simple, pinnate-veined, petiolate. Flowers in capitula, solitary, radiate; calyculi present; receptacle epaleate; ray flowers carpellate; petals 5, basally connate, zygomorphic; disk flowers bisexual; petals 5, basally connate, actinomorphic; stamens 5, epipetalous, the filaments free, the anthers connate; ovary inferior, 2-carpellate, 1-loculate. Fruit a cypsela; pappus present.

A genus of 13 species; North America, West Indies, Mexico, South America, Africa, and Asia. [From the Greek *thymon,* thyme, and *phyllon,* leaf, in reference to the leaf shape.]

Select reference: Strother (2006gg).

Thymophylla tenuiloba (DC.) Small [With narrow leaves.] BRISTLELEAF PRICKLYLEAF.

Hymenatherum tenuilobum de Candolle, Prodr. 5: 642. 1836. *Thymophylla tenuiloba* (de Candolle) Small, Fl. S.E. U.S. 1296, 1341. 1903. *Dyssodia tenuiloba* (de Candolle) B. L. Robinson, Proc. Amer. Acad. Arts 49: 508. 1913.

Decumbent or erect annual herb, to 3 dm; stem striate-angled, glabrous or sparsely hirtellous. Leaves alternate or sometimes opposite at the basal nodes, the blade 12–28 mm overall, 15–30 mm wide, the apex acute or acuminate, the base cuneate, the margin 7- to 15-lobed, the lobes linear or filiform, the upper and lower surfaces glabrous or glabrate, the petiole short, the base clasping. Flowers in capitula, solitary, the peduncle 3–8 cm long, glabrous or hirtellous; calyculi of 3–8 deltate or subulate short bractlets; involucre obconic; phyllaries numerous, lanceolate, 5–7 mm long, unequal, the outer surface glabrous or sparsely hirtellous; receptacle convex, pitted; ray flowers 10–21; corolla yellow-orange, the lamina 4–10 mm long; disk flowers numerous; corolla yellow, 4–6 mm long. Cypsela obpyramidal, 2–4 mm long; pappus of 10–12, 3- to 5-aristate scales 2–4 mm long.

Dry disturbed sites. Rare; Leon County, peninsula. Escaped from cultivation. Massachusetts, South Carolina, Florida, and Alabama west to New Mexico; West Indies, Mexico, Africa, and Asia. Native to Texas and Mexico. Spring–fall.

Tithonia Desf. ex Juss. 1789.

Annuals or perennial herbs, subshrubs, or shrubs. Leaves opposite or alternate, simple, pinnate-veined, petiolate or epetiolate. Flowers in capitula, solitary, radiate; receptacle paleate; ray flowers sterile; petals 5, basally connate, zygomorphic; disk flowers bisexual; petals 5, basally connate, actinomorphic; stamens 5, epetalous, the filaments free, the anthers connate; ovary inferior, 2-carpellate, 1-loculate. Fruit a cypsela; pappus present.

A genus of 14 species; North America, West Indies, Mexico, Central America, South America, Africa, Asia, Australia, and Pacific Islands. [*Tithonus*, prince of Troy of Greek mythology, who is symbolic of old age due to his immortality, in reference to the gray or white indument found in some species.]

Selected reference: La Duke (2006).

1. Outer phyllaries shorter than the inner ones; plant perennial..T. **diversifolia**
1. Outer phyllaries subequaling or longer than the inner ones; plant annual......................T. **rotundifolia**

Tithonia diversifolia (Hemsl.) A. Gray [With leaves of various shapes.] SHRUB SUNFLOWER.

Mirasolia diversifolia Hemsley, Biol. Cent.-Amer., Bot. 2: 168, t. 47. 1881. *Tithonia diversifolia* (Hemsley) A. Gray, Proc. Amer. Acad. Arts 19: 5. 1883. *Urbanisol tagetifolius* (Hemsley) Kuntze, Revis. Gen. Pl. 1: 371. 1891.

Erect perennial herb, subshrub, or shrub, to 2.5(5) m; stem striate-angled, glabrous. Leaves with the blade ovoid or deltate, 0.7–3 cm long, 0.7–2 cm wide, the apex acuminate, the base cuneate or sometimes truncate, the margin crenate-serrate, sometimes 3- or 5-lobed, the upper surface glabrous, the lower surface glabrous or hispid-pilose, usually just on the veins, the petiole 2–6 cm long, the base clasping. Flowers in a solitary capitulum, the peduncle 7–24 cm long, glabrous; involucre hemispheric, 1–2 cm wide; phyllaries numerous in (3)4 series, oblong or ovate, the outer 6–9 mm long, the apex rounded or acute, the outer surface usually glabrous, the inner 10–20 mm long, the apex rounded or acute, the outer surface glabrous; receptacle hemispheric; paleae 10–13 mm long, 3-toothed, the teeth 2–3 mm long, the middle one larger, acute, acuminate, or aristate; ray flowers 7–14; corolla yellow, the lamina 4.5–7 cm long; disk flowers numerous; corolla yellow, 6–8 mm long. Cypsela cuneiform, 4–7 mm long, compressed, sometimes 3–4 angled or biconvex, brown; pappus coroniform with 1–2 subulate scales.

Disturbed sites. Occasional; central and southern peninsula. Escaped from cultivation. Florida and Texas; West Indies, Mexico, Central America, and South America; Africa, Asia, Australia, and Pacific Islands. Native to Texas, Mexico, and Central America. All year.

Tithonia rotundifolia (Mill.) S. F. Blake [With the leaves (3- to 5-lobed ones) rounded in general outline.] CLAVEL DE MUERTO.

Tagetes rotundifolia Miller, Gard. Dict., ed. 8. 1768. *Tithonia rotundifolia* (Miller) S. F. Blake, Contr. Gray Herb. 52: 41. 1917.

Annual herb, to 2(4) m; stem striate-angled, puberulent. Leaves with the blade ovate or deltate, 0.7–2.5 mm long, 0.7–3.5 mm long, the apex acuminate, the base cuneate or sometimes truncate, the margin crenate-serrate, sometimes 3- or 5-lobed, the upper surface glabrous, the lower surface puberulent or glabrous, the petiole 3–8 cm long, the base clasping. Flowers in a solitary capitulum, the peduncle 11–27 cm long, puberulent; involucre hemispheric, 1–2 cm wide; phyllaries numerous in 3–4 series, lanceolate or linear, the outer 1.5–3 cm long, the apex acute, the outer surface puberulent, the inner 1.5–2.5 cm long, the apex rounded or acute, the outer surface usually puberulent; receptacle hemispheric; paleae 12–15 mm long, 3-toothed, the teeth 3–6 mm long, the middle one larger, acute, acuminate, or aristate; ray flowers 8–13; corolla yellow or orange, the lamina 2–3 cm long; disk flowers numerous; corolla yellow or orange, 5–7 mm long. Cypsela cuneiform, 5–7 mm long, compressed, sometimes 3–4 angled or biconvex, brown; pappus coroniform with 1–2 subulate scales.

Disturbed sites. Rare; Leon County. Escaped from cultivation. Florida and Louisiana; West Indies, Mexico, Central America, and South America. Native to Mexico and Central America. Summer–fall.

Tridax L. 1753.

Herbs. Leaves opposite, simple, pinnate-veined, petiolate. Flowers in capitula, solitary, radiate; receptacle paleate; ray flowers carpellate; petals 5, basally connate, zygomorphic; disk flowers bisexual; petals 5, basally connate, actinomorphic; stamens 5, epipetalous, the filaments free, the anthers connate; ovary inferior, 2-carpellate, 1-loculate. Fruit a cypsela; pappus present.

A genus of about 29 species; North America, West Indies, Mexico, Central America, South America, Africa, Asia, Australia, and Pacific Islands. [Three-toothed, in reference to the 3-lobed ray flowers.]

Selected reference: Strother (2006hh).

Tridax procumbens L. [Growing prostrate.] COATBUTTONS.

Tridax procumbens Linnaeus, Sp. Pl. 900. 1753. *Chrysanthemum procumbens* (Linnaeus) Sessé y Lacasta & Moçiño, Fl. Mexic., ed. 2. 190. 1894.

Procumbent or ascending perennial herb, to 4(8) dm; stem striate-angled, hirsute. Leaves with the blade deltate, lanceolate, ovate-lanceolate, or ovate, 1–4(12) cm long, 0.5–2(6) cm wide, the margin often 3-lobed, the ultimate margin coarsely toothed or subentire, the upper and lower surfaces hirsute, the petiole to 1(3) cm long. Flowers in a solitary capitulum, the peduncle 10–20 cm long, hirsute; involucre hemispheric, 5–8 mm wide; phyllaries 11–15 in 2–3 series, lanceolate, the outer surface hirsute; receptacle convex or conic; ray flowers 3–8(13), carpellate; corolla pale yellow, the lamina 2–3 mm long; disk flowers numerous, bisexual; corolla pale yellow, 2–3 mm long. Cypsela obconic or obpyramidal, 2–3 mm long, 3- to 5-angled, densely pilose; pappus numerous plumose, setiform scales 4–5(8) mm long.

Disturbed sites. Frequent; central and southern peninsula, Leon County. Florida and Texas; West Indies, Mexico, Central America, and South America; Africa, Asia, Australia, and Pacific Islands. Native to tropical America. Summer–fall.

Verbesina L. 1753. CROWNBEARD

Herbs. Leaves opposite or alternate, simple, pinnate-veined, petiolate or epetiolate. Flowers in capitula, solitary or in corymbiform, paniculiform, cymiform, or dichasial arrays, radiate or discoid; ray flowers carpellate, sterile, or absent; petals 5 (absent in discoid flowers), basally connate, zygomorphic; disk flowers bisexual; petals 5, basally connate, actinomorphic; stamens 5, epipetalous, the filaments free, the anthers connate; ovary inferior, 2-carpellate, 1-loculate. Fruit a cypsela; pappus present.

A genus of about 300 species; North America, West Indies, Mexico, South America, Europe, Africa, Asia, Australia, and Pacific Islands. [The genus *Verbena,* and *-ina,* to resemble; in reference to the resemblance of its foliage to that of *Verbena* (Verbenaceae).]

Phaethusa Gaertn., 1791; *Pterophyton* Cass., 1818; *Ridan* Adans., 1763; *Ximenesia* Cav., 1793–1794.

Selected reference: Strother (2006ii).

1. Leaves white strigose-canescent or sericeous on the lower surface; plant annual ....**V. encelioides**
1. Leaves other than white strigose-canescent on the lower surface; plant perennial.
   2. Capitula globose; phyllaries deflexed ....**V. alternifolia**
   2. Capitula hemispheric to nearly cylindric; phyllaries erect or slightly spreading (the apex sometimes reflexed).
      3. Leaves not decurrent on the stem.
         4. Capitula radiate, (3)12–15(25), 0.7–1.2 cm wide; leaves sharply serrate nearly to the base ....**V. aristata**
         4. Capitula discoid, solitary (rarely 2–5), 1–2 cm wide; leaves bluntly and coarsely toothed mainly at the middle and above............**V. chapmanii**
      3. Leaves decurrent on the stem (at least the middle and lower ones).
         5. Leaves all or mostly all alternate, the blade usually lobed; ray and disk flowers white............**V. virginica**
         5. Leaves opposite (at least the larger lower and median ones), the upper much reduced and sometimes alternate); ray and disk flowers yellow.
            6. Principal leaves narrowed to a petiolate base, the upper ones not much reduced; inflorescence with numerous capitula............**V. occidentalis**
            6. Principal leaves sessile, the upper ones much reduced below the inflorescence; inflorescence with few capitula............**V. heterophylla**

Verbesina alternifolia (L.) Britton ex Kearney [With alternate leaves.] WINGSTEM.

Coreopsis alternifolia Linnaeus, Sp. Pl. 909. 1753. *Verbesina coreopsis* Michaux, Fl. Bor.-Amer. 2: 134. 1803, nom. illegit. *Verbesina coreopsis* Michaux var. *lutea* Michaux, Fl. Bor.-Amer. 2: 134. 1803, nom. inadmiss. *Actinomeris squarrosa* Nuttall, Gen. N. Amer. Pl. 2: 181. 1818, nom. illegit. *Actinomeris alternifolia* (Linnaeus) de Candolle, Prodr. 5: 575. 1836. *Actinomeris squarrosa* Nuttall var. *alternifolia* (Linnaeus) Torrey & A. Gray, Fl. N. Amer. 2: 335. 1842. *Ridan alternifolia* (Linnaeus) Kuntze, Revis. Gen. Pl. 1: 360. 1891. *Verbesina alternifolia* (Linnaeus) Britton ex Kearney, Bull. Torrey Bot. Club 20: 485. 1893.

Erect perennial herb, to 1(2) m; stem striate-angled, scabrellous. Leaves all or mostly alternate, the proximal sometimes opposite, the blade lanceolate, lanceolate-elliptic, or linear-lanceolate, 1–25 cm long, 2–8 cm wide, reduced upward, the apex acute, the base narrowly cuneate, the margin coarsely toothed or subentire, the upper and lower surfaces scabrellous, the bases cuneate, winged, the wings sometimes decurrent on the stem. Flowers in (3)8–25(50+) capitula in a corymbiform or paniculiform array, globose; involucre saucerlike, 10–12 mm wide; phyllaries 8–12+ in 1(2) series, spatulate, linear-lanceolate, or linear, 3–8+ mm long, spreading or reflexed; receptacle convex; paleae usually navicular; ray flowers (2)6–8+, carpellate or sterile; corolla yellow or orange, the lamina 1.5–2.5+ cm long; disk flowers numerous; corolla yellow or orange, ca. 2 mm long. Cypsela oblanceolate or suborbicular, ca. 5 mm long, slightly flattened, broadly winged, hirtellous or glabrate, dark brown or black; pappus of 2 subulate scales ca. 2 mm long.

Bluff forests and floodplain forests. Rare; Gadsden, Liberty, and Jackson Counties. Massachusetts and New York south to Florida, west to Ontario, Wisconsin, Iowa, Nebraska, Kansas, Oklahoma, and Texas. Fall.

Verbesina aristata (Elliott) A. Heller [Spiny, in reference to the phyllaries, paleae, and pappus awns of the capitulum.] COASTALPLAIN CROWNBEARD.

Helianthus aristata Elliott, Sketch Bot. S. Carolina 2: 428. 1923. *Verbesina aristata* (Elliott) A. Heller, Muhlenbergia 1: 8. 1900. *Pterophyton aristatum* (Elliott) Alexander, in Small, Man. S.E. Fl. 1444, 1509. 1933.

Actinomeris nudicaulis Nuttall, Trans. Amer. Philos. Soc., ser. 2. 7: 364. 1841. *Verbesina nudicaulis* (Nuttall) A. Gray, Proc. Amer. Acad. Arts 19: 12. 1883. TYPE: FLORIDA.

Erect perennial herb, to 1.5 m; stem striate-angled, hirsute. Leaves all or mostly opposite, the distal sometimes alternate, the blade elliptic, 3–6(10) cm long, 1.2–2.8 cm wide, reduced upward, the apex rounded or acute, the base rounded, the margin sharply serrate toothed, the upper and lower surfaces hispidulous-scabrellous, sessile. Flowers in (3)12–25+ capitula in a paniculiform array; involucre hemispheric, 8–10 mm wide; phyllaries numerous in 2–3 series, linear, 2–3+ mm long, erect; receptacle convex; paleae usually navicular; ray flowers (5)11–13, carpellate or sterile; corolla yellow or orange, the lamina (1.5)2.5–3+ cm; disk flowers numerous; corolla yellow or orange, 2–3 mm long. Cypsela subobovate, 4–5 mm long, slightly flattened, narrowly winged, glabrous, purplish black; pappus of 2 subulate scales ca. 1 mm long.

Flatwoods and hammocks. Occasional; northern counties. Georgia, Florida, and Alabama. Spring–summer.

Verbesina chapmanii J. R. Coleman [Commemorates Alvan Wentworth Chapman (1809–1899), American physician and botanist.] CHAPMAN'S CROWNBEARD.

Verbesina chapmanii J. R. Coleman, Rhodora 74: 100. 1972. TYPE: FLORIDA: Liberty Co.: 3 mi. N of Orange, s.d., *McDaniel 4468* (holotype: GH).

Erect perennial herb, to 5(8) dm; stem striate-angled, slightly scabrellous. Leaves all or mostly opposite, the distal sometimes alternate, the blade elliptic, 3–6(10) cm long, 0.8–2.2(3) cm wide, reduced upward, the apex rounded or obtuse, the base cuneate, the margin bluntly and coarsely toothed or subentire, the upper and lower surfaces slightly scabrellous, sessile. Flowers in capitula, solitary or (2)3+ in a dichasial array; involucre hemispheric or turbinate, 8–16 mm wide; phyllaries numerous in 2–3 series, spatulate or linear-lanceolate, 5–9+ mm long, suberect; receptacle convex; paleae usually navicular; ray flowers absent; disk flowers numerous; corolla yellow 2–3 mm long. Cypsela subelliptic, 5–7 mm long, slightly flattened, winged, glabrous or sparsely hirtellous, purplish black; pappus of 2 subulate scales to 0.5 mm long.

Bogs and wet pine flatwoods. Occasional; central panhandle, Walton County. Endemic. Summer–fall.

Verbesina chapmanii is listed as threatened in Florida (Florida Administrative Code, Chapter 5B-40).

Verbesina encelioides (Cav.) Benth. & Hook. f. ex A. Gray [Resembling the genus *Encelia* (Asteraceae).] GOLDEN CROWNBEARD; SKUNK DAISY.

Ximenesia encelioides Cavanilles, Icon. 2: 60, t. 178. 1794. *Ximenesia encelioides* Cavanilles var. *hortensis* de Candolle, Prodr. 5: 627. 1836, nom. inadmiss. *Verbesina encelioides* (Cavanilles) Bentham & Hooker f. ex A. Gray, in S. Watson, Bot. California 1: 350. 1876.

Erect annual herb, to 5(12) dm; stem striate-angled, strigose-scabrellous or sericeous. Leaves all or mostly alternate, the proximal sometimes opposite, the blade deltate-ovate or lanceolate, 3–8(12+) cm long, 2–4(6+) cm wide, the apex acute or acuminate, the base broadly cuneate or truncate, the margin coarsely toothed or subentire, the upper and lower surfaces white strigose-canescent or sericeous, the petiole usually narrowly winged, the base broadly winged and clasping. Flowers in capitula, borne solitary or sometimes 2–3+ in a loose cymiform or corymbiform array; involucre hemispheric or saucerlike, 10–20 mm wide; phyllaries numerous in 1–2 series, ovate-lanceolate, linear-lanceolate, or linear, 6–8+ mm long, suberect or spreading; receptacle convex; paleae usually navicular, linear or filiform; ray flowers (8)12–15, carpellate or sterile; corolla yellow or orange, the lamina 8–10(20) mm long; disk flowers numerous; corolla yellow, 5–6 mm long. Cypsela narrowly obovate, 4–5 mm long, slightly flattened, broadly winged, sparsely strigillose, dark brown or black; pappus of 2 subulate scales 1–2 mm long (absent on the ray cypsela).

Disturbed sites. Occasional; central peninsula, Gadsden County and Monroe County keys. Nearly throughout the United States, West Indies, Mexico and South America; Europe, Africa, Asia, Australia, and Pacific Islands. Native to the western United States and Mexico. Summer.

Verbesina heterophylla (Chapman) A. Gray [With diverse leaf shapes.] DIVERSELEAF CROWNBEARD.

Actinomeris pauciflora Nuttall, Amer. J. Sci. Arts 5: 301. 1822. *Actinomeris squarrosa* Nuttall var. *serrulata* Rafinesque, New Fl. 1: 61. 1836. *Verbesina pauciflora* (Nuttall) Small, Fl. S.E. U.S. 1273, 1340. 1903; non Hemsley, 1881. *Verbesina warei* A. Gray, Poc. Amer. Acad. Arts 19: 12. 1883. *Pterophyton pauciflorum* (Nuttall) Alexander, in Small, Man. S.E. Fl. 1444, 1509. 1933. TYPE: FLORIDA: s.d., *Ware s.n.* (Holotype: PH?).

Actinomeris heterophylla Chapman, Bot. Gaz. 3: 6. 1878. *Verbesina heterophylla* (Chapman) A. Gray, Proc. Amer. Acad. Arts 19: 12. 1883. *Pterophyton heterophyllum* (Chapman) Alexander, in Small, Man. S.E. Fl. 1444, 1509. 1933. TYPE: FLORIDA: s.d., *Chapman s.n.* (Lectotype: NY). Lectotypified by Coleman (1972: 100).

Actinomeris squarrosa Nuttall var. *pauciflora* Rafinesque, New Fl. 1: 61. 1836. TYPE: FLORIDA.

Erect perennial herb, to 1.5 m; stem striate-angled, hispidulous-scabrellous. Leaves all or mostly opposite, the distal sometimes alternate, the blade ovate, elliptic, or lanceolate, 3–7 cm long, 0.8–2.5 cm wide, much reduced upward, the apex obtuse, the base cuneate, winged, decurrent on the stem, the margin toothed, the upper and lower surfaces scabrellous, sessile. Flowers in capitula, solitary or 3–9+ in a loose corymbiform array; involucre hemispheric, 12–15+ mm wide; phyllaries numerous in 2–3 series, narrowly oblong or lanceolate, 4–6+ mm long, suberect; receptacle convex; paleae usually navicular; ray flowers (5)8, carpellate or sterile; corolla yellow or orange, the lamina 1.2–1.5+ cm long; disk flowers numerous; corolla yellow, 2–3 mm long. Cypsela narrow obovate or elliptic, ca. 5 mm long, slightly flattened, winged, glabrous, purplish black; pappus absent or of 2 minute subulate scales.

Flatwoods. Occasional; northern peninsula, Volusia County. Florida and Georgia. Summer.

Verbesina occidentalis (L.) Walter [Western, in reference to its occurring in the western part of eastern North America.] YELLOW CROWNBEARD.

Sigesbeckia occidentalis Linnaeus, Sp. Pl. 900. 1753. *Phaethusa americana* Gaertner, Fruct. Sem. Pl. 2: 425. 1791, nom. illegit. *Verbesina occidentalis* (Linnaeus) Walter, Fl. Carol. 213. 1788. *Verbesina sigesbeckia* Michaux, Fl. Bor.-Amer. 2: 134. 1803, nom. illegit. *Phaethusa borealis* Sprengel, Syst. Veg. 3: 591. 1826, nom. illegit. *Verbesina phaethusa* Cassini, in Cuvier, Dict. Sci. Nat. 59: 143. 1829, nom. illegit. *Phaethusa occidentalis* (Linnaeus) Britton, in Britton & A. Brown, Ill. Fl. N. U.S., ed. 2. 3: 488. 1913.

Erect perennial herb, to 2 m; stem striate-angled, scabrellous. Leaves all or mostly opposite, the distal sometimes alternate, the blade ovate or lanceolate, 6–12(16+) cm long, 3–6(10+) cm wide, the apex acute or acuminate, the base cuneate, the margin coarsely toothed or subentire, the upper and lower surfaces scabrellous, the petiole winged, decurrent on the stem. Flowers in capitula, numerous in a corymbiform or paniculiform array; involucre campanulate or turbinate, 3–5+ mm wide; phyllaries 8–12+ in 2 series, spatulate or oblanceolate, 3–7+ mm long, suberect; receptacle convex; paleae usually navicular; ray flowers (0)1–3(5), carpellate or sterile; corolla yellow or orange, the lamina 1–1.5(2) cm long; disk flowers 8–15; corolla yellow, 3–4 mm long. Cypsela oblanceolate, ca. 5 mm long, slightly flattened, narrowly winged or not winged, strigose, dark brown to black; pappus of 2 subulate scales 3–4 mm long.

Hammocks. Occasional; central and western panhandle. Delaware and Maryland south to Florida, west to Ohio, Kentucky, Missouri, Tennessee, and Mississippi. Fall.

Verbesina virginica L. [Of Virginia.] WHITE CROWNBEARD; FROSTWEED.

Verbesina virginica Linnaeus, Sp. Pl. 901. 1753. *Phaethusa virginica* (Linnaeus) Britton, in Britton & A. Brown, Ill. Fl. N. U.S., ed. 2. 3: 487. 1913.

Sigesbeckia laciniata Poiret, in Lamarck, Encycl. 7: 158. 1806. *Verbesina laciniata* (Poiret) Nuttall, Gen. N. Amer. Pl. 2: 170. 1818. *Verbesina sinuata* Elliott, Sketch Bot. S. Carolina 2: 411. 1823, nom. illegit. *Verbesina virginica* Linnaeus var. *laciniata* (Poiret) A. Gray, Proc. Amer. Acad. Arts 19: 11. 1883. *Phaethusa laciniata* (Poiret) Small, Fl. Miami 195, 200. 1913.

Erect perennial herb, to 2.5 m; stem striate-angled, scabrellous. Leaves all or mostly alternate, the proximal sometimes alternate, the blade ovate-lanceolate, lanceolate-elliptic, or linear-lanceolate, 5–12(22+) cm long, 1–6(12+) cm wide, reduced upward, the apex acute, the base cuneate, winged to the base, the margin sometimes pinnately lobed, coarsely toothed or subentire, sometimes sinuate, the upper and lower surfaces scabrellous or strigillose, sessile, the base decurrent on the stem. Flowers in capitula, numerous in a corymbiform or paniculiform array; involucre obconic or turbinate, 3–5 mm wide; phyllaries 8–12+ in 1–2 series, spatulate or oblanceolate, 3–5(7) mm long, suberect; receptacle convex; paleae usually navicular; ray flowers (1)2–3(7), carpellate or sterile; corolla white, the lamina 3–4(7+); disk flowers numerous; corolla white, 2–3 mm long. Cypsela oblanceolate, 4–5 mm long, slightly flattened, winged, scabrellous, dark brown; pappus of 2 subulate scales 2–3+ mm long.

Open hammocks, stream banks, and coastal hammocks. Common; nearly throughout. Maryland south to Florida, west to Iowa, Kansas, Missouri, Oklahoma, and Texas. Summer–fall.

EXCLUDED TAXON

Verbesina walteri Shinners—Reported for Florida by Small (1903, 1913, as *Actinomeris paniculata* (Walter) Small; 1933, as *Ridan paniculata* (Walter) Small). No Florida specimens seen. Excluded from the state by Cronquist (1980).

Vernonia Schreb., nom. cons. 1791. IRONWEED

Herbs. Leaves alternate, simple, pinnate-veined, petiolate or epetiolate. Flowers in capitula in corymbiform or paniculiform arrays, discoid, bisexual; receptacle epaleate; petals 5, basally connate, actinomorphic; stamens 5, epipetalous, the filaments free, the anthers connate; ovary inferior, 2-carpellate, 1-loculate. Fruit a cypsela; pappus present.

A genus of about 1,000 species; nearly cosmopolitan. [Commemorates William Vernon (ca. 1666–1711), English botanist and entomologist at Cambridge University.]

Selected reference: Strother (2006jj).

1. Leaves mostly basal .... **V. acaulis**
1. Leaves mostly cauline.
   2. Middle cauline leaves linear (less than 1 cm wide).
      3. Upper surface of leaves smooth or nearly so; pappus light yellow .... **V. blodgettii**
      3. Upper surface of leaves scabrous; pappus tawny or purplish .... **V. angustifolia**
   2. Middle cauline leaves lanceolate (1.5 cm wide or wider).
      4. Middle and inner phyllaries long-acuminate to filiform .... **V. noveboracensis**
      4. Middle and inner phyllaries aute or mucronate.
         5. Leaves puberulent on the lower surface; capitula with 10–20(30) flowers .... **V. gigantea**
         5. Leaves tomentose on the lower surface; capitula with 30–60 flowers .... **V. missurica**

Vernonia acaulis (Walter) Gleason [Without a stem, in reference to the mostly basal leaves.] STEMLESS IRONWEED.

Chrysocoma acaulis Walter, Fl. Carol. 196. 1788. *Cacalia acaulis* (Walter) Kuntze, Revis. Gen. Pl. 2: 968. 1891. *Vernonia acaulis* (Walter) Gleason, Bull. New York Bot. Gard. 4: 222. 1906.

Vernonia oligophylla Michaux, Fl. Bor.-Amer. 2: 94. 1803. *Vernonia oligophylla* Michaux, var. *verna* Michaux, Fl. Bor.-Amer. 2: 94. 1803, nom. inadmiss.

Erect perennial herb, to 6(10) dm; stem striate-angled, puberulent, glabrescent. Leaves mostly basal, the blade oblong-elliptic or spatulate, 11–25(30) cm long, (2)6–7 cm wide, the apex acute, obtuse, or rounded, the base cuneate, the margin serrate or dentate, the upper surface scabrellous or glabrate, the lower surface glabrous or glabrate, sometimes hirtellous on the veins, usually gland-dotted, sessile or subsessile. Flowers in capitula in a corymbiform array, the peduncle 0.5–5 cm long; involucre campanulate or hemispheric, 6–7 mm wide; phyllaries numerous in 5–6 series, linear-lanceolate or lanceolate-ovate, 4–9 mm long, the apex subulate or filiform, the margin ciliolate, the outer surface glabrate; receptacle flat or slightly convex; flowers numerous; corolla purplish or pink, 4–6 mm long. Cypsela columnar, ca. 3 mm long, 8- to 10-ribbed, strigillose; pappus of numerous outer scales ca. 1 mm long and numerous inner bristles 5–7 mm long, whitish or stramineous.

Hammocks. Rare; Polk County. Not recently collected. North Carolina south to Florida. Summer.

Vernonia angustifolia Michx. [With narrow leaves.] TALL IRONWEED.

Vernonia angustifolia Michaux, Fl. Bor.-Amer. 2: 94. 1803. *Cacalia angustifolia* (Michaux) Kuntze, Revis. Gen. Pl. 2: 968. 1891, nom. illegit. *Cacalia graminifolia* Kuntze, Revis. Gen. Pl. 2: 968, 1891, nom. illegit. *Vernonia graminifolia* Trelease ex Branner & Coville, Rep. (Annual) Geol. Surv. Arkansas 1888(4): 189. 1891. *Vernonia graminifolia* C. Mohr, Contr. U.S. Natl. Herb. 6: 759. 1901, nom. illegit.; non Gardner, 1847.

Vernonia scaberrima Nuttall, Gen. N. Amer. Pl. 2: 134. 1818. *Vernonia angustifolia* Michaux var. *scaberrima* (Nuttall A. Gray, syn. Fl. N. Amer. 1(2): 91. 1884. *Vernonia angustifolia* subsp. *scaberrima* (Nuttall) S. B. Jones & W. Z. Faust, in Rogerson et al., N. Amer. Fl., ser. 2. 10: 189. 1978.

Vernonia angustifolia Michaux var. *mohrii* S. B. Jones, Rhodora 66: 397. 1964. *Vernonia angustifolia* Michaux subsp. *mohrii* (S. B. Jones) S. B. Jones & W. Z. Faust, in Rogerson et al., N. Amer. Fl., ser. 2. 10: 188. 1878.

Erect perennial herb, to 1 m; stem striate-angled, sparsely appressed puberulent, glabrescent. Leaves mostly cauline, the blade of the midcauline leaves lanceolate, linear, or filiform, 5–12 cm long, 2–4(8) mm wide, the margin entire or subentire, the apex acute, the base cuneate, the upper surface scabrellous, the lower surface with scattered trichomes on the midrib or scabrellous, sometimes gland-dotted, sessile or subsessile. Flowers in capitula in a corymbiform or paniculiform array, the peduncle 0.8–2.5 cm long; involucre subcampanulate or obconic, 4–6(9) mm wide; phyllaries numerous in 5–6 series, lanceolate or ovate-lanceolate, 3–9 mm long, the apex acuminate, the margin ciliolate, the outer surface sparsely scabrellous, glabrescent; receptacle flat or slightly convex; corolla purplish or pink, 4–6 mm long. Cypsela columnar, ca. 3 mm long, 8- to 10-ribbed, strigillose; pappus of numerous outer scales ca. 1 mm long and numerous inner bristles ca. 6 mm long, stramineous or purplish.

Sandhills and dry, open hammocks. Frequent; northern counties, central peninsula. North Carolina south to Florida, west to Louisiana. Summer–fall.

Vernonia angustifolia hybridizes with *V. gigantea* in Florida (=*V.* ×*concinna*) (Jones, 1964).

Vernonia blodgettii Small [Commemorates John Loomis Blodgett (1809–1853), American physician and botanist.] FLORIDA IRONWEED; BLODGETT'S IRONWEED.

Vernonia angustifolia Michaux var. *pumila* Chapman, Bot. Gaz. 3: 5. 1878. TYPE: FLORIDA: Lee Co.: near the Caloosa River.

Vernonia blodgettii Small, Fl. S.E. U.S. 1160, 1338. 1903. TYPE: FLORIDA: Monroe Co.: Big Pine Key, s.d. *Blodgett s.n.* (holotype: NY).

Erect perennial herb, to 3(5) dm; stem striate-angled, glabrous or glabrate. Leaves mostly cauline, the blade linear-lanceolate or linear, 3–7 cm long, 2–8 mm wide, the apex acute, the base cuneate, the margin entire or subentire, the upper surface scabrellous, glabrescent, gland-dotted, the lower surface scabrellous, gland-dotted, sessile or subsessile. Flowers in capitula in a corymbiform or paniculiform array, the peduncle 1.2–3.5 cm long; involucre campanulate or obconic, 6–10 mm wide; phyllaries numerous in 4–5 series, ovate-lanceolate, lanceolate-oblong, or linear-oblong, 3–7 mm long, the apex acute or rounded-apiculate, the margin arachnoid-ciliate, the outer surface pubescent; receptacle flat or slightly convex; corolla purplish or pink, 4–6 mm long. Cypsela columnar, ca. 3 mm long, 8- to 10-ribbed, strigillose; pappus of numerous outer scales ca. 1 mm long and numerous inner bristles 5–7 mm long, stramineous or whitish.

Flatwoods. Occasional; central and southern peninsula. Florida; West Indies. Summer–fall.

Vernonia gigantea (Walter) Trel. ex Branner & Coville [Large, in reference to the plant size.] GIANT IRONWEED.

Chrysocoma gigantea Walter, Fl. Carol. 196. 1788. *Vernonia gigantea* (Walter) Trelease ex Branner & Coville, Rep. (Annual) Geol. Surv. Arkansas 1888(4): 189. 1891. *Cacalia gigantea* (Walter) Kuntze, Revis. Gen. Pl. 1: 324. 1891.

Vernonia altissima Nuttall, Gen. N. Amer. Pl. 2: 134. 1818. *Vernonia fasciculata* Michaux var. *altissima* (Nuttall) Torrey & A. Gray ex Chapman, Fl. South. U.S. 188. 1860. *Vernonia altissima* Nuttall forma *parviflora* A. Gray, Syn. Fl. N. Amer. 1(2): 90. 1884, nom. inadmiss.

Vernonia ovalifolia Torrey & A. Gray, Fl. N. Amer. 2: 59. 1841. *Vernonia gigantea* (Walter) Trelease ex Branner & Coville subsp. *ovalifolia* (Torrey & A. Gray) Urbatsch, Brittonia 24: 237. 1972. *Vernonia gigantea* (Walter) Trelease ex Branner & Coville var. *ovalifolia* (Torrey & A. Gray) D. B. Ward, Phytologia 94: 468. 2012. TYPE: FLORIDA: s.d., *Chapman s.n.* (lectotype: NY). Lectotypified by Urbatsch (1972: 237).

Vernonia oligantha Greene, Pittonia 5: 56. 1902. TYPE: FLORIDA: Manatee Co.: Palmetto, 30 Nov 1901, *Tracy s.n.* (holotype: ?).

Erect perennial herb, to 2(3) m; stem puberulent, glabrescent. Leaves mostly cauline, the blade lanceolate, 12–25 cm long, 2–6 cm wide, the apex acute, the base cuneate, the margin serrate, the upper surface strigillose, glabrescent, the lower surface scabrellous, sometimes glabrescent, sometimes sparsely gland-dotted, the petiole short, winged, sometimes so to the base, sometimes sessile. Flowers in capitula in a corymbiform array, the peduncle 1–12(20) mm long; involucre broadly campanulate or hemispheric, 4–5 mm wide; phyllaries numerous in 4–5 series, ovate-lanceolate or oblong, 2–5 mm long, the apex acute or rounded-apiculate, the margin ciliolate, the outer surface glabrate; receptacle flat or slightly convex; corolla purplish or pink, 4–6 mm long. Cypsela columnar, 3–4 mm long, 8- to 10-ribbed, strigillose; pappus of numerous outer scales ca. 1 mm long and numerous inner bristles 5–6 mm long, purplish or stramineous.

Hammocks and floodplains. Common; northern counties, central peninsula. New York south to Florida, west to Ontario, Michigan, Iowa, Nebraska, Kansas, Oklahoma, and Texas. Summer–fall.

Vernonia gigantea hybridizes with *V. angustifolia* in Florida (=*V.* ×*concinna*) (Jones, 1964).

Vernonia missurica Raf. [Of Missouri.] MISSOURI IRONWEED.

Vernonia missurica Rafinesque, Herb. Raf. 28: 1833.

Erect perennial herb, to 1.2(2) m; stem striate-angled, puberulent. Leaves mostly cauline, the blade elliptic, ovate-lanceolate, or lanceolate, the apex acute, the base cuneate, the margin serrate, the upper surface scabrellous, glabrescent, the lower surface usually puberulent or tomentose, rarely glabrate, gland-dotted, short-petiolate, often winged to the base or sessile. Flowers in capitula in a corymbiform array, the peduncle 0.3–3.5 cm long; involucre broadly campanulate or urceolate, 5–9 mm wide; phyllaries numerous in 6–7 series, lanceolate, linear-oblong, or oblong, 2–7(9) mm long, the apex acute or rounded-apiculate, the margin ciliolate, the outer surface sparsely scabrellous, glabrescent, rarely gland-dotted; receptacle flat or slightly convex; corolla purplish or pink, 4–6 mm long. Cypsela columnar, 3–4 mm long, 8- to 10-ribbed, strigillose; pappus of numerous outer scales ca. 1 mm long and numerous inner bristles 6–8 mm long, stramineous or whitish.

Wet hammocks. Occasional; central and western panhandle. Ontario, Michigan, Kentucky, Tennessee, Georgia, and Florida, west to Iowa, Nebraska, Kansas, Oklahoma, and New Mexico, also Massachusetts. Summer–fall.

Vernonia noveboracensis (L.) Michx. [Of New York.] NEW YORK IRONWEED.

Serratula noveboracensis Linnaeus, Sp. Pl. 818. 1753. *Behen noveboracense* (Linnaeus) Hill, Hort. Kew. 68. 1768. *Cacalia noveboracensis* (Linnaeus) Kuntze, Revis. Gen. Pl. 1: 324. 1891. *Vernonia noveboracensis* (Linnaeus) Michaux, Fl. Bor.-Amer. 2: 95. 1803.

Erect perennial herb, to 1.2(2) m; stem striate-angled, puberulent, glabrescent. Leaves mostly cauline, the blade lanceolate, 9–15(25) cm long, 1.5–4.5(6) cm wide, the apex acute, the base cuneate, the margin serrate, the upper surface scabrellous, often gland-dotted, the lower surface scabrellous or tomentose, gland-dotted, the petiole short, often winged to the base, or sessile. Flowers in capitula in a corymbiform or paniculiform array, the peduncle 0.2–3.5 cm long; involucre subhemispheric, 7–10 mm wide; phyllaries numerous in 4–6 series, lanceolate or oblong, the apex subulate or filiform, the margin ciliolate, the outer surface sparsely tomentose, glabrescent; receptacle flat or slightly convex; corolla purplish or pink, 4–6 mm long. Cypsela columnar, 8- to 10-ribbed, strigillose; pappus of numerous outer scales ca. 1 mm long and numerous inner bristles 6–7 mm long, stramineous.

Floodplain forests and swamps. Rare; northern peninsula west to central panhandle. New

Hampshire, Massachusetts, and New York south to Florida, west to Ohio, Kentucky, Tennessee, and Alabama, also Oklahoma and New Mexico. Fall.

HYBRID

Vernonia ×concinna Gleason (*V. angustifolia* × *V. gigantea*) [Pretty, elegant.]

Vernonia concinna Gleason, Bull. New York Bot. Gard. 4: 225. 1906, pro sp. TYPE: FLORIDA: Lake Co.: vicinity of Eustis, 24 Aug 1894, *Nash 1759* (holotype: NY; isotype: US).

Hammocks. Rare; Clay, Lake, Leon, and Gadsden Counties. Endemic. Summer–fall.

EXCLUDED TAXA

Vernonia glauca (Linnaeus) Willdenow—This more northern species was reported for Florida by Gray (1884, as *V. noveboracensis* (Linnaeus) Michaux var. *latifolia* A. Gray) and Chapman (1897, as V. *noveboracensis* var. *latifolia* A. Gray), the name misapplied to Florida material of *V. noveboracensis*.

Vernonia texana (A. Gray) Small—Reported for Florida by Small (1903, 1913a). Excluded from the state by Small (1933). No Florida specimens seen.

Vernonia ×illinoensis Gleason (*V. gigantea* × *V. missurica*)—Reported for Florida by Faust (1977), in a generalized range map. The parent species occurs sympatrically in Florida, and it is reasonable to expect the hybrid here, but no Florida specimens seen.

Xanthium L. 1753. COCKLEBURR

Herbs. Leaves alternate, simple, pinnipalmate-veined, petiolate. Flowers in capitula in racemiform or spiciform arrays or solitary in the axils, discoid, carpellate or staminate; receptacle paleate or epaleate, petals 5, basally connate, actinomorphic or absent; stamens 5, the filaments connate, the anthers free or connate; ovary inferior, 2-carpellate, 1-loculate. Fruit a cypsela; pappus absent.

A genus of 2 species; nearly cosmopolitan. [From the Greek *Xanthos*, yellow, in reference to an ancient name for the plant, which was used to produce a yellow dye.]

Selected reference: Strother (2006kk).

The taxonomy of *Xanthium* is controversial. Numerous species have been recognized, based primarily on minor differences of fruit morphology. We recognize two species for the genus following Löve and Dansereau (1959), which has generally been adopted by most other workers. However, there is still some question if a third South American species (*X. ambrosioides* Hooker & Arnott) should be recognized as is sometimes done by various workers.

Acanthoxanthium (DC.) Fourr., 1889.

Xanthium strumarium L. [With a cushion-like swelling, in reference to the fruit.] COCKLEBURR.

Xanthium strumarium Linnaeus, Sp. Pl. 987. 1753. *Xanthium vulgare* Lamarck, Fl. Franc. 2: 56. 1779, nom. illegit.

Xanthium canadense Miller, Gard. Dict., ed. 8. 1768. *Xanthium strumarium* Linnaeus var. *canadense* (Miller) Torrey & A. Gray, Fl. N. Amer. 2: 294. 1842.

Xanthium americanum Walter, Fl. Carol. 231. 1788. *Xanthium macrocarpum* de Candolle var. *glabratum* de Candolle, Prodr. 5: 523. 1836. *Xanthium pungens* Wallroth, Beitr. Bot. 227, 231. 1844, nom. illegit. *Xanthium strumarium* Linnaeus var. *glabratum* (de Candolle) Cronquist, Rhodora 47: 403. 1945. *Xanthium glabratum* (de Candolle) Britton, Man. Fl. N. States 912. 1901, nom. illegit.
Xanthium pensylvanicum Wallroth, Beitr. Bot. 228, 236. 1844. *Xanthium pensylvanicum* Wallroth var. *glandulosum* Wallroth, Beitr. Bot. 236. 1844, nom. inadmiss.

Erect annual herb, to 1(2) m; stem striate-angled, sparsely to moderately rough-pubescent with short, stout, broad-based, ascending trichomes, sometimes stipitate-glandular or sessile-glandular. Leaves with the blade broadly ovate or ovate-triangular, 4–12(18) cm long, 3–10(18) cm wide, sometimes palmately 3- to 5-lobed, the apex acute or obtuse, the base cordate, the margin irregularly serrate, the upper and lower surfaces sparsely to moderately roughened with short, broad-based trichomes, sometimes only along the veins, sometimes also stipitate-glandular or sessile-glandular, the petiole 3–10 cm long. Flowers in capitula in a racemiform or spiciform axillary array, sometimes solitary in the axil; proximal capitula carpellate; involucre ellipsoid, 2–5 mm wide at anthesis; phyllaries numerous in 6–12 series, the outer free, the inner proximally connate, the free apex hooked, the whole forming a hard prickly burr; receptacle epaleate; flowers 2; corolla absent; distal capitula staminate; involucre saucer-shaped. 3–5 mm wide; phyllaries 6–16 in 1–2 series, free; receptacle conic or columnar; paleae spatulate, cuneiform, or linear, membranous, distally villous or hirtellous; flowers numerous; corolla white, 1–2 mm long. Cypsela subfusiform, enclosed in an ovoid or ellipsoid, hard, 2-chambered, prickly burr 1–3 cm long.

Disturbed sites. Occasional; nearly throughout. Nearly cosmopolitan. Summer.

EXCLUDED TAXON

Xanthium spinosum Linnaeus—Reported for Florida by Small (1903, 1913a, 1933, as *Acanthoxanthium spinosum* (Linnaeus) Fourreau) and Strother (2006kk). The only vouchering specimen is a purported Chapman collection at MO whose provenance is dubious. Excluded from the state by Cronquist (1980).

Youngia Cass. 1831.

Herbs. Leaves alternate, simple, pinnate-veined, petiolate. Flowers in capitula in corymbiform or paniculiform arrays, liguliform, bisexual; calyculi present; receptacle epaleate; petals 5, basally connate, zygomorphic; stamens 5, epipetalous, the filaments free, the anthers connate; ovary inferior, 2-carpellate, 1-loculate. Fruit a cypsela; pappus present.

A genus of about 36 species; nearly cosmopolitan. [Commemorates Edward Young (1683–1765), English poet and dramatist, and Thomas Young (1773–1829), British physician, physicist, and Egyptologist.]

Selected reference: Spurr (2006).

Youngia japonica (L.) DC. [Of Japan.] ORIENTAL FALSE HAWKSBEARD.

Prenanthes japonica Linnaeus, Mant. Pl. 107. 1767. *Youngia japonica* (Linnaeus) de Candolle, Prodr. 7: 194. 1838. *Crepis japonica* (Linnaeus) Bentham, Fl. Hongk. 194. 1861. *Youngia japonica* (Linnaeus) de Candolle subsp. *genuina* Babcock & Stebbens, Publ. Carnegie Inst. Wash. 484: 98. 1937, nom. inadmiss.

Annual or biennial herb, to 9 dm; stem terete, fistulose. Leaves all or mostly basal, the blade oblong, ovate, or oblanceolate, 3–15(25) cm long, 2–4(6) cm wide, the apex acute or obtuse, the base cuneate, the margin pinnate-lobed, sublyrate, the ultimate margin denticulate, the lateral lobes reduced proximally, the terminal lobe elliptic, ovate, obovate, or oblong-truncate, larger than the laterals, the upper surface glabrous, the lower surfaces glabrous, usually puberulent on the midvein, the petiole 1–10 mm long, glabrous or puberulent, the base often dilated, clasping. Flowers in capitula in a corymbiform or paniculiform array, the peduncle 1–5(15) mm long; calyculi of 3–5 deltate or ovate, membranous bractlets; phyllaries 8 in 1–2 series, lanceolate or linear, 4–6 mm long, subequal, the margin slightly scarious, the apex obtuse to acute, the outer surface glabrous, glabrate, or with appressed trichomes; receptacle flat or convex, slightly pitted; flowers 8–25; corolla yellow, 5–7 mm long. Cypsela subfusiform, 2–3 mm long, compressed, weakly beaked, 11- to 13-ribbed, scabrous on the ribs, reddish brown; pappus of numerous basally connate, white, subequal, barbellate bristles, 3–4 mm long.

Disturbed sites. Common; nearly throughout. Nearly cosmopolitan. Native to Asia. All year.

Zinnia L.

Herbs. Leaves opposite or subopposite, simple, pinnate-veined, epetiolate. Flowers in capitula, solitary, radiate; receptacle paleate; ray flowers carpellate, basally connate, zygomorphic; disk flowers bisexual, basally connate, actinomorphic; stamens 5, epipetalous, the filaments free, the anthers connate; ovary inferior, 2-carpellate, 1-loculate. Fruit a cypsela; present or absent.

A genus of about 17 species; North America, West Indies, Mexico, Central America, and South America, Africa, Asia, Australia, and Pacific Islands.

Selected reference: Smith (2006b).

1. Involucre hemispheric; paleae with a strongly differentiated, fimbriate apex; cypsela cartilaginous-winged; pappus awnless .... **Z. elegans**
1. Involucre cylindric or campanulate; paleae with a slightly differentiated, erose or subentire apex; cypsela wingless; pappus of the disk flowers with a prominent awn .... **Z. peruviana**

Zinnia elegans Jacq. [Elegant.] ELEGANT ZINNIA.

Zinnia elegans Jacquin, Icon. Pl. Rar. 3: 15, t. 589. 1792. *Crassina elegans* (Jacquin) Kuntze, Revis. Gen. Pl. 1: 331. 1891.

Zinnia elegans Sessé y Lacasta & Moçiño, Pl. Nov. Hisp. 142. 1890; non Jacquin, 1792.

Zinnia violacea Cavanilles, Icon. 1: 57, t. 81. 1791, nom. rej. *Zinnia elegans* Jacquin var. *violacea* (Cavanilles) de Candolle, Prodr. 5: 536. 1836.

Erect annual herb, to 1(2) m; stem striate-angled, hirsute, strigose, or scabrous. Leaves with the blade ovate or oblong, 6–10 cm long, 2–6 cm wide, the apex acute, the base rounded or cordate, the margin entire, the upper and lower surfaces scabrellous or glabrate, gland-dotted, sessile. Flowers in a solitary capitulum, the peduncle to 8.5 cm; involucre hemispheric, to 2.5 cm wide;

phyllaries numerous in 3–4 series, obovate, 1–2 cm long, the apex rounded, erose or fimbriate; receptacle conic; paleae obovate or oblong, conduplicate, the apex rounded or obtuse, fimbriate, the outer surface scarious, sparsely strigose, or glabrous; ray flowers 8–numerous; corolla red, yellow, purple or white, the lamina 1–3.5 cm long; disk flowers numerous; corolla yellow, 7–9 mm long. Cypsela 6–10 mm long, 3-angled in the ray flowers, slightly compressed in the disk flowers, cartilaginous-winged, sometimes faintly ribbed, ciliolate; pappus absent.

Disturbed sites. Rare; central peninsula, Madison and Jackson Counties. Escaped from cultivation. Native to Mexico and South America. Summer–fall.

Zinnia peruviana (L.) L. [of Peru.] PERUVIAN ZINNIA.

Chrysogonum peruviana Linnaeus, Sp. Pl. 920. 1753. *Zinnia peruviana* (Linnaeus) Linnaeus, Syst. Nat., ed. 10. 1221. 1759. *Zinnia pauciflora* Linnaeus, Sp. Pl., ed. 2. 1269. 1763. *Lepia pauciflora* Hill, Hort. Kew. 18. 1768, nom. illegit. *Zinnia florida* Salisbury, Prodr. Stirp. Chap. Allerton 205. 1796, nom. illegit. *Crassina peruviana* (Linnaeus) Kuntze, Revis. Gen. Pl. 1: 331. 1891.

Zinnia multiflora Linnaeus, Sp. Pl. ed. 2. 1269. 1763. *Lepia multiflora* (Linnaeus) Hill, Hort. Kew. 18. 1768. *Crassina multiflora* (Linnaeus) Kuntze, Revis. Gen. Pl. 1: 331. 1891.

Zinnia floridana Rafinesque, New Fl. 4: 70. 1838 ("1836"). TYPE: FLORIDA/GEORGIA.

Erect annual herb, to 5(10) dm; stem striate-angled, strigose. Leaves with the blade ovate, elliptic, or broadly lanceolate, 2.5–7 cm long, 0.8–3.5 cm wide, the apex acute or obtuse, the base rounded, truncate, or slightly cordate, the margin entire, the upper and lower surfaces scabrellous, sessile. Flowers in a solitary capitulum, the peduncle 1–5(7) cm long; involucre campanulate, 1–2 cm wide; phyllaries numerous in 3–4 series, obovate or oblong, 1–2 cm long, the apex rounded, entire or erose, sometimes ciliate, the outer surface scarious or glabrous; receptacle conic; paleae obovate or oblong, conduplicate, the apex obtuse, erose or subentire; ray flowers 6–numerous; corolla red or maroon, sometimes yellow, the lamina 0.8–2.5 cm long; disk flowers numerous; corolla yellow, 5–6 mm long. Cypsela 7–10 mm long, 3-angled in the ray flowers, compressed in the disk flowers, ribbed, ciliate; pappus of 1 stout awn, 4–6 mm long.

Dry disturbed sites. Rare; Duval County. Escaped from cultivation. Not recently collected. North America, West Indies, Mexico, Central America, and South America; Africa, Asia, Australia, and Pacific Islands. Native to tropical America. Summer–fall.

EXCLUDED GENUS

Guizotia abyssinica (Linnaeus f.) Cassini—Reported for Florida by Strother (2006x), based on a specimen from Leon County (FSU), collected at a building loading dock in 1997 on the Florida State University campus, where apparently three plants were adventive from a birdseed mix. No other reports since. Unless further records are found, this collection is insufficient for the species to be considered part of the flora.

Literature Cited

Abbott, J. R., C. L. White, and S. B. Davis. 2008. *Praxelis clematidea* (Asteraceae), a genus and species new for the flora of North America. J. Bot. Res. Inst. Texas 2: 621–26.

Ackerfield, J., and J. Wen. 2002. A morphometric analysis of *Hedera* L. (the ivy genus, Araliaceae) and its taxonomic implications. Addansonia, sér. 3. 24: 197–212.

Affolter, J. M. 1985. A monograph of the genus *Lilaeopsis* (Umbelliferae). Syst. Bot. Monogr. 6: 1–140.

Alford, J. D., and L. C. Anderson. 2002. The taxonomy and morphology of *Macranthera flammea* (Orobanchaceae). Sida 20: 189–204.

Anderson, L. C. 1970. Studies on *Bigelowia* (Asteraceae, Compositae): Morphology and taxonomy. Sida 3: 451–65.

Anderson, L. C. 1994. A revision of *Hasteola* (Asteraceae) in the New World. Syst. Bot. 19: 211–19.

Anderson, L. C. 2006a. *Arnoglossum. In:* Flora of North America Editorial Committee. Flora of North America North of Mexico. 20: 622–25. New York/Oxford: Oxford University Press.

Anderson, L. C. 2006b. *Hasteola. In:* Flora of North America Editorial Committee. Flora of North America North of Mexico. 20: 610–11. New York/Oxford: Oxford University Press.

Anderson, L. C., E. L. Bridges, and S. L. Orzell. 1995. New data on distribution and morphology for the rare *Hasteola robertiorum* (Asteraceae). Phytologia 78: 246–48.

Antonelli, A. 2008. Higher level phylogeny and evolutionary trends in Campanulaceae subfam. Lobelioideae: Molecular signal overshadows morphology. Molec. Phylogen. Evol. 46: 1–18.

Applequist, W. L. 2015. A brief review of recent controversies in the taxonomy and nomenclature of *Sambucus nigra* sensu lato. Acta Hort. 1061: 25–33.

Arriagada, J. E. 1998. The genera of Inuleae (Compositae; Asteraceae) in the southeastern United States. Harvard Pap. Bot. 3: 1–48.

Arriagada, J. E., and N. G. Miller. 1997. The genera of Anthemideae (Compositae; Asteraceae) in the southeastern United States. Harvard Pap. Bot. 2: 1–46.

Babcock, E. B. 1947. The genus *Crepis.* Pt. I: The taxonomy, phylogeny, distribution, and evolution of *Crepis.* Pt. II: Systematic treatment. Univ. Calif. Publ. Bot. 21, 22.

Barkley, T. M. 1978. *Senecio. In:* Britton et al., eds. North American Flora. Ser. 2., part 10, 50–139. New York: New York Botanical Garden Press.

Barkley, T. M. 2006a. *Emilia. In:* Flora of North America Editorial Committee. Flora of North America North of Mexico. 20: 605–7. New York/Oxford: Oxford University Press.

Barkley, T. M. 2006b. *Erechtites. In:* Flora of North America Editorial Committee. Flora of North America North of Mexico. 20: 602–4. New York/Oxford: Oxford University Press.

Barkley, T. M. 2006c. *Gynura. In:* Flora of North America Editorial Committee. Flora of North America North of Mexico. 20: 610. New York/Oxford: Oxford University Press.

Barkley, T. M. 2006d. *Pseudogynoxys. In:* Flora of North America Editorial Committee. Flora of North America North of Mexico. 20: 608. New York/Oxford: Oxford University Press.

Barkley, T. M. 2006e. *Senecio. In:* Flora of North America Editorial Committee. Flora of North America North of Mexico. 20: 544–70. New York/Oxford: Oxford University Press.

Barkley, T. M., L. Brouillet, and J. L. Strother. 2006. Asteraceae. *In:* Flora of North America Editorial Committee. Flora of North America North of Mexico. 19: 13–69. New York/Oxford: Oxford University Press.

Bayer, R. J. 2006. *Antennaria. In:* Flora of North America Editorial Committee. Flora of North America North of Mexico. 19: 388–415. New York/Oxford: Oxford University Press.

Bayer, R. J., and G. L. Stebbins. 1993. A synopsis with keys for the genus *Antennaria* (Asteraceae: Inuleae: Gnaphaliinae) in North America. Canad. J. Bot. 71: 1589–1604.

Bell, C. R., and L. Constance. 1957. Chromosome numbers in Umbelliferae. Amer. J. Bot. 44: 565–72.

Bennett, J. R., and S. Mathews. 2006. Phylogeny of the parasitic plant family Orobanchaceae inferred from phytochrome A. Amer. J. Bot. 93: 1039–51.

Bierner, M. W. 2006. *Helenium. In:* Flora of North America Editorial Committee. Flora of North America North of Mexico. 21: 426–35. New York/Oxford: Oxford University Press.

Blake, S. F. 1916. Compositae new and transferred, chiefly Mexican. Proc. Amer. Acad. Arts 51: 515–26.

Blake, S. F. 1921. Revision of the genus *Acanthospermum.* Contr. U.S. Natl. Herb. 20: 383–92.

Bogler, D. J. 2006a. *Crepis. In:* Flora of North America Editorial Committee. Flora of North America North of Mexico. 19: 222–39. New York/Oxford: Oxford University Press.

Bogler, D. J. 2006b. *Hypochaeris. In:* Flora of North America Editorial Committee. Flora of North America North of Mexico. 19: 297–99. New York/Oxford: Oxford University Press.

Bogler, D. J. 2006c. *Lygodesmia. In:* Flora of North America Editorial Committee. Flora of North America North of Mexico. 19: 369–73. New York/Oxford: Oxford University Press.

Bogler, D. J. 2006d. *Nabalus. In:* Flora of North America Editorial Committee. Flora of North America North of Mexico. 19: 264–71. New York/Oxford: Oxford University Press.

Bolli, R. 1994. Revision of the genus *Sambucus.* Dissertationes Botanicae Vol. 223. Berlin, Stuttgart: Cramer.

Botanga, C. J., and M. P. Timko. 2005. Genetic structure and analysis of host and nonhost interactions of *Striga gesnerioides* (witchweed) from central Florida. Phytopathology 95: 1166–73.

Bowden, W. M. 1959. Cytotaxonomy of *Lobelia* L. section *Lobelia.* I. Three diverse species and seven small-flowered species. Canad. J. Genet. Cytol. 1: 49–64.

Bridges, E. L., and S. L. Orzell. 2018. Reassessment of *Pityopsis* section *Graminifoliae* (Small) Semple in peninsular Florida. 32, in Weakley et al. Combinations, rank changes, and nomenclatural comments in the vascular flora of the southeastern United States. III. J. Bot. Res. Inst. Texas 12: 27–67.

Brouillet, L. 2006a. *Eurybia. In:* Flora of North America Editorial Committee. Flora of North America North of Mexico. 20: 365–82. New York/Oxford: Oxford University Press.

Brouillet, L. 2006b. *Oclemena. In:* Flora of North America Editorial Committee. Flora of North America North of Mexico. 20: 78–81. New York/Oxford: Oxford University Press.

Brouillet, L. 2006c. *Taraxacum. In:* Flora of North America Editorial Committee. Flora of North America North of Mexico. 19: 239–52. New York/Oxford: Oxford University Press.

Brouillet, L., J. C. Semple, G. A. Allen, K. L. Chambers, and S. D. Sunberg. 2006. *Symphyotrichum. In:* Flora of North America Editorial Committee. Flora of North America North of Mexico. 20: 465–539. New York/Oxford: Oxford University Press.

Buswell, W. M. 1942. Goldenrods of South Florida. Bull. Univ. Miami 16: 1–14.

Canne-Hilliker, J. M. 2006. *Galinsoga. In:* Flora of North America Editorial Committee. Flora of North America North of Mexico. 21: 180–82. New York/Oxford: Oxford University Press.

Canne-Hilliker, J. M., and J. F. Hays. 2010. Typifications of names in *Agalinis, Gerardia,* and *Tomanthera* (Orobanchaceae). J. Bot. Res. Inst. Texas 4: 677–81.

Chambers, K. L., in B. S. Vuilleumier. 1973. The genera of Lactuceae (Compositae) in the southeastern United States. J. Arnold Arbor. 54: 42–93.

Chambers, K. L., and R. J. O'Kennon. 2006. *Krigia. In:* Flora of North America Editorial Committee. Flora of North America North of Mexico. 19: 362–67. New York/Oxford: Oxford University Press.

Channell, R. B. 1957. A revision of the genus *Marshallia* (Compositae). Contr. Gray Herb. 81: 41–132.

Chapman, A. W. 1860. Flora of the Southern United States. New York: Ivison, Phenney and Co.

Chapman, A. W. 1878. An enumeration of some plants—chiefly from the semi-tropical regions of Florida—which are either new, or which have not hitherto been recorded as belonging to the Flora of the Southern States (continued). Bot. Gaz. 3: 9–12.

Chapman, A. W. 1883. Flora of the Southern United States. ed. 2. New York: Ivison, Blakeman, Taylor and Co.

Chapman, A. W. 1897. Flora of the Southern United States. ed. 3. New York: American Book Co.

Chen, L.-Y., Q.-F. Wang, and S. S. Renner. 2016. East Asian Lobelioideae and ancient divergence of a giant rosette *Lobelia* in Himalayan Bhutan. Taxon 65: 293–304.

Clevinger, J. A. 2006. *Silphium. In:* Flora of North America Editorial Committee. Flora of North America North of Mexico. 21: 77–82. New York/Oxford: Oxford University Press.

Clewell, A. F. 1985. A Guide to the Vascular Plants of the Florida Panhandle. Tallahassee: University Presses of Florida/Florida State University Press.

Clewell, A. F., and J. W. Wooten. 1971. A revision of *Ageratina* (Compositae: Eupatorieae) from Eastern North America. Brittonia 23: 123–43.

Coleman, J. R. 1972. Nomenclatural clarification of two species of *Verbesina* (Compositae) endemic to Florida. Rhodora 97–101.

Correa, M. D., and R. L. Wilbur. 1969. A revision of *Carphephorus* (Compositae-Eupatorieae). J. Elisha Mitchell Sci. Soc. 85: 79–91.

Correll, D. S., and H. B. Correll. 1982. Flora of the Bahama Archipelago. Vaduz: J. Cramer.

Correll, D. S., and M. C. Johnston. 1970. Manual of the Vascular Plants of Texas. Renner: Texas Research Foundation.

Coulter, J. M., and J. N. Rose. 1900. Monograph of the North American Umbelliferae. Contr. U. S. Natl. Herb. 7(1): 9–256.

Cox, P. B., and L. E. Urbatsch. 1994. A taxonomic revision of *Rudbeckia* subg. *Macrocline* (Asteraceae: Heliantheae: Rudbeckiinae). Castanea 59: 300–318.

Cronquist, A. J. 1980. In Radford et al. (eds.) Vascular Flora of the Southeastern United States. Vol. 1: Asteraceae. Chapel Hill: University of North Carolina Press.

DeLaney, K. R., R. P. Wunderlin, and J. C. Semple. 2003. *Chrysopsis delaneyi* (Asteraceae), another new species from peninsular Florida. Bot. Explor. (Florida) 3: 1–37.

DeVore, M. L. 1991. The occurrence of *Acicarpha tribuloides* (Calyceraceae) in eastern North America. Rhodora 93: 26–35.

DeVore, M. L. 1994. Systematic studies of Calyceraceae. PhD dissertation, Ohio State University.

Downie, S. R., and K. E. Denford. 1988. Taxonomy of *Arnica* (Asteraceae) subgenus *Arnica*. Rhodora 90: 245–75.

Easterly, N. W. 1957. A morphological study of *Ptilimnium*. Brittonia 9: 136–45.

Eggers, D. M. 1969. A revision of *Valerianella* in North America. PhD dissertation, Vanderbilt University, Nashville.

Eriksson, T., and M. J. Donoghue. 1997. Phylogenetic relationships of *Sambucus* and *Adoxa* (Adoxoideae, Adoxaceae) based on nuclear ribosomal ITS sequences and preliminary morphological data. Syst. Bot. 22: 555–73.

Faust, Z. 1977. *Vernonia illinoensis* (Compositae): Species or hybrid? Castanea 42: 204–12.

Fehrer, J., B. Gemeinholzer, J. Chrtek, and S. Bröutigam. 2007. Incongruent plastid and nuclear DNA phylogenies reveal ancient intergeneric hybridization in *Pilosella* hawkweeds (*Hieracium*, Cichorieae, Asteraceae). Mol. Phylog. Evol. 42: 347–61.

Fehrer, J., K. Krak, and J. Chrtek. 2009. Intra-individual polymorphism in diploid and apomictic polyploid hawkweeds (*Hieracium*, Lactuceae, Asteraceae): Disentangling phylogenetic signal, reticulation, and noise. BMC Evol. Biol. 2009, 9: 239.

Feist, M.A.E. 2009. Clarifications concerning the nomenclature and taxonomy of *Oxypolis ternate* (Apiaceae). J. Bot. Res. Inst. Texas 3: 661–66.

Feist, M.A.E., S. R. Downie, A. R. Magee, and M. Liu. 2012. Revised generic delimitations for *Oxypolis* and *Ptilimnium* (Apiaceae) based on leaf morphology, comparative fruit anatomy, and phylogenetic analysis of nuclear rDNA ITS and cpDNA *trnQ-trnK* intergenic spacer sequence data. Taxon 61: 402–18.

Ferguson, I. K. 1965. The genera of the Valerianaceae and Dipsacaceae in the southeastern United States. J. Arnold Arbor. 46: 218–31.

Fernald, M. L. 1897. A systematic study of the United States and Mexican species of *Pectis*. Proc. Amer. Acad. Arts 33: 57–86.

Fernald, M. L. 1950. Gray's Manual of Botany. 8th ed. New York: American Book Co.

Florida Exotic Pest Plant Council (FLEPPC). 2017. Florida Exotic Pest Plant Council's 2017 List of Invasive Plant Species. http://www.fleppc.org/list.htm.

Flora of North America Editorial Committee. 2006. Asteraceae. *In:* Flora of North America North of Mexico. Volumes 19–21. New York/Oxford: Oxford University Press.

Gaiser, L. O. 1946. The genus *Liatris* (continued). Rhodora 48: 216–63.

Godfrey, R. K. 1952. *Pluchea*, section Stylimnus, in North America. J. Elisha Mitch. Sci. Soc. 68: 238–71.

Godfrey, R. K. 1988. Trees, Shrubs, and Woody Vines of Northern Florida and Adjacent Georgia and Alabama. Athens: University of Georgia Press.

Godfrey, R. K., and J. W. Wooten. 1981. Aquatic and Wetland Plants of Southeastern United States: Dicotyledons. Athens: University of Georgia Press.

González-Gutiérrez, P. A., and J. Sierra-Calzado. 2004. Aquifoliaceae. *In:* Greuter, W., and R. Rankin Rodríquez, eds. Flora de la Republica de Cuba, Plantas Vasculares, Serie A. 9(1). Ruggell, Liechtenstein: A. R. Gantner Verlag.

Graham, S. A. 1966. The genera of Araliaceae in the southeastern United States. Araliaceae. J. Arnold Arbor. 47: 126–36.

Gray, A. 1884. *Vernonia*. *In:* Synoptical Flora of North America. 1(2): 89–91. New York: Ivison, Blakeman, Taylor, and Company.

Haines, A. 2006. *Euthamia*. *In:* Flora of North America Editorial Committee. Flora of North America North of Mexico. 20: 97–100. New York/Oxford: Oxford University Press.

Hand, M. L., P. Vít, A. Krahulcová, S. D. Johnson, K. Oelkers, H. Siddons, J. Chrtek, J. Fehrer, A. M. Koltunow. 2015. Evolution of apomixis in *Pilosella* and *Hieracium* (Asteraceae) inferred from the conservation of apomixis-linked markers in natural and experimental populations. Heredity 114: 17–28.

Hansen, H. V. 1985. A taxonomic revision of the genus *Gerbera* (Compositae, Mutisieae) sections Gerbera, Parva, Piloselloides (in Africa) and Lasiopus. Opera Bot. 78: 1–36.

Hays, J. F. 2002. *Agalinis* (Scrophulariaceae) of the east Gulf Coastal Plain. MS thesis, University of Louisiana-Monroe.

Hays, J. F. 2010. *Agalinis flexicaulis* sp. nov. (Orobanchaceae: Lamiales), a new species from northeast Florida. J. Bot. Res. Inst. Texas 4: 1–6.

Heiser, C. B., D. M. Smith, S. Clevenger, and W. C. Martin. 1969. The North American sunflowers (*Helianthus*). Mem. Torrey Bot. Club 22(3): 1–218.

Hiroe, M. 1979. Umbelliferae of World. Tokyo: Ariake Book Co.

Holmes, W. C. 2006. *Mikania*. *In:* Flora of North America Editorial Committee. Flora of North America North of Mexico. 21: 545–47. New York/Oxford: Oxford University Press.

Howarth, D. G., M.H.G. Gustafsson, D. A. Baum, and T. J. Motley. 2003. Phylogenetics of the genus *Scaevola* (Goodeniaceae): Implications for dispersal patterns across the Pacific Basin and colonization of the Hawaiian Islands. Amer. J. Bot. 90: 916–23.

Hyatt, P. E. 2006. *Sonchus. In:* Flora of North America Editorial Committee. Flora of North America North of Mexico. 19: 273–76. New York/Oxford: Oxford University Press.

Jackson, R. C. 1960. A revision of the genus *Iva*. Univ. Kansas Sci. Bull. 41: 793–876.

Jansen, R. K. 1985. The systematic of *Acmella* (Asteraceae-Heliantheae). Syst. Bot. Monogr. 8: 1–115.

Johnson, M. F. 1971. A monograph of the genus *Ageratum* L. (Compositae-Eupatorieae). Ann. Missouri Bot. Gard. 58: 6–88.

Jones, A. G., and P. P. Lowry. 1986. Types and selected historic specimens of *Aster* s.l. (Asteraceae) in the Herbarium, Laboratoire de Phanérogamie, Musée National d'Histoire Naturelle, Paris (P). Bull. Mus. Natl. Hist. Nat., B, Adansonia 8: 395.

Jones, G. N. 1940. A monograph of the genus *Symphiocarpos*. J. Arnold Arbor. 46: 201–52.

Jones, S. B. 1964. Taxonomy of the narrow-leaved *Vernonia* of the southeastern United States. Rhodora 66: 382–401.

Karaman-Castro, V., and L. E. Urbatsch. 2006. *Boltonia. In:* Flora of North America Editorial Committee. Flora of North America North of Mexico. 20: 353–57. New York/Oxford: Oxford University Press.

Keener, B. R. 2006. *Balduina. In:* Flora of North America Editorial Committee. Flora of North America North of Mexico. 21: 419–20. New York/Oxford: Oxford University Press.

Keil, D. J. 1986. Synopsis of the Florida species of *Pectis* (Asteraceae). Sida 11: 385–95.

Keil, D. J. 2006a. *Cirsium. In:* Flora of North America Editorial Committee. Flora of North America North of Mexico. 19: 95–164. New York/Oxford: Oxford University Press.

Keil, D. J. 2006b. *Onopordum. In:* Flora of North America Editorial Committee. Flora of North America North of Mexico. 19: 87–88. New York/Oxford: Oxford University Press.

Keil, D. J. 2006c. *Pectis. In:* Flora of North America Editorial Committee. Flora of North America North of Mexico. 21: 222–30. New York/Oxford: Oxford University Press.

Keil, D. J., and J. Ochsmann. 2006. *Centaurea. In:* Flora of North America Editorial Committee. Flora of North America North of Mexico. 19: 181–94. New York/Oxford: Oxford University Press.

Kiger, R. W. 2006. *Cosmos. In:* Flora of North America Editorial Committee. Flora of North America North of Mexico. 20: 353–57. New York/Oxford: Oxford University Press.

Kilian, N. 1997. Revision of *Launaea* Cass. (Compositae, Lactuceae, Sonchinae). Englera 17: 1–478.

Kim, S. C., D. J. Crawford, and R. K. Jansen. 1996. Phylogenetic relationships among the genera of the subtribe Sonchinae (Asteraceae): Evidence from ITS sequences. Syst. Bot. 21: 417–32.

King, R. M., and H. Robinson. 1987. The genera of the Eupatorieae (Asteraceae). Monogr. Syst. Bot. Missouri Bot. Gard. 22: 1–128.

Kirkman, L. K. 1981. Taxonomic revision of *Centratherum* and *Phyllocephalum* (Compositae: Vernoneae). Rhodora 83: 1–24.

Krak, K., P. Caklová, J. Chrtek, and J. Fehrer. 2013. Reconstruction of phylogenetic relationships in a highly reticulate group with deep coalescence and recent speciation (*Hieracium*, Asteraceae). Heredity 110: 138–51.

Kral, R. 1966. Observations on the flora of the southeastern United States with special reference to northern Louisiana. Sida 2: 395–408.

Kral, R., and R. K. Godfrey. 1958. Synopsis of the Florida species of *Cacalia*. Quart. J. Florida Acad. Sci. 21: 193–206.

Kurz, H., and R. K. Godfrey. 1962. Trees of North Florida. Gainesville: University of Florida Press.

La Duke, J. C. 2006. *Tithonia. In:* Flora of North America Editorial Committee. Flora of North America North of Mexico. 21: 138–39. New York/Oxford: Oxford University Press.

Lagomarsino, L. F., A. Antonelli, N. Muchhala, A. Timmermann, S. Mathews, and C. C. Davis. 2014. Phylogeny, classification, and fruit evolution of the species-rich Neotropical bellflowers (Campanulaceae: Lobelioideae). Amer. J. Bot. 101: 2097–112.

Lamont, E. E. 2006a. *Eutrochium. In:* Flora of North America Editorial Committee. Flora of North America North of Mexico. 21: 474–78. New York/Oxford: Oxford University Press.

Lamont, E. E. 2006b. *Garberia. In:* Flora of North America Editorial Committee. Flora of North America North of Mexico. 21: 538–39. New York/Oxford: Oxford University Press.

Lamont, E. E. 2006c. *Sclerolepis. In:* Flora of North America Editorial Committee. Flora of North America North of Mexico. 21: 488–90. New York/Oxford: Oxford University Press.

León de la Luz, J. L., and A. Medel Narváez. 2013. A new species of *Bidens* (Asteraceae: Coreopsidae). Acta Bot. Mex. 103: 119–26.

Liogier, H. A. 1995. Descriptive Flora of Puerto Rico and Adjacent Islands. Spermatophyta-Dicotyledoneae, Volume IV, Melastomataceae to Lentibulariaceae. Río Piedras: Estación de la Universidad de Puerto Rico.

Liogier, H. A., and L. F. 1982. Descriptive Flora of Puerto Rico and Adjacent Islands: A Systematic Synopsis. Río Piedras: Editorial de la Universidad de Puerto Rico.

Long, R. W., and O. Lakela. 1971. A Flora of Tropical Florida. Miami: University of Miami Press.

Löve, D., and P. Dansereau. 1959. Biosystematic study of *Xanthium:* Taxonomic appraisal and ecological studies. Canad. J. Bot. 37: 173–208.

Lowry, P. P., G. M. Plukett, and J. Wen. 2004. Generic relationships in Araliaceae: Looking into the crystal ball. S. African J. Bot. 7: 382–92.

Majeský, L., F. Krahulec, and R. J. Vašut. 2017. How apomictic taxa are treated in current taxonomy: A review. Taxon 66: 1017–40.

Marshall, J. B. 1974. A note on *Conyza sumatrensis* (Retz.) E. Walker. Watsonia 10: 166–67.

Mathias, M. E., and L. Constance. 1941. A synopsis of the North American species of *Eryngium.* Amer. Midl. Naturalist 25: 361–87.

Mathias, M. E., and L. Constance. 1944a. Araliaceae. N. Amer. Fl. 28B: 3–42.

Mathias, M. E., and L. Constance. 1944b. Umbelliferae. N. Amer. Fl. 28B: 43–160.

Mathias, M. E., and L. Constance. 1945. Umbelliferae. N. Amer. Fl. 28B: 161–295.

Mathias, M. E., and L. Constance. 1965. A revision of the genus *Bowlesia* Ruiz & Pav. (Umbelliferae-Hydrocotyloideae) and its relatives. Univ. Calif. Publ. Bot. 38: 1–73.

McAtee, 1956. A Review of the Nearctic *Viburnum.* Chapel Hill, NC: Published by the author.

McVaugh, R. 1943. Campanulales, Campanulaceae, Lobelioideae. N. Amer. Fl. 32A, part 1: 1–134.

McVaugh, R. 1945. The genus *Triodanis* Rafinesque, and its relationships to *Specularia* and *Campanula.* Wrightia 1: 13–52.

Melchert, T. E. 2010. *Bidens.* In: Turner, B. L. (ed.). The comps of Mexico: A systematic account of the family Asteraceae. Chapter 10. Subfamily Coreopsideae. Phytologia Mem. 15: 3–56.

Meyer, F. G. 1951. *Valeriana* in North America and the West Indies. Ann. Missouri Bot. Gard. 38: 377–503.

Middleton, B. A., E. Anemaet, T. E. Quirk, and N. P. Tippery. 2018. *Nymphoides humboldtiana* (Menyanthaceae) in Florida (U.S.A.) verified by DNA data. J. Bot. Res. Inst. Texas 12: 257–63.

Morefield, J. D. 2006. *Filago. In:* Flora of North America Editorial Committee. Flora of North America North of Mexico. 19: 447–49. New York/Oxford: Oxford University Press.

Morgan, D. R., and B. Holland. 2012. Systematics of Symphyotrichinae (Asteraceae: Astereae): Disagreements between two nuclear regions suggest a complex evolutionary history. Syst. Bot. 37: 818–32.

Mulligan, G. A. 1980. The genus *Cicuta* in North America. Canad. J. Bot. 58: 1755–67.

Nesom, G. L. 2001. Taxonomic review of *Chrysogonum* (Asteraceae: Heliantheae). Sida 19: 811–20.

Nesom, G. L. 2006a. *Ageratina. In:* Flora of North America Editorial Committee. Flora of North America North of Mexico. 21: 547–53. New York/Oxford: Oxford University Press.

Nesom, G. L. 2006b. *Ageratum. In:* Flora of North America Editorial Committee. Flora of North America North of Mexico. 21: 481–83. New York/Oxford: Oxford University Press.

Nesom, G. L. 2006c. *Aphanostephus. In:* Flora of North America Editorial Committee. Flora of North America North of Mexico. 20: 351–53. New York/Oxford: Oxford University Press.

Nesom, G. L. 2006d. *Carphephorus. In:* Flora of North America Editorial Committee. Flora of North America North of Mexico. 21: 535–38. New York/Oxford: Oxford University Press.

Nesom, G. L. 2006e. *Chaptalia*. *In:* Flora of North America Editorial Committee. Flora of North America North of Mexico. 19: 78–80. New York/Oxford: Oxford University Press.

Nesom, G. L. 2006f. *Chromolaena*. *In:* Flora of North America Editorial Committee. Flora of North America North of Mexico. 21: 544–45. New York/Oxford: Oxford University Press.

Nesom, G. L. 2006g. *Chrysogonum*. *In:* Flora of North America Editorial Committee. Flora of North America North of Mexico. 21: 74–75. New York/Oxford: Oxford University Press.

Nesom, G. L. 2006h. *Chrysoma*. *In:* Flora of North America Editorial Committee. Flora of North America North of Mexico. 20: 105. New York/Oxford: Oxford University Press.

Nesom, G. L. 2006i. *Croptilon*. *In:* Flora of North America Editorial Committee. Flora of North America North of Mexico. 20: 228–30. New York/Oxford: Oxford University Press.

Nesom, G. L. 2006j. *Erigeron*. *In:* Flora of North America Editorial Committee. Flora of North America North of Mexico. 20: 256–348. New York/Oxford: Oxford University Press.

Nesom, G. L. 2006k. *Facelis*. *In:* Flora of North America Editorial Committee. Flora of North America North of Mexico. 19: 442–43. New York/Oxford: Oxford University Press.

Nesom, G. L. 2006l. *Fleischmannia*. *In:* Flora of North America Editorial Committee. Flora of North America North of Mexico. 21: 540–41. New York/Oxford: Oxford University Press.

Nesom, G. L. 2006m. *Gamochaeta*. *In:* Flora of North America Editorial Committee. Flora of North America North of Mexico. 19: 431–38. New York/Oxford: Oxford University Press.

Nesom, G. L. 2006n. *Hartwrightia*. *In:* Flora of North America Editorial Committee. Flora of North America North of Mexico. 21: 540. New York/Oxford: Oxford University Press.

Nesom, G. L. 2006o. *Ionactis*. *In:* Flora of North America Editorial Committee. Flora of North America North of Mexico. 20: 82–84. New York/Oxford: Oxford University Press.

Nesom, G. L. 2006p. *Koanophyllon*. *In:* Flora of North America Editorial Committee. Flora of North America North of Mexico. 21: 542–43. New York/Oxford: Oxford University Press.

Nesom, G. L. 2006q. *Liatris*. *In:* Flora of North America Editorial Committee. Flora of North America North of Mexico. 21: 512–35. New York/Oxford: Oxford University Press.

Nesom, G. L. 2006r. *Pluchea*. *In:* Flora of North America Editorial Committee. Flora of North America North of Mexico. 19: 478–84. New York/Oxford: Oxford University Press.

Neson, G. L. 2006s. *Pseudognaphalium*. *In:* Flora of North America Editorial Committee. Flora of North America North of Mexico. 19: 415–25. New York/Oxford: Oxford University Press.

Nesom, G. L. 2006t. *Pterocaulon*. *In:* Flora of North America Editorial Committee. Flora of North America North of Mexico. 19: 476–77. New York/Oxford: Oxford University Press.

Nesom, G. L. 2006u. *Rayjacksonia*. *In:* Flora of North America Editorial Committee. Flora of North America North of Mexico. 20: 437–39. New York/Oxford: Oxford University Press.

Nesom, G. L. 2006v. *Sachsia*. *In:* Flora of North America Editorial Committee. Flora of North America North of Mexico. 19: 477–78. New York/Oxford: Oxford University Press.

Nesom, G. L. 2012. Taxonomy of *Aplastrum, Ammoselinum*, and *Spermolepis* (Apiaceae). Phytoneuron 2012-87: 1–49.

Nesom, G. L. 2019. Taxonomic synopsis of *Pityopsis* (Asteraceae). Phytoneuron 2019-1: 1–31.

Nesom G. L., and J. M. Egger. 2014. *Castilleja coccinea* and *C. indivisa* (Orobanchaceae). Phytoneuron 2014-14: 1–7.

Nesom, G. L., and J. M. Stucky. 2004. Taxonomy of the *Liatris pilosa* (*graminifolia*) group (Asteraceae: Eupatorieae). Sida 21: 815–26.

Neves, S. S., and M. F. Watson. 2004. Phylogenetic relationships in *Bupleurum* (Apiaceae) based on nuclear ribosomal DNA ITS sequence data. Ann. Bot. 93: 379–98.

Norton, R. A. 1991. *Bidens alba* (smooth beggar-tick) and *Bidens pilosa* (hairy beggar-tick). *In:* Y.P.S. Bajaj (ed.). Biotechnology in Agriculture and Forestry. Vol. 15. Medicinal and aromatic plants III. Berlin: Springer-Verlag.

Noyes, R. D. 2000. Biogeography and evolution insights on *Erigeron* and allies (Asteraceae) from ITS sequence data. Pl. Syst. Evol. 220: 93–114.

Orzell, S. L., and E. L. Bridges. 2002. Notes on *Carphephorus odoratissimus* (Asteraceae) in peninsular Florida. Sida 20: 559–69.

Panero, J. L., and V. A. Funk. 2002. Toward a phylogenetic subfamilial classification for the Compositae (Asteraceae). Proc. Biol. Soc. Wash. 115: 909–22.

Parker, E. S., and S. B. Jones. 1975. A systematic study of the genus *Balduina* (Compositae, Heliantheae). Brittonia 27: 355–61.

Parks, J. C. 1973. A revision of North American and Caribbean *Melanthera* (Compositae). Rhodora 75: 169–210.

Parks, J. C. 2006. *Melanthera. In:* Flora of North America Editorial Committee. Flora of North America North of Mexico. 21: 123–25. New York/Oxford: Oxford University Press.

Patterson, T. F., and G. L. Nesom. 2006. *Conoclinium. In:* Flora of North America Editorial Committee. Flora of North America North of Mexico. 21: 478–80. New York/Oxford: Oxford University Press.

Pennell, F. W. 1925. The genus *Afzelia:* A taxonomic study in evolution. Proceeding Acad. Nat. Sci. Philadelphia 77: 335–73.

Pennell, F. W. 1928. *Agalinis* and allies in North America. Proc. Acad. Nat. Sci. Philadelphia 80: 339–449.

Pennell, F. W. 1929. *Agalinis* and Allies in North America: II. Proc. Acad. Nat. Sci. Philadelphia 81: 111–249.

Pennell, F. W. 1935. The Scrophulariaceae of eastern temperate North America. Acad. Nat. Sci. Philadelphia Monogr. 1: 1–650.

Philcox, D. 1965. Contributions to the Flora of tropical America: LXXIV. Revision of the New World species of *Buchnera* L. (Scrophulariaceae). Kew Bull. 18: 275–315.

Pinkava, D. J. 1967. Biosystematic study of *Berlandiera* (Compositae). Brittonia 19: 285–98.

Pinkava, D. J. 2006. *Berlandiera. In:* Flora of North America Editorial Committee. Flora of North America North of Mexico. 21: 83–87. New York/Oxford: Oxford University Press.

Plunkett, G. M., P. P. Lowry, D. G. Frodin, and J. Wen. 2005. Phylogeny and geography of *Schefflera:* Pervasive polyphyly in the largest genus of Araliaceae. Ann. Missouri Bot. Gard. 92: 202–24.

Powell, A. M. 1978. Systematics of *Flaveria* (Flaveriinae-Asteraceae). Ann. Missouri Bot. Gard. 65: 590–639.

Pruski, J. F., and G. Sancho. 2006. *Conyza sumatrensis* var. *leiothera* (Compositae: Astereae), a new combination for a common Neotropical weed. Novon 16: 96–101.

Pryer, K. M., and L. R. Phillippe. 1989. A synopsis of the genus *Sanicula* (Apiaceae) in eastern Canada. Canad. J. Bot. 67: 694–707.

Radford, A. E., H. E. Ahles, and C. R. Bell. 1964. Guide to the Vascular Flora of the Carolinas. Chapel Hill: University of North Carolina Press.

Radford, A. E., H. E. Ahles, and C. R. Bell. 1968. Manual of the Vascular Flora of the Carolinas. Chapel Hill: University of North Carolina Press.

Robart, B. W., C. Gladys, T. Frank, and S. Kilpatrick. 2015. Phylogeny and biogeography of North American and Asian *Pedicularis* (Orobanchaceae). Syst. Bot. 40: 229–58.

Rock, H.F.L. 1957. A revision of the vernal species of *Helenium* (Compositae) (continued). Rhodora 59: 168–78.

Rodrigues, A., S. Shaya, T. A. Dickinson, and S. Stefanović. 2013. Morphometric analysis and taxonomic revision of the North American holoparasitic genus *Conopholis* (Orobanchaceae). Syst. Bot. 38: 795–804.

Roquet, C., J.J.A. Sáez, S. Alfonso, M. L. Alarcón, and N. Garcia-Jacas. 2008. Natural delineation, molecular phylogeny and floral evolution in *Campanula*. Syst. Bot. 33: 203–17.

Schilling, E. E. 2006a. *Helianthus. In:* Flora of North America Editorial Committee. Flora of North America North of Mexico. 21: 141–69. New York/Oxford: Oxford University Press.

Schilling, E. E. 2006b. *Phoebanthus. In:* Flora of North America Editorial Committee. Flora of North America North of Mexico. 21: 113–14. New York/Oxford: Oxford University Press.

Schilling, E. E. 2011. Systematics of the *Eupatorium album* complex (Asteraceae) from eastern North America. Syst. Bot. 36: 1088–100.

Schilling, E. E., and K. C. Grubbs. 2016. Systematics of the *Eupatorium mohrii* complex (Asteraceae). Syst. Bot. 41: 787–95.

Schneider, A. C. 2016. Resurrection of the genus *Aphyllon* for New World broomrapes (*Orobanche* s.l., Orobanchaceae). Phytokeys 75: 107–18.

Schubert, M.T.R. 2014. A revision of *Centella* (Apiaceae). PhD dissertation. Rand Afrikaans University, Johannesburg.

Scott, R. W. 2006. *Brickellia. In:* Flora of North America Editorial Committee. Flora of North America North of Mexico. 21: 491–507. New York/Oxford: Oxford University Press.

Semple, J. C. 1978. A revision of the genus *Borrichia* Adans. (Compositae). Annals Missouri Bot. Gard. 65: 681–93.

Semple, J. C. 1981. A revision of the goldenaster genus *Chrysopsis* (Nutt.) Ell. nom. cons. (Compositae-Astereae). Rhodora 83: 323–83.

Semple, J. C. 2003. New names and combinations in goldenrods, *Solidago* (Asteraceae: Astereae). Sida 20: 1605–16.

Semple, J. C. 2006a. *Borrichia. In:* Flora of North America Editorial Committee. Flora of North America North of Mexico. 21: 129–31. New York/Oxford: Oxford University Press.

Semple, J. C. 2006b. *Bradburia. In:* Flora of North America Editorial Committee. Flora of North America North of Mexico. 21: 211–12. New York/Oxford: Oxford University Press.

Semple, J. C. 2006c. *Brintonia. In:* Flora of North America Editorial Committee. Flora of North America North of Mexico. 21: 106. New York/Oxford: Oxford University Press.

Semple, J. C. 2006d. *Chrysopsis. In:* Flora of North America Editorial Committee. Flora of North America North of Mexico. 20: 212–21. New York/Oxford: Oxford University Press.

Semple, J. C. 2006e. *Heterotheca. In:* Flora of North America Editorial Committee. Flora of North America North of Mexico. 20: 230–56. New York/Oxford: Oxford University Press.

Semple, J. C. 2006f. *Pityopsis. In:* Flora of North America Editorial Committee. Flora of North America North of Mexico. 20: 222–28. New York/Oxford: Oxford University Press.

Semple, J. C. 2006g. *Ampelaster. In:* Flora of North America Editorial Committee. Flora of North America North of Mexico. 20: 460. New York/Oxford: Oxford University Press.

Semple, J. C., and F. D. Bowers. 1985. A revision of the goldenaster genus *Pityopsis* Nuttall (Compositae: Astereae). Univ. Waterloo Biol. Ser. 29: 1–34.

Semple. J. C., and J. G. Chmielewski. 2006. *Doellingeria. In:* Flora of North America Editorial Committee. Flora of North America North of Mexico. 20: 43–46. New York/Oxford: Oxford University Press.

Semple, J. C., J. G. Chmielewski, and C. Leeder. 1991. A multivariate morphometric study and revision of *Aster* subg. Doellingeria sect. Triplopappus (Compositae: Astereae): The *Aster umbellatus* complex. Canad. J. Bot. 69: 256–76.

Semple, J. C., and R. E. Cook. 2006. *Solidago. In:* Flora of North America Editorial Committee. Flora of North America North of Mexico. 20: 107–66. New York/Oxford: Oxford University Press.

Semple, J. C., S. B. Heard, and L. Brouillet. 2002. Cultivated and native asters of Ontario (Compositae: Astereae): *Aster* L. (including *Asteromoea* Blume, *Diplactis* Raf. and *Kalimeris* (Cass.) Cass.), *Callistephus* Cass., *Galatella* Cass., *Doellingeria* Nees, Oclemena E. L. Greene, *Eurybia* (Cass.) S. F. Gray, *Canadanthus* Nesom, and *Symphyotrichum* Nees (including *Virgulus* Raf.). Univ. Waterloo Biol. Ser. 41: 1–135.

Semple, J. C., and M. R. Leonard. 2006. *Sericocarpus. In:* Flora of North America Editorial Committee. Flora of North America North of Mexico. 20: 101–5. New York/Oxford: Oxford University Press.

Semple, J., and K. S. Semple. 1977. *Borrichia* ×*cubana* (*B. frutescens* × *arborescens*): Interspecific hybridization in the Florida keys. Syst. Bot. 2: 292–301.

Shan, R. H., and L. Constance. 1951. The genus *Sanicula* in the Old World and the New. Univ. Calif. Publ. Bot. 25: 1–78.

Sherff, E. E. 1936. Revision of the genus *Coreopsis*. Field Mus. Nat. Hist., Bot. Ser. 11: 279–475.

Shinners, L. H. 1946. Revision of the genus *Aphanostephus* DC. Wrightia 1: 95–121.

Shinners, L. H. 1947. Revision of the genus *Krigia* Schreber. Wrightia 1: 187–206.

Shultz, L. M. 2006. *Artemisia*. *In:* Flora of North America Editorial Committee. Flora of North America North of Mexico. 19: 503–34. New York/Oxford: Oxford University Press.

Sieren, D. J. 1981. The taxonomy of the genus *Euthamia*. Rhodora 83: 551–79.

Siripun, K. C., and E. E. Schilling. 2006. *Eupatorium*. *In:* Flora of North America Editorial Committee. Flora of North America North of Mexico. 21: 462–74. New York/Oxford: Oxford University Press.

Small, J. K. 1903. Flora of the Southeastern United States. New York: Published by the author.

Small, J. K. 1913a. Flora of the Southeastern United States. 2nd ed. New York: Published by the author.

Small, J. K. 1913b. Flora of Miami. New York: Published by the author.

Small, J. K. 1913c. Flora of the Florida Keys. New York: Published by the author.

Small, J.K. 1913d. Shrubs of Florida. New York: Published by the author.

Small, J. K. 1933. Manual of the Southeastern Flora. New York: Published by the author.

Smith, A. R. 2006a. *Heliopsis*. *In:* Flora of North America Editorial Committee. Flora of North America North of Mexico. 21: 67–70. New York/Oxford: Oxford University Press.

Smith, A. R. 2006b. *Zinnia*. *In:* Flora of North America Editorial Committee. Flora of North America North of Mexico. 21: 71–74. New York/Oxford: Oxford University Press.

Smith, E. B. 1976. A biosystematic survey of *Coreopsis* in eastern United States and Canada. Sida 6: 123–215.

Spaulding, D. D., and T. W. Barger. 2016. Keys, distribution, and taxonomic notes for the lobelias (*Lobelia,* Campanulaceae) of Alabama and adjacent states. Phytoneuron 2016-76: 1–60.

Spurr, P. L. 2006. *Youngia*. *In:* Flora of North America Editorial Committee. Flora of North America North of Mexico. 19: 255–56. New York/Oxford: Oxford University Press.

Steussy, T. F. 1977. Revision of *Chrysogonum* (Compositae, Heliantheae). Rhodora 79: 190–202.

Steussy, T. F. 1978. A revision of *Lagascea* (Compositae, Heliantheae). Fieldiana, Bot. 38: 75–133.

Strother, J. L. 2006a. *Acanthospermum*. *In:* Flora of North America Editorial Committee. Flora of North America North of Mexico. 21: 36–37. New York/Oxford: Oxford University Press.

Strother, J. L. 2006b. *Acmella*. *In:* Flora of North America Editorial Committee. Flora of North America North of Mexico. 21: 132–33. New York/Oxford: Oxford University Press.

Strother, J. L. 2006c. *Ambrosia*. *In:* Flora of North America Editorial Committee. Flora of North America North of Mexico. 21: 10–18. New York/Oxford: Oxford University Press.

Strother, J. L. 2006d. *Calyptocarpus*. *In:* Flora of North America Editorial Committee. Flora of North America North of Mexico. 21: 133. New York/Oxford: Oxford University Press.

Strother, J. L. 2006e. *Centratherum*. *In:* Flora of North America Editorial Committee. Flora of North America North of Mexico. 19: 206. New York/Oxford: Oxford University Press.

Strother, J. L. 2006f. *Cichorum*. *In:* Flora of North America Editorial Committee. Flora of North America North of Mexico. 19: 221–22. New York/Oxford: Oxford University Press.

Strother, J. L. 2006g. *Conyza*. *In:* Flora of North America Editorial Committee. Flora of North America North of Mexico. 20: 348–50. New York/Oxford: Oxford University Press.

Strother, J. L. 2006h. *Coreopsis*. *In:* Flora of North America Editorial Committee. Flora of North America North of Mexico. 21: 185–98. New York/Oxford: Oxford University Press.

Strother, J. L. *Cyanthillium*. 2006i. *In:* Flora of North America Editorial Committee. Flora of North America North of Mexico. 19: 204–5. New York/Oxford: Oxford University Press.

Strother, J. L. 2006j. *Eclipta*. *In:* Flora of North America Editorial Committee. Flora of North America North of Mexico. 21: 128–29. New York/Oxford: Oxford University Press.

Strother, J. L. 2006k. *Elephantopus*. *In:* Flora of North America Editorial Committee. Flora of North America North of Mexico. 19: 202–3. New York/Oxford: Oxford University Press.

Strother, J. L. 2006l. *Gaillardia*. *In:* Flora of North America Editorial Committee. Flora of North America North of Mexico. 21: 421–26. New York/Oxford: Oxford University Press.

Strother, J. L. 2006m. *Glebionis. In:* Flora of North America Editorial Committee. Flora of North America North of Mexico. 19: 554–55. New York/Oxford: Oxford University Press.

Strother, J. L. 2006n. *Hieracium. In:* Flora of North America Editorial Committee. Flora of North America North of Mexico. 19: 278–94. New York/Oxford: Oxford University Press.

Strother, J. L. 2006o. *Hymenopappus. In:* Flora of North America Editorial Committee. Flora of North America North of Mexico. 21: 309–16. New York/Oxford: Oxford University Press.

Strother, J. L. 2006p. *Iva. In:* Flora of North America Editorial Committee. Flora of North America North of Mexico. 21: 25–28. New York/Oxford: Oxford University Press.

Strother, J. L. 2006q. *Lactuca. In:* Flora of North America Editorial Committee. Flora of North America North of Mexico. 19: 259–63. New York/Oxford: Oxford University Press.

Strother, J. L. 2006r. *Leucanthemum. In:* Flora of North America Editorial Committee. Flora of North America North of Mexico. 19: 557–59. New York/Oxford: Oxford University Press.

Strother, J. L. 2006s. *Melampodium. In:* Flora of North America Editorial Committee. Flora of North America North of Mexico. 21: 34–36. New York/Oxford: Oxford University Press.

Strother, J. L. 2006t. *Palafoxia. In:* Flora of North America Editorial Committee. Flora of North America North of Mexico. 21: 388–91. New York/Oxford: Oxford University Press.

Strother, J. L. 2006u. *Parthenium. In:* Flora of North America Editorial Committee. Flora of North America North of Mexico. 21: 20–22. New York/Oxford: Oxford University Press.

Strother, J. L. 2006v. *Pascalia. In:* Flora of North America Editorial Committee. Flora of North America North of Mexico. 21: 131–32. New York/Oxford: Oxford University Press.

Strother, J. L. 2006w. *Polymnia. In:* Flora of North America Editorial Committee. Flora of North America North of Mexico. 21: 39–40. New York/Oxford: Oxford University Press.

Strother, J. L. 2006x. *Guizotia. In:* Flora of North America Editorial Committee. Flora of North America North of Mexico. 21: 40–41. New York/Oxford: Oxford University Press.

Strother, J. L. 2006y. *Pseudoelephantopus. In:* Flora of North America Editorial Committee. Flora of North America North of Mexico. 19: 204. New York/Oxford: Oxford University Press.

Strother, J. L. 2006z. *Pyrrhopappus. In:* Flora of North America Editorial Committee. Flora of North America North of Mexico. 19: 376–78. New York/Oxford: Oxford University Press.

Strother, J. L. 2006aa. *Smallanthus. In:* Flora of North America Editorial Committee. Flora of North America North of Mexico. 21: 33–34. New York/Oxford: Oxford University Press.

Strother, J. L. 2006bb. *Sphagneticola. In:* Flora of North America Editorial Committee. Flora of North America North of Mexico. 21: 126. New York/Oxford: Oxford University Press.

Strother, J. L. 2006cc. *Stokesia. In:* Flora of North America Editorial Committee. Flora of North America North of Mexico. 19: 201. New York/Oxford: Oxford University Press.

Strother, J. L. 2006dd. *Syndrella. In:* Flora of North America Editorial Committee. Flora of North America North of Mexico. 21: 127. New York/Oxford: Oxford University Press.

Strother, J. L. 2006ee. *Tagetes. In:* Flora of North America Editorial Committee. Flora of North America North of Mexico. 21: 235–36. New York/Oxford: Oxford University Press.

Strother, J. L. 2006ff. *Tetragonotheca. In:* Flora of North America Editorial Committee. Flora of North America North of Mexico. 21: 178–79. New York/Oxford: Oxford University Press.

Strother, J. L. 2006gg. *Thymophylla. In:* Flora of North America Editorial Committee. Flora of North America North of Mexico. 21: 239–45. New York/Oxford: Oxford University Press.

Strother, J. L. 2006hh. *Tridax. In:* Flora of North America Editorial Committee. Flora of North America North of Mexico. 21: 179–80. New York/Oxford: Oxford University Press.

Strother, J. L. 2006ii. *Verbesina. In:* Flora of North America Editorial Committee. Flora of North America North of Mexico. 21: 106–11. New York/Oxford: Oxford University Press.

Strother, J. L. 2006jj. *Vernonia. In:* Flora of North America Editorial Committee. Flora of North America North of Mexico. 19: 206–13. New York/Oxford: Oxford University Press.

Strother, J. L. 2006kk. *Xanthium. In:* Flora of North America Editorial Committee. Flora of North America North of Mexico. 21: 19–20. New York/Oxford: Oxford University Press.

Strother, J. L., and R. R. Weedon. 2006. *Bidens. In:* Flora of North America Editorial Committee. Flora of North America North of Mexico. 21: 205–18. New York/Oxford: Oxford University Press.

Sundberg, S. D., and D. J. Bogler. 2006. *Baccharis. In:* Flora of North America Editorial Committee. Flora of North America North of Mexico. 20: 23–34. New York/Oxford: Oxford University Press.

Thieret, J. W. 1969. Notes on *Epifagus*. Castanea 34: 397–402.

Tippery, N. P., and D. H. Les. 2011. Phylogenetic relationships and morphological evolution in *Nymphoides* (Menyanthaceae). Syst. Bot. 36: 1101–113.

Tippery, N. P., D. H. Les, and E. L. Peredo. 2015. *Nymphoides grayana* (Menyanthaceae) in Florida verified by DNA and morphological data. J. Torrey Bot. Soc. 142: 325–30.

Tomb, A. S. 1980. Taxonomy of *Lygodesmia* (Asteraceae). Syst. Bot. Monogr. 1: 1–51.

Trock, D. K. 2006a. *Achillea. In:* Flora of North America Editorial Committee. Flora of North America North of Mexico. 19: 492–94. New York/Oxford: Oxford University Press.

Trock, D. K. 2006b. *Packera. In:* Flora of North America Editorial Committee. Flora of North America North of Mexico. 20: 570–602. New York/Oxford: Oxford University Press.

Turner, B. L., and M. I. Morris. 1976. Systematics of *Palafoxia* (Asteraceae: Helenieae). Rhodora 78: 567–628.

Urbatsch, L. E. 1972. Systematic studies of the Altissimae and Giganteae groups of the genus *Vernonia* (Compositae). Brittonia 24: 229–38.

Urbatsch, L. E., and P. B. Cox. 2006a. *Ratibida. In:* Flora of North America Editorial Committee. Flora of North America North of Mexico. 21: 60–63. New York/Oxford: Oxford University Press.

Urbatsch, L. E., and P. B. Cox. 2006b. *Rudbeckia. In:* Flora of North America Editorial Committee. Flora of North America North of Mexico. 21: 44–60. New York/Oxford: Oxford University Press.

Urbatsch, L. E., K. M. Neubig, and P. B. Cox. 2006. *Echinacea. In:* Flora of North America Editorial Committee. Flora of North America North of Mexico. 21: 88–92. New York/Oxford: Oxford University Press.

Vaezi, J., and L. Brouillet. 2009. Phylogenetic relationships among diploid species of *Symphyotrichum* (Asteraceae: Astereae) based on two nuclear markers, ITS and GAPGH. Molec. Phylogen. Evol. 51: 540–53.

Ward, D. B. 2004. New combinations in the Florida Flora II. Novon 14: 365–71.

Watson, L. E. 2006a. *Anthemis. In:* Flora of North America Editorial Committee. Flora of North America North of Mexico. 19: 537–38. New York/Oxford: Oxford University Press.

Watson, L. E. 2006b. *Cladanthus. In:* Flora of North America Editorial Committee. Flora of North America North of Mexico. 19: 495. New York/Oxford: Oxford University Press.

Watson, L. E. 2006c. *Marshallia. In:* Flora of North America Editorial Committee. Flora of North America North of Mexico. 21: 456–58. New York/Oxford: Oxford University Press.

Watson, L. E. 2006d. *Soliva. In:* Flora of North America Editorial Committee. Flora of North America North of Mexico. 19: 545–46. New York/Oxford: Oxford University Press.

Wells, J. R. 1965. A taxonomic study of *Polymnia* (Compositae). Brittonia 17: 144–59.

West, E., and L. E. Arnold. 1946. The Native Trees of Florida. Gainesville: University of Florida Press.

Whetstone, R. D., and K. R. Brodeur. 2006. *Launaea. In:* Flora of North America Editorial Committee. Flora of North America North of Mexico. 19: 272. New York/Oxford: Oxford University Press.

Wilhelm, G. S. 1984. Vascular flora of the Pensacola region. PhD dissertation, Southern Illinois University, Carbondale.

Wolf, S. J. 2006. *Arnica. In:* Flora of North America Editorial Committee. Flora of North America North of Mexico. 21: 366–77. New York/Oxford: Oxford University Press.

Wunderlin, R. P. 1982. Guide to the Vascular Plants of Central Florida. Gainesville: University Presses of Florida.

Wunderlin, R. P. 1998. Guide to the Vascular Plants of Florida. Gainesville: University Press of Florida.

Wunderlin, R. P., and B. F. Hansen. 2003. Guide to the Vascular Plants of Florida. 2nd ed. Gainesville: University Press of Florida.

Wunderlin, R. P., and B. F. Hansen. 2011. Guide to the Vascular Plants of Florida. 3rd ed. Gainesville: University Press of Florida.

Wunderlin, R. P., and J. E. Poppleton. 1977. The Florida species of *Ilex* (Aquifoliaceae). Florida Sci. 40: 7–21.

Yarborough, S. C., and A. M. Powell. 2006. *Flaveria. In:* Flora of North America Editorial Committee. Flora of North America North of Mexico. 21: 447–50. New York/Oxford: Oxford University Press.

Zhang, J.-W., Z.-L. Nie, J. Wen, and H. Sun. 2011. Molecular phylogeny and biogeography of three closely related genera, *Soroseris, Stebbinsia,* and *Syncalathium* (Asteraceae, Chichorieae), endemic to the Tibetan Plateau, SW China. Taxon 60: 15–26.

Index to Common Names

Abelia, glossy, 62
largeflower, 62
Achicoria azul, 295
African bushdaisy, 242
African daisy, 259
Alexanders, 113
golden, 114
meadow, 113, 114
Algerian ivy, 70
American bellflower, 35
American bluehearts, 17
American burnweed, 224
American elder, 56
American everlasting, 255
American holly, 32
American squawroot, 19
American wild carrot, 91
Angelica, coastalplain, 81
hairy, 82
Anisescented goldenrod, 368
Ankle-aster, 286
flaxleaf, 286
Annual marshelder, 287
Annual saltmarsh aster, 400
southeastern, 399
southwestern, 387
Annual trampweed, 246
Apalachicola aster, 242
Apalachicola doll's daisy, 163
Apalachicola lobelia, 39
Appalachian gayfeather, 306
Aralia, 74
frosted, 74
Arkansas dozedaisy, 141
Arrowleaf aster, white, 403
Arrowwood, southern, 58
Artichoke, Jerusalem, 278
Aster, 379
annual saltmarsh, 400
Apalachicola, 242
Bahaman, 383
calico, 391
climbing, 383
Dixie, 357
eastern silver, 385
Elliott's, 387
Florida water, 388
Georgia, 389
grassleaf, 239
late purple, 393
perennial saltmarsh, 401
pinebarren, 316
pinewoods, 242
rice button, 386
Savannah, 384
scaleleaf, 382
Short's, 398
Simmonds', 399
skyblue, 392
smooth blue, 390
smooth white oldfield, 396
southeastern annual saltmarsh, 399
southern pine, 240
southern swamp, 241
southwestern annual salt-marsh, 387
Stokes', 379
thistleleaf, 240
toothed whitetop, 356
Walter's, 404
wavyleaf, 402
western silver, 397
white arrowleaf, 403
white oldfield, 394
white panicle, 390
whitetop, 316, 356, 357
willowleaf, 395
Aster family, 115
Asthmaweed, 200
Australian umbrella tree, 75
Aztec marigold, 406

Bahaman aster, 383
Bahama sachsia, 353
Baker's coreopsis, 203
Balduina, purple, 151
Baldwin's eryngo, 94
Ballast eryngo, 95
Balsam groundsel, 320
Barbara's buttons, 308
grassleaf, 308
southern, 310
spoonshape, 309
Bay lobelia, 41
Beachberry, 54
Beach creeping oxeye, 323
Beach false foxglove, 5
Beach naupaka, 54
Beaked cornsalad, 66
Beechdrops, 20
Beggarticks, 154, 155
crowned, 159
devil's, 157
small, 156
smallfruit, 158
smooth, 157
Bellflower, 35
American, 35
Chinsegut, 36
Florida, 36

Robin's, 36
Bellflower family, 34
Bellorita, 226
Big floatingheart, 49
Bigleaf sumpweed, 288
Bishopweed, mock, 102
Bitterweed, 263
Blackeyed Susan, 350
Blackfoot, 310
Blackhaw, rusty, 60
Blackroot, 342
Blacksenna, 24
piedmont, 25
yaupon, 24
Blacksnakeroot, 103
Canadian, 103
clustered, 105
Maryland, 104
Small's, 105
Blanketflower, 250
lanceleaf, 251
Blazing-star, Georgia, 302
Godfrey's, 303
scrub, 302
Blessed thistle, 174
Blodgett's ironweed, 419
Blue aster, smooth, 390
Blueflower eryngo, 95
Bluehearts, 17
American, 17
Bluemink, 135
Blue mistflower, 198
Bluestem goldenrod, 363
Bogbean family, 49
Bogbutton, 354
pink, 354
Boneset, common, 236
false, 167
Mosier's false, 168
rough, 237
Boton de oro, 310
Bowlesia, hoary, 83
Boykin's lobelia, 40
Boynton's false foxglove, 8
Brickellbush, 166
heart-leaf, 167
Bristleleaf chaffhead, 173
Bristleleaf pricklyleaf, 410
Bristly scaleseed, 108
Broombush falsewillow, 149
Broomrape, 15, 22
oneflowered, 15
Broomrape family, 1
Browneyed Susan, 352
Buffalo spinach, 224
Bull thistle, 197
Bullwort, large, 79
Burnweed, 224
American, 224
Burrmarigold, 157
Burrweed, 374
carpet, 376
field, 375
Bushdaisy, African, 242
Bush goldenrod, 182
Bushy seaside oxeye, 165
Butterweed, 319
Button aster, rice, 386
Button eryngo, 96
Button rattlesnakemaster, 96
Buttons, Barbara's, 308
grassleaf Barbara's, 308
southern Barbara's, 310
spoonshape Barbara's, 309
Buttonweed, 375

Calico aster, 391
Calycera family, 55
Camphor daisy, 347
Camphorweed, 280, 332, 334
rosy, 333
stinking, 334
wingstem, 336
Canada goldenrod, 364
Canada lettuce, 292
Canadian blacksnakeroot, 103
Canadian honewort, 89
Canadian horseweed, 200
Canadian licorice-root, 98
Canadian lousewort, 23
Cancerroot, 19
Cankerweed, 315
Cape Sable thoroughwort, 179
Cape Sable whiteweed, 135
Cardinalflower, 40
Caribbean purple everlasting, 254
Carolina desertchicory, 344
Carolina elephantsfoot, 220
Carolina goldenrod, 362
Carolina grasswort, 99
Carolina holly, 27
Carolina woollywhite, 283
Carpet burrweed, 376
Carrot, American wild, 91
wild, 90
Carrot family, 77
Cat'sear, 283
hairy, 285
smallhead, 285
smooth, 284
South American, 284
Tweedie's, 284
Celery, 82
wild, 82
Chaffhead, 170
bristleleaf, 173
coastalplain, 171
hairy, 173
pineland, 170
Chaffseed, 23
Chamomile, 139
corn, 140
stinking, 140
Chapman's crownbeard, 414
Chapman's gayfeather, 298
Chapman's goldenrod, 369
Chattahoochee false foxglove, 12
Cheesewood, 66
Japanese, 67
Taiwanese, 67
Chervil, 85
hairyfruit, 86
spreading, 86
Chicory, 191
Chinsegut bellflower, 36
Chrysanthemum, 295
garland, 259
Cinchweed, 324
Florida, 325
sanddune, 324
spreading, 325
Clasping coneflower, 218
Clasping Venus' looking-glass, 47
Clavel de muerto, 411
Climbing aster, 383
Climbing ginseng, 69
Climbing hempvine, 314
Clustered blacksnakeroot, 105
Clustered yellowtops, 249
Coastalplain angelica, 81

Coastalplain chaffhead, 171
Coastalplain crownbeard, 414
Coastalplain goldenaster, 190
Coastalplain hawkweed, 282
Coastalplain honeycombhead, 151
Coastalplain palafox, 321
Coastalplain tickseed, 204
Coastalplain yellowtops, 248
Coastal ragweed, 137
Coatbuttons, 412
Cockleburr, 421
Common boneset, 236
Common cottonrose, 247
Common dandelion, 408
Common groundsel, 355
Common leopardbane, 142
Common mugwort, 148
Common ragweed, 136
Common sneezeweed, 264
Common sowthistle, 377
Common sunflower, 270
Common winterberry, 33
Common yarrow, 129
Coneflower, 217, 347
  clasping, 218
  cutleaf, 350
  eared, 348
  eastern purple, 218
  grassleaf, 349
  Mohr's, 351
  orange, 349
  pinnate prairie, 346
  prairie, 345
  purple, 218
  shiny, 352
  softhair, 351
  upright prairie, 345
Coralberry, 64
Coral honeysuckle, 63
Coreopsis, Baker's, 203
Coriander, 88
Corn chamomile, 140
Cornel-leaf whitetop, 216
Cornflower, garden, 175
Cornsalad, 65
  beaked, 66
Cosmos, garden, 210
  sulphur, 211
  wild, 211
Cottonrose, 247
  common, 247
Cottonthistle, 316
  Scotch, 317
Cottony goldenaster, 185
Cowbane, 100
  giant water, 111
  Piedmont, 101
  stiff, 100
  water, 110, 111
Cowpea witchweed, 25
Creeping eryngo, 96
Creeping oxeye, 378
  beach, 323
Crested floatingheart, 50
Crownbeard, 412
  Chapman's, 414
  coastalplain, 414
  diverseleaf, 415
  golden, 415
  white, 416
  yellow, 416
Crowned beggarticks, 159
Crown tickseed, 208
Cruise's goldenaster, 186
Cucumberleaf dune sunflower, 273
Cudweed, 253, 339
  Dominican, 339
  elegant, 255
  Heller's, 340
  Jersey, 340
  Pennsylvania, 256
  silvery, 254
  simple-stem, 257
  spoonleaf, 256
  white-cloaked, 255
Cure-for-all, 334
Cutleaf coneflower, 350

Dahoon, 28
  myrtle, 29
Daisy, African, 259
  Apalachicola doll's, 163
  camphor, 347
  doll's, 162
  false, 219
  lazy, 141
  Peruvian, 252
  pineland, 178
  skunk, 415
  smallhead doll's, 163
  Spanish, 263
  strangler, 169
  Transvaal, 258, 259
  white doll's, 163
Daisy fleabane, eastern, 226
Dandelion, 407
  common, 408
Delaney's goldenaster, 183
Delicate everlasting, 254
Dense gayfeather, 304
Desertchicory, 344
  Carolina, 344
Devil's beggarticks, 157
Devil's grandmother, 222
Devil's paintbrush, 328
Devil's walkingstick, 68
Dill, 80
Diverseleaf crownbeard, 415
Dixie aster, 357
Dixie goldenrod, 363
Dogfennel, 232
  weedy, 198
Dog's-tongue, 338
Doll's daisy, 162
  Apalachicola, 163
  smallhead, 163
  white, 163
Dominican cudweed, 339
Downy goldenrod, 370
Downy lobelia, 45
Downy ragged goldenrod, 369
Downy yellow false foxglove, 17
Dozedaisy, 141
  Arkansas, 141
Dress's goldenaster, 189
Dune sunflower, 272
  cucumberleaf, 273
  east coast, 272
  west coast, 273
Dusty miller, 147
Dwarfdandelion, 289
  potato, 290
  Virginia, 291
  weedy, 290
Dwarf schefflera, 76
Dwarf spotflower, 131
Dye-flower, 205
Dyersweed goldenrod, 367

Eared coneflower, 348
Eared goldenrod, 362
Early whitetop fleabane, 229
East coast dune sunflower, 272
Eastern daisy fleabane, 226
Eastern grasswort, 99
Eastern purple coneflower, 218
Eastern silver aster, 385
East Palatka holly, 34
Elder, 56
  American, 56
Elderberry, 56
Elegant cudweed, 255
Elegant zinnia, 423
Elephantsfoot, 220
  Carolina, 220
  smooth, 221
  tall, 221
Elliott's aster, 287
Elliott's goldenrod, 366
Elmleaf goldenrod, 373
English ivy, 70
Entireleaf Indian paintbrush, 18
Eryngo, 92
  Baldwin's, 94
  ballast, 95
  blueflower, 95
  button, 96
  creeping, 96
  fragrant, 93
  scrub, 94
  wedgeleaf, 94
  Everlasting, 253
  American, 255
  Caribbean purple, 254
  delicate, 254
  Pennsylvania, 256
  singlestem, 257
  spoonleaf purple, 256
  sweet, 341

False boneset, 167
  Mosier's, 168
False daisy, 219
Falsefennel, 234
False fleabane, 343
  ladies', 343
False foxglove, 2
  beach, 5
  Boynton's, 8
  Chattahoochee, 12
  downy yellow, 17
  fernleaf yellow, 16
  flaxleaf, 9
  Hampton, 7
  Harper's, 8
  Jackson, 6
  pineland, 5
  Plukenet's, 11
  purple, 12
  saltmarsh, 10
  scaleleaf, 4
  Seminole, 7
  slenderleaf, 14
  smooth yellow, 16
  sprawling, 7
  tenlobe, 11
  threadleaf, 13
  twoline, 9
False goldenaster, 280
False hawksbeard, oriental, 422
False hoarhound, 237
False Indian plantain, 262
False sunflower, 336
  Florida, 327
  pineland, 327
False vanillaleaf, 172
Falsewillow, broombush, 149
  saltwater, 149
False yellowhead, 215
Feay's palafox, 321
Fennel, 97
  sweet, 97
Fernleaf yellow false foxglove, 16
Feverfew, 322
  Santa Maria, 322
Fewflower gayfeather, 303
Fewleaf sunflower, 276
Field burrweed, 375
Field wormwood, 146
Fireweed, 224
Firewheel, 251
Flameflower, 21
Flamevine, Mexican, 341
Flattop goldenrod, 244
  slender, 243
Flatwoods sunflower, 271
Flaxleaf ankle-aster, 286
Flaxleaf false foxglove, 9
Fleabane, 225
  early whitetop, 229
  eastern daisy, 226
  false, 343
  ladies' false, 343
  oakleaf, 228
  Philadelphia, 227
  prairie, 228
  slender, 229
Floatingheart, 49, 51
  big, 49
  crested, 50
  Gray's, 51
  little, 50
  yellow, 52
Floating marshpennywort, 72
Florida bellflower, 36
Florida cinchweed, 325
Florida false sunflower, 327
Florida gayfeather, 302
Florida goldenaster, 184
Florida greeneyes, 154
Florida Indian plantain, 144
Florida ironweed, 419
Florida Keys hempvine, 313
Florida lobelia, 42
Florida paintbrush, 171
Florida shrub thoroughwort, 289
Florida sunflower, 274
Florida tasselflower, 222
Florida tickseed, 204
Florida thoroughwort, 239
Florida valerian, 65
Florida water aster, 388
Florida yellowtops, 248
Flyr's nemesis, 167
Foldear lobelia, 42
Foxglove, beach false, 5
  Boynton's false, 8
  Chattahoochee false, 12
  downy yellow false, 17
  false, 2
  fernleaf yellow false, 16
  flaxleaf false, 9
  Hampton false, 7
  Harper's false, 8
  Jackson false, 6
  pineland false, 5
  Plukenet's false, 11
  purple false, 12
  saltmarsh false, 10

scaleleaf false, 4
Seminole false, 7
slenderleaf false, 14
smooth yellow false, 15
sprawling false, 7
tenlobe false, 11
threadleaf false, 13
twoline false, 9
yellow false, 15
Fragrant eryngo, 93
Fringed sneezeweed, 265
Fringeleaf tickseed, 205
Frosted aralia, 74
Frostweed, 416

Gallant-soldier, 252
Gallberry, 30
large, 29
sweet, 29
Gallfeed, 54
Garber's gayfeather, 300
Garber's scrub stars, 258
Garden cornflower, 175
Garden cosmos, 210
Garland chrysanthemum, 259
Gayfeather, 296
Appalachian, 306
Chapman's, 298
dense, 304
fewflower, 0
Florida, 303
Garber's, 300
Gholson's, 301
Godfrey's, 303
grassleaf, 299
Piedmont, 303
pinkscale, 299
savanna, 304
scaly, 305
shortleaf, 306
slender, 301
tall, 298
Georgia aster, 389
Georgia blazing-star, 302
Georgia holly, 31
Georgia Indian plantain, 145
Georgia tickseed, 208
Gholson's gayfeather, 301
Giant goldenrod, 365
Giant ironweed, 419
Giant ragweed, 138
Giant water cowbane, 111
Gingseng, climbing, 69
Ginseng family, 26
Glade lobelia, 43
Glossy abelia, 62
Godfrey's blazing-star, 303
Godfrey's gayfeather, 303
Godfrey's goldenaster, 184
Golden alexanders, 114
Goldenaster, 165, 182
coastalplain, 190
cottony, 185
Cruise's, 186
DeLaney's, 183
Dress's, 189
false, 280
Florida, 184
Godfrey's, 184
grassleaf, 332
Highlands, 187
Lynn Haven, 187
Maryland, 190
narrowleaf, 189
pineland, 188
scrubland, 191
soft, 166
Golden crownbeard, 415
Goldenmane tickseed, 203
Golden ragwort, 318
Goldenrod, 360
anisescented, 368
bluestem, 363
bush, 182
Canada, 364
Carolina, 362
Chapman's, 369
Dixie, 363
downy, 370
downy ragged, 369
dyersweed, 367
eared, 362
Elliott's, 366
elmleaf, 373
flattop, 244
giant, 365
Leavenworth's, 367
mock, 168
Nuttall's rayless, 162
pinebarren, 364
pineland rayless, 161
rayless, 160
rayless mock, 169
roundleaf, 369
seaside, 371
slender flattop, 243
sweet, 368
twistedleaf, 372
Virginia, 366
wand, 372
woody, 182
wreath, 363
wrinkleleaf, 371
Golden tickseed, 209
Goldentop, 243
Goodenia family, 53
Grandmother, devil's, 222
Grassleaf aster, 239
Grassleaf Barbara's buttons, 308
Grassleaf coneflower, 349
Grassleaf gayfeather, 299
Grassleaf goldenaster, 332
Grassleaf lettuce, 293
Grasswort, 99
Carolina, 99
eastern, 99
Gray's floatingheart, 51
Greater tickseed, 207
Green-and-gold, 181
Greeneyes, 153
Florida, 154
soft, 153
Groundsel, balsam, 320
common, 355
Groundsel tree, 150
Gulf hammock Indian plantain, 262
Gullfeed, 54

Hairy angelica, 81
Hairy cat'sear, 285
Hairy chaffhead, 173
Hairyfruit chervil, 86
Hairyjoint meadowparsnip, 109
Hairy leafcup, 359
Hairy spotflower, 130
Hairy sunflower, 275
Hammockherb, 262
Hammock snakeroot, 133
Hampton false foxglove, 7

Hare's-ear, 84
Harper's false foxglove, 8
Harp onefruit, 261
Hawksbeard, 213
  Oriental false, 422
  smallflower, 213
Hawkweed, 281
  coastalplain, 282
  orange, 328
Heart-leaf brickellbush, 167
Hedgeparsley, 111
  spreading, 112
Heller's cudweed, 340
Hellroot, 22
Hemlock, spotted water, 87
  water, 87
Hemlock waterparsnip, 107
Hempleaf ragwort, 355
Hempvine, 312
  climbing, 314
  Florida Keys, 313
Herbwilliam, 102
Highlands goldenaster, 187
Hispid starburr, 128
Hoarhound, false, 237
Hoary bowlesia, 83
Holly, 26
  American, 32
  Carolina, 27
  East Palatka, 34
  Georgia, 31
  Krug's, 31
  sand, 27
  sarvis, 27
  serviceberry, 27
  scrub, 32
  tawnyberry, 31
  topal, 34
Holly family, 26
Honewort, 88
  Canadian, 89
Honeycombhead, 151
  coastalplain, 151
  oneflower, 152
  purpledisk, 151
Honeysuckle, 62
  coral, 63
  Japanese, 62
  trumpet, 63
Honeysuckle family, 61
Horseweed, 199
  Canadian, 200
  manzanilla, 201
Hummingbird-flower, 21
Hyssopleaf thoroughwort, 233

Indian-current, 64
Indian paintbrush, 18
  entireleaf, 18
Indian plantain, 142
  false, 262
  Florida, 144
  Georgia, 145
  Gulf Hammock, 262
  ovateleaf, 145
  pale, 143
  variableleaf, 144
  white, 143
Inkberry, 30, 54
Ironweed, 214, 417
  Blodgett's, 419
  Florida, 419
  giant, 419
  little, 214
  Missouri, 420
  New York, 420
  stemless, 417
  tall, 418
Ivy, 69
  Algerian, 70
  English, 70
Ivyleaf thoroughwort, 179

Jack-in-the-bush, 180
Jackson false foxglove, 6
Japanese cheesewood, 67
Japanese honeysuckle, 62
Jersey cudweed, 340
Jerusalem artichoke, 278
Joe Pye weed, 245
  sweetscented, 245
Justiceweed, 234

Kidneyleaf rosinweed, 358
King-devil, 328
Knapweed, 174
  spotted, 175
Krug's holly, 31

Lace, Queen Anne's, 90
Ladies' false fleabane, 343
Lakeside sunflower, 271
Lanceleaf blanketflower, 251
Lanceleaf tickseed, 205
Large bullwort, 79
Largeflower abelia, 62
Largeflower tickseed, 205
Large gallberry, 29
Largeleaf marshpennywort, 71, 72
Larkdaisy, 176
Lateflowering thoroughwort, 238
Late purple aster, 393
Lawn marshpennywort, 72
Lazy daisy, 141
Leafcup, 336
  hairy, 359
  Tennessee, 337
Leavenworth's goldenrod, 367
Leavenworth's tickseed, 206
LeConte's thistle, 194
Leopardbane, 142
  common, 142
Lesser snakeroot, 133
Lettuce, 291
  Canada, 292
  grassleaf, 293
  prickly, 293
  woodland, 292
Licorice-root, 98
  Canadian, 98
Lilac tasselflower, 223
Lionsfoot, 315
Little floatingheart, 50
Little ironweed, 214
Lobelia, Apalachicola, 39
  bay, 41
  Boykin's, 40
  downy, 45
  Florida, 42
  foldear, 42
  glade, 43
  McVaugh's, 45
  Nuttall's, 44
  pineland, 43
  shortleaf, 40
  southern, 39, 42
  white, 44
Looking-glass, clasping, 47

small Venus', 46
Venus', 46
Lousewort, 22
Canadian, 23
Low starburr, 128
Lynn Haven goldenaster, 187

Madamfate, 37
Madam gorgon, 55
Manyflower marshpennywort, 73
Manzanilla horseweed, 201
Mapleleaf viburnum, 58
Marigold, 406
Aztec, 406
sweetscented, 407
Marshelder, 287
annual, 287
narrowleaf, 287
Piedmont, 288
seacoast, 288
Marsh parsley, 89
Marshpennywort, 71
floating, 72
largeleaf, 71, 72
lawn, 72
manyflower, 73
whorled, 73
Maryland blacksnakeroot, 104
Maryland goldenaster, 190
Mayweed, 140
McVaugh's lobelia, 45
Meadow alexanders, 113, 114
Meadowparsnip, 109
hairyjoint, 109
purple, 109
Mexican flamevine, 341
Mile-a-minute, 313
Miller, dusty, 147
Missouri ironweed, 420
Mistflower, blue, 198
Mock bishopweed, 102
Mock goldenrod, 168
rayless, 169
Mohr's coneflower, 351
Mohr's thoroughwort, 235
Moshatel family, 56
Mosier's false boneset, 168
Muck sunflower, 277
Mugwort, common, 148
Muster John Henry, 407
Myrtle, sea, 150
Myrtle dahoon, 29

Narrowleaf goldenaster, 189
Narrowleaf marshelder, 287
Narrowleaf silkgrass, 331
Narrowleaf sunflower, 270
Narrowleaf yellowtops, 249
Naupaka, 53
beach, 54
Needles, Spanish, 156, 159
Nemesis, Flyr's, 167
Nerveray, 409
pineland, 409
New York ironweed, 420
Nodeweed, 405
Nuttall's lobelia, 44
Nuttall's rayless goldenrod, 162
Nuttall's thistle, 195

Oakleaf fleabane, 228
Octopus tree, 75
Oldfield aster, smooth white, 396
white, 394
Oneflowered broomrape, 15
Oneflower honeycombhead, 152
Onefruit, 261
harp, 261
Oppositeleaf spotflower, 130
Orange coneflower, 349
Orange hawkweed, 328
Oriental false hawksbeard, 422
Ovateleaf Indian plantain, 145
Oxeye, 279
beach creeping, 323
bushy seaside, 165
creeping, 378
seaside, 164
smooth, 279
tree seaside, 164
Oxeyedaisy, 295, 296

Paintbrush, devil's, 328
entireleaf Indian, 18
Florida, 171
Indian, 18
Palafox, 320
coastalplain, 321
Feay's, 321
Texas, 321
Pale Indian plantain, 143
Paleleaf woodland sunflower, 278
Panicle aster, white, 390
Paraguay starburr, 127
Parsley, 101, 102
marsh, 89
Pennsylvania cudweed, 256
Pennsylvania everlasting, 256
Perennial saltmarsh aster, 401
Peruvian daisy, 252
Peruvian zinnia, 424
Philadelphia fleabane, 227
Piedmont blacksenna, 25
Piedmont cowbane, 101
Piedmont gayfeather, 303
Piedmont marshelder, 288
Pine aster, southern, 240
Pinebarren aster, 316
Pinebarren goldenrod, 364
Pineland chaffhead, 170
Pineland daisy, 178
Pineland false foxglove, 5
Pineland false sunflower, 327
Pineland goldenaster, 188
Pineland lobelia, 43
Pineland nerveray, 409
Pineland purple, 172
Pineland rayless goldenrod, 161
Pineland silkgrass, 329
Pinewoods aster, 242
Pink bogbutton, 354
Pinkscale gayfeather, 299
Pink thoroughwort, 250
Pinnate prairie coneflower, 346
Pittosporum family, 66
Plantain, false Indian, 262
Florida Indian, 144
Georgia Indian, 145
Gulf Hammock Indian, 262
Indian, 142
ovateleaf Indian, 145
pale Indian, 143
Robin's, 227
variableleaf Indian, 144
white Indian, 143
Plukenet's false foxglove, 11
Possumhaw, 30, 59

Potato dwarfdandelion, 290
Prairie coneflower, 345
  pinnate, 346
  upright, 345
Prairie fleabane, 228
Pricklyleaf, 410
  bristleleaf, 410
Prickly lettuce, 293
Purple, pineland, 172
Purple aster, late, 393
Purple balduina, 151
Purple coneflower, 218
  eastern, 218
Purpledisk honeycombhead, 151
Purpledisk sunflower, 271
Purple everlasting, Caribbean, 254
  spoonleaf, 256
Purple false foxglove, 12
Purplehead sneezeweed, 265
Purple meadowparsnip, 109
Purple thistle, 193
Pussytoes, 138, 139

Queen Anne's lace, 90
Queen-devil, 281
Queen-of-the-meadow, 245

Rabbit tobacco, 341
Ragged goldenrod, downy, 269
Ragleaf, redflower, 212
Ragweed, 136
  coastal, 137
  common, 136
  giant, 138
  western, 137
Ragwort, 354
  golden, 318
  hempleaf, 355
  roundleaf, 319
  Small's, 317
Rattlesnakemaster, 93
  button, 96
Rattlesnakeroot, 314
  slender, 314
Rayless goldenrod, 160
  Nuttall's, 162
  pineland, 161
Rayless mock goldenrod, 169
Redflower ragleaf, 212
Resindot sunflower, 277
Rice button aster, 386
Ricepaper plant, 76
Robin's bellflower, 36
Robin's plantain, 227
Rockbell, southern, 48
Romerillo, 155
Rose-rush, 307
Rosinweed, 357
  kidneyleaf, 358
  starry, 358
Rosy camphorweed, 333
Rough boneset, 237
Roughfruit scaleseed, 108
Roundleaf goldenrod, 369
Roundleaf ragwort, 319
Roundleaf thoroughwort, 237
Rusty blackhaw, 60

Sachsia, Bahama, 353
Sagebrush, 146
  white, 147
Saltmarsh aster, annual, 400
  perennial, 401
  southeastern annual, 399
  southwestern annual, 387
Saltmarsh false foxglove, 10
Saltwater falsewillow, 149
Sanddune cinchweed, 324
Sand holly, 29
Santa Maria feverfew, 322
Sarvis holly, 27
Savanna gayfeather, 304
Savannah aster, 384
Savannah sneezeweed, 267
Scaleleaf aster, 382
Scaleleaf false foxglove, 4
Scaleseed, 107
  bristly, 108
  roughfruit, 108
Scaly gayfeather, 305
Schefflera, dwarf, 76
Scotch cottonthistle, 317
Scratchdaisy, 213
  slender, 214
Scrub blazing-star, 302
Scrub eryngo, 94
Scrub holly, 32
Scrubland goldenaster, 191
Scrub stars, Garber's, 258
Seacoast marshelder, 288
Sea myrtle, 150
Seaside goldenrod, 371
Seaside oxeye, 164
  bushy, 165
  tree, 164
Semaphore thoroughwort, 235
Seminole false foxglove, 7
Serviceberry holly, 27
Shaggysoldier, 252
Shepardsneedle, 106
Shiny coneflower, 352
Shortleaf gayfeather, 306
Shortleaf lobelia, 40
Shortleaf sneezeweed, 264
Short's aster, 398
Shrub sunflower, 411
Shrub thoroughwort, Florida, 289
Sickleleaf silkgrass, 330
Silkgrass, 328
  narrowleaf, 331
  pineland, 329
  sickleleaf, 330
  zigzag, 330
Silkleaf, 294
Silver aster, eastern, 385
  western, 397
Silverleaf sunflower, 271
Silverling, 150
Silvery cudweed, 254
Simmond's aster, 399
Simple-stem cudweed, 257
Singlestem everlasting, 257
Skunk daisy, 415
Skyblue aster, 392
Slender flattop goldenrod, 243
Slender fleabane, 339
Slender gayfeather, 301
Slenderleaf false foxglove, 14
Slender rattlesnakeroot, 314
Slender scratchdaisy, 214
Small beggarticks, 156
Smallflower hawksbeard, 213
Smallflower thoroughwort, 238
Smallfruit beggarticks, 158
Smallhead cat'sear, 285
Smallhead doll's daisy, 163
Small-leaf viburnum, 59
Small's blacksnakeroot, 105
Small's ragwort, 317

Small Venus' looking-glass, 46
Small woodland sunflower, 275
Smooth beggarticks, 157
Smooth blue aster, 390
Smooth cat'sear, 284
Smooth elephantsfoot, 221
Smooth oxeye, 279
Smooth white oldfield aster, 396
Smooth yellow false foxglove, 16
Snakeroot, 132
  hammock, 133
  lesser, 133
  white, 132
Sneezeweed, 263
  common, 364
  fringed, 265
  purplehead, 265
  Savannah, 267
  shortleaf, 264
  southeastern, 266
Snowberry, 64
Snowflake, water, 52
Snow squarestem, 311
Snowy white thoroughwort, 236
Soft goldenaster, 166
Soft greeneyes, 153
Softhair coneflower, 351
South American cat'sear, 284
Southeastern annual saltmarsh aster, 399
Southeastern sneezeweed, 266
Southeastern sunflower, 269
Southern arrowwood, 58
Southern Barbara's buttons, 310
Southern lobelia, 39, 42
Southern pine aster, 240
Southern rockbell, 48
Southern swamp aster, 241
Southern whitetop, 217
Southwestern annual saltmarsh aster, 387
Sowthistle, 376
  common, 377
  spiny, 377
Spadeleaf, 85
Spanish daisy, 263
Spanish needles, 156, 159
Spikenard, 68
Spinach, buffalo, 224
Spiny sowthistle, 377
Spiritweed, 95
Spoonleaf cudweed, 256
Spoonleaf purple everlasting, 256
Spoonshape Barbara's buttons, 309
Spotflower, 129
  dwarf, 131
  hairy, 130
  oppositeleaf, 130
Spotted knapweed, 175
Spotted water hemlock, 87
Sprawling false foxglove, 7
Spreading chervil, 86
Spreading cinchweed, 325
Spreading hedgeparsley, 112
Squarehead, 409
Squarestem, 311
  snow, 311
Squawroot, 19
  American, 19
Starburr, 127
  hispid, 128
  low, 128
  Paraguay, 127
Starry rosinweed, 358
Star tickseed, 208
Stemless ironweed, 417
Stiff cowbane, 100
Stiff sunflower, 276
Stinking camphorweed, 334
Stinking chamomile, 140
St. John's Susan, 352
Stokes' aster, 379
Strangler daisy, 169
Sulphur cosmos, 211
Sumpweed, bigleaf, 288
Sunbonnets, 177
  white, 177
  woolly, 178
Sunflower, 268
  common, 270
  cucumberleaf dune, 273
  dune, 272
  east coast dune, 272
  false, 326
  fewleaf, 276
  flatwoods, 271
  Florida, 274
  Florida false, 327
  hairy, 275
  lakeside, 271
  muck, 277
  narrowleaf, 270
  paleleaf woodland, 278
  pineland false, 327
  purpledisk, 271
  resindot, 277
  shrub, 411
  silverleaf, 271
  small woodland, 275
  southeastern, 269
  stiff, 276
  swamp, 270
  variableleaf, 274
  west coast dune, 273
  woodland, 273
Susan, blackeyed, 350
  browneyed, 352
  St. John's, 352
Swamp aster, southern, 241
Swamp sunflower, 270
Swamp thistle, 195
Sweet everlasting, 341
Sweet fennel, 97
Sweet gallberry, 29
Sweet goldenrod, 368
Sweetscent, 335
Sweetscented Joe Pye weed, 245
Sweetscented marigold, 407
Sweet viburnum, 60

Taiwanese cheesewood, 67
Tall elephantsfoot, 221
Tall gayfeather, 298
Tall ironweed, 418
Tall thistle, 193
Tall thoroughwort, 231
Tall tickseed, 209
Tansyaster, 346
Tasselflower, 222
  Florida, 222
  lilac, 223
Tawnyberry holly, 31
Tenlobe false foxglove, 11
Tennessee leafcup, 337
Texas palafox, 321
Texas tickseed, 207
Thistle, 192
  blessed, 174
  bull, 197

LeConte's, 194
Nuttall's, 195
purple, 193
swamp, 195
tall, 193
Virginia, 196
Thistleleaf aster, 240
Thoroughwort, 178, 198, 230, 250, 289
Cape Sable, 179
Florida, 239
Florida shrub, 289
hyssopleaf, 233
ivyleaf, 179
lateflowering, 238
Mohr's, 235
pink, 250
roundleaf, 237
semaphore, 235
smallflower, 238
snowy white, 236
tall, 231
waxy, 234
white, 231
Threadleaf false foxglove, 13
Tickseed, 202
coastalplain, 204
crown, 208
Florida, 204
fringeleaf, 205
Georgia, 208
golden, 209
goldenmane, 203
greater, 207
lanceleaf, 205
largeflower, 205
Leavenworth's, 206
star, 208
tall, 209
Texas, 207
Tobacco, rabbit, 341
woman's, 139
Toothed whitetop aster, 356
Toothpickweed, 80
Topal holly, 31
Trampweed, 246
annual, 246
Transvaal daisy, 258, 259
Tree seaside oxeye, 164
Tropical whiteweed, 134
Trumpet honeysuckle, 63
Tuffybells, 48
Tweedie's cat'sear, 284
Twistedleaf goldenrod, 372
Twoline false foxglove, 9

Umbrella tree, Australian, 75
Upright prairie coneflower, 345

Valerian, 65
Florida, 65
Vanillaleaf, 171, 172
false, 172
Variableleaf Indian plantain, 144
Variableleaf sunflower, 274
Velvetplant, 260
Venuscomb, 106
Venus's looking-glass, 46
clasping, 47
small, 46
Viburnum, mapleleaf, 58
small-leaf, 59
sweet, 60
Walter's, 59
Virginia dwarfdandelion, 291
Virginia goldenrod, 366
Virginia thistle, 196

Walkingstick, devil's, 65
Walter's aster, 404
Walter's viburnum, 59
Wand goldenrod, 372
Water aster, Florida, 388
Water cowbane, 110, 111
giant, 111
Water hemlock, 87
spotted, 87
Waterparsnip, 106
hemlock, 107
Water snowflake, 52
Wavyleaf aster, 402
Waxy thoroughwort, 234
Wedgeleaf eryngo, 94
Weedy dogfennel, 198
Weedy dwarfdandelion, 290
West coast dune sunflower, 273
Western ragweed, 137
Western silver aster, 397
White arrowleaf aster, 403
White-cloaked cudweed, 255
White crownbeard, 416
White doll's daisy, 163
White Indian plantain, 143
White lobelia, 44
Whitenymph, 112
White oldfield aster, 394
smooth, 396
White panicle aster, 390
White sagebrush, 147
White snakeroot, 132
White sunbonnets, 177
White thoroughwort, 231
White thoroughwort, snowy, 236
Whitetop, 216
cornel-leaf, 216
southern, 217
Whitetop aster, 306, 356, 357
toothed, 356
Whitetop fleabane, early, 229
Whiteweed, 134
Cape Sable, 135
tropical, 134
Whorled marshpennywort, 73
Wild carrot, 90
American, 91
Wild celery, 82
Wild cosmos, 211
Willowleaf aster, 395
Wingstem, 413
Wingstem camphorweed, 336
Winterberry, common, 33
Witchweed, 25
cowpea, 25
Woman's tobacco, 139
Woodland lettuce, 292
Woodland sunflower, 273
paleleaf, 278
small, 275
Woody goldenrod, 182
Woolly sunbonnets, 178
Woollywhite, Carolina, 283
Wormwood, field, 146
Wreath goldenrod, 363
Wrinkleleaf goldenrod, 371

Yankeeweed, 232
Yarrow, 129
common, 129
Yaupon, 33
Yellow crownbeard, 416

Yellow false foxglove, 15
    downy, 17
    fernleaf, 16
    smooth, 16
Yellow floatingheart, 52
Yellowhead, false, 215
Yellowtops, 247
    clustered, 249
    coastalplain, 248
    Florida, 248
    narrowleaf, 249
Yerba de San Juan, 325
Youpon blacksenna, 24

Zigzag silkgrass, 330
Zinnia, elegant, 423
    Peruvian, 424

Index to Scientific Names

Accepted scientific names of plants and plant families are in roman type. Synonyms are in *italics*.

Abelia, 62
  ×grandiflora, 62
  rupestris var. *grandiflora*, 62
Acanos spina, 317
Acanthopanax
  aculeatus, 69
  trifoliatus, 69
Acanthospermum, 127
  australe, 127
  hispidum, 128
  humile, 128
    var. *hisidum*, 128
    var. *normale*, 128
  xanthioides, 127
    var. *obtusifolium*, 127
Acanthoxanthium spinosum, 422
Achillea, 129
  millefolium, 129
    subsp. *occidentalis*, 129
    var. *occidentalis*, 129
  occidentalis, 129
Achyrophorus
  chillensis, 284
  radicatus, 285
Acicarpha, 55
  tribuloides, 55
    var. *dentata*, 55
Acmella, 129
  nuttalliana, 130
  occidentalis, 130
  oppositifolia
    var. *oppositifolia*, 131
    var. repens, 130
  pilosa, 130
  pusilla, 131
  repens, 130
Acosta micranthos, 175
Actinomeris
  alternifolia, 413
  heterophylla, 415
  nudicaulis, 414
  paniculata, 417
  pauciflora, 415
  squarrosa, 413
    var. *alternifolia*, 413
    var. *pauciflora*, 415
    var. *serrulata*, 415
Actinospermum, 151
  angustifolium, 151
  uniflorum, 152
Adenimesa atriplicifolia, 143
Adopogon carolinianus, 291
Adoxaceae, 56
Adventina ciliata, 252
Aethusa
  divaricata, 108
  leptophylla, 89
Afzelia, 24
  cassioides, 24
  pectinata, 25
    var. *peninsularis*, 25
Agalinis, 2
  aphylla, 4
  corymbosa, 12
  delicatula, 8
  divaricata, 5
  erecta, 12
  fasciculata, 5
    var. *peninsularis*, 6
  filicaulis, 6
  filifolia, 7
  flava, 16
  flexicaulis, 7
  georgiana, 8
  harperi, 8
  holmiana, 13
  keyensis, 11
  laxa, 9
  linifolia, 9
  maritima
    var. grandiflora, 10
    var. *maritima*, 14
  microphylla, 4
  obtusifolia, 11
  oligophylla, 13
  palustris, 12
  parvifolia, 11
  pedicularia var. *pectinata*, 16
  perennis, 9
  pinetorum, 8
    var. *delicatula*, 8
  plukenetii, 11
  pulchella, 12
  purpurea, 12
    var. *carteri*, 12
  setacea, 13
    var. *parvifolia*, 11
  skinneriana, 14
  spiciflora, 11
  stenophylla, 13
  tenella, 11
  tenuifolia, 14
    var. *leucanthera*, 14
  virginica, 17
Ageratina, 132

altissima, 132
aromatica, 133
var. *incisa*, 133
jucunda, 133
Ageratum, 134
altissimum, 132
conyzioides, 134
hirtum, 134
houstonianum, 135
forma *normale*, 135
var. *typicum*, 135
humile, 134
littorale, 135
forma *album*, 135
forma *setigerum*, 135
maritimum, 135
mexicanum forma *houstonianum*, 135
odoratum, 134
Ageria
cassena, 34
heterophylla, 29
obovata, 29
opaca, 32
palustris, 28
Alitubus millefolium, 129
Ambrosia, 136
artemisiifolia, 136
var. *elatior*, 136
var. *paniculata*, 136
crithmifolia, 137
cumanensis, 138
elata, 136
elatior, 136
var. *artemisiifolia*, 136
glandulosa, 136
hispida, 137
monophylla, 136
paniculata, 136
peruviana, 138
psilostachya, 137
rugelii, 137
trifida, 138
var. *normalis*, 138
Ambrosiaceae, 117
Amellus
aspera, 311
var. *glabriusculus*, 312
var. *normalis*, 311
lanceolatus, 312
niveus, 311
Ammi, 79
capillaceum, 102
divaricatum, 108
diversifolium, 79
var. *latifolium*, 79
elatum, 79
majus, 79
visnaga, 80
Ammiaceae, 77
Ammopursus, 296
ohlingerae, 302
Ampelaster, 380
carolinianus, 383
Anacis tripteris, 209
Anethum, 80
arvense, 80
foeniculum, 97
var. *vulgare*, 97
graveolens, 80
rupestre, 97
Angelica, 81
dentata, 81
hirsuta, 81
spinosa, 68
triquinata, 81
venenosa, 82
villosa, 82
Antennaria, 138
fallax, 139
parlinii subsp. *fallax*, 139
plantaginea, 139
var. *petiolata*, 139
plantaginifolia, 139
solitaria, 139
Anthemis, 139
arvensis, 140
cotula, 140
cotula-foetida, 140
foetida, 140
mixta, 198
repens, 130
Anthriscus arvensis, 112
Apargia chillensis, 284
Aphanostephus, 141
skirrhobasis
var. *skirrhobasis*, 141
var. thalassius, 141
Aphyllon, 15
uniflorum, 15
Apiaceae, 77
Apium, 82
ammi, 79, 90, 115
var. *genuinum*, 90
var. *leptophyllum*, 90
ammi-majus, 79
celleri, 82
cicutifolium, 107
crispum, 102
divaricatum, 108
echinatum, 108
graveolens, 82
leptophyllum, 90
maritimum, 82
petroselinum
var. *crispifolium*, 102
var. *crispum*, 102
visnaga, 80
vulgare, 82
Aplactis paniculata, 182
Apogon
dandelion, 290
gracilis, 290
Aquifoliaceae, 26
Aquilaria pentandra, 67
Araliaceae, 67
Archangelica
dentata, 81
hirsuta, 82
villosa, 82
Archemora
denticulata, 100
rigidior, 100
ternata, 101
Aralia, 68
spinosa, 68
guilfoylei, 74
papyrifer, 76
trifoliata, 69
Arnica, 142
acaulis, 142
claytonia, 142
nudicaulis, 142
Arnoglossum, 142
album, 143
atriplicifolium, 143
diversifolium, 144
floridanum, 144
ovatum, 145
var. *lanceolatum*, 145

plantagineum, 146
sulcatum, 145
Artemisia, 146
campestris
subsp. caudata, 146
var. *caudata*, 146
capillifolia, 232
caudata, 146
forma *typica*, 147
ludoviciana, 147
var. *typica*, 147
officinalis, 148
ruderalis, 148
stelleriana, 147
tenuifolia, 232
vulgaris, 148
subsp. *ludoviciana*, 147
subsp. *typica*, 148
var. *ludoviciana*, 147
var. *normalis*, 148
var. *typica*, 148
Aster
adnatus, 382
agrostifolius, 391
annuus, 226
asperifolius, 402
asperulus, 402
asteroides, 356
bahamensis, 383
baldwinii, 402
bifoliatus, 357
bracei, 401
brachypholis, 396
caesius, 363
camptosorus, 398
canadensis, 200
carolinianus, 383
chapmanii, 384
ciliatus, 397
collinsii, 357
commutatus, 369
concinnus, 390
concolor, 385
var. *devestitus*, 385
var. *simulatus*, 385
conyzoides, 356
coridifolius, 386
diffusus var. *horizontalis*, 391
discoideus, 169
divaricatus, 214, 404
var. *sandwicensis*, 387
divergens var. *pendulus*, 391
diversifolius, 402
dumosus, 386
var. *coridifolius*, 386
var. *gracilentus*, 386
var. *gracilipes*, 386
var. *pergracilis*, 386
var. *subulifolius*, 386
var. *verus*, 386
elliottii, 387
ericoides, 404
forma *villosus*, 394
var. *pilosus*, 394
var. *villosus*, 394
eryngiifolius, 240
exilis
var. *australis*, 387
var. *inconspicuus*, 387
fistulosus, 364
flavovirens, 372
foliolosus
var. *cordifolius*, 386
fontinalis, 388
fragilis var. *brachypholis*, 396
georgianus, 389
gracilipes, 386
gracillimus, 366
graminifolius, 244
grandiflorus, 404
hemisphericus, 240
horizontalis, 391
humilis, 216
inconspicuus, 387
infirmis, 216
laevis var. *concinnus*, 390
lanceolatus, 409
var. *latifolius*, 390
lateriflorus, 391
var. *horizontalis*, 391
var. *pendulus*, 391
var. *spatelliformis*, 391
latissimifolius, 366
var. *serotinus*, 365
leavenworthii, 367
liguliformis, 402
linariifolius, 286
loriformis, 402
marilandicus, 356
mexicanus, 371
miser
var. *pendulus*, 391
var. *vimineus*, 391
mohrii, 402
nudatus, 161
obovatus, 316
odorus, 368
oolentangiense, 392
paludodus, 241
subsp. *hemisphericus*, 240
var. *hemisphericus*, 240
patens, 393
var. *floridanus*, 388
var. *georgianus*, 389
var. *gracilis*, 393
var. *tenuicaulis*, 393
paternus, 356
pauciflosulosus, 182
pediomonus, 240
pendulus, 391
petiolaris, 369
phlogifolius var. *patens*, 393
phyllocephalus, 347
phyllolepis, 397
pilosus, 394
pinifolius, 399
plumosus, 395
praealtus, 395
pratensis, 397
prenanthoides, 405
proteus, 402
puniceus subsp. *elliottii*, 387
pulverulentus, 370
purshianus, 372
racemosus, 396
reticulata, 316
sagittifolius, 405
var. *dissitifolius*, 403, 405
var. *urophyllus*, 403
salicifolius, 395
sandwicensis, 387
scandens, 383
sempervirens, 371
sericeus var. *microphyllus*, 397
sericocarpoides, 217
shortii, 398
var. *camptosorus*, 398
simmondsii, 399
simplex, 405
simulatus, 385

spatelliformis, 391
spectabilis, 242
spinulosus, 242
squamatus, 399
squarrosus, 404
stenactis, 226
strictus, 372
sublitoralis, 366
subulatus, 400, 405
var. *australis*, 387
var. *bahamensis*, 383
var. *elongatus*, 383
var. *euroauster*, 400
var. *ligulatus*, 405
var. *sandwicensis*, 387
sulznerae, 399
tenuicaulis, 393
tenuifolius, 401
var. *aphyllus*, 401
tortifolius, 357
tradescantii, 391
ulmifolius, 373
umbellatus, 217
var. *humilis*, 216
var. *latifolius*, 217
undulatus, 402
var. *asperulus*, 402
var. *diversifolius*, 402
var. *loriformis*, 402
urophyllus, 403
vernus, 229
villosus, 394
vimineus, 391
walteri, 404
Asteraceae, 115
Athanasia
graminifolia, 308
obovata, 309
Atirsita pumila, 94
Aureolaria, 15
flava, 16
subsp. *reticulata*, 16
var. *reticulata*, 16
microcarpa, 17
pectinata, 16
subsp. *floridana*, 16
subsp. *typica*, 16
purpurea, 12
reticulata, 16
tenuifolia, 14
villosa, 16
virginica, 17

Babiron
dichotomum, 108
divaricatum, 108
Baccharis, 148
angustifolia, 149
angustior, 150
cuneifolia, 150
dioica, 149
foetida, 334
glomeruliflora, 150
halimifolia, 150
var. *angustior*, 150
rohriana, 149
sessiliflora, 150
vahlii, 149
Baccharoides punctata, 176
Balduina, 151
angustifolia, 151
atropurpurea, 151
bicolor, 152
lutea, 152
multiflora, 151
uniflora, 152
Batschia nivea, 132
Behen noveboracensis, 420
Benedicta officinalis, 174
Berlandiera, 153
humilis, 154
pumila, 153
subacaulis, 154
tomentosa, 153
×humilis, 154
Berula erecta, 115
Bidens, 154
alba, 155
var. *radiata*, 155
aristosa, 160
artemisiifolia
var. *caudata*, 211
var. *sulphurea*, 211
aurea, 160
bipinnata, 156
caudata, 211
chrysanthemoides, 158
var. *nashi*, 158
coronata, 159
var. *leptophylla*, 158
var. *trichosperma*, 160
var. *typica*, 160
discoidea, 156
frondosa, 157
laevis, 157
leucantha, 155
var. *pilosa*, 159
leucanthema, 155
melanocarpha, 157
mitis, 158
var. *leptophylla*, 158
nashi, 158
nivea, 311
pilosa, 159
forma *discoidea*, 159
forma *leucantha*, 155
forma *radiata*, 155
subvar. *discoidea*, 159
subvar. *radiata*, 155
var. *alba*, 155
var. *bipinnata*, 156
var. *leucantha*, 155
var. *radiata*, 155
sulphurea, 211
trichosperma, 159
tripteris, 209
Bigelowia, 160
nudata, 161
forma *spathulifolia*, 161
forma *virgata*, 161
subsp. australis, 161
subsp. nudata, 161
var. *spathulifolia*, 161
var. *virgata*, 161
nuttallii, 162
virgata, 161
Boltonia, 162
apalachicolensis, 163
asteroides, 163
var. *glastifolia*, 163
diffusa, 163
glastifolia, 163
Borkhausia caroliniana, 344
Borrichia, 164
arborescens, 164
var. *glabrata*, 164
cubana, 165
frutescens, 165
glabrata, 164
×cubana, 165

Bowlesia, 83
incana, 83
forma *crassifolia*, 83
Brachyrhamphus, 295
intybaceus, 295
Bradburia, 165
pilosa, 166
Brassaia actinophylla, 75
Brauneria purpurea, 218
Brickellia, 166
cordifolia, 167
eupatorioides, 167
var. *floridana*, 168
mosieri, 168
Brintonia, 168
discoidea, 169
Buchnera, 17
americana, 17
brevifolia, 18
elongata, 18
var. *obtusa*, 18
floridana, 18
gesnerioides, 25
levicaulis, 18
longifolia, 18
palustris, 18
Buphthalmum
angustifolium, 151
arborescens, 164
frutescens, 165
repens, 378
Bupleurum, 84
lancifolium, 84
rotundifolium, 84

Cacalia
acaulis, 417
angustifolia, 418
atriplicifolia, 143
aurantiaca, 260
cinerea, 215
cordifolia, 313
diversifolia, 144
elliottii, 145
floridana, 144
gigantea, 419
graminifolia, 418
heterophylla, 258
lanceolata, 145
var. *elliottii*, 145
noveboracensis, 420
ovata, 145
plantaginea, 146
sagittata, 223
sonchifolia, 223
sulcata, 145
Calcitrapa
benedicta, 174
lanuginosa, 174
Calea aspera, 311
Calliopsis
basalis, 203
nudata, 208
tinctoria, 209
Calostelma elegans, 299
Calyceraceae, 55
Calyptocarpus, 169
vialis, 160
Campanopsis
linarioides, 48
marginata, 48
Campanula, 35
americana, 35
amplexicaulis, 47
biflora, 46
floridana, 36
linarioides, 48
marginata, 48
perfoliata, 47
robinsiae, 36
Campanulaceae, 34
Campanulastrum, 35
americanum, 35
Caprifoliaceae, 61
Caprifolium
japonicum, 62
sempervirens, 63
Carduaceae, 117
Carduus
altissimus, 193
glaber, 195
horridulus, 193
lanceolatus, 197
lecontei, 194
muticus, 195
nuttallii, 195
pinetorum, 193
repandus, 197
revolutus, 196
smallii, 194
spinosissimus, 193
var. *elliottii*, 193
virginianus, 196
vittatus, 194
vulgaris, 197
Carelia
conyzoides, 134
houstoniana, 135
littorale, 135
maritima, 135
Carota sylvestris, 90
Carphephorus, 170
bellidifolius, 174
carnosus, 170
corymbosus, 171
odoratissimus, 171
var. odoratissimus, 172
var. subtropicanus, 172
paniculatus, 173
pseudoliatris, 173
subtropicanus, 172
tomentosus, 174
Carthamus laevis, 379
Carum
graveolens, 82
majus, 79
petroselinum var. *crispum*, 102
Cassine caroliniana, 27
Castilleja, 18
coccinea, 19
indivisa, 18
Caucalis
arvensis, 112
canadensis, 103
carota, 90
daucus, 90
divaricata, 112
marilandica, 104
Celeri, 82
graveolens, 83
Centaurea, 174
benedicta, 174
calcitrapa, 176
cyanus, 175
maculosa, 176
subsp. *micranthos*, 175
micrantha, 175
micranthos, 175
nigrescens, 176
phrygia, 176

solstitialis, 176
stoebe subsp. micranthos, 175
Centella, 84
asiatica, 85
var. *repanda*, 85
erecta, 85
var. *floridana*, 85
floridana, 85
repanda, 0
forma *floridana*, 85
Centratherum, 176
longispinum, 176
punctatum, 176
Centrospermum xanthioides, 127
Ceratocephalus repens, 130
Chaerophyllum, 85
canadense, 89
floridanum, 86
pecten-veneris, 106
procumbens, 86
var. *tainturieri*, 86
tainturieri, 86
var. *floridanum*, 86
Chamaemelum
arvensis, 140
cotula, 140
foetidum, 140
leucanthemum, 296
millefolium, 129
mixtum, 198
Chaptalia, 177
albicans, 177
dentata, 198
integrifolia, 178
tomentosa, 178
Chlaenobolus
pycnostachyus, 342
undulatus, 342
Chondrilla pulchra, 213
Chondrocarpus sibthorpioides, 72
Chondrophora, 160
nudata, 161
var. *virgata*, 161
virgata, 161
Chromolaena, 178
frustrata, 179
heteroclinia, 180
ivifolia, 179
odorata, 180
rigida, 180
Chrysanthemum
coronarium, 259
leucanthemum, 296
var. *typicum*, 296
var. *vulgare*, 296
pratense, 296
procumbens, 412
segetum, 260
vulgare, 296
Chrysocoma
acaulis, 417
capillifolia, 232
coronopifolia, 232
gigantea, 419
graminifolia, 244
nudata, 161
odoratissima, 172
paniculata, 173
virgata, 161
Chrysogonum, 180
australe, 181
peruviana, 424
virginianum
var. australe, 181
var. *virginianum*, 181
Chrysoma, 181
pauciflosulosa, 182
solidaginoides, 182
Chrysopsis, 182
adenolepis, 329
argentea, 331
aspera, 329
cruiseana, 186
decumbens, 186
delaneyi, 183
divaricata, 214
falcata, 330
flexuosa, 330
floridana, 184
var. *highlandsensis*, 187
gigantea, 187
godfreyii, 184
forma *viridis*, 184
gossypina, 185
forma *decumbens*, 186
forma *trichophylla*, 186
subsp. cruiseana, 186
subsp. gossypina, 186
subsp. hyssopifolia, 186, 187
graminifolia, 331
var. *aspera*, 329
var. *latifolia*, 331
var. *microcephala*, 331
highlandsensis, 187
hyssopifolia, 186
lamarckii, 280
lanuginosa, 187
latifolia, 331
latisquamea, 188
linariifolia, 286
linearifolia, 189
subsp. dressii, 189
subsp. linearifolia, 189
var. *dressii*, 189
mariana, 190
var. *floridana*, 184
microcephala, 331
mixta, 187
nervosa, 331
nuttallii, 166
obovata, 316
oligantha, 332
pilosa, 166
scabra, 280
scabrella, 190
subulata, 191
tracyi, 331
trichophylla, 186
Chrysostemma tripteris, 209
Chthonia
glaucescens, 324
leptocephali, 324
Cicerbita
acuminata, 293
canadensis, 292
elongata, 292
floridana, 292
intybacea, 295
villosa, 293
Cichoriaceae, 117
Cichorum, 191
commune, 191
intybus, 191
var. *caeruleum*, 192
var. *sylvestre*, 192
perenne, 191
var. *sylvestre*, 192
rigidum, 191
sylvestre, 191
Cicuta, 87

bulbifera, 88
curtissii, 87
maculata, 87
var. *curtissii*, 87
mexicana, 87
venenosa, 82
virosa, 88
var. *maculata*, 87
Cicutaria maculata, 87
Cirsium, 192
altissimum, 193
horridulum, 193
forma *elliottii*, 193
var. *elliottii*, 193
var. *megacanthum*, 193
var. *vittatum*, 194
lanceolatum, 197
var. *vulgare*, 197
lecontei, 194
megacanthum, 193
muticum, 195
nuttallii, 195
pinetorum, 193
repandum, 197
revolutum, 196
smallii, 193
forma *purpureum*, 194
virginianum, 196
forma *revolutum*, 196
vittatum, 193
vulgare, 197
Cladanthus, 197
mixtus, 198
Cnicus, 174
altissimus, 193
benedictus, 174
glaber, 195
horridulus, 193
lanceolatus, 197
lecontei, 194
muticus, 195
nuttallii, 195
repandus, 197
spinosissimus, 193
virginianus, 196
Coelestina maritima, 135
Coleosanthes cordifolius, 167
Complaya trilobata, 378
Compositae, 117
Conium maculatum, 115
Conoclinium, 198
coelestinum, 198
dichotomum, 199
flaccidum, 199
nepetifolium, 199
Conopholis, 19
americana, 19
Conophora
atriplicifolia, 143
diversifolia, 144
floridana, 144
ovata, 145
Conopodium canadense, 89
Conradia lecontei, 21
Conyza, 199
asteroides, 356
baccharis, 333
bonariensis, 200
carolinensis, 334
canadensis, 200
var. *pusilla*, 201
cinerea, 215
comphorata, 334
floribunda, 200
halimifolia, 150
laevigata, 201
odorata, 335
parva, 201
purpurascens, 335
pycnostachya, 342
ramosissima, 201
sagittalis, 336
squamata, 399
sumatrensis, 200
var. *floribunda*, 200
Conyzanthus squamatus, 399
Conyzella
canadensis, 200
linifolia, 200
Coreopsis, 202
alba, 155
alternifolia, 413
angustata, 206
angustifolia, 204, 270
aurea, 158, 160
var. *incisa*, 158
var. *leptophylla*, 158
auriculata, 210
bakeri, 203
basalis, 203
var. *typica*, 203
coronata, 159, 208
debilis, 208
delphinifolia var. *rigida*, 207
discoidea, 156
falcata, 204
floridana, 204
gladiata, 204
var. *linifolia*, 207
grandiflora, 205
helianthoides, 204
integrifolia, 205
lanceolata, 205
var. *angustifolia*, 205
var. *glabella*, 205
var. *succisifolia*, 205
leavenworthii, 206
var. *curtissii*, 206
var. *garberi*, 206
var. *lewtonii*, 206
var. *typica*, 206
leucantha, 155
leucanthema, 155
lewtonii, 206
linifolia, 207
longifolia, 204
major, 207
forma *oemleri*, 207
var. *oemleri*, 207
var. *rigida*, 207
var. *stellata*, 207
mitis, 158
nudata, 208
nuecensis, 208
oemleri, 207
palustris, 204
perfoliata, 158
pubescens, 208
var. *debilis*, 208
var. *robusta*, 208
var. *typica*, 208
rigida, 207
senifolia, 207
var. *rigida*, 207
var. *stellata*, 207
stellata, 207
tinctoria, 209
trichosperma, 160
tripteris, 209
verticillata, 210

Coreopsoides lanceolata, 205
Coriandrum, 88
  globosum, 88
  maculatum, 87
  majus, 88
  sativum, 88
    var. *vulgare*, 88
Cosmos, 210
  bipinnatus, 210
    var. *typicus*, 210
  caudatus, 211
  sulphureus, 211
    var. *typicus*, 211
Cotula
  alba, 0219
  mexicana, 376
Crantzia
  carolinensis, 99
  chinensis, 99
  lineata, 99
Crantziola
  carolinensis, 99
  lineata, 99
Crassina
  elegans, 423
  multiflora, 424
  peruviana, 424
Crassocephalum, 212
  crepidioides, 212
  diversifolium var. *crepidioides*, 212
  sonchifolium, 223
Crepis, 213
  foetida, 213
  japonica, 422
  pulchra, 213
    subsp. *typica*, 213
Critamus dauricus, 107
Critonia kuhnia, 167
Croptilon, 213
  divaricatum, 214
Cryptotaenia, 88
  canadensis, 89
Cunigunda
  perfoliata, 236
  purpurea, 245
Cupularia viscosa, 215
Cyanthillium, 214
  cinereum, 214, 215
Cyanus
  arvensis, 175
  cyanus, 175
  segetus, 175
  vulgaris, 175
Cyclospermum, 89
  ammi, 115
  leptophyllum, 89, 90
Cynthia, 290
  dandelion, 290

Dasistoma
  bignoniiflorum, 16
  flavum, 16
  pectinatum, 16
  quercifolium, 17
Dasyanthus conglobatus, 340
Daucus, 90
  carota, 90
    forma *normalis*, 90
    subsp. *mediterraneus*, 90
    subvar. *typicus*, 90
    var. *genuinus*, 90
    var. *sylvestris*, 90
    var. *typicus*, 90
  communis, 90
  divaricatus, 108
  esculentus, 90
  laevis, 80
  pusillus, 91
  scadiophylus, 91
  visnaga, 80
  vulgaris, 90
Deringa canadensis, 89
Didymopanax papyrifer, 76
Diodonta
  coronata, 159
  leptophylla, 158
  mitis, 158
Diomedea
  bidentata, 165
  glabrata, 164
  indentata, 164
Dipligon
  graminifolium, 331
  hyssopifolium, 186
  marianum, 190
  nuttallianum, 166
  scabrellum, 190
  trichophyllum, 186
Diplopappus
  amygdalinus, 217
  annuus, 226
  divaricatus, 214
  dubius, 226
  graminifolius, 331
  lanatus, 186
  linariifolius, 286
  marianus, 190
  obovatus, 316
  trichophyllus, 186
Diplosastera tinctoria, 209
Diplostephium
  linariifolium, 286
  obovatum, 316
Discopleura
  capillacea, 102
  nuttallii, 103
Distreptus spicatus, 338
Dittrichia, 215
  viscosa, 215
Doellingeria, 216
  amygdalina, 217
  humilis, 216
  infirma, 216
  obovata, 316
  reticulata, 316
  sericocarpoides, 217
  umbellata, 217
    subsp. *humilis*, 216
    var. *latifolia*, 217
Doria altissima, 364
Doronicum
  acaule, 142
  nudicaule, 142
  ramosum, 228
Dortmanna
  amoena, 39
  boykinii, 40
  brevifolia, 40
  cardinalis, 41
  feayana, 41
  glandulosa, 43
  nuttallii, 44
  paludosum, 44
  puberula, 45
Dracopis, 217
  amplexicaulis, 218
Drepanophyllum lineare, 107
Dysmicodon perfoliatum, 47

Dysodium divaricatum, 310
Dyssodia tenuiloba, 410

Echinacea, 218
atrorubens, 353
var. *graminifolia*, 349
purpurea, 218
Echinopanax papyrifer, 76
Eclipta, 219
adpressa, 219
alba, 219
forma *erecta*, 219
forma *prostrata*, 219
var. *erecta*, 219
var. *prostrata*, 219
dubia, 219
erecta, 219
longifolia, 219
prostrata, 219
strumosa, 219
Elephantopus, 220
carolinianus, 220
elatus, 221
var. *intermedius*, 221
nudatus, 221
scaber var. *carolinianus*, 220
spicatus, 338
tomentosus, 222
Eleutheranthera divaricata, 310
Eleutherococcus, 69
trifoliatus, 69
Emilia, 222
coccinea, 223
fosbergii, 222
javanica, 223
sonchifolia, 223
var. *javanica*, 223
var. *typica*, 223
Endorima, 151
atropurpurea, 151
uniflora, 252
Endyra, 224
fluctuans, 224
Ennepta
coriacea, 29
myricoides, 30
Epifagus, 20
americana, 20
virginiana, 20
Erechtites, 224
hieraciifolius, 224
var. *glabrescens*, 225
var. *typicus*, 225
Erigeron, 225
annuus, 226
subsp. *strigosus*, 228
var. *ramosus*, 228
var. *typicus*, 226
bellioides, 226
bonariensis, 200
var. *floribundus*, 200
camphoratus, 334
canadensis, 200
var. *pusillus*, 201
carolinianus, 243
expansus, 387
floribundus, 200
laevigatus, 201
linifolius, 200
nervosus, 331
paniculatus, 200
philadelphicus, 227
var. *quercinus*, 228
pulchellus, 227
var. *typicus*, 227
pusillus, 200
quercifolius, 228
ramosus, 228
var. *beyrichii*, 228
ruderale, 200
strigosus, 228
var. beyrichii, 228
var. *typicus*, 228
tenuis, 229
vernus, 229
viscosus, 215
Erinus americanus, 17
Eriocarpum
floridana, 347
megacephalum, 347
phyllocephalum, 347
rubiginosum var. *phyllocephalum*, 347
tracyi, 347
Eriolepis lanceolata, 197
Eryngium, 92
aquaticum, 93
var. *floridanum*, 93
var. *normale*, 93
var. *ravenelii*, 93
var. *synchaetum*, 97
aromaticum, 93
baldwinii, 94
cervatesii, 97
cuneifolium, 94
divaricatum, 95
floridanum, 93
foetidum, 95
gracile, 94
integrifolium, 95
var. *typicum*, 95
longifolium, 97
mettaueri, 93
prostratum, 96
ravenelii, 93
synchaetum, 96
virginianum, 93
yuccifolium, 96
var. *synchaetum*, 96
Erythremia aphylla, 307
Ethulia
bidentis, 248
uniflora, 354
Euonymus tobira, 67
Eupatoriadelphis, 245
fistulosus, 245
purpureus, 245
Eupatoriophalacron album, 219
Eupatorium, 230
ageratoides, 132
album, 231
var. *petaloideum*, 236
var. *typicum*, 231
altissimum, 231
altissimum, 132
anomalum, 239
aromaticum, 133
var. *incisum*, 133
capillifolium, 232
var. *leptophyllum*, 234
catarium, 338
chapmanii, 236
clematideum, 338
coelestinum, 198
compositifolium, 232
connatum, 236
conyzoides, 134, 180
coronopifolium, 232
cuneifolium, 237
var. *semiserratum*, 238

eugenei, 239
fistulosum, 245
foeniculaceum, 232
var. *glabrum*, 234
var. *lateriflorum*, 232
var. *traganthes*, 232
foeniculoides, 232
frustratum, 179
glaucescens, 234
var. *leucolepis*, 234
heteroclinium, 180
hyssopifolium, 233
var. hyssopifolium, 233
var. laciniatum, 233
var. *linearifolium*, 234
var. *tortifolium*, 235
incarnatum, 250
incisum, 133
ivifolium, 179
var. *genuinum*, 179
jucundum, 133
laevigatum, 245
lecheifolium, 233
leptophyllum, 234
leucolepis, 234
linearifolium, 234
maculatum, 246
marrubium, 237
mikanioides, 235
mohrii, 235
odoratum, 180
ovatum, 237
parviflorum, 238
pectinatum, 239
perfoliatum, 236
petaloideum, 236
pilosum, 237
pinnatifidum, 239
pubescens, 237
purpureum, 245
subsp. *angustifolium*, 245
subsp. *typicum*, 245
var. *angustifolium*, 245
recurvans, 235
rotundifolium, 237
subsp. *ovatum*, 237
var. *ovatum*, 237
var. *pubescens*, 237
var. *saundersii*, 237
var. *scabridum*, 237
var. *typicum*, 237
rugosum, 132
scabridum, 237
scandens, 314
semiserratum, 238
serotinum, 238
smithii, 239
speciosum, 299
suaveolens, 133
teucrifolium, 237
var. *verbenifolium*, 237
torreyanum, 233
tortifolium, 235
tracyi, 133
urticifolium, 132
var. *clematideum*, 338
verbenifolium, 237
var. *saundersii*, 237
villosum, 289
×anomalum, 239
×pinnatifidum, 239
Eurybia, 239
chapmanii, 384
eryngiifolia, 240
hemispherica, 240
paludosa, 241
spectabilis, 242
spinulosa, 242
Euryops, 242
chrysanthemoides, 242
Euthamia, 243
caroliniana, 243
fastigiata, 244
graminifolia, 244
var. *hirtipes*, 244
gymnospermoides, 244
hirtipes, 244
leptocephala, 244
microcephala, 243
microphylla, 243
minor, 243
tenuifolia, 243
var. *glutinosa*, 243
Eutrochium, 245
fistulosum, 245
maculatum, 246
purpureum, 245

Facelis, 246
apiculata, 246
retusa, 246
var. *typica*, 246
Falcaria dahurica, 107
Fatsia papyrifer, 76
Ferula
canadensis, 98
graveolens, 80
villosa, 82
Filago, 247
germanica, 247
rotundata, 247
vulgaris, 247
Flamaria coccinea, 21
Flaveria, 247
bidentis, 248
floridana, 248
latifolia, 249
linearis, 249
var. *latifolia*, 249
pinetorum, 248
repanda, 249
tenuifolia, 249
trinervia, 249
×*latifolia*, 249
Fleischmannia, 250
incarnata, 250
Foeniculum, 97
commune, 97
foeniculum, 97
officinale, 97
vulgare, 97
forma *officinale*, 97
Fredia radiata, 66

Gaillardia, 250
aestivalis, 251
var. *flavirens*, 251
amara, 263
bicolor, 251
var. *drummondii*, 251
var. *vulgaris*, 251
drummondii, 251
fimbriata, 267
lanceolata, 251
var. *flavovirens*, 251
lutea, 251
picta, 251
pulchella, 251
forma *picta*, 251
var. *drummondii*, 251

var. *picta*, 251
Galinsoga, 252
ciliata, 252
parviflora var. *quadriradiata*, 252
quadriradiata, 252
Gamochaeta, 253
americana, 257
antillana, 254
argyrinea, 254
chionesthes, 255
coarctata, 255
falcata, 257
pensylvanica, 256
purpurea, 256
simplicicaulis, 257
spicata, 255
subfalcata, 254
Gamolepis chrysanthemoides, 242
Garberia, 257
fruticosa, 258
heterophylla, 258
Georgina bipinnata, 210
Gerardia, 2
afzelia, 24
aphylla, 4
var. *filicaulis*, 6
var. *grandiflora*, 13
cassioides, 24
divaricata, 5
erecta, 12
fasciculata, 5
subsp. *peninsularis*, 6
subsp. *typica*, 6
filicaulis, 6
flammea, 21
flava, 16
var. *reticulata*, 16
georgiana, 8
harperi, 8
holmiana, 13
keyensis, 11
laxa, 9
leucanthera, 14
linifolia, 9
maritima, 14
subsp. *grandiflora*, 10
var. *grandiflora*, 10
var. *major*, 10
mettaueri, 5
var. *clausa*, 5
var. *nuda*, 6
microphylla, 13
nuda, 6
obtusifolia, 11
pinetorum, 8
parvifolia, 11
pectinata, 16
pedicularia var. *pectinata*, 16
plukenetii, 11
var. *microphylla*, 13
pulchella var. *delicatula*, 8
pulcherrima, 12
purpurea, 12
var. *carteri*, 12
var. *fasciculata*, 5
quercifolia, 17
setacea, 13
skinneriana, 14
spiciflora, 11
stenophylla, 13
tenella, 11
tenuifolia, 14
subsp. *leucanthera*, 14
subsp. *typica*, 14
var. *filiformis*, 14
var. *leucanthera*, 14
Gerbera, 258
albicans, 177
jamesonii, 259
jamesonii, 259
walteri, 178
Gifola, 247
germanica, 247
vulgaris, 247
Glebionis, 259
coronaria, 259
segeta, 260
Glyceria repanda, 85
Gnaphalium
americanum, 257
antillanum, 254
coarctatum, 255
dioicum var. *plantaginifolium*, 139
domingense, 339
falcatum, 257
germanicum, 247
helleri, 340
luteoalbum, 340
var. *normalis*, 340
obtusifolium, 341
var. *helleri*, 340
pensylvanicum, 256
peregrinum, 256
plantaginifolium, 139
polycephalum, 340
purpureum, 256
var. *falcatum*, 257
var. *normale*, 256
var. *simplicicaule*, 257
var. *spatulatum*, 256
var. *spicatum*, 255
retusum, 246
simplicicaule, 257
spatulatum, 256
spicatum, 255
suaveolens, 336
subfalcatum, 254
Gonotheca helianthoides, 409
Goodeniaceae, 53
Guizotia abyssinica, 424
Gymnosperma nudatum, 249
Gymnostyles, 374
anthemifolia, 375
chilensis, 375
lusitanica, 376
pterosperma, 375
stolonifera, 375
Gynoxys berlandieri, 342
var. *cordifolia*, 342
Gynura, 260
aurantiaca, 260
crepidioides, 212

Hallomuellera lineata, 99
Haplocarpha, 261
lyrata, 261
Haplopappus
divaricatus, 214
megacephalus, 347
phyllocephalus, 347
subsp. *megacephalus*, 347
subsp. *typicus*, 347
var. *genuinus*, 347
var. *megacephalus*, 347
rubiginosus var. *phyllocephalus*, 347
Harpalyce virgata, 314

Hartwrightia, 261
  floridana, 261
Hasteola, 262
  robertiorum, 262
Hedera, 69
  algeriensis, 70
  arborea, 70
  communis, 70
  diversifolia, 70
  helix, 70
    var. *euhelix*, 70
    var. *europaea*, 70
    var. *typica*, 70
    var. *vulgaris*, 70
  poetarum, 70
  poetica, 70
Hederaceae, 68
Hedypnois taraxacum, 408
Heleastrum
  chapmanii, 240, 384
  hemisphericum, 240
  spinulosum, 242
Helenia
  autumnalis, 264
  decurrens, 264
Heleniastrum
  autumnale, 264
  brevifolium, 264
  fimbriatum, 267
  helenium, 267
  nudiflorum, 265
  parviflorum, 264
  tenuifolium, 263
  vernale, 267
Helenium, 263
  aestivale, 251
  amarum, 263
  anceps, 265
  autumnale, 264
    var. *normale*, 264
    var. *parviflorum*, 264
  brachypodium, 265
  brevifolium, 264
  decurrens, 267
  discovatum, 267
  drummondii, 265
  fimbriatum, 267
  flexuosum, 265
  floridanum, 265
  helenium, 267
  leptopoda, 267
  nudiflorum, 265
  nuttallii, 267
  parviflorum, 264
  pinnatifidum, 266
  puberulum, 266
  quadridentatum, 267
  tenuifolium, 263
  vernale, 267
Helianthella
  grandiflora, 327
  tenuifolius, 327
Helianthus, 268
  agrestis, 269
    forma *almae*, 269
    forma *oswaldii*, 269
  angustifolius, 270
  annuus, 270
    var. *argophyllus*, 271
    var. *debilis*, 272
  argophyllus, 271
  aristata, 414
  atrorubens, 271
    subsp. *heterophyllus*, 274
    subsp. *radula*, 276
    var. *normalis*, 271
  australis, 278
  carnosus, 271
  cucumerifolius, 273
  debilis, 272
    forma *cucumerifolius*, 273
    subsp. cucumerifolius, 273
    subsp. debilis, 272
    subsp. *praecox*, 279
    subsp. *tardiflorus*, 273
    subsp. vestitus, 273
    var. *cucumerifolius*, 273
    var. *tardiflorus*, 293
    var. *vestitus*, 273
  divaricatus, 273
  elongatus, 274
  floridanus, 274
  fulgidus, 349
  giganteus, 279
  glaucus, 279
  heterophyllus, 274
  hirsutus, 275
  hirtus, 350
  laciniatus, 350
  laevis, 157
  microcephalus, 275
  mollis, 279
  montanus, 278
  occidentalis, 276
  parviflorus, 275
  pinnatus, 346
  platycephalus, 270
  praecox, 279
  radula, 276
  resinosus, 277
  simulans, 277
  strumosus, 278
    subsp. *tomentosus*, 278
  tomentosus, 278
  tuberosus, 278
    forma *oswaldiae*, 269
    var. *typicus*, 278
  undulatus, 274
  vestitus, 273
  ×glaucus, 279
Heliopsis, 279
  gracilis, 279
  helianthoides
    var. gracilis, 279
    var. *helianthoides*, 280
    var. *scabra*, 280
  laevis, 157
    var. *gracilis*, 279
  minor, 280
Helosciadium leptophyllum, 89
Heptapleurum arboricola, 76
Heterotheca, 280
  adenolepis, 329
  aspera, 329
  falcata, 330
  flexuosa, 330
  floridana, 184
  gossypina, 186
  graminifolia, 331
    var. *tracyi*, 331
  hyssopifolia, 186
    var. *subulata*, 191
  lamarckii, 280
  latifolia, 280
  latisquamea, 188
  mariana, 190
    subsp. *floridana*, 184
  microcephala, 331
  nervosa, 331
  oligantha, 332

pilosa, 166
scabra, 280
scabrella, 190
subaxillaris, 280
subsp. *latifolia*, 280
var. *latifolia*, 280
var. *procumbens*, 280
trichophylla, 186
Hieracium, 281
argyraenum, 282
asperum, 377
aurantiacum, 328
floridanum, 281, 282
gronovii, 281
javanicum, 223
megacephalon, 282
oleraceum, 377
sonchus, 377
venosum, 282
×*marianum*, 282
Hippia
minuta, 376
stolonifera, 376
Hippobroma, 37
longiflora, 37
Hydrocotyle, 71
ambigua, 74
asiatica, 85
var. *floridana*, 85
australis, 74
bonariensis, 71
var. *tribotrys*, 74
bowlesioides, 72
canbyi, 74
chinensis, 99
erecta, 85
ficaroides, 85
interrupta, 71
lineata, 99
polystachya var. *triradiata*, 74
ranunculoides, 72
forma *genuina*, 72
var. *lobata*, 72
repanda, 85
sarmentosa, 85
sibthorpioides, 72
sinensis, 99
tribotrys, 74
umbellata, 73
var. *ambigua*, 74
var. *bonariensis*, 71
var. *umbellulata*, 73
umbellulata, 73
verticillata, 73
var. triradiata, 74
var. verticillata, 73
vulgaris var. *verticillata*, 73
Hyloseris
caroliniana, 291
virginica, 291
Hymenatherum tenuilobum, 410
Hymenopappus, 282
caroliniensis, 283
integrifolius, 321
scabiosaeus, 283
Hypochaeris, 283
albiflora, 285
brasiliensis
var. *albiflora*, 285
var. *tweediei*, 284
chillensis, 284
glabra, 284
infesta, 285
microcephala var. albiflora, 285
radicata, 285
var. *typica*, 285
var. *vulgaris*, 285
stellata, 285
tweediei, 284

Idianthes pulchra, 213
Ilex, 26
ambigua, 27
var. *coriacea*, 27
var. *monticola*, 34
amelanchier, 27
angustifolia, 28, 29
arenicola, 32
forma *oblanceolata*, 33
forma *sebringensis*, 33
var. *obovata*, 33
var. *paucidens*, 33
var. *transiens*, 33
buswellii, 27
caroliniana, 27, 28
var. *jejuna*, 27
cassena, 34
cassine, 28
subsp. *myrtifolia*, 29
var. *angustifolia*, 29
var. cassine, 28
var. *latifolia*, 28
var. myrtifolia, 29
var. *parvifolia*, 29
coriacea, 28
cumulicola, 32
curtissii, 30
cuthbertii, 30
dahoon, 28
var. *angustifolia*, 29
var. *grandifolia*, 28
var. *laurifolia*, 29
var. *myrtifolia*, 29
var. *parvifolia*, 29
decidua, 30
subsp. *longipes*, 31
var. *curtissii*, 30
var. *longipes*, 31
dubia
forma *chapmaniana*, 27
forma *pseudoambigua*, 27
floridana, 29
glabra, 30
forma *leucocarpa*, 30
krugiana, 31
laurifolia, 29
longipes, 31
lucida, 34
montana, 34
myrtifolia, 29
opaca, 32
forma *subintegra*, 32
subsp. *arenicola*, 32
var. arenicola, 32
var. opaca, 32
var. *subintegra*, 32
var. *xanthocarpa*, 32
pygmaea, 33
var. *subdentata*, 33
ramulosa, 29
verticillata, 33
vomitoria, 33
×*attenuata*, 34
Impla germanicum, 247
Inula
arabica, 343
argentea, 331
divaricata, 214
glandulosa, 190
graminifolia, 331

var. *tenuifolia*, 331
falcata, 330
gossypina, 186
mariana, 190
var. *falcata*, 330
scabra, 280
subaxillaris, 280
trichophylla, 186
viscosa, 215
Ionactis, 286
linariifolia, 286
Isopappus divaricatus, 214
Isotoma longiflora, 37
Iva, 287
angustifolia, 287
annua, 287
asperifolia, 289
cheiranthiifolia, 289
frutescens, 288
imbricata, 288
microcephala, 288
monophylla, 136

Kantemon angustifolium, 63
Kerneria
leucantha, 155
pilosa, 159
var. *discoidea*, 159
var. *radiata*, 155
tetragona, 159
Koanophyllon, 289
villosum, 289
Krigia, 289
caroliniana, 291
var. *leptophylla*, 291
cespitosa, 290
dandelion, 290
leptophylla, 291
oppositifolia, 291
virginica, 291
Kuhnia, 166
albicaulis, 167
eupatorioides, 167
var. *floridana*, 168
var. *gracilis*, 167
glabra, 167
kuhnia, 167
latifolia, 167
mosieri, 168
paniculata, 167
virgata, 167
Kyrstenia
altissima, 132
aromatica, 133
incarnata, 250
jucunda, 133
tracyi, 133

Lacinaria, 296
aspera, 298
chapmanii, 298
chlorolepis, 300
deamiae, 298
earlei, 306
elegans, 299
elegantula, 299
garberi, 300
gracilis, 301
graminifolia, 307
laevigata, 307
laxa, 301
nashii, 300
ohlingerae, 302
pauciflora, 303
scariosa, 307
var. *aspera*, 298
var. *intermedia*, 298
var. *media*, 298
var. *squarrulosa*, 306
secunda, 303
spicata, 304
var. *resinosa*, 305
squarrosa, 305
tenuifolia, 306
Lactuca, 291
acuminata, 293
canadensis, 292
var. *elongata*, 292
var. *typica*, 292
elongata, 292
var. *graminifolia*, 293
var. *integrifolia*, 292
floridana, 292
var. *villosa*, 293
graminifolia, 293
hirsuta, 294
intybacea, 295
oleracea, 377
serriola, 293
villosa, 293
Lagascea, 294
mollis, 294
Lapsana pulchra, 213
Lasallea
adnata, 382
caroliniana, 383
concolor, 385
patens, 393
phyllolepis, 397
sericea subsp. *pratensis*, 397
walteri, 404
Launaea, 295
intybacea, 295
Laurentia longiflora, 37
Leachia lanceolata, 205
Legouzia
biflora, 46
perfoliata, 47
Leighia bicolor, 270
Leioligo
caesia, 363
pulverulenta, 370
Leontodon
carolinianus, 344
chillense, 284
dandelion, 290
erythrospermus, 408
laevigatus, 408
officinalis, 408
taraxacum, 408
subsp. *laevigatus*, 408
subsp. *officinalis*, 408
var. *genuinus*, 408
var. *laevigatus*, 408
var. *vulgare*, 408
tomentosus, 177
vulgaris, 408
Lepachys
columnaris, 345
columnifera, 345
pinnata, 346
pinnatifida, 346
Lepia
multiflora, 424
pauciflora, 424
Lepiactis virgata, 372
Leptamnium virginianum, 20
Leptilon
bonariense, 200
canadense, 200

var. *pusillum*, 200
linifolium, 200
pusillum, 200
Leptocaulis
divaricata, 108
echinata, 108
Leptoclinium fruticosum, 258
Leptopoda
brachypoda, 265
brevifolia, 264
decurrens, 267
fimbriata, 265, 266, 267
floridana, 267
helenioides, 267
helenium, 267
pinnatifida, 266
puberula, 267
var. *pinnatifida*, 266
Leria albicans, 177
Leucacantha cyanus, 175
Leucanthemum, 295
leucanthemum, 296
vulgare, 296
var. *baumgartnerianum*, 296
Liatris, 296
amplexicaulis, 172
aspera, 298
var. *intermedia*, 298
var. *typica*, 298
botrys, 301
chapmanii, 298
var. *longifolia*, 298
corymbosa, 171
earlei, 306
elegans, 299
var. *kralii*, 299
var. *typica*, 299
elegantula, 299
fruticosa, 258
garberi, 300
gholsonii, 301
gracilis, 301
var. *gholsonii*, 301
graminifolia, 307
var. *elegantula*, 299
laevigata, 307
laxa, 301
macrostachya, 304
odoratissima, 172
ohlingerae, 302
paniculata, 173
patens, 302
pauciflora, 303
var. pauciflora, 303
var. secunda, 303
pilosa, 307
var. *gracilis*, 301
provincialis, 303
quadriflora, 307
resinosa, 305
savannensis, 304
scariosa, 307
var. *squarrulosa*, 306
secunda, 303
spicata, 304
var. *macrostachya*, 304
var. *resinosa*, 304
var. *savannensis*, 304
var. *typica*, 305
squarrosa, 305
var. *gracilenta*, 305
var. *typica*, 305
squarrulosa, 306
tenuifolia, 306
var. *laevigata*, 307
var. quadriflora, 307
var. tenuifolia, 306
Ligusticum, 98
barbinode, 109
canadense, 98
foeniculum, 97
Lilaeopsis, 99
attenuata, 100
carolinensis, 99
chinensis, 99
lineata, 99
Limnanthemum
aquaticum, 49
cristatum, 51
grayanum, 51
humboldtianum, 52
indicum, 52
lacunosum, 49
peltatum, 52
trachyspermum, 49
Litrisa, 170
carnosa, 170
Lobelia, 37
amoena, 39
var. *glandulifera*, 42
var. *obtusata*, 42
apalachicolensis, 39
boykinii, 40
brevifolia, 40
cardinalis, 40
var. *meridionalis*, 41
cliffortiana, 46
coccinea, 40
crassiuscula, 43
elongata, 46
feayana, 41
flaccidifolia, 42
floridana, 42
frutescens, 54
georgiana, 42
glandulifera, 42
glandulosa, 43
var. *laevicalyx*, 43
homophylla, 43
longiflora, 37
nudicaulis, 44
nuttallii, 44
paludosa, 44
var. *floridana*, 42
plumieri, 54
puberula, 45
forma *simulans*, 45
var. *glabella*, 45
var. *laeviuscula*, 45
var. *simulans*, 45
rogersii, 45
sericea, 54
var. *typica*, 54
spicata, 46
taccada, 54
xalapensis, 46
Lobeliaceae, 35
Lomaxeta verrucosa, 321
Lonicera, 62
acuminata var. *japonica*, 62
angustifolia, 63
flamea, 63
japonica, 62
var. *typica*, 62
sempervirens, 63
var. *major*, 63
var. *ovata*, 63
symphoricarpos, 64
Lorentea humifusa, 325

Lygodesma, 307
aphylla, 307

Machaeranthera phyllocephala, 347
var. *megacephala*, 347
Macranthera, 21
flammea, 21
fuschioides var. *lecontei*, 21
lecontei, 21
Marsea
bonariensis, 200
canadensis, 201
Marshallia, 308
angustifolia, 308
var. *cyananthera*, 308
graminifolia, 308
subsp. *tenuifolia*, 308
var. *cyananthera*, 308
lanceolata, 309
var. *platyphylla*, 309
obovata, 309
var. *platyphylla*, 309
var. *scaposa*, 309
ramosa, 310
tenuifolia, 308
Maruta
cotula, 140
foetida, 140
vulgaris, 140
Matricaria
asteroides, 163
coronaria, 259
glastifolia, 163
leucanthemum, 296
Melampodium, 310
australe, 127
divaricatum, 310
humile, 128
ovatifolium, 310
Melanthera, 311
amellus var. *subhastata*, 312
angustifolia, 312
var. *subhastata*, 312
aspera, 311
var. *glabriuscula*, 312
var. *subhastata*, 312
brevifolia, 312
deltoidea, 311
hastata, 311
subsp. *lobata*, 311
var. *lobata*, 312
var. *parvifolia*, 312
lanceolata, 312
ligulata, 312
linnaei, 311
lobata, 312
nivea, 311
parvifolia, 312
radiata, 312
trilobata, 311
urticifolia, 311
Menyanthaceae, 47
Menyanthes
cristata, 51
indica, 52
trachysperma, 49
Mesadenia, 142
angustifolia, 145
atriplicifolia, 143
difformis, 144
diversifolia, 144
elliottii, 145
floridana, 144
lanceolata, 145
ovata, 145
sulcata, 145
Mesoligus subulatus, 400
Meum foeniculum, 97
Meyera fluctuans, 224
Microtinus odoratissimus, 60
Mikania, 312
batatifolia, 314
cordifolia, 313
micrantha, 313
forma *typica*, 313
scandens, 314
Mirosolia diversifolia, 411
Mulgedium
acuminatum, 293
floridanum, 292
integrifolium, 292
lyratum, 292
villosum, 293
Mysteron pulchellum, 227
Mylanche virginiana, 20
Myrrhis
canadensis, 89
pecten-veneris, 106
procumbens, 86
Nabalus, 314
albus var. *serpentaria*, 315
autumnalis, 315
fraseri, 315
serpentaria, 315
virgatus, 315
Neactelis strigosa, 276
Nemopanthus ambiguus, 27
Neoceis hieraciifolia, 224
Neurolaena lobata, 336
Neurophyllum longifolium, 101
Nintooa, 62
japonica, 62
Nocco mollis, 294
Northopanax guilfoylei, 74
Nymphoides, 49
aquatica, 49
cordata, 50
cristata, 50, 51
grayana, 50
humboldtiana, 51
indica, 52
lacunosa, 50, 51
peltata, 52

Obeliscaria
columnaris, 345
pinnata, 346
Oclemena, 315
reticulata, 316
Oedera trinervia, 249
Oenanthe
carolinensis, 111
filiformis, 111
rigidius, 100
teretifolia, 111
Oligosporus campestris subsp. *caudatus*, 147
Onopordum, 316
acanthium, 317
Ooclinium, 179
rigidum, 180
Ormenis
bicolor, 198
mixta, 198
Orobanchaceae, 1
Orobanche, 22
americana, 19
minor, 22
uniflora, 15

var. *typica*, 15
virginiana, 20
Osmia, 179
conyzoides, 180
frustrata, 179
heteroclinia, 180
ivifolia, 179
odorata, 180
Osteospermum uvedalia, 359
Othake texana, 321
Oxypolis, 100
caroliniana, 111
denticulata, 100
filiformis, 111
subsp. *greenmanii*, 111
greenmanii, 111
rigidior, 100
ternata, 101

Packera, 317
anonyma, 317, 318
aurea, 318
glabella, 319
obovata, 319
paupercula, 320
tomentosa, 320
Palafoxia, 320
fastigiata, 321
feayii, 321
integrifolia, 321
rosea var. *ambigua*, 322
texana, 321
var. *ambigua*, 322
Paleista procumbens, 319
Paleolaria fastigiata, 321
Panax
aculeatus, 69
papyrifer, 76
quinquefolius, 77
Parthenium, 322
hysterophorus, 322
pinnatifidum, 322
Pascalia, 323
glauca, 323
Pastinaca
anethum, 80
denticulata, 100
graveolens, 80
rigidior, 100
sativa, 115
Pecten veneris, 106
Pectinaria vulgaris, 106
Pectis, 324
ciliaris, 326
floridana, 326
glaucescens, 324
humifusa, 325
leptocephala, 324
lessingii, 324
linearifolia, 325
linifolia, 326
prostrata, 325
×floridana, 326
Pedicularis, 22
canadensis, 23
forma *typica*, 23
var. *dobbsii*, 23
Pentagonia
biflora, 46
perfoliata, 47
Periclymenum sempervirens, 63
var. *latifolium*, 63
Persoonia
angustifolia, 308
lanceolata, 309
Petroselinum, 101
crispum, 102
hortense
forma *crispum*, 102
var. *crispum*, 102
petroselinum var. *crispum*, 102
sativum
subvar. *crispum*, 102
var. *crispum*, 102
Peucedanum
anethum, 80
graveolens, 80
rigidium, 100
teretifolium, 111
ternatum, 101
Phaecasium
lampsanoides, 213
pulchrum, 213
Phaethusa, 413
americana, 416
borealis, 416
laciniata, 416
occidentalis, 416
virginica, 416
Phalacroloma
acutifolium, 226
annuum, 226
beyrichii, 228
Phelypaea biflora, 15
Phenianthus, 62
sempervirens, 63
Phoebanthus, 326
grandiflorus, 327
tenuifolius, 327
Phyteumopsis
angustifolia, 308
lanceolata, 309
Pilosella, 327
aurantiacum, 328
Pimpinella leptophyllum, 89
Pinardia coronaria, 259
Pittosporaceae, 66
Pittosporum, 66
pentandrum, 67
tobira, 67
Pityopsis, 328
adenolepis, 329
aequilifolia, 331
argentea, 331
aspera, 329
var. *adenolepis*, 329
falcata, 330
flexuosa, 330
graminifolia, 331
forma *latifolia*, 331
var. *aequilifolia*, 331
var. *latifolia*, 331
var. *microcephala*, 331
var. *tenuifolia*, 331
var. *tracyi*, 331
latifolia, 331
microcephala, 331
nervosa, 331
var. *tracyi*, 331
oligantha, 332
Placus
odoratus, 335
polycephalus, 353
purpurascens, 335
tenuifolia, 331
tracyi, 331
Pluchea, 332
baccharis, 333
bifons, 336

camphorata, 334
caroliniensis, 334
foetida, 334
var. *imbricata*, 334
imbricata, 334
longifolia, 335
odorata, 335
var. *normalis*, 335
var. *succulenta*, 335
purpurascens, 335
var. *succulenta*, 335
quitoc, 336
rosea, 333
sagittalis, 336
suaveolens, 336
symphytifolia, 336
tenuifolia, 335
Polymnia, 336
canadensis, 337
carnosa, 378
laevigata, 337
tetragonotheca, 409
uvedalia, 359
var. *densipilis*, 359
var. *floridana*, 360
var. *genuina*, 359
Polymniastrum, 359
uvedalia, 359
Polypteris, 320
integrifolia, 321
texana, 321
Polyscias, 74
guilfoylei, 74
Porcellites radicatus, 285
Praxelis, 337
clematidea, 337
Prenanthes
aphylla, 307
autumnalis, 314
fraseri, 315
japonica, 422
javanica, 223
pulchra, 213
serpentaria, 315
virgata, 311
Prinos
ambiguus, 27
confertus, 33
coriaceus, 29
deciduus, 30
glaber, 30
gronovii, 33
verticillata, 33
Prionopsis chapmanii, 240
Prismatocarpus perfoliatus, 47
Pseudelephantopus, 338
spicatus, 338
Pseudognaphalium, 339
domingense, 339
helleri, 340
luteoalbum, 340
obtusifolium, 341
Pseudogynoxys, 341
berlandieri, 342
chenopodioides, 341
Pterocaulon, 342
pycnostachyum, 342
undulatum, 342
virgatum, 343
Pterophyton, 413
aristatum, 414
heterophyllum, 415
pauciflorum, 415
Ptilimnium, 102
capillaceum, 102
nuttallii, 103
Pulicaria, 343
arabica, 343
viscosa, 215
Pyrethrum laucanthemum, 296
Pyrrhopappus, 344
carolinianus, 344
var. *georgianus*, 344
georgianus, 344
multicaulis, 345
pauciflorus, 345

Rapuntium
amoenum, 39
cardinale, 40
coccineum, 40
glandulosum, 43
longiflorum, 37
nuttallianum, 44
paludosum, 44
puberulum, 45
var. *glabellum*, 45
Ratibida, 345
columnaris, 345
columnifera, 345
pinnata, 346
sulcata, 345
Rayjacksonia, 346
phyllocephala, 347
var. *megacephala*, 347
Rhinanthaceae, 1
Rhinanthus virginicus, 17
Ridan, 413
alternifolia, 413
paniculata, 417
Romeria lobelia, 54
Rotantha, 35
floridana, 36
robinsiae, 36
Rothia caroliniensis, 283
Rudbeckia, 347
amplexicaulis, 218
atrorubens, 353
auriculata, 348
bupleuroides, 351
columnaris, 345
columnifera, 345
divergens, 350
floridana, 350
var. *angustifolia*, 350
foliosa, 349
fulgida, 349
var. *auriculata*, 348
var. *spathulata*, 349
glabra, 352
graminifolia, 349
heterophylla, 350
hirta, 350
var. *angustifolia*, 350
var. *floridana*, 350
var. *pulcherrima*, 350
laciniata, 350
var. *heterophylla*, 350
maxima, 353
mohrii, 351
mollis, 351
nitida, 352
var. *longifolia*, 352
pinnata, 346
pinnatiloba, 352
purpurea, 218
radula, 276
serotina forma *pulcherrima*, 350
spathulata, 349

triloba, 352
var. *pinnatiloba*, 352
Russelia flammea, 21

Sachsia, 353
bahamensis, 353
polycephala, 353
Sambucus, 56
bipinnata, 56
canadensis, 56
forma *typica*, 56
subsp. *laciniata*, 56
var. *canadensis*, 57
var. *glabra*, 56
var. *laciniata*, 56
intermedia, 56
nigra
subsp. canadensis, 56
var. *canadensis*, 56
simpsonii, 56
Sanicula, 103
canadensis, 103
var. *floridana*, 104
var. *genuina*, 103
var. *marilandica*, 104
var. *typica*, 103
floridana, 104
gregaria, 105
marilandica, 104
var. *canadensis*, 103
odorata, 105
smallii, 105
triclinaris, 105
triclinium, 105
Sataria linearis, 101
Scaevola, 53
bela-modagam, 54
frutescens, 54
var. *sericea*, 54
lobelia, 54
plumieri, 54
sericea, 54
var. *taccada*, 54
taccada, 54
var. *sericea*, 54
Scandix, 106
pecten, 106
pectenifera, 106
pectiniformis, 106
pecten-veneris, 106
subsp. *eupecten-veneris*, 106
var. *genuina*, 106
procumbens, 86
rostrata, 106
ternata, 89
vulgaris, 106
Schefflera, 75
actinophylla, 75
arboricola, 76
Schwalbea, 23
americana, 23
var. *australis*, 23
australis, 23
Sclerolepis, 354
uniflora, 354
verticillata, 354
Sclerophyllum pulchrum, 213
Selinum
ammoides, 79
anethum, 80
coriandrum, 88
foeniculum, 97
graveolens, 80, 82
leptophyllum, 90
pecten, 106
visnaga, 80
Selloa nudata, 249
Senecio, 354
anonymus, 318
atriplicifolius, 143
aureus, 318
var. *angustifolia*, 318
var. *fastigiatus*, 319
var. *obovatus*, 319
var. *pauperculus*, 320
balsamitae
var. *crawfordii*, 320
var. *pauperculus*, 320
berlandieri, 342
brasiliensis var. tripartitus, 355
cannabinifolius, 355
carolinianus, 319
chenopodioides, 341
confusus, 342
crawfordii, 320
crepidioides, 212
elliottii, 319
glabellus, 319
hieraciifolius, 224
lobatus, 319
lyratus, 319
obovatus, 319
var. *elliottii*, 319
ovatus, 145
pauperculus, 320
subsp. *crawfordii*, 320
var. *crawfordii*, 320
var. *typicus*, 320
rotundifolius, 318
semperflorens, 355
smallii, 318
sonchifolius, 223
tomentosus, 320, 356
tripartitus, 355
vulgaris, 355
Senecioides cinerea, 215
Sericocarpus, 356
acutisquamosus, 357
asteroides, 356
bifoliatus, 357
var. *collinsii*, 357
collinsii, 357
conyzoides, 356
tortifolius, 357
var. *collinsii*, 357
Serinia, 290
cespitosa, 290
oppositifolia, 291
Seriola tweediei, 284
Serratula
cinerea, 215
noveboracensis, 420
speciosa, 299
spicata, 304
squarrosa, 305
Seruneum
glaucum, 323
trilobatum, 378
Seseli
ammi, 90, 115
foeniculum, 97
graveolens, 82
Setachna cyanus, 175
Seymeria, 24
cassioides, 24
pectinata, 25
subsp. *peninsularis*, 25
subsp. *typica*, 25
tenuifolia, 24

Sideranthus
megacephalus, 347
phyllocephalus, 347
Sigesbeckia
laciniata, 416
occidentalis, 416
Silphium, 357
angustum, 358
asteriscus, 358
var. *angustatum*, 358
var. *dentatum*, 358
var. *simpsonii*, 358
compositum, 358
subsp. *ovatifolium*, 359
subsp. *venosum*, 359
var. *michauxii*, 358
var. *ovatifolium*, 358
var. *venosum*, 359
dentatum, 358
var. *angustatum*, 358
gracile, 358
nuttallianum, 154
ovatifolium, 358
pumilum, 153
var. *tomentosum*, 153
reticulatum, 153, 358
scaberrimum, 358
simpsonii, 358
subacaule, 154
tomentosum, 153
trilobatum, 378
venosum, 359
Sison
ammi, 115
aureum, 114
canadensis, 89
capillaceum, 102
divaricatum, 108
majus, 79
trifoliatum, 114
Sitilias, 344
caroliniana, 344
multicaulis, 345
Sium, 105
apium, 82
canadense, 89
cicutifolium, 107
var. *lineare*, 107
denticulatum, 100
floridanum, 107
graveolens, 82
lineare, 107
var. *intermedium*, 107
rigidius, 100
suave, 107
var. *floridanum*, 107
teretifolium, 111
trifoliatum, 114
visnaga, 80
Smallanthus, 359
uvedalia, 359
Smyrnium
aurea, 114
cordatum, 109
Solidago, 260
altissima, 364
amplexicaulis, 362
angustifolia, 372
arguta
subsp. *caroliniana*, 362
var. *arguta*, 373
var. *boottii*, 373
var. caroliniana, 362
aspera, 371
aspericaulis, 364
auriculata, 362
boottii, 374
var. *brachyphylla*, 363
var. *caroliniana*, 362
var. *yadkinensis*, 362
brachyphylla, 363
caesia, 363
var. *zedia*, 363
canadensis
subsp. *altissima*, 364
var. *canadensis*, 374
var. scabra, 364
caroliniana, 243
celtidifolia, 371
chapmanii, 369
chrysopsis, 372
discoidea, 169
edisoniana, 366
elliottii, 366
var. *edisoniana*, 366
var. *typica*, 366
elliptica, 366
fastigiata, 244
fistulosa, 364
flavovirens, 372
gigantea, 365
subsp. *serotina*, 365
var. *leiophylla*, 365
var. *serotina*, 365
gracillima, 366
graminifolia, 244
var. *typica*, 244
hirsutissima, 364
hirtipes, 244
juncea, 374
lanceolata var. *minor*, 243
lateriflora, 391
latissimifolia, 366
leavenworthii, 367
mexicana, 371
michauxii, 243
microcephala, 243
microphylla, 243
milleriana, 369
minor, 243
mirabilis, 366
nashii, 367
nemoralis, 367
subsp. *haleana*, 367
var. *haleana*, 367
var. *typica*, 367
notabilis, 362
odora, 368
forma *inodora*, 368
subsp. *chapmanii*, 369
var. chapmanii, 369
var. *inodora*, 368
var. odora, 368
patula
subsp. strictula, 369
var. *patula*, 374
var. *stricula*ta, 369
pauciflosulosa, 182
petiolaris, 369
petiolata, 374
pilosa, 364
puberula
subsp. *pulverulenta*, 370
var. *puberula*, 374
var. pulverulenta, 370
pulverulenta, 370
rigidiuscula, 374

rugosa
subsp. aspera, 371
subsp. *rugosa*, 374
var. *aspera*, 371
var. *celtidifolia*, 371
salaria, 372
salicina, 369
scabra, 364
sempervirens, 371
subsp. *mexicana*, 371
var. *mexicana*, 371
var. *typica*, 371
serotina, 365
var. *serotina*, 374
var. *gigantea*, 365
speciosa, 374
var. *rigidiuscula*, 374
stricta, 372
subsp. *gracillima*, 366
var. *angustifolia*, 372
tarda, 362
tenuifolia, 243
tortifolia, 372
ulmifolia, 373
virgata, 372
viscosa, 215
yadkinensis, 362
Soliva, 374
anthemifolia, 375
lusitanica, 376
nasturtiifolia, 376
pterosperma, 375
sessilis, 375
stolonifera, 376
Sonchus, 376
acuminatus, 293
asper, 377
forma *typicus*, 377
var. *pungens*, 377
fallax var. *asper*, 377
floridanus, 292
javanicus, 223
oleraceus, 377
var. *asper*, 377
pallidus, 292
Sparganophorus verticillata, 354
Specularia
biflora, 46
perfoliata, 47
Spermolepis, 107
divaricata, 108
echinata, 108
inermis, 108
Sphagneticola, 378
trilobata, 378
Spilanthes
americana, 130
forma *longiinternodiata*, 131
var. *repens*, 130
nuttallii, 130
pusilla, 131
repens, 130
stolonifera var. *pusilla*, 131
Staehelina elegans, 299
Stelmanis scabra, 280
Stemmodontia trilobata, 378
Stenactis
annua, 226
subsp. *strigosa*, 228
beyrichii, 228
ramosa, 228
strigosa, 228
verna, 229
Stokesia, 379
cyanea, 379
laevis, 379
Streblanthus
gracilis, 94
heterophylus, 96
humilis, 94
tenuifolius, 94
Striga, 25
gesnerioides, 25
Symphoricarpos, 64
orbiculatus, 64
parviflorus, 64
symphicarpos, 64
vulgaris, 64
Symphyotrichum, 379
adnatum, 382
bahamense, 383
bracei, 401
carolinianum, 383
chapmanii, 384
concolor, 385
var. *devestitum*, 385
var. *plumosum*, 395
cordifolium, 405
divaricatum, 404
dumosum, 386
var. *pergracilis*, 386
var. *subulifolium*, 386
elliottii, 387
ericoides, 404
expansum, 387
fontinale, 388
georgianum, 389
grandiflorum, 404
kralii, 399
laeve
subsp. *concinnum*, 390
var. concinnum, 390
lanceolatum
var. *lanceolatum*, 404, 405
var. latifolium, 390
lateriflorum, 391
var. *horizontale*, 391
var. *spatelliforme*, 391
oolentangiense, 392
patens, 393
var. *gracile*, 393
pilosum, 394
plumosum, 395
praealtum, 395
pratense, 397
prenanthoides, 405
racemosum, 396
sagittifolium, 405
sericeum var. microphyllum, 397
shortii, 398
simmondsii, 399
simplex, 405
squamatum, 399
subulatum, 400
var. *elongatum*, 383
var. *ligulatum*, 404, 405
var. *parviflorum*, 387
var. *squamatum*, 400
tenuifolium, 401
var. *aphyllum*, 401
tradescantii, 391
undulatum, 402
urophyllum, 403
walteri, 404
Symporia glomerata, 64

Synedrella, 405
nodiflora, 405
vialis, 169
Synstima caroliniana, 27

Tagetes, 406
erecta, 406
lucida, 407
major, 406
minuta, 407
rotundifolia, 411
Tanacetum leucanthemum, 296
var. *pratense*, 296
Taraxacum, 407
caucasicum var. *erythrospermum*, 408
dens-leonis, 408
var. *commune*, 408
var. *laevigatum*, 408
var. *officinale*, 408
erythrospermum, 408
laevigatum, 408
var. *erythrospermum*, 408
var. *normale*, 408
leontodon, 408
officinale, 408
subsp. *vulgare*, 408
var. *erythrospermum*, 408
var. *genuinum*, 408
var. *laevigatum*, 408
var. *pratense*, 408
var. *typicum*, 408
officinarum, 408
palustre var. *vulgare*, 408
taraxacoides var. *laevigatum*, 408
taraxacum, 408
subsp. *genuinum*, 408
subsp. *laevigatum*, 408
var. *erythrospermum*, 408
var. *laevigatum*, 408
vulgare, 408
subsp. *erythrospermum*, 408
var. *genuinum*, 408
Tessenia
canadensis, 200
philadelphica, 227
ramosa, 228
Tetragonotheca, 409
helianthoides, 409
Tetrapanax, 76
papyrifer, 76
Thaspia trifoliata, 109
Thaspium, 109
aureum, 114
var. *cordatum*, 110
var. *trifoliatum*, 110
barbinode, 109
var. *chapmanii*, 109
var. *pinnatifidum*, 109
chapmanii, 109
cordatum, 109
var. *atropurpureum*, 109
trifoliatum, 109, 110
var. *apterum*, 113
var. *flavum*, 110
ziziopsis, 110
Thelechitonia trilobata, 378
Thymophylla, 410
tenuiloba, 410
Thyrsanthema
semiflosculare, 178
tomentosum, 177
Thyrsosma chinensis, 60
Tiedemannia, 110
filiformis, 110
subsp. filiformis, 111
subsp. greenmanii, 111
rigidior, 100
teretifolia, 111
ternata, 101
Tithonia, 410
diversifolia, 411
rotundifolia, 411
Torilis, 111
arvensis, 112
divaricata, 112
japonica, 112
nodosa, 112
Toxopus gymnanthes, 21
Trachysperma
aquaticum, 50
grayanum, 51
humboldtianum, 52
lacunosum, 50
natans, 49
Trachyspermum ammi, 115
Traganthes
compositifolia, 232
eugenei, 239
pectinata, 239
pinnatifida, 239
Tragopogon dandelion, 290
Trattenikia
angustifolia, 308
lanceolata, 309
Trepocarpus, 112
aethusae, 112
Triclinium odoratum, 105
Tridax, 412
procumbens, 412
Trilisa, 170
carnosa, 170
odoratissima, 172
paniculata, 173
subtropicana, 172
Triodanis, 46
biflora, 46
perfoliata, 47
subsp. *biflora*, 46
var. *biflora*, 46
Tripolium
subulatum, 400
var. *boreale*, 400
var. *cubensis*, 387
var. *parviflorum*, 387
Troximon dandelion, 290
Tussilago
albicans, 177
integrifolia, 178

Ucacou nodiflorum, 405
Umbelliferae, 77
Uncasia
alba, 231
altissimum, 231
anomala, 239
cuneifolia, 237
hyssopifolia, 233
lecheifolia, 233
leucolepis, 234
linearifolia, 234
mikanioides, 235
mohrii, 235
perfoliata, 236
petaloidea, 236
pubescens, 237
rotundifolia, 237
scabrida, 237

semiserrata, 238
serotina, 238
torreyana, 233
tortifolia, 235
verbenifolia, 237
Urbanisol tagetifolius, 411

Valeriana, 65
locusta var. *radiata*, 66
radiata, 66
scandens, 65
var. *genuina*, 65
Valerianaceae, 61
Valerianella, 65
radiata, 66
Verbesina, 412
alba, 219
alternifolia, 413
arborescens, 164
aristata, 414
carnosa, 378
chapmanii, 414
coreopsis, 413
var. *lutea*, 413
encelioides, 415
heterophylla, 415
laciniata, 416
nodiflora, 405
nudicaulis, 414
occidentalis, 416
pauciflora, 415
phaethusa, 416
prostrata, 219
sigesbeckia, 416
sinuata, 416
virginica, 416
var. *laciniata*, 416
walteri, 417
warei, 415
Vernonia, 417
acaulis, 417
altissima, 419
forma *parviflora*, 419
angustifolia, 418
subsp. *mohrii*, 418
subsp. *scaberrima*, 418
var. *mohrii*, 418
var. *pumila*, 419
var. *scaberrima*, 418
blodgettii, 419
cinerea, 215
concinna, 421
fasciculata, 215
var. *altissima*, 419
gigantea, 419
subsp. *ovalifolia*, 419
var. *ovalifolia*, 419
glauca, 421
graminifolia, 418
missourica, 420
noveboracensis, 420
var. *latifolia*, 421
oligantha, 419
oligophylla, 418
var. *verna*, 418
ovalifolia, 419
scaberrima, 418
texana, 421
×concinna, 421
×*illinoensis*, 421
Viburnum, 57
acerifolium, 58
var. *densiflorum*, 58
capitatum, 59
cassinoides, 61
var. *nitidum*, 59
corymbosum, 61
densiflorum, 58
dentatum, 58
var. *semitomentosum*, 58
var. *scabrellum*, 58
ferrugineum, 60
involucratum, 58
laurifolium, 59
molle, 61
var. *tomentosum*, 58
nashi, 59
nitidum, 59
nudum, 59
var. *angustifolium*, 59
var. *claytonia*, 59
var. *grandiflorum*, 59
var. *nitidum*, 49
var. *serotinum*, 59
obovatum, 59
odoratissimum, 60
prunifolium, 61
var. *ferrugineum*, 60
rufidulum, 60
var. *floridanum*, 60
rufotomentosum, 60
scabrellum, 58
semitomentosum, 58
tomentosum, 58
Villarsia
aquatica, 49
cordata, 50
cristata, 51
humboldtiana, 52
indica, 52
lacunosa, 49
trachysperma, 49
Virgulus
adnatus, 382
carolinianus, 383
concolor, 385
georgianus, 389
patens, 393
var. *georgianus*, 389
var. *gracilis*, 393
pratensis, 397
walteri, 404
Visnaga
daucoides, 80
vulgaris, 79

Wahlenbergia, 48
linarioides, 48
marginata, 48
Wedelia
carnosa, 378
glauca, 323
trilobata, 378
Willoughbya
cordifolia, 313
heterophylla, 314
micrantha, 313
scandens, 314
var. *normalis*, 314
Winterlia trifolia, 30
Wulffia
angustifolia, 312
deltoidea, 311
hastata, 311
Wylia pecten-veneris, 106

Xanthium, 421
americanum, 422
canadense, 421
glabratum, 422

macrocarpum var. *glabratum*, 422
pensylvanicum, 422
var. *glandulosum*, 422
pungens, 422
spinosum, 422
strumarium, 421
var. *canadense*, 421
var. *glabratum*, 422
vulgare, 421
Ximenesia, 413
encelioides, 415
var. *hortensis*, 415

Youngia, 422
japonica, 422
subsp. *genuina*, 422

Zanthoxylum trifoliatum, 69
Zinnia, 423
elegans, 423
var. *violacea*, 423
elegans, 423
florida, 424
floridana, 424
multiflora, 424
pauciflora, 424
peruviana, 424
violacea, 423
Zizia, 113
aptera, 113
arenicola, 114
aurea, 114
cordata, 109
latifolia, 114
trifoliata, 114

Richard P. Wunderlin, professor emeritus of biology at the University of South Florida, is the author of *Guide to the Vascular Plants of Central Florida* and *Guide to the Vascular Plants of Florida*, and coauthor of *Guide to the Vascular Plants of Florida*, second edition, *Guide to the Vascular Plants of Florida*, third edition, and the *Atlas of Florida Plants* website (www.florida.plantatlas.usf.edu).

Bruce F. Hansen, curator emeritus of biology at the University of South Florida Herbarium, is coauthor of *Guide to the Vascular Plants of Florida*, second edition, *Guide to the Vascular Plants of Florida*, third edition, and the *Atlas of Florida Plants* website (www.florida.plantatlas.usf.edu).

Alan R. Franck, curator at Florida International University, is coauthor of the *Atlas of Florida Plants* website (www.florida.plantatlas.usf.edu).